Scottish Life and Society

Religion

Publications of the European Ethnological
Research Centre

Scottish Life and Society: A Compendium of Scottish Ethnology
(14 Volumes).

Already published:
Volume 3 *Scotland's Buildings*
Volume 9 *The Individual and Community Life*
Volume 11 *Institutions of Scotland: Education*
Volume 14 *Bibliography for Scottish Ethnology*

GENERAL EDITOR:
Alexander Fenton

Scottish Life *and* Society

A COMPENDIUM OF SCOTTISH ETHNOLOGY

RELIGION

Edited by
Colin MacLean
and
Kenneth Veitch

JOHN DONALD
in association with
THE EUROPEAN ETHNOLOGICAL RESEARCH CENTRE
and
THE NATIONAL MUSEUMS OF SCOTLAND

First published in Great Britain 2006 by
John Donald, an imprint of Birlinn Ltd
West Newington House
10 Newington Road
Edinburgh

www.birlinn.co.uk

Copyright © The European Ethnological Research Centre, 2006

ISBN 10: 0 85976 650 0
ISBN 13: 978 0 85976 650 0

All rights reserved

British Library Cataloguing-in-Publication Data
A catalogue record is available on request

Typeset by Carnegie Publishing Ltd, Lancaster
Printed and bound in Great Britain by the Cromwell Press, Trowbridge, Wiltshire

Contents

List of Figures	*viii*
List of Contributors	*xv*
Foreword	*xvii*
Acknowledgments	*xix*
Abbreviations	*xx*
Glossary	*xxi*

INTRODUCTION
Introduction   3
Gordon Graham

PART ONE: CONSOLIDATION
1. Early Christianity in Scotland: The Age of Saints   15
   Alan Macquarrie
2. The Cult of Saints in Medieval Scotland   42
   Alan Macquarrie
3. Religious Life in Scotland in the Later Middle Ages   60
   Audrey-Beth Fitch
4. The Impact of the Reformation in Relation to Church, State and Individual   103
   Francis Lyall

PART TWO: DIVERSITY
5. Presbyterianism   127
   Henry R Sefton
6. Roman Catholics in Scotland: Late Sixteenth to Eighteenth Centuries   143
   John R Watts
7. Roman Catholics in Scotland: Nineteenth and Twentieth Centuries   170
   John F McCaffrey
8. Episcopalians   191
   Allan Maclean
9. Other Christian Groups   235
   James A Whyte
10. The Jews in Scotland   256
    Kenneth Collins

11.	Islam in Scotland After 1945 *Mona Siddiqui*	281
12.	Reflecting World Faiths: BBC Scotland's Experience *Johnston McKay*	295
13.	Reflecting World Faiths: St Mungo Museum of Religious Life and Art *Harry Dunlop and Alison Kelly*	306
14.	Alternative Beliefs and Practices *Steven J Sutcliffe*	313

PART THREE: LANGUAGE, LITERATURE AND MEDIA

15.	The English and Scots Languages in Scottish Religious Life *Graham Tulloch*	335
16.	Gaelic and the Churches *Donald E Meek*	363
17.	Words of Power: Literature, Drama and Religion *Donald Smith*	379
18.	Medium for Godly Messages *Colin MacLean*	393
19.	Marvellous New Trumpets: The Media 1920s–2001 *Colin MacLean*	413

PART FOUR: THE RELIGIOUS COMMUNITY

20.	Ministers and Society in Scotland 1560-c1800 *Ian Whyte*	433
21.	The Church Social *Johnston McKay*	452
22.	Occasions in the Reformed Church *Henry R Sefton*	469
23.	Missions and Missionaries: Home *Frank Bardgett*	479
24.	Missions and Missionaries: Foreign *Andrew C Ross*	517
25.	Church, Law and the Individual in the Twentieth Century *Francis Lyall*	530
26.	From Monochrome to Colour *Elizabeth Henderson*	560

PART FIVE: ARCHITECTURE AND THE ARTS

| 27. | Cathedrals and Churches
James A Whyte | 575 |
| 28. | New Uses for Old Churches and Manses
David Maxwell
with an appendix by *Robin D A Evetts* and *Deborah Mays* | 598 |

29.	Furnishings in the Reformed Church *Henry R Sefton*	616
30.	Music, Church and People *Douglas Galbraith*	629
31.	Christian Art in Scotland *Murdo Macdonald*	652
	Index	681

List of Figures

1.1	Nineteenth-century drawing of the Catstane, Kirkliston, Midlothian, possibly fifth century.	19
1.2	Replica of stone commemorating Ides, Viventius and Mavorius, Christian clergy. Kirkmadrine, Rhinns of Galloway, early sixth century.	21
1.3	Crosses of St John (replica), St Matthew and St Martin, Iona, eighth century.	28
1.4	The 'Apostles Stone', Dunkeld, right side, ninth century.	31
1.5	Back of a cross-slab from Kilbar, Barra, probably eleventh century.	33
1.6	Four monuments showing the spread of Christianity across Scotland.	35
2.1	Eleventh-century bronze crosier head, stripped of its original decoration.	43
2.2	Page from the Winter Propria Sanctorum of the *Aberdeen Breviary* showing the collect and first lesson for the commemoration of St Ethernan (2 December).	45
2.3	Fifteenth-century stone statue of John the Baptist, dredged from the Firth of Forth.	48
2.4	Statue of St Andrew, carved in oak, carrying his cross and a book, *c*1500.	49
2.5	Statue of St Ninian on an altarpiece originally from St Olai Kirke, Helsingør, Denmark.	50
2.6	Painted panel showing Christ with saints. Fowlis Easter collegiate church, second half of the fifteenth century.	51
2.7	Page from the Book of Deer, tenth century, showing Luke the Evangelist.	52
2.8	The opening pages of St Mark's Gospel from Queen Margaret's Gospel Book, eleventh century.	53
2.9	Stone statue of St Duthac from the collegiate church of Tain, Ross and Cromarty, fifteenth century.	54
2.10	Twelfth-century bronze crosier drop found in Loch Shiel, near St Finan's Isle.	55
3.1	Painting depicting the soul of the unrepentant sinner being taken to hell. Fowlis Easter collegiate church, second half of the fifteenth century.	68

3.2	Painting of St Catherine of Alexandria. Fowlis Easter collegiate church, second half of the fifteenth century.	69
3.3	Effigy of George, second Lord Seton, with rosary. Seton collegiate church, early sixteenth century.	70
3.4	Bas-relief on piscina showing devil with bat's wings. Seton collegiate church, early sixteenth century.	73
3.5	Bas-relief of instruments of the Passion. Haddington collegiate church, fifteenth century.	74
3.6	Boss depicting worm in skull. Lady Chapel, Glasgow Cathedral, early sixteenth century.	76
3.7	Bas-relief of deathbed scene, Mary King's Close, Edinburgh, late fifteenth century.	77
3.8	Woodcut of demons surrounding a deathbed. *The Book Intytulid the Art of Good Lywyng and Good Deyng*, 1503.	79
3.9	Bas-relief of Mary leading a layman away from the devil. Rosslyn chapel, mid-fifteenth century.	82
3.10	Woodcut of St Michael and the dragon. *The Golden Legend*, Lyon, 1487.	84
3.11	Woodcut of the real presence in the Eucharist. Arbuthnott Prayer Book, c1482–3.	88
5.1	National Covenant, with signatures of members of the Privy Council, 1638.	129
5.2	Copy of the Solemn League and Covenant, 1643.	131
5.3	Group portrait of the leaders of the Secession of 1737.	132
5.4	Eighteenth-century Bible used at the annual Covenanters' conventicle at Flotterstone, near Edinburgh.	134
5.5	D O Hill, *The Signing of the Deed of Demission* (1866).	135
5.6	Moderator of the Church of Scotland, with ministers, 1850s.	136
5.7	Lochcarron United Free Church congregation worshipping in the manse garden, c1900.	137
5.8	Minister of the United Free Church and his elders, Orkney, c1910.	139
6.1	Traquair House, Innerleithen, Upper Tweedsdale.	144
6.2	Brass crucifix and chalice from a grave in Dumbarton of a seventeenth-century Catholic priest.	148
6.3	Engraving of George Gordon, first Marquis of Huntly (1562–1636), and his wife, Lady Henrietta Stewart.	150
6.4	Bishop Thomas Nicolson (1646–1718), copy of original portrait.	152
6.5	Portrait of Bishop George Hay (1729–1811) in the Scots College, Rome.	156
6.6	The College of Scalan, Upper Banffshire, the new building of 1767.	157

6.7	St Ninian's church, Tynet, Moray.	161
6.8	St Gregory's Preshome, Enzie, Moray, 1788–90.	162
6.9	Aquhorties House, 1799, seminary for the Lowland Vicariate.	163
7.1	Dufftown Roman Catholic chapel, 1824.	171
7.2	Portrait of Charles Petre Eyre, archbishop of Glasgow (1878–1902).	174
7.3	Motherwell Catholic Boys' Brigade, Company I, December 1915.	176
7.4	Scotus Academy, Edinburgh, 1952.	177
7.5	The building of Oban Cathedral, 1932.	180
7.6	Pilgrims at Carfin Grotto, c1947.	182
7.7	Margaret Sinclair.	184
7.8	The Scottish Hierarchy after the Centenary Mass for the consecration of Archbishop John Strain by Pope Pius IX. Basilica of San Lorenzo fuori le Mure, 25 September 1964.	186
7.9	Ticket for Papal Mass, Bellahouston Park, Glasgow, 1 June 1982.	187
8.1	G W Brownlow, *Prison Baptism at Stonehaven* (1864).	193
8.2	The consecration of Bishop Seabury in Aberdeen, 1784.	195
8.3	Forfar Meeting House.	196
8.4	Kilmaveonaig altar vessels.	199
8.5	St Mary's Cathedral, Edinburgh.	207
8.6	A double certificate of Confirmation and Admission to Communion of 1919.	210
8.7	St Andrew's church and rectory, Ardrossan.	212
8.8	Sketch of Auld Sandy Milne.	213
8.9	Principal Alexander Boyd, staff and students at the Episcopal Teacher Training College, Edinburgh, 1928.	215
8.10	The Scottish bishops at Inverness, June 1943.	217
8.11	Armorial sign.	220
8.12	Diocesan Festival Eucharist, St John's Cathedral, Oban, 1979.	222
8.13	Blessing the animals, Holy Trinity Church, Dunoon, Rogation Sunday 1984.	226
8.14	Ordination of the first women priests in the diocese of Argyll and The Isles. Joanna Anderson, warden of the abbey of Iona, and Gillian Orpin, diocesan chaplain. Oban, 26 January 1995.	228
9.1	Portrait of Robert Sandeman.	239
9.2	Interior of Congregational church, Lerwick, Shetland.	241
9.3	Canonmills Baptist church, Edinburgh.	243
9.4	The congregation of Nicolson Square Baptist church, 2004.	245
9.5	A service at St Mark's Unitarian church, Edinburgh.	246
9.6	Salvation Army singers, c1900.	249

9.7	The Salvation Army is noted for its work among the homeless. Glasgow, 2003.	250
9.8	Worship at the CLAN Gathering, St Andrews, 2004.	254
10.1	Portrait of Dr Asher Asher.	258
10.2	Entrance to the Jewish enclosure, Glasgow Necropolis.	260
10.3	Jewish immigrant family, Glasgow, c1910.	261
10.4	Odessa Lodge Friendly Society, c1907.	262
10.5	Exterior of Links' shop (with Yiddish lettering on the windows), Main Street, Gorbals, c1907.	266
10.6	Interior of Garnethill Synagogue, Glasgow, opened 1879.	267
10.7	Jewish soldiers outside South Portland Street Synagogue, World War I.	269
10.8	Sir Myer Galpern, Glasgow's first Jewish Lord Provost (1958).	273
10.9	Jewish Institute Players performing a play.	275
10.10	Jewish Scouts, c1958.	276
11.1	Laying out prayer mats at Pilrig mosque, Edinburgh, 1993.	285
11.2	Edinburgh Central Mosque.	289
11.3	Four mosques join together to celebrate Ramadhan, the Meadows, Edinburgh, 1993.	292
13.1	The Gallery of Religious Art, St Mungo Museum of Religious Art and Life, Glasgow.	308
13.2	Chinese calligraphy at 'Meet Your Neighbour', St Mungo Museum of Religious Art and Life, Glasgow, 2004.	310
14.1	White Buddha in the garden of the Samye Ling Tibetan Centre, Eskdalemuir, Dumfriesshire, 1999.	314
14.2	Green Man, May Day Parade, Stirling Castle Esplanade, 2000.	316
14.3	Celtic Christian pilgrimage, Ninian's Cave, Whithorn, Wigtownshire, 2001.	319
14.4	Wildwood Books, Edinburgh.	322
14.5	The Nature Sanctuary, Findhorn Foundation Community, Moray.	325
15.1	The Ruthwell Cross from Dumfriesshire, with The Dream of the Rood inscribed in runes.	336
15.2	Title-page of the Geneva Bible of 1560.	341
15.3	Title-page of *The Book of Common Prayer*, 1637.	345
15.4	Etching by Wenceslaus Hollar of the riot at St Giles', Edinburgh, 23 July 1637.	347
15.5	A nineteenth-century French image of Morton finding Balfour of Burley reading the Bible.	350
15.6	'The priest-like father reads the sacred page'. An illustration of Burns's 'The Cotter's Saturday Night'.	351
15.7	Title-page of Waddell's translation of the Psalms.	354

18.1	Advertisement for *Herald of Mercy*, in Rev. John Macpherson, *Life and Labours of Duncan Matheson, the Scottish Evangelist*, 1871.	399
18.2	Thomas Chalmers.	400
18.3	Illustration of *Her Day of Service* by Mrs Isabella Fyvie Mayo.	402
18.4	Hugh Miller.	406
18.5	W R Nicoll.	408
19.1	Reverend Dr Ronald Selby Wright.	415
22.1	Silver Communion cups from Leslie church, Aberdeenshire.	470
22.2	Free Church Communion, Plockton, early twentieth century.	471
22.3	John Philip, *Baptism in Scotland* (1850).	473
22.4	Baptismal ewer, Greyfriars, Edinburgh, 1707–8.	473
22.5	Baptismal certificate, United Free Church of Scotland, 1925.	474
22.6	Wedding group, Davidson's Mains parish church, Edinburgh, 1965.	475
22.7	George Rankin (1864–1937), *The Crofters Funeral*.	476
22.8	James Guthrie, *A Highland Funeral* (1881).	477
23.1	Cover page of *The Missionary Magazine* for 1796.	481
23.2	Mission Hall at Halsary, Caithness.	490
23.3	Reverend Thomas Rosie (d. 1862).	496
23.4	Tent Hall postcard c1920, 'A Salvation March'.	498
23.5	Mr John Murchison, Lay Missionary at the Glasgow and West Coast Mission's station at Kylerhea (1921–54).	502
23.6	Mission house and church at Craigton, Strath Halladale, Sutherland.	503
23.7	Scottish Colportage Society's Bible and Gospel Book Van, 1947.	505
23.8	Cover page of the Song Book of the 1955 Billy Graham All-Scotland Crusade.	508
23.9	Congregation of the Bilston Mission, Midlothian, at their jubilee service, 1988.	510
23.10	New Charge church built by the Church of Scotland, 2002–03, for Inverness Inches between a shopping mall and new housing.	511
26.1	Women's Guild, Granton parish church, early 1960s.	562
26.2	Community café, Richmond Craigmillar parish church.	566
27.1	Detail of Pictish stone at Aberlemno.	576
27.2	Iona Abbey.	577
27.3	St Andrews Cathedral.	578
27.4	Leuchars church.	579
27.5	Dunblane Cathedral.	581
27.6	St Columba's church, Burntisland.	584
27.7	Lauder parish church.	585

27.8	Kelso Old parish church.	586
27.9	Pulpit and long table, Plockton parish church.	588
27.10	Plockton parish church.	589
27.11	Kelso North church.	590
27.12	Pulpit and hour glass at St Salvator's, St Andrews.	593
27.13	Bo'ness Roman Catholic church.	594
27.14	The altar in Bo'ness Roman Catholic church.	595
27.15	Chapel at Scottish Churches' House, Dunblane.	596
28.1	Belford Youth Hostel, Edinburgh.	600
28.2	The Tron church, Edinburgh.	601
28.3	Stockbridge church, Edinburgh.	602
28.4	Aberdeen Maritime Museum.	603
28.5	The church of St James', Pollock.	605
28.6	Trinity College Apse, Edinburgh.	606
28.7	The Bedlam Theatre, Edinburgh.	607
28.8	The Episcopal church of the Holy Trinity, Dean Bridge, Edinburgh.	608
28.9	The Queen's Gallery, Edinburgh.	610
28.10	Carrington church.	611
28.11	St Ninian's manse, Leith.	612
29.1	The West Church of St Nicholas, Aberdeen.	616
29.2	A folding stool of the type used in churches in Scotland before and after the Reformation.	618
29.3	Examples of old and new seating, St Giles', Edinburgh.	619
29.4	Etching of the Kilmarnock funeral bell, 1639.	620
29.5	Pulpit (1912), Greyfriars church, Edinburgh.	621
29.6	Nineteenth-century, eagle-style lectern, St Giles', Edinburgh.	623
29.7	Organ (Peter Collins, 1990), Greyfriars church, Edinburgh.	624
29.8	Collection ladles from Auchterless parish church, Aberdeenshire.	626
30.1	Panel of the Dupplin Cross showing David the Psalmist as a Celtic bard with a harp.	631
30.2	Inchcolm Abbey, Firth of Forth.	632
30.3	A Crear McCartney window in St Michael's, Linlithgow.	634
30.4	Initials outside the University Chapel at St Andrews marking the site of the martyrdom of Patrick Hamilton.	635
30.5	Pages of the 1635 Psalter showing the four parts of the Old Hundredth, with the melody in the tenor.	638
30.6	Plaque at the foot of Calton Hill, Edinburgh.	643
30.7	The precentor's pulpit in St Martins church near Perth.	644
30.8	John L Bell of Iona.	648

31.1	Papil Stone, Shetland.	653
31.2	Book of Kells, opening page of St John's Gospel, *c*800.	655
31.3	Detail of the crucifixion painting, Fowlis Easter collegiate church, second half of the fifteenth century.	657
31.4	The Annunciation from the Beaton Panels.	659
31.5	Detail of King David playing his harp from the Dean House Panels.	660
31.6	David Wilkie, *John Knox Dispensing the Sacrament at Calder House*.	663
31.7	William Dyce, *Christ as the Man of Sorrows*.	664
31.8	Phoebe Anna Traquair, *The Progress of the Soul: The Entrance*.	667
31.9	John Duncan, *St Bride*.	668
31.10	Hew Lorimer, *Our Lady of the Isles*, South Uist.	669

List of Contributors

REVEREND DR FRANK BARDGETT
Formerly of the Department of National Mission, The Church of Scotland

DR KENNETH COLLINS
General Medical Practitioner and Chairman of the Scottish Jewish Archives

HARRY DUNLOP
Museum Manager, St Mungo Museum of Religious Life and Art, Glasgow

DR ROBIN D A EVETTS
Historic Scotland Inspectorate

PROFESSOR AUDREY-BETH FITCH †
Department of History, California University of Pennsylvania

REVEREND DOUGLAS GALBRAITH
Office for Worship, Doctrine and Artistic Matters, The Church of Scotland

PROFESSOR GORDON GRAHAM
Henry Luce III Professor of Philosophy and the Arts, Princeton Theological Seminary

REVEREND ELIZABETH HENDERSON
Minister of the Richmond Craigmillar Church, Edinburgh

ALISON KELLY
Curator of World Religions, St Mungo Museum of Religious Life and Art, Glasgow

PROFESSOR FRANCIS LYALL
School of Law, University of Aberdeen

DR JOHN F McCAFFREY †
Formerly of the Department of Scottish History, University of Glasgow

PROFESSOR MURDO MACDONALD
Department of History, University of Dundee

VERY REVEREND ALLAN MACLEAN OF DOCHGARROCH
Canon of St John's Cathedral, Oban

COLIN MACLEAN
Journalist for c30 years, and Publishing Director, Aberdeen University Press, 1979–90

DR ALAN MACQUARRIE
Research Associate, School of Scottish History, University of Strathclyde

REVEREND JOHNSTON McKAY
Formerly Editor of Religious Broadcasting, BBC Scotland, and Clerk to the Presbytery of Ardrossan

DAVID MAXWELL †
Over a period of 27 years, Secretary, Chairman and committee member of the Church of Scotland Committee on Artistic Matters

DR DEBORAH MAYS
Historic Scotland Inspectorate

PROFESSOR DONALD E MEEK
Department of Celtic and Scottish Studies, University of Edinburgh

DR ANDREW C ROSS
School of Divinity, University of Edinburgh

REVEREND DR HENRY R SEFTON
Christ's College, University of Aberdeen

DR MONA SIDDIQUI
Department of Theology and Religious Studies, University of Glasgow

DR DONALD SMITH
Scottish Storytelling Centre, Edinburgh

DR STEVEN J SUTCLIFFE
School of Divinity, University of Edinburgh

PROFESSOR GRAHAM TULLOCH
Department of English, Flinders University, Adelaide

DR JOHN R WATTS
Author and researcher, Addiewell

PROFESSOR IAN WHYTE
Department of Geography, University of Lancaster

VERY REVEREND PROFESSOR JAMES A WHYTE †
Formerly Principal of St Mary's College, University of St Andrews and Moderator of the Church of Scotland

† deceased contributor

Foreword

Anyone embarking on a comparative scientific study of cultures which included Scotland would soon discover that religion has played an important part in Scottish life and society from the earliest times up to the present day. It is therefore not surprising that the European Ethnological Research Centre should include in its *Compendium* series a volume on religion.

Although influenced over the years by developments in other countries, such as Italy, Ireland, Switzerland, the Netherlands and England, religion in Scotland has developed its own character and traditions. The nature of these differences and the reasons for them are an important part of any comparative study, but to undertake such a task on a scientific basis requires a breadth of knowledge which puts it, I think, beyond any single individual to complete satisfactorily. In this volume, therefore, the European Ethnological Research Centre has secured the services of a number of distinguished authorities, who have dealt with aspects of the development of religion in Scotland in a way which together makes up a very complete study. A valuable overview of the volume, and of the subject as a whole, is provided by Professor Gordon Graham of Aberdeen University in an introductory chapter that will be found very helpful by those who wish to study the subject extensively as well as by those who wish to dip into a particular aspect of the subject which engages their interest.

As one who from his earliest days was brought up in a culture in which religion and in particular biblical Christianity had a central place, I believe this volume helps to give an understanding of the relationship between the doctrines and institutions with which I have been familiar and the characteristics of the Scottish people which have played a part in their development. It moreover gives me an insight into the relationship between these doctrines and institutions and the other countries in which similar developments have taken place.

For those of us who are interested in religion in Scotland, it also gives some extremely valuable insights. For example, I believe that the emphasis in post-Reformation Scotland on the ministry of The Word has been an important factor in encouraging activities outside the strict realm of religion, including particularly education but to some extent the law too. I think it is further true that the emphasis on education coupled with the institution of Presbyterian government with the place this gives to the laity in the councils of the Church has meant that there has not been a concentration of doctrinal authority in the hands of the professional ministry that may be discerned elsewhere and that in consequence where matters of interpretation of The Word are concerned there is more room for informed and acute difference of opinion in the Kirk than in many other national Churches in the past. This has no doubt contributed to the number of divisions that have taken place in the Kirk in Scotland over the years. Although the Disruption of 1843 was not primarily a matter of interpretation of The Word, it illustrates the central role that the laity has played in the Kirk. One of the leading exponents of the position adopted by the Disruption-Free Church was a stonemason from Cromarty in the Scottish Highlands whose eloquence and influence as a newspaper editor played a very large part in inspiring and supporting the leadership of the Church in the stand they took against the civil authorities of the time in requiring that the affairs of the Church should not be regulated to the extent that they had been.

In my view, this volume is a distinguished contribution to ethnological study and will repay careful study by all who are interested in Scotland and its culture as well as those who have a particular interest in the development of religion.

Lord MacKay of Clashfern KT PC

Acknowledgments

The editors would like to acknowledge:

    The Scotland Inheritance Fund

    The Trustees of the National Museums of Scotland

    The Russell Trust

Grateful thanks are also due to the following:

    The Scottish Life Archive, National Museums of Scotland, Edinburgh

    The Scottish Catholic Archives, Columba House, Drummond Place, Edinburgh

    The Scottish Jewish Archives, Garnethill Synagogue, Hill Street, Glasgow

Abbreviations

AU	*Annals of Ulster*
BCP	*Book of Common Prayer*, 1662
BL	British Library
DSCHT	*Dictionary of Scottish Church History and Theology* (Cameron, N de S, ed, Edinburgh, 1993)
EERC	European Ethnological Research Centre
HE	Bede's *Ecclesiastical History of the English People*
IR	*Innes Review*
NAS	National Archives of Scotland
NLS	National Library of Scotland
NMS	National Museums of Scotland
PRO	Public Record Office
PSAS	*Proceedings of the Society of Antiquaries of Scotland*
RCAHMS	Royal Commission on the Ancient and Historical Monuments of Scotland
ROSC	*Review of Scottish Culture*
RSCHS	*Records of the Scottish Church History Society*
SEJ	*Scottish Ecclesiastical Journal*
SEM/EEJ	*Stephen's Episcopal Magazine* or *Edinburgh Ecclesiastical Journal*
SHR	*Scottish Historical Review*
SLA	Scottish Life Archive, National Museums of Scotland
SPB	*Scottish Book of Common Prayer*, 1929
SSB	*Scottish Standard Bearer*
SSPCK	Society in Scotland for Propagating Christian Knowledge
VC	Adomnán's *Life of Columba*

Glossary

This selective list contains only those terms found in the text that may aid a non-specialist reader.

Anglican Communion of Churches, a worldwide fellowship of independent Episcopal Churches, in Communion with, and recognised by, the Archbishop of Canterbury, the ecclesiastical head of the Anglican Church.

Apostolic Succession, the belief that bishops are consecrated as successors to, and direct inheritors of, the apostles.

Ascription, the completing of prayer or praise by using the names of the Trinity.

Atonement, the belief that God was reconciled to humankind by Jesus Christ's death.

Banns, a public announcement of an intended marriage.

Beth Din, Jewish law court usually dealing with matters of Kashrut or of personal status such as in granting of divorces. The presiding rabbi is known as the Av (father).

Bred, a small tablet or board serving to receive offerings to an altar, church etc.; an offertory box.

Chassidic, of or relating to the Jewish Chassidim, a sect of Orthodox Jews who follow the Mosaic Law strictly.

Chevra, literally means Society. Often refers to a religious society established as a small place of worship.

Cohen (plural **Cohenim**), Jews of priestly descent who have certain religious duties and privileges related to their status. They are forbidden contact with the dead except at the funeral of their close relatives.

Communion, the intimate relationship of fellowship between God and humankind, instituted by Jesus Christ as a specific task in the life of the Church.

Communion tokens, pieces of metal or card distributed to people to show that they were in good enough standing in the church community to be able to attend a Communion service.

Eid, there are two eids in the Islamic calendar. One comes at the end of the fasting month of Ramadan and the other is celebrated at the end of the annual pilgrimage to Mecca, the Hajj.

Episcopacy, ecclesiastical government by bishops, who are specifically ordained for that purpose.

Evangelical, an approach to belief that gives special prominence to the doctrines of the Fall and the Regeneration and Redemption of humankind, through the work of Jesus Christ, and which is received by free and unmerited grace.

Evensong, in the Episcopal Church a daily evening service of prayer and praise, for clergy and laity, used especially on Sundays.

Eucharist, a name for the Communion service, in origin meaning 'Thanksgiving'.

Fatawa, opinions given by Muslim jurists and persons of recognised religious standing. They are not legally binding but serve to guide the questioner on any issue.

Filioque, the words 'and the son', which were added to the Nicene Creed by the Western Church in the eleventh century.

General Assembly, the supreme court of the Church of Scotland (as of some other Presbyterian churches), having jurisdiction over Synods, Presbyteries and Kirk Sessions.

Hadith, literally means a report or saying. In Islam, the word denotes a saying of the Prophet Muhammad; the sayings are greatly revered and form a basis for the development of Islamic law.

Halal, means legally permissible in Islam. It is used most commonly to refer to meat that has been slaughtered according to the rules of Islamic law.

High Church, in the Episcopal Church, the idea that the Church was specifically founded by Jesus Christ as a group in the closest relationship with him, whose concerns and decisions therefore have priority over those of secular society.

Hijab, literally means screening but is now commonly used to refer to the headscarf worn by Muslim women.

Imam, the person who leads the congregation in prayer.

Kashrut, the state of being kosher.

Kirk session, the governing body of a Presbyterian congregation, consisting of the minister and a number of ruling elders. It was historically responsible for the education of children, the care of the poor, the observance of the Sabbath and the oversight of morals, but by the twentieth century its duties were almost entirely with the life and work of the congregation.

Kollel, an institution for the studying of Talmud by adults who have finished their yeshiva studies.

Kosher, food prepared in accordance with Jewish religious laws.

Liturgy, established and prescribed formularies used in churches at public worship.

Low Church, in the Episcopal Church, the idea that the Church, and its institutions, is totally or largely invented by humankind.

Mattins, in the Episcopal Church a daily morning service of prayer and praise, for clergy and laity, used especially on Sundays.

Mikva, ritual bath used monthly by married women, by converts to Judaism on their acceptance into the faith and by orthodox men before Shabbat and Festivals.

Presbytery, the court standing mid-way between General Assembly (or Synod) and Kirk Session in the hierarchy of courts that constitute the Presbyterian system of Church government. It consists of the ministers and one elder from each church within a district, and its duties include the nomination of candidates for the ministry.

Quietism, the practice of some people in prayer, who by mental passivity, or contemplation, bring the soul into direct union with God.

Recusant, one who refuses to conform to or obey the ordinances of the established Church, applied in Scotland to Roman Catholics who refused to attend the services of the Church of Scotland.

Salat, refers to the five daily prayers which should be observed by every Muslim and which form one of the five pillars of Islam.

Schismatic, one whose action or beliefs separates them fundamentally from the basic tenets of the whole, to such an extent that they cannot be regarded as a part of the whole.

Shabbat, the Jewish Sabbath. Weekly family and community based festival of complete rest from dusk on Friday to sundown on Saturday night.

Shari'a, ideal of Islamic law as contained in the Qur'an and interpreted by the jurists.

Shechita, Jewish method of despatch of kosher animals for food carried out by the ***Shochet***, a highly skilled religious official.

Sunna, the traditional root of Muslim law, based on biographical stories about Muhammad.

Vestry, in the Episcopal Church, the governing body of the secular life of a congregation, of which the rector is usually chairman.

Yeshiva, a religious school, often full-time, for boys over Barmitzvah age (13 years). The principal is known as the Rosh (head).

Introduction

Introduction

GORDON GRAHAM

THE NATURE OF ETHNOLOGY

The dictionary defines ethnology as the 'comparative scientific study of cultures'. The word 'scientific' in this context should be understood in its broad and original meaning, not in the narrow sense it tends to have in contemporary writing. The 'scientific' need not take physics as its model, and in the case of the comparison of cultures most definitely ought not to. There are, at least as yet, no theories of cultural development available to us that have anything like the scope or explanatory power of the theories that can now be called upon to explain the nature and development of the physical universe at both macro and micro levels. Ethnological study is scientific in the sense that it has, or ought to have, the following character – a careful accumulation of empirical facts systematically organised and interpreted in a non-partisan way with no other aim than that of casting these facts in a light genuinely productive of insight and understanding.

This scientific approach to culture is distinctive. We can accumulate facts and organise them for different purposes: to secure among ourselves a stronger sense of cultural identity, for example; to determine collective guilt or innocence for past atrocities; to assess the extent to which some aspect of a culture has progressed (medicine or technology, say). All these are legitimate purposes to which the accumulation of facts and the study of history may be relevant. But they are not scientific. This is not to say that they are unscientific, or in some way dubious. They are, rather, non-scientific, and it is only what has been called 'scientific imperialism' that would take this as a ground for their denigration or dismissal. Our concerns with culture differ, but this in itself does not make any one superior or inferior. How then does the scientific interest in culture differ from others?

In times past, the mark of science was sometimes taken to be identical with the establishment of anything properly called 'facts'. This is the view often known as 'positivism'. At one time in the ascendant, it was subsequently displaced by various versions of 'relativism', which denied the very possibility of 'objective fact'. Impressed by the inescapable necessity of employing concepts that have a history, this doctrine declares the validity of any empirical observation to be relative to the social or personal circumstances of the observer. It is a doctrine that tends even yet to dominate debate in sociological and anthropological circles. But it may be argued that both positivism and relativism have their sights fixed in the wrong place.

The difference between (say) a judicial and an historical inquiry into the past is not that the latter is objective in a way that the former cannot be, still less that the different concepts they use (in so far as they do use different concepts) determine different truth values. Nor does the difference lie in the one being normative while the other is factual. Facts are essential to judicial inquiries; investigative norms are essential in history. The crucial difference lies in their respective purposes. The purpose of a judicial inquiry is to decide what ought to be done if guilt is to be punished and innocence vindicated. The purpose of an historical investigation is to explain how and why what happened, happened.

What then is the purpose of ethnology? One answer is – comparison for its own sake. To know how and why one culture, or one aspect of some culture, differs from place to place is intrinsically interesting. It may also be illuminating by causing us to revise our unexamined presuppositions about the culture we have ourselves inherited. Further, it can reveal to us possibilities of which we were hitherto unaware, novel ways in which human beings organise, celebrate, create or understand themselves. And perhaps most interestingly of all, it may uncover such novelties, not with respect to exotic cultures about which hitherto we knew little, but about cultures with which we believed ourselves to be utterly familiar.

RELIGION IN SCOTLAND

Any particular ethnological study, of course, will attract a variety of readers. To some the subject area will already be familiar, while to others it will be quite new. Accordingly the relative surprise and interest will be different. But it is important for both classes of reader that the scope and focus of the work in question is clear. In this volume, the scope is Scotland past and present, and the focus is religion in all its manifestations. The restriction of scope to Scotland is, in a sense, arbitrary. Cultural phenomena rarely observe political boundaries strictly, and are even less likely to do so when those boundaries are as porous as the boundary between Scotland and the rest of the United Kingdom has been since the Act of Union in 1707. In the case of some phenomena, however, the restriction is more meaningful than it is in others. Religion is one of these. Along with law and education, for a protracted period religion in Scotland had a discernibly different identity from religion in England or Ireland, being Presbyterian rather than Anglican or Catholic. And even before that, its Celtic culture made a difference to its reception of Christianity.

There is then reason to focus on religion in Scotland as a distinctive topic for ethnological study. Here is a phenomenon that we can be sure will differ in at least some respects from religion in other countries. At the same time, there are enough points of contact with religion elsewhere to make comparison with other places instructive. Scotland was not the only Celtic country; Christianity, its religion for most of its history, is a world religion; secularisation and the multiculturalism that came with the end of empire are

European phenomena; and so on. In short, religion in Scotland is a specially suitable topic for ethnological study properly understood.

The scope of the subject, then, is a relatively easy matter to settle. More difficult, perhaps, is the question of method. But even here it is possible to say something useful, provided we steer clear of the narrow agendas of positivism, relativism and the like. First, every cultural phenomenon has its history, so part of the study of culture will have to be historical, an examination of how it arose and developed, some of it expressly concerned with specific periods, as the first four essays in this volume are. At the same time, ethnology must avoid the antiquarian. The past, it has sometimes been said, is another country, but while there are subjects rightly interested in studying the past for its own sake, ethnology studies the past at least in part for the light it throws on the present. There must thus be a kind of dialectical relationship between past and present, though in a volume like this it is often a dialectic that readers rather than writers construct, as they move from essay to essay.

Secondly, any aspect of culture finds expression in at least two different ways – the groups and institutions that make up its structure, and the beliefs and ideas that animate that structure. For this reason it is not merely appropriate but essential for the ethnologist of religion to explore what are now called 'faith communities', or in older terminology, religious denominations – groups socially differentiated by a combination of their beliefs, practices and organisation. The ten essays in the second part of the volume are for the most part written along these lines. The seven essays in the fourth part of the volume take this exploration further by examining the structures and ideas that straddle specific religious divisions and look at the role of, for example, ministers and missionaries in the religion of Scotland.

Thirdly, an adequate understanding of religion requires that its influence on and relationship with other aspects of culture be explored. In some contemporary societies religion may now appear a purely private preoccupation that occupies its own restricted sphere, in the way that stamp collecting or train spotting does. But in so far as this is true, it is quite atypical. In most times and places, religious belief and practice, and their associated organisation, have been deeply embedded in social life more generally. The third and the fifth sections explore some of these relationships at their most marked – the connection between religion and language, and the influence of religion on architecture and the arts.

Fourthly and finally, the phenomenon the ethnologist proposes to study, religion in this case, must not be too narrowly conceived. Better to include borderline cases than run the risk of ignoring important facets. Religious behaviour can show itself in surprising places, but more importantly perhaps, it is not to be assumed that the dominant religion, historically or numerically, is the only significant one. In the case of Scotland, Christianity is of course the major presence. But Judaism and Islam, and now some strands of what is called 'new age', also have a place, albeit a minor one, in the religion of Scotland, and accordingly there are essays here devoted to them.

To cover this ground adequately, twenty-eight contributors have been asked to write over 200,000 words. Faced with this huge volume, and variety, of material, it would be easy to lose sight of any general picture that would hold it all together, any conception within which the various facets of religion in Scotland could be connected and related. The principal purpose of this introduction is to sketch just such a general picture. Of necessity it has two parts – a broad-brush depiction of the shape of the past, and an equally general view of the present. The overall narrative of Scotland's religious past can be structured around two major turning points – the advent of Christianity and the Protestant Reformation. The present may be said to be characterised by a process of secularisation.

CELTIC CHRISTIANITY

The Christian religion came to Scotland with Ninian, tradition holds, around AD 400. The world it entered was importantly different to the world from which it sprang. It is of the greatest importance to understand that Christianity in its earliest phase was not a new religion, but a movement within Judaism, and its gradual emancipation took place as it spread across the Mediterranean, at that time when an imperial world focused upon Rome. With its arrival in Scotland, Christianity encountered a culture quite different to that of the Graeco-Roman world into which it was born. The result, inevitably, was a distinctive variety, whose distinctiveness was strengthened by further missionary endeavours from Ireland, itself a culture at some remove from Rome.

Both Ireland and Scotland were beyond the western limits of the Roman empire, and had cultural histories of their own, about which, however, we know relatively little. The relevance of this to the religion of Scotland, and to its study, is considerable. Christianity is remarkable (in comparison to Hinduism or Islam for example) in its capacity for what is known as 'acculturation': the ability to assume different forms according to the culture in which it is preached. Between English puritanism, Spanish Catholicism, and contemporary Christianity in Africa (say), there are differences so great that one may wonder where precisely the commonality lies that makes them all varieties of the same religion. So too with Scotland. It is essential to observe that Christianity's encounter results in something new, culturally speaking.

It follows that it would be a mistake to think in either of two ways – that the new religion simply 'took over' and effectively suppressed or destroyed the indigenous culture, or that the indigenous culture simply assumed a Christian 'patina' while more or less continuing as it was. The point may be illustrated in a general way by the example of Christmas. There is some reason to believe that with the arrival of Christianity the winter festival of Hogmanay was converted into a celebration of the birth of Christ.

Sometimes people infer from this that Christmas eliminated the celebration of the winter solstice, sometimes that Christmas is 'really' nothing more than an ancient primitive festivity. Both inferences are mistaken. The new religion transformed existing practice, but transformation is not destruction; it retains elements of the thing transformed.

One important implication of this, especially important at the present time perhaps, is that there is no possibility of reaching back into the past and 'recovering' the 'truly indigenous' religious practices of Scotland. This is a danger that protagonists of (somewhat misnamed) 'new age' religions are prone to. But a similar mistake has been more commonly made by enthusiasts for 'Celtic Christianity' who again have sought a 'truly' Scottish version of Christianity, one that predates the Synod of Whitby (664), when the authority of Roman over Celtic patterns of spirituality, liturgy and Church government was asserted in a way that eventually (not immediately) led to the disappearance of the Celtic Church. But the significance of Whitby for the study of Scottish religion is not its being a watershed for a native Scottish Christianity, but a striking instance of the interplay between Christianity as a universal and a local religion.

This interplay resulted eventually in the medieval Scottish Church, which, though it retained some residual Celtic elements, also acquired a European orientation, thanks to the assertion of Rome's supremacy. One of its manifestations was the establishment of the Scottish universities. Starting with St Andrews in 1411, these were founded with the purpose of making local provision for theological and professional education, while at the same time maintaining the connections that Scots students had formed with the universities of Europe, notably the university of Paris. Thus (for example) the Scot Hector Boece was brought back from Paris to be first Principal of the King's College Aberdeen (in 1495).

To a considerable extent, however, the precise nature of this interplay has to be guessed at. Literary and other sources relating to early Celtic culture are limited or obscure, and of the still earlier Pictish culture, almost nothing is known. For the purposes of the ethnology of Scottish religion, the lesson to be learned is one of caution – chiefly caution against a certain sort of romanticism. Here, in fact, we see in sharp relief, the difference between ethnology and other interests in the past; ethnology can never go beyond the careful assembly of what is known, and modesty of interpretation when what is known is limited.

THE PROTESTANT REFORMATION

If relatively little is known of the culture that Christianity encountered when it first arrived in Scotland, almost too much is known about the next major turning point in Scottish religion – the Protestant Reformation. That is to say, the quantity of relevant evidence is vast. Partly this is because the driving force of this huge change was mainly doctrinal rather than, say, political or economic, with the result that a huge number of theological writings were published, and extensive debate took place within a highly organised, nationwide structure – the Church.

The shift from Church to Kirk was significant in a great many ways and is arguably the event from which the culture that can most obviously be called distinctively Scottish arose. There is, however, a *caveat* to be entered. The modern image of Scotland, both at home and abroad, is almost exclusively that of romantic Highlandism, with its familiar icons of bagpipes, tartan, kilt and glen. This inclines us to forget the deep and long-standing division between Lowland and Highland Scotland that has marked most of the country's history, a division intensified by the Reformation. The Kirk was predominantly a Church of the Lowlands, partly because of numbers, literacy and prosperity. It was chiefly in the ancient universities and the principal cities of Scotland that the debate, and battles, about reform took place, and the Highlands and Islands had neither university nor major city. Consequently, Catholicism retained its hold there far longer than it did elsewhere, as did Episcopalianism (whose 'non-juring' adherents refused to swear an oath of allegiance to William of Orange), until the final defeat of the Jacobites in 1745. Ironically, even the current identification of Scotland with the culture of the clans is a product of the Lowlands – the enormous impact of Walter Scott. Ironically, Scott was able to promote the Highland image with such astounding success precisely because politically the Highland clans were a spent force. Nowadays, some parts of the Highlands and Islands (especially Lewis and Harris) are associated with a sternly Calvinist religion, but this is a much later, nineteenth-century product and not a residue of Reformation Scotland.

The move from Church to Kirk has two striking features – the introduction of Presbyterian government and an almost exclusive focus on the religion of the Word. These are both related to but distinct from the fact that the dominant theology was Calvinism. Unquestionably, Calvinist theology has had a profound influence on the history of ideas in Scotland and on some of its finest literary works; Hogg's *Confessions of a Justified Sinner* is possibly the best known but only one of these. Yet it is differing attitudes to Calvinism that can be said to lie at an important division in Scottish Church history between the Puritans and the Moderates. What both had in common, and what consequently proved to be the distinguishing mark of Scottish religion over a long period, was not theological doctrine so much as the form of Church government – Presbyterianism – and the style of Church worship – the ministry of the Word.

Presbyterianism is sometimes regarded as a type of democracy, or at least as democratic in spirit. This is not quite right, but it is certainly true that within Presbyterian government the role of the minister is rather different from that of the priest, and the role of the laity (in the form of ordained elders) correspondingly far more significant. In particular, the Kirk's assemblies, presbyteries and committees (some of them misleadingly named because of their sheer size), constituted *fora* in which the voice of the laity was as strong as that of the clergy, and sometimes stronger. This difference took on even greater significance after the Act of Union when the General Assembly and the Church and Nation Committee were the places where the people and the spirit of Scotland were more directly encountered than they were in Parliament.

It is impossible to assess or perhaps even guess at the full impact of Scotland's Presbyterianism. It is likely, for example, that it stimulated a more literate and reflective laity than in those societies where education was closely connected to the preparation for leadership or the professions, and perhaps for this reason it contributed to what subsequently came to be referred to as 'the democratic intellect' – a culture in which higher education was more easily accessed by the intellectually able, be they rich or poor. It is also plausible to hold that the existence of an ordained and theologically educated laity helps to explain the constant story of splits within the Kirk throughout the post-Reformation period, a rare occurrence in pre-Reformation Catholic Scotland. One of these splits was an event of enormous and continuing consequence – the Disruption of 1843 – whose significance we will shortly consider. Even anti-religious thought and feeling can be seen to be affected by the Presbyterian culture. There is a Scottish manifestation of anti-clericalism (at its most obvious in the novels of Grassic Gibbon), for example, quite different from the anti-clericalism of Catholic countries, and its chief enemy is not Calvinist theology, but the Presbyterian 'minister'. It is worth noting in this context that, while in France protagonists of the eighteenth-century 'Enlightenment', such as Voltaire, were for that very reason anti-clericals, several of the most important contributors to the Scottish Enlightenment (Thomas Reid and Hugh Blair are specially notable examples) were themselves ministers of religion.

The second major aspect of Reformation religion in Scotland lies in its being primarily a religion of 'the Word'. Once again the full ramifications of this are hard to trace, so far reaching and diffuse has its influence been. Notably, and famously, in its war on all elements of 'folk' practice, the religion of the Word led to the abolition of Christmas, and hence by a curious route, the re-establishment of the pagan festival of Hogmanay, with which Scotland subsequently became especially associated. It also led to a distinctive style of church architecture in which places of worship lost all their most striking visual and architectural features to become simple meeting places for the plain hearing and preaching of the Word. In the same spirit, a central place was given to the Bible in worship; with Psalms and Paraphrases (or other parts of Scripture) metrically arranged becoming Scotland's unique contribution to Christian worship in the wider world. The religion of the Word led to a strong and powerful emphasis on the intellectual elements of religion over the affective or the liturgical, and one result was the special relationship of ministers to the universities of Scotland which gave a central place to the education of clergy, and out of which (after seven or eight years) there emerged a 'clerisy' (to use Samuel Taylor Coleridge's term for the intellectual custodians of a tradition) more highly educated in theology and biblical studies than anywhere else in the Christian world.

Any religion of the Word has both strengths and weaknesses. The centrality of the Scriptures leads to a strong emphasis on literacy and academic education more generally. It also places a primary responsibility on clergy to prepare sermons of length and substance that are themselves highly educational. One result is an easy passage between ministry and learning, the transition for instance that the philosopher Thomas Reid and the Hebraeist Robertson Smith made among countless others. However, since the emphasis is on matters intellectual as the route to salvation, there is a tendency for other aspects of human personality to be unengaged – the poetic and the emotive for example. Further, when the theological content of belief is primary, theological differences more readily lead to splits because there is nothing beyond common theological subscription to bind people into a unity. Both these features can be seen to assume a new importance in the religion of Scotland as it faced two challenges common to the rest of Europe – the movement of peoples and the secularisation of culture.

MIGRATION AND SECULARISATION

There is an argument to be made for the view that the single most important event in Scotland's history since the Act of Union was the Disruption of 1843 when the national Church split in two. There had been many splits before, of course, as well as several after, and this one lasted less than one hundred years. Yet the magnitude of this particular division meant that for as long as it lasted there was no one denomination able to speak with the authoritative voice of a national Church. Moreover, it resulted in a massive dissipation of the resources and energies of the Church as its institutional and physical manifestations were systematically replicated across Scotland. In every place rival buildings sprang up, often a mere matter of yards from each other; and in the case of three of the four ancient universities, whole new divinity faculties came into existence. When the split ended, a single national voice proved hard to recover, while the legacy of having far more 'physical plant' than was required proved a constant drain on the re-united Church.

This burden came at a time of declining support and influence. This decline is unquestionably part of the Europe-wide phenomenon of secularisation. Just what the nature, extent and causes of this secularisation are is a difficult and intensely debated issue in both history and sociology. But it is possible to say something about the particular form secularisation has taken in Scotland and the special circumstances prevailing there. As the nineteenth century progressed, the divided national Church faced not just internal rivalries, but competition from a reviving Scottish Episcopal Church. This small denomination was the residue of a pre-Presbyterian Church that had been proscribed after 1745 because, like a few of their counterparts in England, its bishops would not renege on the oath of allegiance they had sworn to the Stewarts. This proscription was lifted in 1792, and in the course of the nineteenth century, the Episcopalians emerged from proscription to capitalise on the Anglo-Catholic revival south of the border. As a result the denomination expanded enormously in terms of both congregations and members. Alongside this, alternative Protestant denominations of an evangelical kind – notably Congregationalist and Baptist – also made strides in the urban areas of nineteenth-century Scotland. And then, in the second half of the century, came the huge influx of Catholics from Ireland, escaping the ravages of famine, and with them an extension of the Roman Catholic Church far beyond the confines of its pre-Reformation outposts in the Highlands and Islands. This expansion of Roman Catholicism was further strengthened by the later revival of other migrant groups, notably the Italians whose ice-cream parlours spread east and north from Glasgow in much the way that Chinese restaurants were later to do.

The Irish, the Italians, the Chinese proved to be just some of the new ethnic groups that came to Scotland, first as a result of the greater levels of migration that cheaper and better transportation made possible, and then as a result of the end of empire. Each of them has altered its religious makeup, some, of course, being more exotic than others. In contemporary Scotland there are Hindus, Sikhs and Muslims, as well as Christians and Jews. Although these late additions have had some visual impact (as, for example, in the construction of the large and impressive mosque in Potterrow, Edinburgh), they have continued to be identified largely with immigrant groups and have not, therefore, presented Christian Scotland with much competition for adherents. Their principal significance lies rather in their impact on the place of religion in public life. Whereas formerly the Kirk can be said to have been the centrepiece of Scottish social structure, this is not possible in a society expressly thought of as multicultural.

Contemporary Scotland is not as culturally heterogeneous as it is officially declared to be, but such cultural differences as there are, are primarily religious. The presence of the practitioners of religions other than Christianity has played an important part in generating a belief in multiculturalism. This has then been taken to sanction a relative relegation of the national Church to the status of one among many. Such relegation is possible only because of a more widespread decline in religious observance, but the two are not identical; when such figures as we have are accurately interpreted, it seems that, even at the height of Victorian church going, a majority of Scots never attended. The difference between then and now cannot be captured in statistics. It is, rather, a change in the social role and status of the Kirk, from central to peripheral, and from the public to the private sphere. *This* change, which is perhaps the greatest point of contrast between past and present, owes something to migration as well as to secularisation.

THE FUTURE OF RELIGION IN SCOTLAND

Arguably the most fruitless of intellectual endeavours is the attempt to predict the future, since human beings have regularly proved spectacularly bad at it. This has never prevented people from engaging in it, however. In the 1980s it was predicted that, given existing rates of decline, the Church of Scotland would cease to exist in 2047. Subsequently this estimate was revised to 2032, an indicator of the Kirk's continued failure to attract new members. Yet we know that the weakest basis for prediction is the projection of existing trends. The complete disappearance of the Kirk seems to me unlikely. More likely is its ceasing to be a national Church in anything but name.

However this may be, it can be said with certainty that a marked feature of contemporary religion in Scotland is its marginalisation from the life of the major institutions. The restored Scottish Parliament decided against formal prayers after the fashion of Westminster; the University of Edinburgh has abandoned the practice of prayers at Graduation; civic visits by the Moderator of the Church of Scotland are not the quasi-State occasions that they were; hundreds of city centre church buildings have been converted to flats, offices, pubs and clubs. None of this rules out religious revival, or the recovery of individual denominations and Churches. But it does seem that the place of religion in the Scotland of the future will be radically different to its place in the past. And this means that, after a time, compendious though this volume is, the ethnology of Scottish religious life will need to be undertaken afresh.

PART ONE
●
Consolidation

1 Early Christianity in Scotland: The Age of Saints

ALAN MACQUARRIE

The origins of Christianity in Scotland are very obscure. We think of the period of conversion, from the fifth to the seventh century, as 'the Age of Saints', in which individuals made more impact than institutions, and charismatic missionaries spread the gospel as solitaries or in small groups. In fact, the picture may be more complex. Our main literary sources, the hagiographical writings of a later age, are notoriously difficult to use.[1] There was an interval of several hundred years between the careers of the individual missionaries and the drawing up of their *vitae* in a form acceptable to medieval churchmen and their lay patrons. The fact that most of the 'saints' whom we will be considering belonged to the late fifth, sixth and early seventh centuries parallels the position in Ireland, where the majority of saints for whom *vitae* were composed lived before 650.[2] It is worth pointing out that likewise Irish learned secular tales set in a period after the mid-seventh century are exceptional.[3] Of the full-length Scottish *vitae* to have survived, the latest chronological location is found in *The Life of St Serf*, which is set in the period of the Pictish king Bridei son of Derilei and St Adomnán (d. 704).[4]

Given such a gap, there is little likelihood of strict historical accuracy in such *vitae*, even if that was what the compilers had been aiming at. But medieval hagiographers were usually more concerned to edify, amaze and entertain than to provide historical data, and their sources were usually fantastic and mythological in any case. Where twelfth-century or later writers had an earlier Celtic *vita* on which to draw, that *vita* usually consisted of a series of miracle stories about the saint which were drawn from a common pool of such stories, many of them based on biblical models from both the Old and New Testaments. These stories are usually located at sites or churches which were claimed as part of the ecclesiastical jurisdiction of the saint's successor, and read at times almost like a collection of the title deeds of the saint's principal church. It has been remarked that in their *vitae* saints are portrayed as figures of an earlier age, heroes whose like is not to be found at the time of writing.[5]

CHRISTIANITY IN ROMAN AND POST-ROMAN NORTH BRITAIN

But the Age of Saints is not the beginning of the story of Christianity in Britain. Some early Christian writers speak of Christianity in Britain: Tertullian, c200, speaks of 'parts of Britain inaccessible to Rome made subject to Christ'; while Origen, c240, asserted that Christianity was a unifying factor in Britain, bringing it closer to Rome.[6] Gildas, writing c540, describes the martyrdom of St Alban, a Roman officer at Verulamium (now St Albans) during the persecution of Diocletian (303–12); he also mentions the martyrdom of Aaron and Julius, citizens of Caerleon, and adds that there were many other graves of martyrs in lands which were later occupied by the Anglo-Saxons.[7] Since doubt has been cast on whether there was any persecution in Britain during the period 303–12, it has been suggested that the martyrdom of St Alban may belong rather to the reign of Septimius Severus (193–211) or some other time later in the third century.[8]

In the fourth century the picture becomes clearer. In 314, within a year of Constantine's edict of toleration, three British bishops attended the Council of Arles, accompanied by lesser clergy. Clearly a church hierarchy already existed in Britain. According to Athanasius, British clergy attended the councils of Nicaea (325) and Sardica (342–3). Three British bishops attended the Council of Rimini in 359.[9]

Archaeology provides some clues which help to confirm and fill out this picture. In Roman villas, mosaics have been found which could bear a Christian interpretation. A silver hoard of Christian vessels was buried at Durobrivae (Water Newton, Cambs) in the Nene valley some time before 350. From the same period come the fine Christian friezes in a villa chapel at Lullingstone in Kent, and floor mosaics depicting Christ in majesty at Hinton St Mary.[10] The cumulative evidence suggests 'an educated and wealthy Christian society in the Romano-British countryside in the second half of the fourth century'.[11]

It was this society which produced Roman Britain's only noteworthy theologian, Pelagius. Pelagius was born somewhere in Britain c360 and died probably in the 420s having lived most of his life outside his native land. We do not know how far his British origins influenced Pelagius' thinking; he emerges as a thoroughly Mediterranean scholar and thinker, perhaps more heavily influenced by Origen than anyone else.[12]

It is important to remember, however, that most of the land which we now call Scotland had never been part of the Roman province, and was only remotely affected by Roman civilisation. The Flavian conquests under Agricola, who fought against a Caledonian confederation, were largely abandoned within a few years, and Hadrian commissioned his stone wall along the Tyne–Solway line c120. The Antonine decision to recommission Agricola's Forth–Clyde frontier with a permanent wall c140 was also short-lived, although the northern Wall may have been reoccupied several times during the second century.[13] By the end of this century we hear of a tribe called the Maeatae, dwelling close to this Wall, raiding into Roman Britain with the help of the Caledonii, who dwelt beyond them. Septimius Severus (193–211) campaigned more extensively in Scotland than anyone since Agricola, but in the end he decided to recommission the Hadrianic frontier. Thereafter there was peace between Roman Britain and the tribes beyond the Wall for most of the third century.

We know little about the tribes living beyond Hadrian's Wall. Ptolemy's *Geography* locates four tribes in the area between the two walls: the Novantae in the south-west (Galloway), the Selgovae in the Southern Uplands, the Damnonii in Strathclyde, and the Votadini in Lothian. Beyond the northern Wall he names a large number of tribal groupings, giving the Caledonii a prominent and central place. By the time of Severus the northern tribes were divided into two great confederations, Maeatae (Dumyat in Clackmannanshire was presumably one of their strongholds) and Caledonii (who have left several place-names in the central Highlands, which is where Ptolemy's map locates them). By the end of the third century we hear for the first time of 'Picti' and of the 'Caledonians and other Picts', and in the 360s there is a reference to the northern tribes being divided into two groups, Dicalydonae and Verturiones, who again attacked Hadrian's Wall. The Romans had apparently recruited native scouts whom they called *areani* ('field-dwellers' or perhaps 'desert-dwellers') 'to penetrate deep into enemy territory and give our commanders warning of the movements of border peoples'; but these 'were convicted of taking bribes to betray our army to the enemy' and had to be disbanded.[14]

About this time archaeology detects changes in the nature of the defence of the Wall, with regular legionary camps giving way to fortified villages held by militiamen and their families.[15] Evidence for Christianity from this more militarised northern zone remains thin. But most of the fourth-century emperors had encouraged Christianity among the army, and the two usurpers from the period who were stationed in Britain, Magnus Maximus (383–8) and Constantine III (407–10), commanded the garrison on the Wall and were Christian themselves. A few Christian symbols have been found at military sites in northern England, but before the fifth century the evidence from the north is very slight.

By *c*400, if not earlier, Hadrian's Wall appears to have been abandoned, and Stilicho's 'Pictish War' of 400–2 was the last attempt to hold the frontier.[16] Soon after came the usurpation of Constantine III, the great barbarian invasion across the Rhine in 406–7, and the sack of Rome by the Goths in 410. Britain was instructed by the emperor to look to its own defence. There seems to have been a rapid deterioration in British culture, with coinage for trafficking going out of use within about a generation; the province passed into the hands of local 'tyrants' who carved it up and ruled their tribal areas from hill-forts.

The fifth and sixth centuries have been described as 'lost centuries' lying between the end of the Roman occupation (*c*410) and the documented histories of the barbarian kingdoms of Great Britain (*c*550).[17] Written sources for this period are very sparse, being confined mainly to the writings of Constantius, Gildas and Patrick, the later compilation known as 'Nennius', and still later British legends and genealogies.

Constantius' *Life of Germanus* describes how the heretical teaching of Pelagius became so influential in his native country (which he had left long before) that Germanus of Auxerre had to come to Britain c429 to preach against it and restore the Britons to Christian orthodoxy. He was met by a sophisticated and Romanised aristocracy at Verulamium led by a man 'of tribunician power', and there is a suggestion of a considerable amount of factious feuding among the Britons. About ten years later he revisited Britain and was again preaching to the British aristocracy against Pelagianism when a report of a raid by Picts and Saxons reached them; Germanus allegedly led the faint-hearted Britons into battle and gave them fresh courage by teaching them the battle-cry 'Alleluia', by which they gained the victory.[18] The contrast between the relative peace of Germanus' first visit in the late 420s and the unsettled conditions of his second are perhaps suggestive of changed times. Also by the time of his second visit (perhaps c440) the Saxons seemed to be well established and the Picts from the north were raiding deep into formerly Roman Britain. But there is no suggestion that either Saxons or Picts were influenced by Roman civilisation or by Christianity.

The evidence of St Patrick's writings suggests that some kind of Romanised administration and Christian society persisted in Britain in Patrick's youth, and that by his old age even barbarian warlords on the periphery of formerly Romanised Britain pretended to some attachment to Christianity and Roman citizenship: 'I do not call you my fellow-citizens,' he writes, 'or fellow-citizens of the holy Romans, but fellow-citizens of demons' – a taunt which would have had no point if they had not aspired to be thought of as Romans.[19] The difficulty of using the evidence is that the dates of Patrick's life and career, his place of origin and the extent of his mission, are the subject of lively controversy, in which there is no general agreement. What is accepted is that he was a Romanised Briton who spent part of his adolescence as a captive in Ireland and was later active as a missionary in Ireland some time in the fifth century (certainly not later than the conversion of the Franks c496).[20]

From early in the fifth century, however, comes one very important piece of archaeological evidence from the north: the Traprain Hoard. This is a collection of silver plate weighing over 117 kg (53 lb), containing pieces of more than 150 different objects. It also contains several coins, which point to a date for the burial of the hoard of around AD 410–25, or possibly slightly later. Some of the silver vessels had been cut up and the hoard was tightly packed in a pit in the ground, as if the silver was being prepared to be melted down, and then rapidly buried. The burial of the semi-dismantled hoard in a shallow pit is indicative of unsettled times.[21]

A number of interpretations is possible. The hoard could have been booty brought back by raiders from across the Wall. Or it could have been domestic silver used as bullion to pay native mercenaries or allies of the

Figure 1.1 Nineteenth-century drawing of the Catstane, Kirkliston, Midlothian, possibly fifth century. The inscription has been read: *IN OC TUMULO IACIT VETTA F[ILIA] VICTI* (In this tomb lies Vetta daughter of Victus). Courtesy of the Trustees of the National Museums of Scotland, NMS SLA C14251.

imperial province. Or it could have been the domestic silver of a powerful and wealthy chief of the Votadini (the tribe located in this area by Ptolemy in the second century), reflecting his culture and taste, which had been broken up and buried during troubled times. The archaeology of the hill itself, however, does not solve this question; neither do Christian symbols on some of the items necessarily reflect the taste and religion of the Votadinian chief of Traprain *c*425. Some of the objects, moreover, have overtly pagan symbolism. They do show, however, that Christianity was within reach of the lands north of the Wall by this time, and within a generation or so we should expect to be finding more concrete signs of Christianity.

In a broad sweep through an area to the west and south-west of Traprain we find a group of some half-dozen funerary monuments spaced south to north from Chesterholm on the Wall to the Catstane at Kirkliston (Fig. 1.1); it has been argued that 'these are native memorials' and range from the fifth to the seventh century.[22] The inscriptions of Peebles and Yarrowkirk presumably indicate the presence of Christianity among the people whom Ptolemy calls the Selgovae. That of the Catstane at Kirkliston with its cemetery is evidence for Christianity among his Votadini, quite possibly in the fifth century; and the ambiguous evidence of the Traprain Hoard may possibly, but does not necessarily, point in the same direction. But the most impressive epigraphic and archaeological, as well as literary, evidence for early Christianity comes from Galloway, the south-west of Scotland, where Ptolemy places the tribe which he calls the Novantae.

A number of years ago a memorial stone was found at the small cathedral town of Whithorn in the southern Machars of Galloway, bearing an inscription in Latin in Roman capitals which can be translated: 'We praise thee, O Lord. Latinus, 35 years of age, and his daughter, four years of age. Here the grandson[s] of Barrova[n]dus made [this] monument.' On epigraphic grounds this inscription has been dated to the mid- to late-fifth century, although there is no universal agreement about its date. It is thus possibly even earlier than a group of monuments that have been found at a similar site at Kirkmadrine in the Rhinns of Galloway, which commemorate Christian clergy: Ides, Viventius, Mavorius, Florentius, and a deacon (or possibly subdeacon) called Ventidius (Fig. 1.2). At Whithorn there exists another stone, probably slightly later in date (seventh or even early eighth century), which proclaims that it is the '[Cross] of the *locus* of St Peter the Apostle'.[23]

Archaeological evidence is coming increasingly to suggest that the site of Whithorn began as a greenfield site settled by incomers *c* AD 500. The incomers had access to some of the technology of the Roman Empire; their place of origin is uncertain, but Romanised Britain or northern Gaul seems most likely.[24] The origins of Whithorn seem therefore to belong to the respite period *c*490–540, when there was a modest revival of British fortunes and a recollection of the past glories of the imperial province. This perhaps suggests that the settlers at Whithorn possibly were not fleeing from external expansionist pressures, but were themselves expanding confidently from secure bases. It may be that this was a period of Christian missionary work among the peripheral barbarians. Bede denounces the Britons for their failure to convert the Anglo-Saxons, but this may simply reflect his prejudice. Patrick, whose career was unknown to Bede, lived probably a generation or more before Ninian, but he laid foundations on which others were to build. We know of Vinnio or Vindobarr, venerated in Ireland as St Finnian, a British missionary active in Ireland in the first half of the sixth century, who corresponded with Gildas and who taught St Columba.[25] Gildas denounces the clergy of his own day as lazy and luxurious, while excepting 'a very few good pastors'; but he speaks of widely travelled, foreign-educated presbyters as well. Such clergymen would have been aware of the newly fashionable cult of St Martin in Gaul, which spread to Rome during the pontificate of Symmachus (498–514), and later to England and Ireland.

ST NINIAN

Geographically, Whithorn does not look like a base for missionary activity among the Picts; with Kirkmadrine, it points rather towards Ireland. When

Figure 1.2 Replica of the stone commemorating Ides, Viventius and Mavorius, Christian clergy. Kirkmadrine, Rhinns of Galloway, early sixth century.

we move from the anonymous world of archaeology to that of named Christian missionaries such as Ninian, we are moving into the 'Age of Saints' *per se*. It is to St Ninian that we must now turn.

A number of years ago Professor A A M Duncan attempted to reconstruct Bede's sources of information for northern events, and in particular for St Ninian and the conversion of the Picts.[26] This highly original work has had few followers. Although Professor Duncan demonstrated (among other things) that Bede's chief source of information about the Picts was a letter running under the name of Nechton king of the Picts to Abbot

Ceolfrith of Jarrow, this point has been ignored by most scholars, who argue that Bede's only source about St Ninian was Pehthelm, bishop of Whithorn.[27] Pehthelm, they argue, had access to a Celtic *vita* of Ninian drawn up at Whithorn *c*550–650.[28] Although one can certainly make a case for a *vita* underlying the eighth-century *Miracula Nynie Episcopi* and the twelfth-century *Vita Niniani*, there is no reason for believing it could be as early as the sixth century. It has been shown that the earliest *vitae* of Celtic saints are Cogitosus' *vita* of St Brigit of Kildare, Cumméne's *Liber de Virtutibus Sancti Columbe* and Muirchu's *vita* of St Patrick.[29] Not until the last quarter of the seventh century was the fashion of writing *vitae* of founders well established among the Celtic peoples. So the early Celtic *vita* of Ninian, assuming it existed, is unlikely to be much earlier than *c*700. Like all Celtic *vitae*, it will have been written long after the death of its subject.

If it may be doubted that such a *vita* could be so early, one must also doubt that it could have been Bede's source. Let us remember what Bede tells us, and the context in which he tells it. His words are:

> In the year 565 ... there came from Ireland a priest and abbot named Columba, a monk by his habit and his way of life, coming to preach the word of God to the provinces of the northern Picts, that is, those who are divided from their southern regions by a range of rough mountains. The Southern Picts, however, who live on this side of these mountains, had, so they claim (*ut perhibent*), received the word of faith a long time before and abandoned the error of idolatry, through the preaching of the word to them by Nynia, a most holy and reverend bishop, a man of the British race, who was regularly instructed at Rome in the faith and mysteries of truth. His episcopal see is celebrated for its church and the name of St Martin, where his body rests with those of many saints. The English race now [*c*730] holds it. This place, belonging to the province of Bernicia, is commonly called *ad Candidam Casam* ('At the White House'), because he built there a church of stone, of a kind unknown among the Britons. Columba came to Britain while Bridius *filius* Meilochon, a very powerful king, was ruling over the Picts ...[30]

Bede is little concerned in this passage to tell us about the foundation of Whithorn. His concern is the conversion of the Picts; it is for this reason only that he introduces Ninian as a missionary independent of Columba, and he only mentions *Candida Casa* in passing as the site of Ninian's see and place of his burial.

The bias and emphasis in this passage is obviously Pictish; the spelling of Bridei's name, *Bridius filius Meilochon*, is very close to that of the 'Poppleton' MS (*Breidei filius Mailcon*), which appears to conserve Pictish spellings mostly free from the influence of Gaelic orthography[31] (cf. Adomnán's *Brudeus, Bruideus* – never with a patronymic – and *Annals of Ulster's* [*AU*] *Bruide mac Maelcon*). He is flatteringly described as 'very powerful', and he is credited with granting Iona to Columba, even though what appears to be Iona's own account of its foundation, found in the *AU* sa 574, credited the grant rather to Conall mac Comgaill, king of the Dál Riata.

So why does Bede mention Ninian at all? Not because Pehthelm of Whithorn had told him a contradictory story about the conversion of the Picts, which he inserted into his otherwise 'Pictish' source, but rather because the Picts themselves included a mention of Ninian in their letter to Ceolfrith. In fact, Bede appears to attribute his information about the conversion of the Southern Picts to the Southern Picts themselves: 'The Southern Picts, so they claim, had long before abandoned the error of idolatry and received the true faith through the preaching of the word to them by Bishop Ninian ...'.[32] The mention of Ninian does not look like an interpolation in or contradiction of an otherwise Pictish account of the conversion of the Picts. It in fact agrees in broad terms with Adomnán's account of Columba's dealings with the Picts, which involve him travelling up the Great Glen to King Bridei's hall somewhere near Inverness.[33] As far as Adomnán was concerned, the Miathi who dwelt around the head of the Firth of Forth were *barbari*.[34] Neither Adomnán nor the Picts seem to have been aware of Columban activity between Forth and Mounth (something of the sort may be obscurely suggested in the *Amra Coluim Chille,* although this is uncertain).[35] There is ample evidence, however, in place-names and archaeology for British missionary activity among the Picts south of the Mounth, and it would be incredible that the Picts themselves had forgotten this or for some other reason failed to mention it in their letter to Ceolfrith.[36] The name of Nynia, Ninian, is so obscure that it can hardly be a Pictish invention; it looks like an attempt to reproduce the British name Ninniaw (later Latinised as Nennius). (The possibility that it could be a confusion or misreading of the name Uinnio or Finnio will be considered below.)

But although there can be little doubt that the Picts told Ceolfrith that they had been converted by a British bishop called Nynia, it is unlikely that it was the Picts who told him that Ninian was buried *ad Candidam Casam*, at Whithorn, in a church dedicated to St Martin, where he was surrounded by the burials of many saints. This does look more like a story from Whithorn itself, and as such must have come to Bede from Pehthelm. 'The English race now holds his episcopal see', writes Bede; 'it belongs to the province of Bernicia'. This clearly is not information from the Picts. Professor Duncan points out that Bede's alliterative Latin *ad Candidam Casam* appears to derive from Old English *æt ðæm hwitan ærne,* 'at the white house', at Whithorn.[37] Presumably this represents an English translation of a British name, possibly involving the Celtic element **loukos,* 'white', cognate with Latin *luceo, lux,* English *light,* etc. Ptolemy's *Geography* records that there was a place in Galloway called *Loukopia* (or something similar), but its identification with Whithorn, though attractive, must remain uncertain.[38]

So although Bede learned of Ninian's missionary activity among the Picts from the Picts themselves, presumably in a written source containing the name *Nynia,* he also knew of his association with Whithorn, apparently from an English source. The use of two sources explains the apparent contradiction between the statement in *HE* iii, 4 that Ninian had been bishop of Whithorn and the statement in v, 23 that Pehthelm was the first bishop of Whithorn.

How did Bede get this information from Pehthelm? Pehthelm is mentioned several times in Bede's writings. In *HE* v, 13, Bede recounts a long story which took place in the time of Cenred king of Mercia (704–9), which he tells 'as I learned it from Bishop Pehthelm'. Elsewhere (v, 18) Bede records miracles which occurred at the place where Bishop Hædde of Wessex had died (c705), which 'reverend Bishop Pehthelm used to relate'. Bede tells us that Pehthelm was for a long time deacon and monk with Hædde's successor Aldhelm, bishop of Wessex at Sherborne. He appears to have come from southern England; Professor MacQueen has demonstrated that his name, despite a striking coincidence, probably has nothing to do with Picts.[39] Although Bede promises that he will say more about Pehthelm, he in fact only mentions him once more briefly, in *HE* v, 23, where he is enumerating the bishoprics of Northumbria as they stood in 731; Pehthelm has recently (*nuper*) become first bishop of *Candida Casa*, because of an increase among the faithful there. In the first two cases, Bede is repeating stories which he has heard told by Pehthelm; in the third, he does not tell us how he came by the information. It is likely, however, that he learned this *via* a messenger from Whithorn. The erection (or revival) of a bishopric there under an English bishop would have been welcome news to the churches of Northumbria.[40] Certainly there is nothing to indicate that Bede's information about the revived bishopric of Whithorn came from a written source.

So some time around 731 Bede learned from Pehthelm about the new bishopric *æt ðæm hwitan ærne* (at the white house). This he turned into elegant alliterative Latin as *ad Candidam Casam*. This is now Whithorn in Galloway. Presumably he also learned from Pehthelm that there was a church at Whithorn dedicated to St Martin and that it was the burial place of a saint called Nynia, who rested there 'together with many saints'. Nynia was associated with the building of a stone church there. Pehthelm may not have told Bede that Nynia was a bishop, since Bede later assumed that Pehthelm himself was the first bishop at Whithorn.

Bede identified this Nynia, builder of a stone church at Whithorn, with a British bishop of the same name of whom he had learned in a letter from Nechton king of the Picts to the abbot of Jarrow. This letter claimed that Nynia was a bishop, regularly trained at Rome, who had converted the Southern Picts to Christianity 'a long time before' Columba's visit to the Northern Picts.

Was Bede correct to identify the British bishop Nynia, preaching to the southern Picts 'a long time before' Columba, with Nynia, the builder of a stone church dedicated to St Martin at Whithorn? The answer is that we will probably never know for certain, but we are allowed to advance probabilities. The first point is that it would probably be unwise to multiply unnecessarily persons of the same name, roughly contemporary with one another; if Bede knew from two sources of a person called Nynia operating in northern Britain c500–50, we do not need to assume two separate persons called Nynia operating in the same area and period. The fact that Whithorn looks towards the Irish Sea, to Man, to the Rhinns and to Ireland, need not distract us; Dark Age travel and communication was possible by land as well as by water. We see from the careers of many of the figures who appear in Bede's *HE* that wide travel was commonplace, and that men often spent different parts of their careers in different places. Gildas, describing the clergy of exactly this period, says that they delight 'to cross the seas and travel through wide lands ... then they return home'.[41] It would be characteristic of the age for one man to have spent some time travelling among the Picts preaching and baptising and to have ended his days across the mountains beside the Western Sea facing towards Ireland. The most obvious example of such a career is that of Columba. Nynia or Ninian, assuming there was just one person of the name, could provide a close parallel.

Is there a possibility that Ninian's name, *Nynia*, is the result of scribal confusion or error? In the first half of the sixth century (the period to which, I have argued elsewhere,[42] we should assign Ninian) there was an important British bishop and missionary active in Ireland called probably *Uindobarros* or (in a shortened form) *Uinnio*.[43] In Ireland he came to be known as Finnio, and he may underlie the saints called Finnian who were venerated at Clonard and Moville, and also St Finbar of Cork.[44] Some later Irish traditions link St Finnian of Moville with Whithorn.[45] Given the ease of confusion between *u* and *n* in written sources, is it possible that St Nynia or Ninian of Whithorn is another manifestation of the same personage?[46]

This is a question which I discussed in print a number of years ago, and which has more recently been revived by others.[47] I think this is a possibility which would have to be taken very seriously if Bede had only had one written source for St Ninian, in which an original **uinia(u)* (or something similar) could have been misread as *nynia* or *ninia*. But we have seen that Bede's information about Ninian did not come from a single source. He had a letter from King Nechton to the abbot of Jarrow which explained, among other things, why the Picts did not venerate Columba as their only apostle, but also remembered a British bishop called Nynia; and he had an account, probably a verbal message, from Pehthelm of Whithorn, telling him that somebody of the same name had built a stone church dedicated to St Martin at Whithorn, where he was buried together with many saints. A single written source (such as Nechton's letter) could possibly have led to such a scribal error; a corroborating source, probably verbal, makes it virtually impossible.

At the time of the Anglian occupation of the Machars the population would have been British speaking. Possibly, as has been argued,[48] a *vita* of Ninian was already in existence, which the Anglian monks versified as the *Miracula Nyniae Episcopi* and of which a version was later turned into Latin prose as the *Vita Niniani* attributed to Ailred of Rievaulx.[49] Bede did not have access to this *vita*, but he spoke to people who had access to the traditions of Ninian preserved at Whithorn. These people will have pronounced the name correctly (the attested British name *Ninniaw*), and it will have appeared as *Niniauus* or something similar in the putative Whithorn *vita*, giving rise to the twelfth-century spelling *Ninianus* and our modern name Ninian. Bede's spelling *Nynia* was presumably found in Nechton's letter to Ceolfrith. He identified this person, probably correctly, with the person buried at Whithorn whom he had heard about from Pehthelm.

It is noteworthy that Ninian's connection with Whithorn was not widely known when Bede was writing. Pehthelm was the recipient of a letter *c*735 from Boniface, bishop of Mainz, which shows that he was regarded as an authority on canon law. The letter contained gifts for the bishop's church, as was customary at the time, but no mention is made of St Ninian; evidently Boniface had not heard of the saint buried at Whithorn.[50] This contrasts with Ninian's fame some fifty or sixty years later, when Alcuin wrote to the monks of Whithorn about their patron, and sent a gift specifically to his shrine.[51] Ninian owes his fame to Bede's account of him, and Bede is our only real source for St Ninian. Bede's interest in Ninian is above all as the missionary to the Southern Picts. Elaborate hypotheses about Ninian's identity and activities which ignore these facts cannot be convincing.

ST COLUMBA

The best known figure of the 'Age of Saints' in Scotland is undoubtedly St Columba. Although we possess a eulogy of the saint written shortly after his death, it is obscure and difficult to assess.[52] Most of our knowledge about Columba comes from the *Life* written by Adomnán, abbot of Iona, about 700, a century after the saint's death.[53] Additional information is provided by Bede, writing some thirty years after Adomnán;[54] and by the collections of Irish annals, which seem to embody material compiled at Iona down to the mid-eighth century.[55] Adomnán had access to at least one earlier written source, a book of *The Virtues of St Columba* written by his predecessor Abbot Cumméne (657–69), but the bulk of his narrative is probably based on oral tradition collected among the monks of Iona during the time of Abbot Ségéne (*c*624–52).[56] Later hagiography tells us much about how later generations viewed Columba, as a powerful miracle worker and ascetic, but tells us less of historical value for his life.[57]

Little is known about Columba's career in Ireland. He was born probably in 521, into the family called the Cenél Conaill, which ruled over much of County Donegal (the name is still preserved in the barony of Tirconnell).[58] At an early age Columba was fostered to a priest, and in his youth he studied under some of the greatest ecclesiastical teachers in Ireland at the time, including St Finnian.[59] It is uncertain whether he began his career as a founder of monasteries before he came to Scotland. Bede states that the Columban monastery of Durrow (County Offaly) was founded before Columba left Ireland, but Adomnán implies that it was founded, and the buildings were in process of being built, after Columba settled at Iona.[60] It seems that Columba's visit to Clonmacnoise during the abbacy of Ailither (c586–99) was connected with the foundation of Durrow.[61] Although there was later a Columban monastery at Derry, the date of its foundation is uncertain. Since Iona 'held the pre-eminence' over all other Columban foundations, it seems likely that Iona was in fact the earliest of Columba's houses.

A turning point in Columba's life seems to have come in 561, the year of the battle of Cul Dreimne. This battle was fought between branches of the royal house of the Uí Néill, of which Columba's own Cenél Conaill was part, and Columba was held responsible for instigating the battle and for the bloodshed involved.[62] The details of the battle and its causes are very obscure; but it seems to have resulted in considerable unpopularity for Columba, for in the following year (562) he was excommunicated by a synod of Irish clergy meeting at Tailtiú (now Teltown, County Meath).[63] Adomnán, perhaps not surprisingly, is rather coy about these events. Although Columba's excommunication was not long-lasting, and he was defended at the synod by St Brendan of Birr, it was probably these events which decided Columba to leave Ireland. So in 563, accompanied by twelve companions, he sailed away from Ireland to the court of Conall mac Comgaill, king of the Dál Riata in Scotland.[64]

From this story Columba emerges as a political exile, as much, if not more than, as a pilgrim saint. And although the Dál Riata were not related to the Uí Néill, they shared a common Gaelic language and culture. Adomnán's comment, 'he sailed away from Ireland to be a pilgrim',[65] is expanded in the *Secunda Praefatio* to read 'he sailed away from Ireland to Britain, wishing to be a pilgrim for Christ'.[66]

Later tradition has it that Columba sailed directly to Iona, and landed at Port na Curaich on the south side of the island. But this appears to be contradicted by Adomnán's statement that on his first arrival in Britain Columba resided with King Conall,[67] and the Irish annals state that it was Conall who gave him Iona for the foundation of his monastery.[68] Iona was probably established as Columba's home and monastic centre soon after his arrival in Scotland.

Adomnán has little to say about Columba's relationship with Conall after his initial (probably brief) residence at court. Perhaps these early years in Scotland were spent in work at Iona and on the foundation of other island monasteries in the southern Hebrides. Columba's secular prominence may date from the death of Conall in 574, when he was involved in legitimising the succession of Conall's cousin, Aedán mac Gabráin, by a service of laying on of hands at Iona.[69]

This ceremony is one of the earliest recorded instances of a barbarian king being inaugurated in a Christian ritual, and as such the story deserves special attention. Not surprisingly, Adomnán's account has led to some incredulity.[70] There is no suggestion in Adomnán's account, however, that Columba took part in Aedán's election. Adomnán merely states that Aedán came to Iona and met Columba there; Columba accepted his kingship as a *fait accompli*, and 'laying his hand upon his head he ordained and blessed

Figure 1.3 Crosses of St John (replica), St Matthew, and St Martin, Iona, eighth century. Courtesy of the Trustees of the National Museums of Scotland, NMS SLA C14252.

him'. There is no suggestion, either, that Aedán did not undergo the normal secular inauguration of a Gaelic king, which probably would have involved oath-taking and acclamation, the recitation of the king's genealogy and the setting of his foot in a sacred carved footprint (one is still visible at Dunadd in Knapdale; another, fancifully associated with St Columba, is extant near Southend in Kintyre). It cannot be determined whether his 'ordination' by Columba preceded or followed such a ceremony. The point is, however, that Columba does not emerge as a kingmaker; the allegation that Adomnán promotes 'an inflated interpretation of the rights and powers of Iona abbots with regard to kings' overstates the claims that Adomnán is in fact making.[71]

Subsequently, Columba and Aedán seem to have co-operated closely on a number of occasions. At some point in his reign (the Annals give 575, but a date in the early 590s would fit the circumstances better), Aedán took Columba with him to a meeting with Aed mac Ainmirech, king of the northern Uí Néill (and thus Columba's kinsman), at Druim Cett near Limavady in northern Ireland, at which Columba acted as Aedán's adviser in matters concerning the relationship of the two kings.[72] It has been shown that much of the later legend which surrounds the Convention obscures the fact that it was a secular gathering, a meeting of kings to discuss matters of taxation and military service.[73] In fact the main purpose of the meeting may well have been to conclude a military alliance between the Northern Uí Néill and the Dál Riata against the Ulaid, one of the most powerful Ulster peoples. Columba's role will have been to act as a mediator between King Aedán and his cousin Aed mac Ainmirech of the Northern Uí Néill.[74]

Aedán's activities may have affected Columba's relationships with kings in Scotland at the time as well. On at least one occasion he visited Bridei, king of the Northern Picts, who had a royal fortress near the River Ness; the hilltop of Craig Phatric beside Inverness has been suggested as the most likely location.[75] Adomnán records a number of miracles which occurred at King Bridei's court and elsewhere in Pictish lands.[76] During the same visit Columba asked for safe-conduct for the pilgrim monk Cormac, seeking a hermitage in the northern ocean; he asked Bridei to instruct the sub-king of the Orkney islands, who was at his court and of whom Bridei held hostages, not to molest Cormac should he come into his territories.[77]

The reference to the Orkney islands is of interest, because we know that one of King Aedán's most ambitious recorded military exploits was a raid on these islands c580.[78] It is tempting to see Columba's visit to Bridei's court as a diplomatic mission in connection with that event. Embedded within the narrative is mention of another purpose, the release of at least one Gaelic female captive. Adomnán himself engaged in redemptive activity of this kind,[79] so it is disappointing that he does not say more about Columba engaging in such work. There are hints in Adomnán's narrative about unsettled conditions in Glen More which might have involved slave-raiding,[80] and it is easy to conceive Aedán's attack on the Orkneys as retaliation for such a raid. The evidence would be consistent with Columba making a diplomatic visit to Bridei's court near Inverness, perhaps c580, in connection with raiding and counter-raiding and the redemption of captives. Although Adomnán does mention a small number of Pictish converts made by Columba (e.g. stating that he converted and baptised a Pictish household in the region of Urquhart when he was travelling beside Loch Ness),[81] he does not speak of a large-scale conversion of Picts such as is described by Bede. Most crucially, he does not mention the conversion and baptism of Bridei himself. So it must be doubted if the conversion of the Picts took place on any large scale until after Columba's death, or if his visit or visits into Pictish territory were intended as missionary journeys.[82] If this was in fact a diplomatic visit, as appears more likely, it follows that King Aedán was able to call on Columba's services for work of this kind, and this tells us something about the relationship between king and abbot.

On another occasion, Adomnán describes how Columba received a secret message from Rhydderch Hael, king of the Britons of Dumbarton, described as a friend of the saint, through one of his own monks.[83] Rhydderch sought to know whether he would die by the hand of his enemies, and Columba assured him that he would die in peace in his own house. The Britons of Strathclyde were Christians by this time, which may explain why Rhydderch is called Columba's friend. But there are known instances of hostility between Aedán and Rhydderch in this period, such as a Welsh 'triad' which describes how Aedán by an 'unrestrained ravaging ... came to the court of Rhydderch Hael at Dumbarton; he left neither food nor drink nor beast alive'.[84] It is possible either that Columba was assuring Rhydderch that Aedán would not repeat an earlier raid against him, or else that Aedán raided into Strathclyde in spite of Columba's assurances to the contrary. There is a lost tale *Orgain Sratha Cluada* (The Slaughter of Strathclyde), which appears in learned lists of Irish tales, which may refer to the same incident.[85]

Many of Adomnán's stories about his dealings with aristocrats and laymen of lesser rank concern acts of kindness rewarded or slights punished. But Adomnán also speaks at length of the monastic life on Iona, its buildings, its monks and its many visitors. The great bulk of the Life of Columba is a series of short anecdotes about Columba's words and activities among his monks – an intimate portrait of the ideal abbot.

The longest and most moving chapter of Adomnán's Life of Columba is his description of the saint's death. As well as its touching description of Columba's last days and hours, this chapter also sheds light on monastic life at Iona, so it is worth considering in some detail. Adomnán describes how, in the month of May, some weeks after Easter, Columba was taken in a cart to visit the monks who were at work in the western parts of the island. He told them that his end was drawing near, and blessed them and the island; then his cart was brought back to the monastic enclosure. On the following Sunday, he saw a vision of angels while celebrating the Eucharist in the monastic chapel. On the Saturday after that, he and his personal attendant Diarmait went out for a short walk, but Columba's age (he was about 76) prevented him going further than the nearest farm-buildings. These he blessed, while he told the sorrowing Diarmait that he expected to die that same night. On the way back to the monastery he sat down to rest at a point which was later commemorated by the placing of a wooden cross set in a mill-stone, which still stood in Adomnán's time. While he sat resting, he was approached by one of the monastery's horses, 'an obedient servant who was accustomed to carry the milk-vessels between the cow-pasture and the monastery'. The horse placed its head in the saint's bosom and seemed to weep, as if it knew that its master would soon be taken from it. Diarmait wanted to drive the beast away, but Columba would not allow this; rather he allowed it to nuzzle against him, before 'he blessed his servant the horse, as it turned sadly away from him'. Then he climbed a small hill overlooking the monastery (possibly Sgurr an Fhithich, just west of the abbey), and gave his monastery his last blessing: 'On this place, small and mean though it be, not only the kings of the *Scoti* with their peoples (ie, the Gael of Scotland and Ireland), but also the rulers of barbarous and foreign nations with their subjects, will bestow great and especial honour'.[86]

Then he returned to his writing-hut to continue the copying of a psalter on which he had been working. He continued this activity up to the time of Vespers, the office which initiated the Lord's Day; and after attending Vespers in the chapel he retired to his own lodgings to sleep. Adomnán tells us that right up to his death Columba slept on the bare ground, using a stone for a pillow. When the bell rang for the midnight office, Columba hurried to the chapel ahead of the monks; and they, coming in carrying lamps to light the dark chapel, found him dying in front of the altar. He raised his hand

Figure 1.4 The 'Apostles Stone', Dunkeld, ninth century, right side. The main figure possibly represents St Columba. © Crown Copyright: RCAHMS NO 04 SW 1.

feebly in a last benediction, and expired. The date of his death was Sunday 9 June 597.[87]

An assessment of Columba's life and career is not easy. Adomnán is more concerned to demonstrate Columba's power and God's favour towards him than to set out the events of his life in chronological order or assess his secular influence and importance. There can be no doubt that Columba did have secular influence, as a high-born aristocrat who moved easily in the company of kings and princes. But it would be misguided and cynical to view him only as a clerical manipulator of secular politics. He was concerned equally with the spiritual well-being of his monks and with the pilgrims and penitents who visited him. It is hinted by Adomnán that sick people came to Iona seeking remedies for physical ailments, and that these were supplied by Columba's monks;[88] and there are constant references to the copying of books, especially the Bible, at Iona. The monastic life at Iona also consisted of manual labour, in the fields and on building projects. Columba, however, seems not to have insisted on too rigorous a programme of labour; he once expressed disapproval of the way his monks at Durrow were being over-burdened with work.[89] Finally, there are references to the regular monastic offices, including Vespers (the evening office) and the midnight office, and the celebration of the Eucharist on Sundays and holy days. Wednesday was normally observed as a fast day at Columba's monasteries, but this could be relaxed if there were important visitors.[90]

It is perhaps impossible to get any closer to Columba. Subsequent tradition and hagiography have obscured him as much as they have revealed him, and tell us as much about the later reputation of Iona as they do about Columba himself. This process had already started long before Adomnán wrote a century after his founder's death, and it is a credit to Adomnán's honesty and singleness of purpose that we know as much about Columba as we do.

ST KENTIGERN

Probably a near contemporary of Columba was St Kentigern or Mungo of Glasgow. There is an obit given c612 (= c614) in the Welsh Annals. We possess a complete twelfth-century *vita* by Jocelin of Furness and a fragment of another, which have a very complex source history. Whether or not we can say very much about the career of Kentigern himself is questionable. His name is British, but the stories of his conception in Lothian and his birth and education at Culross are clearly unhistorical. He is claimed to have been bishop of Strathclyde during the reign of Rhydderch Hael. It is not asserted that he was the first bishop of Strathclyde, and there is some slight evidence suggestive of Christianity in the area earlier.[91] He probably died c614 and was buried at Glasgow, which he is alleged to have founded. Other churches associated with him are Cadder, and possibly Hoddom. This is about the sum of our historical knowledge.

Kentigern's hagiography, and the disentangling of the various threads

Figure 1.5 Back of a cross-slab from Kilbar, Barra, probably eleventh century. The runic inscription has been read: After Thorgerth, Steinar's daughter, this cross was raised. Courtesy of the Trustees of the National Museums of Scotland, NMS XIB. 102.

in its makeup, help to provide some evidence about the development of the church of Glasgow, which is otherwise very obscure. The earliest stratum in Jocelin's *vita* could belong to a period about a century after the saint's death, a time when founders' *vitae* were very much in fashion. Of the later additions, some could be as early as the tenth century, others may be as late as the twelfth century. Some appear to come from a *vita* drawn up in a Gaelic style, replete with Irish-style miracles and ascetic feats, and concerned to connect Kentigern with saints of Scotia north of the Forth like Columba and Serf.

The only independently attested 'fact' of Kentigern's career is the date of his death, c614, in the *Annales Cambriae*. These seem to have reached their present form soon after 950, although by this time they have long since ceased to carry northern material. The northern material is closely related to the northern genealogies, which extend down to the 870s. There is reason to believe that some northern material could have been transferred into Gwynedd c890. These arguments contain an element of speculation. What is important is that Kentigern is a real historical figure, and that a small number of facts about him can be teased out of the web of twelfth-century and later hagiography.

OTHER EARLY SAINTS

We can assign an approximate date to Kentigern because of the independent date in the Welsh Annals. For many saints, we have no chronological data at all. We have a twelfth-century *vita* of St Serf, associating him with many places in south-west Fife, Clackmannanshire, and adjacent regions, but without corroboration we cannot be very confident about the chronological associations claimed, with St Adomnán (d. 704) and King Bridei son of Derilei (d. c706). It is difficult, indeed, to go much further than to say that a churchman, possibly Pictish, called Serf was active in founding churches in that area at some time during the 'Age of Saints'.[92]

The same is true of many other saints, for some of whom our only written record is a brief office in the *Aberdeen Breviary* or a note in a Calendar. We know of St Baya and Maura, associated with the Cumbraes and adjacent mainland of Cunninghame, St Blane of Kingarth on Bute, St Conval of Inchinnan, St Devenic of Banchory, St Drostan of Aberdour and Deer in Buchan, St Duthac of Tain, St Ethernan (who appears disguised as St Adrian of the Isle of May), St Fillan of Loch Earn, Glendochart and Strathfillan, St Kessog of Luss and elsewhere on Loch Lomond, St Machar of Aberdeen, St Machutus of Lesmahagow, St Mirren of Paisley, and many others.[93] The topographical associations may be correct, so we know where they probably lived; but we have no way of knowing when they lived. The feast-dates in calendars may in many cases reflect a memory of the dates on which a saint died; but if in many cases we know on what day a saint died, in very few cases do we know in what year. There are exceptions, where we have corroborative data from a chronicle or other source: Palladius (431), Kentigern (614), Donnan of Eigg (617), Boniface or Cuiretán of Rosemarkie (present at the Synod of Birr, 697), Adomnán of Iona (704), Maelrubai of Applecross (722) and Baldred of Tynninghame (756) are examples.[94]

There are also many examples of dedications and kirks associated with saints who probably or certainly did not visit the location in question. St Ninian has dedications as far north as the Shetland Isles, but his missionary influence does not appear to have spread beyond the Mounth. A legend associates St Laurence of Canterbury with the kirk of Conveth in Mearns (whence the modern name Laurencekirk), but although the dedication to St Laurence of Canterbury is not in question, it is inconceivable that Laurence actually visited the place.[95] St Finnbarr is associated with Dornoch and Barra, but the legends of St Finnbarr in the *Aberdeen Breviary* are drawn from an Irish *betha* of a Munster saint.[96] Shadowy traditions link St Constantine with Kilchousland in Kintyre and the ancient site of Govan on the Clyde, but the identity of the saint in question and the authenticity of the associations is a matter of pure guesswork.[97] Which of the dozens of

Figure 1.6 Four monuments showing the spread of Christianity across Scotland. (Clockwise from top left): (a) Kildalton Cross, Islay, eighth century; (b) Pictish Cross-slab, Aberlemno, Angus, eighth century; (c) Cross-shaft, Govan, Strathclyde, c900; (d) Sueno's Stone, Forres, Moray, probably tenth century. Photographs by the author, except 1.6d which is © Crown Copyright: Reproduced courtesy of Historic Scotland.

saints called Colmán or Colmoc is commemorated at Inchmahome is beyond speculation.

And so it is for many saints: obscure names like Fergus, Fotin, Kenneth, Kentigerna, Kevoca, Machan, Madoc, Maioca, Marnoc, Medan, Monan, Regulus, Ternan, Triduana, Winnoc and Winnin may not all be lost beyond all hope of recovery, but some of them certainly are, and others are very difficult to recover. The reasons behind the spread of dedications are only slowly becoming understood. From other sources we can detect a process in several stages, with the spread of Christianity among the Picts from the British province in the fifth century, from Whithorn in the sixth, from Iona in the seventh, from Northumbria in the eighth, and from the West again in the ninth. We would expect dedications to reflect this complex and confusing spread, and so we find.

This chapter, however, has been devoted to the impact of the saints; the later development of their cult is another story. What can be said with certainty is that by the end of the 'Age of Saints', early in the eighth century, there was no part of the land we now call Scotland where the symbol of the Christian cross was not familiar, accepted and understood. That was a great achievement.

NOTES

1. Ó Riain, 1982, 146–59.
2. Sharpe, 1991, 9–10.
3. Mac Cana, 1980, 99–101; Byrne, 1973, 105.
4. Macquarrie, 1993a, 122–52.
5. Sharpe, 1991, 9.
6. Tertullian, *Adversus Iudaeos*; Origen, *Homilies on Ezekiel*, col. 698.
7. Gildas, 1978, 10–11.
8. Todd, 1981, 10–11.
9. Mansi, 1759, cols. 463ff.; Athanasius, *Historia Arianorum*, cols. 725–6; Athanasius, *Apologia contra Arianos*, cols. 249–50; Sulpicius Severus, *Chronica*, 94.
10. Todd, 1981, 228–9; Johnson, 1980, 166–9.
11. Frend, 1968, 42.
12. Rees, 1988; Evans, 1968.
13. Keppie, 1990; Breeze, 1979.
14. See Macbain, 1911, 29ff.; Dio Cassius quoted in Mann and Penman, 1978, 28–9; Ammianus Marcellinus quoted in Mann and Penman, 1978, 44, 46. It has been suggested that *areani* could be a misreading for *arcani* (spies).
15. Duncan, 1975, 28.
16. Claudian. On the Consulship of Stilicho. In *Monumenta Germaniae Historica*, X, Berlin, 1892; and Platmauer quoted in Mann and Penman, 1978, 47–8.
17. Campbell, 1982, 20ff.
18. Constantius, 1920, 247ff.
19. Patrick's letters are edited in Bieler, 1961, and Hood, 1978.
20. Among the most important of the many works are: Bury, 1905; MacNeill, 1964; O'Rahilly, 1942; Carney, 1961; Mohrmann, 1961; Binchey, 1962; Hanson, 1968; Thomas, 1981; Sharpe, 1982; Dumville, 1993; see also Macquarrie, 1997, cap. 2.

21. Curle, 1923; *The Treasure of Traprain* (NMS Information Sheet, no. 7, 1980); Jobey, 1976, 191–204; Close–Brooks, 1983, 206–23.
22. Thomas, 1992, 2–3. They are recorded in Macalister, 1945, nos. 510–11, 514–15, and some later discoveries.
23. Allen and Anderson, 1903, 496–7; Macalister, 1945, nos. 519, 520; Thomas, 1992, 3–11, 19; Campbell, 1982, 21; Wall, 1965, 208–11.
24. Hill, 1997.
25. Dumville, 1984, 207–14; Sharpe, 1984, 196–201.
26. Duncan, 1981, 1–42.
27. MacQueen, 1989, 2–3, 11.
28. MacQueen, 1989, 11.
29. Macquarrie, 1997, 230; Sharpe, 1991, 8–17.
30. Colgrave, B and Mynors, R A B. *Bede's Ecclesiastical History of the English People*, Oxford, 1969 [hereafter *HE*], iii, 4.
31. Anderson, 1973, 235–60.
32. *HE*, 223. Their translation obscures this point by rendering *ut perhibent* as a passive, 'so it is said'. But Bede's active verb refers back to the main subject, *australes Picti*. See also MacQueen, 1989, 12; but this does not take us very far. Bede uses the verb *perhibere* (to claim, assert) some twenty-one times in *HE*; on eight occasions it is passive ('it is said', etc), on thirteen occasions an active form is used. In all but two of these thirteen cases, the active verb can be related to an expressed subject. On only two occasions (ii, 2 and iii, 5), the subject of *ut perhibent* is not expressed, so a passive meaning ('so it is said') may be understood; but that is not the case here. See Jones, 1929, sv *Perhibeo*, etc.
33. Anderson, A O and M O. *Adomnán's Life of Columba*, 2nd edn, Edinburgh, 1991 [hereafter *VC* (Vita Columbae)], i, 37; ii, 33, 35, 42.
34. *VC*, i, 8.
35. Stokes, 1899, 31–55, 132–83, 248–89, 400–37; Stokes, 1900, 133–6. Most recently edited in Clancy and Márkus, 1995, 96–128. It may be that the *tuatha Toí* mentioned in the *Amra* were the peoples dwelling on Ptolomy's *Tuesis aestuarium*, the mouth of the Spey. But see Clancy and Márkus, 1995, 118–9; Watson, 1986, 50–1.
36. Barrow, 1983, 1–15.
37. Duncan, 1981, 28–30.
38. Watson, 1986, 33–4.
39. MacQueen, 1989, 40.
40. It is noteworthy that immediately after this mention of Pehthelm in *HE*, v, 23, Bede goes on to describe peaceful relations between Northumbria and the Picts.
41. Gildas, *De Excidio Britonum*, cap. 67.
42. Macquarrie, 1987, 1–25; Macquarrie, 1997, cap. 3.
43. Mentioned in *VC* as *Findbarrus* (i, 1; ii, 1); *Vinniauus* (ii, 1) *Finnio* (acc. *Finnionem*) (iii, 4). Mentioned by Columbanus as *Vennianus auctor*, correspondent of Gildas (*Monumenta Germaniae Historica*, I, *Epistolae*, III, 156ff.).
44. Ó Riain, 1982.
45. Bernard and Atkinson, 1898, I, 22; Plummer, 1910, II, 60, 263.
46. I have previously alluded to this possibility in Macquarrie, 1997, cap. 3 n. 88, p. 73.
47. Macquarrie, 1997, 65, 73, n. 88; Clancy, 2001, 1–28.
48. MacQueen, 1989, 4ff.
49. See MacQueen, 1989, 4 and n. 7, p. 125, on the likelihood of this attribution.
50. *Monumenta Germaniae Historica*, I, *Epistolae*, I, 282–3.
51. *Monumenta Germaniae Historica*, IV, *Poetae*, 942–62.
52. *Amra Coluim Cille*, see n. 35 above.
53. Picard, J M. The Purpose of Adomnán's Vita Columbae, *Peritia*, 1 (1982), 160–77, at 167ff.

54. *HE*; also Plummer, 1896.
55. Bannerman, 1974, 9–26. The main collections of annals are: *Annals of Ulster*, Mac Airt and Mac Níocaill, 1983 (hereafter *AU*); *Annals of Tigernach*, Stokes, 1896, (hereafter *AT*); these are the most important. Also noteworthy are: *Chronicon Scotorum*, (an abbreviation of 'Tigernach') ed. Hennessy, W M, Rolls Series, 1866; *Annals of Innisfallen*, ed. Mac Airt, S, Dublin, 1951; *Annals of Clonmacnoise* (an English translation of a set of annals related to 'Tigernach' and *Chronicon Scotorum*), ed. Murphy, D, 1896; *Annals of the Four Masters*, ed. O'Donovan, J, 1856. In citing *AU* entries, I have followed the usual practice of adding 1 to the later *Anno Domini* dates in the MS for this period.
56. *VC*, iii, 5; see also i, 1; i, 2; i, 3; ii, 4; iii, 19. The 1st edn (1961) has a valuable introduction which has been shortened in the 2nd edn. On Adomnán's use of oral tradition, see Ó Riain, 1982, and Sharpe, 1995.
57. For the Irish Life of Columba and the development of Columban hagiography, see Herbert, 1988, 211–88.
58. *VC*, Second Preface; see *VC*, pp. xxviii ff.
59. *VC*, i, 1; ii, 1; iii, 4. See also Dumville, 1984, 207–14. On the multiplication of saints of the same name, see Ó Riain, 1982, 147–55.
60. *HE*, iii, 4; *VC*, i, 3; iii, 15.
61. *VC*, i, 3; Edwards, 1986, 31.
62. *AU*, sa 561.
63. *VC*, iii, 3.
64. *VC*, iii, 3; i, 7; see Bannerman, 1974, 128.
65. *VC*, i, 7.
66. *VC*, Second Preface.
67. *VC*, i, 7.
68. *AU*, sa 574: 'Conall mac Comgaill ... obtulit insolam Iae Columbe Cille'; Iona's own record of its foundation.
69. *VC*, iii, 5.
70. Enright, 1985, 86–7.
71. Enright, 1985, 84.
72. *VC*, i, 10–11, 49; Bannerman, 1974, 157–70; Byrne, 1973, 110–11.
73. Bannerman, 1974, 157–70.
74. Byrne, 1973, 111.
75. Craig Phatric was the site of an Iron-Age hill-fort, with traces of secondary occupation during the Pictish period; Ritchie, 1989, 44; Alcock, 1984, 23; Alcock, 1987, 82f. Bridei's fort was 'not far from the River Ness, though far enough for an urgent message to have been carried on horseback'; so the site of Inverness Castle is probably ruled out. See *VC*, p. xxxiv.
76. *VC*, ii, 36. It has been pointed out that some of these stories are said to have taken place *trans Dorsum Britanniae* (i.e. across Drumalban) and others are located *in Provincia Pictorum*. Adomnán may have been working from two narratives which recorded visits to the Inverness area, but that does not in itself require more than one visit. The editors of *VC* (1st edn, pp. 81–2) doubt that Columba made more than one visit to Bridei's territories; and see also Bieler, 1963, 184. On the other hand, the words *in prima fatigatione itineris* could be taken to imply subsequent journeys; and Bridei is said to have honoured Columba *ex ea in posterum die ... suae omnibus vitae reliquis diebus*, which might imply that they met subsequently. The evidence is not conclusive either way. See my review of *VC* (2nd edn), Macquarrie, 1993b, 213–5.
77. *VC*, ii, 43.
78. *AU*, sa 580, 581. The entry, however, is in Irish (*'Fecht Orc la hAedhán mac Gabráin'*), so it may be doubtful if it belongs to the earliest stratum of the Annals, which seem to have been in Latin.

79. *AU*, sa 687.
80. *VC*, i, 28.
81. *VC*, ii, 32; iii, 14.
82. *HE*, iii, 4.
83. *VC*, i, 8.
84. Bromwich, 1978, 147–8; Bannerman, 1974, 86–90; Duncan, 1981, 16–19.
85. See Macquarrie, 1997, 108–9.
86. *VC*, iii, 23. Adomnán, of course, was writing with hindsight.
87. *VC*, iii, 23; See Anderson, 1922, I, 103.
88. *VC*, i, 27.
89. *VC*, i, 29.
90. *VC*, i, 26.
91. See Macquarrie, 1997, cap. 5
92. Macquarrie, 1993a.
93. *Breviarium Aberdonense*.
94. Palladius is mentioned in the *Chronica* of Prosper of Aquitaine; Kentigern in the *Annales Cambriae*; Donnan, Adomnán and Maelrubai in the Irish Annals; Cuiretán in the records of the Synod of Birr known as *Cain Adomnáin*; and Baldred in the *Historia Regum* of Symeon of Durham.
95. Macquarrie, 1996, 95–109. The speculative suggestion by Clancy, 1999, 83–8, that the dedication might have been to St Laurence the Martyr is highly unlikely. It would be common to expect a dedication to transfer from an obscure regional saint to a popular universal one, but not the other way round. Besides, the fragment shows that the dedication to St Laurence of Canterbury was already fixed by the eleventh century.
96. In Plummer, 1922.
97. Ritchie, 1994, esp. 27–32.

BIBLIOGRAPHY

Alcock, L. A survey of Pictish settlement archaeology. In Friell, J G P and Watson, W G, eds. *Pictish Studies*, Oxford, 1984, 7–41.
Alcock, L. Pictish studies: Present and future. In Small, A, ed. *The Picts: A New Look at Old Problems*, Dundee, 1987, 80–92.
Allen, J R and Anderson, J. *Early Christian Monuments of Scotland*, Edinburgh, 1903.
Anderson, A O. *Early Sources of Scottish History, AD 500–1286*, Edinburgh, 1922.
Anderson, A O and M O. *Adomnán's Life of Columba*, 2nd edn, Edinburgh, 1991.
Anderson, M O. *Kings and Kingship in Early Scotland*, Edinburgh, 1973; 2nd edn, Edinburgh, 1980.
Athanasius. *Apologia Contra Arianos*. In Migne, J-P. *Patrologiae Cursus Completus, Series Graeca*, 161 vols., Paris, 1857–68, XXV, col. 248–252.
Athanasius, *Historia Arianorum*. In Migne, J–P. *Patrologiae Cursus Completus, Series Graeca*, 161 vols., Paris, 1857–68, XXV, col. 1327–1332.
Bannerman, J W M. *Studies in the History of Dalriada*, Edinburgh, 1974.
Barrow, G W S. The childhood of Scottish Christianity: A note on some place–name evidence, *Scottish Studies*, 27 (1983), 1–15.
Bernard, J H and Atkinson, R H. *The Irish Liber Hymnorum*, London, 1898.
Bieler, L. *Libri Epistolarum S Patricii Episcopi*, Dublin, 1961.
Bieler, L. Review of *Adomnán's Life of Columba*, *Irish Historical Studies*, 13 (1963), 175–84.
Binchey, D A. St Patrick and his biographers, ancient and modern, *Studia Hibernica*, 2 (1962), 7–173.
Breeze, D J. *Roman Scotland: A Guide to the Visible Remains*, Newcastle, 1979.

Breviarium Aberdonense, Edinburgh, 1509–10, reprinted Edinburgh, 1854.
Bromwich, R. *Trioedd Ynys Prydein*, Cardiff 1961; 2nd edn, Cardiff, 1978.
Bury, J B. *The Life of St Patrick*, London, 1905.
Byrne, F J. *Irish Kings and High–Kings*, London, 1973.
Campbell, J, et al. *The Anglo–Saxons*, Oxford, 1982.
Carney, J. *The Problem of St Patrick*, Dublin, 1961.
Clancy, T O. The real St Ninian, *IR*, 52 (2001), 1–28.
Clancy, T O and Márkus, G. *Iona: The Earliest Poetry of a Celtic Monastery*, Edinburgh, 1995.
Clancy, T O. The foundation legend of Laurencekirk revisited, *IR*, 50 (1999), 83–8.
Close–Brooks, J. Dr Bersu's excavations at Traprain Law, 1947. In O'Connor, A and Clarke, D V, eds. *From the Stone Age to the 'Forty–Five: Studies Presented to R B K Stevenson*, Edinburgh, 1983, 206–23.
Colgrave, B and Mynors, R A B, eds. *Bede's Ecclesiastical History of the English People*, Oxford, 1969.
Constantius. *Vita Germani Episcopi*, Monumenta Germaniae Historica Scriptores Rerum Merovingicarum, Berlin, 1920.
Curle, A O. *The Treasure of Traprain: A Scottish Hoard of Roman Silver Plate*, Glasgow, 1923.
Dumville, D. *Saint Patrick, AD 493–1993*, Woodbridge, 1993.
Dumville, D. Gildas and Uinniau. In Lapidge, M and Dumville, D, eds. *Gildas: New Approaches*, Woodbridge, 1984, 207–14.
Duncan, A A M. *Scotland: The Making of the Kingdom*, Edinburgh, 1975.
Duncan, A A M. Bede, Iona and the Picts. In Davis, R H C and Wallace–Hadrill, J M, eds. *The Writing of History in the Middle Ages: Essays Presented to Richard William Southern*, Oxford, 1981, 1–42.
Edwards, N. The South Cross, Clonmacnoise. In Higgitt, J, ed. *Early Medieval Sculpture in Britain and Ireland*, Oxford, 1986, 23–48.
Enright, M J. Royal succession and abbatial prerogative in Adomnán's *Vita Columbae*, *Peritia*, 4 (1985), 83–103.
Evans, R F, ed. *Pelagius: Inquiries and reappraisals*, London, 1968.
Frend, W H C. The Christianising of Roman Britain. In Barley, M W, and Hanson, R P C, eds. *Christianity in Britain, 300–700*, Leicester, 1968, 37–49.
Gildas, *De Excidio Britonum*. In Winterbottom, M, ed. *The Ruin of Britain and other Works*, London and Chichester, 1978.
Hanson, R P C. *St Patrick*, Oxford, 1968.
Herbert, M. *Iona, Kells and Derry*, Oxford, 1988.
Hill, P. *Whithorn and St Ninian: The excavation of a monastic town, 1984–91*, Stroud, 1997.
Hood, A B E. *St Patrick: His Writings and Muirchu's 'Life'*, Chichester, 1978.
Jobey, G. Traprain Law: A summary. In Harding, D W, ed. *Hillforts: Later prehistoric earthworks in Britain and Ireland*, London, 1976, 191–204.
Johnson, P. *Romano-British Mosaics*, Princes Risborough, 1982.
Jones, P F. *Concordance of the Historia Ecclesiastica of Bede*, Cambridge, Mass, 1929.
Keppie, L. *Scotland's Roman Remains*, Edinburgh, 1990.
Mac Airt, S and Mac Níocaill, G, eds. *Annals of Ulster*, Dublin, 1983.
Macalister, R A S. *Corpus Inscriptionum Insularum Celticarum*, I, Dublin, 1945.
Macbain, A. *Etymology of the Principal Gaelic National Names, Personal Names and Surnames, to which is added a Disquisition on Ptolemy's Geography of Scotland*, Stirling, 1911.
Mac Cana, P. *The Learned Tales of Medieval Ireland*, Dublin, 1980.
MacNeill, E. *St Patrick*, 1934, reprinted Dublin, 1964.
Macquarrie, A. The date of St Ninian's mission: A reappraisal, *RSCHS*, 23 (1987), 1–25.
Macquarrie, A. *Vita Sancti Servani*: The Life of St Serf, *IR*, 44 (1993a), 122–52.
Macquarrie, A. Review of *Adomnán's Life of Columba* (2nd edn), *SHR*, 72 (1993b), 213–5.
Macquarrie, A. An eleventh–century account of the foundation legend of Laurencekirk, and of Queen Margaret's Pilgrimage there, *IR*, 47 (1996), 95–109.

Macquarrie, A. *The Saints of Scotland*, Edinburgh, 1997.
MacQueen, J. *St Nynia*, 2nd edn, Edinburgh, 1989.
Mann, J C and Penman, R G, eds. *Literary Sources for Roman Britain*, London, 1978.
Mansi, J D. *Sacrorum Conciliorum Nova et Amplissima Collectio*, Venice, 1759.
Mohrmann, C. *The Latin of St Patrick*, Dublin, 1961.
Monumenta Germaniae Historica, I, Epistolae, III, *Epistolae Merowingici et Karolini Aevi*, Berlin, 1892.
Monumenta Germaniae Historica, VII, Constantius, *Vita Sancti Germani*, Berlin, 1920.
Monumenta Germaniae Historica, IV, *Poetae Latini Aevi Carolini*, Berlin, 1923.
O'Rahilly, T F. *The Two Patricks*, Dublin, 1942.
Ó Riain, P. Towards a methodology in early Irish hagiography, *Peritia*, 1 (1982), 146–59.
Origen. *Homilies on Ezekiel*. In Migne, J-P. *Patrologiae Cursus Completus, Series Graeca*, 161 vols., Paris, 1857–68, XIII.
Plummer, C. *Venerabilis Baedae Opera Historica*, Oxford, 1896.
Plummer, C. *Vitae Sanctorum Hiberniae*, Oxford, 1910.
Plummer, C. *Bethada Náem nÉrenn: Irish Saints' Lives*, Oxford, 1922.
Rees, B R. *Pelagius: A Reluctant Heretic*, Woodbridge, 1988.
Ritchie, A. *Picts*, Edinburgh, 1989.
Ritchie, A, ed. *Govan and its Early Medieval Sculpture*, Stroud, 1994.
Sharpe, R. St Patrick and the See of Armagh, *Cambridge Medieval Celtic Studies*, 4 (1982), 33–59.
Sharpe, R. Gildas as a father of the Church. In Lapidge, M and Dumville, D, eds. *Gildas: New approaches*, Woodbridge, 1984, 193–205.
Sharpe, R. *Medieval Irish Saints' Lives*, Oxford, 1991.
Sharpe, R. *Adomnán of Iona, Life of St Columba*, Harmondsworth, 1995.
Stokes, W. The Bodleian Amra Choluimb Chille, *Revue Celtique*, 20 (1899), 31–55, 132–83, 248–89, 400–37; 21 (1900), 133–6.
Stokes, W. Annals of Tigernach, *Revue Celtique*, 17 (1896), 6–33, 119–263, 337–420.
Sulpicius Severus, *Chronica*. In *Corpus Scriptorum Ecclesiasticorum Latinorum*, Vienna, 1866–, I, col. 3–105.
Tertullian. *Adversus Iudaeos*. In *Corpus Christianorum, Series Latina*, Turnhout, 1953–, II, *Tertulliani Opera*, II, col. 1337–1398.
Thomas, A C. *Christianity in Roman Britain*, London, 1981.
Thomas, A C. *Whithorn's Christian Beginnings*, Whithorn, 1992.
Todd, M. *Roman Britain: 55 BC–AD 400*, Brighton, 1981.
Wall, J. Christian Evidences in the Roman period, *Archaeologia Aeliana*, 4th ser. 43 (1965), 208–11.
Watson, W J. *The History of the Celtic Place-Names of Scotland*, Edinburgh, 1926, reprinted Edinburgh, 1986.

2 The Cult of Saints in Medieval Scotland

ALAN MACQUARRIE

Like all of medieval Christendom, Scotland participated in the cult of saints. Modern theologians would explain the phenomenon of the cult of saints by saying that as the medieval conception of God became more remote and transcendent, the saints were necessary as intermediaries, bridging the gulf between the divine majesty and the weakness and frailty of man. Thus popular attitudes to saints are sometimes ambiguous, with saints being both fearsome wonder-workers and objects of homely affection. Many of them have, in addition to their formal names, an affectionate pet-name or hypocoristic, often beginning with *mo*, 'my'. The most famous example is the patron saint of Glasgow, whose formal name Kentigern < *Cyndyern* < *Conthigirn* < **Cunotigernos*, 'lord of the hounds', but whose pet-name Mungo < *mo (n)chú*, 'my little dog'.[1]

We will consider briefly three aspects of the cult of saints as it was manifested in medieval Scotland. The first is hagiography, writings about saints, mainly their lives or *vitae*. The second is the presence of saints in art. Finally, something will be said about the popular expression of devotion to saints through pilgrimage.

HAGIOGRAPHY: SAINTS' LIVES

The ambiguous character of saints is everywhere manifest in hagiography. Often saints were portrayed as having very human failings, such as anger or laziness: St Fillan caused a goose to peck out the eye of a novice whom he caught spying on him, while St Kentigern indolently neglected the fire which burned in St Serf's monastery at Culross. But their human frailties were always made good because they were particular recipients of divine favour: Fillan regretted his wrath and healed the blinded novice, while Kentigern relit his master's fire through prayer and spontaneous combustion.[2] Saints' legends sometimes seem to have an ambiguous attitude to moral goodness, on the one hand excusing conduct which might be regarded as morally suspect while on the other stressing their heroes' virtues.

An example of this can be seen in some of the stories of kindness to animals with which saints' lives abound. In a rural society of bare subsistence, a fairly unsentimental attitude to animals might seem to be the

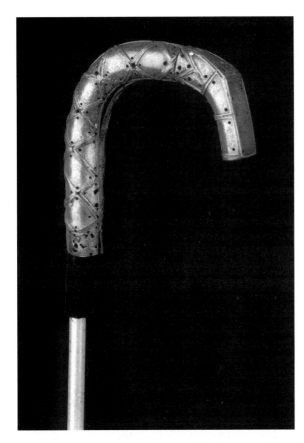

Figure 2.1 Eleventh-century bronze crosier head, stripped of its original decoration. It was housed inside the Coigrich of St Fillan. Courtesy of the Trustees of the National Museums of Scotland, NMS H.KC.2.

norm. Animals were sources of labour and food; where they provided neither, they were either ignored or, if dangerous or competitive, destroyed. Saints could miraculously provide animals for food, as Columba did for a poor peasant by blessing a wooden stake for him; in his journeying among the Picts, Columba repulsed both a savage boar which charged him and a fierce water-monster which threatened to attack one of his monks.[3] St Machar turned a wild boar to stone simply because it trod down his cornfield.[4] St Fillan showed a slightly milder attitude to animals when a wolf slew one of his draught-animals: he ordered the wolf to submit to the yoke until its task was done, then released it with an injunction never to harm another creature ever again.[5]

The famous story of Columba's hospitality to a migrant crane or heron which landed on the western shore of Iona and was cared for there by one of his monks for three days is traditionally held up as an example of a saint showing sentimental kindness to animals. This view has more recently been criticised by some who fear that modern sentimentality may be carried too far. They point out that Columba justified his kindness because the bird came from his own native Donegal, and therefore belonged to his kin. But there is

THE CULT OF SAINTS • 43

no escaping the point of the story, or Adomnán's own stated reason for including it in Columba's *Life*: it is, he says, 'a matter of less moment, but so beautiful that it cannot, I think, be passed over in silence'. He tells the story, he says, because it is *iucunda* (beautiful, lovely – a pretty story). Columba's explanation for his kindness to the bird, that it came from the land of his fathers, may be regarded as a secondary apology to his monk for sparing the bird the pot.[6]

Indeed, in the wider context of hagiography in general Columba's story fits well with attitudes to animals elsewhere. In the *Book of Armagh* there is a story describing how Patrick and his companions went for a walk on the hill of Armagh, where they found a hind with a newborn fawn at the spot where they proposed to build a church. Patrick would not allow his companions to kill the animals, but lovingly lifted the fawn in his arms and carried it off the hill while its mother obediently followed him; once they were thus gently removed, building work could begin.[7]

This consideration for animals who might have proved a useful source of food reflects the austerities which are often attributed to saints: many of them eat little or no meat, and they reprove gluttons who eat too much of it. To slaughter a crane or a deer when one had it at one's mercy would be an earthly reaction; to spare it, for whatever excuse, was otherworldly.

A special bond with animals, sympathetic or telepathic, is common in hagiography, and symbolises the saint's closeness to God's creation. St Kentigern can yoke a stag and a wolf to the plough, like St Fillan mentioned above. He can also yoke two untamed oxen to a funeral cart and let them draw it unbidden to a suitable site for building his church.[8] This miracle is lifted from a story relating to Patrick's burial in the *Book of Armagh*, but the model is ultimately biblical, deriving from the story of the return of the Ark of the Covenant from Philistia to Israel in I Samuel 6. The *Life of Columba* tells a touching story of Columba saying farewell to one of Iona's horses on the last day of his life.[9]

The earliest insular hagiography – of the seventh and eighth centuries – tends to be moralistic and edifying in tone. The period of the first Viking raids, from about 790 until well into the ninth century, is rather less productive of hagiography.[10] From about 900 onwards there is a new outburst of hagiographic activity, but it is in some ways different in character. For one thing, it contains works in Irish as well as in Latin, there are mixed 'Hiberno-Latin' lives, and the Latin tends to be of poorer grammatical and literary quality. Also lives now are less edifying and moralistic, with saints becoming amoral wonder-workers who seem to enjoy going around scaring people. Although biblical models are still to be found, the hagiography of this period shows more influence of secular literature.

Some saints' lives read like a collection of title deeds, with a saint travelling from place to place working miracles in each, and each place being then claimed as part of the saint's jurisdiction. The twelfth-century *Vita Sancti Servani* is the best Scottish example of such a life, with its hero visiting

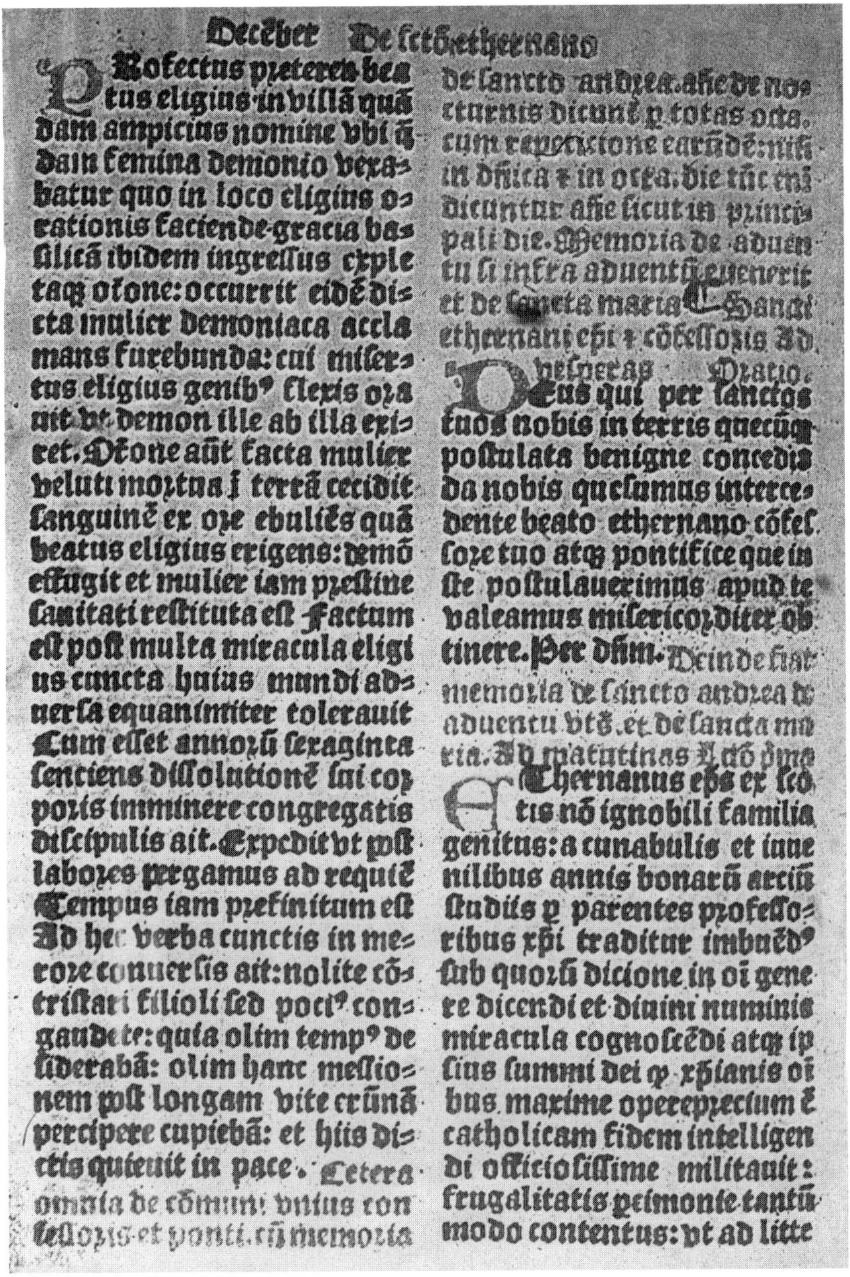

December De scō ethernāno

Rofectus preterea beatus eligius in villā quādam ampicius nomine vbi quadam femina demonio vexabatur quo in loco eligius orationis faciende gracia basilicā ibidem ingressus expletaq̃ orōne: occurrit eidē dicta mulier demoniaca acclamans furebunda: cui misertus eligius genib9 flexis orauit vt demon ille ab illa exiret. Oratione aūt facta mulier veluti mortua ĩ terrā cecidit sanguine ex ore ebulies qua beatus eligius erigens: demō effugit et mulier iam prestine sanitati restituta est. Factum est post multa miracula eligius cuncta huius mundi aduersa equanimiter tolerauit Cum esset annorū seraginta sentiens dissolutionē sui corporis imminere congregatis discipulis ait. Expedit vt post labores pergamus ad requiē Tempus iam prefinitum est Ad hec verba cunctis in merore conuersis ait: nolite cōtristari filioli sed pocius congaudete: quia olim temp9 deliberabā: olim hanc messionem post longam vite erūnā percipere cupiebā: et hiis dictis quieuit in pace. Cetera omnia de cōmuni vnius confessoris et ponti. cū memoria

de sancto andrea. ase de nocturnis dicunt p totas octa. cum repeticione earūdē: nisi in dnica ⁊ in octa. die tūc eni dicuntur ase sicut in principali die. Memoria de aduentu si infra aduentū euenerit et de sancta maria ⁂ Sancti ethernani epi ⁊ cōfessoris Ad hesperas Oratio.

Deus qui per sanctos tuos nobis in terris quecūq̃ postulata benigne concedis da nobis que sumus intercedente beato ethernano cōfessore tuo atq̃ pontifice que iuste postulauerimus apud te valeamus misericorditer obtinere. Per dn̄m. Deinde fiat memoria de sancto andrea ⁊ aduentu vt s. et de sancta maria. Ad matutinas Lcō imē

Ethernanus eps ex scetis nō ignobili familia genitus: a cunabulis et iuuenilibus annis bonarū arciū studiis p parentes professoribus xp̄i traditur imbuēd9 sub quorū dicione in oī genere dicendi et diuini numinis miracula cognosceēdi atq̃ ipsius summi dei q̃ xp̄ianis oībus. maxime opereprecium ē catholicam fidem intelligendi officiosissime militauit: frugalitatis primonte tantūmodo contentus: vt ad litte

Figure 2.2 Page from the Winter Propria Sanctorum of the *Aberdeen Breviary* showing the collect and first lesson for the commemoration of St Ethernan (2 December). Courtesy of the Trustees of the National Library of Scotland, NLS F.6. f. 5., pars hiemalis, 2 December.

Inchkeith, Kinneil, Culross, Lochleven, Dysart, Tullybody, Tillycoultry, Alva, Airthrey and Dunning, and performing miracles in most of these places. This neatly delineates the *parochia* of St Serf centred on Culross.[11] This tendency is not new to the post-Viking period, however; examples go back to Tirchán's *Memoir* of St Patrick in the *Book of Armagh*.[12]

In some cases the miracles become fantastic to the point of hysteria: St Laurence of Canterbury, in his fragmentary *vita* from Laurencekirk embedded in an otherwise sober Canterbury life, summons down fire from heaven on his enemies (twice), walks dryshod on water, raises the dead, and causes a spring to well up in dry ground, all in the space of less than two pages.[13]

In the twelfth century, saints' lives once again become edifying and increasingly moralistic. This is evident in the several *vitae*, abbreviated, fragmentary or complete, of St Kentigern, which culminate in the *Vita Kentigerni* of Jocelin of Furness of c1180. This spares no opportunity to moralise and reprove in the most austere Cistercian fashion. Now the extinguishing of St Serf's fire results not from Kentigern's indolence but from the wicked envy of his companions, who 'secretly extinguished all the fire within the habitations of the monastery' and then try to put the blame on him. When St Serf's pet robin is killed by the novices while playing, and they again try to place the blame on Kentigern, he 'kept himself entirely apart from the affair' before healing the bird by prayer.[14] This pure and moral boy contrasts with the indolent and careless youth of the earlier versions, and reflects the changing values in the twelfth-century Scottish Church in the age of monastic reform.

In a different category comes a small group of *vitae* of eleventh- and twelfth-century 'royal saints'. Chief among these, and probably setting the pattern for others, is the *Vita Margaretae Reginae* by Thurgot, prior of Durham.[15] This survives in two versions. The longer version was written 1104x1107, but the shorter version may be an earlier draft composed very shortly after the queen's death in 1093. The fuller version was addressed to Queen Matilda, Margaret's daughter, who married Henry I of England and whose brothers Edgar (1097–1107), Alexander (1107–24) and David (1124–53) were successively kings of Scots.

Margaret's youngest son, David, whose generosity to the religious orders became proverbial, was never canonised as a saint, though many contemporaries allude to his saintly character. Ailred, later abbot of Rievaulx, occupied an important place in King David's household before his conversion to the religious life, and after the king's death he wrote a *Lamentatio* in hagiographic style, which he dedicated to the future King Henry II of England (therefore dateable 1153x1154).[16]

A third *vita* of a 'royal saint' is the *Vita Waldeui* of Jocelin of Furness, concerning the life of Abbot Waldef of Melrose (d. 1159). Waldef was the son of Earl Simon I de St Lis or Senlis, and King David's stepson. His career and tenure of office as abbot of Melrose may in fact have been less distinguished than is suggested in Jocelin's *Vita*, but the *Vita* certainly supported the pretensions to sanctity of the Scottish royal house in the same way that the *Vita Margaretae* and the *Lamentatio* for King David did.[17]

On the whole, the hagiographic legacy of Scotland is slender between these twelfth-century productions and the end of the fifteenth century. Medieval inventories of the books of Scottish cathedrals make references to *Legenda Sanctorum*, but most of these have been lost, or survive only as fragments. A rare survival is a vernacular verse collection of saints' lives, formerly attributed to John Barbour, 'The Legends of the Saints'. Among its apostles, evangelists, virgins, confessors and martyrs, mostly drawn from the *Legenda Aurea* and the *Specula* of Vincent of Beauvais, are lives of St Machar of Aberdeen and St Ninian of Whithorn.[18] The latter appears to be drawn from Ailred's Latin *Vita*. The attribution to John Barbour, archdeacon of Aberdeen and author of *The Brus* (c1375), is no longer widely accepted. Other nationalistic chroniclers of the later Middle Ages such as John of Fordun (writing c1385), Andrew of Wyntoun, (c1410), and Walter Bower (c1440) made much of earlier Scottish saints, using hagiographical materials which are now lost.[19] But Bower in particular may have been conscious of the relative shortage of writings about Scottish saints, since he felt the need to expropriate large numbers of Irish saints into his huge *Scotichronicon* and pass them off as Scots. This was made possible because the words *Scotia* and *Scotus* were still used to mean 'Ireland' and 'Irish' in the twelfth century. This has greatly annoyed some modern Irish scholars; but Bower can perhaps be forgiven for his expropriations, since the hagiographic record of his own country is relatively sparse.

About half a century after Bower wrote, the task of giving Scotland a large-scale national hagiography was taken in hand by William Elphinstone, bishop of Aberdeen. His *Breviarium Aberdonense* (*Aberdeen Breviary*), published in Edinburgh in 1509/1510, is the most important collection of Scottish saints' lives (in the form of short *lectiones* for their feast-days) as well as having the distinction of being Scotland's first printed book. For many saints the *Aberdeen Breviary* provides our only information. For others it presents variants on traditions which are recorded elsewhere. It is by far our greatest treasury of Scottish hagiography. In spite of its late date, the *Breviary* is important because it records many earlier traditions which have otherwise been lost. Preliminary research suggests that where the *Breviary* compilers had earlier materials on which to draw, these are often of twelfth-century or even earlier date.[20]

THE SAINTS IN ART

Because of the comprehensive nature of the Reformation in Scotland, there has been great loss of images of saints in the form of statues of wood or stone, stained glass, paintings, and manuscript illuminations. But examples of most of these do survive.[21]

There are relatively few representations of saints on Early Christian carved stones in Scotland. Some of the early high crosses of the Iona school

Figure 2.3 Fifteenth-century stone statue of John the Baptist, dredged from the Firth of Forth. Courtesy of the Trustees of the National Museums of Scotland, NMS H.KG.6.

Figure 2.4 Statue of St Andrew, carved in oak, carrying his cross and a book, c1500. Courtesy of the Trustees of the National Museums of Scotland, NMS H.KL.188.

Figure 2.5 Statue of St Ninian on an altarpiece originally from St Olai Kirke, Helsingør, Denmark. Courtesy of the National Museum of Denmark, inventory no. CDXXXVI.

show the Virgin and Child with angels, in a way which is reminiscent of the Virgin and Child page in gospel books, such as the Book of Kells. A cross-slab at Dunkeld shows on one face the twelve Apostles standing as witnesses to the miracle of loaves and fishes, and on a side panel a large figure of a monk with a halo or cowl round his head (Fig. 1.4); this is possibly the earliest representation of St Columba in stone.[22]

Stone statues survive in a number of Scottish churches from the later medieval period.[23] Melrose Abbey has a number of statues of saints, including the Blessed Virgin and others which are less easy to identify. At Lincluden there are a number of surviving statues of saints from the reredos. These include St Paul, St John the Evangelist, St James the Less, St Thomas, and a female, probably St Brigid, patron of the Douglases. At Linlithgow there is a statue of St Michael slaying the dragon. Glasgow Cathedral has a fine late medieval choir-screen with altar bases on either side bearing the arms and initials of Robert Blackadder, bishop and archbishop of Glasgow (1483–1508). There are statues of saints, probably the apostles, on the front of the altar bases. In the Blackadder Aisle at Glasgow there is a carving in the vault showing the body of St Fergus on his funeral cart being drawn by oxen. At Paisley Abbey there is a carved stone frieze in the St Mirin chapel,

Figure 2.6 Painted panel showing Christ with saints. The figure on the far right represents St Ninian. Fowlis Easter collegiate church, second half of the fifteenth century.

showing scenes from the life of St Mirin.[24] There is a fifteenth-century statue of St Magnus in Kirkwall Cathedral.[25] A number of churches have surviving sculpture scenes of the Virgin as part of an Annunciation scene.

Scotland's greatest surviving treasury of late medieval sculpture is at Roslin (or Rosslyn) chapel in Midlothian, but only a small proportion of the total shows images of saints. The apostles are shown on one of the arches in the south aisle; there are also statues of apostles on a band outside one of the windows on the north side. St Christopher and St Sebastian stand above the capitals of the west doorways of the choir.[26]

Very few wooden statues have survived. A timber statue of a bishop, probably St Ninian, was found in a bog close to Whithorn Priory. An oak statue of St Andrew carrying his cross has survived from St Andrews Cathedral (Fig. 2.4); probably it was once one of many statues in the choir.[27] A remarkable statue of St Ninian, without the manacles which he sometimes carries, is on an altarpiece now in the National Museum of Denmark in Copenhagen, originally from St Olai Kirke in Helsingør (Fig. 2.5). This reflects the devotion to their saints which the Scots took with them overseas.

Hardly any medieval glass has survived in Scotland in any state of completeness: the glass in the Hammermen's Chapel in the Cowgate of Edinburgh is the most important survival, but it is of heraldry rather than hagiology. Some glass from Holyrood Abbey has been incorporated in a window in the Palace of Holyroodhouse. Two windows of French medieval glass have been inserted into Douglas Kirk when it was restored in modern times by the earl of Home, and they include St Paul, the Virgin Mary, and the four Evangelists. In Victorian times it again became acceptable to put coloured glass into churches, and many churches now have fine representations of saints in their windows.

Very few paintings have survived in Scottish churches. The most remarkable series is the set of paintings on oak panels in Fowlis Easter church near Dundee. In addition to a large painting of the crucifixion and other scenes from the life of Christ (Fig. 31.3), there is a series of panels showing Christ and a number of saints, which probably came originally from the rood screen of the collegiate church (Fig. 2.6).[28] These have been

Figure 2.7 Page from the Book of Deer, tenth century, showing Luke the Evangelist. By permission of the Syndics of Cambridge University Library, CUL MS Ii.6.32, f.29v.

identified, by the symbols which they carry, as St Catherine (Fig. 3.2), St Matthias, St Thomas, St Simon, St John, St Peter, St Anthony, St James the Less, St Paul, and St Ninian.[29] St Ninian is shown as a bishop holding a manacle, with the lion of Scotland at his foot. Ninian is usually represented

Figure 2.8 The opening pages of St Mark's Gospel from Queen Margaret's Gospel Book, eleventh century. Courtesy of the Bodleian Library, University of Oxford, MS. Lat. Liturgy. f. 5, incipit page.

holding chains to symbolise his attribute as a freer of captives, although this image does not correspond exactly with any episode in the twelfth-century *Vita Niniani* or earlier hagiography.

A good number of medieval manuscripts survive from Scotland, and some of these have images of saints. Some early gospel books, such as the Book of Deer (tenth century, now in Cambridge) and Queen Margaret's Gospel Books (eleventh century, now in the Bodleian Library, Oxford), have illuminated evangelist pages at the beginning of each gospel (Figs. 2.7, 2.8). Some later books have illuminated pages with images of saints. Good examples are found in a series of liturgical manuscripts written for Arbuthnott church by Mr James Sibbald; these include pictures of St Ternan and the Mass of St Gregory. There is a fine image of St Ninian, with his manacles, in the Salisbury Book of Hours, now in Edinburgh.[30]

PILGRIMAGE

There were several important pilgrimage centres in Scotland, and the following account is not intended to be comprehensive.[31] Whithorn held the relics of St Ninian, and seems to have been a major shrine and centre of

Figure 2.9 Stone statue of St Duthac from the collegiate church of Tain, Ross and Cromarty, fifteenth century. It represents St Duthac as a bishop, standing on a calf. Courtesy of the Trustees of the National Museums of Scotland, NMS H.KG.139.

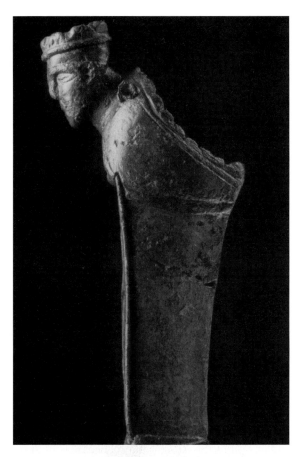

Figure 2.10 Twelfth-century bronze crosier drop found in Loch Shiel, near St Finan's Isle. Courtesy of the Trustees of the National Museums of Scotland, NMS H.1993.634.

pilgrimage from c500. By Bede's time, c730, Whithorn was famed for its church dedicated to St Martin and for the relics of St Ninian and 'many saints'. The fame of the church, but not of Ninian's relics, had reached Germany by about the same time. By the end of the eighth century the power and sanctity of Ninian's relics were known on the Continent. The Anglian minster seems to have been an ambitious timber structure designed to accommodate pilgrim traffic.[32] In the twelfth century the bishopric of Whithorn was revived and its cathedral rebuilt in stone by the lords of Galloway, and fragments of the twelfth-century cathedral are still visible. Later a gothic crossing, transepts and aisled chancel were built over a crypt which housed the tomb of St Ninian. Other places sacred to Ninian were venerated in the locality: his chapel at Isle of Whithorn was rebuilt in the thirteenth century beside a harbour which welcomed pilgrims from England, Ireland and Man; his cave was round the coast on the shore at Physgill. Whithorn numbers Robert I and James IV among its royal pilgrims.

Iona is second in age to Whithorn as a place of sanctity, but thanks to modern restoration its buildings are the most complete pilgrimage site in

Scotland. All that survives from Columban and immediately post-Columban times is the substantial perimeter earthwork or *vallum* at the north-west corner of the enclosure. From the eighth and ninth centuries there are stone high crosses, one of which survives complete and four in fragments. A small stone chapel of the tenth or eleventh century, 'St Columba's shrine', is believed to mark the spot where Columba died in 597; it was visited by King Magnus Barelegs in 1098. In the twelfth century a new chapel was built in the *Reilig Odhrain* burial ground, and in the following centuries the church was rebuilt in a succession of gothic styles following its conversion to Benedictine monasticism in 1203. At one point a large south transept was projected but never completed, probably to accommodate pilgrim traffic around the chancel.[33] Columba's relics have a very complex history, much of it remote from Iona itself; but even without them Iona retained and still retains its attractiveness to visitors and pilgrims.[34]

There were places of pilgrimage in the north as well. At Tain there are three medieval buildings: a small church by the seashore, probably the original site of St Duthac's church; a modest twelfth-century parish kirk in the town itself; and beside it a large late-medieval collegiate church, which reflects Tain's popularity as a place of pilgrimage by the fifteenth century. King James IV visited Tain in 1503.

The cathedral at Kirkwall housed the relics of St Magnus, the earl of Orkney martyred in 1116. His cult developed quickly after his death, and the core of the present cathedral – the crossing and transepts and the west end of the choir and eastern bays of the nave – belong to the late twelfth century. In the thirteenth century the east and west ends were extended as pilgrim traffic increased.

Glasgow Cathedral housed the shrine of St Kentigern in what is the Scottish mainland's most complete surviving medieval cathedral. Owing to the fall in the ground at the east end of the church, the chancel was built on two storeys, with the saint's tomb and holy well in the lower church and his shrine behind the high altar in the upper choir. The design is intended to facilitate the circulation of pilgrims on both levels. As at Iona, a large south transept was projected but never completed; the vaulting of the lower stage was eventually completed by Bishop Robert Blackadder (d. 1508), but the upper storey of the transept was never built. Part of the shrine, or its base, survives; it is on display in the lower church.[35]

Probably the only medieval building in Scotland that was more magnificent than Glasgow Cathedral was the great cathedral of St Andrews, which was comparable in scale with English or continental cathedrals. There was a religious community on the site from the eighth century, which claimed to hold certain bones of St Andrew the Apostle. By the late eleventh century this had become such a place of pilgrimage that Queen Margaret endowed the pilgrims' ferry at Queensferry; in the twelfth century the earls of Fife supplemented this with another ferry to the east at Earlsferry. Until the late twelfth century the main building was St Rule's church with its tall tower (110 feet), but from 1160 onwards it was replaced with a huge stone-vaulted church to hold the bones of the Apostle, which underwent successive modifications throughout the medieval period.[36]

Dunfermline was the burial place of Queen Margaret and her husband Malcolm Canmore (both d. 1093). Margaret's grandson David I built the magnificent Romanesque nave of the abbey, modelled on Durham Cathedral, and his thirteenth-century descendants rebuilt the choir and east end as a shrine to their saintly ancestor, in emulation of Edward III's rebuilding of Westminster Abbey to house the relics of Edward the Confessor. The crossing and choir have been swept away and replaced by the present parish church, but the foundations of the eastern chapel and the base of St Margaret's shrine have survived.[37] St Margaret's relics survived the Reformation and were taken to France, but they did not survive the French Revolution.

These are just a few examples of Scotland's most important centres of pilgrimage. There were many others. At Restalrig, the blind came to St Triduana's chapel to seek sight from a virgin who had blinded herself to avoid marriage to a prince. Whitekirk in East Lothian had a well dedicated to the Virgin, who took terrible vengeance on English soldiers who pillaged her shrine. At Kilpatrick on the Clyde near Dumbarton, reputed birthplace of St Patrick, pilgrims came in the twelfth century to a well-endowed hospice to visit the saint's healing well. Just across the river at Inchinnan was an *ymago* of St Convall which could heal physical deformity. Throughout Scotland there were similar wells, shrines and images with healing properties of varying degrees of wealth and popularity.

Although the reformers attempted to sweep all this away in the sixteenth century, there is evidence that popular devotion died hard: throughout the 1580s the presbytery of Stirling was still rebuking many 'pepill that passis in pilgrimage to Chrystis woll and usis gret idolatrie or superstitioun thairat'.[38] But gradually the shrines were deserted and the cult of saints faded. Owing to the level of destruction of buildings, books and furnishings, it is difficult now to imagine how widespread the cult of saints once was in Scotland.

NOTES

1. Watson, 1986, 169.
2. *Breviarium Aberdonense*, Propria Sanctorum, 9 January; *Vita Kentigerni* in Forbes, 1874.
3. Anderson, A O and M O. *Adomnán's Life of Columba*, 2nd edn, Oxford, 1991 [hereafter, *VC*], ii, 37; ii, 26; ii, 27.
4. *Breviarium Aberdonense*, 12 November.
5. *Breviarium Aberdonense*, 9 January.
6. *VC*, i, 48; Meek, 1999, 253–70, at 263.
7. Hood, 1978, 25.
8. *Vita Kentigerni*, 20, 9.
9. *VC*, iii, 23.
10. Sharpe, 1991, cap. 1.
11. Macquarrie, 1993, 122–52.
12. Bieler, 1979.

13. Macquarrie, 1996a, 95–109.
14. *Vita Kentigerni*, 5, 6.
15. *Vita Margaretae Reginae* in Hodgson Hynde, 1868.
16. Skene, 1871, v, 35ff.
17. *Acta Sanctorum, Augusti*, i (1733), 248–76.
18. Metcalfe, 1887–92.
19. Amours, 1903–14; Watt, 1987–1998.
20. Boyle, 1981, 59–82; Macquarrie, 1996b, 31–54.
21. Descriptions can be found in MacGibbon and Ross, 1896–7; Cruden, 1960.
22. Macquarrie, 2004, 77.
23. Coltart, 1936, 55–65 and *passim*.
24. Yeoman, 1999, 30–2.
25. Yeoman, 1999, 94.
26. Coltart, 1936, 62.
27. Yeoman, 1999, 42, 68.
28. See also Chapter 31, Christian Art in Scotland.
29. Coltart, 1936, 208–10; Fawcett, 1985, 58–9; Fawcett, 2002, 292–3.
30. McRoberts, 1953; Dowden, 1910, 96, 98; Yeoman, 1999, 34.
31. See Yeoman, 1999. Statements in the following section which are not otherwise referenced are mostly from Yeoman, 1999.
32. Hill, 1997.
33. RCAHMS, 1982.
34. Herbert, 1988; Bannerman, 1993, 14–47.
35. Fawcett, 1998.
36. McRoberts, 1976.
37. Fawcett, 1994, 26–9.
38. Kirk, 1981.

BIBLIOGRAPHY

Acta Sanctorum, Augusti, i (1733).
Amours, F J. *Androw of Wyntoun's Orygynale Cronykil of Scotland*, Scottish Text Society, 1st ser. 50, 53–4, 56–7, 63, Edinburgh, 1903–14.
Anderson, A O and M O. *Adomnán's Life of Columba*, 2nd edn, Oxford, 1991.
Bannerman, J W M. *Comarba Coluim Chille* and the relics of Columba, IR, 44 (1993), 14–47.
Bieler, L. *The Patrician Texts in the Book of Armagh*, Dublin, 1979.
Boyle, A. Notes on Scottish Saints, IR, 32 (1981), 59–82.
Breviarium Aberdonense, Edinburgh, 1509–10; reprinted Edinburgh, 1854.
Coltart, J S. *Scottish Church Architecture*, London, 1936.
Cruden, S. *Scottish Abbeys*, Edinburgh, 1960.
Dowden, J. *The Medieval Church in Scotland*, Glasgow, 1910.
Fawcett, R, ed. *Medieval Art and Architecture in the Diocese of Glasgow*, British Archaeological Association, 1998.
Fawcett, R. *Scottish Medieval Churches*, Edinburgh, 1985.
Fawcett, R. *Scottish Abbeys and Priories*, London, 1994.
Fawcett, R. *Scottish Medieval Churches: Architecture and furnishings*, Stroud, 2002.
Forbes, A P. *Lives of SS Ninian and Kentigern*, Edinburgh, 1874.
Herbert, M. *Iona, Kells and Derry*, Oxford, 1988.
Hill, P. *Whithorn and St Ninian*, Stroud, 1997.
Hodgson Hynde, J. *Symeonis Dunelmensis Opera et Collectanea*, Durham, 1868.
Hood, A B E. *St Patrick: His writings and Muirchu's Life*, Chichester, 1978.
Kirk, J. *Stirling Presbytery Records*, Edinburgh, 1981.

MacGibbon, D and Ross, T. *The Ecclesiastical Architecture of Scotland*, Edinburgh, 1896–7.
Macquarrie, A. *Vita Sancti Servani*: the life of St Serf, *IR*, 44 (1993), 122–52.
Macquarrie, A. An eleventh-century account of the foundation legend of Laurencekirk, and of Queen Margaret's pilgrimage there, *IR*, 47 (1996a), 95–109.
Macquarrie, A. Lives of Scottish saints in the Aberdeen Breviary: Some problems of sources for Strathclyde saints, *RSCHS*, 26 (1996b), 31–54.
Macquarrie, A. *Medieval Scotland: Kingship and nation*, Stroud, 2004.
McRoberts, D. *Catalogue of Scottish Medieval Liturgical Books and Fragments*, Glasgow, 1953.
McRoberts, D. *The Medieval Church of St Andrews*, Glasgow, 1976.
Meek, D E. Between faith and folklore: Twentieth-century images and interpretations of Columba. In Broun, D and Clancy, T O, eds. *Spes Scotorum: Hope of Scots: Saint Columba, Iona and Scotland*, Edinburgh, 1999, 253–70.
Metcalfe, W M. *Legends of the Saints*, Scottish Text Society, 1st ser. 13, 18, 23, 25, Edinburgh, 1887–92.
RCAHMS, *Argyll: An inventory of the Monuments: Iona*, Edinburgh, 1982.
Sharpe, R. *Medieval Irish Saints' Lives*, Oxford, 1991.
Skene, W F. *Johannis de Fordun Chronica Gentis Scottorum*, Edinburgh, 1871, v.
Watt, D E R, et al. *Scotichronicon by Walter Bower*, Aberdeen, 1987–98.
Watson, W J. *The History of the Celtic Place-names of Scotland*, Edinburgh, 1926; reprinted Edinburgh, 1986.
Yeoman, P. *Pilgrimage in Medieval Scotland*, London, 1999.

3 Religious Life in Scotland in the Later Middle Ages

AUDREY-BETH FITCH

INTRODUCTION

Explorations of religious life in late medieval Scotland reveal that certain religious practices and beliefs owed their origins to the pagan past. One early Victorian writer claimed that, in his own day, cows were buried alive as a cure for mental illness and epilepsy.[1] Such reservations aside, most late medieval Scots were Christians whose beliefs and practices can be explored through written and material evidence.

During the sixteenth century, reformers characterised traditional Roman Catholic Christianity as mechanistic. They argued that its rituals were designed to improve spiritual standing without emotionally or spiritually engaging individual Christians, or explaining the theological basis of the faith. Yet the faith of late medieval Scottish Christians was a personalised one, and theological concepts informed people's attitudes and actions. Scots believed implicitly in the reality of the Day of Judgement, heaven, hell and purgatory, and understood that God, Mary, Jesus, the saints and the devil had an impact on life before and after death. God judged people; Jesus, Mary and the saints aided them; and the devil tried to lead them astray. Learning about the afterlife and its supernatural beings contributed to a successful life and prepared people for death spiritually.

This study asserts that many religious beliefs and practices crossed class, occupational and gender lines, yet studies of religion often distinguish between 'popular' and 'élite' religion. For Robert Scribner 'popular beliefs' were those cherished by the 'mass of the people' (e.g. cults of saints and relics) rather than the 'religious élite'. He argued that distinguishing 'beliefs' from 'practices' might not be useful or even possible when defining popular belief. Religious beliefs may have informed religious practice, such as the founding of perpetual prayers and masses for the dead, but lay people rarely operated from a clear conceptual framework.[2] I would argue, however, that Scots did act on the basis of internalised religious concepts, and that male and female laity and clergy from different classes had a common religious understanding. Yet Scribner reminds us that practices should not be explained primarily through theology, for popular devotion was more fluid, less structured and less wedded to officially sanctioned, theologically nuanced religious interpretations than orthodox belief. Instead, popular religious understanding was developed in an ad hoc manner through exposure to religious imagery in secular as well as spiritual contexts. People were reminded of their relationship with the supernatural through paintings, carvings, statues, relics, vestments and altar ornament decorations, as well as popular and priestly preaching, the liturgy, devotional and secular literature and drama, and saintly tales told in church. The reported efficacy and popularity of certain shrines also influenced people's ideas. The secular was sacralised when people met religious imagery in daily activities. For example, a small three-dimensional carved image of Mary standing with the baby Jesus in her arms decorated a mid-sixteenth-century oak cabinet,[3] and an image of Mary and the baby Jesus was carved on Banff's market cross.[4]

Enthusiastic destruction during the post-Reformation period makes evaluating many sources difficult. Nevertheless, written evidence from foundation charters, theological tracts, poetry, plays and burgh council and craft and merchant guild records provides a framework for understanding popular religion in later medieval Scotland. It is unfortunate that many liturgical and devotional works were destroyed, and access to material evidence is even more problematic. Certain artefacts survive, but often we learn about material evidence through descriptions in church inventories or other listings of altar ornaments and vestments, books, statues, paintings and furnishings. Yet the material evidence which does survive tells us a great deal about the environment of faith in which pre-Reformation Scots lived and worked, their religious ideas, and the activities they believed essential to their spiritual welfare. Indeed, without reference to the material context of faith in the pre-Reformation period, a large piece of the puzzle is missing. As Carl Jung has opined: 'Without images you could not even speak of divine experiences. You would be completely inarticulate ... Since it is a matter of an ineffable experience the image is indispensable ... God approaches man in the form of symbols.'[5]

Owing to the uneven survival of written records and material artefacts and low levels of literacy, the portrait of Scottish religious faith and action which usually emerges is a clerical product and treats primarily urban élites, rural lairds and magnates, and the monarchy. In burghs, masters are the easiest to identify, and the religious outlook of apprentices, maidservants and common labourers can only be surmised. Only occasionally are we given a hint of rural attitudes, apart from those of lairds and magnates. For example, in the early sixteenth century Alexander Mylne, canon of Dunkeld and abbot of Cambuskenneth (1474–1548), commended Dunkeld's clergy for rooting out sin and heresy in the rural population. In particular, he praised Sir Thomas Greig, prebendary of Alight. A layman, pretending to be dumb, had convinced his fellow Scots that he could tell the past and future through signs and nods. Greig convinced the man to admit that the devil had inspired him to do this.[6]

Comparing trends in different parts of the country can offer a regional perspective on attitudes and activities. The north east, for example, has been regarded as a place where traditional religion was particularly strong. According to kirk sessions and the General Assembly, all classes in the north east suffered from the 'unhappy seed of popery'.[7] Yet the uneven survival of evidence makes generalisations about regional differences of dubious value. The relative wealth of material evidence which survives in the north east, for instance, may reflect the power of a landed élite able to protect religious imagery in homes and churches rather than a region whose inhabitants were more firmly committed to traditional practices than other people in Scotland. Certainly Bishop Gavin Dunbar (c1455–1532) was not complacent about the region's adherence to Catholicism. Despite the reforming zeal of Bishop Elphinstone (1431–1514) and his subordinates, in 1525 Dunbar reported to the king that Aberdeenshire was afflicted by Lutheran doctrines.[8]

At a national level there are a variety of sources to suggest religious values, but these are most helpful for the Scottish élite. In 1525 the parliament published its first act against heresy, and in 1532 James V (1513–42) declared to the parliament that it should defend the Catholic faith and Church institutions as willingly as its ancestors had accepted the faith. Defence of traditional religion meant defence of God, Church and Pope, and punishment of heresy in all its manifestations.[9] One element of traditional religion which needed defending was devotion to the Blessed Virgin Mary. In 1540 a parliamentary statute insisted on worship and honour of Mary. Her intercession with God, the Son and the Holy Ghost on behalf of the whole country would safeguard the kingdom and the succession, welfare and prosperity of James V, his queen Mary of Guise (1515–60) and their successors. By helping to maintain peace, unity and concord between the king and other Christian princes, Scotland would resist the enemies of Catholicism and remain constant in the faith.[10] To some modern commentators this act might suggest lack of support for the cult of the Virgin in Scotland, at least at the grass-roots level. Yet overall support for Mary was strong. John Gau reported in 1533 in 'The Richt Way to the Kingdome of Hevine' that many people believed that salvation came through prayers and service to Mary.[11] Marian prayer and pilgrimage were valued as a route to personal sanctification, and collective spiritual welfare was enhanced through new foundations and re-dedications. For example, in 1543, at the behest of parishioners and local lairds and clerics, the parish church of Cullen, Banffshire, became the collegiate church of St Mary.[12] Collective support for Mary could also be expressed by purchasing and publicly displaying Marian paintings or statues: in 1526 the goldsmiths of Edinburgh sent a guild member to Flanders to obtain an image of Our Lady of Loretto.[13] Christocentric reformers had to work hard to dislodge Mary from the hearts and minds of the populace; the act of 1540 merely testifies to the iconoclasm and anti-Marian rhetoric being employed to this end.

Kirk session and General Assembly records reveal the outlook of the Protestant religious élite after the Reformation. They also suggest that traditional religious beliefs and practices survived the Reformation. People suffered prosecution in order to take part in Catholic Eucharistic, baptismal and funeral rites.[14] Pre-Reformation Scots were not unthinking adherents of traditional religion; many believed deeply in the value of certain rituals.

THE CONSTRUCTION AND EXPRESSION OF RELIGIOUS IDEAS BY GROUPS AND INDIVIDUALS

Since most pre-Reformation Scots were illiterate, they educated themselves about matters of the faith by keeping their eyes and ears open. They learned about the nature of sin and the afterlife, the role of saints and the holy family, and routes to spiritual improvement from fellow lay people (e.g. family, friends, guild brothers, burgh council members, poets, playwrights), the clergy (e.g. friars, parish priests and craft chaplains), and a physical environment laden with religious symbolism (e.g. art, architecture, religious pageants and plays). Although churchmen and their activities are often used as a measure of religious activity, secular institutions such as parliaments, the monarchy, burgh councils, and craft and merchant guilds helped to construct and reinforce Scottish religious conceptions. Parliamentary acts protected and promoted traditional religion by punishing 'heresy', and monarchs confirmed the religious foundations of burgesses, lairds and magnates through the Great Seal and acted as pious exemplars for their subjects. In 1511, for example, James IV (1473–1513) confirmed a chaplainry foundation made one year earlier by Elizabeth Mason, widow of John Scrimgeour, burgess of Dundee.[15] Royal support for traditional church activities included purchasing books for liturgical use, commissioning religious art, and donating money on feast days. James V donated a great antiphonal book to the chapel in Holyroodhouse in 1538/9,[16] James III (1452–88) commissioned an altarpiece in the late 1470s for his mother's collegiate foundation of Trinity College, Edinburgh,[17] and James III and Queen Margaret (c1457–86) made an offering of 27s. 6d. in Edinburgh Castle on St Margaret's Day, 1473.[18] In their private capacity, monarchs purchased books for private devotion and attended church regularly. James IV commissioned a lavishly illustrated book of hours for himself and his wife Margaret Tudor (1489–1541),[19] and the meeting of James V and Peter Swave, emissary of Christian, duke of Holstein, took place near the altar in Jedburgh's monastery church in 1535.[20]

Scots followed the royal lead. It was most likely a glover who, in 1557, donated a painting of St Bartholomew for the craft to use as an altarpiece at their altar in Perth's burgh church.[21] Other people purchased volumes for private devotion and public ritual, albeit ones less extravagantly illustrated. For example, Sir Robert Arbuthnott of Arbuthnott (d. 1506) commissioned a prayer book (completed by the laird's priest-in-residence c1482–3), and the baxters of Dundee purchased a missal in 1486 for use at their craft altar in the parish church of St Mary's.[22]

Burgh councils defended traditional religion. They took charge of prosaic matters such as repair of the church roof, and made sure that weekly offerings were collected. They also helped to ensure that high standards of service were maintained by funding and supervising regular church services in the choir as well as altar, chaplainry and obit (perpetual anniversary masses for the dead) foundations. In 1495, Aberdeen burgh council confirmed that forty-five burgh dwellers had made loans to help pay for work on the church roof,[23] and in 1521 the burgh council levied a fine of 10s. for collectors who did not take up the weekly bred silver collection. By 1537 the problem of non-collection had obviously worsened, for the fine was raised to the sum equivalent to that collected the previous Sunday.[24] Stirling burgh council made similar arrangements with respect to the 'Rood bred'. On 7 September 1556 the fine for non-collection was the sum equivalent to that collected the previous or following Sunday.[25]

Paying stipends to supplement or fully finance clerical positions gave burgh councils the right to insist that standards of service remain high. In 1536/7 Ayr burgh council paid stipends to several chaplains in St John's church.[26] In 1517, a burgess assigned the burgh of Peebles a supervisory role in his altar and chaplainry foundation,[27] and much earlier in 1496 burgess George of Spalding assigned annual rents to Dundee burgh council to support two obit foundations for himself and his wife.[28] In 1505, Wat Strathin received a 20s. stipend from the burgh of Aberdeen to sing and help with divine service in the choir and church of St Nicholas. In addition, eight burgh-dwellers were to supply him with food and drink.[29] In 1528/9 the burgh council set up a system to monitor the performance of burgh council appointees. On 15 January 1529 Thomas Menzies of Pitfodels, bailie David Anderson, William Rolland, and Master Andrew Tulideff were appointed to collect monies owed to the burgh's singers and to find two of the best choir chaplains to monitor their fellow singers. Each quarter, a fault book would be read and fines levied against the singers' quarterly stipends. Fines would be levied for absenteeism, chattering during services or leaving one's post before the end of the service, and would amount to 2d. on weekdays, 4d. on holy days and 8d. on principal feasts.[30] In 1533, the burgh met and 'all in ane voce' fired all the singers in the choir except Sir Andrew Coupar, an old and valued employee. Clerical misconduct, directed at God and the burgh, would not be tolerated.[31] In 1556, the burgh council appointed Master Andrew Gray as a chaplain in the choir because he was learned and qualified, but he was reminded that good attendance and personal residence were required; the council fully intended to enforce the will of the founder.[32]

Merchant and craft guilds honoured patron saints, ensured perpetual prayers for dead members, and organised the religious life of guild members. Guilds also supported traditional religion by taking part in burgh processions (e.g. Edinburgh hammermen, Corpus Christi celebrations).[33] As the dominant force on burgh councils, merchants in particular helped determine burgh-dwellers' religious experiences. Guilds built and equipped altars and founded and maintained perpetual chaplainries, utilising dues and fines levied against masters for producing shoddy goods or disregarding trade rules. In its seal of cause of 20 October 1531, the master tailors of Edinburgh assigned their weekly penny to St Anne's altar in St Giles' church.

Funds maintained the fabric, supported daily services, and purchased altar ornaments.[34] In 1527 the walker craft of Dundee founded a chaplainry in honour of St Mark at St Michael's altar in their parish church.[35] In 1528 the Glasgow weavers' seal of cause stipulated that the weekly fee for master craftsmen would be one penny, and for apprentices, servants and waged workers a halfpenny. Glasgow's cordiners followed suit in 1559.[36] In 1543 the Perth hammermen craft fined masters who took more than 15 days to hand over their weekly offering,[37] and in 1523 the baxters of Edinburgh fined members one pound of wax to the altar of St Cuthbert for selling substandard or rotten goods.[38] In 1533 the skinners and furriers levied a fine of one pound of wax to their altar of St Christopher for taking another man's apprentice or servant.[39]

Popular devotion was expressed collectively as well as individually.[40] It is untrue that individual devotion revealed 'true' piety whereas participation in public rituals signified mere conformity. Nor were the 'common folk' unconcerned with the behaviour of magnates and kings. In the late Middle Ages the community had everything to do with the religious fortunes of the individual; people's spiritual fortunes were affected by the behaviour of social superiors. Writers warned the élite that their activities endangered the spiritual welfare of the collectivity.[41] In *The Complaynt of Scotland* (1546), Robert Wedderburn (c1515–57), vicar of Dundee, railed at his countrymen, particularly lairds and magnates, for fighting with each other and preying on the poor and labouring classes. God's acceptance and mercy would not come to those who held drawn swords toward the innocent. He urged them to obey God so He would end the wars, plagues and famines besetting Scotland.[42]

Founders of obits and chaplainries requested prayers for benefactors as well as ancestors, the 'faithful [Christian] dead', immediate family members and themselves. Kings might also be named, for they represented the state in a spiritual as well as secular sense. Walter Chapman of Ewirland, burgess of Edinburgh and 'familiaris servitoris' to James IV, named the king in his chaplainry foundation of 1513,[43] and in 1529 Thomas Neill (d. 1533), burgess of Ayr, and Agnes Wishart his wife asked priests at their obit mass to exhort the people to pray for the welfare of the king's grace.[44] In 1543 Philip Gibson's obit foundation requested prayers for James V as well as Mary, Queen of Scots (1542–87).[45] A king's bad behaviour or immorality could bring God's wrath upon the whole nation. The author of *The Book of Pluscarden* claimed the plague of 1362 was God's punishment for King David II's sins, which included keeping a mistress.[46] Other Scots contributed to Scotland's difficulties, such as merchants and craftsmen who cheated customers.[47] Robert Henryson (c1430–1500), a popular and vocal writer,[48] roundly criticised all manner of Scots who led lives of sexual excess and self-gratification.[49] He lambasted élites as well as the middling sort for failing to turn from sin and prepare for Judgement. He considered the violent plague outbreaks of his own day to be God's judgement on a sinful people.[50]

Public rituals had spiritual value for the individual, and emphasised social harmony. For example, public penitential rites taught that forgiveness was central to community reconciliation, and reconciliation with one's neighbour preceded reconciliation with God. In 1554 Edinburgh burgh council punished candlemaker William Young for breaking craft statutes. He was ordered to walk to the parish church at the time of High Mass carrying a candle for St Giles' light. Once in the church he was to ask forgiveness of the bailies and provost. The latter would then grant forgiveness on behalf of himself and the community, and in so doing restore the relationship between Young and burgh society. Young rejected the burgh council's claim to represent the entire community, accusing it of oppressing burgh-dwellers,[51] but his was the minority opinion; others accepted that it was the responsibility of the secular leadership to maintain unity so that the burgh might continue to function as a religious community. The vernacular 'Ane Godlie Exhortatioun' (The Twapenny Faith) (1559), which Archbishop John Hamilton of St Andrews (1512–71) intended to be read aloud before Communion, reminded parishioners that they were a Christian community. It required that neighbours reconcile before Communion or risk damnation.[52]

A WEEK IN THE LIFE OF THE AVERAGE BURGH DWELLER

For the average person, religious life was far more than attendance at High Mass once a week. People's religious sensibility was moulded on a daily basis through conversation, participation in a variety of activities, and contemplation of religious art and architecture. The weekly activities of a hypothetical busy and prosperous burgh-dweller included several interactions with the sacred. Although expected to attend Sunday's High Mass, he would not have been required to take the sacrament more than once a year, usually at Easter. However, our burgh-dweller paid heed to poet William Dunbar (c1460–1514), and worried that he would forget to confess some sins if he took Communion so infrequently.[53] Untutored in Latin, during High Mass he contemplated a rood screen painting depicting Jesus' death on the Cross (e.g. Fowlis Easter collegiate church, c1480), and meditated on Jesus' suffering and its meaning for salvation. He took special note of the angel tenderly carrying the repentant sinner's soul to heaven, and the black winged creature clutching the unrepentant sinner's soul in its claws (Fig. 3.1). Sobered, he vowed to spend more time in prayer and meditation, resist more strongly the seven deadly sins, and stop complaining about his weekly payments to the craft altar.[54] The statue of the Blessed Virgin Mary at the high altar (e.g. St Andrews Cathedral) brought home to him the compassionate nature and intercessory power of Jesus' mother, and embroidered images of saints on the priest's vestments and altar cloths (e.g. King's College, Aberdeen)[55] reminded him that he should follow their devout and moral example.

In late medieval Scotland there was ample opportunity for burgh dwellers to express devotion to saints, and they were happy to do so. Certainly poet and courtier Sir David Lindsay's (?1486–1555) aversion to the cult of saints put him in the minority.[56] Returning from Somerset's foray into Scotland in 1548, Londoner William Patten wrote to patron and privy councillor Sir William Paget that Scots were devoted to their 'pardon beades' and rosaries, that they went on pilgrimages, worshipped idols, gave oblations and offerings to saints, and seemed to have 'charmes for every diseas' and 'suffrages for every sore'.[57] Saintly imagery abounded. Paintings, statues, carvings, vestments, ornaments, woodcuts, prayers and offerings expressed devotion to saints and reminded people of saintly miracles.[58] Not long ago near Glasgow a small statute of St Eloy was found buried in the mud. It may well have reposed on the altar of the hammermen craft for whom St Eloy was the patron.[59] Behind the altar in a chapel in Paisley abbey a retable honoured St Mirin's life and miracles.[60] Saints transferred to supplicants merits won through suffering, thereby assisting people before and after death. Martyr saints were particularly powerful intercessors, martyrdom being the fullest expression of devotion to God. Healing was believed to take place at a variety of holy wells and shrines, including St Triduana's well in Restalrig, the Cross of Crail, and the shrine of Loretto near Musselburgh.[61]

Devotional books were full of woodcuts of saints; each image conveyed meaning by highlighting one aspect of the text.[62] The prayer book, albeit unusually lavish, of Robert Blackadder, archbishop of Glasgow (c1445–1508), contained an illustration on almost every page. One image depicted St Catherine of Alexandria holding a wheel, the symbol of her martyrdom, whilst the text praised her as 'beloved virgin and advocate with God in [our] struggle'.[63] In a large painting in Fowlis Easter church she rammed a large sword through the head of the man who had tried to take her virginity (Fig. 3.2).[64] To women in particular this bespoke a strong-minded woman who understood the spiritual power of sexual purity, was utterly devoted to Jesus, and would act courageously on behalf of her supplicants. It also reminded female viewers that God valued spiritual commitment and sexual self-restraint.

At some point in the week our burgh dweller offered prayers at the obit service of a fellow burgess. The foundation charter stipulated that services be held at the altar of St Barbara, a saint whose popularity was increasing in the decades leading up to the Reformation. The founder wished to honour God, the Blessed Virgin Mary and St Barbara, and benefit the souls of himself, his wife, parents, benefactors and the faithful dead. To support the obit financially in perpetuity, his friend had donated annual rents from various properties. Understanding that badly executed services angered God, he had carefully outlined the procedures to be followed by clerical celebrants. On the eve of St Ninian, as our burgh dweller made his way to the church, a crier invited the living to pray for the dead. The curate and chaplains of the choir sang the obsequies of the dead at the altar of St Barbara, then sang a mass the next day, followed by a trental of masses the following week. To ensure that his gravestone was not overlooked, the founder had insisted that prayers be said there and a 'table' placed on it with a pall and lights. No one in the burgh could have missed the obit, for bells were rung on the Eve of St Ninian as well as St Ninian's Day (e.g. St Nicholas church, Aberdeen, 1509).[65] After the obit-mass, our burgh dweller looked with compassion on the poor to whom an 'obit dole' was being paid in compensation for their prayers.[66] Apart from extra prayers for the founder's soul, such generosity to the poor counted as a good deed.[67] On his way out of the church, our burgh dweller spent a few moments contemplating the rosary clasped in the hands of a deceased local magnate carved in effigy (e.g. Lord Seton, Seton collegiate church (Fig. 3.3)).[68] He wondered if the man had been grateful for the Blessed Virgin Mary's company on the frightening journey to Judgement.

Later in the week our burgh dweller attended a special saint's day mass, listened to the life story of the saint, and made an offering (e.g. James IV's offering at St Catherine's mass, Linlithgow, 1490).[69] Unfortunately his parish priest did not use the Scots tongue when reciting the common prayers preceding the litany of the saints, or he might have learned even more about the saint.[70] In common with other popular saints, she had been martyred in pagan times. Our burgh-dweller contemplated an image of the saint in the church (e.g. St Catherine of Alexandria, Fowlis Easter collegiate church and St Salvator's College, St Andrews)[71] as well as a relic of her tomb (Glasgow Cathedral)[72] displayed in a monstrance on her altar. He doubted his capacity to suffer so grievously in defence of the faith. Perhaps a generous donation

Figure 3.1 Painting depicting the soul of the unrepentant sinner being taken to hell. Fowlis Easter collegiate church, second half of the fifteenth century. Photograph by the author.

Figure 3.2 Painting of St Catherine of Alexandria. Fowlis Easter collegiate church, second half of the fifteenth century. Photograph by the author.

Figure 3.3 Effigy of George, second Lord Seton, with rosary. Seton collegiate church, early sixteenth century. Photograph by the author.

of 'lightsilver' to keep a candle burning in her honour would do instead (e.g. St Barbara's light, Kirk o' Field church, Edinburgh)?[73] Our burgh dweller felt lucky to live in a Christian country where prayer, meditation on the Passion, church ritual and the example and intercession of the saints helped people achieve spiritual worthiness. Above all, Jesus' Crucifixion gave him hope for salvation, and he trusted that the prayers and masses of the living would raise him from purgatory after he died. He made a mental note to make arrangements for an obit once he had a moment to spare (e.g. obit at altar of St Gabriel the Archangel, St Giles' church, Edinburgh, 1523).[74]

The death of a fellow guild member took our burgh-dweller to a funeral, where he contemplated a retable behind the craft altar (e.g. two Marys at the foot of the Cross, Paisley abbey),[75] and was reminded of the devotion of Jesus' family and friends. He also participated in a Corpus Christi pageant where the biblical narrative was re-told (e.g. Dundee, Lanark, Edinburgh, Perth).[76] In the procession, the confraternity of the Holy Blood carried a banner which reminded him not only of Jesus' suffering on his behalf, but also His willingness to listen to human concerns. Along with a depiction of the Crucifixion, Jesus looked down reassuringly at a layman (e.g. Fetternear banner belonging to the confraternity of the Holy Blood, St Giles' church, Edinburgh, after 1516).[77]

Apart from His role as sacrifice, Jesus was a kind friend, brother and beloved son of Mary. He had suffered on the Cross for humanity's sake, listened to its prayers,[78] helped it turn from sin,[79] and mediated with God on its behalf. The devotion and grief displayed by family and friends at the

Crucifixion reminded Scots that Jesus had formed loving relationships on earth, that His separation from the living was as heart-breaking as theirs would be, and that His self-sacrifice proved He loved humanity. In devotional collection BL Arundel MS 285, compiled about 1540, eleven of the seventeen woodcuts depicted Mary and Jesus together. In nine of these images Jesus was an infant. In one representation Mary looked lovingly down at Jesus' upturned face and in another He clasped her around her neck.[80] He was portrayed sitting in his grandmother's (St Anne's) lap while leaning toward his mother to investigate what she was holding in her hand, and rambunctiously jumping from Mary to St Anne.[81] Images in Archbishop Blackadder's prayer book encouraged the viewer to value Jesus' human sacrifice. There were several images of Him growing up as well as suffering on the Cross. In one particularly touching woodcut, a delighted baby Jesus stood in his walker.[82] Dedications of religious foundations also reminded people of Jesus' humanity. In 1522, David Fowrross, burgess of Haddington, founded a chaplainry at his altar of the Three Kings and the Blessed Virgin Mary,[83] and when the community of Peebles erected their collegiate church in 1541 they dedicated one prebend to St Mary in Childbirth.[84]

In the decades before the Reformation, Mary increasingly was depicted as a grief-stricken, middle-aged mother at the Crucifixion, and Jesus as her suffering young son, vulnerable, bloody and sad. This reflected people's growing fear of Judgement and hell and their reliance on Mary's intercession and Jesus' sacrifice to gain them access to heaven. Mary of Guise's Book of Hours included a 'Devotion to the Passion' and illustration of the Five Wounds,[85] and in Fowlis Easter collegiate church fragments of a Deposition painting portrayed Mary holding her dead son and a large canvas depicted Jesus' followers stricken by grief at the foot of the Cross. BL Arundel MS 285 depicted Jesus wearing the crown of thorns and holding up His pierced hands across from the prose prayer 'Remembrance of the Passion',[86] and across from Walter Kennedy's 'The Passion of Christ'[87] there was an image of Jesus being scourged. Just before the devotional poem the 'Golden Latany' in BL Arundel MS 285 He was portrayed bearing the Cross while blood flowed from His wounds.[88] Portrayals of Jesus suffering encouraged Scots to empathise with Him as they meditated on His Passion; meditation helped people focus on spiritual matters and displayed gratitude for Jesus' sacrifice. Distinguishing between God and Jesus in art and architecture also helped people empathise with Jesus' human suffering. Divine judge, creator father, punisher of sin and bringer of death, God often was depicted as a stern, grey-bearded middle-aged man (e.g. image of particular Judgement in Arbuthnott prayer book, image in Archbishop Blackadder's prayer book, and boss, Glasgow Cathedral).[89] Jesus, on the other hand, was depicted as a pleasant, comely young man (e.g. head of Jesus, St Andrews Cathedral). The Arbuthnott prayer book made the distinction stark. A powerful, stern, greying king on a throne held a small, thin, pathetic young Jesus across his lap.[90] Other images of the young Jesus were more powerful. The Fowlis Easter Trinity panel painting portrayed Jesus as young and personable, triumphant in glory as the risen lord, not the vulnerable victim of the Crucifixion. In the *Catechism* of 1552, it was a young Jesus who arrived to judge cowering bodies emerging from graves at the general Day of Judgement.[91] Yet as both triumphant lord of heaven and human victim it was the young Jesus to whom the laity gravitated, not the grey-bearded God.

After a long period of meditation on the Passion, and careful reflection on the magnitude of human sinfulness, our burgh dweller was ready to help stage his craft guild's play about the Creation and the Fall. His task was to purchase 'blood' and equipment for the devil and his minions (e.g. Perth hammermen).[92] He was sobered by the Crucifixion play staged by another craft. The demonstrable suffering of Jesus, and Mary's grief, reminded him of humanity's obligation to meditate on the Passion and thank Mary for her compassion and sacrifice. He clicked his rosary beads and murmured Ave Marias on his way home from the pageant. Later in the week he attended an 'allegorical pageant' or 'tableaux vivants' of the virtues (e.g. Edinburgh from 1503). Contemplating the personified virtues and their corresponding vices,[93] he felt renewed determination to live a moral life. He even considered buying a printed prayer book or collection of saints' lives, perhaps an inexpensive one with only a few illustrations (e.g. *Legenda Sanctorum*, Lyons, 1554).[94]

A few days later he attended the baptism of his new baby, and gazed at scenes of the Passion carved in bas-relief on the baptismal font (e.g. Seton collegiate church). Images of sinful Adam and Eve on the offering plate reminded him of Jesus' crucial role in human salvation (e.g. Fowlis Easter collegiate church), and the carved devil with bat's wings on the piscina offered a sharp warning that the devil was ever-present (e.g. Seton collegiate church (Fig. 3.4)). He rushed off to confess to his parish priest, overcome by an urge to cleanse himself of sin. Later that evening he had a long conversation over dinner with the craft chaplain (e.g. surgeons and barbers, cordiners, and skinners and furriers of Edinburgh)[95] about the state of the church roof, and the reports of clerical absenteeism and sloppy procedure at the altar where his mother-in-law's obit would be celebrated the following week (e.g. Perth altar foundation, 1504).[96]

Our burgh-dweller dashed off on pilgrimage to a nearby saint's shrine to pray for his wife, who was lying ill from a fever after the birth of their baby (e.g. shrine of St Mary of Loretto, Musselburgh).[97] He decided to cheer her up by buying her a pendant depicting Jesus on the Cross. It would reassure her that Jesus, God's beloved Son, was watching over her (e.g. pendants in the shape of a crucifix worn by George, 5th Lord Seton (c1530–85), and Mary, Queen of Scots).[98] He also founded an altar in his parish church and dedicated it to his wife's name saint (e.g. altar of St Barbara the Virgin, Perth, 1525),[99] hoping to gain the saint's prayers for his wife's speedy recovery. At home, he prayed to Jesus, Mary and St Margaret of Antioch. St Margaret might be willing to continue to intercede for his wife. After all, she had ensured that his wife and baby survived childbirth after his wife prayed to her at a new altar in the parish church (e.g. altar of St

Figure 3.4 Bas-relief on piscina showing devil with bat's wings. Seton collegiate church, early sixteenth century. Photograph by the author.

Thomas the Apostle and St Margaret the Virgin, Dundee, 1521).[100] His wife had decided to seek St Margaret's aid after listening to the saint's *vita* in church:

> ... quat-evir woman It be
> that in tyme of byrth callis on me,
> grant that woman but [ony] wath
> deliveryt be & the barne bath.[101]

However, if his ailing wife did not recover, he hoped that the Blessed Virgin Mary would give her three days advance warning of her death.[102]

Each day on the way to his stall in the high street our burgh-dweller passed an image of Mary and the baby Jesus carved on the market cross (e.g. Banff).[103] He genuflected and offered a quick prayer for his wife's recovery. Passing a collegiate church and royal palace, where instruments of the Passion had been carved in bas-relief, he was reminded of his debt to Jesus (e.g. Haddington collegiate church (Fig. 3.5), Falkland Palace). On the exterior of the Lady Chapel of the local cathedral, a bas-relief of a dragon reminded him that the devil was omnipresent (e.g. Glasgow Cathedral). He resolved to work harder to resist temptation, but given his enjoyment of sex and other sinful pleasures, it would be difficult to avoid a long stint in purgatory. He made a mental note to speak with church officials about founding another perpetual chaplainry for himself and his wife, so that after they died the living would continue to pray for their souls (e.g. chaplainry, altar of St Barbara, church of the Blessed Virgin Mary, Leith, 1499).[104] Finally,

Figure 3.5 Bas-relief of instruments of the Passion. Haddington collegiate church, fifteenth century. Photograph by the author.

he promised God, Mary and the saints that he would spend more time during religious services thinking about matters of the spirit and less about what was going on at work, where he was having trouble with a new apprentice.

CONCERNS ABOUT DEATH, JUDGEMENT AND HELL

Our burgh dweller's religious concerns and experiences were shared by people all over Scotland. People increasingly felt unworthy of heaven and susceptible to evil both within and without. They did their best to cultivate personal morality, engage in church rituals, do good deeds and hold firm in the face of death, but looked anxiously to the saints, Jesus and Mary for intercession. They also wanted the living to pray and celebrate church rituals on their behalf after death so that they could escape purgatory more quickly and ascend to heaven.

In the hundred years before the Reformation, death was 'harde, perelus, ande rycht horreble',[105] and its time and cause uncertain; infant and maternal mortality was high, plague devastated town and country, agricultural accidents often turned fatal, and wars and famines took a heavy toll.[106] Art offered poignant reminders of death and its indifference to earthly status. In Rosslyn chapel (Midlothian) a fifteenth-century foundation of the Sinclairs of Rosslyn, a bas-relief of the Dance of Death was carved near the main altar in the Lady Chapel. Each figure represented a different occupation and was accompanied by a skeleton. A ceiling boss in the early sixteenth-century Lady Chapel built by Archbishop Blackadder had a skull with a worm on it (Fig. 3.6), and Mary, Queen of Scots, gave her companion Mary Seton a silver watch with the Fall on one side and a skull on the other, reminding her of humanity's sinful nature and the inevitability of death.[107] A 'cart' carrying a corpse was painted on the ceiling at the entrance to Glasgow Cathedral's Lady aisle, and a woodblock of a horse-drawn cart carrying human remains appeared in *The Crafte to Lyue Well and to Dye Well* (1505), an English rendering of a 1503 Scots version of the French *Tractatus de Arte Bene Vivendi et Bene Moriendi*. A man and boy accompanied the horse-drawn cart[108] but, as the author of 'The Thre Prestis of Peblis how thai%tald thar talis' reminded his audience, family and friends departed at the graveside. Only 'almos deid and cheritie' accompanied people into the afterlife to 'slake the Judges [God's] ire'. That is, good deeds were the best means of gaining God's mercy at the particular Day of Judgement.[109]

Choice of burial site (lair) was an important consideration for Scots wishing to reduce their time in purgatory and reach heaven quickly. From poor craftsmen to kings, people wanted to be buried in consecrated ground.[110] A sampling of several hundred testaments from 1545 to 1559 in the dioceses of Glasgow, St Andrews and Dunblane suggests that testators wished to remain with the same spiritual community in life and death. That is, they wanted to be buried in their parish church or churchyard.[111] This was equally true of earlier generations. In 1506, for example, Sir David Sinclair of Swynbrocht donated money, goods and religious objects to a variety of churches, but was buried in his local church of St Magnus of Tingwall, Shetland.[112] Sometimes people were buried alongside relatives,[113] or in a site provided by a craft or merchant guild for members and their families.[114] But most people chose sites as close to centres of supernatural power as they could afford. Burial near a patron saint's altar brought his or her intercession, as did burial near altars dedicated to Mary. Sites near the high altar and in the choir were the most popular,[115] and thus the most costly.[116] To jog the memory of the living and ensure that anniversary prayers and masses took place at the correct location, candles or ornamentation could be placed at the tomb or a flat gravestone erected. As part of an obit foundation, Janet Paterson of Edinburgh spent 7s. on four candles to illuminate the tombs of her parents (1523),[117] and in 1456 Alexander Sutherland of Dunbeath requested a flat gravestone to identify his grave.[118] A series of Dundee burgh and court books from the 1520s and 1550s recorded numerous payments for burial sites in the parish church. On 30 September 1521, the burgh stated that the fee for burial in the choir was 10 merks, and that no one would be buried there unless his friends first paid up.[119] Yet the burgh sometimes used burial sites to reward good service. In 1516, common clerks Robert Seres, elder, and Robert Seres, younger, were given free burial sites and bell-ringing at their funerals as a reward for past and future service to the burgh.[120] The burgh council eventually decided to subsidise the poor, declaring on 5 October 1556

Figure 3.6 Boss depicting worm in skull. Lady Chapel, Glasgow Cathedral, early sixteenth century. Photograph by the author.

that people who could not afford a burial site would not be charged.[121]

At death, body and soul separated, and the soul rose to the particular Judgement, after which it departed for heaven, hell or purgatory, usually the last. The body was buried, awaiting reunion with the soul at the general Day of Judgement. After the general Day of Judgement people were sent to heaven or hell, purgatory no longer being an option. Visual images helped men and women understand death and the judgement process, and provided a focus for meditation and prayer. In Sir Robert Arbuthnott of Arbuthnott's prayer book, an illustration depicted a woman's death and the particular judgement. The woman's body lay on a bed while a priest sprinkled her with holy water in the sacrament of extreme unction. As she was being cleansed spiritually, her husband and two children solemnly leaned over her bedside. Above the woman's prone body two angels held her soul aloft in a white sheet. This image depicted the moment of death when body and soul separated. The body remained on earth and the soul was carried upward to a solemn, white-bearded God sitting in Judgement. By depicting her soul as a vulnerable, white-skinned baby, viewers were reminded that death returned people to their original state. Souls lost all knowledge, prestige, power and goods, the protection of church ritual, and even emotional ties with family and friends.[122] A fifteenth-century bas-relief of a similar deathbed scene depicted adult family members and a priest clustered anxiously around the bedside (Fig. 3.7). Also present were three children, each in varying states of distress. One very small child huddled near the head of the bed, a slightly older one knelt at the foot of the bed, holding on to his mother's feet with his hands, and another child, his back to the bed, leaned on a small desk with his head in his hand.[123] These images were familiar, poignant reminders of the wrenching loss many families experienced. It was common, for example, to lose young women in childbirth.

Deathbed images reminded people that they might die suddenly, and that they should take steps to prepare themselves for the afterlife by 'dying well', or rather, preparing for death throughout life. Individual activities that helped people 'die well' included prayer to God, Mary, Jesus and the saints, and meditation on the Passion. It was also important to adhere to orthodox beliefs, make religious foundations and donations, and participate in a variety of church rituals, particularly penance and the Eucharist. All these activities helped people live ethically according to the Ten Commandments, develop a strong personal morality, and shift their attention from sex, family and worldly entertainments.[124] Preaching aided spiritual development by focusing people's minds on spiritual matters and educating them on the theological basis of the faith. Consequently Scotland's provincial council in 1549 promised to increase preaching.[125] Preaching might be done by priests, bishops or specially appointed preachers,[126] or by friars, who were popular

Figure 3.7 Bas-relief of deathbed scene, Mary King's Close, Edinburgh, late fifteenth century. Courtesy of the Trustees of the National Museums of Scotland, NMS E/0079.

in Scotland[127] despite reformers like George Wishart describing them as 'sergeantis of Sathan, and deceavaris of the soules of men'.[128] The *Catechism* warned priestly advisors that people surrendered to sexual and other worldly pleasures and to the devil. He urged advisors to convince lay people to submit to the discipline, correction and penance decided upon by priests during confession.[129]

Despite the warnings of writers and artists, most people began preparations for death near the end of their lives. Consequently advisory texts often focused on the final hours of life. Demons swirled around deathbeds tempting the dying in devotional works such as *The Book Intytulid the Art of Good Lywyng and Good Deyng* (1503) (Fig. 3.8). Yet alongside these demons, priests administered the last rites, and angels, Jesus, Mary and the saints waited patiently. Whatever happened in these last moments, the Church, Holy Family and the saints would help people fight temptation and maintain their faith in God.[130] Extreme unction, the restoration of ill-gotten goods, donations to the Church, the foundation of perpetual prayers and obits and last-minute confessions, all might be part of those final hours. In Gavin Douglas' allegory 'The Palice of Honour' (c1530), the protagonist waited in purgatory before going on a long journey to heaven to face a king (God) who valued good works rather than people's status on earth.[131] When Nicholas Baxter, burgess of Aberdeen, founded an obit and a weekly mass of the Five Wounds at the altar of the Holy Cross in the nave of his parish church, he claimed that he wished to

> anticipate by works of mercy that day of last judgement (which to those who have done good works in this life shall be full of joy, but full of anxiety and sadness ever-during to those who do the opposite), so that the things which we have sown on earth we may be able to reap with multiplied fruit in heaven ...[132]

Yet *The Craft of Deyng*, written in the mid-fifteenth century, instructed priests to advise parishioners that God's mercy came not from works but through faithful love for God, contrition, and a determination to die in the sacraments of the Church.[133] By and large people were urged to trust in God and resist despair,[134] rejoicing in death as the natural ending of life and the beginning of eternal joy with the Father.[135] Executors of testaments could, of course, reassure the dying that what was left undone would be accomplished a short time after their death. Executors disposed of testators' goods, and settled debts to benefit testators' souls. They founded obits or other religious services, made donations to hospitals or schools, and otherwise improved the testators' standing in purgatory.[136] Testators in mid-sixteenth-century Glasgow, St Andrews and Dunblane dioceses invariably issued a final threat to executors, reminding them that they would be answerable for their actions to the 'high judge' (God) on the Day of Judgement.[137]

Controlling sexual urges was of tremendous concern to commentators, for sexual sins were a primary reason for being denied access to heaven. Hell became associated with punishment for sexual excess, and the devil with

Figure 3.8 Woodcut of demons surrounding a deathbed. *The Book Intytulid the Art of Good Lywyng and Good Deyng*, 1503. Courtesy of the Master and Fellows of Emmanuel College, Cambridge, Emm. Coll. Lib. S9.2.51.

sexual temptation. In Rosslyn chapel's choir, finished in 1484, a bas-relief of the seven deadly sins placed entwined lovers next to the devil, who raked them in to hell. Women in particular needed to cultivate sexual purity.[138] Robert Henryson urged women to wear chastity like an undergarment, close to the skin.[139] Sir David Lindsay was convinced that most female inhabitants of hell were being punished for sexual crimes for which they had not expressed contrition or made confession.[140] As noted above, a painting in Fowlis Easter collegiate church depicted St Catherine ramming a large sword through the head of a man who had threatened her virginity. The image sent a clear message: sexuality was evil, and denial of it gave women high status in heaven. On a spare page in the Acts of the Parliaments published in 1540, the woodcut depicting heaven contained married women and chaste widows, although virgins were the most numerous. The more women restrained themselves sexually, the better off they were spiritually.[141] One option was to lead normal lives as wives and mothers and then enter convents or join the confraternity affiliated with the Franciscan order with their husbands and take vows of chastity.[142]

The painting of the Crucifixion above the rood screen in Fowlis Easter collegiate church reminded people that it was not safe to rely entirely on the actions of the living for protection from hell (Fig. 31.3). Carrying out the penitential sacrament of contrition, confession and satisfaction was essential to success in the afterlife. In the painting, the naked, baby-like soul of the unrepentant criminal crucified with Jesus was carried immediately towards hell; it was given no second chance in purgatory. Yet even if people made sufficient satisfaction for sin before death to be sent to purgatory, there was still a chance they would be sent to hell after the general Day of Judgement. In hell people suffered greatly, for soul and body had reunited by that time.[143] Bodily pain was easy to imagine for people who lived in an era with limited medical treatment. To frighten the living into repentance, and to encourage them to found prayers and masses for the dead, art and literature emphasised the derisive cruelty of the devil and his fiends, the pains and torments experienced by sinners, and the strength and loathsome nature of the devil. In devotional collection Harleian MS 6919 (c1550), William of Touris' 'The Contemplacioun of Synnaris' depicted the devil as a scaly, goat-like creature on a throne, delightedly supervising the torture of souls.[144] Text and image worked together to convince people of the terrible, endless torments of that 'dirk dungeoun' and 'perpetuall presoun'.[145] The devil was often inhuman and ugly, ugliness being associated with evil.[146] The winged fiend in the Fowlis Easter painting breathed fire through its open jaws as it grasped the unrepentant sinner's soul in its claws, and dragons were carved on ceiling bosses and an exterior wall of Archbishop Blackadder's Lady Chapel. In Rosslyn chapel's north wall the devil was portrayed as a human with the ears and horns of a goat. Ugliness of character often was equated with dark skin, leading to descriptions of the devil as 'blak as pik'.[147] The demon or devil who mingled with the crowd in the Fowlis Easter's Crucifixion painting had dark skin and white flashing teeth as well as horns. Possibly he was meant to appear Moorish; certainly William Dunbar's devil was a Muslim in 'The Dance of the Sevin Deidly Synnis'.[148] In keeping with this idea of the devil as 'infidel', or fallen angel, the devil in pageants was dressed to symbolise all that was evil and against God.[149] In Rosslyn chapel the devil appeared upside down, bound with a rope. In 1518, the devil, his man and three tormenters joined Adam and Eve for the Corpus Christi play of Perth's hammermen.[150] By 1553 the craft was focusing even more heavily on the war between good and evil. The cast had begun to include a 'big' devil, his chapman and a serpent to represent evil and the devil, and two angels to represent goodness and God's will. Payments for blood and torture implements suggest scenes of violence and/or torture in hell. By vicariously experiencing hell's pains, people learned which sins were punishable in hell. By 1553, the only saint remaining in the cast was St Eloy, the craft patron, a rather mild-looking individual if the craft statue which survives is any indication.[151] People were coming to rely less heavily on saintly intercession and saintly merits to achieve salvation. Increasingly they took personal responsibility for the eradication of sin,[152] and threw themselves on the mercy of Jesus and Mary, the most powerful intercessors available to them.

GAINING THE AID OF MARY, JESUS AND THE SAINTS

As people cultivated personal relationships with supernatural beings, particularly the Blessed Virgin Mary and Jesus, devotional books came in handy. They inspired and guided private devotion, and could be used during long Latin services. Books of hours were popular as they included vernacular translations of parts of the Breviary (e.g. Office of the Virgin, penitential psalms, litany, office of the dead and suffrages of the saints), and thus helped people understand church services. The rosary or prayers of the 'Ave Maria' (Office of the Virgin) was the most popular, and could be recited in church in a low voice.[153] The *Catechism* encouraged lay devotion to the rosary, deeming it the most effective method of approaching Mary and improving spiritual standing.[154] In 1484 Abbot George Shaw built a wall enclosing Paisley abbey's lands. Along the wall were niches for statues of saints. One inscription emphasised Mary's critical role in the workings of divine grace: 'Go not this way unless you have said *Ave Maria* / Let him be always a wanderer who will not say *Ave* to thee'.[155]

Mary helped the living lead a moral life, encouraging them through her example to eschew sin. She also counselled Jesus to purge people of sin, forgive them their sins, and allow them into heaven.[156] She brought them news of imminent death,[157] and when death came she fought the devil and stood by them, hustling them off to the care of the angels.[158] A bas-relief in Rosslyn chapel depicted Mary shepherding a man away from a chagrined Satan, her arm flung protectively over the man's shoulders (Fig. 3.9). Mary was often depicted as a powerful queen in heaven. She stood next to Jesus in a Resurrection painting in Fowlis Easter collegiate church, and appeared as a crowned queen of heaven in the Arbuthnott prayer book and the illustration of the afterlife in the Acts of the Parliaments of 1540.[159] In the arms of the royal burgh of Cullen she was arrayed in the full trappings of state. A crowned Mary sat on a gold throne-like chair, holding the baby Jesus and a sceptre surmounted by a gold fleur-de-lys.[160] Mary's power was allied to compassion. She comforted and guided people on the road to heaven, where she had the ear of the king (God) and interceded for them.[161] 'Suppos all sanctis our synfall prayere contempne / Thyne eres ar' ay opyn at our reqwest'.[162]

Marian and Passion images were often entwined. This reinforced a popular notion that Mary was co-redeemer with Jesus,[163] reminded people that mother and son were committed to humanity's spiritual welfare, and encouraged viewers to empathise with their emotional and physical suffering. Jesus and Mary's monograms, the Marian symbols of lilies, roses, daisies, stars and moons, and Tree of Jesse, Annunciation and rosary images were depicted alongside instruments of the Passion. The font in St John's church, Aberdeen, bore a crowned Gothic 'M' along with rose and Passion symbolism (c1525),[164] and a Missal binding produced in 1506 was decorated with a Tree of Jesse as well as instruments of the Passion.[165] Fowlis Easter's sacrament house placed Jesus' head, flanked by angels holding the pillar of

Figure 3.9 Bas-relief of Mary leading a layman away from the devil. Rosslyn chapel, mid-fifteenth century. Courtesy of the Rosslyn Chapel Trust.

scourging and the Cross, below a bas-relief of the Annunciation. Medieval stained glass windows in St Bride's church in Douglas, Lanarkshire, had Tree of Jesse imagery,[166] and oak panels carved for Cardinal Beaton in the mid-sixteenth century placed Mary and the baby Jesus at the top of the Tree of Jesse with the apostles on branches below.[167] Thus eternal life and assistance came through Mary and her son Jesus, and only secondarily through the prayers of the apostles. On a missal binding from St Nicholas church, Aberdeen, a Tree of Jesse rested below an inscription invoking Mary's aid,[168] echoing a reference to Mary as the tree of life in 'The Lang Rosair' in BL Arundel MS 285.[169]

Easter was a time to meditate on Jesus' sacrifice but also to remember Mary's emotional suffering and consequent intercessory power. During Lent the collegiate church of Crail hung a veil of red worsted cloth over Mary's image,[170] and in the Arbuthnott prayer book and Fowlis Easter Crucifixion painting Mary stood grieving at the foot of the Cross (Fig. 31.3). In BL Arundel MS 285 she knelt on the ground, stabbed through with swords, and in another image she held a vulnerable, dead Jesus across her lap, swords stabbing into her body to represent a mother's grief.[171] The illustration of the afterlife in the Acts of the Parliaments of 1540 made her intercessory power very clear. A circlet of roses surrounded heaven. Below heaven, souls in the flames of purgatory stretched their hands upward while various figures knelt on earth, clasping rosaries and praying to Mary. Superimposed upon the heavenly scene, within the circlet of roses, Jesus hung on a Cross festooned with roses.[172]

Saints apart from Mary were advocates in the struggle for moral purity and divine acceptance. Individuals founded and augmented masses and chaplainries in their honour, and donated ornaments, paintings, statues, altar cloths, monies and wax to their altars and chaplainries to gain their favour. For example, William Kemp and his wife donated a silver chalice to St James' altar in Haddington parish church and had their 'mark' inscribed on it. John of Crumme and his wife donated a silver rood chalice to the Rood altar in the same church. It was decorated with a crucifix in the foot and inscribed with their names.[173] St Michael the Archangel was a popular saint, often depicted attacking the devil. As the devil represented humanity's susceptibility to sin as well as evil forces in the world, this image reassured people not only that St Michael was a powerful saint, but also that, in general, the forces of good won out over evil. On the seal of Patrick, bishop of Moray, St Michael stood tall as he drove a sword through a human devil at his feet,[174] and in the prayer book of Archbishop Blackadder an armoured St Michael on horseback speared a dragon.[175] On the tomb of Alexander Macleod of Dunvegan on Harris, a winged St Michael fought hand-to-hand combat with a bestial devil,[176] and in a 1487 copy of the popular *The Golden Legend* by Jacobus de Voragine, a winged St Michael in full armour wielded a sword over a devil who was part human and part animal (Fig. 3.10).[177] Another powerful saintly advocate was St John the Evangelist. The prayer 'O Intemerata' described him as a blessed and familiar friend of Jesus, and enjoined supplicants to pray to St John and Mary to show them their sins and intercede on their behalf on the Day of Judgement.[178] Visual prominence reminded people of St John's value to humanity. A panel painting in Guthrie collegiate church displayed Mary and St John on their knees interceding for sinners at the general Day of Judgement,[179] and Jesus, Mary and St John were portrayed in stained glass windows in St Bride's church in Douglas.[180] St Salvator's collegiate church in St Andrews owned a silver cross for use in processions which was gilded

Figure 3.10 Woodcut of St Michael and the dragon. *The Golden Legend*, Lyon, 1487. Courtesy of the Trustees of the National Library of Scotland, NLS Inc.289.5, f. 3v.

with images of Mary and St John, and images of the Trinity, Mary and St John decorated the provost's cap.[181]

THE USES OF CHURCH RITUAL

Individual efforts to cultivate personal morality and relationships with supernatural beings, or to otherwise improve spiritual standing, were encouraged and valued. However, most religious experiences occurred within a group context, whether in royal, magnate and lairdly households, burgh and parish communities, merchant and craft guilds or families. Communal experiences also incorporated individual faith experiences. For example, people read devotional literature and recited the rosary during church services. In 'The Tua Mariit Wemen and the Wedo', William Dunbar mentioned women taking prayer books to church.[182]

There is plenty of evidence that celebration of the Eucharist was an important element of religious life in royal, magnate and lairdly households. Malcolm, Lord Fleming, reflected a widely held view when he declared that, if celebrated properly, the Mass could 'snatch the souls of the faithful departed from the pains of purgatory, and bring them to the full enjoyment of blessed glory'. In the foundation charter of the collegiate church of Biggar (1546), he made provision for the quick dismissal of clerics who performed poorly, assigned fines for absenteeism at services, and posted a register of services in the college as a guideline for clerical staff.[183] Monarchs moved Eucharistic equipment when they changed residences. For example, on Corpus Christi Day in 1513, James IV paid 13s. to carter Alexander Ramsay to carry the chapel gear from Stirling back to Edinburgh.[184] Several magnates and lairds in the late fifteenth and early sixteenth centuries requested the right to have portable altars to facilitate household worship. William, third Lord Erroll, Constable of Scotland (d. 1507) and his wife Elizabeth Leslie (d. 1511) requested the right to a portable altar, claiming that they did so out of sincere and fervent devotion. In 1483 Sixtus IV granted their request, making it possible for them to have priestly functions, masses and other divine offices celebrated in their own and their family's domestic quarters. That the lords of Erroll believed in the efficacy of the Eucharistic rite is borne out by the records of the General Assembly of the Scottish Church. In 1592 the Assembly reported that the 'idolatrie of the Mess' was celebrated in the house of the earl of Erroll in Logiealmond, Perthshire, and Slains, Aberdeenshire, and that he otherwise subverted the 'true religioun'.[185] Also citing sincere devotion to the Church and personal religious fervour, in 1490 Marion Scrimgeour and her husband Robert Arbuthnott of Arbuthnott requested the right to a portable altar to facilitate regular celebration of mass and other divine services.[186] In 1502 the papacy allowed William Graham of Kincardine, Archibald Edmonston of Duntreath, Stirlingshire, and Patrick Home of Polwarth, Berwickshire, to have portable altars. This facilitated celebration of the Eucharist in the presence of families and household servants whenever they wished, even under interdict.[187] A late medieval portable altar from Coldingham Priory, Berwickshire, is part of the collection of the National Museums of Scotland in Edinburgh. It is a small, flat, thin, square piece of sandstone marked with five crosses, one in each corner and one in the centre; the central cross is carved with a crown of thorns, a reminder of the historical suffering of Jesus re-enacted in the Eucharist.[188] This altar could easily have accompanied a peripatetic lairdly, magnate or royal household. As we have no record of household attendance at religious services, we must assume that members of households agreed with the religious preoccupations of their employers. Post-Reformation evidence does suggest that households functioned as Christian communities. In 1587, for example, the General Assembly lambasted the laird of Leslie for holding public Mass in his household with 'two idoles above the altar'.[189]

Many collective religious experiences were enjoyed in rural and urban parishes. For example, people joined together to watch and participate in processions, religious pageants and plays. Sir Richard Maitland of Lethington remarked that Scots expressed their devotion to God 'outwardly' through processions.[190] In April of 1534 Haddington's crafts consented of their own free will to produce pageants on Corpus Christi Day,[191] and Edinburgh's burgesses carried a wooden image of burgh patron St Giles in the St Giles' Day procession in 1552. Sir David Lindsay criticised both laity and clergy for placing their trust in religious images and activities, particularly those associated with the saints. He agreed that images of saints helped teach the illiterate, and accepted the clerical argument that knowledge of saints' sufferings could fortify the faith. However, he believed that the prayers, offerings, processions and pilgrimages popular with his contemporaries too easily became the basis for idol worship.[192]

Parishioners congregated for baptisms, marriages and funerals, and church services focused on celebration of the Eucharist. Holy water cleansed Original Sin, as people were reminded in poetry[193] and devotional works as well as the liturgy and preaching. In *The Book Intytulid the Art of Good Lywyng and Good Deyng* a priest grasped a Bible in one hand and in the other a scoop of water. He held it poised over the head of a naked baby who was being held by his father and mother over the baptismal font.[194] Holy water also healed and defended against the evils of the devil and one's neighbours, as people were reminded in Seton collegiate church, where bat wings and the face and ears of the devil were carved on the piscina. Water associated with shrines or relics was particularly effective as a healing agent. In 1500, Alexander Mylne, bishop of Dunkeld, reported that a bone of Saint Columba dipped in water had helped cure parishioners of the plague.[195]

Funerals prepared the deceased spiritually for the afterlife. Not all people were willing to trust executors to make funeral arrangements. For example, Katherine Carmichael, lady of Cambusnethan, whose testament was drawn up in Kilmarnock in 1552, gave £20 toward funeral expenses at the parish church of St Aidan.[196] Margaret Hutcheson of 'Russall's Mill', St Andrews diocese, left 30s. to Sir Walter Airth, no doubt for funeral services. This amount was a much greater percentage of her net worth (£4 9s. 4d.) than that left by many Scots.[197] Margaret Fullarton (d. 1550), wife of John White and inhabitant of Irvine, made her husband and son executors, but stipulated that 5s. be paid to Irvine's vicar, 6s. 8d. to the friars minor of Ayr, and 2s. to each person offering prayers for the dead at her funeral.[198] Friars and monks included donors in regular prayers, although those best served made large donations or offered continuing support through assignment of annual rents from properties. The Preceptory of St Anthony's, Leith, clarified the matter in 1526, stating that each Sunday in perpetuity it would pray for people who gave annual rents, or augmented religious services through a foundation, or had in other ways given substantially of their goods to the building, reparation or maintenance of their abbey and hospital.[199]

People in late medieval Scotland supported a number of church rituals, but none were as effective in cleansing sin as the Eucharist.[200] In Dunbar's words, 'Thy Haly Supper for my syn recompence, / And of my gilt the holy satisfactioun'.[201] The late fifteenth-century theologian John of Ireland argued that the Passion of Jesus was the spoke upon which the wheel of salvation turned. The Eucharist brought participants closer to heaven by re-enacting the historical Passion, and made other sacraments effective by bringing God's mercy and grace.[202] The real presence of Jesus in the Eucharist ensured the sacrament's value as recompense for human sin. Viewers were reminded of this in a bas-relief in Rosslyn chapel which depicted a young Jesus holding the Eucharistic chalice. In an illustration in Sir Robert Arbuthnott of Arbuthnott's prayer book, the Eucharistic chalice rested on an altar before which three people knelt, two of them clerics (Fig. 3.11). Above the chalice a young, bearded Jesus rose from his tomb. Blood dripped from the crown of thorns and the nail holes in his hands. He pressed the gash where Longinus' spear had pierced him, causing the blood from his side to ripple down into the chalice on the altar.[203] Jesus' role as human sacrifice in the Mass could not have been more clearly expressed. In 1551 Elizabeth Gordon and her spouse Alexander Ogilvy of Deskford and Findlater built a sacrament house in Deskford, Banffshire, to house and protect the consecrated Host. Its inscription, dating from 1554, confirmed that Gordon and Ogilvy believed that the bread and wine (body and blood) of the Eucharist were the source of salvation.[204] The inscription, along with a bas-relief of two angels raising a monstrance, reminded viewers that Jesus was made really present through the elevation of the bread in the Mass. Seton's fifteenth-century octagonal font linked the power of the Crucifixion to the sacrament of baptism by depicting the Crucifixion and the events leading up to it in bas-relief on its exterior.

Jesus had suffered at the Crucifixion, and the process continued in the sacrifice of the altar (Eucharist). People were reminded of Jesus' suffering through artistic representations of Crucifixion implements and the sites on Jesus' body where He had suffered the greatest pain. Common 'instruments of the Passion' were nails, a crown of thorns, whip, ladder, spear and pillar,

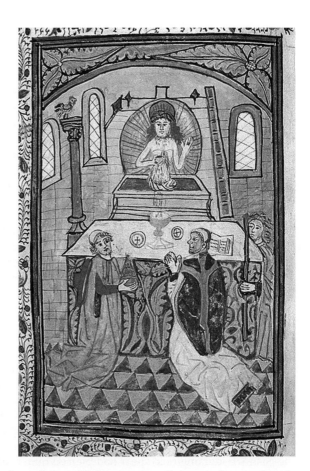

Figure 3.11 Woodcut of the real presence in the Eucharist. Arbuthnott Prayer Book, c1482–3. Courtesy of Renfrewshire Council.

as well as a heart, hands and feet. Nails, crowns of thorns, hearts, hands and feet were carved in bas-relief in Haddington and Seton collegiate churches and Glasgow Cathedral. The symbols appeared on a shield above the central pillar of the west doorway at St Mary's, Haddington, on corbels of buttresses on the south and north exterior walls of Seton, and on ceiling bosses in Archbishop Blackadder's Lady aisle, Glasgow. Passion symbolism even decorated primarily secular buildings, providing daily reminders of Jesus' suffering. During James V's reign, angels carrying instruments of the Passion were carved on consoles below niches for saints along the exterior wall of Falkland Palace's chapel wing. The Fetternear Banner owned by Edinburgh's confraternity of the Holy Blood was decorated with various instruments of the Passion. Such processional banners reminded onlookers of Jesus' sacrifice on their behalf.[205]

Apart from the weekly High Mass,[206] private or 'low' masses improved the spiritual standing of absent founders and other named souls. In principle, prayers and masses benefited all 'neighbours', enemies as well as friends,[207] but in practice they were intended to benefit named souls, usually those of the founder(s), his or her immediate relatives, benefactors and the faithful dead. Low masses were inexpensive, requiring few paid participants, and could be dedicated to saints, Mary or the Passion itself. The increasing emphasis on Jesus' historical and Eucharistic suffering, and real presence in the Mass, led to the foundation of masses of the 'Passion', 'Name of Jesus', 'Five Wounds' of Jesus,[208] 'body of Christ' and 'Holy Blood'. On 8 September 1519, Alexander Malison, burgess of Aberdeen, and his son Gilbert founded a mass of the Five Wounds in St Nicholas church on Fridays, a day traditionally reserved for commemorative masses of the Passion.[209] The mass was to be celebrated at the altar of St John the Evangelist by the curate and choir chaplains,[210] an unsurprising choice given St John's power as intercessor on the Day of Judgement.

Dedicating obits, chaplainries, altars, chapels or aisles and prebends to Jesus in His Passion brought the full power of the Crucifixion to the aid of donors and named souls. They also gained the prayers of other supernatural beings associated with the foundation (e.g. chaplainry of St Catherine at the altar of the Holy Rood in the aisle of the Blessed Virgin Mary). Obits offered annual prayers and masses to help souls suffering in purgatory. On 25 January 1497, Robert James, citizen of Brechin, founded an obit which included a mass of the Name of Jesus to be celebrated on the 40 weekdays following the obit.[211] In 1551 Marion Bruce, wife of burgess John Gray, founded an obit at the altar of Corpus Christi (body of Christ) in the parish church of St Michael, Linlithgow, for the souls of chaplain and curate Sir Henry Louk, who in 1540 appeared in the records as chaplain of the altar of the Blessed Virgin Mary and curate of Linlithgow.[212] In 1556 Peter Newlands also founded an obit at this altar, but did so for the souls of the late James Naismith and Elizabeth Louk his spouse.[213] In 1527 John Murray founded an obit and daily mass at St Salvator's altar in St Nicholas church, Aberdeen. Two years later his executor Patrick Gordon of Methlick, Aberdeenshire, founded a trental of masses for Murray's soul.[214] In 1493 Alexander Donaldson founded a chaplainry at his new altar of the Holy Cross in the monastery church of Jedburgh,[215] and in 1523 John, Lord Hay of Yester, founded a chaplainry at the altar of the Holy Cross in the collegiate church of Bothans.[216] In 1508, Ade Hepburn of Cragis, one of the 'familiaris' of James IV, founded a chaplainry at the altar of the Holy Wounds of Jesus in the cathedral church of Aberdeen, the altar having been founded by Master Alexander Cabell, rector of Banchory and canon of Aberdeen.[217] In 1506 an altar of the Holy Rood of Christ was founded by burgess William Cunningham in the aisle of the Blessed Virgin Mary in the parish church of Dumfries.[218] In St Giles' church, Edinburgh's merchants founded a lavish aisle in honour of the Holy Blood and placed a bas-relief in the chapel decorated with a crown of thorns and heart. The cult of the Holy Blood, favoured by Scottish merchants, emphasised Jesus' self-sacrifice. The 'Holy Blood' was the drops of blood shed from Jesus' side, the same blood which flowed from Jesus' side into the altar chalice in the Arbuthnott prayer book.[219] Cullen (1543) and Peebles (1541), primarily lay collegiate foundations, both had prebends dedicated to the Holy Cross. By the Reformation, St Giles', Edinburgh, had prebends dedicated to the Holy Rood, St Salvator and the Holy Blood, and St Mary in the Fields, Edinburgh, had a prebend dedicated to the Name of Jesus.[220]

Many prayers and masses were slated to commence after the death of the founder when souls needed assistance in purgatory. Although Jesus' Crucifixion had made complete satisfaction for 'actual' human sin, the *Catechism* argued that people continued to sin. Scots could even be held collectively responsible for the sins of individuals (e.g. the king). The sacrament of penance helped the living make satisfaction for 'dreggis of syn', depending on their quantity and quality,[221] but most people believed it unlikely they would be free of sin at death. Complete satisfaction for sin came after death, and readied body and soul for reunion at the general Day of Judgement.[222] In the decades before the Reformation there was an increased reliance on lay prayer to lift souls from purgatory; founders assigned substantial sums for lay prayers, whether as part of daily mass, obit, or chaplainry foundations. Soliciting lay prayers as well as clerical prayers and masses pleased God as a good deed and reminded Him that Jesus' historical Passion had secured human salvation. It also gained the intercession of Mary, Jesus and the saints.[223] The poor, elderly or infirm were the most willing to offer prayers, for they received monies and/or food and drink in return for attending obit services and praying for the deceased. On 19 December 1492, Andrew Moubray, burgess of Edinburgh, founded an obit at his newly founded altar of St Ninian in St Giles', Edinburgh, and ordered that money be distributed to paupers during the obit. He wanted the foundation charter to be read at the altar after Mass in the presence of the people and the church dean to ensure that the obit was carried out according to his very specific instructions.[224] Founders of obits at the parish church of Ayr assigned increasingly large sums to poor people praying at religious services. In 1507, Adam Wishart (d. 1507) decided that 20 per cent of the funds for his father John's obit would be distributed to the poor after the obit-mass.[225] Several years later John Brown (d. 1524) and his wife Mariote Petheid assigned 43 per cent of obit monies to the poor, more than that paid to clerical celebrants. The foundation charter declared that helping the poor was part of the reason for making the foundation. But Brown, like other Scots, wanted more than the poor to pray for him. He asked that church bells be rung to call Ayr residents to his obit-mass. In this manner the 'people' would be summoned 'to put forth prayers, and intercede with God, and his spotless mother, for the souls of the said John and Mariote'.[226] The laity came to believe that the quantity as well as quality of prayers mattered. Consequently, wealthy Scots as well as parish communities founded collegiate churches to ensure daily celebration of masses for souls (e.g. Biggar (1546) and Peebles (1541), respectively). Yet even annual obit-masses brought a substantial number of prayers, if only once each year. Belief in the value of lay prayer continued beyond the Reformation. According to burgess Robert Birre, in October of 1566 Edinburgh received word that the Queen was 'deadly sick'; she asked that the bells be rung so 'all the people' could go to church to pray for her.[227]

Besides prayers and masses, founding schools and hospitals raised the spiritual stock of souls in purgatory.[228] Benefactors could also gain the favour of God, Mary and the saints by making donations to religious houses. In 1532 the hospital of St Mary Wynd, Edinburgh, was singled out for assistance by John Levington and his wife Christian Lamb. On 9 September they donated an annual rent to the widows or bedeswomen of the hospital and their successors. In return, the women were to keep a lamp burning all year in front of the statue of the Virgin Mary in the chapel.[229]

CONCLUSION

Far from leaving it up to the clergy to get them to heaven, late medieval Scots took personal responsibility for spiritual improvement. They carried in their mind's eye an image of the afterlife and the supernatural beings which peopled it, and chose individual and communal activities to ensure success after death. Propitiation of saints, Mary, Jesus and God through personal prayer and church ritual, meditation on the Passion, careful choice of burial site, and 'dying well' were ways they built spiritual capital, hoping to avoid hell, reduce time in purgatory, and gain access to heaven. In particular, they strongly supported the Eucharist and tried to restrain their sexuality. As life was hard and God's will absolute, death could come at any time. Artists and writers urged people to fear God and Judgement, resist sin, learn about the virtues of the Passion and the sacraments, and die in conformity with orthodox belief.[230] It was a tall order, but Scots were willing and able to act on their own behalf, and knew what was needed to reach the eternal joy of heaven.[231]

NOTES

1. Simpson, 1863, 216.
2. Scribner, 1981, 95. Cf. Scribner, R. Magic and the formation of Protestant popular culture in Germany, *Popular Culture in Question* (conference), University of Essex, April, 1991.
3. The cabinet possibly belonged to Mary, Queen of Scots, in NMS, Edinburgh.
4. Urquhart, 1973, 79.
5. Jung, 1988, 268.
6. Mylne, 1915, 328.
7. Thomson, 1839–45, ii, 716–7 and 724 (Session of 6 February 1587), iii, 829–30 (Session of 5 January 1592), 949 (Session of 31 March 1596), and 1047–9 (Session of 27 July 1608). Cf. McLennan, B. The Reformation in the burgh of Aberdeen, *Northern Scotland*, 2 (1976–7), 128–9 and 133–8.
8. Letter from James V to his sheriff in Aberdeen, referring to reports of heresy by Gavin Dunbar. In Stuart, 1844, i, 110–1.
9. Leslie, 1888–95, ii, 226–7.
10. Thomson and Innes, 1814–44, ii, 370.
11. Gau, J. The Richt Way to the Kingdome of Hevine. In Laing, 1855, iii, 357.

12. Alexander Ogilvy of Findlater, Alexander Dick, archdeacon of Glasgow, John Duff of Muldavit, the bailies, councillors and community of Cullen, and the parishioners of the church founded the collegiate church. Cowan and Easson, 1976, 218.
13. Minutes of the Edinburgh goldsmith craft (1520s), NAS GD1/482/1, f. 7r.
14. e.g. Thomson, 1839–45, i, 784 (Session of 4 August 1590). Cf. Fleming, 1889–90, i, xvi–xvii.
15. Chaplainry at the altar of St Thomas. Balfour Paul and Dickson, 1877–1913, ii, 3650, p. 789.
16. Balfour Paul and Dickson, 1877–1913, vii, 132.
17. Macmillan, 1990, 18.
18. Balfour Paul and Dickson, 1877–1913, i, 64.
19. Österreichische Nationalbibliothek, Book of Hours of James IV and Margaret Tudor, Codex 1897, *passim*.
20. e.g. Brown, 1891, 57.
21. McRoberts, 1959, 283.
22. Maxwell, 1891, 30.
23. Stuart, 1844, i, 56–7.
24. Cooper, 1888–92, ii, 354; Stuart, 1844, i, 151.
25. Renwick, 1887, 68.
26. Pryde, 1937, 27.
27. Chambers, 1872, 43–5.
28. Charter of obit foundations by George of Spalding, burgess of Dundee, NAS GD76/152.
29. Stuart, 1844, i, 75; Cooper, 1888–92, ii, 343. The providers of sustenance were the 'gudmen' of the burgh. Cooper, 1888–92, ii, 354 (1521).
30. Cooper, 1888–92, ii, 360.
31. Stuart, 1844, i, 143.
32. Stuart, 1844, i, 301.
33. Mill, 1927, 72.
34. Marwick, 1871, 55.
35. Balfour Paul, et al., 1882–1914, iii, 435, pp. 96–8.
36. Marwick, J D and Renwick, R, eds. *Charters and Other Documents Relating to the City of Glasgow AD 1175–1649*, 3 vols., Scottish Burgh Records Society, 1897, i, p. lxiv, and Campbell, W. *History of the Incorporation of Cordiners in Glasgow*, Glasgow, 1883, i, 249.
37. Hunt, 1889, 50.
38. Marwick, 1869, 214–15.
39. Marwick, 1871, 62.
40. Scribner, 1981, 95.
41. e.g. Henryson, The Taill of the Scheip and the Doig. In *Poems*, 1974, ll. 1258–1320, pp. 39–40.
42. Wedderburn, 1979, 131, 133 and 135–6.
43. Balfour Paul and Dickson, 1877–1913, ii, 3872, p. 844.
44. Paterson, 1848, 24 and 36–7. Note that he wished a weekly mass to be celebrated at the Holy Blood altar and his obit to be celebrated at the altar of the Holy Rood, both dedications emphasizing the suffering humanity of Jesus.
45. Bryce, 1909, ii, 41–2.
46. Skene, 1877–80, i, 232.
47. Sir David Lindsay, The Dreme of Schir David Lyndesay. In *Works*, 1931, i, ll. 309 and 314–15, p. 13.
48. Henryson was first published in 1508. Aldis, 1904, n. p.
49. Henryson, The Taill how this foirsaid Tod maid his Confessioun to Freir Wolf Waitskaith. In *Poems*, 1974, ll. 782–5, p. 24 and ll. 793–5, p. 25.

50. Henryson, Ane Prayer for the Pest. In *Poems*, ll. 6 and 25–6, p. 137, and ll. 36–40, p. 138.
51. Marwick, 1871, 202. Cf. In 1505 the 'body of the towne' was declared to be represented by 'the aldirman, ballieis, ande diuerss of the counsale and communite', *viz* the secular élite. Stuart, 1841–52, 35.
52. Hamilton, J. Ane Godlie Exhortatioun (1559). Patrick, 1907, 189–90.
53. Dunbar, The maner of passyng to confessioun. In Bennett, 1955, ll. 19–21, p. 257, ll. 22–8 and 43–54, p. 258, and ll. 60–70, p. 259.
54. e.g. In 1543 Perth hammermen fined masters who took more than fifteen days to hand over their weekly offering. Hunt, 1889, 50. In 1531 the weekly penny of Edinburgh's master tailors was spent on St Anne's altar in St Giles'. It supported daily services, maintained the fabric, and bought altar ornaments, in seal of cause of 20 October 1531, Marwick, 1871, 55.
55. Eeles, 1956, 33, 36 and 65.
56. Simpson, 1863, 215; Lindsay, The Monarche (Ane Dialogue betuix Experience and ane Courteour, Off the Miserabyll Estait of the Warld). In *Works*, 1931, i, ll. 2280–3 and 2307, p. 267, ll. 2311–22, p. 268, and ll. 2387–413, pp. 270–1.
57. William Patten, The expedicion into Scotlande of ... prince Edward, Duke of Somerset, London, 1548, NLS AdvMS 28/3/12, fos. 57v and 61r–v. Numerous licences were issued by the Crown to lairds and magnates who wished to go on pilgrimage overseas. For example, in 1506 the Earl of Crawford received a licence to go to Amiens on pilgrimage. Livingstone, *et al.*, 1908–82, i, 1251, p. 181.
58. Richardson, 1928, 202.
59. Museum of Religious Life and Art, Glasgow. Cf. Lindsay, The Monarche. In *Works*, 1931, i, ll. 2367–8, p. 269.
60. Paisley Abbey, Renfrewshire.
61. Ross, 1909, 334, and Lindsay, The Monarche. In *Works*, 1931, i, l. 2359–62 and 2366, p. 269, and ll. 2685–92, p. 279.
62. Chrisman, 1982, 103–6. Cf. Scottish devotional work c1540, BL Arundel MS 285, included 17 woodcuts.
63. Blackadder's prayer book, NLS MS 10271, fos. 98r–v and 99r–v.
64. Fowlis Easter collegiate church, Angus.
65. Cooper, 1888–92, ii, 97–9. Usually the services in the choir (*viz Placebo, Dirige*, and nine lessons in chant) took place on the date of death, and the next day the obit–mass was celebrated at a saint's altar. Cf. Obit of John Cochren of Ayr, d. 1526, in *The Obit Book of the Church of St. John the Baptist*, Ayr, 53; Darlington, 1967, 2.
66. Foundation of an obit by Janet Paterson of Edinburgh, on 1 June 1523, for her parents, plus a donation of 32s. to the deserving poor. Balfour Paul, *et al.*, 1882–1914, iii, 234, pp. 52–3.
67. John of Ireland, Of Penance and Confession. In Asloan, 1925, i, 75, cited in Durkan, 1962, 116.
68. Seton collegiate church, East Lothian.
69. 25 November offering of 18s. Balfour Paul and Dickson, 1877–1913, i, 132.
70. Not until 1559/60 did Scotland's provincial council require priests to use the vernacular to facilitate lay understanding of and devotion to the saints, in General Statutes of 1558–9. Patrick, 1907, 157.
71. Fowlis Easter collegiate church, Angus, and Register of Vestments, Jewels, and Books for the Choir, etc., Belonging to the College of St Salvator in the University of St Andrews, circa AD MCCCCL. In Macdonald and Dennistoun, 1843, iii, 204.
72. Dowden, 1898–9, 298.
73. On St Barbara's Day in December of 1511, James IV donated 14s. to St Barbara's light in the church of Kirk o' Field, Edinburgh. Balfour Paul and Dickson, 1877–1913, iv, 180.

74. Balfour Paul et al., 1882–1914, iii, 234, pp. 52–3.
75. Normally Jesus would have hung on the Cross above them, but Mary Magdalene and the Virgin Mary were positioned to suggest that they held up the Cross. This expressed female nurturing as well as support for Jesus' self-sacrifice. Paisley Abbey, Renfrewshire.
76. Mill, 1927, 70–2 and 92.
77. Fetternear Banner, NMS, Edinburgh; Bawcutt, 1976, 9.
78. Henryson, The Thre Deid-Pollis. In *Poems*, 1974, ll. 52–61 and 64, p. 135.
79. The Jesus Psalter. In Bennett, 1955, ll. 145–6, p. 199 and ll. 130 and 135, p. 198.
80. Scottish devotional work c1540, BL Arundel MS 285, fos. 182r and 185r.
81. Scottish devotional work c1540, BL Arundel MS 285, fos. 190v and 208v.
82. Blackadder's prayer book, NLS MS 10271, f. 33v.
83. Altar and chaplainry donation by David Fowrros, burgess of Haddington (1522), NLS RH1/2/339.
84. Cowan and Easson, 1976, 224.
85. Donaldson, 1990, 54.
86. Scottish devotional work c1540, BL Arundel MS 285, f. 125v.
87. This image originally appeared in *Die Negen Couden*, a devotional work published at Antwerp c1505 by Heinrich Eckert Van Homberch, in Scottish devotional work c1540, BL Arundel MS 285, f. 5; Bennett, 1955, xxxii–xxxiii and facing p. 7.
88. Scottish devotional work c1540, BL Arundel MS 285, f. 116v.
89. Arbuthnott prayer book, Paisley Museum, Renfrewshire, unfoliated; Blackadder's prayer book, NLS MS 10271, f. 44r.
90. Arbuthnott prayer book, Paisley Museum, Renfrewshire, unfoliated. Cf. Seal of Patrick, bishop of Moray, in which God is seated with Jesus on the Cross between his knees. Letters of presentation from Patrick, bishop of Moray to Sir James Cuthbert of the chaplainry of St John the Baptist, 1550, NAS GD103/1/44.
91. Hamilton, 1552, f. xcir.
92. e.g. Perth hammermen's costs for their Corpus Christi pageant from 1518 to 1553, Mill, 1970, xlix, 146–7; Hunt, 1889, 3 and 18–19.
93. Mill, 1927, 80–2.
94. e.g. A copy of the *Legends of the Saints* was owned by John Greenlaw of Haddington, vicar of Keith Humbie (d. 1566). de Voragine, 1554.
95. In Edinburgh several guilds required that master craftsmen take turns supplying the craft chaplain with food and drink. This arrangement can be documented for the surgeons and barbers (1515), cordiners (1536) and skinners and furriers (1533), the last-named guild fining masters 8d. per day for failing to fulfil their obligations. Marwick, 1871, 62, 79 and 103. Food was offered as a reward to Aberdeen chaplain Andrew Lame on 3 May 1533. The burgh council awarded him one meal per week in seven different households. On Sundays he ate with provost Gilbert Menzies, on Mondays with Thomas Menzies, on Tuesdays with David Anderson, on Wednesdays with Andrew Cullane, on Thursdays with Watt Cullen, on Fridays with John Black, and on Saturdays with Andrew Durty. Stuart, 1844, 148–9.
96. Altar foundation by Robert Clark, burgess of Perth, NAS GD79/4/140. Clark dedicated the altar to St Serf the Confessor. Absenteeism led to dismissal.
97. Leslie, 1888–95, ii, 253.
98. NMS.
99. e.g. Alexander Tyrie, burgess of Perth, founded this altar. Lawson, 1847, 74.
100. David Rollok of Bello founded this altar. Dundee burgh and head court books i (1454 and 1520–5), NAS RH2/8/46, pp. 89–90.
101. Metcalfe, 1896, ii, ll. 664–7, p. 66. Sir David Lindsay verified that women called on St Margaret to support them in childbirth. Lindsay, The Monarche. In *Works*, 1931, i, ll. 2379–80, p. 269.

102. Rubric to the 'Orisouns'. In Bennett, 1955, ll. 1–5, p. 279.
103. Urquhart, 1973, 79.
104. Foundation of Gilbert Edmonston for himself and his wife, Elizabeth Craufurd, and other souls. *Registrum Magni Sigilli Regum Scotorum*, ii, 2496, 531.
105. The Craft of Deyng. In Lumby, 1870, 1.
106. Sir David Lindsay warned people that death could come suddenly, and if it were the general Day of Judgement it would be too late to make recompense for sin. Lindsay, The Monarche. In *Works*, 1931, i, ll. 5530–3, p. 362 and ll. 5538–45 and 5550–3, p. 363. William Dunbar wrote of the certainty of illness and death and the need to make preparations to ensure success in the afterlife. Dunbar, Lament for the Makaris. In *Poems*, 1932, ll. 9–12, p. 20 and ll. 93–100, pp. 22–3.
107. *The Glasgow Herald* (23 July, 1957), 5, and the testament of Mary Seton (b. 1541), to whom Mary, Queen of Scots gave the watch. In her testament of 1602, Seton gave the watch to executor Mr Anthony of Beauchesne, priest and subchanter in the church of Rheims, France. Seton, 1896, ii, 962–3. Watches and clocks were symbols of mortality, reminding people of the passage of time and the inevitability of death.
108. Lewington, T. The Crafte to Lyue Well and to Dye Well (1505). In Johnstone and Robertson, 1929, 8 and facing 9. The 1503 Scottish original was attributed to Alexander Barclay by Johnstone and Robertson, but the author was Thomas Lewington. The Scottish translation was entitled *The Book Intytulid the Art of Good Lywyng and Good Deyng*.
109. Robb, 1920, ll. 1301–10, 1316–20, p. 54.
110. e.g. In 1508 James IV paid 28s. to the German who was to make his 'lair' or burial place, in Balfour Paul and Dickson, 1877–1913, iv, 132.
111. Glasgow commissariat register of testaments 1547–55, NAS CC9/7/1 (unfoliated); St Andrews commissariat register of testaments 1549–51, NAS CC20/4/1 (unfoliated); Dunblane commissariat register of testaments 1539–47 and 1553–58, NAS CC8/8/1A (unfoliated), *passim*.
112. For example, Sir David left velvet coats to the high altar of the cathedral church of Orkney and the 'Cross Kirk' of Dunrossness, Shetland, two nobles to the Holy Cross in Stanebruch, the golden chain given him by the King of Denmark to the cathedral church of Roeskilde, Denmark, and two nobles and *The Buk of Gud Maneris* to Sir Magnus Harrode, indicating that at least some Scottish magnates read advisory works. The Testament of Sir David Synclar of Swynbrocht Knycht. In Laing, 1855, iii, 107–10. Cf. *The Porteous of Noblenes*. In Asloan, 1925.
113. Aries, 1981, 76. Cf. Plat Pays region near Lyon, France. In Lorcin, 1972, 292.
114. Aries, 1981, 76.
115. McKay, 1962, 111.
116. Dundee burgh and head court book, i (1454 and 1520–5), NAS RH2/8/46, and Dundee burgh and head court books, ii (1550–3) and iii (1555–7), NAS RH2/8/47, *passim*.
117. *Registrum Magni Sigilli Regum Scotorum*, iii, 234, 53.
118. The Testament of Alexander Sutherland of Dunbeath, at Rosslyn castle, 15 November 1456. Laing, 1855, iii, 96.
119. Dundee burgh and head court book, i (1454 and 1520–5), NAS RH2/8/46, f. 46b, p. 62.
120. Dundee burgh and head court book, i (1454 and 1520–5), NAS RH2/8/46, f. 168b, p. 267.
121. Dundee burgh and head court book, iii (1555–7), NAS RH2/8/47, f. 72b, p. 182.
122. Arbuthnott prayer book, Paisley Museum, Renfrewshire, unfoliated.
123. Mary King's Close, Edinburgh, late fifteenth century, in NMS.
124. Hamilton, 1882, xii, clv and clix; Dunbar, Of the Warldis Vanitie. In *Poems*, 1932, ll. 4–5, p. 150, and 'The Craft of Deyng', 5–6.

125. General Statutes of 1549. Patrick, 1907, 192–3, p. 101.
126. e.g. Edict anent preaching within the diocese of Glasgow. Romanes, 1917, iii, 168–9. The library of preacher John Watson, canon of Aberdeen, contained a number of volumes useful for sermons, such as Hughes de Vinac's *Sermones super evangelia, tempore hyemali* (Paris), John Royard's *Homiliae in omnes epistolas dominicales* (Paris, 1544) and a Roman Breviary (Lyons, 1546). In Durkan and Ross, 1961, 157.
127. Lindsay, The Monarche. In *Works*, 1931, ll. 2597–600, p. 276, ll. 2509–708, pp. 273–9, and *passim*.
128. These words were attributed to George Wishart, who was referring to two Franciscans listening to his sermon in Inveresk, Midlothian, in 1545. John Knox, History of the Reformation in Scotland. In Laing, 1846, i, 136.
129. Hamilton, 1882, clv and clix.
130. Lewington, 1503, 27(I_1) (Extreme unction), 31(O_2) (Temptation to lack of faith), 32 (O_4) (Be secure of faith), 33(O_6) (Temptation to mis-hope or despair), 34(P_2v) (Do not despair). The English version of the Scottish translation, which resides in Emmanuel College, Cambridge, is *The Crafte to Lyue Well and to Dye Well* (1505), attributed to Alexander Barclay in Johnstone and Robertson, 1929, 8–9. However, F H Stubbings, Hon. Keeper of Rare Books, Emmanuel College, Cambridge, has identified the Aberdonian writer as Thomas Lewington (letter of 3 May 1991). Cf. *Short Title Catalogue of English Books before 1641*, 791.
131. Douglas, The Palice of Honour. In Douglas, 1967, 123.
132. Cooper, 1888–92, ii, 160.
133. Cooper, 1888–92, ii, 3.
134. Hamilton, 1882, clix.
135. The Craft of Deyng, 1–2.
136. Hamilton, 1882, The tabil, iii; Sheehan, 1963, 262–3; and Engelmann, 1903, 83.
137. Cf. Glasgow commissariat register of testaments 1547–1555, NAS CC9/7/1 (unfoliated); St Andrews commissariat register of testaments 1549–1551, NAS CC20/4/1 (unfoliated); Dunblane commissariat register of testaments 1539–1547 and 1553–1558, NAS CC8/8/1A (unfoliated).
138. e.g. Hali Meidenhad. In Millett and Wogan–Browne, 1990, 21–3.
139. Henryson, The Garmont of Gud Ladeis. In *Poems*, 1974, ll. 9–10, p. 125.
140. Lindsay, The Dreme of Schir David Lyndesay. In *Works*, 1931, i, ll. 267–87, pp. 12–13.
141. *The New Acts and Constitutionis of parliament made be Iames the Fift kyng of Scottis*, 1540, f. 27v.
142. Ladies of Seton, Glenbervie and Bass founded the Dominican convent of St Catherine of Siena in 1517. Cowan and Easson, 1976, 152. Marion Scrimgeour and Robert Arbuthnott of Arbuthnott joined the confraternities of the Franciscans and St John of Jerusalem in 1487. Bryce, 1909, ii, 263–4.
143. Lindsay, The Monarche. In *Works*, 1931, i, ll. 6008–21, p. 376; ll. 5488–9, p. 361l; ll. 5583–602, pp. 364–5; and ll. 5614–39, pp. 365–6.
144. The illustration is of the early sixteenth-century Flemish or German school. William of Touris, The Contemplacioun of Synnaris. In Bennett, 1955, Harleian MS 6919 illustration, facing p. 143.
145. In Bennett, 1955, Harleian MS 6919, ll. 1193–200, p. 144.
146. Literary examples include Dunbar, The Dance of the Sevin Deidly Synnis. In *Poems*, 1932, l. 82, p. 122; Dunbar, On the Resurrection of Christ. In *Poems*, 1932, l. 5, p. 159; Dunbar, Ane Ballat of Our Lady. In *Poems*, 1932, pp. 160–2; and The Lang Rosair. In Bennett, 1955, ll. 69–72, p. 324.
147. Dunbar, The Devillis Inquest. In *Poems*, 1932, l. 106, 79.
148. Dunbar, The Dance of the Sevin Deidly Synnis. In *Poems*, 1932, l. 6, p. 120.
149. Noddings, 1989, 90; and Mill, 1927, 84. The devil featured in the 'annual riding' honouring the patron saint of Perth's baxters. Mill, 1927, 74.

150. Mill, 1970, 146–7; Hunt, 1889, 3.
151. Museum of Religious Life and Art, Glasgow.
152. Hunt, 1889, 18–9, 78–9; Mill, 1970, 146–7.
153. Aston, 1984, 122–3.
154. Hamilton, 1882, clxxxxiii–clxxxxvi.
155. Malden, 1993, 21.
156. Walter Kennedy, Closter of crist riche Recent flour delyss. In Asloan, 1925, ii, ll. 9–14, p. 273, and ll. 65–70, p. 275.
157. Rubric to the 'Orisouns'. In Bennett, 1955, ll. 1–5, 279.
158. e.g. Kennedy, Closter of crist riche Recent flour delyss. In Asloan, 1925, ii, ll. 57–64, p. 274–5. Cf. Dunbar, Ane Ballat of Our Lady. In *Poems*, 1932, ll. 29–30, p. 161.
159. Arbuthnott prayer book, Paisley Museum, Renfrewshire, unfoliated; *The New Acts and Constitutionis of parliament made be Iames the Fift kyng of Scottis, 1540*, f. 27v.
160. Urquhart, 1973, 80, and Mackinlay, 1910, i, 87.
161. Dunbar, Ane Ballat of Our Lady. In *Poems*, 1932, ll. 11–12 and 37–8, p. 161; and ll. 47–8, 51–2 and 67–8, p. 162.
162. Walter Kennedy, part two of a series of poems 'Ballatis of Our Ladye'. In Asloan, 1925, ii, ll. 43–4, 274.
163. Cf. View of St Bernard of Clairvaux that Mary gained humanity access to divine favour. Bauckham, 1989, 156–7. In the prayer 'O Clementissime', included in BL Arundel MS 285 and written by St Bernard, Mary is asked to grant eternal life, in O Clementissime. In Bennett, 1955, l. 98, p. 281.
164. Bennett, 1955, 123.
165. Carter, 1956–7, 126.
166. Virgin Mary and child and King Hezekiah, according to Eeles, probably from a window representing the Tree of Jesse. Eeles, 1914–5, 87.
167. NMS.
168. Eeles, 1898–9, 441.
169. The Lang Rosair. In Bennett, 1955, l. 291, p. 332.
170. Chartulary of Crail collegiate church (1528), NLS AdvMS 34/4/6, f. 108v. Cf. The Cross of Crail was a pilgrimage destination, in Lindsay, The Monarche. In *Works*, 1931, i, ll. 2385–6.
171. Arbuthnott prayer book, Paisley Museum, Renfrewshire, unfoliated; Fowlis Easter collegiate church, Angus; and Scottish devotional work c1540, BL Arundel MS 285, fos. 178v and 204v.
172. *The New Acts and Constitutionis of parliament made be Iames the Fift kyng of Scottis, 1540*, f. 27v.
173. Haddington burgh court books, NAS GD1/413/1, p. 106.
174. Letters of presentation from Patrick, Bishop of Moray to Sir James Cuthbert of the chaplainry of St John the Baptist, 1550, NAS GD103/1/44.
175. Blackadder's prayer book, NLS MS 10271, f. 92r.
176. Bas–relief on tomb of Alexander Macleod of Dunvegan, St Clement's church, Rodil, Harris.
177. Legenda Sanctorum of Jacobus de Voragine (1487), BL Inc. 289.5, no. cxl Sig. v3v.
178. O Intemerata. In Bennett, 1955, ll. 193–4, p. 284–5.
179. Guthrie panel painting of Judgement, NMS.
180. Eeles, 1914–15, 87.
181. Register of Vestments, Jewels, and Books for the Choir, etc., Belonging to the College of St Salvator in the University of St Andrews, circa AD MCCCCL. In Macdonald, Dennistoun, 1843, 202 and 205.
182. Dunbar, The Tua Mariit Wemen and the Wedo. In *Poems*, 1932, ll. 410–17, p. 94 and ll. 422–35 and 451–75, p. 95–6.

183. Rutherford, 1946, 28–9 and 31–3.
184. Balfour Paul and Dickson, 1877–1913, iv, 412.
185. Grant of portable altar to William, third Lord Errol and his wife Elizabeth Leslie (1483), NAS RH2/7/5, 1443–4; Cheney, 1978, 38; *Acts and Proceedings of the General Assemblies of the Kirk of Scotland, from the Year MDLX* (*The Booke of the Universall Kirk of Scotland*), ii, 829–30 (Session of 5 January, 1592).
186. Grant of portable altar to Robert Arbuthnott of Arbuthnott (1490), NAS GD1/2/294.
187. Grant of portable altars to William Graham of Kincardine, Archibald Edmonston of Duntreath, Stirlingshire, and Patrick Home of Polwarth, Berwickshire, NAS GD103/2/46.
188. NMS.
189. *Acts and Proceedings of the General Assemblies of the Kirk of Scotland, from the Year MDLX* (*The Booke of the Universall Kirk of Scotland*), ii, 717 (Session of 6 February 1587).
190. Sir Richard Maitland of Lethington, Of the Wynning of Calice (1558). In *Poems*, 1830, 10.
191. Haddington burgh court books, NAS GD1/143.
192. Lindsay, The Monarche. In *Works*, 1931, i, ll. 2313–48, pp. 268–9, and ll. 2501–40, pp. 273–4, i, ll. 2359–96, pp. 269–70, ll. 2509–2540, pp. 273–4, and ll. 2677–700, pp. 278–9. Cf. Robert Lindesay of Pitscottie, citing Patrick Hamilton's view that prayers to saints dishonoured Jesus. Robert Lindesay of Pitscottie. *The Cronicles of Scotland*, ed. John Graham Dalyell, 2 vols., Edinburgh, 1814, ii, 316.
193. Dunbar, The Tabill of Confession. In *Poems*, 1932, l. 45, p. 164.
194. Lewington, 1503, 23(I_1).
195. Mylne, Lives of the Bishops. In Hannay, 1915, 313. Cf. Water from St Mary's well, half a mile from Cullen collegiate church, was used in church services. Mackinlay, 1910, i, 87.
196. Glasgow commissariat register of testaments 1547–55, NAS CC9/7/1 (unfoliated).
197. St Andrews commissariat register of testaments 1549–51, NAS CC20/4/1 (unfoliated).
198. Glasgow commissariat register of testaments 1547–55, NAS CC9/7/1 (unfoliated).
199. An Obituary from the Rental Book of the Preceptory of St Anthony, near Leith M.D.XXVI. In Laing, 1855, ii, 299.
200. Brilioth, 1930, 84.
201. Dunbar, The Tabill of Confession. In *Poems*, 1932, ll. 43–4, p. 164.
202. John of Ireland, Of Penance and Confession. In Asloan, 1925, i, 7–8.
203. Arbuthnott prayer book, Paisley Museum, Renfrewshire, unfoliated.
204. Inscription was 'Caro mea vere est cib et sangvis me vere e pot q manvcat mea carnet bibit mev sangvine vive i eternvm'. MacGibbon and Ross, 1896, iii, 404–5 and 406.
205. NMS.
206. Dunbar, The Tabill of Confession. In *Poems*, 1932, l. 84, p. 165.
207. *Poems*, 1932, xxxviii and xli.
208. e.g. Rolland Blackadder, subdean of Glasgow, founded a chaplainry at the altar of Saints Nicholas and John in Glasgow cathedral, and established a Friday mass of the Five Wounds of Christ. The editor believes that the foundation was made near the end of Blackadder's life, which may have been in 1541. Renwick, 1896, ii, 111 and 103.
209. Cf. Malcolm, Lord Fleming's foundation of the collegiate church of Biggar reserved Thursday for the Mass of the 'Body of Christ' and Friday for the Mass of the 'Five Wounds of our Lord'. Provost John Steinson's 'Register of Services'. In Rutherford, 1946, 34.
210. Cooper, 1888–92, ii, 135.

211. Innes and Chalmers, 1856, ii, 138–9.
212. Angus, 1931, i, 124, p. 23 and p. 47.
213. Obit foundation at the altar of Corporis Christi by Peter Newlands, NAS GD76/4/59.
214. Cooper, 1888–92, ii, 136 and 143.
215. Donaldson, 1952, 665, p. 149.
216. Harvey and Macleod, 1930, 404, p. 127.
217. Harvey and Macleod, 1930, 3196, p. 683.
218. *Registrum Magni Sigilli Regum Scotorum*, ii, 3010, p. 641.
219. Wall, 1910, 151; Arbuthnott prayer book, Paisley Museum, Renfrewshire, unfoliated.
220. Cowan and Easson, 1976 , 218, 220 and 224.
221. Hamilton, 1882, clv–clvi.
222. Lindsay, The Monarche. In *Works*, 1931, i, ll. 487–501, p. 213 and ll. 5586–611, pp. 364–5.
223. Cf. Hamilton, 1882, cciii.
224. Balfour Paul, et al., 1882–1914, ii, 2120, p. 448.
225. *The Obit Book of the Church of St. John the Baptist, Ayr*, 52.
226. *The Obit Book of the Church of St. John the Baptist, Ayr*, 58.
227. Diary of Robert Birre, NLS AdvMS 33/7/28, 5.
228. Hamilton, 1882, cc.
229. Durkan, 1985, 438, p. 143.
230. Kennedy, Honour with Age. In MacQueen, Scott, 1989, 99–100; Ireland, Of Penance and Confession. In Asloan, 1925, i, 3 and 5–6.
231. James III wrote to his brother-in-law about the death of Queen Margaret, declaring that she had left earth's miseries to enter an eternal life of joy. Letter from James III to John, king of Denmark and Norway, NAS RH2/8/35, 26–7.

BIBLIOGRAPHY

Primary sources – MS

Altar foundation by Robert Clark, burgess of Perth, NAS GD79/4/140.
Blackadder's prayer book, NLS MS 10271.
Charter of obit foundations by George of Spalding, burgess of Dundee, NAS GD76/152.
Chartulary of Crail collegiate church, 1528, NLS AdvMS 34/4/6.
Diary of Robert Birre, NLS AdvMS 33/7/28.
Dunblane commissariat register of testaments, 1539–47 and 1553–58, NAS CC8/8/1A.
Dundee burgh and head court books, 1 (1454 and 1520–5), NAS RH2/8/46.
Dundee burgh and head court books, 2 (1550–3) and 3 (1555–7), NAS RH2/8/47.
Glasgow commissariat register of testaments, 1547–55, NAS CC9/7/1.
Grant of portable altar to Robert Arbuthnott of Arbuthnott, 1490, NAS 1/2/294.
Grant of portable altar to William, third Lord Errol and his wife Elizabeth Leslie, NAS 1483, RH2/7/5, 1443–4.
Grant of portable altars to William Graham of Kincardine, Archibald Edmonston of Duntreath, Stirlingshire, and Patrick Home of Polwarth, Berwickshire, NAS GD103/2/46.
Haddington burgh court books, NAS GD1/143.
Legenda Sanctorum of Jacobus de Voragine, NLS Inc. 289.5, no. cxl Sig. v3v.
Letter from James III to John, king of Denmark and Norway, NAS RH2/8/35, 26–7.
Letters of presentation from Patrick, Bishop of Moray to Sir James Cuthbert of the chaplainry of St John the Baptist, 1550, NAS GD103/1/44.
Minutes of the Edinburgh goldsmith craft, 1520s, NAS GD1/482/1.
Obit foundation at the altar of Corporis Christi by Peter Newlands, NAS GD76/4/59.

Österreichische Nationalbibliothek, Book of Hours of James IV and Margaret Tudor, Codex 1897.
Paisley Museum, Arbuthnott prayer book, 1482–3.
St Andrews commissariat register of testaments, 1549–51, NAS CC20/4/1.
Scottish devotional work, c1540, BL Arundel MS 285.
Testament of Dame Catherine Lauder, 1515, NAS GD12/80.
William Patten, The expedicion into Scotlande of ... prince Edward, Duke of Somerset, London, 1548, NLS AdvMS 28/3/12.

Primary sources – printed

Angus, W, ed. *Protocol Book of Sir Robert Rollok, 1534–1552*, Edinburgh, 1931.
Asloan, J. *The Asloan Manuscript. A Miscellany in Prose and Verse*, ed. W A Craigie, 2 vols., Edinburgh, 1923–25.
Balfour Paul, J and Dickson, T, ed. *Compota Thesaurariorum Regum Scotorum. Accounts of the Lord High Treasurer of Scotland*, 11 vols., Edinburgh, 1877–1913.
Balfour Paul, J, et al., eds. *Registrum Magni Sigilli Regum Scotorum. The Register of the Great Seal of Scotland*, 11 vols., Edinburgh, 1882–1914.
Bennett, J A W, ed. *Devotional Pieces in Verse and Prose*, Edinburgh, 1955.
Brown, P H, ed. *Early Travellers in Scotland*, Edinburgh, 1891.
Bryce, W M, ed. *The Scottish Grey Friars. Documents*, 2 vols., Edinburgh, 1909.
Chambers, W, ed. *Charters and Documents Relating to the Burgh of Peebles AD 1165–1710*, Edinburgh, 1872.
Cooper, J, ed. *Cartularium Ecclesiae Sancti Nicholai Aberdonensis*, 2 vols., Aberdeen, 1888–92.
Darlington, I, ed. *London Consistory Court Wills 1492–1547*, London, 1967.
de Voragine, J. *Legenda Sanctorum*, Lyons, 1554.
Donaldson, G, ed. *Protocol Book of James Young 1485–1515*, Edinburgh, 1952.
Douglas, G. *The Shorter Poems of Gavin Douglas*, ed. Priscilla Bawcutt, Edinburgh, 1967.
Dunbar, W. *The Poems of William Dunbar*, ed. William Mackay Mackenzie, Edinburgh, 1932.
Durkan, J, ed. *Protocol Book of John Foular, 1528–1534*, Edinburgh, 1985.
Eeles, F C, ed. *King's College Chapel Aberdeen. Its Fittings, Ornaments and Ceremonial in the Sixteenth Century*, Edinburgh, 1956.
Fleming, D H, ed. *Register of the Minister, Elders and Deacons of the Christian Congregation of St. Andrews*, 2 vols. Edinburgh, 1889–90.
Fraser, W. *Memorials of the Earls of Haddington. Correspondence and Charters*, 2 vols., Edinburgh, 1889.
Hamer, D, ed. *The Works of Sir David Lindsay of the Mount 1490–1555. Text of the Poems*, Edinburgh, 1931.
Hamilton, J. *The Catechisme*, 1552.
Hamilton, J. *The Catechism set forth by Archbishop Hamilton Printed at St. Andrews –1551 together with The Two-Penny Faith 1559*, ed. A F Mitchell, Edinburgh, 1882.
Hannay, R K, ed. *Rentale Dunkeldense*, Edinburgh, 1915.
Harvey, C H and Macleod, J, eds. *Calendar of Writs Preserved at Yester House 1166–1625*, Edinburgh, 1930.
Henryson, R. *Poems*, ed. and intro. Charles Elliott, 2nd edn, Oxford, 1974.
Hunt, C, ed. *The Perth Hammermen Book (1518 to 1568)*, 1889.
Innes, C and Chalmers, P, eds. *Registrum Episcopatus Brechinensis. Registrum. Appendix Cartarum*, 2 vols., Edinburgh, 1856.
Johnstone, J F K and Robertson, A W, eds. *Bibliographia Aberdonensis Being an Account of Books Relating to or Printed in the Shires of Aberdeen, Banff, Kincardine or Written by Natives or Residents or by Officers, Graduates or Alumni of the Universities of Aberdeen 1472–1640*, Aberdeen, 1929.
Laing, D, ed. *The Works of John Knox*, Edinburgh, 1846.
Laing, D, ed. *The Bannatyne Miscellany*, Edinburgh, 1855.

Lawson, J P. *The Book of Perth: An Illustration of the Moral and Ecclesiastical State of Scotland Before and After the Reformation*, Edinburgh, 1847.
Leslie, J. *The Historie of Scotland*, trans. Father James Dalrymple (1596), ed. E G Cody, 2 vols., Edinburgh, 1888–95.
Lewington, T. *The Book Intytulid the Art of Good Lywyng and Good Deyng*, Paris, 1503.
Lindesay of Pitscottie, R. *The Cronicles of Scotland*, ed. J G Dalyell, 2 vols., Edinburgh, 1814.
Livingstone, M, et al., eds. *Registrum Secreti Sigilli Regum Scotorum. The Register of the Privy Seal of Scotland*, 8 vols., Edinburgh, 1908–82.
Lumby, J R, ed. *Ratis Raving, and Other Moral and Religious Pieces, in Prose and Verse*, London, 1870.
MacQueen, J and Scott, T, eds. *The Oxford Book of Scottish Verse*, Oxford, 1989.
Maitland of Lethington, R. *The Poems of Sir Richard Maitland of Lethingtoun, Knight*, ed. J Bain, Edinburgh, 1830.
Marwick, J D, ed. *Extracts from the Records of the Burgh of Edinburgh AD 1403–1528*, Edinburgh, 1869.
Marwick, J D, ed. *Extracts from the Records of the Burgh of Edinburgh AD 1528–1557*, Edinburgh, 1871.
Metcalfe, W M, ed. *Legends of the Saints*, 3 vols., Edinburgh, 1896.
Millett, B and Wogan–Browne, J, eds. *Medieval English Prose for Women. Selections from the Katherine Group and Ancrene Wisse*, Oxford, 1990.
Mylne, Alexander. *Vitae Dunkeldensis Ecclesiae Episcoporum*, ed. T Thomson, Edinburgh, 1823.
The New Acts and Constitutionis of parliament made be Iames the Fift kyng of Scottis, 1540, Edinburgh, 1541.
Paterson, J, ed. *The Obit Book of the Church of St. John the Baptist, Ayr*, Edinburgh, 1848.
Patrick, D, ed. *Statutes of the Scottish Church 1225–1559*, Edinburgh, 1907.
Pryde, G S, ed. *Ayr Burgh Accounts 1534–1624*, Edinburgh, 1937.
Renwick. R, ed. *Extracts from the Records of the Royal Burgh of Stirling AD 1519–1666*, Glasgow, 1887.
Renwick, R, ed. *First Protocol Book of William Hegait, 1555–60*, 2 vols., Glasgow, 1896.
Robb, T D, ed. *The Thre Prestis of Peblis how thai tald thar talis*, Edinburgh, 1920.
Romanes, C S, ed. *Selections from the Records of the Regality of Melrose and from the Manuscripts of the Earl of Haddington, 1547–1706*, Edinburgh, 1917.
Rutherford, D S. *Biggar St. Mary's. A Medieval College Kirk*, Biggar, 1946.
Skene, F J H, ed. *The Book of Pluscarden*, 2 vols., Edinburgh, 1877–80.
Stuart, J, ed. *Extracts from the Council Register of the Burgh of Aberdeen 1398–1570*, Aberdeen, 1844.
Stuart, J, ed. *The Miscellany of the Spalding Club*, 5 vols., Aberdeen, 1841–52.
Thomson, T, ed. *Acts and Proceedings of the General Assemblies of the Kirk of Scotland, from the Year MDLX (The Booke of the Universall Kirk of Scotland)*, 3 vols., Edinburgh, 1839–45.
Thomson, T and Innes, C, eds. *The Acts of the Parliaments of Scotland*, 11 vols., Edinburgh, 1814–44.
Wedderburn, R. *The Complaynt of Scotland (c. 1550)*, Edinburgh, 1979.

Secondary sources
Aldis, H G. *A List of Books Printed in Scotland Before 1700*, Edinburgh, 1904.
Aries, P. *The Hour of Our Death*, trans. Helen Weaver, London, 1981.
Aston, M. *Lollards and Reformers. Images and Literacy in Late Medieval Religion*, London, 1984.
Bauckham, R. The origins and growth of western Mariology. In Wright, D F, ed. *Chosen by God: Mary in evangelical perspective*, London, 1989, 141–60.
Bawcutt, P. *Gavin Douglas. A Critical Study*, Edinburgh, 1976.

Brilioth, Y. *Eucharistic Faith and Practice. Evangelical and Catholic*, trans. A G Hebert, London, 1930.
Carter, C. The Arma Christi in Scotland, *PSAS*, 80 (1956–7), 116–29.
Cheney, C R, ed. *Handbook of Dates for Students of English History*, London, 1978.
Chrisman, M. *Lay Culture, Learned Culture. Books and Social Change in Strasbourg, 1480–1599*, New Haven, 1982.
Cowan, I B and Easson, D E. *Medieval Religious Houses Scotland*, London, 1976.
Donaldson, G. *Faith of the Scots*, London, 1990.
Dowden, J. The inventory of ornaments, jewels, relicks, vestments, service books, etc. belonging to the cathedral church of Glasgow in 1432, illustrated from various sources, and more particularly from the inventories of the cathedral of Aberdeen, *PSAS*, 33 (1898–9), 280–329.
Durkan, J and Ross, A, eds. *Early Scottish Libraries*, Glasgow, 1961.
Durkan, J. Care of the poor: Pre-Reformation hospitals. In McRoberts, D, ed. *Essays on the Scottish Reformation 1513–1625*, Glasgow, 1962, 116–28.
Eeles, F C. Notes on a missal formerly used in St. Nicholas, Aberdeen, *PSAS*, 33 (1898–9), 440–60.
Eeles, F C. Mediaeval stained glass recently recovered from the ruins of Holyrood abbey church, *PSAS*, 49 (1914–15), 81–91.
Engelmann, J. *Les Testaments Coutumiers au XVe Siècle*, Geneva, 1903.
Galbraith, J. The Middle Ages. In Forrester, D and Murray, D, eds. *Studies in the History of Worship in Scotland*, Edinburgh, 1984.
Jung, C G. Jung and Religious Belief. In Hull, R F C, ed. *Psychology and Western Religion*, London, 1988.
Lorcin, M–T. Les clauses religieuses dans les testaments du plat pays lyonnais aux XIVe et XVe siècles, *Le Moyen Age. Revue d'Histoire et de Philologie*, 78, 2 (1972), 287–323.
MacGibbon, D and Ross, T. *The Ecclesiastical Architecture of Scotland from the Earliest Christian Times to the Seventeenth Century*, 3 vols., Edinburgh, 1896.
McKay, D. Parish life in Scotland, 1500–1560. In McRoberts, D, ed. *Essays on the Scottish Reformation 1513–1625*, Glasgow, 1962, 85–115.
Mackinlay, J M. *Ancient Church Dedications in Scotland*, 2 vols., Edinburgh, 1910.
Macmillan, D. *Scottish Art 1460–1990*, Edinburgh, 1990.
McRoberts, D. A sixteenth century picture of St. Bartholomew from Perth, *IR*, 10 (1959), 281–6.
Malden, J. *The Abbey and Monastery of Paisley*, Paisley, 1993.
Maxwell, A. *Old Dundee, Ecclesiastical Burghal and Social, Prior to the Reformation*, Edinburgh, 1891.
Mill, A J. *Mediaeval Plays in Scotland*, Edinburgh, 1927.
Mill, A J. The Perth hammermen's play. A Scottish Garden of Eden, *SHR*, 49 (1970), 146–53.
Noddings, N. *Women and Evil*, Berkeley, 1989.
Richardson, J. Fragments of altar retables of late mediaeval date in Scotland, *PSAS*, 62 (1928), 197–224.
Ross, T. Restalrig and the well of St. Triduana, *SHR*, 6 (1909), 334–5.
Rubin, M. *Corpus Christi. The Eucharist in Late Medieval Culture*, Cambridge, 1991.
Scribner, R. *For the Sake of Simple Folk. Popular Propaganda for the German Reformation*, Cambridge, 1981.
Seton, G. *A History of the Family of Seton*, 2 vols., Edinburgh, 1896.
Sheehan, M. *The Will in Medieval England*, Toronto, 1963.
Simpson, J Y. Notes on some Scottish magical charm-stones, or curing-stones, *PSAS*, 4 (1863), 211–24.
Urquhart, R M. *Scottish Burgh and County Heraldry*, London, 1973.
Wall, J C. *Relics of the Passion*, London, 1910.

4 The Impact of the Reformation in Relation to Church, State and Individual

FRANCIS LYALL

INTRODUCTION

We tend to speak of 'the Reformation' as if it was something accomplished swiftly. It was not. The harbingers of the Reformation are to be found decades before Luther's Ninety-five Theses,[1] and the political, economic and theological struggles of the Reformation go on for decades thereafter. Indeed, notwithstanding the vogue for ecumenism, it could be argued that the Reformation, and the Counter-Reformation, are not yet done. More conservatively the Reformation in Scotland arguably started soon after the ideas of Luther came to the kingdom in the early sixteenth century, and did not find substantial completion until its end with the commitment to Presbyterianism. But even then, until the defences of the Church were ground into the arrangements for the Union of the Parliaments of England and Scotland in 1707, a degree of uncertainty persisted. In the twentieth-first century it is fair to say that those defences now appear weakened, as much because of internal rot as through external attack – but that is the matter of a subsequent chapter.

In Britain a clear distinction can be made between England and Scotland, for in the sixteenth century they were separate kingdoms, and the progress of the Reformation was different in each. While in England there was indeed a theological element to what happened, the political element and the personal wishes of Henry VIII had a major role to play, and, notwithstanding the restoration during the reign of Mary, the legal breach with Rome was accomplished earlier, albeit less thoroughly.

In Scotland, the theological element, spearheaded by ministers of religion, was stronger and more focused, although it would be wrong thereby to under-estimate the political element in the success of the Scottish Reformation. In England the Reformation was, to a degree, top-down, with the result that the English Church was less thoroughly reformed. What was at stake in England was 'who is the head of the Church?' rather than the belief system that the Church espoused. Indeed, although under the Submission of the Clergy Act 1533 (25 Hen. 8 c. 19) a commission of 32 was provided for to revise the Canons of the English Church, that commission may never have been constituted, and due attention was not given to matters of belief until later.[2] By contrast, the Scottish reformers prepared, and the Scottish parliament adopted, the Scots Confession within days of the decision to abolish the papal jurisdiction. That difference between the Reformation in the two kingdoms should be remembered. It clearly demonstrates the balance between politics and theology in the two kingdoms both at the higher and at the lower levels of their respective societies.

AMBIT

My task in this chapter is to give an overview of the Reformation in Scotland, as it affected the Church, the individual, the organs and the organisation of the State. In order to do this we must review the pre-Reformation situation, and also extend the discussion into the seventeenth century, for it was not until the legal position of the Church of Scotland as Presbyterian was secured in the arrangements for the Union of the Parliaments in 1707 that things were sufficiently settled.

I would also say that I write as a lawyer, rather than as a historian.[3] My sources are to a considerable extent legal, including the statutes of the Scottish parliament, and the cases of the Court of Session, the highest court in Scotland at the time,[4] so far as they are available to me. Contemporaneously, we have also John Knox's own account of *The History of the Reformation in Scotland*,[5] as well as the *First* and *Second Book of Discipline*, of which more anon,[6] and the *Buk of the Universall Kirk*.[7] Of course, since then many scholars have laboured, the majority from more or less obvious religious positions. Most have written from a reformed perspective, some passionately so, but Roman Catholic scholars have also been active, for example, in a fascinating set of articles by the Scottish Catholic Historical Association that appeared in the *Innes Review* in 1959, which have also been published in a single volume.[8] Recent legal scholarship has taken a detached view.[9]

PRE-REFORMATION SCOTLAND

Scotland was a poor country, with significant differences between its different parts. The Lowlands and coastal districts were economically better off than the Highlands. Major seaports engaged in trade and fishing. The burghs and their hinterlands operated a money economy, while rural areas depended on what amounted often to subsistence farming. Agriculture did not improve until the fifteenth century, and the fuller introduction of a feudal system of landholding in that century introduced the relationship of feudal superior and vassal more widely than it had been known, with a consequent stratification of society.[10]

In pre-Reformation Scotland, the links between Church and State were considerable. The system under which a defined territorial area was allocated for spiritual purposes begins with the responsibilities of bishops early in the history of the Church, but the further subdivision of a bishop's territory into parishes, in which a particular priest had a responsibility for bringing the ordinances of religion, also made sense. Exactly when parishes came to Scotland is unknown, but obviously once the Roman Catholic version of Christianity was dominant it was that Church's parish structure that was effective. Many modern Church of Scotland parishes can be traced back well before the Reformation, particularly in rural areas.[11] By the time of the Scottish Reformation in 1560 all Scotland had, at least on paper, been divided into parishes each with at least one church, and the inhabitants of each area were, as it were, allocated to that parish and church, for the upkeep of which they were responsible. But the constitution of a parish for 'spiritual purposes' also meant considerations beyond prayers and church services. The Church also accepted responsibility for education and for welfare, albeit the discharge of that responsibility was erratic.[12] This caused disquiet, not to say dissension, which was exacerbated by the fact that the Church owned anything up to one half of the land, and of that one half belonged to the monasteries.[13] However, by the later fifteenth and early sixteenth centuries, through a series of devices, the Crown had obtained what amounted to the right to present to major Church appointments. Many of the abbeys and priories were effectively in the hands of relatives of nobility,[14] who were without interest in the spiritual functions of their offices, and the personal lives of such were a common scandal.[15]

THE REFORMATION

Apart from its religious element, there were political and economic factors at play in the Scottish Reformation. The question of the French alliance was involved, as were property matters. The English interest was that the French alliance should be dissolved. In Scotland some who were against the role that France was playing in the kingdom, particularly during the regency of Mary of Guise, the queen mother, sided with the reformers. Others saw opportunity of getting their hands on the vast wealth and property of the Church. It would be unfair, however, to displace the religious factor as the prime mover in what happened.

The general history of the Reformation in Scotland is well-known and we need not rehearse its progress in detail. Essentially, it was a moderate and gradual matter,[16] the final break coming forty-three years after Luther's Theses. During that period there were halts and starts. Some in the then Church were conscious of its imperfections. The General Statutes of the Provincial Council of the Prelates and Clergy of the Realm of Scotland (i.e. of the then Roman Catholic Church in Scotland), held in 1549, 1551-2 and 1558-9, show disquiet and a move towards the redress of error and wrongs,[17] but matters had by then gone too far. For decades, not to say centuries, Scottish scholars had been studying on the Continent, and some at least had imbibed the new wine of the Reformation, and brought it home.[18] Indeed, some Catholic priests were themselves preaching the new doctrine, and were later to pass over into the ministry of the new Church. But some of the ecclesiastical reaction to the new (or old)[19] ideas was fierce: Patrick Hamilton, a young preacher, was burned as a heretic in St Andrews in 1528 by Cardinal James Beaton. Eighteen years later, in 1546, religious tension was heightened by the martyrdom of George Wishart by Cardinal David Beaton, nephew of the aforementioned James. This led to the assassination of David Beaton at his castle in St Andrews, the siege of that castle by the forces of the Crown assisted by French troops, its reduction, and the imprisonment of its defenders, including John Knox who had ministered to them.

The contacts with the Continent which Scotland had built up over the years were crucial in the development of the religious ideas of the Scottish Reformation. The connection to Holland was important, as was the influence of Geneva, and Calvin's theology, which was fully opposed to Rome, became crucial. No compromise with the Roman form of Church government, or with its doctrines, was to be contemplated. The result was that at the end of the process there was a new Church in Scotland, albeit one which would have claimed, in doctrine at least, to be the true original Church of New Testament times re-established.

INSTITUTIONAL ASPECTS

The origins of the present institutional Church of Scotland can be traced back to December 1557. Various persons bound themselves in what is now known as the First Covenant to maintain, nourish and defend the whole congregation of Christ against the attacks of the 'congregation of Satan', which they expressly renounced.[20] Even then, despite the inflammatory language, there might have been hope for tolerance and a more peaceful progress of reform. However, in 1559 a riot in Perth caused the Lords of the Congregation to take up arms in their defence. The Crown sought to suppress this insurrection, but Elizabeth of England intervened to support the rebels. With the death of Mary of Guise, the queen regent, the French forces (which were the mainstay of the Crown) were willing to come to terms with the English. The withdrawal of both French and English troops was subsequently agreed by the Treaty of Edinburgh.[21] Concessions attached to the treaty provided that Mary, now fully queen of Scots, would take no reprisals against the rebels, and that the Convention of Estates should meet in August 1560 to name commissioners to proceed to France to discuss the religious questions that lay at the base of the rebellion.

When that parliament met, however, it did so without having been summoned by the Crown, and without its normal balance of composition. These points may have helped Mary not to assent to what the parliament did, for although its business had apparently been established by the concessions annexed to the Treaty of Edinburgh, it was petitioned to go further than just appoint persons to discuss theological questions. It was asked to abolish popery. This was achieved by three acts: the Papal Jurisdiction Act

1560 c. 2,[22] which abolished papal jurisdiction; the Act 1560 c. 3,[23] which abolished idolatry; and the Act 1560 c. 4,[24] which proscribed the Mass. The matter of the doctrines of the Church was referred to a committee of six ministers including John Knox with instructions to prepare a statement of the new belief, and in four days a draft of what came to be known as the Scots Confession was produced.[25] This Confession was adopted by the parliament by the Confession of Faith Ratification Act 1560, c. 1.

These four acts are extremely important. There is some argument as to their lawfulness, however, as Mary refused her consent. It may seem idle from a twenty-first-century perspective to deprive the acts of their legal force on this technicality, but the Reformation at this stage was defective. Nothing was done at the statutory level to settle the system of government of the Church, or to deal with matters of property. The same small committee of ministers that had produced the Confession of Faith was asked by the Privy Council to draw up a constitution for the Church. They came back with *The First Book of Discipline*.[26] There is controversy about the *Book*, doubt being possible as to its exact text, and as to when certain elements of it were added.[27]

The *First Book of Discipline* outlined a form of Church government based on a modified episcopal system, in which the superintendents (?=bishops) did not have an episcopal jurisdiction within the meaning normal to systems of episcopal polity.[28] The First Head of the *Book* dealt with doctrine, the Second Head with the sacraments, and the Third with the abolition of idolatry. The *Book* then turned to ministers and their lawful election (Fourth Head) and the provision to be made for them (Fifth Head). These Heads, with the additional duties given to the superintendents, went a long way towards establishing what the polity of the Church might be. In addition the Fifth Head anent the Provision for Ministers, also dealt with 'the Distribution of the rents and possessions justly appertaining to the Church', and the Sixth Head discussed the 'Rents and Patrimonie of the Church', with, expressly as a temporary measure, ministers being restricted to receiving the 'thirds of benefices'. But in their financial demands the reformers had gone on to more contentious ground. More was to follow, with unnumbered, and possibly inserted sections in Heads Five and Six on the provision by the 'godly Magistrates' of schools within every parish, of schools to teach grammar and Latin in the larger towns, and a college within every town with a superintendent where 'the arts, Logick and Rhetorick, together with the tongues' would be taught by 'sufficient maisters for whom honest stipends must be appointed'. Fathers would be compelled to 'bring up their children in learning and vertue', without exception for rich or poor. Inspectors (the minister and the elders and the rest of the 'learned men') in every town would quarterly secure that the youth had thereby profited. Further, for those with aptitude, what we would now call 'higher education' was to be available. Universities would be provided in St Andrews, in Glasgow and in Aberdeen, so that education might be taken to the highest level possible. The syllabus would differ in the three proposed institutions, but included variously mathematics, economics, ethics and physics, municipal and Roman law, Hebrew, divinity, and so on. Each university would have a rector and a principal, together with regents. The stipend of each of these was indicated. To meet expenses, each university would be provided with temporal assets including land, from the rents and revenues of which both salaries and upkeep would be provided, although some charge was to be made on those attending the universities, if they were sufficiently well self- or family-financed. Bursaries were to be available for needy students, seventy-two in St Andrews, and forty-eight each in Glasgow and Aberdeen.

Not surprisingly, parliament did not approve the *First Book of Discipline*. The financial and economic upheaval consequent upon the redistribution of assets which would have been required was unacceptable, no matter how desirable the ends sought. It must also be recalled that, notwithstanding the technicalities, much of the property of the Roman Catholic Church was effectively in the hands of the nobility. The result was that the endowments of the Church were left virtually intact, and in the hands of those already drawing on them. Further, as the *Book* was not accepted, the previous Church polity was not affected at this time, though ministers of the former Church were effectively forbidden by the Reformation statutes from preaching the older doctrines. The abbeys and monasteries and other religious houses were not, as in England, suppressed or dissolved. Rather these institutions were allowed to die along with their personnel. In fact some of the Roman Catholic ministers remained in post in their parishes, though those agreeing with the new doctrines moved over to the new Church.[29] (This indeed could be seen happening earlier, as many of the reformers had been Roman Catholic priests). But the 'transferees' did not take their stipend or other emoluments with them. That meant that ministers of the new faith were alimented only by the thirds of benefices, a payment from Church lands previously made to the Crown.

The situation was, therefore, curious. The old Church continued in existence with its estates and revenues, but with its ministers forbidden to preach or celebrate its sacraments, while the new Church, including those former Roman Catholic priests who had 'reformed', struggled to establish its polity with insufficient resources. The reformers seem at this time to have played a waiting game, there being little point in pressing the take-over of the institutional Church. That made sense, for at the time the Roman Catholic Church was closely allied to Queen Mary, who still had French links. Her abdication in 1567 changed all that, the regent on behalf of the infant James VI, the earl of Moray, being a Protestant.

The 1567 parliament repealed all former acts in support of the papacy, these being all acts 'made against God his word, and maintenance of Idolatrie in ony times bypast' (1567 c. 4). What might not agree with God's word would be anything contrary to the 1560 Confession of Faith, and therefore to assist the citizenry that Confession was annexed to the statute, and thereby once more engrossed in the register of parliament. The 1567 parliament also ratified various other of the acts of 1560, thus curing any defect in their original enactment through Mary's failure to accept them. Thus the Papal Jurisdiction Act 1560 c. 2, was ratified by the Act 1567 c. 3 (c. 2 12mo), 'Anent the abolishing of the Pape, and his usurped authoritie', and the proscription of the Mass by the Act 1560 c. 4 was confirmed by the Act 1567 c. 5 (c. 3 12mo).

Thus far the 1567 parliament had in effect repeated its predecessor, but three further steps were taken properly to establish the new Church. Of these the Church Act 1567 c. 6 and the Church Jurisdiction Act 1567 c. 12 are fundamental to the history and existence of the present Church of Scotland at least as a human institution.

The Church Act 1567 c. 6 [30] 'Anent the trew and haly Kirk and of tham declarit not to be of the same', defined that Kirk to consist of the ministers that agreed properly as to doctrine and sacrament, and the people of the realm that professed Christ 'as he is now offerit and do communicat with the haly sacramentis as in the reformit kirkis of this realm' as publicly administered according to the Confession of Faith. Further, by its definition, all others were no Church. That was made starkly clear by the Church Jurisdiction Act 1567 c. 12 which narrated that the Crown declared and granted jurisdiction to the Church which preached 'the trew word of Jesus Christ, correctioun of maneris and administratioun of haly Sacramentis ...'. The act further declared that 'thair is na uther face of Kirk nor uther face of Religion than is presentlie be the favour of God establisheit within this Realme And that their be na uther iurisdiction ecclesiasticall acknawlegeit within this Realme uther than quhilk is and salbe within the same Kirk or that quhilk flowis therefra concerning the premissis ...'. Despite such language, the Act 1567 c. 12 was not satisfactory to the reformers who appreciated that by its terms it might appear that the jurisdiction of the Church was at the gift of the Crown. However, no remedy of that point was forthcoming from the civil authorities.

The third innovation in the work of the 1567 parliament on Reformation matters was the Coronation Oath Act c. 8. This, requiring all future monarchs to swear to protect the true Church as established and to root out all those opposed to its teaching, remains in force, and will probably present a point of discussion at the next succession to the throne. Suffice it to say here that at the time it was of more political than legal interest, although its breach was cited as the first head of complaint in the Claim of Right against James VII, 1689, c. 28.[31]

By these acts, then, the Scottish Reformation was accomplished in its institutional sense through the definition of the new Protestant Church, and the condemnation of all others. The matter of its exact governance was left silent, and for the next quarter of a century the polity of that Church seems to have oscillated between a quasi-Presbyterian and a quasi-episcopal system, the role and powers of the superintendents appearing much to depend on the personality of those so serving, and on the reaction of their ecclesiastical brethren to them. This was not uniform throughout the realm.

The failure of the 1560 parliament to enact or otherwise adopt the ideas if not the terms of the *First Book of Discipline* continued to cause problems. Also arising from that failure, the finances of the new institution remained a difficulty, interacting with the question of ecclesiastical polity. As in 1560, parliament was not willing to adopt the *First Book of Discipline*, and steps were taken not so much to disendow the Roman Catholic Church as slowly to take over its benefices as the persons holding them passed on to a greater reward. Thus the Act 1567 c. 10 provided that the patronage of benefices should remain with their existing patrons, but that the reformed Church, through its superintendents, should be the judges of the qualifications of a new presentee when the matter arose. True Protestant ministers would therefore in due course take over. In the meantime the Act 1567 c. 7 laid down that 'thirds of benefices' should, as the Sixth Head of the *First Book of Discipline* had indicated, be paid to ministers for their maintenance '... ay and quhill the Kirk come to the full possession of their proper patrimonie, quilk is the teindes ...'.

The intention, therefore, was that benefices would be taken over, and that payment of ministers through teinds, a charge on land, would be introduced.[32] The latter was to come into proper being in the next century, but the matter of benefices dragged. Thus under the Concordat of Leith of 1572, episcopal sees falling vacant were to be filled by Crown appointment, subject to examination of the qualifications of the presentee by the Church. That was open to abuse, and was used by the Crown to assert practical control of the Church in that the bishop's oath of office recognised the royal supremacy in matters religious.[33] Even so, also in 1572, by its Act c. 3 (c. 46 12mo) a new parliament enacted testing legislation, requiring the subscription of the Confession of Faith by all ministers 'so that the adversaries of Christ's Evangell sall not injoy the Patrimonie of the Kirk'. This act made possible the deposition of ministers who refused to conform to the Confession, and was effective in part. It was also, however, used by the Crown once more to exact an acknowledgement of the royal supremacy. Taking the acts of 1567 and 1572 together, the benefices formerly belonging to ministers of the Roman Catholic Church were slowly taken over by the new institution. But nothing was done to recover properties which had already passed under secular control, nor to deal with the property of abbeys and monasteries, institutions which had not been taken over by the Kirk. Thus, in the region from which I write, north-east Scotland, the abbey of Deer was seized by the Earl Marischal. That he was later, in 1593, to found a university,[34] partly financed from the funds of the abbey of Deer, does not fully expiate the transfer of that wealth from the ecclesiastical realm to the secular.[35] In any event, the gradual removal of Church property into secular hands continued, the Crown itself taking care to benefit. The Act 1587 c. 8 (c. 29 12mo) 'Annexation of the Temporalities of benefices to the Croun' dealt with lands which had not yet passed into secular hands. It preserved for the life of existing beneficiaries tithes, tacks etc., but otherwise allowed the Crown to deal with that property. The Act 1592 c. 13 (c. 121 12mo) provided that transfers to secular hands after the 1587 act were valid only if made to lords of parliament, and the Act 1592 c. 44 (c. 131 12mo) required the compilation of lists of rents etc. of Church properties so that the 1587 Annexation Act could be properly implemented. Steps were thereafter taken to see that property or income therefrom was not further diverted.[36]

It may be gathered from the foregoing that by the 1570s a quasi-episcopal form of Church government was operating throughout much of Scotland, that that polity was being used by the Crown in measure to control the Kirk, that some among the noble families had used various devices to benefit themselves from Church funds, and that many others of the nobility were using the 'Reformation of religion' to divert Church funds into their coffers on a more permanent and absolute basis. By contrast, the Kirk and its ministers were subject to grave financial limitations, as even the rudimentary provisions as had been made by statute were not being properly implemented. That was the situation found by Andrew Melville when he returned to Scotland from Geneva in 1574. To meet this he formulated a new Book, the *Second Book of Discipline*,[37] which was approved by the General Assembly of the Kirk in 1578, another document essential for an understanding of the development of the Kirk.[38]

The *Second Book of Discipline* condemned episcopacy owing to its possible abuse, and set out clearly and cogently the basic elements of what is the normal form of conciliar Church government in Presbyterianism. It stated, rather than argued, the independence of the Church from civil authority, Church powers and authority being derived from God. It also argued that the Church had the right and duty to instruct the civil authority in matters of conscience and religion, and indeed the right of the Kirk to tell the civil arm how it should exercise its own proper authority in accordance with the Word of God. This came close to arguing for a theocracy, and, not surprisingly, was avoided and indeed attacked by the Crown as being subversive of the Crown's own authority and powers.

One year later, as noted above, by the Act 1579 c. 6 (c. 68 12mo), parliament re-enacted the Church Act of 1560, and by the Act c. 7 (c. 69 12mo) did the same for the Church Jurisdiction Act 1567 c. 12. In 1581 the king and his household subscribed the Negative Confession,[39] denying all doctrines contrary to or incompatible with the Scots Confession of 1560. Despite this, subsequent legislation of 1584, known as the Black Acts, sought to re-impose episcopacy, denounced Presbyterianism and placed all authority, including ecclesiastical authority, under the Crown. But by 1590 the Presbyterian party was once more in the ascendant, resulting in the Act 1592 c. 8 (c. 116 12mo), which established the Presbyterian government of the Kirk.

The General Assembly Act 1592 c. 8 (c. 116), sometimes referred to as the Great Charter of the Church, is the statutory foundation of the Presbyterian government of the Church of Scotland. In it the former acts regarding the Protestant faith were ratified and approved, with the exception of the Black Acts of 1584. Drawing on some of the phraseology of the *Second Book of Discipline*, the government of the Church through general assembly, synods, presbyteries and kirk sessions was also approved. From the Kirk perspective, there were defects, notably in the preservation of patronage of the right of presentation to ministerial charges (appointments), and that the calling of General Assemblies was left a matter for the Crown. James VI used this latter power to call Assemblies at places less under the control of 'conviction Protestants' (to use a modern term). After his accession to the English throne in 1603 he was able in a few years to secure the re-creation of bishops, and through them to control the Church. Thereafter the struggle between Presbyterianism and episcopacy was to darken the seventeenth century, but that takes us well beyond the scope of this chapter. Suffice it to say that Presbyterianism was to be fully re-established, and protected in the terms of the Union negotiations and agreements that led to the Union of the Parliaments in 1707.

The other point that should perhaps be made is that there was not a strong Counter-Reformation movement in Scotland at this time. On the Continent, wars were to be fought, but within Scotland, separated from the Continent by distance, as well as by Protestant England, the Catholic cause did not present a unified front but rather survived as pockets of the 'old faith'.[40] While these were protected by powerful Catholic families in certain parts of Scotland, they did not unite, or rise up. Their protection and continued toleration meant that enclaves of Catholicism remained, for example in remoter Aberdeenshire and in the southern Hebrides, but this did not seriously affect the progress of the Kirk. It is even the case that the Kirk was prepared to debate rather than take action.

THE STATE

A simple point may be made to start with: the word 'State' in the title of this section is misconceived if it is intended to have a legal content. There was no Scottish 'State' at the time of the Reformation, convenient as it may be to use the jargon: in fact even at present there is no British 'State' as a legal entity. In United Kingdom law the term the 'State' is incorrect. We live in a kingdom, under a monarchy, and all government takes place in the name of and in exercise of the powers of that monarchy and the institutions which it has created. Of course the question then arises whether the Church is a creation of the monarchy, and it can be seen that during the settlement of Church polity at the time of the Reformation, the Crown occasionally did try to indicate or assert such control. That, however, was clearly ruled out by the eventual legislation in the matter, the Church Jurisdiction Act 1579 c. 7, repeating the Act 1567 c. 12, affirming the Confession, which held in specific terms that only Jesus Christ was the head of the Church.[41] Further, the *Second Book of Discipline* had indicated it as a duty of the Kirk to correct the civil arm if it got things wrong. In short, implicit in the Reformation, there was built in to the constitutional arrangements in Scotland that the Kirk might be a source of opposition to the Crown. This indeed it proved to be during the years of the Reformation, in the reign of Mary Queen of Scots, and to an extent in that of her son and his heirs. James VI, who had been tutored by

George Buchanan, the author of *De Iure Regni apud Scotos* (1579),[42] was under notice. The struggles of the seventeenth century were to prove the importance of opposition, which, albeit political in effect, had its basis in a theology and a system of Church government where the 'priesthood of all believers' was the base.[43] The happier side of all this from the point of view of the Kirk, was that the Crown and its officers could serve as the 'godly magistrate', a concept clearly to be found in John Calvin and others.[44] Although pleased to see itself as the godly magistrate, the Crown did not always share the Kirk's view as to the content of the conduct which that godliness required. Whether this sat well with the granting of a vote in parliament to the holders of bishoprics, abbacies or other prelatic tenure by the Act 1597 c. 2 (c. 235 12mo), may be doubted.

The other matter to be noted is that for many purposes the parish came to be considered as the unit of local government. As we will see below, the parish was the unit for such matters as the administration and securing of poor relief, and of education. It was also the unit used for the registration of titles to land in the General Register of Sasines, the public record of landholding in Scotland that dates from 1617,[45] and only now, some four centuries later, is being replaced by the system of Registration of Title under the Land Registration (Scotland) Act 1979, c. 33.

THE INDIVIDUAL

Of course all that has been previously discussed affected individuals. While the actual effect might vary throughout Scotland, the willingness of the new Kirk to continue in the tradition, or aspirations, of its predecessors could not be gainsaid. Thus the pre-Reformation parliament of 1551 had passed an act 'Anent them that perturbis the Kirk, in time of divine service' (c. 8, c. 17 12mo), another 'Anent them that knawis themselves under process of cursing' (Church censure – in the ultimate, excommunication) and turn up at church service notwithstanding (c. 9, c. 18 12mo), one 'Anent them that maries twa wives or husbandes' (c. 11, c. 19 12mo), and finally, one 'Anent Adulterers' (c. 12, c. 20 12mo). Four years later the parliament of 1555 required that 'na maner of person be chusen Robert Hude, Little John, Abbotis of Un-reason, Queenis of Maij nouther in Burgh nor to landwart, in ony time to cum', and specified punishment for 'ony Provost, Baillies, Councell or Communitie' which might select persons to play these roles (1555 c. 40, c. 61 12mo).

The reformers continued in those tracks. The parliament of 1563 passed legislation 'Anent Adulterie' (c. 10, c. 74 12mo) prescribing the death penalty for both parties, while also preserving the right of the 'innocent party' to sue for divorce. The 1567 parliament dealt with fornication (c. 4, c. 13 12mo), incest (c. 15, c. 14 12mo) which incorporated parts of Leviticus 18, and degrees of consanguinity (c. 16, c. 15 12mo), that of 1592 with usury (c. 56, c. 140 12mo – 'against unlawful conditions in contracts') and the 1594 parliament dealt with Parricide (c. 30, c. 224 12mo). But short of statutory criminality, it was not for nothing that the part of the Church Jurisdiction Act 1567 c. 12 which I previously quoted granted jurisdiction to the Church which preached 'correctioun of maneris'. While Scotland was well short of that experience of the twentieth century which we may call 'government by Ayatollah' or by 'Taliban', the reformed Kirk was to take its responsibility for the morality of the people with increasing seriousness over the next decades.[46] The agency for such supervision was the parish, through the minister and kirk session, albeit that it was more than a generation before every parish had a session.[47]

As has already been noted, the territory of Scotland had been divided into parishes before the onset of the Reformation. The Act 1581 c. 3 (c. 100 12mo) called for the provision of ministers for all parishes and for stipends for them. They were to be responsible for the moral as well as the spiritual and physical welfare of their parishioners. How far this went in the years immediately after the Reformation is moot. Wormald[48] and Donaldson[49] among others indicate that the process was slow. Wormald notes that in many ways the real struggle of the times was between the secular Protestant leaders and the ministers of the new Church. More effort was to go into that than into the supervision of the laity until a date later than the immediate compass of this chapter.[50] The major exception to that is the matter of witchcraft. The Act 1563 c. 73 c. 9 (c. 73 12mo) ' Anentis Witchcraftes' proscribed witchcraft, sorcery and necromancy under pain of death. What precisely put this matter ahead of that of adultery on the parliamentary agenda is a matter of conjecture. Certainly witchcraft had been a concern for almost one hundred years since the Bull of Innocent VIII, *Summis desiderentes affectibus* of 9 December 1484 had appointed Heinrich Kramer and Jacob Sprenger as inquisitors of the phenomenon.[51] Certainly, Exodus 22:18 provides that witches are to be exterminated, and the 1563 parliament was following the normal pattern of the time. But it is notable that later the Crown was to be itself involved, James VI publishing *Daemonologie* in 1597, though perhaps that book marked the high point of Crown interest and whether it was triggered by the particular case of the North Berwick witches,[52] or was basically an attack on Reginald Scot, is moot.[53]

In the long term, more important was the gradual implementation of the ideas of the *First Book of Discipline* as to education, which were to have substantial effects on the development of this small and rather insignificant territory perched on the far north-west of Europe. Education had previously been of concern,[54] but the proposals of the *Book* showed a greater interest, and included detailed and country-wide proposals. Education was becoming a right and a duty, rather than a privilege. While the *Book* itself was not adopted by parliament, as Duncan Shaw shows, the General Assembly of the Church took the matter of education seriously.[55] Writing some four centuries later, I am grateful for that new directing of interest, from which, at many removes, I have benefited. But the effect throughout the country was patchy.[56] There are elements of Head V of the *Book* which indicate Kirk control of education. Scholars might go to a university if able, but also only with a certificate from one's minister and school teacher as to 'learning, docility, age and parentage'. It may also be noted that at this time parish schools were not the only schools available. In various burghs grammar schools and 'sang schools' existed, and had in some instances for centuries. These were now subjected to presbytery inspection, the Act 1567 c. 11 requiring that all teachers had to be Kirk approved, and prohibiting the private instruction of the youth.[57] Further, following a resolution of the General Assembly in 1578, the Act 1579 c. 9 (c. 71 12mo) provided that the pains of barratry were to be inflicted on those who sent their children abroad for their education without royal permission.

What was taught in the parochial schools was, of course, also supervised. Catechetical instruction was important,[58] but so was the availability of suitable reading material. The reformers carried over into their practice the control of printing and the censorship of materials to be published that had previously been instituted at the behest of the Roman Catholic Church.[59] Heretical books were not to be printed, and any such imported into the kingdom confiscated.[60] The Act 1587 c. 2 (c. 23 12mo) provided for the punish- ment of adversaries of the 'true religion', and those who should seek to persuade others to decline from the profession of the true religion. The circulation of books or letters arguing against the reformed faith was also prohibited by the act, and this was underlined by the next statute, Act 1587 c. 4 (c. 25 12mo) dealing with those who trafficked in, or imported, 'erroneous' books.

Formally, as noted above, the question of welfare had been a concern of the pre-Reformation Church, although implementation of any sort of duty thereanent seems to have been sporadic. After the Reformation parishes continued in measure to care for their sick. However, the control and operation of the hospitals never formally passed to the Kirk, and in fact seems to have become a responsibility of the Crown. By the Act 1579 c. 17 (c. 74 12mo) a major attempt was made to deal with such questions. The lord chancellor was instructed to make inquiry into the foundation documents of all hospitals, with a view to returning them to their original purpose which was considered as being to serve the 'aged, impotent and pure [poor]'.[61] The poor were instructed to return to the parishes of their birth or of their ordinary residence of at least seven years, where they would be entered on a register. Thereafter, in both burgh and in landward parishes the assistance to individuals would be provided by a tax on all the inhabitants within the parish, without exception. There was also provision for allocating recipients to their appropriate parishes, and for transfer between them, severe punishment being meted out on those who refused to obey directions. There is little by way of direct evidence how well this system worked, but the Act 1592 c. 159[62] anent the hospital in Edinburgh may show how such matters were dealt with.[63] Later this duty, at least in landward parishes, was passed to the kirk sessions, and a substantial body of law developed dealing with the administration of poor relief.[64]

Correlative to the relief of the poor and the ill, concern was shown as to undue or conspicuous consumption. Sumptuary laws, as the legislation on such matters is known, were common throughout Europe in the fifteenth and sixteenth centuries, as matters of class distinction, morality and the preservation of guild monopolies. Scotland was no different,[65] but it is interesting to find such questions still of concern to the post-Reformation arbiters of propriety, though perhaps their motives were slightly different.[66] Thus the first part of the Act 1581 c. 18 (c. 113 12mo) is 'Against the excess of costly clothing', setting out a scale of income and status appropriate for particular styles of dress, and this was repeated in other admonitions.[67]

The other matter which the Act 1579 c. 17 (c. 74 12mo) dealt with, after the question of poor relief, was that of 'strang and idle beggaris' and 'vagabondis'. These were to be deterred by stocks and irons, the possibility of a whipping, and having their ears pierced by a hot iron. 'Ordinary' beggars might ply their trade only within the parish of their residence.[68] Particular other kinds of itinerant beggars were made the subject of further provision. Those who used subtle, crafty and unlawful games such as juggling or fast-and-loose were to be punished. Gypsies, the 'idle peopill calling themselves *Aegyptians*, or any uther, that feinzes [feigns] them to have knawledge of Charming, Prophecie, or uthers abused sciences, quairby they perswade the peopill that they can tell their weirdes, deathes and fortunes, and sik uther phantasticall imaginations' were also to be dealt with. So indeed were all minstrels, singers and storytellers who were not in the special service of lords or burghs. Intriguingly, vagabond scholars of the universities of St Andrews, Aberdeen and Glasgow, not licensed by their university to take alms, were to be fined, as were seamen who neglected their calling. From a modern perspective these provisions have a social justification. They were perhaps more influenced, however, by the Pauline exhortation to work for one's own keep (II Thess. 3: 10), and the biblical admonition against trafficking with spirits (Exod. 22: 18; cf. I Sam. 28). In any event by the Act 1592 c. 69 (c. 149 12mo) the civil authorities in the rest of Scotland were instructed to put the 1574 and similar acts into force in their areas as Edinburgh did, and, by the Act 1597 c. 39 (c. 272 12mo), the power to enforce these acts was transferred to kirk sessions in landward areas (i.e. not within the burghs).

In the realm of spiritual welfare, of course, Sunday was important.[69] As in some other matters, the interest of the Roman Catholic Church and that of the reformers coincided. The first legislation on the matter in Scotland is the Act 1503 c. 28 (c. 83 12mo), which extended beyond Sundays, dealing with fairs held on all holy days or within kirk yards. By the Act 1551 c. 8 (c. 17 21mo), those who disturbed Sunday worship were to be fined. After the Reformation, however, there is increased legislative activity as to Sunday observance, there being down to the Union of the Parliaments a whole series of statutes on various questions of Sunday observance.[70] The need for this legislation may indicate that it tended to fail. Kirk censures by the kirk session may have been more effective as to personal habits.[71]

However, one long-lasting effect in the law was that it was held that, while bargains made on a Sunday are not illegal, judicial proceedings on a Sunday or the execution of a civil warrant (except one preventing a debtor from fleeing the jurisdiction) were unlawful and ineffective.[72] Sunday was said to be a *'dies non'*.

The kirk session was, of course, a court of the Church and obviously would affect socially and otherwise any individual dealt with by it. It is important, however, to note that one result of the Reformation was that the courts of the Church of Scotland were and are recognised as courts of the realm, and not just the internal judicatures of a private association or club.[73] But that meant that the relationship of the Kirk and the civil court system(s) had to be dealt with, and that at a time when the relationship between Church and State in such matters was widely under question.

Before the Reformation, Church courts, the Officials Courts,[74] exercising jurisdiction on behalf of the bishops, dealt with matters which had an ecclesiastical element through being matters of faith and morals. Matrimonial causes clearly fell into that category. Matters of wills (testaments) and succession also depended on relationships defined by Church law, and so fell under their jurisdiction. But, because in contrast to other courts the Church courts were well staffed with able professional lawyers, other matters tended to be brought to them. Thus arguably questions of obligations, particularly contract, could be dealt with, since the performance or non-performance of a contract, being based on promise, had a spiritual element to it. In these cases the sentence of the ecclesiastical tribunal was accompanied with a warning that failure to obey would involve 'cursing', that is ecclesiastical censure and ultimately exclusion from the means of grace through excommunication. Although, by the time of the Reformation, the lowered prestige of the Church was having its effect, that threat still operated on many minds.

There was, however, by this time an alternative: a civil legal system. Building on sporadic previous arrangements and on an Italian model, the College of Justice had been created in 1532 by the Act 1532 c. 2 as a permanent civil court. As originally constituted, seven of the fourteen ordinary senators of the College of Justice were ecclesiastics, as was its president. But, by the time of the Reformation, all was not well. The Act 1579 c. 38 (c. 93 12mo) refers to unqualified persons being appointed to the Session, and to bribery and corruption either direct, or through the wives and family of judges. Accordingly the 1579 Act provided that the Crown should only present for appointment 'ane man that fearis GOD', well-read, practical and knowledgeable in the law, who acts with expedition and dispatch, and having a sufficient living within his own resources. Further, new appointees should be examined as to their knowledge by the existing Lords of Session, who should refuse those found lacking. Finally, the requirement that the president of the College be an ecclesiastic was dispensed with. Five years later by the Act 1584 c. 6 (c. 133 12mo) ministers of religion were forbidden from holding office as judges, advocates, clerks or notaries (except in the making of a testament)[75] although it took time for these reforms to take effect as the existing judicial incumbents demitted office or died off.

These steps did not entirely solve problems of jurisdiction. The remit of the Church courts had come under question with the abolition of papal jurisdiction by the Act 1560 c. 2, but it was not entirely clear whether that legislation abolished papal jurisdiction only, or whether the Church courts also had had the ground cut from under them. In fact various of the Officials continued to deal with matters for some years, their staff now serving the new superintendents in consistorial matters. In 1563, however, the Privy Council set up a Commissary Court in Edinburgh to adjudge consistorial questions for mid-Scotland. Local commissaries elsewhere retained their roles for some years, and then were replaced by a more regular commissary system, based on the old boundaries of the bishoprics, and staffed by persons nominated by the College of Justice.[76]

The other question is that of the law to be applied by the courts where there was a religious element or connotation. Simply put, the reformers and those of their persuasion had a hearty dislike of the Canon Law which the Roman Catholic Church had developed over the centuries. One result is that there is little trace of Canon Law in much of the Law of Scotland general, save perhaps in matrimonial law. There Scotland, having separated from Roman Catholic Church before the abolition of irregular marriage by the Council of Trent in 1563, retained those forms of marriage, marriage by declaration *de praesenti* and by promise *subsequente copula* being abolished only by the Marriage (Scotland) Act 1939 c. 34. Marriage by 'habit and repute' remains possible.[77] That said, the traces of Canon Law remaining in both the civil and ecclesiastical law are intriguing, if they are traces, and not merely the recognition of the desirability of a particular rule.[78] On the other side of matrimony, the reformers made divorce, as opposed to separation 'from bed and board' (*a vinculo et thoro*) possible, though there is some evidence that this was already being permitted for adultery. This was now, however, made quite clear and general. Divorce for adultery was contemplated by Head IX of the *First Book of Discipline*,[79] and for desertion was recognised by the Act 1573 c. 1 (c. 55 12mo).[80]

So, we may leave the Reformation in Scotland and its effects on both the State and the individual. In a later chapter we will turn to the modern position.

NOTES

1. Kidd, 1911, Part I, Lutheran, Doc. no. 11, 20–6.
2. The breach with Rome came in 1533, the much criticised unofficial *Ten Articles* in 1536. The Ecclesiastical Canons Act 1535 (27 Hen 8 c. 15) and the Canon Law Act 1543 (35 Hen 8 c. 16) might have allowed progress, but again nothing was officially done. The 1533 and 1543 Acts did provide that the existing Canons should continue in force and as a result 'belief' was not affected. The *Thirty-nine Articles of Religion* come in 1563, but in final form only in 1571. See Hill, M. *Ecclesiastical Law*, 2nd edn, Oxford, 2001, 6–7; *The Canon Law of the Church of England*, the Report of the

Archbishop's Commission on Canon Law, London, 1947; Gibson, E C S. *The Thirty-nine Articles*, 6th edn, London, 1908.
3. On the whole matter see Lyall, 1980.
4. The Court of Session was created in 1532 by the Act 1532 c. 2.
5. Dickinson, 1949.
6. Cameron, 1972; Dickinson, 1949, vol. 2, App. VIII, 280–324.
7. *The Book of the Universal Kirk*, Thomson, T, ed., Bannatyne and Maitland Clubs, 3 vols., Edinburgh, 1839–45; *The Booke of the Universall Kirk of Scotland*, Peterkin, A, ed., Edinburgh, 1839. For ecclesiastical laws, cartularies and records, see: Maxwell, L F and Maxwell, W H. *Scottish Law to 1956*, vol. 5 of *A Legal Bibliography of the British Commonwealth of Nations*, London 1957, 33–8; for synod, presbytery and kirk session records see, Grant, F J. Presbyterian Court Records. In *An Introduction to the Sources and Literature of Scots Law*, Stair Society, vol. 1, Edinburgh, 1936, 144–62.
8. McRoberts, 1962.
9. See Cairns, J W. Historical introduction. In Reid, K and Zimmerman, R, eds. *A History of Private Law in Scotland*, Oxford, 2000, vol. 1, *Introduction and Property*, 14–184, particularly, sec. V, 'From the Reformation to the Restoration', at 74–105. See also Walker, 1995, particularly for its sources, and Ch. 9 'The Church', 248–319.
10. Wormald, 1981, 3–72.
11. *Origines Parochiales Scotiae*, 2 vols., Bannatyne Club, 1851; Cowan, I B. The development of the parochial system in medieval Scotland, *SHR*, 40 (1972), 43–55; Cowan, I B. *The Parishes of Medieval Scotland*, Scottish Record Society, vol. 93, Edinburgh, 1967.
12. Durkan, J. Care of the poor: Pre-Reformation hospitals. In McRoberts, 1962, 116–28; Durkan, J. Education in the century of the Reformation. In McRoberts, 1962, 145–68; McKay, D. Parish life in Scotland, 1500–1560. In McRoberts, 1962, 85–115; Wormald, 1981, 75–94; Dilworth, M. *Scottish Monasteries in the Late Middle Ages*, Edinburgh, 1995, 57–74.
13. Dilworth, 1995, at 57, citing Grant, I F. *The Social and Economic Development of Scotland before 1603*, Edinburgh, 1930, at 223, while noting some caution with Grant's figures.
14. Various 'commendators' of such property appear in cases reported in Morison, W M. *The Decisions of the Court of Session in the form of a Dictionary, to 1808*, 21 vols., Edinburgh, 1811–16, title 'Kirk Patrimony'.
15. Sir David Lindsay's *Ane Satyre of the Thrie Estaits* is the example most commonly referred to: see, Dickinson, 1949, who reviews much of the material. See also the *Statutes of the Scottish Church: 1225–1559*, Scottish History Society, vol. 54, 1907.
16. Donaldson, 1960; Burleigh, 1960, 153–257. Documents are collected in Kidd, 1911, Part II, Reformed, sec. B.4 Scotland, Docs. no. 334–51, 686–715.
17. *Statutes of the Scottish Church*, 84–19; cf. Donaldson, 1960, 1–28.
18. The Act 1525 c. 4 prohibited the import of Lutheran books: Kidd, 1911, Doc. no. 335, p. 689. The Act 1551 c. 26 (c. 27 12mo) required that printers print no book or other material without having a licence for the publication from the Church.
19. It is an error, and a misleading one, to categorise the Protestant movement as 'protest', though one can see how that etymology can be constructed. Rather the concept is of *protestatio* – a call for a return to the earlier position, the original doctrine of the Church. Only later did questions of polity arise.
20. Kidd, 1911, Doc. no. 344, 696–7.
21. Kidd, 1911, Doc. no. 347, 699–70.
22. Act 1560 c. 2: Dickinson, 1949, I, 341. It should be explained that certain of these old Scots Acts, still being extant, were 'baptised' with 'short titles' by the Statute Law Revision (Scotland) Act 1964 s. 2, sch. 2. Acts repealed prior to the 1964 Act lack short titles, but an indication of their content, together with their Record and

12mo edn. numbering, is given in the Chronological Table of the Acts of the Parliament of Scotland 1424–1707, in Part II of the *Chronological Table of the Statutes*, published regularly by HMSO, London. I cite the Record edition, with the 12mo added where different. Sir Thomas Murray of Glendook's *The Laws and Acts of Parliament made by King James the First, Second Third, Fourth, Fifth, Queen Mary King James the Sixth, Charles the Second who presently Reigns*, Edinburgh, 1681, in the A5 version, occasionally cites chapter numbers different from the 12mo edn.

23. This Act annulled previous Acts of the Scottish parliament which were neither in agreement with the Word of God, nor the new 1660 Confession of Faith.
24. See also Dickinson, 1949, I, 340.
25. Henderson, 1960; Dickinson, 1949, II, App. VI, 257–72.
26. Cameron, 1972; Dickinson, 1949, II, App. VIII, 280–324.
27. Cameron, 1972, 3–77.
28. Cameron, 1972, 115–18.
29. Donaldson, G. The Parish Clergy and the Reformation. In McRoberts, 1962, 129–44.
30. The Church Act 1567 c. 6 was defectively printed, but was re-enacted with due explanation as the Act 1579 c. 6 (c. 68 12mo).
31. Technically this was an Act of the Convention of [the Scottish] Estates, rather than an Act of the Scots parliament. Whatever, in contradistinction to the language of its English equivalent, the Claim of Right concluded that for various specified reasons James VII had 'forfaulted [forfeited] the right to the Croune and the throne is become vacant'.
32. On teinds see Connell, Sir J. *A Treatise on the Law of Scotland respecting Tithes, and the Stipends of the Parochial Clergy*, 2nd edn, 2 vols., Edinburgh, 1830; Buchanan, W. *Treatise on the Law of Scotland on the subject of Teinds or Tithes*, Edinburgh, 1862, cf. Gibson, A J H. *Stipend in the Church of Scotland*, Edinburgh, 1961.
33. Dickinson, 1968, 12; *The Book of the Universal Kirk*, Bannatyne and Maitland Clubs, 3 vols., Edinburgh, 1839–45, vol. I, 220.
34. Now, since 1860, Marischal College of the University of Aberdeen, under the Universities (Scotland) Act 1858, 21 & 22 Vict. c. 83.
35. Henderson, G D. *The Founding of Marischal College, Aberdeen*, Aberdeen, 1947.
36. See generally Walker, 1995, 770–2.
37. Kirk, 1980. The 'Introduction', 1–157, deals informatively with the antecedents and tenor of the *Book*.
38. Henderson, 1954; Kennedy, 1960; MacGregor, 1926.
39. Dickinson, 1968, III, 32–5.
40. For the post-Reformation history of Scottish Catholicism down to the restoration of the hierarchy in 1878, see Anson, P. *Underground Catholicism in Scotland, 1622–1878*, Montrose, 1970; and Chapters 6 and 7 in this volume.
41. Henderson, 1960, c. XVI.
42. Buchanan, G. *The Art and Science of Government among the Scots*, McNeil, D H, trans., Glasgow, 1964: also translated as *The Powers of the Crown in Scotland*, Arrowood, C E, Austin, TX, 1949. According to Irvine Smith, the Act 1584 c. 9 (c. 134 12mo, against treason), ratified by the Act 1585 c. 1 (c. 10 12mo, against leasing making), was directed against Buchanan's book, and extended the crime of 'leasing making' (sc. slander of public officials) to include all speeches and sermons, public or private, which meddled in public affairs; Irvine Smith, J. The Transition to the Modern Law, 1532–1660. In *An Introduction to Scottish Legal History*, Stair Society, vol. 20, Edinburgh, 1958, 25–43 at 41.
43. The democratic imperative can be traced to this, at least as one source of that immensely important concept of the desirably proprieties of civil government.
44. Calvin, J. *Institutes of the Christian Religion*, 1536, ed., trans. and annotated, Battles, J F, London, 1975, Ch. VI, 'Christian freedom, ecclesiastical power and political

administration', 176–226; *Calvin: Institutes of the Christian Religion*, 1559 edn (the final edn), Library of Christian Classics, vol. 20, McNeill, J T, ed., Battles, F L, trans., 2 vols., Philadelphia, 1960, Book IV. Ch. XX, vol. 2, 1485–1521. Cf. material noted and bibliographed in the latter version. Cf. also Buchanan's *De Iure Regni*, 1579, just referred to, and, from a later date, Samuel Rutherford's *Lex, Rex or The Law and the Prince; A Dispute for the Just Prerogative of King and People*, 1644, rep. Sprinkle Publications, Harrisonburg, VA, 1980.
45. Act 1617 c. 16.
46. Walker, 1995, 474–6.
47. *Selections from the Records of the Kirk Session, Presbytery and Synod of Aberdeen*, Spalding Club, Aberdeen, 1846; Kirk, J, ed. *Stirling Presbytery Records, 1581–1587*, Scottish History Society, Edinburgh, 1981; Edgar, A. *Old Church Life in Scotland: Lectures on Kirk Session and Presbytery Records*, Paisley, first series, 1885; second series, 1886.
48. Wormald, 1981, 122–39.
49. Donaldson, 1960, 76–101.
50. Cf. Cowan, 1982, 159–205.
51. Kramer and Sprenger wrote the *Malleus Maleficarum*, (Summer, M. trans., London, 1928) the authoritative text of the time on witchfinding. The *Malleus* retains its interest in its discussion of proof of criminality and matters of the proprieties of the laws of evidence.
52. Normand, L and Roberts, G. *Witchcraft in Early Modern Scotland: James VI's Demonology and the North Berwick Witches*, Exeter, 2000.
53. Scot, R. *The Discoverie of Witchcraft*, 1584, Summers, M, ed. and intro., London, 1930: New York, 1970. Cf. Wormald, 1981, 168–70. On witchcraft in Scotland see also Walker, 1995, 477–83, with citations to cases in Pitcairn, R. *Ancient Criminal Trials in Scotland, 1488–1730*, 3 vols., Edinburgh, 1829–33; Larner, C. *Enemies of God: The witch–hunt in Scotland*, London, 1981; Macdonald, S. *The Witches of Fife: Witch–hunting in a Scottish Shire, 1560–1710*, East Linton, 2002.
54. Thus the Act 1496 c. 1 (c. 3 12mo) provided under a penalty of £29 Scots, that all barons and freeholders of substance should send their eldest sons to grammar school from the age of six or nine until they had perfect Latin, and after that for three years to the schools of art and law so that they understood the laws. See also Durkan, J. Education in the Century of the Reformation. In McRoberts, 1962, 145–68; Scotland, J. *The History of Scottish Education: From the Beginning to 1872*, 2 vols., London, 1969, vol. I.
55. Shaw, 1964, 180–202.; cf. Scotland, 1969; Grant, J. *The History of the Burgh Schools in Scotland*, London, 1876.
56. Wormald, 1981, 177–90.
57. Scotland, 1969, vol. I; Grant, 1876. See also the background facts in *The Presbytery of Elgin v The Magistrates and Town Council of Elgin* (1861) 23 D.287, a saga that wended its way through the courts for ten years.
58. Shaw, 1964, 200–2.
59. Shaw, 1964, 220–30.
60. Shaw, 1964, 220. It may be noted the General Assembly of the Reformed Kirk (or the commission thereof that made specific decisions) operated the Act 1551, c. 26 (c. 27 12mo) – a pre-Reformation statute.
61. In fairness we may note that a similar inspection for the same purpose was instructed by the Act 1424 c. 2 (c. 27 12mo).
62. Not printed in Record edition as not being a public and general statute.
63. Shaw, 1964, 203–5.
64. On the subsequent Poor Law, see Dunlop, A. *Poor Laws of Scotland*, Edinburgh, 1825, 1828, 1854; Lamond, R P. *The Scottish Poor Laws*, Edinburgh, 1870, 1892.

The administration of poor relief became a governmental function through parochial boards in Victorian times. It may be noted that the parish remained the appropriate district for the administrative purposes, though the matter had been removed from Kirk Sessions.

65. Cf. the Acts 1429 cc. 8, 9 and 10 (c. 118 12mo) 'Anent the persones that sall weare claithes of Silke and Furringes' (sc. fur trimming), and the Act 1457 c. 13 (c. 70 12mo), 'Of coastly clothing'.
66. On the 'Sumptuary laws', see Irvine Smith, J and MacDonald, I. *An Introduction to Scottish Legal History*, Stair Society, vol. 20, Edinburgh, 1958, 280–301; Walker, 1995, 649.
67. For example, the Act 1587 c. 16 (c. 36 12mo) required members of the Estates (i.e. of parliament) to be modest in their dress, the penalty for any breach being exclusion from parliament house.
68. In the fifteenth century the principle was established that someone might be licensed to beg within the local authority area, if otherwise unable to find work: Act 1424 c. 25 12mo (not in Record edn).
69. Head IX of the *First Book of Discipline* provided that on Sunday before noon the Word should be preached, the sacraments celebrated and marriages solemnised. After noon was for the catechising of the children in public. Other days might also have sermon periods, during which in the 'notable towns' no–one should be at work.
70. The bulk of the Sunday Observance Acts so far as they had not been expressly repealed, were held to have fallen into desuetude in their criminal aspect by Lord Mackay in *Brown* v. *Magistrates of Edinburgh* 1931 S.L.T.456. As enumerated by Lord Mackay the Acts were 1503 c. 28 (c. 83 12mo), 1579 c. 8 (c. 20), 1592 c. 17 (c. 124), 1593 c. 6 (c. 163), 1594 c. 8 (c. 201), 1661 c. 18 (c. 181), 1661 c. 338 (c. 38), 1663 c. 43 (c. 19), 1672 c. 58 (c. 22), 1690 c. 7 (c. 5), 1693 c. 64 (c. 40), 1696 c. 31 (c. 31) and 1700 c. 12 (1701 c. 11). See also the list in 'Time' cited below n. 72 para 817 n. 13. The Statute Law Revision (Scotland) Act 1964 c. 80 eliminated most of the remainder. The 'clustering' of the legislation is noteworthy, and in major part reflects contemporaneous civil and ecclesiastical turmoil and struggle. Latterly Justices of the Peace were under a duty to enforce the Acts, but they lacked penal powers for the purpose.
71. Walker, 1995, 649–50.
72. The present legal position is more moot: see 'Time' in the *Stair Memorial Encyclopaedia of the Laws of Scotland*, vol. 22, para. 817.
73. *Presbytery of Lews* v. *Fraser* (1874) 1 R. 888.
74. Walton, F P. The Courts of the Officials and the Commissary Courts, 1512–1830. *An Introduction to the Sources and Literature of Scots Law*, Stair Society, vol. 1, Edinburgh, 1936, 133–53.
75. Not now. The Requirements of Writing (Scotland) Act c. 7, s. 9(6) redefined those who can execute a deed on behalf of someone blind or unable to write. Ministers do not come within the new definition.
76. Shaw, 1964, 206–19; Donaldson, G. The Church Courts. In *An Introduction to Scottish Legal History*, Stair Society, vol. 20, Edinburgh, 1958, 363–73; Scanlan, J D. Husband and Wife: Pre-Reformation Canon Law of Marriage of the Officials' Court. In *An Introduction to Scottish Legal History*, Stair Society, vol. 20, Edinburgh, 1958, 69–81; Ireland, R D. Husband and Wife: (a) Post-Reformation Canon Law of Marriage of the Commissaries' Courts and (b) Modern Common and Statute Law. In *An Introduction to Scottish Legal History*, Stair Society, vol. 20, Edinburgh, 1958, 82–9; Ireland, R D. Husband and Wife: Divorce, Nullity of Marriage and Separation. In *An Introduction to Scottish Legal History*, Stair Society, vol. 20, Edinburgh, 1958, 90–8; cf. Irvine Smith, 1958.

77. A decree of the Court of Session is necessary in the matter for such a marriage to have civil effect. The procedure is likely to be abolished in 2006.
78. Smith, D B. Canon Law. In *An Introductory Survey of the Sources and Literature of Scots Law*, Stair Society, vol. 1, Edinburgh, 1936, 183–92; Cf. Gardner, J C. The influence of the Law of Moses. In *An Introductory Survey of the Sources and Literature of Scots Law*, Stair Society, vol. 1, Edinburgh, 1936, 235–40. Cf. also Miller, W G. The Canon Law and Scottish Presbyterianism, *J. of Jurisprudence*, 32 (1888), 113–27. The long–standing rule of Scots Law that permitted reference in certain cases to the 'writ *or oath*' (emphasis added) of a defender, looks suspiciously as if it relied on a religious sanction as well as criminality by contempt of court. The rule is obsolescent since the coming into force of s. 14 of the Requirements of Writing (Scotland) Act 1995.
79. As noted above, adultery was subject to penalty – even death – by the Act 1563 c. 10 (c. 74 12mo), but divorce seems to have been more usual. The Act 1592 c. 11 (c. 119) stripped an adulterous woman of her rights to dispose of her property to the prejudice of her lawful heirs. The Act 1600 c. 29 (c. 20 12mo) prohibited the marriage of adulterers with the person with whom in the relevant divorce proceedings they had been adjudged to have committed adultery. The practice soon developed of omitting the name of the 'other party' from the decree of divorce.
80. The Act 1573 c. 1 (c. 55 12mo) narrates that divorce for desertion had been lawful since August 1560 – the date of the Reformation in Scotland.

BIBLIOGRAPHY

Burleigh, J H S. *A Church History of Scotland*, Oxford, 1960.
Cameron, J K, ed. *The First Book of Discipline*, Edinburgh, 1972.
Cowan, I B. *The Scottish Reformation: Church and Society in Sixteenth Century Scotland*, London, 1982.
Dickinson, W C, ed. *John Knox's History of the Reformation in Scotland*, 2 vols., London, 1949.
Dickinson, W C, et al. *Source Book of Scottish History*, Edinburgh, 1968.
Donaldson, G. *The Scottish Reformation*, Cambridge, 1960.
Henderson, G D. *Presbyterianism*, Aberdeen, 1954.
Henderson, G D, ed. Bulloch, J, trans. *The Scots Confession 1560*, Edinburgh, 1960.
Kennedy, J. *Presbyterian Authority and Discipline*, Edinburgh, 1960.
Kidd, B J, ed. *Documents Illustrative of the Continental Reformation*, Oxford, 1911.
Kirk, J, ed. *The Second Book of Discipline*, Edinburgh, 1980.
Lyall, F. *Of Presbyters and Kings: Church and State in the Law of Scotland*, Aberdeen, 1980.
MacGregor, J G. *The Scottish Presbyterian Polity*, Edinburgh, 1926.
McRoberts, D, ed. *Essays on the Scottish Reformation, 1513–1625*, Glasgow, 1962.
Shaw, D. *The General Assemblies of the Church of Scotland, 1560–1660*, Edinburgh, 1964.
Walker, D M. *A Legal History of Scotland*, vol. 3, *The Sixteenth Century*, Edinburgh, 1995.
Wormald, J. *Court, Kirk and Community: Scotland 1470–1625*, Edinburgh, 1981.

PART TWO
●
Diversity

5 Presbyterianism

HENRY R SEFTON

Presbyterianism is not native to Scotland but has come to be regarded as characteristic of Scottish church life. It has been both a unifying and divisive factor in the life of the community. It is basically a conciliar form of government which refuses to entrust individuals with ultimate authority and therefore ought to have a unifying effect on society. It has been divisive because of the relationship between Church and king in Scotland.

The origins of Presbyterianism have been traced to the reform movements in Strasbourg, Geneva and Frankfurt.[1] The structures set up there related mainly to cities. In Scotland the system had to be adapted to cover a whole country. Interestingly enough, the first traces of Presbyterianism in Scotland are to be found in cities like St Andrews and Dundee. The latter has been described as 'the Geneva of Scotland'.[2] There is a widespread impression that Presbyterianism is government by presbyteries, but presbyteries were slow to emerge in Scotland and were not fully recognised until 1592. Presbyterianism is properly government by presbyters and is to be contrasted with Episcopalianism which is government by bishops. Much of Scottish life in the sixteenth and seventeenth centuries was dominated by the struggle between the two forms of government. To this day 'bishop' is an emotive word in Scottish Presbyterian Churches.

The Scottish reformers were at pains to clarify the characteristics of the 'trew Kirk'. In the *Confession of Faith* of 1560, now known as the *Scots Confession*, they declared that the notes of the true Church were as follows:

> The notes therefore of the trew Kirk of God we beleeve, confesse and avow to be First, the trew preaching of the Worde of God, into the quhilk God hes revealed himselfe unto us, as the writings of the Prophets and Apostles dois declair. Secundly, the right administration of the Sacraments of Christ Jesus quhilk man be annexed to the word and promise of God, to confirm and seale the same in our hearts. Last, Ecclesiastical discipline uprightlie ministred, as Goddis Worde prescribes, whereby vice is repressed, and vertew nurished.[3]

The first two of these notes are to be found in the writings of Continental reformers. The Scots are distinctive in their emphasis on discipline as a mark of the true Church. This helps to explain their stubborn refusal to contemplate any form of Church government which seemed to them inconsistent with the word of God and the strictness of the personal conduct required of Church members.

Before 1560, the conventional date of the Scottish Reformation, there were elections of elders and deacons from the communities of Ayr, Dundee, Perth and St Andrews and in some cases of a minister.[4] Thus the Church court which came to be known as the kirk session was the first to be set up in Scotland. This of course was the court which impinged most directly on the average member of the community for it was concerned with the regulation of individual conduct.

The legislation of the Reformation Parliament of 1560 made no provision for Church organisation but in December of that year there was a gathering of ministers and elders which has been recognised as the first General Assembly. This gathering approved a book of policy which has come to be known as *The First Book of Discipline*, but this document did not receive any authority from Parliament. It did receive qualified approval from the Privy Council with the condition that the bishops, abbots and other benefice holders should be allowed to receive their revenues for life as long as they made provision for the maintenance of the ministry of word and sacraments. This of course limited very severely the ambitious plans in the book for education and the care of the poor. It was in order to tap these revenues that the General Assembly agreed in 1572 to the appointment of titular bishops. Unfortunately the revenues did not come the way of the Church but passed into various secular hands and this did much to discredit the office of bishop.

The First Book of Discipline is rather vague on Church organisation beyond the local Congregation but it is very specific in matters of discipline. The seventh 'head' deals with ecclesiastical discipline and details the offences to be dealt with by the 'civill sword'. These are offences like blasphemy, adultery, murder and perjury and 'other crimes capitall, worthy of death'.

> But drunkenness, excesse be it in apparel, or be it in eating or drinking, fornication, oppressing of the poore by exactions, deceiving of them in buying and selling by wrang met or measure, wanton words and licentious living tending to slander, doe openly appertaine to the kirk of God to punish them, as God's word commands.[5]

The procedures for dealing with offenders are given as private and public admonition and in extreme cases excommunication. This latter involved social ostracism not only locally but throughout the realm.

There is no mention of witchcraft in the *First Book of Discipline*, but in 1563 a comprehensive act regarding 'witchcraft sorcery and necromancy' was passed by the Scottish Parliament. These were all forbidden under pain of death, and the judges of the realm were ordered to deal rigorously with those who practised these. All three practices are described in the act as superstition, a term which in this context has theological overtones. Such practices had had credence heretofore but now such empty nonsense is to be made to cease. It has been suggested by P G Maxwell-Stuart that the

Figure 5.1 National Covenant, with signatures of members of the Privy Council, 1638. Courtesy of the Trustees of the National Museums of Scotland, NMS H.OA21.

Witchcraft Act was aimed not only at magical practices, which had been condemned by the medieval Church, but also at surviving Catholic rituals and devotions which had survived but which were regarded as 'superstition' by the reformers.[6] The implementation of the act was entrusted to the judges and sheriffs but they had the enthusiastic support of the Church courts. In his book *The Witches of Fife*, Stuart Macdonald suggests that the Church's support of the prosecution of witches was part of the quest for a

Godly society. This meant eradicating evil in all its forms, and witchcraft was one of these forms. 'The church was driven by its vision of what society could be.'[7]

Rosalind Mitchison and Leah Leneman have pointed out that some 80 per cent of the victims of witchcraft accusations and trials were women and they suggest that the gender element was significant in the motivation of such events.[8] The numbers of those accused and executed have been much discussed by recent scholars and while there is a downward trend the figures are still considerable. Christina Larner estimated that there were 1,337 executions for witchcraft in Scotland.[9]

The last execution for witchcraft was that of Janet Horne in Dornoch in 1727. The act was repealed in 1735 but the Kirk Session of Kinneff investigated a case of witchcraft in 1739.[10] The last execution for blasphemy was that of Thomas Aikenhead in 1697. While the sentence was handed down by the High Court of Justiciary in terms of a parliamentary act against blasphemy renewed in 1695, the ministers of Edinburgh could have obtained a reprieve from the Privy Council had they chosen to do so. Instead they approved and encouraged the implementation of the sentence.

The close co-operation between civil and Church courts was substantially modified in 1712 by the passing of the Toleration Act by the Parliament of Great Britain. This act was primarily for the relief of Episcopalians but it included a section in which it was declared that 'no civil pain or forfeiture or disability whatsoever' should be incurred by proceedings in a Church judicatory. Civil magistrates were prohibited from compelling accused persons to appear before Church judicatories and from giving effect to Church sentences. The Blasphemy Act was repealed in 1812. Despite the repeal of supportive civil acts ecclesiastical discipline continued to be exercised particularly in cases of ante-nuptial fornication and adultery. The records of kirk sessions in the eighteenth century bear witness to their diligence in this. Offenders were required to make several appearances on the place of repentance before being formally absolved. An appearance of this kind is recorded as late as 1884 in the Free Church of Lochcarron.

The parliamentary 'Act for abolisheing of the actis contrair the trew religion' passed in 1592 (c. 8–116) is regarded as the Magna Carta of Presbyterianism. In the Articles Declaratory of the Church of Scotland of 1921 it is described as having ratified the liberties of the Church. The act recognised the lawfulness of Church government by general assembly, synods, presbyteries and kirk sessions. General Assemblies were to be held at the appointment of the king and in the presence of the king or his commissioner. Throughout the seventeenth century, the General Assembly was to prove the most vulnerable court. It was frequently subverted by royal intervention and for several years was in abeyance. The office of bishop had never been formally abolished by Parliament and from 1600 appointments were made by the king to all 13 of the mediaeval dioceses. Despite this, the Presbyterian structures beneath the General Assembly remained in place though the jurisdiction of the courts was infringed by the establishment of

Figure 5.2 Copy of the Solemn League and Covenant, 1643. Courtesy of the Trustees of the National Museums of Scotland, NMS H.OA51.

courts of High Commission under the control of the archbishops of St Andrews and Glasgow.[11]

Bishops were regarded by Presbyterians as not only contrary to Holy Scripture but also a means of interference in the affairs of the Church by the king who in their eyes was merely a member of the Church and not its head. The bishops had an important role as one of the three Estates of Parliament and consequently in the Committee of Articles, which controlled the business of Parliament. When they visited their dioceses they would preside as moderator of synod, presbytery and even kirk session. Thus for much of the seventeenth century there was an amalgamation of Presbyterianism and Episcopalianism which worked reasonably well in some parts of the country but was bitterly opposed in others, notably in the south-west.

Presbyterian opposition to bishops and royal interference in Church affairs was given expression in the National Covenant of 1638 and the Solemn League and Covenant of 1643. The latter envisaged the extension of Presbyterianism to England and Ireland but immediately meant the participation of commissioners from Scotland in the Westminster Assembly of Divines summoned by the English Parliament. This assembly produced the classic documents of Presbyterianism:

> *The Confession of Faith*, usually known as *The Westminster Confession*;
> *The Larger Catechism*;
> *The Shorter Catechism*;

The Directory for the Publick Worship of God;
The Form of Presbyterial Church Government.

All of these documents were approved by the General Assembly of the Church of Scotland 1645 to 1648. *The Directory* was approved and established by act of the Estates of Parliament in 1645 and *The Confession of Faith* was ratified and established by acts of Parliament in 1649 and 1690 as 'the public and avowed confession of the Church of Scotland'. The Westminster documents did not gain similar acceptance in England or Ireland to the disappointment and indeed annoyance of the Scots.

The act of 1690 (c. 7–5) not only ratified *The Confession of Faith* but also settled Presbyterian Church government by kirk sessions, presbyteries, provincial synods and general assemblies as 'the only government of Christ's church within this Kingdome'. The position of the Church of Scotland and Presbyterian Church government was further strengthened by the Act of Security of 1706 (c. 6–6). This latter act provided for an oath to be taken on accession by the Sovereign to maintain Presbyterian Church government in Scotland.

Prelacy or Episcopal Church government was specifically abolished by the Scottish Parliament in 1689 but in 1712 the Parliament of Great Britain passed an act 'to prevent the Disturbing of those of the Episcopal Communion in that part of Great Britain called Scotland in the Exercise of their Religious Worship and in the use of the Liturgy of the Church of

Figure 5.3 Group portrait of the leaders of the Secession of 1737. Courtesy of the Trustees of the National Museums of Scotland, NMS SLA C12288.

England'. This act is usually referred to as the Toleration Act and gives legal recognition to worship and Church government other than that of the Established Church. Thereafter Church and State are never even in theory coterminous. Dissent became a normal and legal part of Church life in Scotland. Presbyterianism remains the predominant form of Church government, but it is no longer only the preserve of the Established Church.

The 1690 settlement was not unanimously accepted even by Presbyterians. Extreme Covenanters, who came to be known as Cameronians, strongly objected to a settlement based on the will of king and nation rather than on the divine right of Presbyterianism and on the perpetual binding obligation of the Covenants of 1638 and 1643. This was a lay protest as all three Cameronian ministers conformed. Only after two ministers adhered to them was it possible to set up the Reformed Presbytery in 1743. In 1988, the Reformed Presbyterian Church had five congregations and about 150 members, and continues to this day.

What is usually termed the First Secession had occurred in 1733 with the formation of The Associate Presbytery in 1733. The Seceders were in secession from 'the prevailing Party' rather than the Church of Scotland. They objected not only to the leniency shown to divinity professors whom they regarded as heretics and to the severity shown to those at the opposite end of the theological spectrum, but also to the neglect of the Covenants. But the immediate cause of the Secession was the lukewarm attitude of the General Assembly towards patronage.

Patronage was of long standing in the Church of Scotland and consisted of the right of certain corporate bodies, such as town councils and colleges, and individuals, normally local landowners, to present ministers to parishes without reference to the wishes of the parishioners. Curiously enough, this had not been a burning issue in the sixteenth and seventeenth centuries but it was to be a major cause of the divisions of Presbyterianism in the eighteenth and nineteenth centuries. The Presbytery of Relief was founded in 1761 'for the relief of Christians oppressed in their Christian privileges'. The primary relief which the Relief Presbytery was determined to provide was liberty for congregations to elect their own pastors, an aim shared with the Seceders of 1733. Patronage was also a major issue in the Disruption of 1843 when about a third of the ministers of the Church of Scotland left to form the Free Church of Scotland.

It can, however, be argued that the basic cause of Presbyterian divisions has been the relationship of the Presbyterian Churches with the State. The breach which occurred among the first Seceders in 1747 was over the attitude to be taken to the Burgess Oath. This was an oath required of citizens of Edinburgh, Glasgow and Perth endorsing the religion professed in the realm. Did this imply approval of the establishment in the Church of Scotland? Those who thought it did condemned the oath and came to be known as Antiburghers. Those who were prepared to take the oath came to be known as Burghers. The Antiburghers set up the General Associate Synod and claimed to be the authentic continuation of the Seceder body.

Figure 5.4 Eighteenth-century Bible used at the annual Covenanters' conventicle at Flotterstone, near Edinburgh. Courtesy of the Trustees of the National Museums of Scotland, NMS SLA C15726.

Up to 1761 it was held by all Presbyterians that it was the right and duty of the 'Civil Magistrate', that is to say the government, to support the Church and not interfere in its spiritual functions. The Relief Presbytery and later Synod gradually came to the conclusion that the Church should be free of State support or interference. The Church should be an entirely voluntary body. In 1799 New Light came upon the Associate (Burgher) Synod with regard to the relationship of Church and State. This permitted individual ministers and elders to have their own views in this matter. A minority disagreed with this and left to form the Associate (Burgher) Presbytery. For the sake of clarity the majority are known as New Lights and the minority as Old Lights or Auld Lichts. The General Associate (Antiburgher) Synod adopted a similar New Light stance in 1804 and two years later a small minority formed an Auld Licht body called the Constitutional Associate Presbytery.

The two New Light bodies came together in 1820 to form the United Associate Synod of the Secession Church and this in turn united with the Relief Synod in 1847 to create the United Presbyterian Church. This union was born out of strenuous efforts to promote a policy of voluntaryism in relations between Church and State. Despite this, the Free Church was at pains to declare that its members were not voluntaries and that although they were quitting a vitiated establishment they would rejoice to return to a pure one. This divergence of view from that of the United Presbyterian Church prevented union until 1900. The abolition of patronage in 1874 by act of Parliament did not appease either Church and they co-operated in calls for the disestablishment of the Church of Scotland.

Strangers visiting Scotland in the eighteenth and early nineteenth centuries would have found it difficult to distinguish between the separated Presbyterian Churches. They all had a hierarchy of courts, kirk session, presbytery and, in most cases, synod, but only the Established Church had a general assembly. There was no attempt to cover the whole country and the strength of all the seceding Churches was to be found in the central belt of Scotland.

The Disruption of 1843 saw the setting up of an alternative or Free Church of Scotland aiming to cover the entire country. It had the full range of courts, including the General Assembly, and set up churches in nearly every parish. This meant that many congregations were not self-sufficient, and means had to be found to support them since they had no income from the teinds payable by heritors of the parishes. A central 'sustentation fund' was established to which congregations contributed according to their ability and from which their ministers received what was called an 'equal dividend'. This meant that the Free Church was more centralised than the other Churches in Scotland. The centralising tendency increased until it culminated with the commissioning of a large office building at 121 George Street, Edinburgh, in 1909, now regarded by many as the 'headquarters' of the reunited Church of Scotland. The Assembly Hall was built by the Free Church in 1858–9 and is now the normal meeting place of the Church of Scotland General Assembly.

During the nineteenth century the classic Presbyterian structure of courts was supplemented by the creation of other bodies. Hitherto the kirk session had been responsible for all local parish activity with the assistance from time to time of deacons who were responsible for collecting

Figure 5.5 D O Hill, *The Signing of the Deed of Demission* (1866).

Figure 5.6 Moderator of the Church of Scotland, with ministers, 1850s. Courtesy of the Trustees of the National Museums of Scotland, NMS SLA 58/60/22.

charitable funds and distributing to the poor under the supervision of the Session. The Free Church set up a body known as the deacons' court, consisting of minister, elders and elected deacons, to deal with the financial affairs of the congregation. The United Presbyterian Church went further and elected boards of managers whose members were quite separate from the kirk session to manage the temporal affairs of the congregation. Although responsibility for care of the poor and the provision of schools were taken away from the Churches, the courts of the Presbyterian Churches delegated more and more work to committees. In 1892 the *Handbook Of Scottish Church Defence* listed seven major committees and twelve lesser committees appointed by the General Assembly of the Church of Scotland which it described as 'the permanent executive for carrying on the general work of the Church beyond that proper to the local church courts'.

The Church of Scotland *Year Book 2003/2004* lists twenty-seven General Assembly boards and committees. Several of these boards have major committees and sub-committees so that the central organisation of the Church is highly complex as well as highly centralised. Dr Andrew Herron suggested that the development of committees was a major reason for the perceived redundancy and abolition of synods in the Church of Scotland in 1992. Only the Free Church of Scotland now has the four-fold structure of Presbyterian courts. The synod is the supreme court of the Free Presbyterian Church with the same powers as a general assembly.

Presbyterianism is often described as a democratic form of Church government. This is a misunderstanding for it is government not by the people but by ministers and elders and, in the case of the Church of Scotland since 1992, deacons. Deacons in this context are full-time Church workers formerly known as lay missionaries or deaconesses. In the Church of Scotland it has been the practice that only ministers can preside as moderator of a kirk session but the General Assembly of 2003 adopted an overture which permits elders who have undergone a course of training to preside at kirk session meetings. In the United Free Church, elders can act as interim moderator in a vacancy in the pastoral charge. Elders and deacons are eligible to be elected as moderator of a presbytery or of the General Assembly and several presbyteries have elected elders as moderators. In 2003 Dr Alison Elliot, an elder in Edinburgh, was nominated as moderator of the General Assembly of 2004 and so became the first woman as well as the first elder since the sixteenth century to be nominated for the office.

For most of its history Presbyterianism has been male dominated and women have been excluded from participation in the courts of the Church. John Knox inveighed against 'the monstrous regiment [rule] of women'. It is the more remarkable that women were among the prominent supporters of Presbyterianism during the seventeenth century and some of them were martyrs in the cause of the Covenants. In the eighteenth century Lady Glenorchy built chapels in Edinburgh and Strathfillan in Perthshire and

Figure 5.7 Lochcarron United Free Church congregation worshipping in the manse garden, which they did for a whole summer until the manse as well as the church was seized by the Free Church, c1900. Courtesy of the Trustees of the National Museums of Scotland, NMS SLA C.897.

remained a loyal Presbyterian despite the persuasions of John Wesley. The Woman's Guild of the Church of Scotland was founded in 1887 'to unite the women of the Church in the dedication of their lives to the Lord Jesus Christ through Worship, Fellowship and Service'. The Guild developed a parallel organisation of Presbyterial councils, central committee and annual meeting. The following year the General Assembly decided to revive the office of deaconess in the Church of Scotland. Women played a prominent part in the overseas missionary activity which was a notable feature of the life of all three major Presbyterian Churches in the nineteenth century. Many became missionaries and fulfilled leadership roles denied to them in the home Churches. Mary Slessor was a pioneer and established missionary stations in Okoyong, Itu and Arochuku in eastern Nigeria. In 1898 she was appointed vice consul in Okoyong, thus becoming the first woman magistrate in the British Empire.

In 1919 the General Assembly of the United Free Church passed an act declaring that the office of deacon was tenable by men ordained for life or by men and women appointed for three years. Ordination of women even to a managerial post like that of deacon was still unacceptable. Shortly after the union of the United Free Church and the Church of Scotland the question of the ordination of women was again raised by a petition to the General Assembly of 1931. The matter was remitted to a committee which recommended that women should be eligible for ordination as deacons and elders. The Assembly of 1935 agreed only to the ordination of women as deacons. Ordination of women to the eldership was agreed by the Assembly of 1966.

The first General Assembly of the continuing minority of the United Free Church resolved that 'any member of the Church in full communion shall be eligible to hold any office within the United Free Church'. Elizabeth Barr was ordained as minister of Auchterarder United Free Church in 1935 and in 1960 became the first woman to be moderator of the General Assembly of any British Presbyterian Church. It was not until 1968 that the Church of Scotland General Assembly decided that women were eligible for ordination to the ministry on the same terms and conditions as men. In the following year Catherine McConnachie was ordained by the presbytery of Aberdeen. The other Scottish Presbyterian Churches do not admit women to the eldership or ministry. In the Church of Scotland a significant number of women have been ordained as ministers and inducted to parishes. They are also eligible to be moderator of presbytery and the General Assembly.

Despite the reunions of the nineteenth and twentieth centuries there are still seven different Presbyterian Churches in Scotland. Some of these have already been noted. In 1839 the Auld Licht Burghers joined the Church of Scotland but most of them went out in the Disruption of 1843 to join the Free Church. A majority of the Auld Licht Anti-Burghers joined the Free Church in 1852 but a minority continued under the name of Original Secession Church until 1956 when they acceded to the Church of Scotland. In 1876 almost all the Cameronians joined the Free Church but a minority has continued as the Reformed Presbyterian Church.

Figure 5.8 Minister of the United Free Church and his elders, Orkney, c1910. Courtesy of the Trustees of the National Museums of Scotland, NMS SLA 50/39/25A.

The division leading to the formation of the Free Presbyterian Church is unique in the history of Presbyterian secession in that it hinged on a specifically doctrinal issue – the terms of subscription to the Westminster Confession of Faith. The General Assembly of the Free Church had in 1892 passed a Declaratory Act permitting 'diversity of opinion ... on such points in the Confession as do not enter into the substance of the Reformed Faith'. Two ministers and a substantial number of elders, members and adherents of the Free Church objected to the act as detrimental to the distinctive Calvinism of the Confession and formed the Free Presbyterian Church at a meeting in August 1893 at Portree in Skye. A division in the Free Presbyterian Church occurred in 1989 on a matter of discipline. A minister and an elder were suspended by the synod because of their involvement in the worship of other Churches. The case of the elder attracted wide attention as he was Lord High Chancellor of Great Britain (Lord Mackay of Clashfern). Fourteen ministers and about thirty elders protested against the decision as an infringement of the right of private judgement 'in matters relating to the application of the Christian faith to daily living'. When the protest was not received they seceded to form the Associated Presbyterian Churches.

Discipline was also at the heart of the most recent Presbyterian division. Despite the acquittal of Professor Donald Macleod in the Sheriff Court at Edinburgh, a minority in the Free Church were not satisfied with the verdict and considered that ecclesiastical censure was necessary. When they failed to persuade the General Assembly they seceded to form the Free Church (Continuing) in 2000.

The seven Presbyterian Churches existing in 2005 were therefore:

The Church of Scotland;
The Reformed Presbyterian Church of Scotland;
The Free Church of Scotland;
The Free Presbyterian Church of Scotland;
The United Free Church of Scotland;
The Associated Presbyterian Churches of Scotland and Canada;
The Free Church of Scotland (Continuing).

The seven Presbyterian Churches are similar in government but vary in ethos and lifestyle. It is significant that the New Lights and the Old Lights have never been reunited. The New Lights were part of the United Presbyterian Church while the Old Lights joined the Church of Scotland and the Free Church. While the United Presbyterian Church in its entirety entered the union with the Free Church in 1900 its voluntary stance on Church and State is continued in the present-day United Free Church. All seven Churches regard *The Westminster Confession* as their Principal Subordinate Standard of Faith but in the Church of Scotland the *Confession* is little known and not much heeded. The other Presbyterian Churches give greater allegiance but many of their members are more familiar with the Shorter Catechism than with the Confession itself. The Church of Scotland is often described as 'a broad Church' and certainly has more variety of belief and practice than the others which are fairly homogeneous as well as being very much smaller. The Church of Scotland is larger than all the others combined but carries a great many 'passengers'. Forty-two per cent of the population of Scotland claim to belong to the Church of Scotland, but most of them never attend church and do not appear on any membership rolls. The official figure according to the *Church of Scotland Year Book 2003–4* is 571,698 communicants.

The relationship of the Church of Scotland with the State is widely regarded in Scotland to be the ideal solution of the former tensions, but the settlement recognised in the Church of Scotland Act 1921 has been explicitly rejected as a model for the Church of England.[12] It could be said that the Church of Scotland is now recognised by law rather than established by law and the 1921 Act provides for the possible recognition of other Churches unlike previous legislation.

Attached to the 1921 Act there are 'Articles Declaratory of the Constitution of the Church of Scotland in Matters Spiritual'. These articles were drawn up and agreed by both the Church of Scotland and the United Free Church and paved the way for the union of the former with the great majority of the latter in 1929. The seventh of these articles 'recognises the obligation to seek and promote union with other Churches in which it (the Church of Scotland) finds the Word to be purely preached, the sacraments administered according to Christ's ordinance, and discipline rightly exercised'. This recalls the marks of the true Church as set out in the *Scots Confession* of 1560. Since the union of 1929 there has been only one union with a Presbyterian Church – that of 1956 with the Original Secession Church. As the 50th anniversary of the 1929 union approached, a committee was set up by the General Assembly to explore the possibility of union with one or more of the other Presbyterian Churches. The replies of the other Churches to the approaches of the Church of Scotland varied in tone but all were in the negative.[13] In the March 2003 issue of the magazine *Life and Work* Professor Andrew McGowan urged the Church of Scotland to seek reunion with other Presbyterian Churches with which it has so much in common rather than a union which would involve accepting bishops. Accepting bishops, he suggests, would mean throwing away its Presbyterian heritage.

An American scholar, Professor Margo Todd, has paid tribute to the profound influence Presbyterianism has had on Scotland: 'It was the kirk that wrought the profound cultural change that has given us a Scotland characterised (and caricatured) by abstemious self-restraint, sober but affective piety, unrelenting sabbatarianism, highly visible and rigorous social discipline, and a militant conviction of the rightness of the Calvinist cause.' This cultural transformation was achieved because its agents were local, the kirk sessions.[14]

Professor Todd writes of early modern Scotland. An observer of contemporary Scotland might come to rather different conclusions. Discipline is no longer administered by kirk sessions. Shopping and the opening of licensed premises on Sunday are common in mainland Scotland

and not unknown in the islands. Piety when visible is often charismatic rather than sober. Indulgence in drink and drugs have replaced abstemious self-restraint. Most Presbyterians have little knowledge of Calvin.

Despite all this, Presbyterianism is still part of the fabric of Scottish life but this may be more apparent to visitors and foreigners than to the Scots themselves.

NOTES

1. Macgregor, 1926, 19–61.
2. Baxter, 1960, 3.
3. Henderson, 1937, 75.
4. Cameron, 1972, 6.
5. Cameron, 1972, 167f.
6. Maxwell–Stuart, 2001, 35–45.
7. Macdonald, 2002, 197.
8. Mitchison and Leneman, 1989, 50.
9. Larner, 1983, 38f.
10. Clark, 1929, 165.
11. Foster, 1975, 47–9, 188–9.
12. Archbishops' Commission, 1970, 65f.
13. The writer was a member of the committee.
14. Todd, 2002, 402f.

BIBLIOGRAPHY

Archbishops' Commission. *Church and State*, Church Information Office, London, 1970.
Baxter, J H. *Dundee and the Reformation*, Dundee, 1960.
Cameron, J K, ed. *The First Book of Discipline*, Edinburgh, 1972.
Clark, I M. *A History of Church Discipline in Scotland*, Aberdeen, 1929.
The Confession of Faith, Edinburgh, 1948.
Foster, W R. *Bishop and Presbytery – The Church of Scotland 1661–1688*, London, 1958.
Foster, W R. *The Church before the Covenants – The Church of Scotland 1596–1638*, Edinburgh, 1975.
Henderson, G D. *Scots Confession, 1560, and Negative Confession, 1581*, Edinburgh, 1937.
Henderson, G D. *Presbyterianism*, Aberdeen, 1954.
Henderson, G D, ed, Bulloch, J, trans. *Scots Confession 1560*, Edinburgh, 1960.
Herron, A. *The Law and Practice of the Kirk*, Glasgow, 1995.
Johnston, C N. *Handbook of Scottish Church Defence*, Edinburgh, 1892.
Larner, C. *Enemies of God – The Witch–hunt in Scotland*, Oxford, 1983.
Lyall, F. *Of Presbyters and Kings. Church and State in the Law of Scotland*, Aberdeen, 1980.
Macdonald, S. *The Witches of Fife: Witch-hunting in a Scottish shire, 1560–1710*, East Linton, 2002.
Macgregor, Janet G. *The Scottish Presbyterian Polity*, Edinburgh, 1926.
Maxwell–Stuart, P G. *Satan's Conspiracy – Magic and Witchcraft in Sixteenth Century Scotland*, East Linton, 2001.
Mitchison, R and Leneman, L. *Sexuality and Social Control – Scotland 1660–1780*, Oxford, 1989.
Todd, M. *The Culture of Protestantism in Early Modern Scotland*, New Haven and London, 2002.
Watson, G. *Bothwell and the Witches*, London, 1975.

6 Roman Catholics in Scotland: Late Sixteenth to Eighteenth Centuries

JOHN R WATTS

'The devout children of worthy men offered sacrifice in secret'
The Book of Wisdom, 18:8

'Remember how our fathers were put to the test, to prove that they truly honoured God'
Judith, 8:21b

At the seizure of power by the reformers in 1560 the Catholic Church in Scotland experienced swift and almost total collapse. A report smuggled to Rome just twenty years afterwards told of the loss of nearly all its priests. Many had apostatised, whether from conviction or self-seeking, and accepted posts in the new religion; others, though not abandoning their faith, had married; a number had found lay work in the service of the gentry; others were reduced to begging; only a handful were still travelling the country and administering the sacraments.[1] The Church had also lost virtually all its property, much of which, being unsuitable for Protestant worship, had already been demolished or allowed to decay.[2] When in 1603 the last trace of its organisation and authority structure died with the death in exile of its only remaining pre-Reformation bishop and Rome decided to discontinue the Scottish hierarchy, its eclipse seemed complete.

There were nonetheless still many who considered themselves Catholics at the end of the sixteenth century. It is estimated that perhaps a third of the nobility and some 240 knights and gentlemen still adhered to the old faith,[3] though the number was diminishing yearly, and their houses served as a rallying point for their servants and tenants and occasionally as a refuge for priests.

In the south, small Catholic communities survived at Terregles and Traquair (Fig. 6.1) where they enjoyed the protection of the Herries, Maxwell and Stuart families, at New Abbey on the Solway where two long-serving priests continued to provide the sacraments for a hundred years after the Reformation, and in a few other centres.[4] In the Gordon lands of the northeast, Catholics remained numerous in the Enzie, in Strathbogie, and in parts of upper Banffshire. But these were oases: in most places elsewhere the Catholic faith was dying on the branch. As the old generation passed on,

Figure 6.1 Traquair House, Innerleithen, Upper Tweedsdale. Drawing by Pádraig Watts.

with almost no priests to minister to the new, and the Kirk strong, Catholicism was becoming no more that a second-hand memory.

The picture was different in the western Highlands and Islands. For the space of perhaps forty years there were no priests whatever there. But the Reformed Church had not yet reached many of the remoter areas either, particularly where the clan chiefs had not embraced Protestantism, and in that vacuum the people clung as best they could to the old practices. They had a unique and rich oral tradition of spirituality to fall back on, and without instruction or the sacraments were keeping the faith alive by old remembered prayers and formulae. A simple people of long memory who lived close to the supernatural, intensely conservative and self-contained, they held on to the old and that which was local, and instinctively resisted anything new introduced by strangers.[5]

The first priests to reach the western Highlands and the Islands were Gaelic-speaking Irish Franciscans, who began their work in earnest in the 1620s. Though few in number they enjoyed spectacular success, baptising thousands of all ages.[6] They referred to the adult baptisms as 'conversions', though in many cases the people can hardly have been converts as we would use the term, but rather formal initiates into a faith that they already belonged to by desire.

Jesuit priests returned to Scotland shortly after the Reformation and maintained a precarious presence, but not until after 1617 were they able to build their numbers and create an organised and coherent mission. Though trained abroad they were all Scots. They confined their work almost entirely to the south, the central belt and the east, usually attaching themselves to the houses of the Catholic gentry where they acted as family chaplains and which they used as bases for work further afield.

In 1622 Pope Gregory XV founded the Congregation *de Propaganda Fide* to support and co-ordinate the Church's evangelical work in the 'mission territories', which included the Protestant states of Europe. The Scottish Mission was established and its secular clergy placed in the direct organisational, spiritual and financial care of Rome. Thus by coincidence the Franciscans, Jesuits and seculars all began or stepped up their work at about the same time, and if the first two decades of the seventeenth century marked the nadir of Catholicism's fortunes in Scotland, the 1620s signalled the beginning of its recovery.

What emerged was a Church unchanged in its essentials, but in its ways utterly new. It had fallen headlong from a position of power, privilege and wealth to a position of none, and had been driven from the centre of society to the very periphery. Its laity were now but a tiny remnant, its priests pariahs. It was a new, lean Church in which only the deeply committed remained, and these 'so steadfast that they seemed to have inherited the fervour of the primitive Christians'.[7] If the old had often not been admirable, the new would be a Church that, looking back today, we cannot but admire. It was a painful time, but at least it was the time for a new start and a new role. Adversity was bringing its own unique opportunity, for priests and people, to live the Beatitudes and the call to walk humbly with their God.

Having gained power and complete ascendancy the reformers were determined to use the force of law and every economic, social and ideological sanction at their disposal to eliminate the Catholic minority and forestall any return to the old faith. The Reformation Parliament of 1560 immediately introduced legislation not only outlawing the celebration of and attendance at Mass (on pain of confiscation of property for the first offence, banishment for the second, and death for the third), but also requiring attendance at Protestant public worship and reception of the sacrament. Convictions for attending Mass in fact usually proved difficult for lack of evidence, unless on the word of spies or apostates. But failure to attend the kirk was an easy matter to prove, and those convicted faced excommunication which, if it mattered little to them on religious grounds, had devastating social effects, particularly for persons of property. A missionary priest writing in 1610 explained how the process worked:

The system of persecution followed is first to excommunicate the Catholic landowner, who afterwards is either banished or imprisoned or deprived of all his possessions. If his wife is a Catholic she is excommunicated, and her husband is not allowed to keep her in his home. And if the husband is a heretic [Protestant] the ministers and magistrates allow him to take another wife according to the laws of the country. Should the children or servants be Catholics they are likewise excommunicated, and the parents are obliged to turn them out of the house.[8]

Those excommunicated might find their names posted on the court-house door, the community forbidden to converse with them, their crops trampled by the military and their cattle driven away.[9] They might even be charged with high treason.

Such legislation was bound to have a massive and immediate impact, despite all the efforts of the few missionaries to hold the line. 'Very few are willing to take such great risks, and all but one or two here and there comply with the demands of the law,' one Jesuit admitted frankly; 'They tell us that it is easy for us to give advice, because we have nothing to lose ... and if the worst came to the worst we can always go back to our college [abroad].'[10]

The laws penalised Scottish Catholics even more severely than their co-religionists in England and Ireland,[11] and over time there was a ceaseless effort to fine-tune them. They were reissued and amplified no less than twenty-one times in the century between the Union of Crowns and the Union of Parliaments, a statistic that indicates both the State's obsession with the problem and its determination to make the laws more effective.[12]

Nonetheless, Scottish Catholics were not called to face the extremes of persecution suffered for a while in England and Wales. Only one, the Jesuit John Ogilvie hanged at Glasgow Cross in 1615, is officially recognised as a martyr, and only a handful more died as a direct result of maltreatment in captivity. More typical were the hundreds who faced ruin, the thousands who suffered deprivation. For them the prospect was not death on the gibbet but a lifetime of attrition.

Nor was the persecution uniform. The laws were implemented unevenly, more closely to the letter in some years than others, largely according to the exigencies of national politics and the power manoeuvres of Crown and Parliament.[13] There were even times, in some parts of the country, when life for Catholics was almost normal.

The Kirk's hand was strengthened with the Revolution Settlement of 1688-9, but after the Union of Parliaments in 1707 it could not always assume that its concerns would loom so large in London as they had in Edinburgh. Nonetheless, the Penal Laws as they applied to Scotland were maintained throughout the eighteenth century, and even consolidated. Two features of the later legislation proved particularly difficult for Catholics since both were introduced to bypass the problem of securing hard evidence against them. The first were those laws[14] which in effect put the onus upon them to prove their innocence by forswearing their religion, refusal to do so being taken as an admission of guilt. The second, which applied to the clergy, were those acts[15] which required merely that a man be 'held and repute' to be a priest in order to be convicted. Technically, these and the rest of the formidable battery of accumulated legislation remained on the statute book until the last decade of the eighteenth century.

The most obvious impact of the Penal Laws upon the little remnant Church was to drive it underground and force it to devise strategies for practising in secret. Only 'low' Masses were now possible, without music, incense or any ceremonial that might attract attention, and with the homily often preached beforehand to minimise the time that the priest would need to remain vested. Local communities adopted recognised signals alerting them to a forthcoming Mass – in one town a fiddler going round the streets, in another a figure in a white sheet appearing in the bushes at night.[16] It has even been suggested that the *piobaireachd* tune 'The Little Spree' was in fact composed as a covert message for the same purpose. At times of greatest harassment, Mass was often held just after midnight on Sunday morning, to allow the congregation to assemble and return home under cover of dark. According to a report from the 1660s, services were always held in private houses in the Lowlands, but in some parts of the Highlands and the Isles, where Catholics were in a majority and protected by their remoteness, they could safely be held in the open. The survival of numerous place-names – *Port na h-Aifrinne* (the harbour bay of the Mass), *Clach an t-Sagairt* (the [altar] stone of the priest), etc. and several Mass stones themselves, suggests that the practice was indeed widespread.[17]

Baptisms and weddings, which should have been joyful family occasions, had to be low-key and hurried affairs. Well into the seventeenth century some Catholic parents still preferred to have their babies baptised quietly in the now ruined pre-Reformation churches. And there must have been many marriages like that held near Dumfries in 1634, which took place in a field at night with only the priest and four witnesses present, and which was only preserved on record because the lighted candles led to the arrest, trial and conviction of the groom.[18]

The priests themselves rarely appeared in public unless in laymen's clothes, particularly when they were on the road. One whom we know of always travelled in the guise of a peasant. Even in the relative safety of the north-west the presence of garrisons forced the priests to appear in Highland habit and armour, 'soe that strangers could not easily distinguish them from the countrey people'.[19]

It was when on the road that they were in greatest danger since the Mass kit and breviary they carried could easily betray their identity (Fig. 6.2). At least one of the religious orders working in Scotland in the seventeenth century received special papal dispensation to substitute the rosary, psalms and other prayers for the normal Daily Office, since these could be learned

Figure 6.2 Brass crucifix and chalice from a grave in Dumbarton of a seventeenth-century Catholic priest. Of continental origin, the crucifix was probably adapted from an altar or processional cross. Courtesy of the Trustees of the National Museums of Scotland, NMS H.KE17, H.KJ239.

by heart and allowed them to travel without the breviary.[20] Priests could not avoid spending long periods on the road, since until 1700 none had a fixed station to work in but was expected to move from place to place wherever the need was seen to be greatest.

To avoid detection many posed as itinerant tradesmen. 'What disguises have I not worn,' wrote one, consciously echoing St Paul, 'what arts have I not professed! Now master, now servant, now musician, now painter, now brass-worker, now clock-maker, now physician, I have endeavoured to be all to all, that I might save all.'[21] In the mid-seventeenth century the leader of the Scottish Mission actually recommended to *Propaganda* that all student-priests should be taught a trade or profession as part of their seminary training, which upon their return to Scotland they could use as a smoke screen for their own safety and that of the families accommodating them, and as a passport into the homes of Protestants.[22]

The priests routinely camouflaged their identity by adopting aliases, which they also used in correspondence as part of an elaborate code for disguising people, places and addresses. Alexander Smith, for instance, was known as 'Mr Short' on account of his diminutive stature, and Alexander

Leslie as 'Hardboots' in recognition of the miles he covered for the Church; the Benedictines were 'Crows' by their black habits, and the Jesuits 'Birlies' because according to their enemies they would twist to whatever position suited them; the work of the Mission was 'trade', the priests themselves were 'labourers', and those in training 'prentices'. Since suspect letters were always likely to be opened, those bound for a 'Catholic' destination abroad were given a 'Protestant' address – 'Amsterdam' for Paris and 'Hamburg' for Rome.

The regular priests[23] and the Jesuits at least enjoyed the organisation and financial support of their orders, and could look forward to retirement in a mother house abroad. The seculars, however, had no such comforts. Financially, they relied entirely on funding from *Propaganda*, since it was accepted that their congregations, who were of course obliged to contribute to their local Kirk parishes, should not be burdened with paying for them also. But *Propaganda*'s grant for this purpose had been set at 500 Crowns in 1653 when there were five missioners to pay, and was not increased for 80 years despite inflation and a threefold growth in the workforce.[24]

As a result the priests were frequently destitute, barely able to eat, and unable to replace vestments worn out and mouldering with age. Several times the bishops advised Rome of their 'half-famished' state, their utter exhaustion and demoralisation and their 'daily threats to quit the Mission'.[25] Similar complaints and threats also recur as a *leitmotif* in the correspondence of the priests themselves, and the number who did in fact abandon the Church over the period was not inconsiderable.[26]

Others succumbed to their life of hardship. A striking number died within a few years of ordination, while ill-health rendered many quite unfit for work, exacerbating the shortage that already existed. For a sacramental Church there could be no greater impediment to recovery and growth than a lack of priests.[27] Yet only by growing could it hope to produce more and so ease the burden on those it had. Its most pressing dilemma throughout the Penal era was a seemingly endless 'clergy crisis', from which it only finally began to emerge as the eighteenth century came to a close.

In their hard and often short lives the priests lived like their flocks, and among them. And as such they earned their reverence, in a way that a more comfortable life-style never could have. 'The people will not pass me without going down on their knees,' wrote one; and another recalled, 'They will not converse with me until they have first knelt for my blessing.' Rarely perhaps were priests and people closer.

The laity were also suffering under the laws, of course. The great landowners had most to lose. George Gordon, first Marquis of Huntly (Fig. 6.3), was imprisoned no less than eight times for his faith; he was convicted of high treason, banished, and saw his castle burned; and his last term in jail, when in his mid-seventies, brought on his death in 1636. Many men of standing sought an accommodation with the State, maintaining their position by subscribing publicly to the doctrines of the Reformation and attending

Protestant worship, while continuing to practise their Catholic faith in private.[28]

The system in fact actively encouraged nominalism. The Kirk must have been aware of this, and such 'conversions' can hardly have afforded it real satisfaction, though they gave the required public message. By its very nature we can never know the extent of crypto-Catholicism, of course. But it was prevalent enough to prompt fierce debate within the Mission: was outward conformity to Protestantism, with inner 'mental reservation', morally justifiable in extreme circumstances? Some priests were willing to permit it. And what of converts to Catholicism? To avoid ruin they often delayed their change of faith or kept it secret. When MacDonald of Keppoch publicly embraced Catholicism in middle age in 1728 he revealed that he had secretly been a Catholic 'in judgment' for 30 years, though a practising Protestant for all that time.[29] The circumstances of conversion could generate heated dispute among missionaries. Was it right to receive converts into the Church, as some priests did, while allowing them to continue Protestant practice? Many thought not: as one put it, 'Such proceeding is no conversion, bot delusion to the scandal of constant Catholiques who hath soferet.'[30]

The Church continued to suffer a gradual loss of its leading laity throughout the period, particularly in times of fiercest persecution. Thus, at the time of the Commonwealth, the marquis of Douglas and the earls of Errol, Caithness, Sutherland and Winton all apostatised within a few years

Figure 6.3 Engraving of George Gordon, first Marquis of Huntly (1562–1636), and his wife, Lady Henrietta Stewart. Courtesy of the National Galleries of Scotland, SP II 35.1.

of one another. And as the letter reporting their loss noted, 'the defection of a single powerful noble always brought with it the fall of very many besides himself'.[31] His house was no longer available as a refuge and base for the clergy. And his servants and tenants, who depended on him for their homes and livelihoods, enjoyed his protection and feared his displeasure, often followed his lead.

The trickle of apostasy might have been much greater but for the tenacious faith and often heroic resistance of many. It gradually took its toll, however, and the loss of numerous men of property, and the ruin of others, left the Church poor and without influence.[32]

The report of William Ballentine to *Propaganda* in 1660 begins like Caesar's *Gallic War* – 'The kingdom of Scotland is divided into two parts.' He was referring to the Highland–Lowland divide, which affected every aspect of life, including the spiritual. The Scottish Mission was in effect two missions that ran along separate paths each barely aware of the other's movements. Highland Catholicism was in outlook, language and practice far more akin to Irish than Lowland Catholicism. All recent missionary work in the Highlands had been in the care of Irishmen and, after Ballentine's report, the Western Isles were actually placed for a short time under the jurisdiction of the archbishop of Armagh.

This duality, as well as the autonomy of the regular priests, which at times set them at cross purposes with the seculars, and the lack of any overall strategy among the latter, all argued for the appointment of a single leader with episcopal authority to bring coherence to the Mission. Until Scotland had a bishop, it was argued, 'there could be neither order nor decency in the Church'. The request was first put to *Propaganda* in 1631 and was repeated regularly. But Rome procrastinated and made no move for more than sixty years.[33]

Finally, in 1694, the pope ratified the appointment of Thomas Nicolson (Fig. 6.4) as bishop and Vicar Apostolic. As the title implied, he was Rome's representative under the direct authority of *Propaganda*, and Scotland remained a 'Mission' – not until 1878 would it be granted autonomous standing with the restoration of the hierarchy. But it at least came under a single authority, to whom all priests were bound. Bishop Nicolson's *Statutes* of 1700 were to be its blueprint for most of the eighteenth century. His appointment marked the beginning of the modern Catholic Church in Scotland.

Catholic devotion has always had both a communal and an individual dimension. But the former was now severely restricted. The public face – the festivals, processions and open-air liturgies still so characteristic of Catholic countries today – entirely disappeared. Laws were even passed forbidding the naming of feast days and compelling the people to work on them. Public pilgrimages ceased and the Catholic minority had to be content with clandestine private visits to the old holy places. We know from William Ballentine's report that they were still visiting the well of Our Lady of Grace

Figure 6.4 Bishop Thomas Nicolson (1646–1718), copy of original portrait. Courtesy of Blairs Museum, Aberdeenshire.

in Banffshire a hundred years after the Reformation, for example. 'Whether I went by day or night,' he wrote, 'I never saw it without some pilgrims kneeling in prayer. Yet there is nothing to be seen but a heap of stones and a clear spring. The ministers have often had it blocked up, but such is the devotion of the people that they clear it out again each time.'[34] At the other end of the country incisions on the walls of St Ninian's Cave near Whithorn show that it too was still a gathering place for prayer, continuing an unbroken tradition of a thousand years.[35]

For the most part, however, the feasts of the Church were now celebrated at home. Catholicism had become an indoor religion. It was within doors that the faithful fulfilled the obligations placed upon them, obligations considerably more demanding than those we are used to today. Mass attendance was compulsory not only on Sundays but on thirteen designated Holy Days of Obligation in the year. The rules of fasting and abstinence applied throughout the forty days of Lent and on twenty-three Ember Days (vigils of major feasts), while other traditional fasts such as those of St Ninian and St Francis were often practised voluntarily.

Driven in upon themselves, Catholics found strength in family devotions. A handbook of family prayer compiled in the late seventeenth century affords us an insight into the form that these must often have taken. It recommended that the family and servants gather every morning for a prayer of thanks for safekeeping in the night, a passage from Scripture, recital of the Angelus, and a blessing by the father of the house, interspersed with readings from a compilation such as the *Manual of Prayers* or Austin's *Devotions*. Evening prayers might consist of a private examination of conscience for the day, a litany of the saints, the Angelus and a blessing, with a reading from a standard work of piety such as *The Imitation of Christ* or St Francis de Sales' *Introduction to the Devout Life*.[36]

The rosary seems to have grown in popularity in penal times, being ideally suited to family prayer and private meditation and as a teaching aid for children. When Elizabeth Howard married the fourth Marquis of Huntly in 1676 she started a rosary confraternity in the household. Her action serves to remind us that the role of wives and mothers must often have been crucial in upholding family devotions, though it is hardly ever recorded in the writings of the time. We can safely assume, indeed, that women played an irreplaceable part in ensuring the continuity of an embattled faith from one generation to the next, especially in the case of mixed marriages or where under the laws of property the husband had abandoned or concealed his religion.[37]

Catholics not only found their civil rights denied in law; they also suffered the vilification of their faith by the Scottish public. The sixteenth-century reformers were unequivocal in their judgement on Catholicism: it was the Congregation of Satan and idolatry, a tyranny and an abomination to be cleansed from the land. Much of this rhetoric was echoed by their successors, and its constant repetition from pulpit and printing press gradually created among the people an *idée fixe*, an axiomatic perception of Catholicism that prevailed throughout the seventeenth century and well into the eighteenth.

To religious obloquy was added political suspicion. The Catholic Church's efforts to win back its old place by political means in the early years after the Reformation inevitably caused its followers to be tarred indiscriminately with disloyalty to Crown and government. Thereafter, the formal instruction issued by *Propaganda* in 1659, that Mission Churches must avoid involvement in domestic politics and accept the government of the day,[38] was largely ignored in Scotland for more than a century. Catholicism's association with Montrose's campaign and the Stuart risings merely confirmed the suspicion that at heart it was an enemy of the Crown, and was used to justify and prolong the Penal Laws. The fact that it was financed by Rome and that its priests, though Scotsmen themselves, were trained in the lands of Britain's traditional enemies, seemed the final proof that it was in reality a fifth column within the State.[39]

Naturally, the threat of the 'enemy' was nearly always exaggerated. Pamphlets and reports warned of 'multitudes' of priests roaming the country and of 'great numbers' of boys sent abroad every year to complete their priestly training, where in fact there were all too few of either. It was even put about that the priests were paid a commission by the pope for every Protestant converted.[40] Perhaps such claims were sincerely believed by those who made them; certainly they were effective in playing on popular fears.

Over time the drip-feed of propaganda left its mark on Catholics and Protestants alike. It confirmed Catholics as a people on the margin and the first on whom suspicion fell. 'The fault of the guilty is calumniously laid upon the innocent, and every man's crime is visited upon our heads,' wrote the leader of the Jesuit mission in 1670.[41] When the nation needed a scapegoat the Catholics were there, and if they could not strictly be called a people of the ghetto – if only because they were never actually confined within designated areas[42] – they were certainly the victims of a sporadic moral Inquisition.

Their treatment forged in them a dogged survival instinct, a sense of identity and mutual dependence, and a special closeness between priest and laity that left an indelible mark. But it also created an introversion. The key Christian tasks of being a prophetic voice and a leaven with the larger society were denied to Scottish Catholicism, and for two centuries it was forced to play the role, one might almost say, of a chaplaincy rather than a fully fledged Church.

For Protestants, the propaganda provided a minority to despise, an opportunity particularly attractive to those on the lowest rungs of society who had little else to look down on. Not surprisingly, therefore, it revealed itself most commonly and spectacularly in the periodic violence of the urban mob. Theirs was not the spontaneous expression of popular will and 'righteousness' that has sometimes been claimed, but rather an eruption of latent fears and hatreds manipulated by others whenever their vested interests were under threat. Those fears were genuinely held, nonetheless, as was their determination to hold on to a 'freedom' genuinely believed in, of which the laws against Catholics seemed to stand as both guarantee and symbol.

In every respect – the subjugation of a section of society by law, the attitudes of a majority that believed itself threatened, even the scriptural 'justification' of the system – the treatment of Catholics amounted to a form of apartheid. And as in modern forms of apartheid, it was a situation in which the perpetrators suffered as well as the victims. Hatred impoverished the haters, and Scottish society itself was weakened since one of its limbs was not permitted to function. If the material destruction of the Reformation brought immediate and irreplaceable cultural loss, the subsequent marginalisation of Scottish Catholics resulted in two centuries of lost talent and goodwill.

Though we have records of the depredations of the mob at times of crisis, we know little of the day-to-day relations between Protestants and Catholics in normal times. There is no reason to assume that they were always hostile. Even among the two clergies one may come upon scattered references to neighbourliness, friendships and even ecumenical endeavour, especially towards the end of the period. Occasionally too there are signs of openness, warmth and protectiveness on the part of the Protestant laity for Catholic neighbours, with whom they had after all grown up and often had family ties. We know, for example, that it was the Protestant gentlemen of Glenlivet who protected the local Catholic seminary from attack by extremists in the 1720s; and half a century later we hear of a Presbyterian minister, a personal friend of Bishop Hay, aiding the establishment of an emigrant Catholic community in Nova Scotia; and it would be good to think that many similar examples of Christian charity occurred that have gone unrecorded. Inevitably, the contemporary writings of each side tended to lay emphasis upon the other as enemy, but in doing so they may in fact have left us a picture at some distance from the whole truth.

Though the great majority of the Scottish people wholeheartedly adopted the new religion, they did not entirely abandon the old practices and spirituality, especially in the remoter areas. When William Ballentine reported on the crowds visiting the well of Our Lady of Grace, he was not referring to Catholics only. And at the end of the seventeenth century the people of the Protestant Hebridean islands were still observing the major feast days, and Orcadians venerating pre-Reformation chapels, in defiance of the teaching of the Kirk.[43]

Time and again in Mission correspondence one comes upon assertions of a continued interest in and curiosity concerning Catholicism, and of the plentiful 'harvest' of converts to be gathered if only the labourers were greater in number, and though such reports certainly had their own agenda there is no reason to doubt their general truth. Small numbers did in fact come into the Church yearly, even in times of greatest persecution, despite the risks and sacrifices involved. Particularly striking were the number who then proceeded to the priesthood, and the proportion who became key figures in their adopted religion. William Ballentine, Thomas Nicolson and George Hay (the undisputed Catholic leader of the late eighteenth century, Fig. 6.5), were all converts, as were numerous others who worked under them. They brought to the service of the Church diverse talents, and the

Figure 6.5 Portrait of Bishop George Hay (1729–1811) in the Scots College, Rome. Courtesy of Mgr. J McIntyre, Baillieston.

benefits of an education that would never have been open to them had they been born into it.

Education in fact posed a delicate problem for Scottish Catholics in the penal era. During that time the vision of the Reformation Parliament that every parish should have its own school was gradually realised, and to these were added more than 300 schools established in the Highlands and Islands by the Society in Scotland for the Propagation of Christian Knowledge. In all these schools the pupils were taught the tenets of *The Westminster Confession* and were normally required to attend Protestant worship on the Sabbath. The grammar schools laid comparable obligations on their students, while at the universities rather similar expectations obtained as a condition of both attendance and graduation. Catholic schools were proscribed by law, and Catholic colleges were non-existent. Thus the effective parental choice for Catholic families was Protestant education or no education.

Naturally there were ways of side-stepping the system. Several of the grammar schools had Catholics on their staff (though this too was forbidden in law) who quietly excused the Catholic pupils from catechism and prayers.[44] In a few Highland strongholds the Church opened its own schools from time to time, but their existence was precarious and their educational standards very mixed. The one on Barra was so poor and remote that parents reputedly vowed that they would sooner send their children to school in the Indies. In the Gordon lands of Banffshire the Kirk reported a rash of small schools in the early eighteenth century, run by women who had themselves been instructed by the local priests, but again these were short-lived.[45] Forty years later an old woman was running a school in the Catholic Braes of Glenlivet, where the children were taught reading but not writing since she could not write herself. In the city, 'dancing classes' in a hired room might hide the existence of a tiny covert Catholic school.

Such establishments served the local children as an alternative to the parish schools. Most of the Church's meagre resources, however, were directed towards serving the propertied class, and attempting to match the curriculum and standards of the grammar schools. Typical was the school at Arisaig, where 'above thirty scholars off the best Gentlemens children of the Highlands' were enrolled in the early eighteenth century.[46] In making this their priority the bishops were being realistic. They recognised the importance of a loyal, informed laity of the gentleman class, both for leadership and as the principal traditional source of vocations to the priesthood. But the fact that they could rarely offer a grammar school education to anyone else merely perpetuated the existing situation.

In 1716 they opened a small, clandestine seminary at Scalan (Fig. 6.6) in the remote uplands of Banffshire which, in defiance of the law and against all odds, was to survive until the very last year of the century.[47] Its main purpose was to train boys for the priesthood, and in its time some seventy were in fact ordained, but it also accepted fee-paying lay students and over

Figure 6.6 The College of Scalan, Upper Banffshire, the new building of 1767. Photograph by the author.

the years provided an excellent general education to a number of boys, particularly from the north-east. Similar ventures in the west Highlands also admitted lay boarders along with those training for the priesthood, but they were beset with financial and other problems, and none came near to matching the success of Scalan.[48]

For families of means there was also the possibility of schooling abroad. The seminaries of Rome, Paris, Douai and Madrid had originally been founded partly to provide secular, university-level education for Scottish youth denied it in their own country, and they continued to welcome lay students throughout the period. In Madrid the lay students were always in the majority and often comprised the entire student body; at Douai the Jesuits maintained a lay preparatory school alongside their senior seminary.[49]

In addition to these schools, English Benedictines, Dominicans and Jesuits had established boarding schools in northern France and Belgium after fleeing their homeland at the Reformation, to which Scottish boys as well as their own countrymen were admitted.[50] The Scottish Parliament attempted to close this loophole by passing laws forbidding parents to send their sons abroad for their education.[51] Though they failed to eliminate the practice entirely, they reduced it to a trickle. Nor was there, of course, any comparable provision for girls.[52]

For one reason or another, therefore, Catholic education was always beyond the reach of all but a tiny minority. For the rest, much depended upon the resolution of the parents. The more devout preferred no schooling at all. As one Highland bishop explained:

> Our youth have no opportunity of being taught unless they be sent to the Prespiterian Masters which our people look upon to be much the saime as to complyment them to the Devil, therfore they think it much better for their Children to have religion without learning, than learning without religion.[53]

Others took a more pragmatic view. Thus when the Scottish Society for the Propagation of Christian Knowledge (SSPCK) opened a school in Glenlivet in 1713 half of those who enrolled were Catholics. When their bishop heard that their textbooks included *The Dangers of Popery*, *Protestant Resolution* and *Funeral of the Mass* and tried to persuade the parents to withdraw them, they stoutly resisted until he agreed to open a Catholic school in the neighbourhood.[54]

Over the years many hundreds of young Catholics attended the parish, SSPCK and grammar schools, where their exposure to a consistent religious message, combined with the natural reaction of children (which every parent has experienced) that 'teacher knows best', must have done much to weaken the standing of their Church. Bishop Nicolson saw for himself the damage being done in the Highlands, where many Catholic gentlemen sent their sons to the Protestant colleges for the liberal education considered necessary for persons of their station. The boys 'had imbibed not only learning but heresy there,' he warned *Propaganda*, 'which they communicated to their friends and dependants, wherever there were no missionaries to stand out against them.'[55] One of the main tasks of the missionary priests, in fact, as his words remind us, was to repair the harm done to the Catholic faith by the Scottish education system.

The exclusion of Catholics from the higher levels of education, their frequent self-exclusion from the lower, and the shortcomings of their Church's own provision, created over time a sub-group generally worse educated than the society at large. This, with their generally greater poverty, must surely have compounded a self-perception of inferiority, almost as a concomitant of their baptism.

How then did they learn their own faith? For most, the homily of the priest on his far-flung circuit of visitations provided their sole opportunity. Beyond this they had few resources. Not until the very end of the period was the Bible made available in an English Catholic version. Before that, some perhaps used the Latin Vulgate or the French Douai text, and it was claimed that many in the north-east used the Protestant Bible.[56] But unguided private reading of Scripture was not a general practice among Catholics at this time. A papal warning against it, issued in 1713 in the wake of the 'excesses' of Jansenist biblical exegesis, had prompted Scotland's bishops to discourage it through much of the eighteenth century.[57]

Books of apologetics and controversy were available and were certainly circulated among the clergy. But catechisms, which would have been of the greatest practical use for priests and laity alike, were scarcely to be had until the mid-eighteenth century. Earlier versions had been withdrawn as doctrinally unsound, and it was only in 1750 that Bishop Smith's catechism was finally approved after fierce opposition at home and close scrutiny in Rome.[58]

When the bishop for the Highlands received his small consignment he expected them to suffice for years to come, considering how few of his flock could read them.[59] In his part of the country, at least, illiteracy remained the greatest obstacle to self-instruction and ensured that the voice of the priest must remain the chief means of handing on the faith. But this only confirmed the people's inclination to rely on their priests at every turn, and made for a laity who, though often deeply pious, were rarely deeply knowledgeable about religion, and thus vulnerable to those of other persuasions who were.

Some of the practical difficulties of an underground faith – problems of Mass attendance, premises, vestments, etc. – have already been touched on; but there were numerous others. One recurrent problem concerned the blessing of the oils used at baptism and confirmation. Before the appointment of Thomas Nicolson as Vicar Apostolic, the oils had to be sent from London, since the blessing could only be performed by a bishop, and even thereafter the time and form of the ceremony were strictly regulated. Stringent rules applied indeed for every liturgical act, defining the vestments and vessels to be used, the exact wording of prayers, and so on. There is no doubt that the formalism of the Catholic Church in such matters compounded the difficulties imposed by its enemies. The Tridentine Church simply had no adequate 'underground' strategy. How useful, for example, would today's lay Eucharistic ministers have been in those times! How much easier if priests and laity had not had to live under self-imposed restrictions that seemed immovable at the time but have since been removed without any problem.

There are, of course, certain rules of the Church which, since principle is closely involved, are not open to change. The rubric remains firm, for example, in requiring that Communion hosts be made of wheaten bread. However, since wheat was rarely grown in the Highlands before 1800, the flour had to be imported for use in the Highland Vicariate. In the 1760s the bishop had it sent from Edinburgh annually, because he could not be sure that what passed for wheat flour in the Highland markets was genuine.[60] Such scrupulous adherence to the rubric, which the Church demanded, heaped further burden on those already burdened and it was not unknown for the faithful to be without the Eucharist for half a year while they waited for hosts.

Difficulties of this kind were not created by the authorities, and it is important to remember that while the Church's worship was undoubtedly, as Dilworth asserts, 'conditioned by the pressure exerted by its adversaries',[61] it would be unjust to lay all of its hardships at their door. Some it created itself. And many were simply a problem of the times. In an age when carriageways rarely extended far beyond the towns, it was the remoteness of most Catholic communities that more than anything determined their daily lives and the practice of their faith. When William Harrison was appointed priest of Morar in the 1730s he encountered the problems at first hand. 'We being remote from all posts, we can have no news here more than we were not in the world,' he told a colleague in a letter that had to be borne across country by a team of runners. 'My knee was my writing table and gun pouther my inck,' he added in excuse for his handwriting. He was not even certain that he had in fact celebrated Easter on the correct date that year.[62] And his was by no means the most isolated station on the Mission.

If Mr Harrison had been abreast of news from the outside world he must have noticed the changes beginning to overtake Scottish society at this date, changes in economic conditions and in social, political and religious attitudes that were about to impact upon his Church. The old passions were cooling. The winds of religious controversy that had swept Europe for two centuries were abating. The last witch-burning fire had been extinguished and the civic authorities were beginning to lose their relish for implementing the anti-Catholic laws. Though they would re-impose them vigorously in the decade following the 1745 rebellion, progressively thereafter they would become a dead letter, an anachronism in what, by the 1770s, was becoming a fundamentally secularist State. By that date a government committed to justice and moderation was in the process of granting freedom of worship in England, Ireland, and its new Catholic colonies recently wrested from the French. The case of Scottish Catholics for similar rights was unanswerable.

And at last the political obstacles to granting them were also coming down. The '45 generation was passing on and, following a pointed lead from Rome, the bishops of the Scottish Mission were at pains to distance themselves from the Jacobite cause, whose threat was in any case evaporating. Their Church was finally disengaging itself from domestic politics and emerging as an independent entity whose agenda was solely spiritual.[63]

Bishop Hay, its acknowledged leader, was not slow to claim for Scotland what was being ceded elsewhere in Britain. In doing so he not only appealed to the 'benevolent principles of humanity' of the Enlightened age, but also pointed to the demonstrable loyalty of his own people (exemplified by the bravery of those who had enlisted for the French and American wars) and Catholicism's growing and positive contribution to Scottish society. His own philosophical writings, the achievements of Alexander Geddes in biblical scholarship and the contributions of Bishop John Geddes to the *Encyclopaedia Britannica*, along with the bishop's personal friendships with several of Edinburgh's leading *literati*, had brought them into the fellowship of the republic of letters, and their Church at least closer to the mainstream.[64]

Catholicism was emerging from the underground. The penal laws were now an absurdity and their repeal could only benefit all parties. The need for secrecy was gone. Where hitherto Mass houses had been camouflaged as domestic or farm buildings (Fig. 6.7), the Mission now began to

Figure 6.7 St Ninian's church, Tynet, Moray. Originally a cottage, it was extended in 1755 to look like a sheepcote, with slit windows (later enlarged). Photograph by the author.

Figure 6.8 St Gregory's Preshome, Enzie, Moray, 1788–90. Photograph by the author.

erect buildings that were quite openly and unmistakably churches (Fig. 6.8). In 1793 the Scottish Catholic Relief Act was passed in Parliament. It did not cancel existing legislation, but relieved Catholics of the penalties attached to it.[65] It did not cede them full emancipation or citizenship, but it gave them what they had not known for more than two hundred years, freedom of worship.

The Church that entered this fresh era and prepared for the new century was very different from that which had begun the old. It was now administered in two vicariates, Highland and Lowland, a structure that better matched its spiritual and linguistic duality. Its priests now received at least their initial training in purpose-built seminaries in Scotland (Fig. 6.9). Prudent management had brought the Church to something approaching solvency. Its actual membership had not grown, certainly, and may actually have fallen below 2 per cent of the national population.[66] And its rural demography, having crystallised in the mid-seventeenth century, had remained largely unchanged,[67] with great tracts of the country still entirely void of Catholics and only theoretically within its pastoral structure (when Bishop Geddes visited Orkney in 1790, no-one was sure to which vicariate it belonged!). But whereas at the beginning of the century some fourteen aristocratic houses and many lesser lairds had been Catholic, by its end almost all had abandoned the faith or been lost through emigration.[68] In the late 1780s Bishop Hay noted that this traditional source of religious vocations had entirely dried up, and that even the gentleman-farmer class, previously a rich source also, was in decline. His main hope now lay with the tenants and small farmers.[69]

Interestingly, he still did not look to the towns, which hitherto had given the Mission neither priests nor lay leaders. In the eyes of his generation the Church was still essentially rural. But its demography was changing. He had himself witnessed the arrival of Catholic Highlanders on the streets of Edinburgh in the famine of 1771–3, and had opened a Gaelic chapel for them there. Now, following a second famine within a decade, more were streaming into the Lowland towns. In their old homes they had been penniless, but most had had their plot of land; in the Lowlands they took the most menial work and became part of the new proletariat. In 1786 Bishop Geddes discovered that there were now some 200 Catholics in the Glasgow–Greenock area, mainly Highlanders and Irish, and warned that they could no longer be properly served from Edinburgh.[70] Meantime, for every Highland family finding its way to the burgeoning industrial towns, many more were leaving for the New World. Perhaps up to 4,000 left Catholic Knoydart, Glen Garry and Morar, and elsewhere the trend was similar if less spectacular. Where at the beginning of the century Highland Catholics had outnumbered Lowland by at least 5:1, by its end they were in the minority.

Thus for one reason or another Scottish Catholicism was taking on a new appearance, far more urban and (though the phrase would not have been used) working-class. And the trend would continue with the coming of the Irish, a trickle that would soon become a flood, transposing the Church's

Figure 6.9 Aquhorties House, 1799, seminary for the Lowland Vicariate. Photograph by the author.

epicentre to the west central belt and standing the old demography on its head.

The nineteenth century would witness the full maturation of this 'new' Church – urban, working-class, largely immigrant, numerous – with a persona quite different from the old. Hard questions would then be asked of a leadership accustomed to caring for something small, native, rural and clandestine. And yet, as we have seen, certain features that one associates with Scottish Catholicism in the nineteenth century – the solidarity, the privacy amounting to introversion, the indispensability of the priest and his closeness to the people, the general lack of involvement in national affairs – were already to be found (though perhaps for somewhat different reasons) in the earlier native Church. Such features have proved remarkably long-lived, for traces of them may be detected still upon the religious landscape of Scotland.

But when we look back at that native Church, these are not the qualities that make the deepest and most lasting impression. The overriding qualities of Catholicism in the penal era, those that most define its character and constitute its most precious legacy and 'message' for today, are the ones that enabled it to survive and put out frail new shoots – its firmness in faith and tenacity in hope, its strength in weakness, and its readiness to suffer and endure.

NOTES

1. Report of the Jesuit Fr Robert Abercrombie; Latin text and translation in Anderson, W J. Report of Robert Abercrombie SJ in the Year 1580, *IR*, 7 (1956), 27–59.
2. See McRoberts, D. Material destruction caused by the Scottish Reformation. In McRoberts, 1962, 415ff. The author notes that within a generation of their confiscation nearly all the larger churches and many of the former parish churches, particularly in the east and central Lowlands, lay in ruin.
3. Hay, 1929, 73 fn; Forbes Leith, 1885, 291 fn and 374; both citing contemporary sources. See also Sanderson, M H. Catholic recusancy in Scotland in the sixteenth century, *IR*, 21 (1970), 87–107.
4. For Terregles and Traquair, see Blundell, O. *Ancient Catholic Houses of Scotland*, Edinburgh and London, 1907, chs. 3, 4 and 7, 8 respectively.
5. See Report of Cardinal Rospigliosi, 1669, *Propaganda* Archives Acta 1669, f. 462, cited in Bellesheim, 1890, IV, 85 fn; and MacLean, D. Catholicism in the Highlands and Isles 1560–1680, *IR*, 3 (1952), 5–13, at 7.
6. Giblin, C, ed. *Irish Franciscan Mission to Scotland 1619–1646*, Dublin, 1964. The work comprises Latin documents, with translations, from the Roman archives, largely the reports and correspondence of the missionaries themselves. Though they first arrived in 1619 the main work was undertaken by four priests from 1626–37. Their claims to have brought some 10,000 into the Church have been questioned but may well be fairly accurate.
7. Memorial of Patrick Anderson SJ, early seventeenth century (in Latin), trans. in Forbes Leith, 1885, 294.
8. Letter Thomas Abercromby SJ to Fr Claudio Acquaviva, General of the Society, 30.10.1610 (in Latin) trans. in Forbes Leith, 1885, 290f.
9. See, for example, Annual Letter of the Superior of the Jesuit Mission in Scotland,

1629, Stonyhurst MS (in Latin) trans. in Forbes Leith, 1909, I, 47ff.
10. Letter Alexander MacQuhirrie to Fr Claudio Acquaviva, 7.9.1604 in Forbes Leith, 1885, 278.
11. In England attendance at Protestant worship could be bypassed by payment of a fine. In Ireland the great number of Catholics made this and some other legislation unworkable.
12. MacInnes, A I. Catholic recusancy and the penal laws, 1603–1707, *RSCHS*, 33.1 (1987), 27–63. MacInnes' detailed account gives references to the *Acts of the Parliaments of Scotland* throughout.
13. Thus, for instance, in the early years of the reign of James VI and I, when he was seeking to restore episcopacy in Scotland, he launched a new persecution of Catholics as a sop to the Presbyterian party; but at the end of his reign, when negotiating the marriage of his son Charles to Henrietta Maria of France, he signed an agreement guaranteeing all Catholics freedom of worship and ordered an end to their harassment in Scotland. Similar vicissitudes of fortune for Catholics marked the reigns of his successors.
14. Several such laws were consolidated in the Act of Abjuration (9 Geo I c. 24) of 1723: An Act to oblige all Persons, being Papists, in that part of Great Britain called Scotland refusing or neglecting to take the Oaths … to Register their Names and Real Estates, etc.
15. Notably, the oft-invoked Act for Preventing the Grouth of Popery, Acta Parliamentorum Gulielmi, 1700, *APS*, X, 215ff.
16. Cf. Halloran, B M. Jesuits in eighteenth century Scotland, *IR*, 52 (2001), 80–100, at 97.
17. Report of Alexander Winster to *Propaganda*, 1688, in Bellesheim, 1890, IV, 116ff. For two Mass stones still surviving in the Glen Spean–Glen Roy district, see MacDonell, A and McRoberts, D. The Mass Stones of Lochaber, *IR*, 17 (1966), 71–81.
18. Privy Council Records of the meetings of 27 Feb, 3 June and 3 July 1634, in Hume Brown, P, ed. *Register of the Privy Council of Scotland*, 2nd series, V, Edinburgh, 1904, 260, 263, 293 and 606; a summary also in Chambers, R. *Domestic Annals of Scotland*, Edinburgh and London, 1858, II, 72f. 6. The groom was sentenced to imprisonment. The bride was already in Dumfries gaol, with fourteen other women, for hearing Mass. Six of her cell-mates, refusing to promise that the offence would not be repeated, were returned to confinement in the Edinburgh tollbooth at their own expense.
19. Army Intelligence Report, Lochaber, no date but c1702, CH 1/2/29, 569, National Archives of Scotland. We cannot assume, however, that the conditions described in such reports necessarily applied throughout the period.
20. The dispensation had been granted to the Irish Dominicans by Pius V – cf. Ross, A. Dominicans in Scotland in the seventeenth century, *IR*, 23 (1972), 40–75, at 58. In his report to *Propaganda* of 1681 Alexander Leslie recommended that the dispensation be extended to all priests. Alexander Leslie *Report*, 4.3.1681 (in Latin), English translation SM 2/9/3, 32, Scottish Catholic Archives. This document, which provides a detailed demographic break-down and responses to more than 80 specific questions posed by *Propaganda* regarding the state and needs of the Church in Scotland, is an invaluable source for the period.
21. Report of Fr John Innes SJ to the General of the Society, 1702, in Forbes Leith, 1905, II, 195.
22. Report of the Prefect–Apostolic of the Scottish Mission to *Propaganda*. See Anderson, W J. Prefect Ballentine's report c1660, part 2, (Latin text and translation), *IR*, 8 (1957), 112–21.
23. That is, priests of the regular orders. The four main orders working in Scotland were the Franciscans, Dominicans, Benedictines and Vincentians.
24. Cf. Hay, M. Too little too late, *IR*, 6 (1955), 19–21; MacMillan, J. 'The root of all evil'?

Money and the Scottish Catholic Mission in the eighteenth century, *Studies in Church History*, 24 (1987), 267–82.
25. For example, Report of Bishops Gordon and Wallace to *Propaganda*, 1730, (in Latin) trans. in Bellesheim, 1890, IV, appendix XII, 383ff.
26. See Szechi, D. Defending the True Faith: Kirk, State and Catholic missioners in Scotland, 1653–1755, *Catholic Historical Review*, 82 (1996), 397–411, at 406. Szechi calculates that during this period 14 to 16 priests apostatised, 10.5 per cent of the total workforce (or 8.5 per cent if one discounts those who later returned). One or two of those he includes are dubious cases, however.
27. Priests were necessary for administering the sacraments of Penance ('Confession'), Holy Eucharist ('Communion'), Extreme Unction ('the Sacrament of the Sick'), and in normal circumstances Baptism (though lay persons might baptise in cases of necessity). Their presence was also required for the sacrament of Matrimony, though strictly speaking the couple conferred this sacrament upon each other. Only a bishop could confer the sacraments of Confirmation and Holy Orders.
28. A notable early example was the chancellor Lord Alexander Seton in the first decade of the seventeenth century. Cf. letter Fr James Seton SJ to the General of the Society, 30.9.1605, quoted in Forbes Leith, 1885, 279. Apparently, most of the leading Catholics of the day followed his example.
29. Letter of James Gilchrist, 4.4.1729, General Assembly Papers, CH 1/2/59, 83v, NAS.
30. Letter of Fr Christie SJ, 27.2.1653, cited in Hay, 1929, 195ff.
31. Alexander Winster Report, *loc. cit.*
32. Towards the end of the period William Robertson, principal of the university of Edinburgh, calculated that less than twenty Catholics in Scotland possessed £100 p.a. in land, and that they had virtually no influence in society – Report of General Assembly debate in *Scots Magazine*, 41 (1779), 413.
33. *Propaganda* had compromised by the appointment of a Prefect–Apostolic from 1653, but his powers were strictly limited.
34. Anderson, 1957, 111.
35. McRoberts, D. *The Pilgrimage to Whithorn*, Glasgow, nd, 7.
36. Cf. Anderson, W J. Catholic family worship on Deeside in 1691, *IR*, 18 (1967), 151–6. The handbook was Robert Strachan's *The Manner of Prayer in a Family Both Morning and Evening* (1691). The *Manual of Prayer* was first published in Rouen in 1583, and ran to many editions before being superseded by the *New Manual* in 1688. Francis de Sales' *Introduction to the Devout Life* had been available in English translation since 1613.
37. The one attempt to treat the theme to date is: Roberts, A. The role of women in Scottish Catholic revival, *SHR*, 70 (1991), 128–50. His manful trawling of the literature for references, anecdotes and incidents reveals, apart from anything else, the sparseness of available evidence.
38. Propaganda *Instructiones*, 1659, in Seredi, ed. *Codex Juris Canonici*, VII, 4463, which ordered Mission leaders to instruct the faithful to obey their rulers even if 'hostile', to avoid criticism of them, and to 'suffer in silence'.
39. Cf. e.g. James Gordon *Narratio* to *Propaganda*, 1703 (in Latin), copy SCA SM 3/8, para 25.
40. For priests: *A Seasonable Warning by the Commission of the General Assembly Concerning the Dangers of Popery*, Edinburgh, 1713. For students: Representation by the Committee of the Commission of the General Assembly of the Church of Scotland Anent the Growth of Popery in the North, 1722, NLS MS 3430, 239f. For commission: Memorial Concerning the Growth of Popery in Diverse Places in the North and Islands of Scotland, Causes thereof and Some Remedies Humbly Proposed, General Assembly, 1718, NLS MS 3430, f. 220.

41. Annual Letter, Jesuit Mission, 1670, (in Latin) trans. in Forbes Leith, 1909, II, 20.
42. There were occasions at times of national emergency when the movement of Catholics was restricted. For example, when William III's position was threatened in 1690 all adult British Catholics were ordered to remain within five miles of their homes. Proclamation for the Confinement of Popish Recusants within Five Miles of their respective Dwellings, 17.9.1690, copy NAS GD 103/2/237. A similar ruling was made at the time of an expected Jacobite landing in 1708. Proclamation of Queen Anne (no title), 1708, copy NAS RH 14/569.
43. Martin, M. *A Description of the Western Islands of Scotland circa 1695*, 1703, Stirling, 1934 edn, 108 re Lewis, 264 re Gigha, 271f re Jura, 280 re Colonsay, 310 & 443 re St Kilda, 372f re Orkney. For examples of persistence of old practices in the north east, see McPherson, J. *Primitive Beliefs in the Northeast of Scotland*, London, 1929, *passim*.
44. For example, the schools at Fochabers and Strathbogie in the early eighteenth century. In these areas, where Catholics were numerous, teachers abetted the parents in keeping the pupils from catechism and prayers. Inspectors were sent in to ensure that they complied with the law. Minutes of synod meetings over a period of two years show that the problem continued. Synod of Moray Minutes of meetings 26.5.1708, 1.6.1709, and 30.5.1710, NAS CH 2/271/4, 11, 199, 224, 263f.
45. Particular condescendance of some grievances from the encresce of Poperie, and the Insolence of popish priests & Jesuits, 19.5.1714, NLS MS 976, f. 143. In the later seventeenth and eighteenth centuries schools were opened in Glen Garry and Arisaig, and on Barra, Canna and South Uist.
46. Letter Thomas Innes to William Leslie, 13.10.1701, SCA, Blairs Letters.
47. For a detailed account, see Watts, 1999. Scalan replaced the small seminary opened on Eilean Bàn, Loch Morar in 1714, which had closed following the failure of the 1715 rebellion. It originally served the whole Mission, but with the creation of separate Highland and Lowland vicariates in 1732 it became the seminary of the Lowland vicariate. In normal times its roll typically varied between c5–12, of whom perhaps a third would be lay students; their ages ranged from c9–24, the majority being teenagers. Fees for lay students were £6 pa, raised in 1779 to £8.
48. Seminaries were opened to serve the Highland vicariate at Eilean Bàn (1732–38), Gaotal in Arisaig (1738–46), Glenfinnan (1768–70), Buorblach in North Morar (1770–79) and Samalaman in Moidart (1783–1803).
49. For the original role of the Scots College Rome see Dilworth, M, chap. 1 of McCluskey, R, ed. *The Scots College Rome 1600–2000*, Edinburgh, 2000, 20. For Madrid see Taylor, M. *The Scots College in Spain*, Valladolid, 1971.
50. Cf. Barnes, A S. *The Catholic Schools of England*, London, 1926: for English College at Douai (1579), 58ff; for St Gregory's Douai (Benedictine, 1611), 78ff; for St Omer's nr Calais (Jesuit, 1592), 66ff; for Dominican school at Bornhem (1673), 90ff; for St Laurence's Dieuleward (Benedictine, 1669), 85ff.
51. Act Anent nobilmen who sendis thair sones oute of the cuntrie, Acta Parl. Jacobi VI, 1609, *APS*, IV, 406; Ane Act anent the childrene of noble men and otheris remaning in seminaryis of popishe religioun beyond sea and aganis Jesuitis and messe priestis, Acta Parl. Caroli I, 1625, *APS*, V, 177.
52. From the time of its establishment in York in 1686 by sisters of the Institute of the Blessed Virgin Mary, the Bar Convent provided a school for daughters of the English gentry. At least 34 of the girls who enrolled before 1800 were Scots – information supplied by Sr Gregory IBVM, archivist. Her booklet *History of the Bar Convent*, York, 2001, offers a fascinating picture of convent and school life in the late seventeenth and eighteenth centuries. For a more general account of education opportunities for Catholics, see Dealy, Sr M B. *Catholic Schools in Scotland*, Washington, 1945.

53. Letter Bishop Hugh MacDonald to Peter Grant (Scots Agent in Rome), 20.6.1740, SCA, Blairs Letters.
54. Cf. Register of Society's Schools, NAS GD 95/9/1, microfilm. For the bishop's intervention, Additional Memorial Concerning the Growth of Popery, General Assembly, 1720, NLS MS 68, f. 31.
55. Bishop Nicolson Report to *Propaganda*, 1701 (in Latin) trans. in Bellesheim, 1890, IV, Appendix VIII, 371ff, para 16.
56. Bishop Hay's English edition finally appeared in 1796, though parts, edited by Bishop John Geddes and Gallus Robertson, had been available several years earlier. Hay's was essentially the eighteenth-century translation of the Douai Vulgate made by Bishop Challoner, the leader of the Catholic Church in England – see Anderson, W J. Fr Gallus Robertson's edition of the New Testament, 1792, *IR*, 17 (1966), 48–59. For 'most' Catholic families of the north-east owning and using the Protestant English Bible – Anderson, 1966, 51f.
57. The papal injunction rejecting the assumption that 'reading of sacred Scripture is for all' had appeared in the Bull *Unigenitus Dei Filius* of Clement XI, published specifically as a condemnation of the Jansenist heresy.
58. Gordon, 1867, 9f.
59. Letter Bishop Hugh MacDonald to Bishop Smith, 26.11.1760, SCA, Blairs Letters.
60. Letter Bishop John MacDonald to Procurator of the Mission, 27.10.1767, SCA Blairs Letters.
61. Dilworth, M. Roman Catholic worship, chap. 8 of Forrester, D and Murray, D, eds. *Studies in the History of Worship in Scotland*, Edinburgh, 1984, 113.
62. Letter William Harrison to Robert Gordon, 1.4.1738, SCA, Blairs Letters.
63. The key moment in Rome had been the refusal of the Vatican, on the death of James III, to recognise Charles Edward Stuart as the rightful heir. If one looks for a particular moment of *peripeteia* within the Mission itself, it was perhaps the instruction given to the faithful in the bishops' pastoral letter of January 1780 to join Protestants in praying for the royal family on a Fast Day appointed by the Kirk (3 Feb 1780). Significantly, their pastoral letter followed close upon the death of the last uncompromising Jacobite bishop, John MacDonald.
64. Cf. Goldie, M. The Scottish Catholic Enlightenment, *Journal of British Studies*, 30 (1991), 20–62.
65. Act Requiring a certain Form of Oath of Abjuration, and Declaration, from His Majesty's Subjects professing the Roman Catholick Religion, in that Part of Great Britain called Scotland, 33 Geo III c 44, 3.6.1793, extending to Scotland the Act of 1791 (An Act to Relieve, upon Conditions and under Restrictions, the Persons therein described from certain Penalties and Disabilities to which Papists or those Professing the Popish Religion are by Law subject, 31 Geo III c. 32), which had applied to all parts of Britain except Scotland. It relieved Catholics from the penalties of the Act of 1700, dependent upon their taking an oath of allegiance to the Crown and disclaiming any temporal claims of the papacy. It allowed them to worship freely and to enjoy real or personal property, but it did not permit them to take posts as teachers or governors in schools, private houses or universities.
66. The Catholic population has been estimated at c30,000 (1.8 per cent of the national population) in 1800 – cf. Darragh, J. The Catholic population of Scotland since the year 1680, *IR*, 4 (1953), 49–59. For the 1700 figure estimates are more problematic, but it may have been c24,000 (2.4 per cent of the national population).
67. The religious affiliations of the clans had become permanent by the end of the Montrose campaign; since then the position in the Highlands had changed little, the Catholic Church making some gains particularly in Lochaber and Kintail in the eighteenth century, and dying out in some remote areas no longer reached by priests after c1700.

68. For gentry losses, Johnson, 1983, 43. Most had occurred through marriage, entry into the British army, or a general desire to integrate into the establishment.
69. Bishop Hay, Reflections on the Present State of this District [the Lowland vicariate] with Regard to Providing and Admitting Boys to the Seminary, MS, 9.3.1787, SCA CS 1/1/11.
70. Letter Bishop Geddes to Bishop Hay, 30.11.1786, SCA, Blairs Letters.

BIBLIOGRAPHY

Anderson, W J. Prefect Ballentine's report c1660, part 2 (Latin text and translation), *IR*, 8 (1957), 39–66; 99–129.
Anson, P F. *Underground Catholicism in Scotland 1622–1878*, Montrose, 1970.
Bellesheim, A, trans. Hunter Blair, D O. *History of the Catholic Church of Scotland*, Edinburgh and London, 1890, vols. III, IV.
Blundell, O. *The Catholic Highlands of Scotland*, 2 vols., Edinburgh and London, 1909, 1917.
Bossy, J. *The English Catholic Community 1570–1850*, London, 1975.
Forbes Leith, W, ed. *Narratives of Scottish Catholics Under Mary Stuart and James VI*, Edinburgh, 1885.
Forbes Leith, W, ed. *Memoirs of Scottish Catholics During the XVIIth and XVIIIth Centuries*, 2 vols., London, 1909.
Gordon, J W S. *Journal and Appendix to Scotichronicon and Monasticon*, Glasgow, 1867.
Hay, M V. *The Blairs Papers (1603–1660)*, London and Edinburgh, 1929.
Johnson, C. *Developments in the Roman Catholic Church in Scotland 1789–1829*, Edinburgh, 1983.
McRoberts, D, ed. *Essays on the Scottish Reformation 1513–1625*, Glasgow, 1962.
Watts, J. *Scalan: the Forbidden College 1716–1799*, East Linton, 1999.
Watts, J. *Hugh MacDonald, Highlander, Jacobite and Bishop*, Edinburgh, 2002.

7 Roman Catholics in Scotland: Nineteenth and Twentieth Centuries

JOHN F McCAFFREY

Any consideration of the position of Roman Catholics in modern Scotland starts from the basic fact that they do not fit easily into a society so profoundly marked by the Reformation as Scotland has been. Consequently, they have had to tread warily in their search for a place in the Scotland which has developed since 1800. This explains the anonymity and inward-looking aspects which have characterised their lives as a community in Scotland, as well as the periodic misunderstandings and antagonisms which they have experienced throughout their period of growth over the past two centuries. Clinging to a faith opposed to the nature of the society around them has the potential for friction, and this has been exacerbated by the Irish contribution to the growth of Scottish Catholicism in this period, as well as the association of this growth with the brutal processes of industrialisation which shattered the old Scotland forever. This period has been one of self-discovery for them, as indeed it has been for Scottish society as a whole.

At the same time, it has to be recognised that while at the level of worship and community they have lived as an enclosed, defensive group, at another, they too have been profoundly affected by the material changes which have transformed Scottish society since 1800. Despite their self-contained nature, the demands of daily living have also forced them to look outwards through their involvement in the activities of the wider community. Linkages through marriage and social intercourse have always existed. They have shared many of the economic and living conditions of their fellow citizens. They have had to participate in common struggles to maintain family decency in one-room houses, or for better working conditions through voluntary and political movements. The growing middle class of Catholics in professions and commerce today, just like the steelworkers, miners, craftsmen and labourers before them, have a constant and direct contact with and knowledge of their fellow Scots through their daily work with them in environments outwith the Catholic community.

The heterogeneity of the Catholic community has also to be recognised. Before Irish Catholics came to swell their numbers, the existing stock of Scots Catholics may have been numerically small at some 30,000 in 1800, but they had a significant presence in the Gaelic-speaking west Highlands, in the rural north-east around Banffshire and in the older towns like

Edinburgh and Aberdeen. These Scots Catholics had a strong sense of identity derived from the experiences of the past two and a half centuries of Scottish history. Their centre of gravity was already shifting from the north to the central belt of Scotland stretching from Dundee in the east to Ayrshire in the west and, above all, to Glasgow and the Clyde Valley. These were areas where they were hitherto largely unknown and consequently lacked the resources in churches, priests and neighbourly contacts which had been so carefully built up in the old centres. Here they were joined by a new stream of Irish Catholics, mainly from Ulster, which became a torrent in the Famine years of the 1840s and 1850s. Whereas Scottish Catholics had learned to live with the Reformation by keeping a low profile and displaying loyalty to the State, their new co-religionists brought with them an equally sharp sense of their past in which an ingrained suspicion of British government was seen as a badge of belief. From the start of the century, therefore, until the more settled period after the restoration of the hierarchy in 1878, the two streams of old and new co-existed in a potentially uneasy alliance in which strategies and leadership were liable to strains and challenges, not only from the new social and economic developments, but also from national differences. Despite the overwhelming superiority in numbers it never became an entirely 'Irish' church. What prevented this and instead created a new amalgam was, first, the continuing existence of an indigenous Catholic group which remained a significant, if necessarily small, element in its

Figure 7.1 Dufftown Roman Catholic chapel, 1824. Courtesy of the Scottish Catholic Archives, GB 0240 Photographic Collection.

expansion and, second, the fact that clerical control remained firmly in Scottish hands for most of the period. Complaints as late as the 1860s from the Irish Catholics about the favourable treatment being given to Scots Catholics, lay as well as clerical, show that the latter's presence and influence in the growing urban population should not be ignored. Only from the later 1840s and the 1850s did priests from Ireland begin to play a more continuous and settled part in building up the Church's expansion in the new industrial areas. The bulk of the priests serving in Clydeside in this period came from the old Catholic areas of the north. The bishops and senior clergy who controlled the course of Catholic life in Scotland (with exceptions in Glasgow like the English aristocrat Charles Petre Eyre, or John Maguire, the son of immigrants but very much schooled in the Eyre mould of Britishness) continued to come from the old traditional north-eastern and Highland areas until well into the twentieth century. As late as 1963 the major see of Glasgow was occupied by a Gaelic speaker from Invernesshire, Archbishop Donald Campbell (1945–63); that of Edinburgh was occupied by Cardinal Gordon Gray (1951–93), born in Leith, but with a Banffshire family background.[1]

As the countryside was rationalised by the shedding of less cost-efficient labour, and as towns sucked in workers to occupy the jobs being created, first in textiles and then, after 1830, increasingly in heavy industries, the problem facing the Catholic community for much of the rest of the century was how to keep up with a constantly changing situation. Nor did it help that the country's infrastructure of housing, schools, urban amenities, etc., was also struggling to keep up with demand, resulting in a tradition of poor housing and disease which bore down particularly hard on the lowest ranks wherein most Catholics were situated. The main challenge facing Catholics, therefore, from the outset was how to achieve an ordered parochial life in which the exigencies of daily existence would benefit from the light of religious truth and practice. The size of the task is shown by the following figures. Between 1800 and 1851 there was a staggering five-fold increase in the number of Catholics: numbers doubled to around 70,000 in 1827 (with 25,000 of these now in the Glasgow region), then more than doubled again by 1851 to 146,000 (the majority of these, 100,000, again in the Western District). However, the number of chapels barely doubled in this period, from 40 in 1800 to 86 by 1851, with the majority of them located in the old areas of the north-east and the Highlands. Similarly, the number of priests to serve these enormous numbers grew only by a factor of two and a half, from 45 in 1800 to 118 by 1851, with the great majority of these again in the north and north-west. Inevitably, the greatest pressures were felt in the expanding towns. In 1808 the priest in charge of Glasgow counted some 4,000 Catholics there and in Paisley, served by himself and only one other priest. By 1838, when there were at least 43,000–44,000 Catholics in Glasgow and its environs, the number of priests to serve them had grown to a mere five. At the same time, Edinburgh's 14,000 Catholics were being served by only four priests. Given this and the existence of only two chapels in each city, the fact that each was well attended says something about the religious faith of this new, constantly shifting population. On the other side of the ledger, it also accounts for the inevitable leakages.[2]

The history of the Catholic community, however, was not simply a reflection of what was going on in Glasgow. In 1827 the old Highland and Lowland Vicariates were divided into a Northern, Eastern and Western District, each under its own vicar-apostolic. The Catholic life in each reflected the strongly regional character of the nation. The Northern District included all the area north from Dornie in Kintail through the eastern Grampians to Aberdeen. Here the pace of economic change allowed for a more gradual adjustment. This District witnessed a period of steady growth from an already secure foundation of 14 chapels in 1800 to 26 by 1851, with a similar increase in priests from 15 to 27. This was an area where traditional relationships remained good (as exemplified by the esteem in which the veteran Aberdeen priest, Charles Gordon [1795–1855] was held by all – his biography written by an admiring Presbyterian). It was able to export priests to supply shortages in Dundee and Glasgow, to expand to meet new demands in the fishing ports, and to build the imposing St Mary's (later the cathedral church) in central Aberdeen.[3] Development in the Eastern District was more fraught – a struggle to obtain priests and build churches and schools to serve the growing numbers of Irish Catholics flooding into the linen and jute mills of Dundee. Edinburgh with its 14,000 Catholics by the 1830s was under pressure too. However, in Bishop James Gillis (1838–64) it had a flamboyant Scot who was willing to take a chance on the future, introducing orders of nuns to expand education, providing welfare schemes for his flock, and engaging in literary controversy with his fellow citizens and co-operating with them in cross-denominational educational ventures like the United Industrial School.[4] The Western District was the most complex. In its Gaelic northern section, covering the Hebrides, Moidart, Arisaig, Knoydart and the Great Glen, the period before the Famine of the mid-1840s to 1850s was one of apparent quiet consolidation which masked a mounting subsistence crisis resulting in both depopulation in some areas and congestion in others. Its 14 chapels and priests remained static up to 1850. Emigrants increasingly made their way south or overseas to Canada or Australia. The Knoydart mission was affected worst of all by the Clearances, its Catholic population reduced from around 1,000 in 1851 to about 70 by 1854, when it was merged with Morar. The area from Ayrshire down to Wigtownshire, with its contrasting economy of small farms, small-scale fisheries and scattered textile, mining and iron villages made great demands on its clergy in terms of the distances to be covered and the poverty and transient nature of its newcomers. By the 1840s only Ayr, Dalry, Kilmarnock, Newton Stewart and Campbeltown had chapels.[5]

It was in the Glasgow and Clyde area and the hinterlands of Lanarkshire with its burgeoning mines and ironworks, however, that the greatest strains were being felt. In Lanarkshire permanent chapels only began to be established in the 1840s in Airdrie, Coatbridge and Hamilton,

Figure 7.2 Portrait of Charles Petre Eyre, archbishop of Glasgow (1878–1902). Courtesy of SCA, GB 0240 Photographic Collection.

the last-named mission covering a vast area from Strathaven to Holytown. Before then priests from Glasgow had travelled out periodically to say Mass in rented premises. Successive vicars-apostolic had to judge how far the present situation would eventually stabilise or whether existing pockets of activity would disappear through emigration or further relocation to where new industrial development offered a chance of a living.[6] Bishops like Andrew Scott (1832–45) and John Murdoch (1845–65) tried to develop a strategy of tentative consolidation, encouraging their flocks to be good citizens, gradually bringing in Irish priests where they thought the local community could support them, and discouraging the diversion of resources and efforts to political movements led by Daniel O'Connell. Scott had some excuse for his anxiety. He had been warned that unless he reined in the political activists in his flock, Glasgow industrialists might withdraw their support for his community ventures and for the Glasgow Catholic Schools Society which they had funded since 1817. Murdoch was equally suspicious of such public activity, expressing the hope in 1848 that the Young Irelanders 'would get a skinfull of bullets'! Although he received stalwart support from legendary Irish priests like Michael Condon, he was reluctant to rely too heavily on them regarding them as lacking in commitment to the Scottish Mission. Most Catholics remained supportive of their religious leaders but they wanted a greater commitment to their future and were unwilling to drop their support for social and political movements that promised some improvement in their condition. Some found outlets in lay initiatives like temperance movements, a development which managed to upset the more even tenor of Edinburgh Catholicism when an indignant Bishop Gillis found himself being lectured by members of his flock.[7]

The 1850s marked a turning point, making it essential that these tensions be brought to some sort of a resolution. The thousands fleeing the Irish potato famine after 1845 not only brought the numbers in existing centres to breaking point, but increased the Catholic presence further through the central belt into Stirlingshire, the Lothians, across the Forth and into the Borders. At the same time the restoration of the English hierarchy in 1850 encouraged a new mood of militant Protestantism. The easier relationships of the early century faded. There was a sharper edge now to the social, institutional and political activities which affected the Catholic community. Bishop Murdoch and his few priests found themselves ministering to the sick and dying in coffin-like closes and even in the open-air as famine fever struck their flocks.[8] The growing number of Irish priests now found themselves under greater pressure to establish new missions. Claims began to arise that once these had been put on a good footing they were then transferred to Scottish clergy, so that 'where Paddy sows, Sandy reaps'. The Western District became riven by dissension by the 1860s. In 1864 a number of Irish priests petitioned Rome for an Irish bishop, and in 1866 James Lynch, rector of the Irish College in Paris was appointed co-adjutor vicar-apostolic to the Western District with right of succession to Bishop Gray. The Scottish clergy protested that this was delivering the Scottish Catholic community (the vast bulk of them anyway) into the hands of Fenian rebels. This was a strange claim, given the antipathy of Lynch and his mentor, Cardinal Cullen, to Irish revolutionary secret societies. Rome sent Archbishop Manning of Westminster to investigate. He in turn recommended that a neutral outsider be brought in to reconcile the two sides, and in 1869 Charles Petre Eyre, vicar-general of Hexham and Newcastle, was made apostolic administrator of the Western District, all this obviously as a preliminary to the restoration of a territorial hierarchy in Scotland which was finally realised in 1878.

There was more to this episode than national and political differences and petty squabbling. The Irish had perhaps become over-sensitised by the strains of expansion in a strange land over the past decades, but the native Scottish clergy showed that they too could exchange insults and stoke up resentments. These quarrels were in fact a sign of vibrancy and growth and an attempt to adjust more effectively to the nineteenth-century world of industrialisation, urbanisation and modernisation.[9] Despite the trauma of the famine, the political structures which Roman Catholics had developed for their social progress were still vibrant enough to expand in the 1860s and 1870s. Priests like the Rev. Peter Forbes and leading lay Catholics like Robert Monteith of Carstairs had been instrumental in introducing new orders to help build up the educational and social welfare institutions of the community – the Marists in 1858 and the Jesuits in 1859, as well as Franciscan nuns in 1847 and Daughters of Charity in 1860. Parish missions to inspire the faithful to perseverance and more frequent reception of the sacraments became a feature from the 1850s. What the more far-seeing priests and the Irish memorialists wanted was a commitment to the future through greater devotional enrichment and parochial development. There was truth in the

Figure 7.3 Motherwell Catholic Boys' Brigade, Company I, December 1915. Courtesy of SCA, GB 0240 Photographic Collection.

charge that the vicars-apostolic were too cautious and unwilling to change the way things had always been done. Most of them had been good appointments and they had all continued to work hard like ordinary missioners after their appointment. But there was very little national co-ordination, and decisions taken by their small coterie, often linked by family ties, left them open to criticism. Nor were they now running a small concern as in 1800. The *Catholic Directories* of the period show that there was the beginnings of a 'lace curtain' as opposed to a 'shanty' element among their flock. One of the lay Irish complainers in the 1860s was bothered not about the nationality of the upper clergy, but about the lack of business accountability in handling the quite substantial loans he had been making for the upkeep of the churches in the Western District. Sometimes church services would be adorned by visiting performers from Italy with renderings of motets and baroque masses. Eyre, from his background, was able to inaugurate a period of consolidation bringing the Church more into the mainstream of the English-speaking Roman Catholic world. The vicars-apostolic had derived their powers from Rome; the occupants of the territorial sees after 1878 were able to act more on their own authority which in turn meant that the parishes they were responsible for enjoyed a more uniform treatment. Now Edinburgh was turned into an archbishopric with suffragan sees in Dunkeld (Dundee), Aberdeen, Argyll and Galloway, with an archbishopric also for the extensive see of Glasgow (eventually to have its own suffragans in 1947 in Paisley and Motherwell). Probably too many dioceses had been formed for they were not all equal in resources, but it did make sense to recognise the distinct nature of Gaelic Catholicism in the west and separate the south-west region from Glasgow.

One of Eyre's major acts was to establish a diocesan seminary in 1874 to ensure a supply of priests suitable for the needs of his district, and indeed the ensuing period saw a remarkable flourishing not only in parochial life and in spirituality, but also in the communal organisations which raised the self-esteem of the by now more settled and growing Catholic communities in Scotland. In 1878 they numbered about 330,000, two-thirds of them in the archdiocese of Glasgow, and had grown to about 513,000 by the eve of World War I, making them about 11 per cent of the total Scottish population. With more churches and priests, attendances at the obligatory Sunday Mass became more regular. A liturgical sequence of May Devotions, June Sacred Heart Devotions, and remembrances of the Holy Souls in November helped to bond local communities and their priests. A glance at the *Catholic Directories* in the 1890s and 1900s reveals an array of guilds for young men and women, of St Vincent de Paul welfare societies, Catholic Young Men's Societies and branches of the temperance League of the Cross (particularly dear to Eyre and to Manning, who took a keen interest in what was going on in Scotland). By 1890, St Patrick's in Edinburgh's Cowgate, for instance, not only had the usual array of men's and women's devotional societies, but also a yearly Sick and Burial Society for the families administered by its members, a Penny Savings Bank, a Total Abstinence Society with about 5,000 members, and a library and reading room containing about 2,600 volumes

Figure 7.4 Scotus Academy, Edinburgh, 1952. Courtesy of SCA, GB 0240 Photographic Collection.

'largely taken advantage of '. A Catholic Choral Society was re-instituted in Glasgow in 1897. Parochial summer outings to the countryside or coast, organised games, etc., were all displays of what the community could achieve.[10]

Of course, some of this was inward-looking and unadventurous, an effort to protect their flocks from the contamination of the world around them or to prevent proselytism and leakage. And efforts to promote temperance were needed in a community where drink was, and has remained, a cancer. Not all of those born into the group remained within it or could withstand the pressures to conform to the atmosphere of clerical control which much of this culture implied. It was a culture which can also be criticised for stifling independence of thought and limiting action in the industrial and political spheres to what was cautiously acceptable to the parish priest, who could often be a fearsome figure of authority.[11] However, it could also be argued that efforts to promote domestic virtue were essential in a community too often bounded by the dark horizons of slum housing and the daily threats to life from underground mining and blast furnaces. It was a Catholic culture which, despite the presence of men like the third Marquess of Bute, lacked the expansiveness provided in England by a wider recusant base and the intellectual force of the Oxford converts. However, in other respects it was very much the Catholicism typical of the rest of the English-speaking world. It resembled the deferential clericalism of Ireland, but the sacrifices made by their pastors gave them a status and respect not often found in Catholic Europe. It was a context which produced exceptional individuals like John Wheatley and James Connolly. Wheatley's career was nurtured in St Bridget's parish, Baillieston. Burned in effigy in 1912 when he went outside the community to challenge his parish priest over Labour politics, the ostracism which would have been permanent in Ireland died down once the initial flare-up had subsided. And for all the charges of narrowness and anti-intellectualism which could be levelled against this Catholic culture, it did have a cohesiveness and a social ethos often in advance of that found elsewhere in Britain. Church leaders such as Archbishops Eyre and Maguire in Glasgow, or MacDonald in Edinburgh, were willing to see Catholics playing an active part in the public life of the country and working for the good of all its people.[12]

It should be remembered, of course, that much of this was typical only in the industrial areas, reaching its fullest expression in cities like Glasgow and Dundee. In the smaller Catholic communities of the north-east and west, devotional life was more restrained and the parochial atmosphere less intense, partly due to their more scattered nature and also because of their more traditional lifestyles and culture. Nor were the industrial areas as monolithic in their parochial culture as is often imagined. In the 1890s new groups provided further variety to the Catholic community as Lithuanians and Italians began to settle in Scotland in some numbers. Though never numerically preponderant, the Lithuanians nevertheless played a significant part in building up the parochial structure in the mining areas of central

Lanarkshire and north Ayrshire, while the Italians developed as a community of small businessmen, principally in the catering trades, throughout the coastal and inland towns. Both groups had to struggle to maintain their identities and the Lithuanians were not initially welcomed when competing with Irish workers for jobs. The Italians, in particular, with their individualistic business interests, shared little in the educational, economic and political aspirations of the Scoto-Irish in the central belt. All of these aspects have to be kept in mind when assessing the identity of Catholics in modern Scotland.[13]

The cultural, institutional and political influences which prevailed in Scotland meant, however, that the Catholic community had to develop within a quite distinctive national context. An example of this was the Scottish Education Act of 1872 which made school attendance compulsory for all. This forced Scottish Catholics to act in a way unimagined elsewhere in the British Isles. Largely because their educational presence and permanence was still problematic in the 1860s, the 1872 Act had made very little provision for a group which was to become by the 1880s and 1890s a major element in Scottish life, particularly in its industrial heartland. Because there were no guarantees of religious freedom to satisfy them, the bishops decided to maintain their own schools, a policy which placed tremendous burdens on Catholics (despite government grants) since they had to support their own schools as well as pay rates for the public board schools. This had two consequences. They had to organise their voting power, utilising the proportional voting system set up in 1872, to ensure the return of Catholic representatives in order to have a say in the policy of the local school boards. This tied in with their growing participation in Scottish politics especially when the parliamentary franchise was extended in 1884–5. Secondly and more directly, it led to the development of a large Catholic educational sector which further reinforced their identity and group pride.

Developing schools that met national standards proved a hard task which had to be tackled principally in the towns. In the Highlands and north-east, school boards proved accommodating in catering for the religious needs of their smaller Catholic population, but this was not the case in Glasgow or Edinburgh or in the smaller industrial towns. Three-quarters of all Catholic schoolchildren were located in the relatively poor Catholic community of Glasgow. Children tended to leave early, attendance was irregular, there was an over-reliance on pupil-teachers and, despite heroic efforts by the religious orders, there was always a deficiency of college-trained staff. There was a constant financial struggle to keep up with the higher standards in buildings and curriculum increasingly demanded by the State, particularly when a post-elementary provision was required by the raising of the leaving age to fourteen years in 1901. Much of the material and emotional resources in Catholic parishes were invested in this educational crusade, and any sign of improvement or praise from the Scotch Education Department's inspectors was seized on and savoured. Notwithstanding the enormous barriers facing a community which was economically at the foot of the socio-

Figure 7.5 The building of Oban Cathedral, 1932. Courtesy of SCA, GB 0240 Photographic Collection.

economic scale, the years up to 1914 did see the creation of a working system of elementary education which represented a minor social revolution when the lifestyles of the 1870s and 1900s are compared. Between 1872 and 1918 the number of schools had grown from 65 to 226 and the attendance had gone from a pitiful 12,000 to around 94,000, equivalent to the national norm. In 1894 Dowanhill College for the training of teachers had been set up and was beginning to bear fruit in the higher quality of women teachers emerging. (A second training college was established at Craiglockhart in Edinburgh in 1920.) Already before 1918, informal soundings were being undertaken to see how the Catholic schools could be brought within the national system. The Education Act of 1918 transferred the upkeep of their schools and teachers' salaries to the State, the teachers appointed having to be approved by the Church authorities as to religious belief and character. In one sense, these guarantees for Catholic schools as part of the public system constituted an acknowledgement of their place in modern Scotland. The feeling that the 1918 Act had brought the by now largely Scoto-Irish Catholic community into the life of the nation while preserving its ethos lies behind its defensive reactions in the twentieth century to any proposals to change what is seen as something of a constitutional symbol. It is worth noting, too, that the bargain was not all on the one side. The bishops lost direct control of their schools and had to trust that the public authorities would act generously in response.[14]

Catholics were also becoming more involved in modern Scottish life through politics. Here again there were variations. In Edinburgh and among the Scottish aristocracy there was a strong tradition of Conservatism tinged with Scottish sentiment. Northerners and Highlanders were as likely to follow a good local personality as a party, the second only becoming a well-drilled political interest group during the 'Crofters' War' of the 1880s and 1890s. It was in the main centres of industrial Scotland that Catholics were most united in support of the Irish Home Rule movement. Through this they forged links with the dominant Liberal party and by the 1880s they were a formidable, radical force whose support had to be courted. Generally they followed Gladstone, but their support was always conditional as in 1885 when they voted against Liberal candidates on Parnell's instructions to bring pressure to bear on Gladstone to give ground to Irish demands. Although this pattern, and the growing self-confidence it engendered by the consideration given to them by the Liberals, continued up to World War I, urban Catholics were beginning to share with their fellow Scots not only the same accents but the same interests in jobs, wage rates and living conditions. Home Rule always had more than a constitutional aspect for them: it was also seen as something which held a general promise of social progress. It is no surprise, therefore, that in the years up to 1914 increasing contacts were made with the growing Labour movement. The leader of the Glasgow Irish, John Ferguson, and his friend and ally, the Irish Land Reformer Michael Davitt, were very influential in the direction Catholic politics would take. Ferguson was a founding member with Keir Hardie of the first ever Labour Party set up in Britain, the 1888 Scottish Labour Party. Davitt was an important figure in the contemporary crofters' struggle for land rights. While the official nationalist line was to condemn these as deviations harmful to the main Irish-Catholic goals, economic interests in Scotland were always pulling Catholics in other, leftward directions. In the 1890s and 1900s there was a lively debate within Catholic circles on social issues anticipating much that was to be found in Leo XIII's 1891 encyclical, *Rerum Novarum*. There was a growing trade union activity among Catholic workers in the west of Scotland before 1914; John Wheatley, a faithful Catholic from the east end of Glasgow who was to become a minister in the first Labour government, was moving out of his base in local Irish politics to try to persuade his co-religionists that Catholicism and socialism could be compatible. Despite much clerical opposition, the bishops kept a low profile in all this, their only overt political act being to continue the ban on the Ancient Order of Hibernians as a secret subversive society up to 1909. These moves to the left may have represented minority interests, but collectively they do indicate that the Catholic vote in the west of Scotland already had the potential to go to Labour once the franchise had been extended in 1918 and Ireland began to fade from the British political scene. What was happening here was a transition from the mid-century exclusive inner parochial concerns of Catholics to involvement with the wider political and class movements emerging in the surrounding society.[15]

The two world wars of the twentieth century were defining moments for Catholics as for others, shaking people out of their old certainties and undermining the economic forces which had hitherto been so powerful in shaping their community. The second war further enriched the mixture of Scottish Catholicism by introducing a large number of Polish soldiers who had escaped here to fight against Bolshevism and Nazism. Before 1914 the bulk of Catholics had lived restricted lives but in an atmosphere of economic expansion and opportunity. In the 1920s and 1930s they were hit by prolonged economic depression, unemployment, low wages and lack of opportunity. The worst effects of this economic stagnation were most strongly felt in those areas where Catholics were most numerous: mining villages and steel towns, shipbuilding and heavy engineering centres. Housing and health conditions, which had always been bad in central Scotland, worsened. There was a change from the confidence of the nineteenth century. Numbers of Catholics remained high at 607,900 in 1931 (over 12 per cent of the Scottish population), with 450,000 of them in the Glasgow diocese, but many of them sought relief in emigration. In the Glasgow area, a strenuous effort was made by the new archbishop, Donald Mackintosh, who had spent most of his life as a professor at the Scots College in Rome, to establish a sounder financial basis, but one gets the impression that this was a period of marking time in parochial life.[16] There were challenges, too, in the shape of a sustained attack on the Catholic presence in the west of Scotland from within the Church of Scotland, summarised in its 1923 report entitled *The Menace of the Irish Race to Our Scottish Nationality*. This was due to a number of factors: the febrile post-war atmosphere; the sudden collapse of industrial security and the progressive consensus which had sustained it; the 1918 Act recognising Catholic educational claims; and the sudden emergence of a strong Labour Party among the newly enfranchised Catholic

Figure 7.6 Pilgrims at Carfin Grotto, c1947. Courtesy of SCA, GB 0240 Photographic Collection.

masses in the west challenging the status quo. It should also be said that the report was the work of a party group which had lost its influence within the Church of Scotland by 1939 and was no means representative of mainstream Presbyterian opinion, some of which spoke out bravely against it. But it was accompanied in the 1930s by outbreaks of political anti-Catholicism at town council elections led by demagogues like Alexander Ratcliffe in Glasgow and John Cormack in Edinburgh. Orange and green factionalism had always been a feature of industrial Scottish life but what occurred now – the serious riots of 1935 in Edinburgh when a civic reception for the Catholic Young Men's Society was disrupted and organised mobs attacked the Catholic Eucharistic Congress – was something new.[17]

The technicolour nature of these events should not obscure the fact that the real problems Catholics had to deal with in the twentieth century were the changing political and scientific perspectives constantly being opened up both nationally and internationally. The rise of dictators and the failures of capitalism in the early 1930s, when up to half of the men in a Clydeside town could be on the dole, meant a rethinking of the fundamentals of the faith and everyday life and conditions. In a superficial sense the response to this might be traced in the support that the bulk of Catholics gave to the emerging Labour Party but, insofar as this represented a Catholic commitment as opposed to a general class interest, the reality is more complex. The coalition of interests that are political parties meant that Catholics had to choose from a range of policies none of which might be totally satisfactory in a religious sense. Bodies like the Catholic Union tended instead to concentrate on specific Catholic interests where voting pressure could be brought to bear, as in the *ad hoc* Education Authorities while these lasted up until 1929. Otherwise relations between the Church and Labour remained tentative, with the former deeply suspicious of its secularist social engineering strand or its support for left-wing anti-clerical parties in Europe. Contrary to popular belief, until the 1960s Catholic representation at Westminster or in local government was meagre in proportion to its electoral support.[18] More effort was made in challenging Communist influence in industry through encouraging members of parochial groups like the Catholic Young Men's Society to play a more active role in their trade unions. The Catholic Truth Society and the Catholic Social Guild engaged in study sessions on how the Church's social teaching might be implemented. These meetings, and visits by speakers such as Frank Sheed or Vincent McNabb, provided Catholics with some idea of the wider world of European Christian Democracy and the liturgical basis needed to sustain it. Although many might be untouched by these developments, the Dominican Anthony Ross recalls in his memoirs how unemployed miners in the village of Whiteriggs, near Airdrie, were able to argue cogently about politics and economics and Catholic social teaching. The demand for graduates to staff Catholic schools increased the number of Catholics in the Scottish universities to over 500 by the 1930s. This created another forum for the widening of social and political horizons under the leadership of Catholic university chaplains like W E Brown at Glasgow, or the Dominicans at

Edinburgh, where the archbishop Joseph McDonald was open to these new initiatives. There were tensions, however, within these developments: the ambivalence of the Church to authoritarian regimes in Italy and Spain could lead to heated disagreements and alienation.[19]

There was light and shade too in the ethos of the Catholic community up to the middle 1950s. There is no doubt of the extent of spirituality. Numbers in the churches on weekdays as well as Sundays, the range of devotional exercises so well attended, the increasing reception of Holy Communion in accordance with the precepts of Pius X, all suggest that there was a deep prayer life among many ordinary people. Something of this spirit can be sensed in the erection of the remarkable Marian shrine at Carfin, which was built by unemployed miners in 1923, and the growing numbers who visited it on pilgrimage. It can be seen, too, in the saintly life of Margaret Sinclair (1900–25), an ordinary working girl in central Edinburgh whose cause for beatification was begun in 1952 and who was declared Venerable in 1978. There was a great burst of activity in the founding of new parishes after 1945, and in this period the total of priests and vocations reached their peak. There was, however, an awareness of a steady leakage in spite of all the efforts to stem it. The links between priests and community through the schools lessened, especially with the expansion by the 1950s and 1960s of secondary schools covering wide areas. The movement of population to new housing schemes and jobs, and the development of a range of

Figure 7.7 Margaret Sinclair. Courtesy of SCA, GB 0240 Photographic Collection.

social welfare services after 1945 reduced the comprehensiveness of the old parochial structure. Some religious practices tended towards formalism with a stress on legalism rather than the deepening of Christian formation, and developments in religious teaching depending on the old penny catechism could not compare with the secular learning available in the cinema and press as well as the school. A clerical style which was often authoritarian and discouraging of lay initiatives, sat uneasily with the expectations raised by the better times which came after the ending of World War II.[20]

A more hopeful side to this emerged in a greater willingness to assess these trends. Patrick MacGill had already broken the ground in his explorations of the dark side of immigrant Catholic life in novels like *Children of the Dead End* (1914) and *The Rat Pit* (1915). Local writer Edward Gaitens in *The Dance of the Apprentices* (1948), and playwright Paul Vincent Carroll, a founding member of James Bridie's Citizens' Theatre, followed similar paths. The Marian Players, a dramatic group from Garnethill, contributed to the development of this world of the imagination. Convert writers like Bruce Marshall, Compton Mackenzie and George Scott-Moncrieff encouraged a fresh look at the Scottish condition. Some like Carroll and Gaitens may be regarded as minor voices today but they were attempting to make sense of the world they saw around them and their work, for all its imperfections, provides sharp insights into contemporary reality as well as being infused throughout with a sense of hope. The early novels of A J Cronin, too, often go beyond the stereotypes of popular Catholicity in examining the mixed Catholic–Protestant inheritance which has been the experience of a significant number in the west of Scotland more often than is acknowledged. Further efforts in creating a new intellectual base came in 1950 with the Jesuit-inspired *Mercat Cross* and the *Innes Review*, the latter the product of a remarkable band of young historians, clerical and lay, of the Scottish Catholic Historical Association.[21]

In retrospect, one can see this as a process striving, in common with other parts of the universal Church, for a new balance in the modern world, a series of tensions between old and new which found its realisation in the vision provided by the Second Vatican Council (1962–5). Conclusions drawn in the 1970s or 1980s as to the latter's effects on Scottish Catholics keep changing. It is obvious, however, that the institutional Church in Scotland was unprepared for these new directions. The attitude of the Scottish bishops seems to have been one of passivity, with their first efforts directed more at controlling the extent of change by trying to keep a tight rein on the news emerging from the Council than in a positive response to the proposed changes. There were sharp differences of opinion, for instance, between Archbishop Scanlan and the expanding university constituency of his diocese over how its university chaplaincy should be directed.[22] Elsewhere, extremes hit the headlines, one wing retreating to Latin Mass Societies, the other not only throwing out the old missals but the Church statues of ancient hallowed association as well. The main liturgical reforms of the vernacular Mass celebrated facing the people were patchily implemented. These upsets

Figure 7.8 The Scottish Hierarchy after the Centenary Mass for the consecration of Archbishop John Strain by Pope Pius IX. Basilica of San Lorenzo fuori le Mure, 25 September 1964. Courtesy of SCA, GB 0240 Photographic Collection.

seemed to be accompanied not only by a lessening of lay observance, but also by defections from the priesthood. In time, however, younger and more forward-looking and enterprising bishops have launched a series of pastoral initiatives aimed at explaining the changes in heart that the liturgical changes imply. Today, even parishioners of the older generation would be loath to go back to the old ways, while those in their twenties can scarcely credit that services were conducted in Latin as recently as the 1960s. Inter-Church relations have improved as well. Archbishop Thomas Winning, for example, addressed the General Assembly of the Church of Scotland in 1975, while the visit by Pope John Paul II to Scotland in 1982 generated much good will on all sides.[23] At parish level too there have been interdenominational efforts to reach out to the surrounding society. Catholics have been taught to measure themselves not so much as an élite but as a group with a responsibility to all their fellow men and women. The reaction to Cardinal Winning's death suggests that there has been a fundamental shift in the position of Catholics in a Scotland which is currently experiencing radical changes in its institutions and culture.

Some of this has come from the educational opportunities which have expanded since the 1970s allowing Catholics to become, like the rest of Scotland, a more varied society with an increasingly affluent, professional middle class. With around 14 per cent of the population and probably the largest Christian grouping they are now a substantial element in contemporary Scottish life.[24] Their contribution to the arts has also been significant – in musical life, for instance, through composers such as Thomas Wilson and James MacMillan. It is difficult for the historian to judge the significance of these trends. The immediate situation is one of flux and perhaps more the province of the sociologist or psychologist. Still, a knowledge of history provides some sense of perspective in which a few tentative observations might be made. One obvious question is whether the material advance of Catholics in Scotland has been matched by a similar development in their intellectual and spiritual vitality. Insofar as this can be measured objectively, in one sense the answer is yes. Catholic congregations today tend to be made up of outward-looking and deeply committed Christians. Faith for them consists not only in external worship but in Scripture-based activities on behalf of the local, the national and international community. However, the numbers attending the central act of their faith, the Sunday Eucharist, are, at around 33 per cent, back to the levels of the 1830s when there were few chapels to serve them and numbers outstripped resources.[25] Greater lay participation has sometimes resulted more from the shortage of priests than any positive vision of how a more fruitful co-operation between priests and laity might be constructed. There is still a feeling that often the first time Catholics hear of a problem is when they read about it in the secular press. Some high-profile cases of abuse which have afflicted the Catholic Church, in common with other welfare institutions throughout Britain, as well as the defection of a bishop, might have been better handled with a more

Figure 7.9 Ticket for Papal Mass, Bellahouston Park, Glasgow, 1 June 1982. Courtesy of SCA, GB 0240 GD 19/.

open and trusting approach. The Catholic educational system has made great advances but the quality of religious formation among its products is still questionable. Many seem to possess little sense or knowledge of the Church's past and are unaware of the richness of the heritage which has been passed on to them. Also, with greater affluence and social position, there is sometimes associated a reassertion of Irishness outmoded even in today's modern Ireland. While there has been widespread discussion of religious bigotry against Catholics in recent years, more thought needs to be given to what the Catholic community itself might do to lessen this by the leadership and example it could give to others. By making a greater input into contemporary discussions on the tensions affecting Scottish society, for instance as regards violence in sport, it would then be facing up to the changed role it now plays in the public life of Scotland, and the responsibilities which go with this.

On the other hand, many of the problems lamented today are ones which have been experienced by Catholics in Scotland in the nineteenth century. A chart of the numbers of priests would show that they reached their highest numbers between 1900 and 1960.[26] These were the days when parishes consisted of four or five clergy on a constant round of home visits. But this 'golden age' has not been typical of the Catholic experience over the last two hundred years. For most of this time resources have been stretched, and today's parishes of one solitary priest are in some ways a return to the early days. The challenges of the secularism and hedonism of contemporary society and the amorality of a supposedly neutral biomedical science have taken the place of the despair and materialism of the industrial society of the nineteenth century. The decline in vocations can be traced to the later 1950s before the period of change inaugurated by the Second Vatican Council. There were deficiencies, too, in the nineteenth century which had to be made good by priests from Ireland or Belgium and Holland. The Catholic community which we see today with all its diversities, its resolved and unresolved problems, its achievements and failures, is a living community which has been formed by the history of the past two centuries in Scotland and has also been part of the making of that history.

NOTES

1. For an overall view see Handley, 1945 and 1947; Darragh, 1986; McCaffrey, J F. Roman Catholics in Scotland in the nineteenth and twentieth centuries, *RSCHS*, 21 (1983), 275–300.
2. Darragh, J. The Catholic population of Scotland since the year 1680, *IR*, 4 (1953), 50, 56–8; Forbes, F and Anderson, W J. Clergy lists of the Highland District, 1732–1828, *IR*, 17 (1966), 129–84; Johnson, C. Secular clergy of the Lowland District 1732–1829, *IR*, 34 (1983), 66–87; Johnson, C. Scottish secular clergy, 1830–1878: the Northern and Eastern Districts: the Western District, *IR*, 40 (1989), 24–68, 106–52; *Royal Commission on Religious Instruction Scotland*, PP, 1837–8, vol. 21, 174, vol. 32, 92–5.
3. Johnson, 1989a, 25; Stark, J. *Priest Gordon of Aberdeen*, Aberdeen, 1909.
4. Johnson, 1989a, 26–7; Anson, 1970, 253–7, 265–71, 278–87; Mackie, P. The foundation of the United Industrial School, *IR*, 39 (1988), 133–50.
5. Johnson, 1989b, 106–9.

6. Mitchell, M J. The establishment and early years of the Hamilton Mission. In Devine, 1995, 1–30.
7. Mitchell, 1998, 127–9, 256; Aspinwall, B. The Irish abroad: Michael Condon in Scotland, 1845–1878. In Sheils, W J and Wood, D, eds. *The Churches, Ireland and the Irish*, (Studies in Church History, volume 25), Oxford, 1989, 279–97; McCaffrey, J F. The stewardship of resources: Financial strategies of Roman Catholics in the Glasgow district, 1800–70. In Sheils, W J and Wood, D, eds. *The Church and Wealth* (Studies in Church History, volume 24), Oxford, 1987, 359–70; McCaffrey, J F. Irish immigrants and radical movements in the west of Scotland in the early nineteenth century, *IR*, 39 (1988), 49, 53, 60.
8. McCaffrey, J F. Reactions in Scotland to the Irish Famine. In Brown, S J and Newlands, G, eds. *Scottish Christianity in the Modern World*, Edinburgh, 2000, 155–75.
9. In addition to Handley, 1947, 59–92 and McRoberts, 1979, 11–23, see McCaffrey, 1983, 278–90 and Aspinwall, B. Scottish and Irish clergy ministering to immigrants, 1830–1878, *IR*, 47 (1996), 45–68, for a more up-to-date analysis.
10. Aspinwall, B. The formation of the Catholic community in the west of Scotland, *IR*, 33 (1982), 44–52; Aspinwall, B. Children of the Dead End: The formation of the Archdiocese of Glasgow 1815–1914, *IR*, 43 (1992), 119–44; McCaffrey, J F. The Roman Catholic Church in the 1890s: Retrospect and prospect, *RSCHS*, 25 (1995), 426–41.
11. For these aspects see Walker, W M. Irish immigrants in Scotland: Their priests, politics and parochial life, *Historical Journal*, 15 (1972), 649–67.
12. Wood, I S. John Wheatley, the Irish and the Labour movement in Scotland, *IR*, 31 (1980), 71–85.
13. O'Donnell, E. 'To keep our fathers' faith ...': Lithuanian immigrant religious aspirations and the policy of west of Scotland Catholic clergy, 1889–1914, *IR*, 49 (1998), 168–83; Aspinwall, B. Baptism, marriages and Lithuanians; or 'Ghetto? What ghetto?', *IR*, 51 (2000), 55–67; Colpi, T. The Italian community in Scotland: Senza un campanile?, *IR*, 44 (1993), 153–67; Aspinwall, B. The ties that bind and loose: The Catholic community in Galloway, 1800–1998, *RSCHS*, 29 (1999), 70–106.
14. Treble, J H. The development of Roman Catholic education in Scotland 1878–1978. In McRoberts, 1979, 111–23; Treble, J H. The working of the 1918 Education Act in Glasgow Archdiocese, *IR*, 31 (1980), 27–44; Fitzpatrick, T A. *Catholic Secondary Education in South–West Scotland Before 1972*, Aberdeen, 1986, 25–53.
15. McCaffrey, J F. Politics and the Catholic community since 1878. In McRoberts, 1979, 140–55; Smyth, J J. *Labour in Glasgow 1896–1936: Socialism, suffrage and sectarianism*, East Linton, 2000, 125–45.
16. Gallagher, 1987, 104–28 passim; Ross, A. The development of the Scottish Catholic community 1878–1978. In McRoberts, 1979, 43–50.
17. Brown, S J. 'Outside the Covenant': The Scottish Presbyterian churches and Irish immigration, 1922–1938, *IR*, 42 (1991), 19–45.
18. McCaffrey, 1979, 151–2; Smyth, 2000, 145–54; Knox, W, ed. *Scottish Labour Leaders 1918–39, a Biographical Dictionary*, Edinburgh, 1984, 29–30, calculates that 14.3 per cent of Labour leaders were Catholics as against 30.9 per cent from the Church of Scotland.
19. Ross, A. *The Root of the Matter: Boyhood, Manhood and God*, Edinburgh, 1989, 122, 137–42, 150–1; Gallagher, 1987, 206–22; Fitzpatrick, T A. The Catholic Social Guild: Fr Leo O'Hea, SJ and the west of Scotland connection, *IR*, 50 (1999), 127–38.
20. Ross, in McRoberts, 1979, 50–5; Gourlay, T. Catholic schooling in Scotland since 1918, *IR*, 41 (1990), 122–4.
21. Edwards, O D. The Catholic press in Scotland since the restoration of the Hierarchy. In McRoberts, 1979, 178–82; Reilly, P. Catholics and Scottish literature 1878–1978. In McRoberts, 1979, 183–203.

22. Gallagher, 1987, 279–83.
23. Belton, 2000, 52–8, 61–4, 91–3, 124–42.
24. Some of the implications of this are analysed by the various contributors to Boyle and Lynch, 1998, and Devine, 2000.
25. It is difficult to get a consistent figure for Mass attendance but those available all agree on the decline since the 1960s. The statistician, James Darragh, in The Catholic population of Scotland 1878–1977, in McRoberts, 1979, 220, cites among several surveys a figure of 64 per cent for 1967. The sociologists Iain R Paterson (Devine, 2000, 219) and Steve Bruce (*IR*, 43 (1992), 152) estimate a fall from around two–thirds in 1959 to about one–third in 1997. This is reflected in the figures given by Bishop Devine for the diocese of Motherwell (*A Pilgrim People*, 1998, 125) of a fall from almost 89,000 at Sunday Mass in 1981 to just over 62,000 in 1996.
26. Darragh, 1953, 221, 239.

BIBLIOGRAPHY

Anon. *The Boyhood of a Priest*, London, 1939.
Anson, P F. *The Catholic Church in Modern Scotland*, London, 1937.
Anson, P F. *Underground Catholicism in Scotland 1622–1878*, Montrose, 1970.
Belton, V. *Cardinal Thomas Winning An Authorised Biography*, Dublin, 2000.
Boyle, R and Lynch, P, eds. *Out of the Ghetto? The Catholic Community in Modern Scotland*, Edinburgh, 1998.
Brown, C. *Religion and Society in Scotland since 1707*, Edinburgh, 1997.
Bruce, S, et al. *Sectarianism in Scotland*, Edinburgh, 2004.
Canning, B J. *Irish–Born Secular Priests in Scotland 1829–1979*, Inverness, 1979.
Catholic Directories for Scotland, 1829 to date.
Darragh, J. The Catholic population of Scotland since the year 1680, *IR*, 4 (1953), 49–59.
Darragh, J. *The Catholic Hierarchy of Scotland, a Biographical List, 1653–1985*, Glasgow, 1986.
Devine, J. *A Pilgrim People. Diocese of Motherwell 1948–1998*, Glasgow, 1998.
Devine, T M, ed. *Irish Immigrants and Scottish Society in the Nineteenth and Twentieth Centuries*, Edinburgh, 1991.
Devine, T M, ed. *St. Mary's Hamilton: A Social History 1846–1996*, Edinburgh, 1995.
Devine, T M, ed. *Scotland's Shame? Bigotry and Sectarianism in Modern Scotland*, Edinburgh, 2000.
Gallagher, T. *Glasgow: The Uneasy Peace*, Manchester, 1987.
Handley, J E. *The Irish in Scotland 1798–1845*, Cork, 1945.
Handley, J E. *The Irish in Modern Scotland*, Cork, 1947.
Johnson, C. *Developments in the Roman Catholic Church in Scotland 1789–1829*, Edinburgh, 1983.
Johnson, C. Scottish Secular Clergy, 1830–1878: the northern and eastern districts, *IR*, 40 (1989a), 24–68.
Johnson, C. Scottish Secular Clergy, 1830–1878: the western district, *IR*, 40 (1989b), 49–59.
Johnson, C. *Scottish Secular Clergy 1879–1989*, Edinburgh, 1991.
McRoberts, D, ed. *Modern Scottish Catholicism 1878–1978*, Glasgow, 1979.
McCluskey, R, ed. *The See of Ninian, a History of the Medieval Diocese of Whithorn and the Diocese of Galloway in Modern Times*, Ayr, 1997.
Mathews, E. *Counting My Blessings*, London, 1989.
Mathews, E. *Counting My Joys*, Girvan, 1990.
Mitchell, M J. *The Irish in the West of Scotland 1797–1848. Trade Unions, Strikes and Political Movements*, Edinburgh, 1998.
Turnbull, M T R B. *Cardinal Gordon Joseph Gray*, Edinburgh, 1994.

8 Episcopalians

ALLAN MACLEAN

INTRODUCTION

The Scottish Episcopal Church has a long tradition of religious practice and thought, drawing inspiration from pre-Reformation times, the Reformation era, times of persecution, the influence of Anglicanism of English and Irish varieties, and more recently through ecumenism from other denominations.

The Episcopal Church's adherence to bishops as being essential to a Church's existence and validity is usually considered to be its principal tenet, but the use of a prayer book and a liturgical service has also been regarded as its distinctive characteristic. More recently its alliance with worldwide Anglicanism has become, for some, its *raison d'être*.

1690–1788 TIME OF PERSECUTION

Church Affairs
The re-establishment of episcopacy in the Restoration Settlement of 1661/2 and the toleration of Presbyterianism as a separate ecclesiastical body were reversed in 1688/90. After a time of uncertainty in the relationship between the adherents of the two systems, including the passing of an Act of Parliament associated with the Treaty of Union securing Presbyterianism, and the attempt by the presbytery of Edinburgh to curtail the ministration of an episcopally ordained minister, the Toleration Act of 1712 was intended to stabilise the position. To take advantage of the act, clergy had to be licensed by taking the oaths of allegiance and abjuration, and pray during services for the queen's majesty, the Princess Sophia of Hanover and all her family. This started the split in the Episcopalian community that was to last for a century.

In 1715 the overt Jacobitism of the non-juring Episcopal clergy, who had not qualified under the 1712 act, led to the imposition of penal statutes on these clergy, which included all the Scottish bishops. In 1745/6, the Government was well aware of the continuing connection between the remaining Episcopal congregations and support for the Jacobites. The resulting destruction of the 'meeting houses', whether or not the clergy were Jacobite, the imprisonment of the clergy and the execution of a clergyman who took part in the Rising[1] led to far more stringent penal statutes, affecting for the first time the laity as well as the clergy. Furthermore, from 1748, to qualify the clergy had to have been ordained by an English or Irish bishop. Scottish Episcopal clergy could thus no longer qualify, even if they wished to.

The number of Scottish Episcopal congregations continued to decrease; the remaining congregations being in distinct geographical areas. Some whole Episcopal congregations qualified, but generally congregations split, and a portion was licensed as a 'qualified congregation'. These qualified congregations were known as 'English chapels', which was perhaps the origin of this name being attached to all Episcopal congregations. Despite having no Episcopal oversight, they may have had some form of organisation for in 1760 the qualified clergy of various counties sent an address of congratulation to King George III.[2] The popularity of these congregations was both because the laity wished to free themselves from the taint of Jacobitism, and also because those attending non-qualified congregations could not vote or take up any civil position, be it in parliament, the armed services, the justiciary of the peace, customs and excise, or teaching in a parochial school.

In 1784 some of the bishops of the Episcopal Church, as a free and independent Episcopal Church, consecrated Samuel Seabury as a bishop for the United States of America, and this passed into the folk memory of the Episcopal Church as one of its most significant acts.

Denominational Characteristics
The two ecclesiastical bodies of Episcopalians, non-juror and qualified, often called respectively 'Scottish' and 'English', were quite independent of each other.

The non-juring Scottish Episcopal Church was for most of the century quite clear about its Jacobite leanings. Its principal characteristic was a great interest in liturgical matters, and various versions of the liturgy were developed. The Scottish Liturgy became the focus of great loyalty among the remaining Episcopalians.

In the areas of Episcopalian adherence, including Edinburgh, the congregations represented a cross-section of society. Elsewhere the Episcopalians were often the landed families, but they might well conform to the English congregations or the Established Church, save when an opportunity for Episcopal worship came to them. Bishop Forbes' journals of 1762 and 1770 show the large number who were confirmed on those occasions, not just in the Episcopalian heartlands.[3]

The qualified or English congregations were anxious to be recognised as legal, and not Jacobite. They looked to conformity with the Church of England and used the *Book of Common Prayer, 1662* [BCP].

Theology, Ideology and Thought
The Scottish Episcopal clergy and laity were committed to Jacobitism because to them the Stuart monarchy was a divine gift, which must be accepted, and the element in political theology which distinguished Episcopalians was an almost mystical, sacred, theory of monarchy.[4] This was

Figure 8.1 G W Brownlow, *Prison Baptism at Stonehaven* (1864), depicts the Rev. John Troup performing a baptism through the bars of Stonehaven jail, where he was imprisoned in 1748–9. The attendant fisherfolk were modelled on current members of the congregation. It is now at St Paul's Cathedral, Dundee, but it was reproduced as an oliograph in about 1934 and distributed to all congregations, who displayed it prominently, and it became an 'icon' of the history and integrity of the Episcopal Church.

characterised by a sense of intense loyalty,[5] which showed itself in a general outlook of passive resistance. They wished to be lawful, following the law of the land as far they could, and beyond that suffering from the consequences, but as Bruce Lenman has termed it, 'learning, social order, the aristocratic graces, were all identified with Episcopal religion and Jacobite politics'.[6] In 1715 they rallied to the Chevalier, the Old Pretender, but in 1745 many were more wary. As the hopes of Jacobitism dwindled they transferred their loyalty to the Hanoverians, as *de facto* rulers, and the ideology of the Episcopal clergy gradually moved away from overt Jacobitism, remaining in a theology of resignation, given-ness, and passive resistance. There may be a connection with the Hutchinsonianism that was taken up by many of the clergy.[7]

The reason why various parts of Scotland remained fervently Episcopalian is not clear, though it is sometimes thought to be due to 'intense conservatism'.[8] There was no question of feeling an ecclesiastical responsibility for the whole country (it was difficult enough to find the resources to minister to those who adhered to episcopacy), but the Episcopalians consistently called themselves 'The Church of Scotland'.

Many sermons emphasised the differences between Episcopalians and the 'schismatics', i.e. followers of Presbyterianism. Baptism in this period by a Presbyterian was considered by most Episcopal clergy to be inadequate, and was sometimes administered again, usually conditionally, before confirmation.

In theological thought they were interested in primitive and apostolic authority as a background for their stand on Church government, and they were in no sense Calvinist, finding their inspiration from the Caroline Divines. There was also an extraordinary current of religious mysticism which linked the thought of some north-east Episcopalians, especially Dr George Garden, to the quietism of Madame Guyon and Madame Bourignon.[9]

The qualified congregations were more moderate in their theology, but were similarly not Calvinist, using the Anglican *BCP* as their standard, but, since they did not have any bishops, being reliant on English or Irish bishops, they were in similar circumstances to many overseas Anglican congregations, though the latter came under the jurisdiction of the bishop of London.

Clergy
The last Episcopal incumbent within the Established Church died in 1728,[10] but the bishops, their senior bishop being termed 'Primus' as convenor of the College of Bishops, had continued to ordain clergy and to choose new bishops. However, there was a progressively more serious situation about the lack of men coming forward for ordination. The number of Gaelic-speaking clergy was very small, though there were catechists who ministered in the Highlands.[11]

The Episcopal clergy were the most significant single group of men creating and transmitting articulate Jacobite ideology.[12] The background of the clergy is difficult to establish, though many seem to have come from farming families of the north-east, but Bishop Campbell (1669–1744) and Bishop Rattray (1684–1743) came from landed families. There being no or little pay, as well as penal restrictions, many clergy took up positions as tutors in Episcopalian households. Tristram Clarke refers to 'the unassuming and down-to-earth character which circumstances forced on many of [the] bishops ... Both the bishops and their clergy frequently appear in the records in the character of men of dignified but impoverished good breeding, who gladly eschewed the trappings of the establishment in their Christian witness.'[13]

John Skinner (1721–1807), the poet 'Tullochgorum' admired by Robert Burns, who was the incumbent of Longside, Aberdeenshire for sixty-five years, was the best-known clergyman of the period.

The antecedents of the qualified clergy show that many were from the same farming background as the Episcopal clergy. By the end of the century many Scots were being ordained in England for qualified congregations in

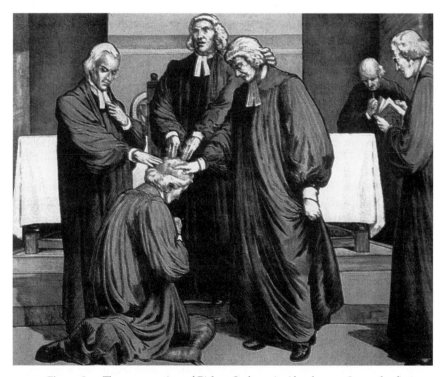

Figure 8.2 The consecration of Bishop Seabury in Aberdeen, 1784, as the first bishop in an independent Episcopal Church in the United States of America is usually considered to be the foundation of the Anglican Communion of Churches. This depiction by Walter J Morgan (1934), like the Stonehaven baptism painting (Fig. 8.1), was distributed to all congregations and became an 'icon' of the Episcopal Church's crucial role in the history of the Anglican Church.

Figure 8.3 Forfar Meeting House. The original chapel in Forfar had been closed in 1746, and this meeting house at 131 High Street was used by the congregation while the Penal Laws were rigorously enforced. Holes were pierced in the 'trance' or lobby for the purpose of enabling those in attendance to take part in the service, while the officiating clergyman stood either in the lobby or on the top of the stair. In 1770 a new chapel was built where 'we have got our principal folks [including the Strathmore family from Glamis Castle] well placed'.

Scotland. James Traill, the son of a Presbyterian minister at Montrose, became bishop of Down and Connor (1765–83) and ordained a Scot, in Scotland, for a qualified congregation.[14]

Worship and Ritual
Initially the services differed little from those of the Presbyterians, except perhaps the introduction of the Lord's Prayer and the ascription. Quite quickly, however, some congregations began to use parts of the *BCP*, and the qualified congregations used the *BCP* from the start. The Scottish Episcopalians were much concerned with the liturgy, both in the words of the rite and some aspects of ritual. The Scottish prayer book of 1637, including the Scottish Liturgy ('Laud's Liturgy') was reprinted in 1712. Various versions of the Scottish Liturgy were made, including a principal text, introduced in 1764.[15] These liturgies were called 'wee bookies'. There was considerable controversy about the 'usages', details of ritual practice, and with time more and more congregations accepted them.

In some places, including some qualified congregations, communion tokens were issued[16] and there was a special prayer of dismissal at the end of each rail-full of communicants, both said to be relics from the days before the Revolution.[17] There is also evidence that in some places the long tables of contemporary Presbyterianism were in use.[18]

The clergy in both traditions wore a black gown for liturgical purposes. Catechists or lay leaders took the services when ordained clergy were not available, while the laity are known to have read the *BCP* service at home, if they could not get to church.

The 'meeting houses', as both Scottish and English chapels were termed, were similar to Presbyterian churches, but probably had a small sanctuary for a communion table. All the non-jurors' buildings were destroyed in 1746, and owing to the new penal laws, which did not allow more than five people to be present (including the minister), a variety of buildings were erected where people could hear over the walls, and through the doors, while the clergyman was ministering to the allotted four people. George Innes, Perth, in 1747 had to preach six or seven times on a Sunday to observe the law, and others record ten or twelve relays.[19] With time, more usual buildings could be erected, though they were usually very simply built and furnished.[20]

The qualified congregations built some fine buildings, after the pattern of American Episcopal churches, of which the design of the Musselburgh chapel has been successfully recreated.[21]

Congregational and Individual Characteristics
Five distinct groups of Episcopal congregations, apart from those in Edinburgh, can be identified: the north-east rural communities; the north-east fishing communities; the Gaelic areas of Appin and Lochaber; around Inverness; and Highland Perthshire.

The rites and customs of the people in the congregations were little different from their Presbyterian neighbours. Some Episcopal congregations maintained the practice of a kirk session, in which the elders sat with the minister and meted out punishment to offenders,[22] while qualified congregations were ruled by a board of managers, who took wide powers on themselves in the absence of Episcopal authority.

Fear of imprisonment was a very real threat, and in 1750 George Innes noted that he was being watched every day by the authorities and, as late as 1756, he complained of his difficulties with soldiers attending his services in disguise, dressed in women's clothes.[23] In 1764 a new non-qualified chapel was built in Leith, and was immediately attended by a group of soldiers, so that Bishop Forbes, the clergyman, thought it wise to go away for a few months.[24] As late as 1770 the Peterhead and Lonmay chapels were closed down by the authorities. Improvisation was often the practice and Bishop Torry (1763–1856) described how, when a young priest at Arradoul in Enzie in the 1780s, he had been forced to celebrate Holy Communion on the table of a farm kitchen, hastily scoured and prepared for the occasion.[25]

Among other characteristics of an Episcopalian's life were baptism in the house, and confirmation by a bishop, though the latter was not considered so vital in the qualified tradition, where in any case it was not always possible to secure a bishop's ministrations. Marriage by a Scottish Episcopal clergyman was illegal, and in 1755 John Connachar, 'of irreproachable character and of considerable attainments', was tried, found guilty and transported.[26] However, towards the end of the century Episcopal clergy were taking marriages, without hassle from the authorities. Qualified clergy were allowed to perform marriages, provided the banns were called at the Established Church, and the entry made in the parish register. However, the qualified congregations often kept their own records.[27] The burial of the dead marked a distinction from the Presbyterians, for Episcopalians had prayers round the coffin in the house and often at the graveside.[28]

The most influential layman of the period was the banker, Sir William Forbes (1739–1806), and the best-known were 'the Gentle Lochiel' (c1700–48) and James Stewart of the Glen (c1705–52).[29]

1788–1845 TIME OF RECONSTRUCTION

Church Affairs

The death of Prince Charles Edward in 1788 was the opportunity for the Scottish Episcopal Church, most of whose clergy had long since given up being strictly Jacobite, to start negotiations for the repeal of the Penal Laws and the consequent reunion with the qualified congregations. The resulting Relief Act of 1792 required an abjuration oath (i.e. that the Stuart pretenders had never, even in the past, had any claim to the throne), and Bishop Skinner, though no Jacobite, was not prepared to take this retrospective oath and nor were the other clergy.[30] The act also barred any clergyman in Scottish orders from taking a service in England.

In 1804 a Convention was held at which the Episcopal clergy accepted the 39 *Articles of Religion* of the Church of England. This enabled the qualified chapels to join the Scottish Episcopal Church and come under the jurisdiction of the Scottish bishops. The reunion was marked by an English clergyman of a formerly qualified congregation, Daniel Sandford, becoming bishop of Edinburgh in 1806.

This was a time of consolidation and, with the reunion, congregations now effectively covered most of Scotland, except the south west, with the beginnings of new congregations in Paisley and Glasgow to cater for the large number of Anglicans who were settling there. Despite this, the concentration of charges was in the north east and in Edinburgh. A General Synod in 1838 issued a Code of Canons and set up the Scottish Episcopal Church Society and an associated Episcopal Fund for the sustentation of poor parishes. There was a clause for school provision.

A sign of the acceptance of the Episcopal Church was the audience that the bishops had with King George IV during his visit to Scotland in 1822, along with the gift of a *Regium Donum* to support clergy in poor congregations.

Some evangelical clergy and congregations could not accept the straitjacket of only using the *BCP* services, so in 1843 they founded the 'Church of England in Scotland', often referred to as the 'English Episcopal Church'.[31]

Figure 8.4 Kilmaveonaig altar vessels. In the back row, eighteenth-century pewter vessels, such as were usual in most congregations. These were generally replaced by embellished silver vessels in the late nineteenth century, the chalice and paten shown here dating from 1898.

Denominational Characteristics
The consolidation and rebuilding of churches and congregations involved much expenditure.[32] Services from the *BCP* became the norm which distinguished all the churches and gave them a close identity.

Some people felt that it was a time of almost total inaction and the bare discharge of ordinary duty,[33] but others believed it to be a period in which the Church was gaining reputation and respect: 'When she was established, she was hated; when she was persecuted, she was brought low; now she is neither established nor persecuted, and she is becoming every day more respected.'[34]

Congregations were not concerned in initiatives among the general poor (considering this to be the work of the Establishment) or in proselytising, but did feel a responsibility for ministering to Anglicans wherever they might be. The foundation of the Gaelic Society was a sign of this, as was the gathering of congregations in Greenock, Kirkcaldy, Paisley and elsewhere. In 1839 it was estimated that there were 10,000 Anglicans in Glasgow and its suburbs: 'Of these a large proportion are miserably poor, without the means, and what is worse, without the inclination, of supplying themselves with spiritual instruction.'[35]

There was a certain amount of interest in foreign missions in Edinburgh in the 1820s. A branch association of the Church Missionary Society had been founded in 1818, which by 1846 had raised upwards of £10,000. The basing in Edinburgh of the Scottish Episcopal Church Society was a sign that the centre of the Church was moving from Aberdeen and the north-east and that it was developing a more national identity.

The number of Episcopalians increased, some in Edinburgh perhaps drawn by the associations of belonging to an anglicised Church, but mostly due to having encountered Anglican services elsewhere.

> The increase of Episcopalians arises from many causes, some of which may be briefly stated. Families who have resided for any length of time in England or in the colonies, and gentlemen of the army, return home confirmed Episcopalians. Episcopalians intermarrying with those of other denominations, almost invariably bring them over to the Church. The constant influx of English families in Edinburgh, either for business or the education of their children, the number of English governesses, and of English servants of both sexes – all these retain their attachment to Episcopacy and the sublime Liturgy of the English Church, which distance from home only tends to increase.[36]

Theology, Ideology and Thought
The prevalent attitude in Episcopalian theology was an almost complete indifference to Calvinism, and in the north-east among the former non-juring congregations a 'high church' theology continued.

The Church's duty was believed to be to minister to all Episcopalians, wherever they were. 'Nothing was more desirable than that we should bring within the fold of the church the thousands of Episcopalians in Glasgow who ought to belong to us', said Bishop Russell in 1845.[37] Of the Gaelic congregations Dean Ramsay said 'Here are our own flesh and blood, men who lift up the same holy hands, breathe the same air, are of the same household of faith, join in the same sacraments and use the same sacred liturgy as ourselves.'[38]

Bishop Skinner in his negotiations for the Relief Act made it clear that it was the Church's duty to uphold the government. This was not inconsistent with the Jacobite past, and, since Jacobitism was dead, the Episcopalians could be no threat to the present establishment. The ideology was one of political submission and acceptance, Episcopalians making good deferential citizens while children were taught to be content with their lot in the world. The general climate was one of non-involvement in society. '[The clergy] shrank from observation. The Bishops never allowed, if they could help it, mention of an Episcopal act to appear in the newspapers; afraid that outsiders should come to know that there were live Bishops in Scotland.'[39]

A strong theme was anti-Popery, as shown when the meagre resources of the Gaelic Society were used for circulating in Gaelic Faber's *Facts and Assertions respecting Popery* and Blanco White's *Poor Man's preservative against Popery*.[40]

Evangelical enthusiasm was slow in appearing in the Episcopal Church, and found little room. It was largely an importation from England. The Duchess of Gordon moved from the Episcopal Church, which she had supported at Fochabers, to the Free Church once it was formed,[41] but most Episcopalian evangelicals joined the 'English Episcopal' congregations.

In this period there was a general warming of relationships with the Presbyterian clergy. Concerning the apostolic succession, it was said of the Episcopal clergy, 'they believed that they themselves did right, but they did not speculate much as to how far their dissident neighbours did wrong ... Time and Biblical criticism had dissipated prejudices and broken down barriers'.[42]

A critic in 1833 observed that the laity were attached to their pastors and well acquainted with the distinctive parts of their creed, but he noticed that the majority in Scotland, while prepared to accept any innovating sect, had a kind of antipathy towards the Episcopal Church. The reasons for this, he believed, were that the Episcopal clergy held that the principles of their Church were unassailable, and they acted on the supposition that the truth would maintain its own ground.[43]

Clergy
Episcopalian clergy had been recruited from among the farming families of the north-east, and in 1810 Miss Panton of Fraserburgh left her money to found a college for the training of clergy. Many candidates, however, were advised to seek ordination in England because of the continuing disqualification of those ordained by a Scottish bishop being allowed to minister in England.

In 1833 the bishops and clergy were found to be 'men of piety, learning and moral worth, and faithful in the discharge of their duties', but it was pointed out that the standard of education of the Episcopal clergy was not as high as the education and learning of the laity demanded. There is a good description of a country clergyman at Muthill.[44] Alexander Jolly while ministering in Turriff 1777–84, found himself very bored with the farming talk of his fellow clergy.[45] The smallness of the stipends did not encourage people to enter the ministry, though the clergy in some cities were people of means. There was a special difficulty in recruiting Gaelic-speaking clergy.[46]

Among the clergy who were central to the thought of the Church were Bishop John Skinner (1744–1816) and Dean Ramsay (1793–1872). E B Ramsay, dean of Edinburgh, was a crucial figure in the resurgence of the Episcopal Church. Connected with St John's, Edinburgh, for nearly 50 years, rallying the congregation which became the most generous in the Episcopal Church, he was instrumental in many initiatives in the wider Church. It was through his influence that the Edinburgh congregations and the Episcopal Church began to find their new role after the years of persecution. His *Reminiscences of Scottish Life and Character* went through twenty-two editions in his own lifetime.

Worship and Ritual
The services were based on the *BCP*, but with some flexibility. However, Canon XXVIII in 1838, forbidding the use of any service not in *BCP*, was the reason for the evangelicals to withdraw, as the bishop of Edinburgh began forbidding weekday prayer meetings if a *BCP* service was not used. This prohibition was an innovation. The Scottish Liturgy was the usual form of

Communion service among the congregations of the north-east and Highlands. Officially, by act of Parliament, each congregation had to pray for the monarch and royal family by name (in case they were praying for the wrong royal family), and the doors of all churches had to remain unlocked during public worship.

In the country areas Holy Communion was celebrated five times a year, namely Christmas, Easter and Whitsunday, and the first Sundays in August (for Lammas) and October (for Michaelmas and Harvest Thanksgiving). Later, the first Sunday in Lent was added.[47] Reservation of the sacrament, so that it could be taken to the housebound, has often been regarded as a particular characteristic of the Episcopal Church, but William Walker records that most people preferred to have a separate celebration in the house.[48]

Until about 1819, the clergy for services wore a black gown, black scarf and white linen bands. From that date the surplice or 'white goon' was used by all the clergy, though not for preaching. This innovation was due to English influence. The ritual, particularly in the country areas, was of the simplest kind. People, long accustomed to it, were rather suspicious and intolerant of anything different.[49]

From early in the nineteenth century, the design of church buildings, especially in the towns and cities, tended to follow the English precedent of an aisled or arcaded hall plan, with a central aisle, and by the 1830s the sanctuaries were becoming deeper, though they were not yet fully chancels. Most, but not all, of the new churches were in the Gothic style.

Congregational and Individual Characteristics
The Episcopalian heartlands of the eighteenth century were still distinct communities, but now there were congregations in most burghs and in some industrial towns. These latter were usually of Church of Ireland origin and this showed in a dislike of any sort of ritual or ceremonial, with a green altar frontal taken to be a sign of Irish Catholic tendency.[50]

The fisherfolk of the east coast were a distinctive group, well recorded for their congregational characteristics:[51] 'The fisherwomen reigned supreme over every fisherman, whether husband, or son, or brother, or father. The fish found in the boats they regarded as their own unquestioned and unquestionable property, and with their administration they would brook no interference.'[52]

The spirituality of the farming community of the north-east can be gauged from the survival of the diary of a farm servant (born 1799), written in 1830. He was a member of the congregation at Meiklefolla, his religious education being the work of the Rev. George Innes, based on a catechism of 1779, bound up with *Some instructions proper for Young Persons before they are Confirmed* (Edinburgh, 1779). The diary records the sermons of the day and the application to the writer's own case. He spent some Sunday afternoons in 'meditation and reading' but sometimes 'very unbecomingly' or 'thoughtlessly employed'. One sermon was recorded as 'no high station in life should make us think ourselves above doing every act of kindness and love to our fellow-creature, especially when the welfare of his soul is concerned', to which the diarist responded 'When I look back on my conduct through the past month, I find I have trifled away time which I ought to have spent in gratitude for numberless mercies which I have daily received.' He also prayed 'O Giver of life, who fitted man for society and who hast enjoined us to love one another, direct and enable me to find such a companion through life, that we may promote each other's temporal and eternal welfare.'[53]

There is a very full description of the congregation at Inverness and its traits: its popularity with all people, including the tinkers, who came for baptism and marriage, the tips for the pew-opener and caretaker of the building, the children's class held round the railing of the communion table every Sunday after the afternoon service.[54]

For the Gaelic-speaking congregations,[55] the Gaelic Episcopal Society was formed in 1832. Dean Ramsay was its secretary and he appears to have been responsible for raising the funds, but the society was not well supported. At that date there were five ministers in the Gaelic-speaking areas. One reported that an average Sunday collection was 3d. or 4d., from a congregation of 80 or 100. The Society decided to apply its funds not only to the education of students for the ministry, but to the regular maintenance of settled ministers. 'The case is rendered so urgent, by the rapid ebbing away of the Episcopal community in these districts, that any measures to prevent the extinction of the Episcopal Church, to be successful, must be adopted speedily.'[56] In 1837, they had three schools in action, at North Kessock, Muir of Ord and Ballachulish, attended by about 300 children. The Society also sent out Gaelic Bibles and prayer books and gave donations to some clergy for pastoral visitations, but they did not have finance to help extend church provision.

In 1833 the *Ecclesiastical Journal* stated:

> In the Highland districts, where Episcopacy was once prevalent, a lingering affection for it still remains. It discovers itself in a very peculiar way, namely in an apparent disregard of all religious feeling. This is easily accounted for: The inhabitants have no opportunity of exercising that form of worship which they have so often heard their fathers praise, nor of becoming members of that church whose reputation has descended from father to son, and they appear willing wholly to forgo the benefits of religion, rather than formally renounce that which their ancestors admired and practised.[57]

There were, however, two attempts to found congregations for the Gaelic-speaking Episcopalians in Glasgow, but both failed. It was stated that away from home they did not mind where they went to church, provided there was a good sermon. Apparently they also felt that, whatever the Episcopal Church did, it always let down its Gaelic members, and there is evidence of a failure by the Church authorities to realise that the Gaelic language was itself crucial to their spirituality. The English language was used at Appin whenever an English monoglot speaker was in the congregation, even though this excluded most of the congregation.[58]

The Episcopal Church was known as the Church where the clergy baptised all the children, whatever their origin or the faults of the parents. However at St John's, Edinburgh, in 1831, the vestry resolved that, for fear of incurring a charge of proselytism, only the children of members of the congregation and visitors to Edinburgh should be baptised.[59] Baptisms were still most often conducted at home, and this was to cause the mission clergyman, D'Orsey, some concern when establishing his congregation at Anderston, since it was contrary to his view of church discipline.[60]

F C Eeles has recorded some of the old marriage customs among the north-east Episcopalians, including the custom of the bride and bridegroom being each attended by a maiden and a young man, and the placing of the wedding ring on the thumb and other fingers in succession, which Dr Eeles equates with the old pre-Reformation ceremony. The great occasion for ceremony was the 'kirkin' or first appearance in church of the newly married couple. At Muchalls, the old style marriage survived among the fisherfolk until the end of the nineteenth century, with the peculiarity of the bridegroom's young man, carrying a staff (about 3 feet long) made of some white wood, with the bark peeled off, and a bunch of blue ribbons tied to the top of it. In the evening, the ribbons or streamers were tied round the right arm of the bridegroom, who wore them that night.[61]

The *BCP* service of 'Churching of Women', about which there were many superstitions in England, was used, but not as a separate service. Dr Eeles records that it was interpolated into Mattins. In Buchan there were similar superstitions to those in England.[62]

Another custom among Episcopalians was seeking a bishop's blessing. At Fraserburgh it was still usual in the 1830s for some members of the bishop's congregation to remain after the service in order to receive the bishop's blessing, though Bishop Jolly may have been the last of the bishops whose blessing was much sought after.[63]

The distinctive feature of the Episcopalian funerals in the north-east was the 'chestings', the evening before the funeral, in the presence of a few women. A lighted candle was placed beside the corpse and a plate of salt was placed on the breast. This evening service (usually from the *BCP*) was distinctive because the Presbyterians had no service at a funeral, and their ministers did not attend funerals. The saying of prayers at the graveside was unknown (hence the prayers the night before), and its introduction by Episcopalians caused a great stir in country areas. Often Episcopalians in referring to the dead would attach 'God sain them' to their names.[64]

Some landed families who lived at a distance from an Episcopal church were Presbyterian in the countryside and Episcopalian in Edinburgh. Among the Church's noted lay people were Sir Walter Scott (though also a Presbyterian elder), Sir Henry Raeburn, the banker Sir William Forbes (1773–1838) and his brother, the judge Lord Medwyn (1776–1854).

1845–1920 TIME OF EXPANSION

Church Affairs

This period saw a new confidence, huge growth in numbers and full integration into the Anglican Communion of Churches, which stretched throughout the British Empire, and through the United States to many other places.

The new confidence came in part from the influence of the Oxford or Tractarian Movement, which emphasised the unique nature of the Church in its life and rituals, but there was considerable unease at some of the more extreme teachings of the movement. Many English tractarians looked to the Scottish Episcopal Church as a paradigm of what an Anglican Church should be like, not least because of its recent times of persecution and its freedom from the State. To counter these teachings, there was a trend towards conformity with the Church of England. This was shown in the downgrading of the Scottish Liturgy, which led to the lifting by Parliament in 1864 of the ban on clergymen ordained by a Scottish bishop being allowed to minister or hold livings in England.

The first recorded use of the term 'The Anglican Communion' appears in the Episcopal Synod Minute Book of 1850[65] and in 1867 the Scottish bishops were invited to the first Lambeth Conference. Following the ideas of that conference, in 1876 the Episcopal Church reorganised its government as an Episcopal Synod, a Provincial Synod and a Representative Church Council (including lay representatives from every congregation).

The fourteen ancient dioceses were rearranged into seven dioceses, five being unions. There was a concerted move to found congregations wherever Anglicans were living, in a geographical system covering the whole country. By 1870 there were also eleven congregations of the evangelical English Episcopalians, who, 1877–87, paid their own independent bishop, Bishop Beckles. However they were already seeking reunion with the Episcopal Church, made possible in their eyes by the down-grading of the Scottish Liturgy.[66]

The publication in 1884 of the seminal *Scottish Communion Office, 1764* led to a renewed interest in the Scottish Liturgy and in 1911 a revised Scottish Liturgy was issued. Celebrations were also held in 1884 to mark the centenary of the consecration of Bishop Seabury, bolstering the existing folk memory of this unique Scottish Episcopalian event.

Denominational Characteristics

Congregations were regarded as the centre of ministry, with all the associated activities to keep a congregation together, both of an ecclesiastical and of a social nature. In particular the foundation of schools, linked with the congregations, was seen as a vital way of providing a Christian and social service both to the congregation and to other families in the community.[67] Through the schools many people came into contact with the Episcopal Church. School buildings were often used for social gatherings but church halls were built for most congregations. In the country, graveyards were created round the church buildings.

Following the Anglican pattern, cathedrals were regarded as centres of diocesan mission and each diocese built or created a cathedral. St Mary's Cathedral, Edinburgh (Fig. 8.5), in particular, became a focus for diocesan and Scottish unity, with a large staff of clergy and a full choir and choir school. Some extra support came from central resources, including the religious orders, social service agency and educational help for which a teacher training college was founded.

Trinity College, Glenalmond, was founded in 1847, in line with similar foundations in England and Ireland. It consisted of a school, governed by Church teaching and principles, for the middle class and landowners, and a theological college, for students of all backgrounds. The college's foundation was supported financially from all corners and ranks of the Church, but the theological training was later moved back to Edinburgh. Other private schools with an Episcopalian foundation were also started. At Aberlour in Banffshire, an Episcopalian orphanage was established in 1875, though not restricted to Episcopalian orphans. Its work was supported by congregations across the country and became a permanent interest.

Home mission was regarded as claiming one's own, and any suggestion that the Episcopal Church was not a Church of the poor was countered.[68] However, it was true that some churches which were intended for poor Anglicans were not always attended by the poor, having been taken over by the wealthy and middle classes. This phenomenon was to be seen in Anderston, Paisley, Pollockshields and elsewhere in the west of Scotland. St Columba's-by-the-Castle, Edinburgh, reported in 1898 that it was now what its founders had intended it to be, the church of the working classes, as was Galashiels; but many other missions, opened in the poorer areas, did not last long.[69] Although some wealthy congregations sponsored 'mission' congregations, pew rents, though they were being given up from the 1890s, were the usual way of financing a congregation.

The development of large choirs and sung services helped to keep the interest and commitment of many, but there was a feeling that they did not always encourage new members, who were unfamiliar with the special form of the Anglican services. Sometimes the sung services were considered to be 'popish intrusions' or giving the character of a cathedral to simple congregational worship. In the west of Scotland, the Orange influence was strong within the Episcopal Church and Irish influence was also seen in a lack of ritual and ceremonial in the working-class congregations there. Anglicisation was seen in the change of terminology from 'Incumbent' to 'Rector', as also from 'Parsonage' to 'Rectory'.

Theology, Ideology and Thought
The Anglo-Catholic wing of the Oxford Movement believed that they had found, in the old non-juring tradition of the Episcopal Church, the theological principles which it was trying to encourage throughout the

Figure 8.5 St Mary's Cathedral, Edinburgh, was built and endowed from the legacy of the Walker family, grandchildren of an Aberdeenshire Episcopalian clergyman, and opened in 1879. Designed by Sir Gilbert Scott, the leading architect of the day, the cathedral's size and beauty showed the prestige to which the Episcopal Church aspired.

Church. However, the teaching on the Eucharist by their representative, Bishop Forbes of Brechin, brought about an action before the Episcopal synod in 1860 and much embarrassment among the more moderate Church members.

Bishop Wordsworth was anxious for reunion with the Presbyterians and Bishop Ewing believed the way towards this was by the Church conforming to and then joining the Church of England. Like Bishop Russell, Ewing took up the views of John Macleod Campbell on the Atonement, which echoed the views of Thomas Erskine of Linlathen, the Episcopal Church's greatest theological thinker.[70]

Respectability and acceptance were still the key ideas throughout the period and the scene was set in 1843 when the Duke of Buccleuch said 'This church is not an aggressive church. It seeks not to encroach upon the rights and privileges of others; its only object is to seek out the scattered members of our communion and to give them those benefits and privileges which they have a right to demand.'[71] Despite this, Episcopalian landlords were not necessarily favourable to the Free Church,[72] and later, at a time when there was a threat of the disestablishment of the Church of Scotland, it was claimed that in the Episcopal Church they would find 'their fastest friends'.[73] However, the tractarian views of many in the Church, especially among the clergy, were not always consistent with this view and were seen in a new narrowness, with an emphasis on defining who was within the Church. Presbyterian clergy were often regarded as not being 'validly' ordained, and this inevitably led to a hardening of attitudes.

Attempts to make the Episcopal Church more acceptable in the community at large were seen in an appeal to history, though it did not overcome prejudice. At least the Episcopalians should have been better informed, since much energy was expended on retelling the stories of the days of persecution and of the indigenous tradition in the north east and Highlands. A depiction of the baptisms from the prison window at Stonehaven jail was painted in 1865 (Fig 8.1). Scholarship was seen in the collections of the Spottiswoode Club and the largely Episcopalian Spalding Clubs. The work on liturgy by Bishop Dowden led to the revision of the Scottish Liturgy. F C Eeles' *Traditional Ceremonial and Customs connected with the Scottish Liturgy* of 1910 was among the Alcuin Club collections, whose objective was to find an indigenous source for Anglican ritual and ceremonial.

Not surprisingly considering the origins of so many of its members, the emphasis on the Scottishness of the Church was countered by a clear Unionist stand, particularly among the Orange element. Anti-popery continued to be an important element in the Church's thinking. In 1898 a Woman's Conference was held in connection with the meeting of the Representative Church Council, the one recorded speech being on over indulgence to children, which met with various criticisms.[74]

Clergy
The period was marked by the importing of English clergy to fill senior positions in Scotland. While it is sometimes difficult to distinguish whether a person was a Scot or not, some of the bishops had no Scottish antecedents. However they were of a similar background and upbringing to many of their English counterparts, among whom were many Scots, including Archbishops Tait and Davidson.

Few able Scots who went to English universities and theological colleges returned to Scotland and the biographies of the clergy in the *Scottish Standard Bearer* show a clear dichotomy between the children of farmers from the north east and those with English experience. Many of the deans were of the former category and never reached the episcopate. Exceptions to the appointment of non-Scots include Bishop Anthony Mitchell of Aberdeen and Orkney, who was raised in humble circumstances in Aberdeenshire; and the general feeling in the Church was 'let the large majority of our clergy be Scotsmen in every respect'.[75]

The need for Gaelic-speaking clergy was often made, but the period showed that the number did increase relatively and most of the Gaelic charges were filled with Gaelic speakers through the later nineteenth century. There was, of course, no reason to think that Gaelic speakers wished to remain in the Highlands and moves by them were recorded to Burntisland and Dunbar.

Among the scholarly clergy were Bishop Dowden (1840–1910), Bishop Mitchell (1868–1917) and Bishop Maclean (1858–1943), as well as historians and scholars John Pratt (1798–1869), J F S Gordon (1821–1904), William Temple (1827–1906) and G H Forbes (1848–75).

Worship and Ritual
Apart from the varying amount of use of the Scottish Liturgy, the *BCP* was used for all services throughout the period, and in 1912 *Common Prayer (Scotland)* was published, which included the Scottish Liturgy of 1911.

The pattern of worship changed at the beginning of the period, when the long *BCP* morning service, used in most congregations, was shortened. There was a variety of practice, though St Peter's, Peterhead, was not unusual in 1878 when, on a Sunday, there was Holy Communion at 8am, except on the first Sunday of the month, when it came after the main service of the day, 11am Mattins, which always included a sermon. Evensong with sermon was at 6pm. In more and more congregations the psalms, canticles and litany were sung. Congregational hymns became normal long before the Presbyterians used them.[76]

Gradually more ceremonial was introduced, with black scarves being replaced by coloured stoles and even Eucharistic vestments in some places. Lighted candles on the altar became normal, except in the west of Scotland, where Irish influence prevailed. Many churches began to follow a Scottish version of medieval ritual, similar to that which was increasingly approved in the Church of England.

The vessels used for Holy Communion[77] were replaced with more valuable versions, often including jewels, donated by pious ladies. Among the finest of such objects was the 'Scottish national chalice' at St Mary's Cathedral, Edinburgh, which was designed by Omar Ramsden and included a figure of St Andrew in the stem.

Between 1840 and 1860 eighty-eight new churches were built, with many more to follow, and church design took a completely new style with the start of the Ecclesiological Movement and its emphasis that the building was a 'sermon in stone'. Each feature had a theological or ceremonial reason, from the raised altar in a separate chancel to the stone building itself being the rock of faith, through which the pulpit steps often led. John Henderson, and his simple country medieval recreations, was followed by Sir Robert Rowand Anderson, Sir Robert Lorimer and the ubiquitous Alexander Ross of Inverness, all four Episcopalians.[78] Sir Ninian Comper (1864–1960), also a Scottish Episcopalian and committed churchman, did some fine work in Scotland, based on Scottish medieval precedent.

Congregational and Individual Characteristics
A difference between the congregations in the cities and urban areas, the small burghs and the rural communities existed through the whole of this period, but in practice the former two had become the *de facto* church of English and Irish settlers.

It has been suggested that among some congregations there was a 'hereditary fondness' for the Episcopal Church,[79] allied to feelings of intense loyalty. There are many references to 'hereditary Episcopalians' and they were held in great respect in some quarters, though their resistance to change

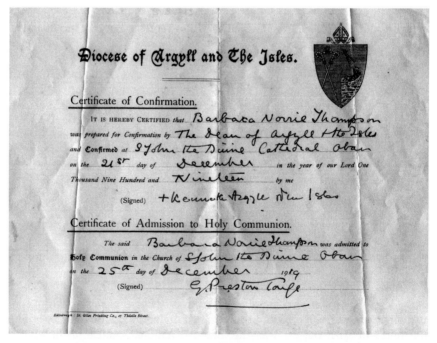

Figure 8.6 A double certificate of Confirmation and Admission to Communion of 1919, signed by the bishop, using his ecclesiastical-style signature. It shows that the candidate was prepared by the incumbent and admitted on Christmas Day. The candidate was 14 years old.

was a characteristic that was often mentioned. Dean Hatt complained to Bishop Forbes that the fisherfolk at Muchalls were so conservative and obstinate that, do what he would, they could not be persuaded to come to Holy Communion more than three times a year. 'And pray,' said the bishop, 'what was it that kept them loyal to the Church but that same conservatism and obstinacy'.[80] However, it was also said that the dislike of changes to the services came from fear that they were anglicising influences.

Although vestries no longer acted as moral watchdogs with disciplinary powers, the clergy, under the influence of the Oxford Movement, were regarded as the guardians of spiritual and moral conduct, giving advice towards leading a better or more serious life. They were very much the leaders of their church communities. General visiting of the whole congregation by the clergy was an expectation;[81] the tradition of pastoral care by the clergy being strong.

The church buildings gradually became the centre for most liturgical occasions, baptisms, marriages and funerals, though in country areas some marriages were still performed in private houses and funerals were often in the house.[82] The buildings began to be felt to be sanctuaries, to be kept apart, beautified and visited. The churches were considered as second homes and the hall was the focus of many of the social activities of the congregations, while the school made another strong link.

For some the idea of belonging to the 'English Church' gave them a feeling of security and many incomers did not shrink from calling it by that name. Others felt that this was a disloyal term and a positive hindrance in country areas. The urban congregations, especially in Glasgow, would have had no reason not to think of themselves as English or Irish chaplaincies. There are frequent references to these congregations being designed for the mass of working people, and there are many instances of the mission charges being thronged with working people, though it was also said that they had no feelings of loyalty to Anglicanism, merely going where the hand-outs were best. It is a fallacy, however, to think that all the English settlers were Anglican, or that they would not find a congenial home within the Presbyterian Churches. The confirmation register at the working-class congregation, St Columba's-by-the-Castle, Edinburgh, shows that a large proportion had not been baptised there, many not even baptised as Episcopalians. However, whether or not the working people regularly attended church, many came for baptism, and St Andrew's, Glasgow, had a baptismal list of thousands each year.[83]

Abstinence societies are recorded in the urban congregations, and antipopery continued as an influence throughout this period, the Episcopal Church being firmly in the Protestant fold.[84]

The fishing communities kept their separate existence, and Rowan Strong has suggested that their numbers have been underestimated.[85] Canon William Low has recorded reminiscences of his time at Cruden, 1870–80, including a description of Jake Alexander, 'Provost of the Bullers', a fine seaman and a skilful fisherman, with a kindly feeling for gulls, 'tammies'

Figure 8.7 St Andrew's church and rectory, Ardrossan, was opened in 1883 and completed in 1894. This view shows the sort of buildings that most Episcopal congregations would hope to have. Built under the patronage of the earl of Eglinton, missions were started at Dalry, Irvine, Stevenston and West Kilbride.

(puffins) and all sea creatures, and a member of St James'. 'Like so many inhabitants of the old fishing villages, he was a staunch Episcopalian, and was never half-hearted or backward as defender of the faith.'[86] The old ways continued at Muchalls too, a congregation that has been well documented, where the funerals were 'walking' funerals and the coffin was carried for 5 miles from Skateraw to Cowie, on poles, at arm's length, the first hearse being used in 1870, and then only because the men were away at the herring fishing.[87]

As well as the resident fishing communities at Peterhead and Torry, the congregations in Orkney and Lewis did pastoral work among the fishing people who went there in a seasonal capacity. However, in Stornoway it was also said that the fishermen came to the Episcopal church because it was the only church with services in English.[88]

There was a continuing feeling of exclusiveness among the Highland Episcopalians, but in most places, especially the Black Isle, numbers were much reduced. In 1847 at Duror of Appin it was reported that there was a congregation of several thousands and that Episcopalians formed half of the entire population.[89] Churches and schools were built and embellished in these traditionally Episcopalian areas, usually due to the generosity of certain individuals. It could be that the loss of adherents was due to the arrival of the railway and the natural movement of people to the towns. However, there may have been a more underlying problem involving the playing down of the Gaelic language, the natural medium for worship of the inhabitants, even when they were English speaking; a situation of which the Presbyterians were able to take advantage.[90] 'Abrach' noticed that the new edition of the Gaelic prayer books had English printed alongside Gaelic and saw this as a sinister step towards the eradication of the Gaelic language.[91]

A special characteristic of the Episcopal congregations was the keeping of all records, perhaps to establish exactly 'who was in and who was out' of the Church, though the statutory requirement to keep registers in the Church of England may have been an influence. Registers of each service, including weekdays, are a feature of many congregational archives. Records at Meiklefolla, Aberdeenshire, include the names and destinations of emigrants among the congregation.[92]

Familiarity with the Scriptures was a usual trait of the era for all denominations, but Canon Low wrote of a parishioner, Auld Sandy Milne, 'He knew how to use his Prayer Book at home, and his Bible and hymn book too. I used to provide him with a *Churchman's Almanack*, to make it easier for him.'[93] Apart from knowing the Prayer Book, what characterised the Episcopalians more than anything else were their celebrations of Christmas, which if it was a weekday involved working people going to a service before work started, and of Easter and the other Church festivals.

Although there were few Episcopalians among the leaders of the working people, several of the Episcopalian landowners were members of Parliament. However, William Ewart Gladstone (1809–98), Prime Minister, always set great store by his connections with the Scottish Episcopal Church, and he entered into many of its theological controversies, pressing the lay claims, as well as supporting every venture, not least the foundation of Trinity College, Glenalmond. Canon Malcolm MacColl (1831–1907), the

Figure 8.8 Sketch of Auld Sandy Milne. Canon Low described ministering to Sandy Milne (d. 1885), a genial and peace-loving old seafarer, who had in his youth gone out with the whalers, and now pottered around the rocks, wearing a short blue jacket, trousers carefully turned up at the foot, and a scarlet Kilmarnock night cap. He was probably typical of the older generation of Episcopalian fisherman at Cruden, when he stated, in the face of the elements, 'There's a Han' aboon a''. From *SSB*, May 1896.

noted Christian Socialist, and protégé of Gladstone, was also an Episcopalian, while there are countless examples of Episcopalians serving on school and poor boards. In 1896, when the Trades Union Congress was held in Edinburgh, seven special sermons, through the agency of the Christian Social Union, were preached in Episcopal churches, 1,800 people attending St Mary's Cathedral to hear about 'equality of social justice, equality of social opportunity'.[94]

The Episcopal Church in this period was able to take its place alongside the three Presbyterian Churches with roughly equal numbers of congregations and clergy to them, as seen in Post Office directories, especially in the countryside and Edinburgh. This gave the Episcopal Church a bigger presence than it later had. Among its noted lay-people, apart from the aristocrats, were the theologian Thomas Erskine of Linlathen (1788–1870), the novelist Catherine Sinclair (1800–64), the historian W F Skene (1809–92), and the scientist Lord Kelvin (1824–1907).

1920–65 TIME OF CONFIDENCE

Church Affairs
In this period, the number of committed Church people, as revealed by the communicant numbers, rose to a peak, but the number of adherents fell dramatically.

More involvement by the laity in the running of the Church was a continuing theme, culminating in their admission to the Provincial Synod in 1961. In the Representative Church Council the number of laity was roughly half of the 650 members, each congregation being directly represented.

The incorporation of the Episcopal schools into the State system in 1918 led over the next thirty years to a drastic reduction in their number and influence as they were closed or united with neighbouring non-denominational schools.

Links with Anglican dioceses around the world developed, including two dioceses in India and Africa, Chanda and Kaffraria, with which there were special links and responsibilities. However, many overseas dioceses were becoming independent, and provinces were being formed and made self-governing. Chanda became part of the united Church of North India, which united several denominations and caused some doctrinal problems among Episcopalians.

The Episcopal Church was involved in various ecumenical dialogues, initially as an adjunct to the Church of England/Church of Scotland talks, but later as full partners. Most of the remaining members of the Catholic Apostolic congregations joined the Episcopal Church at this time, and brought a sense of commitment with their allegiance.

Denominational Characteristics
This was a time of confidence, although some of the seeds of future decline and change were beginning to appear.

Figure 8.9 Principal Alexander Boyd, staff and students at the Episcopal Teacher Training College, Dalry House, Edinburgh, 1928, in which year 'secular' training was transferred to Moray House, Edinburgh.

The Church felt at unity within itself, with little evidence of disagreement on ecclesiastical matters, and it was drawn together by the production of the *Scottish Book of Common Prayer* (*SPB*) in 1929. The book effectively brought together traditionalists, who valued the Scottish Liturgy, and the anglicisers, who looked to England, where a very similar new Prayer Book had been prepared.

Although many nominal Episcopalians were disappearing from the records, the Church still felt a duty to provide ministrations for all Anglicans in Scotland. This was beginning to become expensive, as the Clergy Sustentation Fund had to contribute in increasing amounts towards keeping clergy in places that could not pay the full stipend.

There were two specific appeals for money in the 1920s and 1940s. The former led to the building of ten churches, but the latter was less successful, and at some point after World War II the Episcopal Church lost its momentum for building churches in the new suburbs, and those that were built were rarely lasting. However, those built in new towns, East Kilbride (1956), Cumbernauld (1962) and Glenrothes (1962), all flourished, but did not pick up all the displaced Episcopalian adherents who were moved to these towns. However, individuals moving from one small town to another were usually commended to the clergyman in the new community, who grafted them on to the new congregation. There was also a steady flow of Anglicans coming into Scotland, who found a home in the Episcopal Church, and this covered over the dramatic loss in loosely attached adherents. Several families joined the Episcopal Church through the mixed marriage of Roman Catholics and Protestants, each partner feeling comfortable enough with the services and community. This phenomenon had been noted in the nineteenth century too and was more common in Scotland, where the Episcopal Church was not the predominant denomination, as compared with England.

The principal characteristic of the period was the running-down and closure of Church schools. Roman Catholic and Episcopal schools were brought into the State system in 1918, with safeguards (which in the case of the latter proved illusory) to preserve their sectarian character. The denominational schools were important missionary agencies in giving an identity for the children as well as building up a large number of adherents, who, even if they did not attend church, looked to the Episcopal Church for their experience of the spiritual and knew its regular cycle of the Church's year, its feasts and fasts. In 1918 there were 85 schools and in 1950 there were 33, most of which closed soon afterwards. Dalry House, the teacher training institute, closed in 1934 after 2,000 Episcopal teachers had passed through it over the years.[95]

Sunday schools were also on the decline, but a forward looking project, 'Sunday School by post', centred in rural Aberdeenshire, ensured teaching for both children in small congregations and those who could not get to any church.

A social service board was set up in 1919, which was later called the Social Responsibility committee and among the projects supported were St Mary's Mothercraft Centre in Dundee and several eventide homes.

Theology, Ideology and Thought
The prevailing Anglican theology concerned the involvement of Christians in society, and for some this meant through party politics. Certain individuals, including clergy, took an openly political stance, among whom were John McBain, an early member of the Labour Party and a Glasgow councillor, and George Henderson (1921–96), Labour Provost of Fort William (later Bishop of Argyll and the Isles, and Primus). At the same time, Scottish Episcopalian Sir Alec Douglas Home was Prime Minister.

The Episcopal Church had hoped for some sort of union with the Established Church, but the union between the Established Church and the United Free Church in 1929 left the Episcopal Church rather on its own. This was especially felt in the subsequent unity talks between the Church of Scotland and the Church of England in 1932–4 and 1949–51. The Episcopal Church was brought in to the talks in 1954 and the subsequent report of 1957, which became known as 'The Bishops' Report' resulted in the vigour of a hostile press, which considered bishops in any form, but especially an Anglican form, to undermine Scotland's identity and integrity, being launched against the Episcopal Church.[96]

There was a certain defiance within the Episcopal Church after 1929 and this was seen in the stand taken by the Episcopal Church in 1960, the

Figure 8.10 The Scottish bishops, robed in medieval-style vestments, at the consecration of Piers Holt Wilson as bishop of Moray, Ross and Caithness, Inverness, June 1943.

fourth centenary of the Reformation. A document was printed, 'issued with the approval of the Scottish Bishops'. After making a short statement about the need for a Reformation, it mentioned the tragedy of the resulting divisions and concluded 'After careful consideration the Scottish Bishops have decided that our right course is to refrain from participation in the official celebrations of the events that took place in 1560, because such participation would seem to indicate a denial of or indifference to those very elements of historic Christianity and the undivided Church which we exist to preserve and uphold and under God to restore.'[97]

A further indication of the assertiveness of the Episcopal Church was its distribution to each congregation of copies of paintings of the baptisms at Stonehaven jail during the time of persecution (Fig. 8.1) and of the consecration of Bishop Seabury (Fig. 8.2). At the same period, a hanging sign was designed to mark each Episcopal church (Fig. 8.11), so that it could be recognised as a presence in Scotland. The sign contained heraldry and mitres with the motto 'Evangelical Truth and Apostolic Order'. For those who knew, they made a further statement, since some of the diocesan coats of arms were those from the pre-Revolution Established Church, which the Episcopal Church had legally inherited.

Despite this confidence, the Episcopal Church could not rid itself of the view stated by the historian Agnes Mure Mackenzie, herself an ardent Episcopalian, and which she attributed to a Presbyterian, 'that the Episcopal Church is an English exotic brought in by the laird with his background of an English public school, or by the laird's English wife, and supported mainly by people who hope the laird will ask them to dinner'.[98]

Clergy
The clergy were largely trained at the Episcopal Theological College in Edinburgh, but some who graduated at English universities were trained in England. As in the previous period, many Episcopal clergy were freemasons, which enabled them to fraternise with many of the community who were not members of their congregations.

Among the notable clergy were the brothers John and Alexander McBain (both d. 1936), natives of Rathen, Aberdeenshire, whose ministry at Christ Church, Bridgeton, Glasgow, lasted thirty-seven years, both being Christian Socialists and Labour councillors. John McBain said of his congregation 'There is no such completely proletarian mass of people anywhere in these islands.' While Malcolm MacColl (latterly rector of St John's, Ballachulish, d. 1946) was involved with Sinn Fein and was probably in Ireland during World War II renewing contacts.[99]

Among the bishops of note were Arthur John Maclean (1858–1943), a scholar and liturgist, and Duncan Macinnes (d. 1970), a Gaelic speaker from Glencoe, who showed enormous bravery as a chaplain and prisoner of war.

Worship and Ritual
The 1929 *SPB* became the norm for most services. It contained the revised Scottish Liturgy, though the *BCP* liturgy was also used in some places. From the 1950s the Eucharist was the usual main service on one Sunday a month, the others having Mattins, and at the same date there was a dramatic decline in the number of people attending Evensong.

An importation at the beginning of this period was Harvest Thanksgiving, a popular service in the Church of England. Occurring in about September, it allowed the church building to be decorated and a festival to be celebrated. Christmas cribs and Easter gardens were introduced too. Some congregations from the 1930s had an Easter Eve service of blessing of fire and lighting an Easter candle, but this did not come into more general use until the 1960s. Presbyterians were known to attend Episcopal services on Good Friday in the nineteenth century, and the Three Hours Service, which was introduced fairly generally in the cities and towns from the 1920s, proved popular with many outwith the congregation.

Eucharistic vestments came into general use and altar candles were nearly always lit, except in a few 'low church' congregations. The distribution of palm crosses, an Anglican devotion, was generally introduced to mark a person's commitment to Holy Week. It became a custom for Episcopalians to hand them round both to those who could not get to church and to Presbyterian neighbours.

Congregational and Individual Characteristics
The principal characteristic of Episcopal congregations in this period was an aloofness that largely came from an upbringing of certainty in Church and society. This may have been taught in school or Sunday school, in confirmation preparation or heard from the pulpit. *St Columba's Companion to the Scottish Liturgy* was often given to confirmation candidates, in which they were asked to pray 'for those separated from the Church, especially Presbyterians'. It was shown in particular when many pilgrims went to Iona in 1963 for the 1,400th anniversary of the arrival of St Columba. The occasion was purely Episcopalian and boats came from England and Ireland, along with the Archbishop of Canterbury. There was no concession made to any other Church.

At the same time there was a general knowledge of the Episcopal Church's place in history. The books of Agnes Mure Mackenzie and M E M Donaldson were favourites. The latter wrote *Scotland's Suppressed History*, 'intended as an Eye-opener for Everyman', a version of Scottish history seen from the point of view of the Episcopal Church. At Galashiels, the Episcopal rector wrote in 1942: 'St Peter's Church has stood for and worked for the true catholic and apostolic faith of Scotland for eighty-eight years. Without condemning others, we have held fast to the Faith once delivered through the apostolic succession of Bishops, Priests and Deacons.'[100] In Compton Mackenzie's *Hunting the Fairies*, Aeneas Lamont described himself as a 'Scottish Episcopalian Catholic, like St Columba',[101] and this would have been the view of many Episcopalians.

St Saviour's Child Garden was founded in 1906, with close links to Canon Laurie and Old St Paul's congregation in the Old Town of Edinburgh, and lasted until 1977. It accepted children from the age of five, from some of the poorest families, its history being written as *The Life History of a Slum Child*.[102] Many tinkers and travelling people belonged 'by heredity' to the Episcopal Church.

A congregation generally knew the hierarchy of the members and the different roles of the sexes. Thus at Fort William in the 1940s the two churchwardens were the laird, Lochiel, and the station master. A similar position existed in Oban in the 1990s. Women were often not allowed in the sanctuary of the church; though they served on some vestries from the 1920s. At Oban, when the church was raised to cathedral status in 1920, the women, whose voices were needed to supplement the men, had to sit apart from the men, unrobed, behind the choir stalls. They were not admitted to the choir at St John's, Edinburgh, until 1927.[103]

The clergy were important figures in the community, especially for church members. In the 1920s, Dean Alexander Macinnes at Glencoe (d. 1933) would search out the young men of his congregation who were

Figure 8.11 This armorial sign was produced to draw attention to the presence of the Episcopal Church, and was erected near every church building. The idea stemmed from Provost Shannon's visit to the United States in the early 1960s, as part of a Bishop Seabury exchange, when he saw that Episcopal churches there were marked by a corporate-style welcome sign.

attending a Saturday night dance in Lent and send them home.[104] This was perhaps only possible in a community like Glencoe, but the keeping of church discipline was urged by the clergy in their magazines and no doubt from the pulpit. At Galashiels, during World War II, the rector complained that a large number of young people had made their Communion at the 11am Eucharist. This is 'meant to be a service of Worship and Thanksgiving for all and of Communion only for those who through age or infirmity or for other reasons are not able to come earlier. For young people to come at such a late hour to meet their Lord, even though they have prepared themselves with prayer and fasting, does seem like laziness, and sloth or slackness is one of the seven deadly sins.'[105] However, the bishop might exercise a harsher discipline. In the 1940s the verger at Fort William, though very devout, was excommunicated for alcoholism.

In 1950 St Margaret's, Newlands, south Glasgow, had 2,200 adherents, including 600 communicants. In that year there had been 45 baptisms, 21 confirmations, 16 marriages and 20 funerals. There were three clergy, two mission congregations and over 25 monthly or weekly organisations. Its magazine included news of the Episcopal Church's Home Mission Campaign for church sites and buildings, and there were details of the BBC's religious programmes by the month, with an article 'Why go to church when you can listen to good services on the BBC?'. Party politics were never mentioned in the rector's letters, which were always of a spiritual nature, though he did once mention that 'democracy tends to lower the standard of excellence to the average requirements'. The magazine included an article in 1951 titled 'The enemy at our gate', which began 'Communism, no matter what fair words may be used to cloak its true meaning, is a blasphemous doctrine and a perverse way of life'. In December of the same year there was an article against gambling: 'The very idea is repulsive ... Betting on pools shows utter selfishness and a lack of knowing how to help many really deserving cases and institutions which need our sympathy.'[106]

In the early 1960s, congregations were encouraged to participate in stewardship campaigns, and members were asked to contribute regularly what they could towards church funds. It was said that, if the congregation did this, there would be no occasion to have any other fund-raising activity. However, fund raising was an activity that many members enjoyed and it kept their attachment to the Church, as well as allowing other people in the community to show solidarity by supporting the Episcopal congregation.

The church building remained a home for its members, where an individual's significant moments were marked: baptism, confirmation, marriage and funeral. Booklets were distributed at these occasions and the confirmation class in particular was a time for receiving a Rule of Life. Most congregations had a branch of the Church Women's Mission Association, which gave a focus for the ubiquitous 'work parties' of women, and Mothers' Unions were established in a few congregations. In 1941, St Peter's, Galashiels, also had a Boys' Brigade, a Lifeboy team, a badminton club, a Girls' Guild, a Mission Society, a table tennis group, the Guild of St Agnes and a group that arranged socials.[107]

Where there was an Episcopalian school, Episcopalians were regarded as perfectly normal citizens and in no sense alien; children brought up in an Episcopalian atmosphere did not consider themselves to be unusual. George Henderson remarked that in Oban in the 1930s the Episcopalians did not think of any particular connection with the Church of England, or

Figure 8.12 Diocesan Festival Eucharist, St John's Cathedral, Oban, 1979, with the bishop (at the altar) as celebrant and Richard Holloway (later to be Bishop of Edinburgh and Primus) as preacher. The music was provided by the brass ensemble of Dunoon Grammar School and by guitars of the Fisher Folk, part of the USA Community of Celebration, at that time based at the Cathedral of the Isles, Cumbrae. Photograph by Eric Walton, Oban.

Anglicanism, but Richard Holloway records the difficulties of belonging to the 'English Church' in the west of Scotland in the 1940s. Knowledge of the Church's year was probably the one distinctive characteristic of the ordinary Episcopalian, which included the collecting of palm crosses and the attendance at a Christmas service, often at midnight on Christmas Eve.[108]

The Strathmore family were staunch Episcopalians and it was well known that Lady Elizabeth Bowes-Lyon, later Queen Elizabeth (1900–2002), had been confirmed at St John's, Forfar, as one of the local congregation. The genealogist, Neil, 10th Duke of Argyll (1872–1949), was an ardent Episcopalian, who built, in part himself, the stone bell tower at All Saint's, Inveraray. Dr I F Grant (1887–1983), the author of *Highland Folk Ways* and the founder of *Am Fasgadh*, the museum of Highland Life at Kingussie, was also an Episcopalian.

1965–2000 TIME OF RADICAL CHANGE

Church Affairs

In this period, numbers began to fall, though not as noticeably as in other Churches in Scotland or in the Church of England. This trend, combined with increases in clergy stipends and expenses, in congregational and building maintenance costs, and the general financial changes of the 1970s, led to far greater calls being made on central funding. At the same time the certainties of the Episcopal Church were challenged on all sides.

The Episcopal Church became the object of some hostile attacks, which it took to heart. During the unity talks of 1957, the press had already rallied the nation against ecumenical co-operation in a series of onslaughts. The interference of an American priest and bishop in the ecumenical scene in Edinburgh brought further attacks on Episcopalian discipline, as did Ian Henderson's *Power without Glory* (1967), which likened the Anglican Churches, and the Scottish Episcopal Church in particular, to the remaining imperialism of the British Empire.

The Episcopal Church was also challenged by a series of bishops, who exerted considerable influence: Kenneth Carey (1908–79) came from Cambridge, having written *The Historic Episcopate*, which many believed undermined the importance of episcopacy; Michael Hare Duke (b. 1925), from a background of pastoral psychology, questioned the purpose of the Church and its traditions; Alastair Haggart (1915–98) challenged the Church to be rigorously intellectual in its approach; and Richard Holloway (b. 1933) raised questions about the basis of belief.

In 1977, the requirement of clergy to accept the *39 Articles of Religion* was dropped, as they were no longer considered a standard of faith. This left the *SPB* alone as the standard. Women were allowed to be ordained as deacons in 1986 and as priests in 1994.

A critical report showed that there were more clergy in proportion to laity in the Scottish Episcopal Church than in any other Anglican Church in the UK and a decision was made to decrease the number of clergy as quickly as possible, introducing for a time a 'tartan curtain', which prevented any clergyman from outwith Scotland taking up a paid position in the Episcopal Church. In 1991, compulsory retirement at a certain age was introduced. However, the introduction of non-stipendiary ministry effectively increased the number of clergy far beyond previous ratios, mostly in the larger congregations. This increase became even more pronounced with the ordination of women, many of whom had been waiting for some time for the decision to allow their ordination.

In 1982 a General Synod was inaugurated, in line with other Anglican Churches, to govern the Church. It was given almost total powers and its membership of about 165, being elected by the dioceses, where a similar regime was introduced, saw the end of direct representation of each congregation. With control over doctrine as well as finance, the synod could inaugurate many changes that small congregations might have wished otherwise, including the virtual abolition of sustentation of stipends and the concentration of finances in certain specific areas, such as a few large urban housing estates.

The influence of the Anglican Communion continued to increase and the proliferation of Anglican Churches and provinces around the world meant that the Scottish Episcopal Church became one of many, and numerically one of the smallest. At the same time the Episcopal Church entered into a relationship of interchangeability with some Lutheran Churches, in the 'Porvoo Agreement', thus aligning itself as part of a larger European and Western union of 'middle of the road' Churches, but diluting the doctrine of the Apostolic Succession. On a different ecumenical tack, the synod voted to abolish the 'filioque' clause in the Nicene Creed, bringing it into line with the Orthodox Churches and more primitive practice, but ecumenical relationships in this quarter were soured by the ordination of women.

Denominational Characteristics
The principal characteristic of the period was the erosion of those former certainties in theology, ministry and liturgy, which had been the special mark of the Episcopal Church for centuries. The changing pattern of thought was seen in a new and more liberal attitude to the Bible and to tradition as sources of authority, and this was reflected in new attitudes to ministry, sexuality and marriage discipline.

A campaign of renewal in 1964, 'To Serve Thee Better', had been unable to write a clear-cut statement of belief and a request for a revised catechism or statement of faith was not pursued since 'so much of the traditional theology and teaching of the Church had been, and was still being, re-appraised and expressed in new forms, that any statement was likely to become "dated" within a month of its appearance'.[109]

The ordained ministry came under special scrutiny, and the theology of the greater involvement of the laity in congregational ministry, as well as in the councils and decision-making of the Church, went hand-in-hand with a reappraisal of the 'validity' of non-episcopally ordained ministers in other Churches. The ecumenical parish of Livingston and Local Ecumenical

Projects were signs of these changing ideas, which led to the practice of interchange of ministries in various places. Non-stipendiary clergy ('worker priests') and women's ordination were very visible alterations in the life of many parishes.

The general uniformity of liturgical practice, based on the *BCP*, which had been the distinguishing feature of the Anglican Communion, was completely overturned. The Eucharist became the norm for most congregations on Sunday mornings. This presented problems when children or visitors from other Churches were present, as they appeared to be marginalised. With a new emphasis on baptism being the entry into the universal Church, confirmation was no longer regarded as essential to the taking up of church membership. Children were admitted to Communion before confirmation, and members of other Churches were welcome without special invitation both to receive Communion and to the ecclesiastical structures.

A reaction to the fundamental changes, which in particular the advent of the General Synod had allowed, was seen in the Truth and Unity Movement, the Prayer Book Society (Scotland), and probably in the foundation of the Gaelic Society of the Scottish Episcopal Church.

Marriage discipline was altered when the marriage of people with a former spouse still living was allowed, under the bishop's discretion. This was an acknowledgement of the move in society which had seen divorce become far more acceptable.

Two institutions underwent dramatic changes. The Theological College, with its residential side for ordinands and those clergy wishing a 'refresher course', its historic library and its full-time staff of specialists, was sold and replaced by a Theological Institute, which provided largely part-time training courses for students all over the country. The Aberlour orphanage closed as an institution and the trustees, playing down its connection with the Episcopal Church, opened small units across the country in conjunction with local government.

Ecclesiastically, the Episcopal Church witnessed an influx of evangelicals, particularly in the Edinburgh diocese, with the dividing and planting elsewhere of their successful and popular congregations.

Theology, Ideology and Thought
The theology of the Church and ministry went through profound change during this period, questioning the fundamental basis on which the tradition had developed. The resulting uncertainty led to much experimentation and many *ad-hoc* arrangements, as if the success of any scheme or the acceptability of an idea was a sign of its truthfulness.

The theory that the Episcopal Church had a duty to minister to all Anglicans wherever they were in Scotland, and that it should be present in every large community, disappeared, as the existence and integrity of other Churches and their ministry were acknowledged.

It was emphasised that ministry was for the whole congregation and that the ordained ministry had specific tasks within that body. The concomitant of this was that ministry should not be restricted by gender, despite the witness of tradition down the centuries, and the ordination of women was regarded as a turning point, not just in discipline, but as an acknowledgement of the manner in which Bible and tradition were regarded. This theology of inclusiveness led to a questioning of many other traditional tenets, such as those concerned with sexuality, while marriage discipline was relaxed in line with current secular thought.

The Church's relationship with society was also re-examined and money redirected to the presence of spiritually aware and prayerful people, lay or ordained, in places of deprivation. As a token example, money was collected in a 'Community Fund' and distributed to people in real and often sudden need.

The changing ideas were seen in liturgical reforms. The Scottish Liturgy was given a mild but significant revision in 1970, making it more in line with services in other Churches, which was regarded very favourably by Gordon Donaldson as 'Hopes fulfilled'.[110] This was followed by the evolution of a completely new liturgy, which not only used modern language, but also had a new theological emphasis. Donaldson recorded that the element of sacrifice and atonement had been deliberately excluded.[111] The pax became a physical action, which many regarded as the outward sign of the new liturgical theology.

While synodical government had allowed many of these ideas to be expressed and supported, a group within the Church held to the traditional views and an example of their uneasiness was Arthur Hodgkinson's *Whither the Episcopal Church?*

Figure 8.13 Blessing the animals, Holy Trinity Church, Dunoon, Rogation Sunday 1984, a service that proved popular in a rural congregation.

Clergy
The fundamental changes in forms of ministry resulted in clergy being ordained who had a wide experience of society and life.

In 1975 only two of the bishops were Scots born, while in 1995 all the bishops were Scots and none had been to Oxford or Cambridge universities. This may reflect the changes in the way in which bishops were chosen and the vision of the sort of people that it was considered were needed. In 1974, the college of bishops rejected the elected nominee of the diocese of Glasgow and Galloway as not fitting in with the way that the bishops felt that the Church was moving, though his election was confirmed when the diocese voted for him again.

Bishop Alastair Haggart's intellectual views were considered highly both in the Churches in Scotland and overseas. Bishop Richard Holloway's doctrinal views hit the headlines, inspiring many in all Churches, but sometimes embarrassing and annoying both traditionally minded and evangelical Episcopalians. Canon Kenyon Wright had a high profile at the time when the Scottish Parliament was being established. Though not ordained, Professor Donald Mackinnon (1913–94) was an academic philosopher of great influence.

Worship and Ritual
In line with the rest of the Anglican Communion, the *BCP*, and *SPB* version, lost its unifying pre-eminence and at the same time the Eucharist became the normal morning service; apart from some traditionalist congregations, where Mattins still existed. The exceptions to this generality were the non-Eucharistic 'family services' and the 'healing services', which were introduced, far removed from the liturgy of the *BCP*.

The Scottish Liturgy of 1970 allowed a certain flexibility in language and approach, and the 1982 Scottish Liturgy was consciously designed to have as few instructions as possible, allowing the involvement of the laity in virtually all parts of the service, including the prayer of consecration and the absolution. A new lectionary, shared with many other Churches in Scotland and world-wide, and a revised Church calendar lost much of the individuality of the *SPB*. New texts were also prepared for other services.

The interiors of church buildings were often rearranged so that the altar became free-standing and the celebrant could stand behind it. This was more in line with Roman Catholic than Presbyterian precedent. Traditional ceremonial and ritual largely disappeared and kneeling for prayer became almost obsolete, hassocks being colourfully embroidered as a decoration rather than for use.

Congregational and Individual Characteristics
In the period, loyalty, either to Anglican upbringing or discipline, was no longer enough to ensure that people remained Episcopalian, especially when they moved house, itself an increasingly frequent occurrence. For many people, the congregation, whatever its denomination, and its relationship to

Figure 8.14 Ordination of the first women priests in the diocese of Argyll and The Isles. Joanna Anderson, warden of the abbey of Iona, and Gillian Orpin, diocesan chaplain. Oban, 26 January 1995.

the general community became the draw, while a survey of Episcopalians carried out in the 1980s revealed that it was the liturgy and style of worship that drew them to the Episcopal Church.[112]

In Oban in the 1990s, a third of the Episcopal congregation of 700 (representing nearly 10 per cent of the population, and including adherents) were local, including Appin 'hereditary Episcopalians' and those educated at the Episcopal school (closed 1997/8), a third came from elsewhere in Scotland, and a third were English incomers. Several of the last two were not Episcopalian in origin. At St Mary's Cathedral, Edinburgh, the huge number of adherents noted in the 1960s and 1970s, living, due to rehousing, all over the city and formerly members of the three 'mission congregations', completely disappeared. In 2000, the cathedral congregation with 1,098 communicant members, 49 extra adherents and an average Sunday congregation of 500[113] consisted largely of families drawn by the musical excellence and the congregational community.

Congregations were no longer the 'second home' for people, and general visiting by the clergy of their congregation, apart from those in need, was abandoned as a normal practice, though older members of a congregation still expected it. Little social life centred on the church community. Most people who attended church did so for a service and might remain for coffee fellowship afterwards; only the very committed were involved in house groups and other courses. On weekdays, church buildings were either kept locked or used as centres for the wider community, with some churches becoming multi-purpose. In many ways the lack of congregational social events meant that members became more involved in events in the wider community. Christian Aid Collection and the World Day of Prayer were occasions that stretched across the Churches. Ecumenical co-operation was shown in clergy from other Churches being seen far more frequently in Episcopal churches and the clergy of the different Churches often knew each other better than their congregations realised.

Baptism was not automatically sought, confirmation largely disappeared, and a church marriage (or remarriage), if asked for, was usually in church. The custom of cremation rather than burial resulted in very few funerals being held in church, except where there was a graveyard, and funerals were generally regarded as services of thanksgiving, rather than prayer for the dead and bereaved. However, at Ballachulish in the 1990s the old custom continued of daffodils being placed on all the graves at Easter, as tokens of the Resurrection.

An ignorance of doctrine and the Bible became a characteristic of congregations (as of society at large), with little enthusiasm for spending time to remedy this. For some older people, who thought that they knew certainties, the changes of doctrine and practice dissolved their loyalty to their Church. Despite this, most congregations had an inner group of committed members who wished to learn more, and undertook Bible studies, house groups, 'cursillo' and 'alpha' courses.

Political concern for the poor and unemployed had become a challenge for the Episcopal Church in the 1960s, which put resources into keeping an Episcopalian presence in the east end of Glasgow, the Gorbals and Govan. Political involvement of the laity, however, was considered by many as a duty. Commitment to rent crusades, housing trusts, nuclear disarmament and 'peace, justice and the integrity of creation' were regarded as ways in which the Church and its members could be involved.

The 'Partners in Mission' consultation in 1982 gave the advice that the Church as a body, and its individual members, had forgotten their past and needed 'to re-learn their story' and so be assured of their identity. This was echoed by Bishop Richard Holloway, who said that the Scottish Episcopal Church needed to act as if, and believe that, it had not 'dropped out of the sky *de novo* in 1980'.[114] It had forgotten its story and past and was rapidly becoming a club for like-minded people. In a book of a 100 Scots' 'personal reflections on Scottish identity today', twelve are probably Episcopalian, but only two mention the fact as having any bearing on their Scottishness.[115]

Among the laity who should be mentioned are Lord Kilbrandon (1906–89), whose Report on Marriage pointed out that a marriage without the intention of lifelong commitment could not be deemed to be a marriage; Professor Gordon Donaldson (1913–93), Historiographer Royal and liturgist, who was drawn to the Episcopal Church by the strength of the truth that it proclaimed;[116] and Hamish Henderson (1919–2002), the collector of oral tradition, who was brought up in humble conditions as an Episcopalian and always believed that the Episcopal Church was historically 'the Scottish Church'.[117]

The Episcopal Church at the end of the twentieth century was generally regarded, both inside and outside its borders, as the 'Scottish branch of Anglicanism', or perhaps as the 'Anglican version of Scottish Christianity'. The characteristics of Episcopalians were a willingness not always to go with the majority, or even with the prevailing view. Circumstance of upbringing may have placed them in the Episcopal Church, but for most members their continued adherence was due to choice.

ABBREVIATIONS

BCP   Book of Common Prayer, 1662
SEJ   Scottish Ecclesiastical Journal
SEM/EEJ   Stephen's Episcopal Magazine or Edinburgh Ecclesiastical Journal
SPB   Scottish Book of Common Prayer, 1929
SSB   Scottish Standard Bearer

NOTES

1. Rev. Robert Lyon was executed at Carlisle, 28 October 1746. His theological views are contained in a letter to his mother and sisters and in his 'Last and Dying Speech', Forbes, 1895–6, 3–21.
2. Lawson, 1843, 313.
3. Craven, 1886, 249–55, 327–32, 341–8.
4. Nockles, 1994, 50–3. The attachment to the Stewart monarchy, as well as the Act for securing Presbyterian Church government as a condition of the Treaty of Union, resulted in a clear anti-Union stance in the period after 1707. Pittock, 1995, 88–110, where it is remarked on p. 97 that 'Episcopalians were political nationalists and ecclesiastical unionists, while the established Presbyterian Kirk was the reverse'.
5. Dr Archibald Cameron's education, quoted in Lenman, 1980, 27.
6. Lenman, 1980, 223.
7. White, 1982, 164.
8. Lenman, 1980, 129; Donaldson, 1966.
9. Lenman, 1980, 129–31; Henderson, 1934.
10. Rev. John Burnett at Monymusk. Bertie, 2000, 531.
11. Craven, 1886, 103; Craven, 1907, 256–7.
12. Lenman, 1982, 36.
13. Clarke, 1987, 15.
14. Skinner, 1818, 173.
15. Dowden, 1922, 11.
16. Eeles, 1910, 30–2; Logie, 1904, 81–2.
17. Dowden, 1922, 225–6.
18. e.g., Muthill congregation. Shepherd, 1907, 19–20.
19. Clarke, 1987, 18; Ranken, c1900, 54.
20. e.g., description of 'English chapel', Inverness, by James Boswell, on his tour with Dr Johnson, 29 August 1773: 'The altar was bare fir table, with a coarse stool for kneeling on, covered with a piece of thick sail–cloth doubled, by way of cushion.' Chapman, 1924, 241. The various versions in the development of the design are recorded by Dean Hatt in SSB (April 1897), 87–8. Further details in Eeles, 1910, 14–18.

21. Smith, 1966, 18–19.
22. e.g., Longside 1727–42, quoted in Clarke, 1987, 21.
23. Clarke, 1987, 18.
24. Craven, 1886, 24–30.
25. Ranken, c1900, 55.
26. Goldie, 1976, 104; Lawson, 1843, 305–8.
27. Clarke, 1987, 17.
28. Ranken, c1900, 58–9.
29. 'I die an unworthy member of the Episcopal Church of Scotland, as established before the Revolution'. Mackay, 1931, 297.
30. Mather, 1977, 540–72.
31. Ford, 1997.
32. The cost of 31 new churches at £78,000, enumerated in Lawson, 1843, 485–514.
33. Skinner, Forfar, circular letter to the Bishops and Clergy in Scotland, NAS CH12/12/1153.
34. SEM/EEJ, I, 264.
35. Rev. Robert Montgomery, St Jude's, Glasgow, quoted in Lawson, 1843, 431.
36. SEM/EEJ, I, 264.
37. Quoted in Strong, 2002, 169.
38. SEM/EEJ, III, 222.
39. Ranken, c1900, 57–8.
40. SEM/EEJ, I, 22.
41. Strong, 2002, 279.
42. Walker, 1904, 123, 127.
43. SEM/EEJ, I, 220–7.
44. SEM/EEJ, I, 220–7; Reminiscences of the Rev. Alexander Cruickshank of Muthill, 1783–1834, SSB, (June 1892), 113.
45. Walker, 1878, 32.
46. Strong, 2002, 84–5, 105.
47. Walker, 1904, 117. Dean Hall describes the significance of these dates in 'Some traditional elements in the Scottish Church', SSB (May 1897), 119.
48. Eeles, 1910, 85; Hatt, SSB (May 1897), 119; Walker, 1904, 122.
49. Walker, 1904, 115; Ranken, c1900, 58.
50. In Dundee. Humphrey, 1896, 36.
51. Humphrey, 1896; Hill, 1956.
52. Humphrey, 1896, 19.
53. SSB (Feb 1895), 39–40. There is a description of the spirituality of a farming family, when the son, aged 19, died of scarlet fever in 1849, written by his sister, Isabella Davidson, a dressmaker at Hatton, near Cruden, in SSB (May and June 1904), 99–101, 128–30.
54. Anderson, 1885, 103–17.
55. Strong, 2002, 70–137.
56. SEM/EEJ, III, 224.
57. SEM/EEJ, I, 223.
58. Strong, 2002, 115/6, 120.
59. Balfour–Melville, 1959, 16.
60. Strong, 2002, 171–2.
61. Eeles, 1910, 124–32.
62. Eeles, 1910, 136–37.
63. Walker, 1878, 64–5; Ramsay, n.d., 66.
64. Humphrey, 1896, 2; Ranken, c1900, 58–9; SSB (May 1897), 118.
65. Luscombe, 1996, 7.
66. Strong, 2002, 219, 230, 235.

67. A description of the founding and life of the school at Inverurie is in 'The Piskie's School' Low, 1904, 105–13.
68. Strong, 2002, 227.
69. *SSB* (Feb 1898); Galashiels was a 'splendid type of working man's church', *SSB* (Nov 1895); Strong, 2002, 195–9.
70. Walker, 1904, 164.
71. Episcopal Church Society, *Report* (1843), App 54.
72. For example, Sir James Riddell of Ardnamurchan did not allow a site at Strontian and a boat was used as a floating church by the Free Church.
73. Ranken, c1900, 66–7.
74. Strong, 2002, 181–2. 'Women's work in the Church', gives the speech by Lady Frederick Cavendish and the responses. *SSB* (Jan 1899), 4–6.
75. Quoted in Strong, 2002, 290.
76. Maclean, 1996, 112–3, 121; Luscombe, 1996, 96–7.
77. For earlier plate, Eeles, 1910, 25–28.
78. Maclean, 1997, 48–57.
79. Strong, 2002, 110.
80. Perry, 1939, 156.
81. Dean Nicolson at St Salvador's, Dundee, visited every family in his congregation twice a year, over and above special visitations. There is a description of his methods in 'Personal recollections of two Scottish Deans', *SSB* (Nov 1899), 234.
82. Description of a house funeral, with an open coffin, at Cruden, is in Low, 1904, 171–2.
83. Strong, 2002, 199–200; Clarke, 1987, 20.
84. *SSB*, (Dec 1899), 272–4; Strong, 2002, 180–2.
85. Strong, 2002, 65.
86. Low, 1904, 145–53.
87. Hill, 1956, 80–1; Church work among fisher-folk, 1 Old-time customs among the east coast fishers, *SSB* (May 1900), 105–9.
88. Church work among fisher-folk, Stornoway, Peterhead, Torry, Buckie, *SSB* (1900), 128–30, 152–5, 174–7, 240–3, 257–9; Strong, 2002, 118.
89. Edinburgh Church Building Society, *Report* (1847).
90. Strong, 2002, ch. 7 and p. 132.
91. Probably James Aberigh-Mackay (1820–1908), Gaelic–speaking incumbent of St John's Inverness, ordained in the USA, and runner-up in the election for the bishopric of Moray in 1851; *SEJ* (Oct 1856), 157–8, in which there is also other correspondence about 'Highland Charges'.
92. Clarke, 1987, 18.
93. Low, 1904, 43.
94. *SSB* (Oct 1896), 232.
95. White, 1998, 82–9; White, 1996/7, 49–56; Luscombe, 1996, 104.
96. Luscombe, 1996, 119.
97. The Scottish Episcopal Church, 1960.
98. Quoted in White, 1998, 121.
99. White, 1998, 53, 145.
100. *St Peter's Magazine*, (August 1942).
101. Mackenzie, 1949, 126.
102. From 1960 it was run by the local authority, due to increasing costs. McCallum, 1997, 15–16.
103. Balfour-Melville, 1959, 41.
104. Personal conversation with James Christie, Oban, (1917–96).
105. *St Peter's Magazine* (April 1940).
106. *St Margaret's Magazine*, 44 (1952), 188; 43 (1951), 111, 32, 180–1.
107. From *St Peter's Magazine*.

108. Angus, 2001; personal conversation with George Henderson; Holloway, in Devine and Logue, 2002, 100–1.
109. Luscombe, 1996, 140.
110. Donaldson, 1975; Luscombe, 1996, 85–6; Maclean, 1996, 123–4. The first version, called 'the Grey Book', was issued in 1966.
111. Donaldson, 1990, 140; Donaldson, 1991, 5–6.
112. Partners in Mission consultation, 1982.
113. General Synod, *Report* 19 (2001), 76.
114. Personal conversation.
115. Richard Holloway and Michael Russell MSP in Devine and Logue, 2002.
116. Personal conversation.
117. Obituary, *Scotsman*, 11 March 2002.

BIBLIOGRAPHY

Anderson, I. *Inverness before Railways*, Inverness, 1885.
Angus, R J. *An Episcopalian Household*, for St Columba's-by-the-Castle, Edinburgh, 2001.
Balfour–Melville, E W M. *A Short History of the Church of St John the Evangelist*, Edinburgh, 1959.
Bertie, D M. *Scottish Episcopal Clergy 1689–2000*, Edinburgh, 2000.
Chapman, R W, ed. Johnson, Samuel. *Journey to the Western Islands of Scotland*, and Boswell, James. *Journal of a Tour to the Hebrides with Samuel Johnson LLD*, Oxford, 1924.
Clarke, T. The Episcopal Chest, the Jolly Kist and other Episcopalian Records, *Scottish Records Association*, Conference Report 7 (March 1987), 13–21.
Cooper, W. The Ministry in the Scottish Episcopal Church in the eighteenth century, *SEC Review* 2.2 (Winter 1993/4), 34–44; 3.1 (Summer 1994), 25–31.
Craven, J B, ed. *Journals of Episcopal Visitations of the Right Rev. Robert Forbes MA*, London, 1886.
Craven, J B. *Records of the Dioceses of Argyll and The Isles 1560–1860*, Kirkwall, 1907.
Devine, T and Logue, P, eds. *Being Scottish. Personal Reflections on Scottish Identity Today*, Edinburgh, 2002.
Donaldson, G. Hopes fulfilled: the 'Grey Book', *Liturgical Review* 5:2 (Nov 1975), 21–8.
Donaldson, G. Scotland's conservative north in the sixteenth and seventeenth centuries, *Transactions of the Royal Historical Society*, fifth series 16 (1966), 65–79.
Donaldson, G. *The 'Blue Book' and All That*, privately printed, 1991.
Donaldson, G. *The Faith of the Scots*, London, 1990.
Donaldson, M E M. *Scotland's Suppressed History*, London, 1935.
Dowden, J. *The Scottish Communion Office, 1764*, Oxford, 1922.
Edinburgh Church Building Society annual reports.
Eeles, F C. *Traditional Ceremonial and Customs Connected with the Scottish Liturgy* (Alcuin Club Collections XVII), London, 1910.
Episcopal Chest, National Archives of Scotland, NAS MS CH12/12.
Episcopal Church Society annual reports.
Farquhar, G T S. *Three Bishops of Dunkeld, 1743–1808*, Perth, 1915.
Forbes, R. *The Lyon in Mourning* (Scottish History Society, First Series, XX–XXII), Edinburgh, 1895–6.
Ford, D. D T K Drummond and the foundation of St Thomas's Church, *Book of the Old Edinburgh Club*, New Series 4 (1997), 51–68.
Goldie, F. *A Short History of the Episcopal Church in Scotland*, Revised edition, Edinburgh, 1976.
Henderson, G D, ed. *Mystics of the North–East*, (Third Spalding Club) Aberdeen, 1934.
Henderson, I. *Power without Glory, A Study in Ecumenical Politics*, London, 1967.
Hill, J P. *The Episcopal Chapel at Muchalls*, London, 1956.

Hodgkinson, A. *Whither the Episcopal Church?*, Aberdeen, c1970.
Humphrey, W. *Recollections of Scottish Episcopalianism*, London, 1896.
Lawson, J P. *History of the Scottish Episcopal Church from the Revolution to the Present Time*, Edinburgh, 1843.
Lenman, B. *The Jacobite Risings in Britain 1689–1746*, London, 1980.
Lenman, B. The Scottish Episcopal Clergy and the Ideology of Jacobitism. In Cruickshanks, E, ed. *Ideology and Conspiracy: Aspects of Jacobitism, 1689–1759*, Edinburgh, 1982.
Logie, I. *Notes on Episcopacy in Arbroath, 1596–1904*, Arbroath, 1904.
Low, W L. *Vignettes from a Parson's Album*, Dumfries, 1904.
Luscombe, E. *The Scottish Episcopal Church in the Twentieth Century*, Edinburgh, 1996.
Luscombe, E. *Steps to Freedom. Laurencekirk 1804*, Edinburgh, 2004.
McCallum, R I. Historical notes on Chessel's Court: St Saviour's Child Garden, *Book of the Old Edinburgh Club*, New Series 4 (1997), 15–16.
Mackay, D N. *Trial of James Stewart (The Appin Murder)*, Edinburgh, 2nd edn, 1931.
Mackenzie, C. *Hunting the Fairies*, London, 1949.
Maclean, A. Episcopal worship in the nineteenth and twentieth centuries. In Forrester, D and Murray, D, eds. *Studies in the History of Worship in Scotland*, 2nd edn 1996, 107–25.
Maclean, A. The Scottish Episcopal Church and the Ecclesiological Movement 1840–60, *Architectural Heritage*, VIII Caledonia Gothica (1997), 47–59.
Mather, F C. Church, Parliament and Penal Laws, some Anglo–Scottish interactions in the eighteenth century, *English Historical Review*, 92 (July 1977), 540–72.
Nockles, P. *The Oxford Movement in Context: Anglican High–Churchmanship 1760–1857*, Cambridge, 1994.
Perry, W. *Alexander Penrose Forbes, Bishop of Brechin, the Scottish Pusey*, Edinburgh, 1939.
Pittock, M. *The Myth of the Jacobite Clans*, Edinburgh, 1995.
Ramsay, E B. *Reminiscences of Scottish Life and Character*, Edinburgh, 22nd edn, [Foulis] n.d.
Ranken, A. *Our Scottish Episcopacy, being Sketches of the History of the Church of Scotland from the period of the Reformation*, Edinburgh, 3rd edn, c1900.
St Margaret's Magazine, Newlands, Glasgow.
St Peter's Church Monthly Magazine, Galashiels.
Scottish Episcopal Church. *The Scottish Episcopal Church and the Scottish Reformation, issued with the Approval of the Scottish Bishops*, Edinburgh, 1960.
Scottish Episcopal Church, General Synod annual reports.
Scottish Ecclesiastical Journal
Scottish Standard Bearer
Shepherd, J. *Episcopacy in Strathearn: A History of the Church at Muthill from Earliest Times to the Present Day*, Dumfries, 1907.
Skinner, J. *Annals of Scottish Episcopacy, from the Year 1788 to the Year 1816 Inclusive*, Edinburgh, 1818.
Smith, G. *St Peter's Musselburgh, A Short History of Episcopacy in Musselburgh from 1560 to 1966*, Edinburgh, 1966.
Stephen's Episcopal Magazine or *Edinburgh Ecclesiastical Journal*
Strong, R. *Episcopalianism in Nineteenth–Century Scotland, Religious Responses to a Modernising Society*, Oxford, 2002.
Walker, W. *Reminiscences, Academic, Ecclesiastical and Scholastic, Aberdeen in the nineteenth century until now*, Aberdeen, 1904.
Walker, W. *The Life of the Right Reverend Alexander Jolly, DD, Bishop of Moray*, 2nd edn, Edinburgh, 1878.
White, G. Hutchinsonianism in eighteenth–century Scotland, *RSCHS*, 21 (1982), 157–69.
White, G. Schools in Scotland: some SEC input, *SEC Review*, 5:2 (Winter 1996/7), 49–56.
White, G. *The Scottish Episcopal Church, A New History*, Edinburgh, 1998.

9 Other Christian Groups

JAMES A WHYTE

Each of the three words of the title of this chapter requires definition. 'Other' in this context simply means other than Presbyterian, Episcopalian and Roman Catholic, which have been dealt with in separate chapters. But what is meant by 'group'? A group, in the wider sense, is any gathering of more than two persons, usually held together by a special interest. It could be a congregational prayer-group, or an association or society within a denomination, or one which draws its members from different Church traditions – for example, a renewal group or an ecumenical association. One person may belong to many of these. On the other hand, the word group may describe what we normally call a denomination, or a sect. Unlike societies or associations, denominations are mutually exclusive, in that it is unusual, if not impossible, for anyone to belong to more than one at one time. However friendly and ecumenical the denominations are, you cannot be at the same time a member of the Methodist Church and a member of the Church of Scotland. People may move in and out of denominations, because of changing conviction, or from convenience. Methodists coming to Scotland may find no Methodist congregation within reach. They may join the parish church, which will usually welcome them. Then on moving to a city, or returning to England, they may revert to Methodism again. But, unless there is a congregation which itself straddles the two denominations, they cannot be a Methodist and Presbyterian at one and the same time.

The definition of Christian is, surprisingly, the most difficult. The denominations generally, however deeply they may disagree on doctrine or practice, recognise one another as Christian, as branches of Christ's Church. Ecumenical instruments, such as ACTS (Action of Churches Together in Scotland) bring the main Churches together in co-operation. But how far can recognition extend? Some would hesitate to include Unitarians, who deny the doctrine of the Trinity, as 'Christian'. There are many who would not wish to include groups such as the Mormons and the Jehovah's Witnesses, who define themselves over against the main Christian tradition. Since the purpose of this chapter is to describe rather than to judge, these groups will be included, as will, as far as space and information allow, the multitude of small evangelical and charismatic groups in Scotland today.

These groups are commonly known as 'the sects'. The defining characteristic of a sect (the word means 'division') is not, however, its size (it may be small or large), but its intense exclusiveness, the belief that we, and only we, have the truth, that we alone are the true Christians, we are the true Church. That claim may seem incongruous when made by a tiny group of believers in a dingy mission hall, but it is no different when it is made from the centre of a large and highly organised body of believers across the world. Some of the sects which are small in Scotland are of considerable size worldwide. In some of their attitudes the largest Christian bodies in the world (the Orthodox Church and the Roman Catholic Church) are, or have been, sects, claiming that they, and they alone, are the one true Church.

The situation is constantly changing. Groups which have made exclusive claims may find themselves drawn into co-operation with other Christians, and into ecumenical activity, and in so doing move from the attitude of 'sect' to that of 'denomination'. The smaller evangelical groups and the newer charismatic groups tend to form very loose associations which can readily change. New associations may come into being and then dissolve. It is very difficult to give an accurate picture of such groups, or one that will stand the test of time. It is well-nigh impossible to enumerate them all. But though there is such a variety of Christian groups, in Scotland as elsewhere, their numbers remain small, and even taken together would be only a small fraction of the main Churches: Presbyterian, Roman Catholic and Episcopalian.

Those who believe strongly in the unity of the Christian Church should perhaps be dismayed by the existence of so many groups and sects. Those who believe in the diversity of gifts may find in some of these groups a witness to some aspect of the truth that would otherwise be forgotten, and may consider that no one human group or Church can ever express all of 'the unsearchable riches of Christ'. We may thank God for the principled non-conformist.

The thrawnness of the Scot and his love of argument may make Scotland fruitful soil for religious division. While it is not the purpose of this paper to judge between the denominations and the sects, but to describe them fairly, it is impossible to ignore the difference between those which breathe a spirit of Christian charity and those whose main characteristic is their hostility to other Christian groups.

The writer has not attempted to deal with New Age religious groups. A group which draws its inspiration from the sayings of a dead American Indian chief to a medium may be admirably peaceful, charitable and beneficial in its activities, but it cannot be termed 'Christian' in its beliefs, and many of the New Age groups would not wish that appellation.

Most of the larger Churches, while accepting the authority of the Bible, believe that it must be interpreted and applied in the light of modern critical scholarship. Most of the sects, and mission halls, and some individuals and groups within the larger denominations are fundamentalist, i.e. they hold strongly to the inerrancy of Scripture and reject a critical approach.

Scotland has always been open to influence from outside. The influences that led to the Reformation of religion in 1560 were from the Continent of Europe – Lutheranism in Germany and Calvinism in Geneva. Contact and correspondence with the Continent never ceased, but in the seventeenth century the influences towards change, acceptable and unacceptable, were from the kingdom of England and from Cromwell's Commonwealth. Some of these, for example the Anabaptists, stemmed from the radical wing of the Continental Reformation. In the nineteenth and twentieth centuries the influences and importations were almost all from America. But even Scottish religion was not so rigid that it did not generate its own movements of change, some of which were exported in their turn. It was a surprise to a Scot visiting a theological seminary in South Carolina to see on the wall above the president's chair a portrait of Ebenezer Erskine, the leader of the Original Secession from the Church of Scotland in 1733 – more honoured there than in his native land.

Scots often conveniently forget that their country was conquered and occupied by the Commonwealth army in the 1650s. That occupation and forced union with England brought both Independents and Anabaptists, and their army chaplains, into Scotland. It also brought the Quakers. All three groups – Independents or Congregationalists, Baptists and Quakers – can trace their origins to the Commonwealth period. None, however, except perhaps the Quakers, survived the Commonwealth, and it is to later events that they owe their present existence in Scotland.

THE RELIGIOUS SOCIETY OF FRIENDS (QUAKERS)

The Religious Society of Friends today is known for its pacifist witness, for its gentle and practical Christianity, and for its strong social conscience. Its meetings for worship are characterised by silence and inward recollection, and waiting for the spirit. There is, as Quakers would admit, an expected progression, an unwritten order, within the meeting, but it is an order that can be broken or changed as the spirit moves. The membership today is, for the most part, educated, middle-class, and professional. It is undogmatic, and demands no credal subscription from its members, but tolerates a wide variety of views. It has no ministry and observes no sacraments, though some Quakers would hold that every daily meal is the communion with the Lord. Peaceableness would seem to many the chief, and most attractive, quality of the Quakers.

Yet in its origins the movement was far from peaceable. Its attitudes were aggressive and, far from being silent, its meetings were marked by fierce denunciations of the 'steeple-houses', as George Fox, its founder, contemptuously called the Churches. The nickname Quaker, which they gladly accepted, testifies to the highly emotional (and presumably quite noisy!) character of these revivalist gatherings. From 1653 on, bands of Quaker missionaries were entering Scotland, and some Friends who had retired from Cromwell's army were settling there. George Fox himself visited Scotland in 1657. The movement made converts in Aberdeen, Edinburgh, Inverness and the west of Scotland. There were probably always those who were irked by the dogmatic Calvinism of Scottish Presbyterianism, and looked for an emphasis on Christ-like living rather than on correct belief. Yet, at its height, the converts were few. The radical criticism of the Churches provoked at times opposition and persecution, which Quakers indeed expected, and reported to the 'Meeting for Sufferings' in London. But the Kirk was too much concerned with its own sufferings in the Covenanting period to give much attention to the Quakers, who remained few and became fewer.

After the revolution of 1689 they became more pietistic. They abandoned in time the archaism in dress and speech which had been their witness to simplicity and honesty, without deserting these virtues themselves. They had never abandoned their witness to the importance of the 'inner light', the testimony of the Spirit within, and the Christ-like life, as against credal conformity. Their meetings, always small, reflect a quiet spirit but now also a passionate concern for peace and for justice in the world. They play their part in ecumenical activity among the Churches, even though their rejection of sacraments makes it difficult for some to recognise them.

They remain few in number. There are fewer than twenty Quaker meetings in Scotland, and few of these would have more than twenty members. Yet their influence, and the respect in which they are held, is out of all proportion to their numbers.

CONGREGATIONALISTS (INDEPENDENTS)

Robert Browne, who is credited with being the founder of English Independency, visited Scotland in 1583, an exile from persecution in England. He was at first charitably received in Dundee and St Andrews, and then went to Edinburgh. His views were not so charitably received when he debated with the ministers about the nature of the Church and about sponsors in baptism. Browne's view was that of the 'gathered Church'. 'The Kingdom of Heaven is to be begun not by whole parishes, but by the worthiest only, be they never so few.'[1] The Kirk was committed to the view of the national Church, the parish Church, where everyone in the parish was under the care of the Kirk. Browne's view was that the 'Church' was properly, and only, the local, gathered, congregation. For the Presbyterians 'the haill Kirk' was the Church of Scotland, meeting in General Assembly. They were diametrically opposed. Browne was also opposed to 'nocent ceremonies' in worship, among which were the reciting of the Lord's Prayer and the use of sponsors in baptism. Browne's visit bore little fruit in Scotland, but in time the Commonwealth army brought many Brownists, as the Independents were often called. Prominent figures, such as John Owen, ministered to the Commonwealth troops. Yet when these troops returned to England, the Independent cause vanished for a time from the Scottish scene.

Mention may be made here of an interesting Scottish movement. John Glas, a son of the manse, was born in Auchtermuchty in 1695, and in 1719 was inducted as minister to the parish of Tealing, north of Dundee. He came

Figure 9.1 Portrait of Robert Sandeman.

to criticise the Kirk's theology and its establishment, and began to gather a group of like-minded people. In 1730 he was deposed by the General Assembly, and set up an independent congregation in Dundee. The octagonal building (now part of the halls of St Andrew's parish church) was long known as the 'kail kirk' because, along with weekly communion, Glas restored the practice of the *agape*, a common meal eaten between the services. At one time there were Glasite churches in Dundee, Edinburgh, Galashiels, Kirkcaldy, Montrose, Paisley and Perth. In 1790 they had 1,000 members, and more adherents. Glasites advocated infant baptism and seem to have felt no obligation to evangelise or proselytize. The Glasite Church became virtually extinct in 1999 with the death of its last elder. Their fine building in Barony Street, Edinburgh, is happily preserved.

In 1765 Archibald McLean left the Glasgow Glasite church and formed, with Robert Carmichael, the first Baptist church in Edinburgh (now Bristo church). This Glasite-influenced body, usually called the **Scotch Baptists**, was also known as the **Sandemanians** after Glas' son-in-law, Robert Sandeman (Fig. 9.1). Scottish Baptists today see this group as the originators of the Baptist cause in Scotland.

Scottish Congregationalism today owes its origins to a revivalist movement at the end of the eighteenth century. Revivalism had not been unknown in Presbyterian Scotland in the late seventeenth and early eighteenth centuries. A strong, emotional preaching could awaken people to a sense of their spiritual need and bring them to a new commitment and religious enthusiasm. In the seventeenth century David Dickson of Irvine was one whose name was associated with such revival. For the most part these movements were held within Presbyterianism, and were sometimes associated with the Communion seasons. The advent of a warm evangelicalism, in the 'praying societies', was joined to the popular dislike of patronage, which denied the people their right to call a minister, and led to the Presbyterian Secessions from 1733 on. Information also came from America of the 'surprising work of God' which had accompanied the preaching of the Calvinist Jonathan Edwards in Northampton, Massachusetts, in 1734–5. Edwards had been preaching on hell and damnation, and in such vivid terms that his congregation were struck with extreme terror. The desire for such an 'awakening' was widespread. Cambuslang saw a famous revival in 1742.

Edwards had been surprised by the reaction to his preaching. What surprised him became a technique with later revivalists. The preaching of sin and the dreadful consequences of sin aroused in the listeners an extreme fear of death and hell, before they were offered, in conversion, a way of escape from damnation and hell-fire, through submission and the acceptance of grace. This could lead to an evangelical concern for those who had not yet experienced such a conversion, but might also divide the Church into the saved and the unsaved. Ordinary church life and preaching often seemed unexciting after such revivals, and this led to an appetite for further revivals.

The brothers, Robert and James Haldane, were the sons of a landed proprietor. Robert had a brief naval career before settling on his estate. James, the younger, was for about ten years in the service of the East India Company. Both were converted in 1795 through the influence of an English Independent minister, David Bogue. They maintained their allegiance to the Church of Scotland, but felt called to the work of evangelism. From 1797 James engaged in preaching tours of Scotland and, with the support of his brother Robert, began, in 1798, the Society for Propagating the Gospel at Home, training catechists and missionary preachers. When James toured, ministers of an evangelical persuasion would often invite him into their pulpits to preach. On such occasions he did not hesitate to criticise the neighbouring ministers who were not evangelicals. After many complaints at this divisive behaviour, the General Assembly in 1799 passed an act debarring from the pulpits all who were not ministers, probationers for the ministry, or divinity students. This led the Haldanes to leave the Church of Scotland and to form Independent Congregations.

In 1808 the Haldanes themselves accepted the Baptist view that only believers should be baptised, and some churches followed them (see below, Baptists). Those who continued formed in 1812 a Union of Congregational Churches, which then numbered 55, growing to 96 by the year 1896. In that year the Congregational Union of Scotland was formed by a union with the churches of the Evangelical Union.

Figure 9.2 Interior of Congregational church, Lerwick, Shetland. Courtesy of Reverend Alan Catterall.

This owed its origins to James Morison who, in 1841, was deposed from the ministry of the United Secession Church for heretical views, which included the belief in the universality of God's love, the universality of Christ's atonement and the universality of the work of the Holy Spirit. He had been, and continued to be, influenced by the writings of the American revivalist Charles Finney, and used his methods of preaching. In 1843 he and three other deposed ministers (among them his father) formed the Evangelical Union, rejecting Calvinism, which stressed God's election and predestination to faith, and accepting Arminianism, which stressed faith as the free choice of the individual. The congregations of the Union kept the Presbyterian system of elders and kirk session. They were joined by some from the Congregational Union, who retained their system of government by the church meeting. The Evangelical Union had considerable success among the working classes, and by 1896, when it agreed to merge with the Congregational Union of Scotland, it numbered nearly ninety congregations.

Congregational churches have been found mainly in the towns and cities. There are social reasons for this. Independency required some independence of life and thought. Rural life was dependent on the laird, and had not the same freedom or resources. The shifting of population after World War II has affected Congregationalism as it has the other Churches, with congregations scattered, and no longer living near the church buildings.

Congregationalism has in modern times been socially active and concerned. It takes a critical view of Scripture, and its theology is liberal evangelical. It was the first of the Scottish denominations to ordain women (Vera Finlay, 1929). The Congregational Union played a full part in the ecumenical activities which emerged in Scotland in the twentieth century, in Councils of Churches, nationally and locally, and in quests for union. In 1970, after ten years of negotiation, proposals for union between the Congregational Union and the Church of Scotland proved unacceptable to both sides. The disproportion in size between the two made it seem like an absorption of the Union by the Church of Scotland. The Congregational Union continued to play its part in Multilateral Conversations. In 1988 a proposal to join the United Reformed Church failed to gain the necessary 75 per cent majority. Ecumenical considerations led to the decision, in 1994, to transform the union into the Scottish Congregational Church. In classic Congregationalism the word 'Church' applies properly only to the local congregation, and 27 churches withdrew at that point to maintain the independent principle. They joined the Congregational Federation, formed by those churches of the Congregational Union of England which, for similar reasons, did not enter into the United Reformed Church in 1972. In 2000 the Scottish Congregational Church had 56 churches and 6,697 members. The decision was made that year to follow six congregations of the Churches of Christ into the United Reformed Church. Ward Chapel, Dundee, dissented.

CHURCHES OF CHRIST

In 1808 Alexander Campbell, an Irishman studying in Glasgow, was much influenced by Greville Ewing, one of the founders of Scottish Congregationalism. He went to America and founded the denomination later known as the Disciples or Churches of Christ. Based on an understanding of the New Testament Church, they practised believer's baptism and weekly Communion, and held to the independence of the local congregation. Campbellite congregations began to appear in Scotland from the 1830s on, and the denomination flourished, largely recruiting from Scotch Baptists in the towns and in the fishing areas of Banff. Their theology of baptism, and of the work of the Holy Spirit, kept them separate from the Scottish Baptists. By the end of World War I there were 49 churches, and by 1933 the total membership was 2,809. Thereafter membership declined to 725 in 1979.

UNITED REFORMED CHURCH

In 1981 the Churches of Christ in Great Britain resolved to join the United Reformed Church, which had been formed in 1972 by a union of English Presbyterians and a large part of the Congregational Union of England. Six congregations of the Churches of Christ in Scotland, with a total membership of 257, joined the union. The Reformed Church of England and Wales thus became the United Reformed Church of the United Kingdom, with a tiny presence in Scotland. The accession, in 2000, of the majority of the Scottish Congregational has strengthened that presence, and there is now a Scottish Synod of the United Reformed Church.

BAPTISTS (ANABAPTISTS)

Among the soldiers of Cromwell's army of occupation in the 1650s were Baptists, and among the chaplains, Baptist ministers. They were not slow to debate and to proselytise, taking advantage of the toleration allowed by the Cromwell government. In garrison towns, such as Cupar, Fife, there was public debate between the parish minister and the Baptist chaplain. The Baptist church at Hexham in 1652 sent a missionary to Leith, but he, having left the society, was replaced in 1653. There is an account of public meetings at Bonnington Mill in 1653, when some people were baptised (or re-baptised) in the Water of Leith as believers. Groups existed in Edinburgh, Leith and Perth, and some parish ministers were said to have maintained Anabaptist beliefs. But Baptists became suspect and out of favour for political reasons, and when the Commonwealth troops departed, the Baptist cause in Scotland withered.

Like the Independents, the Baptists believe that the Church is a company of true confessing Christians, and they hold to the autonomy of the local congregation. Unlike the other Independents they repudiate the baptism of infants. From the fact that recorded baptisms in the New Testament are of converts and it is not clear whether the 'households' who were baptised included infants, they conclude that baptism should be on confession of faith after a conversion experience – i.e. 'believers' baptism', and they do not hesitate to baptise as believers those who have been

Figure 9.3 Canonmills Baptist church, Edinburgh. Photograph by David Veitch.

baptised in infancy. Hence, from the Reformation on, they were known as Anabaptists, those who baptise again. (One of the 'vestiges of the true kirk' which the Scots reformers discerned in the medieval Church was baptism, and no reformer repudiated his own baptism). The Baptists themselves do not regard their actions as re-baptism, since for them the baptism of an infant is not scriptural baptism. The classic Baptist position also requires baptism to be by immersion, and not by the pouring or sprinkling of water on the head. As with Presbyterians, Baptists do not believe that baptism is necessary for salvation.

The Baptist view of the Bible is generally conservative or fundamentalist. Baptists are strongly evangelical, and in the past have not hesitated to seek converts from other Churches. Today they play a cautious part in ecumenical activity, but would find it difficult to enter into union with any Church that practises infant baptism or that would compromise the independence of the local congregation.

These problems aside, however, the Baptist position is clearly attractive on one reading of the New Testament, and Baptist views emerged in Scotland in the eighteenth century. As noted, Scottish Baptists today see the Sandemanians (see above) as the originators of the Baptist cause in Scotland. Other small congregations were formed in the eighteenth century, at Keiss in Caithness (under Sir William Sinclair) and in the Lowlands, but a new start came in 1808 when James and Robert Haldane and some of the churches from their revival adopted the Baptist position. In 1869 the Haldanite churches united with the Scotch Baptists and the churches of the English Baptist Order (such as the Charlotte Baptist Chapel in Edinburgh) to form the Baptist Union of Scotland. This body now consists of 171 churches, with about 14,000 members. Statistics show a slow decline in numbers.

Most of the churches in the Union are strict Baptists, but there are a small number of open Baptist churches, which, while they themselves practise believers' baptism, admit to their membership, without re-baptism, those who have been baptised in infancy. Notable among these are Hillhead Baptist church in Glasgow, and Canonmills (formerly Dublin Street) in Edinburgh (Fig. 9.3).

METHODISTS

Beginning in 1751, John Wesley paid no fewer than twenty-two visits to Scotland and made some influential friends, but for a number of reasons the Methodist cause did not flourish north of the border. Methodism, like most revival movements, was Arminian in theology – i.e. it stressed the freedom of the individual to choose or reject the gospel. In Scotland, unlike in eighteenth-century England, the parish system was still effective, and the Kirk, through its education system, still had a hold on the minds and hearts of the people. The structure of Methodism, with conference, synods and districts, resembled Presbyterianism with its general assembly, synods and presbyteries, but whereas the Kirk had an educated and qualified minister in

Figure 9.4 The congregation of Nicolson Square Methodist church, 2004. Courtesy of Reverend Paul King.

every parish, the Methodist ministers were 'travelling preachers', who went round a wide circuit, most of the services being conducted by 'local preachers', who at first had no special education or training. These differences have become less sharp with the passage of time, but for that reason Methodist witness is less distinct from that of the parish Church. Generally, Methodists accept the results of modern scholarship, and take a critical, rather than a fundamentalist, view of the Bible. Following the lead of the Methodist Conference of Great Britain, the Scottish Synod of the Methodist Church plays an active part in ecumenical activity, in Councils of Churches, and in union conversations. A scheme of union with the Church of Scotland failed to gain Methodist approval in 1979, largely because of the discrepancy in size between the two bodies. Methodists took part in the Multilateral Conversations in Scotland, and participated in 2001 in SCIFU – the Scottish Churches Initiative for Union.

The Methodist Church has been active in social concern, and its Central Halls, for example in Edinburgh and in Leith, have been centres of social outreach. The Church is strongest in Moray and the central belt, and also in Shetland, where it has been the largest denomination after the Church of Scotland. Like other Churches, its numbers declined in the late twentieth century. In 2000 there were just under 4,000 members in mainland Scotland and another 500 or more in Shetland.

UNITARIANS

The Doctrine of the Trinity, that the one God is Father, Son and Holy Spirit, has been the subject of dispute since the early Christian centuries. The simplest way out of the tight-rope walking that is Trinitarian orthodoxy is to deny the doctrine completely, as at the time of the Reformation did Faustus

Socinus, who denied that Christ is God, save in the sense that the Father shared his power with him after the Ascension. The Church he formed was crushed, but his writings had continued influence. In 1704 the General Assembly of the Church of Scotland felt it necessary to condemn Socinianism, which suggests that some ministers were attracted to this position. Separate influences within Scotland itself (from such as James Fraser of Brea, Neil Douglas, Henry Williamson and George Harris) and from England (William Christie and Thomas Palmer) led to the establishment of Unitarian churches in Aberdeen, Dundee, Edinburgh, Glasgow and Montrose. The Universalist Church of America also sent a mission to Scotland during the nineteenth century.

Unitarians do not subscribe to creeds, but their belief generally is that God is one and that his love is available to all (hence they are sometimes called Universalists). They have an optimistic view of human nature, denying the doctrine of original sin. They tend to be cheerfully well-disposed to others, and have a strong social conscience. Each congregation is independent, but there is a General Assembly of Unitarian and Free Christian Churches in Britain, and a Scottish Unitarian Association within it. Most of the churches baptise infants and celebrate the Lord's Supper quarterly. The ministry is well trained theologically, and is open to women as well as men. At the beginning of the twenty-first century the Unitarians had four congregations in Scotland (in the four older cities), with a total membership of about 500.

Figure 9.5 Many Unitarian churches have a rich tradition of vocal and instrumental music. A service at St Mark's Unitarian church, Edinburgh. Courtesy of Reverend Andrew Hill.

CHRISTIAN (PLYMOUTH) BRETHREN

This movement owes its origins to J N Darby, then a curate in the Church of Ireland. A group which formed around him began, in 1827, to seek a fellowship more in keeping with the New Testament than the existing Churches seemed to be. The first meeting in England was at Plymouth, by which name they are popularly known, though today they prefer to be known as Christian Brethren.

They appeared in Scotland in 1838, and seem to have gathered groups that had seceded from the earlier evangelical Churches, such as the Evangelical Union and the Baptists. Though they have been prone to split into opposing factions (Open Brethren, Exclusive Brethren, the Church of God) these all have certain principles in common: a fundamentalist view of Scripture, lay leadership (no clergy), believers' baptism by immersion, a weekly celebration of the Lord's Supper, biblical teaching and exposition (with a particular fondness for interpretation of prophecy and the end of the world). Women are kept in a subservient place. The elder or senior brother may sometimes wield power such as no ordained minister has. Each meeting or congregation is known as an 'Assembly', and these are independent of one another. The Brethren are strongly evangelistic, admitting converts on the basis of saving experience rather than profession of faith.

The Open Brethren are among the most numerous, and increased greatly during a time of revival in 1859–60, though some of their gains were at the expense of other evangelical groups, such as the Baptists. Their work took root in the cities and industrial towns, and among the fishing communities of the east coast. In 1933 they numbered about 30,000 but during the second half of the century there was a decline, both in membership and in the number of their assemblies.

The Open Brethren always practised open Communion (a point of contention between them and other groups) and, while still holding strongly to their principles, have an open and brotherly attitude to other Christians. The question 'Are you saved?' is easily parried 'Yes, are you?', which leads to a long and happy account of the saving experience of the original questioner. The Open Brethren have produced some fine scholars, notably F F Bruce.

The Exclusive Brethren, in a split going back to 1848 in England, reject open Communion. Since they had been called out to reject the heresies of the Churches, they must have no fellowship with them. In 1908 there was a further split between a more moderate Glanton meeting and a more rigorous London meeting. An extreme view of exclusiveness has led to publicised cases where father and son were unable to have a meal or even a cup of tea together because they belonged to different sections of the brethren.

The Churches of God broke from the Open Brethren in 1892, arguing the need for separation, and the admission to Communion of only those who had been baptised by immersion as believers. They in turn split in 1904 on the question of independency or the centralised authority of 'overseers'.

These groups, with their tendency to in-fighting, are now very small in numbers in comparison with the Open Brethren.

THE MISSION HALLS

The evangelistic revivals of the nineteenth century led to the establishment of missions and mission halls in the cities, e.g. the Glasgow City Mission and Carrubbers Close Mission in Edinburgh. These were established, and are supported, by Church members, mostly laymen, influenced by the work of Charles Finnie, Dwight L Moodie and others. They were a form of outreach to the poor and were not intended to become denominations, but inevitably they become the Church for those who have found here their first and only experience of Christian fellowship. They are characterised by a traditional and warm evangelicalism.

CHURCH OF THE NAZARENE

In 1906 George Sharpe, a Scot who had been ordained in the Methodist Episcopal Church in the United States, was forced to resign from the pastorate of Parkhead Congregational Church in Glasgow because of doctrinal divisions. With his supporters he formed an independent congregation, and by 1915 there were ten congregations of the Pentecostal Church of Scotland. In that year they united with the Pentecostal Church of the Nazarene in America, and in 1919 the word Pentecostal was dropped to avoid association with speaking in tongues, which the Nazarene Church believes belonged only to the first disciples.

The Church combines a Methodist structure with an emphasis on the independence of the local congregation. It includes both infant and believers' baptism. Its doctrine claims roots in Methodism, and stresses Wesley's belief in entire sanctification. It is strongly fundamentalist in its view of the scripture, and evangelical in its outreach at home and abroad. The discipline of members includes the prohibition of strong drink and tobacco, extravagance in dress and attendance at theatres.

Worldwide, the Church of the Nazarene has over a million members. At the beginning of the twenty-first century it had nineteen congregations in Scotland, with a membership of 2,006.

THE SALVATION ARMY

The Salvation Army was begun by William Booth in 1865, as a Christian Mission to the social and religious needs of 'darkest England'. In 1878 the Mission became The Salvation Army. Seeing itself as 'a militant arm of the Church of Christ' it is organised in military terms, with officers in rank from Soldier to General, its congregations known as Corps (the military parallels being delightfully inexact), its buildings as Citadels, and its characteristic uniform. A strict military discipline is enforced. No other Church is more authoritarian in its structure. Its doctrine is fundamentalist and evangelical, with a belief in Scriptural Holiness or Entire Sanctification. The sacraments of baptism and the Lord's Supper are not observed. Because the Army's work has been mainly in areas of deprivation and degradation, the new recruit is obliged to abstain from alcoholic drink.

Booth's original aim was not to found another Church, but to link his converts to the existing Churches. It was perhaps inevitable that this did not happen, and the Salvation Army became, in fact, a Church. It now works worldwide with over a million members in 108 countries. Its aim is 'to preach the gospel of Jesus Christ and meet the human needs in his name without discrimination'. It came to Scotland in 1879, but there had been earlier contacts between William Booth and the Edinburgh Christian Mission. In the late nineteenth century it grew and developed, and benefited from spiritual revivals in some fishing communities in the 1920s. In the second part of the twentieth century, in common with other Churches, it has seen a decline in numbers. It now has 4,590 members in ninety-eight corps in Scotland.

It also has thirty-two social and community centres. Since the 1890s the Army has been noted for its programmes of social care, which it develops in response to changing social needs – for example, work among addicts and the homeless. Because of its social work, and the speed of its response to disasters, the Army has great respect among the public at large,

Figure 9.6 Salvation Army singers, c1900. Courtesy of the Trustees of the National Museums of Scotland, NMS SLA 58/23/1.

Figure 9.7 The Salvation Army is noted for its work among the homeless. Glasgow, 2003. Courtesy of the Salvation Army.

and gathers much of its resources from appeals to the Christian conscience. The picture of the Salvation Army Band and the diminutive Salvation Army girl bravely selling *Warcry* in the pubs is fixed affectionately in the public mind.

The Salvation Army is now ecumenical, willing to participate in councils of Churches and to co-operate in other ways.

PENTECOSTAL AND CHARISMATIC CHURCHES

Assemblies of God
Worldwide, the Assemblies of God is probably the largest of the Pentecostal Churches, with a membership of several millions. When it organised in 1924 there were four churches in Scotland. In 1992 there were 26 groups in Scotland, and in 2001 there were 50. Although growing, its membership is still not more than 3,000.

The local churches are self-governing, but there is a quarterly meeting of pastors and delegates, and an executive committee to take care of business. Baptism is of believers, and by immersion. Communion is celebrated weekly. The view of scripture is conservative, and the theology conservative evangelical. The Assemblies of God are not exclusive, but co-operate with other Pentecostal groups and with renewal groups within the Presbyterian and Episcopal Churches in a desire for Pentecostal revival within all the Churches.

Elim Pentecostal Churches
In 1915 Principal George Jeffreys, a Welsh evangelist, began, in Ireland, a movement of evangelism based on the 'foursquare' faith in Salvation through Christ, Divine Healing, the Baptism of the Holy Spirit, and the Second Coming. The Elim churches hold a fundamentalist view of the Bible. Baptism is of believers, by immersion, and Communion is weekly. They are not exclusive, and co-operate readily with other evangelical and renewal groups.

Modern Charismatic Groups
The Pentecostal Churches almost all have some connection to the 'speaking in tongues' at 312 Azuza Street, Los Angeles, in 1906. The modern charismatic movement is within the same stream, but emerges in Britain, first in England, then in Scotland about 1960. Michael Harper and Tom Smail were two who gave leadership in person and by their publications. Certain ministers, in the Church of Scotland, Church of England and other 'mainstream' Churches, came to believe that this was the way of renewal for the Church, and the Scottish Churches Renewal Movement came into being. From the mid-seventies new charismatic groups or Churches began to appear. These form very loose and changing associations: for example, the Gate Fellowship in Dundee fostered the infant Eden Christian Fellowship in St Andrews, but

Figure 9.8 Worship at the CLAN Gathering, St Andrews, 2004. Photograph by Iain Craig. Courtesy of New Wine Scotland.

encouraged it to independence. The CLAN Gathering (Christians Linked Across the Nation), held annually at St Andrews, brings together up to 3,000 people, from as many as 400 congregations and groups, including ministers and others from the Church of Scotland, the Episcopal Church and the Baptist Union as well as the charismatic fellowships. The desire is to be a blessing rather than an irritant to the Churches. Speaking in tongues is not particularly stressed in these fellowships, and they are realistic about vetting those who participate in charismatic worship, in order not to be dominated by cranks. They are inclusive in theology.

Christian Science
Founded by Mary Baker Eddy, who had herself had an experience of spiritual healing, the Christian Science Association was reorganised as the First Church of Christ Scientist in Boston, Massachusetts, in 1892. It is now a worldwide movement, and came to Scotland through two sisters, Miss C Lilias Ramsay and Miss E Mary Ramsay, of Leith, who had met Mrs Eddy in 1899 in the United States. An impressive Romanesque church was built in Inverleith Terrace in 1910. The second church, in Young Street, is now the only Christian Science Church in the city. A branch must maintain a reading-room, with Christian Science publications, open to the public. In Scotland there are churches and reading-rooms in Edinburgh and Glasgow, societies and reading-rooms in Helensburgh and Perth, and a society in Dumfries.

Christian Science is a form of idealism, for which God is the only reality, and matter and evil are alike illusions. Spiritual healing is practised, and there are recognised practitioners. The Bible and Mary Baker Eddy's *Science and Health with a Key to the Scriptures* are equally authoritative. The spirit of the movement is unlike some other modern American cults in that its exclusiveness is gentle and friendly and not aggressive and hostile to others.

The Mormons, the Seventh-Day Adventists, and Jehovah's Witnesses
All of these groups originated in the United States in the nineteenth century, and their presence in Scotland is as a result of their foreign mission work here. Many households in Scotland can recall having a visit, often after 11am on a Sunday morning. In the case of the Jehovah's Witnesses it could be a man or a woman, often with a young child, who offers to help you read and understand the Bible. They are usually Scots, not American, and only after questioning do they reveal that they are Jehovah's Witnesses. In the case of the Mormons they are usually two very well-scrubbed and tidy young men who have given a year or more of their lives to do this mission work, voluntarily and unpaid, in Scotland. The Seventh-Day Adventists may come bearing literature to prove that it was the Council of Laodicea (or, in the case of Scotland, Queen Margaret) who moved the Sabbath from the scriptural Saturday to Sunday. Each group is quite exclusive, has copious literature, co-operates with no other body, and does not hesitate to seek its converts from them.

The Mormons (Church of Jesus Christ of the Latter-Day Saints) were founded by Joseph Smith in New York in 1830 after the uncovering of a supposed translation of *The Book of Mormon*, which claims to be a record of peoples living in America between 600 BC and AD 400. Mormons hold a fundamentalist view of the Bible, but regard the *Book of Mormon* as a supplementary, and equally authoritative, revelation. Following faith and repentance, baptism is administered for the remission of sins, and the gift of the Holy Spirit is by the laying-on of hands. The Lord's Supper is celebrated at every service, water being used instead of wine. Strong beliefs in the pre-existence of the soul, and in a complicated set of stages in the life after death have led to two unusual practices. Plural marriage, practised by some, was in order to reduce the queue of souls waiting to come into this world, and baptism for the dead was to benefit unbaptised ancestors. The first is now illegal and no longer practised; the second, always carefully controlled, has given rise to the Mormon interest in genealogy, and the photocopying of records, which has made them more readily available to others. Mormons inculcate in their members a healthy diet and lifestyle, with abstinence from alcohol, tobacco, tea and coffee. As a result they claim a good record in health and social independence.

The Mormon Church is highly organised, ordering every aspect of the life of its members. It came early to Scotland. Among the early converts were two Scots, Alexander Wright and Samuel Mulliner, in Ontario. They returned to Scotland in 1839 to begin missionary work here, with some success. Modern missionary work has mainly been by young American males, backed by American money, and the allure, for the young, of a place to play basketball. How many of the young stay to believe and propagate the doctrines of the Church is not clear, but the Church now claims 2,300 members in Scotland and is determined to increase that number.

The Seventh-Day Adventists are a product of religious revival in America. A group led by William Miller broke up after the failure of his prediction that the second coming of Christ would be in 1844. Continuing Miller's belief in the second advent, but prudently declining to set a date for its final achievement, the Seventh-Day Adventists were able to continue, and are now a highly organised and aggressively missionary Church across the world. Their extreme fundamentalism led them to hold the seventh day of the week as the Sabbath (a view shared with some Baptists in America). They hold a doctrine (which Calvin had been at pains to refute in the sixteenth century) of the Sleep of the Souls – i.e. that the dead sleep until the general resurrection. They are exclusive, and co-operate with no one. Baptism is by immersion on profession of faith, and Communion is celebrated quarterly, preceded by a ceremony of foot-washing. Ministers are ordained after a six-year course, followed by a time of probation. In the absence of a minister, an elder may take the Communion service. Women are allowed to be Bible workers.

Like the Mormons, the Seventh-Day Adventists place much emphasis on health. Alcohol and tobacco are banned, the use of drugs in medicine is discouraged, and nature cures and health foods are manufactured and produced by the Church. Again, like the Mormons, this Church seeks to order every aspect of the life of its members. Missionary work began in Scotland in 1886, and the first congregation was established in 1901–2. There are nine congregations in Scotland with a membership of under 300, but there is a health centre and nursing home, and a retirement home run by the Church.

Jehovah's Witnesses is the preferred name of a body which has been known as Millennial Dawn, International Bible Students Association, and Watch Tower Bible and Tract Society. Founded in 1874 by Charles T Russell, Pittsburgh, it was led after his death in 1916 by Judge Rutherford. It denounced all the Churches under the control of the Devil, and set forth remarkable doctrines, which it claims are based on the Bible. Before his birth, Jesus was the Archangel Michael, and he and his apostles are the latest in a line of witnesses to Jehovah. The doctrine of the Trinity is thus rejected. There are teachings about the end of the world (the Second Coming took place in 1914) and the future life, but the prolific literature of this movement is not noted for its consistency. Its congregations are called 'companies', and its buildings 'Kingdom Halls'. Baptism is of believers by immersion, and the Lord's Supper is celebrated once a year. The moral demands are strict, and Witnesses refuse military service. They also reject abortion, blood transfusions, tobacco and drugs.

The group is highly organised and authoritarian, and has the efficiency of a well-run business – which, in part, it is. It came to Scotland in 1890, and continues to be active, particularly in proselytising from door to door. It has companies throughout Scotland, and about 9,000 members.

The Cheerful Eccentrics
In the changing kaleidoscope which is religious life within the nation, one encounters not only the grim side of religion, the closed sects, convinced that they alone are right and everyone else is going to hell, but also the cheerful side, the tiny groups which do their own thing with great good will to all. Some of these have already been mentioned, but they are probably more than can easily be numbered. One such is the Ancient Orthodox Church in Milton of Campsie. The Most Revd Dr Ronald T Pengilley Harwood, now, alas, somewhat frail, is clear that the aim of this small body is not to be the true Church, but to be an expression of it, in orthodox doctrine and gentle spirituality, and in unity, not rivalry, with fellow Christians. Such bodies, however unusual their nomenclature or their rituals, may, in their own way, enrich the tapestry of religious life.

Strangers Within our Gates
There are many people living in Scotland whose native language is not English. Like the Scot abroad they seek worship in their native tongue. Thus there are German congregations in Edinburgh and in Glasgow; there is in Glasgow a cathedral of the Orthodox Church, and in Kirkcaldy a

congregation of the Coptic Orthodox Church, which ministers to Coptic Christians all over Scotland. There is a Scottish-Asian Christian Fellowship, begun in 1998, which includes members of Indian, Pakistani and Scottish origin, from many different Churches, including the Roman Catholic. Services are held monthly, and on special occasions, in a Free Church of Scotland hall in Glasgow, but the Fellowship does not aim to be a substitute for participation in the local church.

It is good to see imaginative responses to the new religious needs that develop in our pluralist society, and to be able to end this chapter on such a positive note.

NOTE

1. Willison, G F. *The Pilgrim Reader*, Garden City, NJ, 1953, 7.

BIBLIOGRAPHY

Brierly, P and Macdonald, F, eds. *Prospects for Scotland 2000*, Edinburgh, 1995.
Brierly, P, ed. *UK Christian Handbook*, Millennium Edition, London, 2000.
Cameron, N M de S, ed. *Dictionary of Scottish Church History and Theology*, Edinburgh, 1993.
Thomson, D P, ed. *The Scottish Churches' Handbook*, Dunfermline, 1933.

10 The Jews in Scotland

KENNETH COLLINS

Jewish history in Scotland extends back little more than about two centuries with communities being established in Edinburgh in 1816 and in Glasgow in 1823. However, there is evidence of Jewish life in Edinburgh from the end of the seventeenth century, and in the medieval period there are indications that Jews, usually based in England, helped finance royalty, noblemen and the Church.

MEDIEVAL HISTORY

There is no record of Jews settled in Scotland before the expulsion of the Jews from England in 1290. However, from the last years of the twelfth century there are records of financial transactions between Scots and Jews. Robert de Quincy of Edinburgh came to an arrangement in 1171 with the monks of Newbattle Abbey to enable him to repay the sum of £80 borrowed from a Jew called Abraham.[1] A few years later there was a regulation from the bishop of Glasgow, passed between 1181 and 1187, prohibiting 'Churchmen pledging their benefices for money borrowed from the Jews or other usurers'.[2]

Abraham is unlikely to have lived in Scotland and it is more probable that he was based in York, or perhaps in Lincoln, the home of Aaron of Lincoln (d. 1185), who lent the considerable sum of £2,776 to William I, king of Scots.[3] Medieval Scotland did not have the population or urban life likely to attract a Jewish community and possibly lacked the necessary royal protection to ensure its safety.

Jews, usually converts, first appear in Scotland in the middle of the seventeenth century. The first and third occupants of the Chair of Hebrew and Oriental Languages at the University of Edinburgh, Julius Conrad Otto and Alexander Amadeus, both converted Jews, were appointed in 1641 and 1679, respectively.[4] Otto and Amadeus were appointed before the official resettlement of Jews in England in 1656, at a time when there were no Jews in Scotland. At that period, academic appointments could be held only by professing Christians.

THE SEVENTEENTH AND EIGHTEENTH CENTURIES

In the seventeenth and early eighteenth centuries it was necessary to obtain permission from the town council to reside and work in Edinburgh. In

September 1691, the council approved the petition of a Jew named David Brown, a relative of the *Haham*, the religious leader of the Sephardi Jews in London, to live and trade in Edinburgh as it was '... denied that there are lawes in this Kingdom restraining Jewes fra liberty of trade'.[5] Brown, or Pardo (the names were interchangeable at this time), is thus the first professing Jew identified in Scotland. A few other Jews appear in Edinburgh documents of the period. Moses Mosias was admitted to the roll of burgesses in 1698, while Theodore Marine and Isaac Queen (?Cohen) were given permission to trade in 1700 and 1717, respectively.

Other Jews living in Edinburgh during the eighteenth century have been identified from local and national records. Yet, despite the presence of a few individual Jews in the city, there was still no settled Jewish community. However, from the middle of the eighteenth century Jewish links with Scotland were growing. This was mostly due to the increasing reputation of Scottish medical schools. Edinburgh Medical School, which did not have religious tests for entry to study or graduate, in particular attracted a steady stream of Jewish students to the city.[6]

The attachment of Jews to the medical profession dates back to antiquity. In Scotland, however, the conditions for their entry into the modern medical profession only occurred from the middle of the eighteenth century. The first Jewish graduate in Scotland was Jacob de Castro Sarmento (1691–1762), who was awarded an MD by Marischal College, Aberdeen, in July 1739.[7] At that time the colleges in Aberdeen awarded medical degrees to established practitioners on the affidavits of their colleagues. Though such a procedure was open to abuse, it enabled a regular number of overseas Jewish doctors, fifteen in number between 1739 and 1824, to obtain British medical qualifications.

Jewish medical undergraduates first appear at the University of Edinburgh in 1767 and in 1779 New York-born Joseph Hart Myers became the first Jewish physician to have graduated in Britain after a period of regular undergraduate study.[8] During the last decades of the eighteenth century and the first years of the nineteenth century there was hardly a year in which there was not a Jewish student enrolled at the University of Edinburgh. As professional options were limited for Jewish doctors without money or patronage, few of these first Jewish graduates based their medical careers in Scotland and many emigrated to places like Australia and the West Indies.

JEWISH IMMIGRATION

Jewish communities were formally established in Edinburgh in 1816 and in Glasgow in 1823.[9] These communities remained small at first and it was not till the end of the nineteenth century that there was larger-scale Jewish immigration to Scotland, when the Glasgow Jewish community came to outstrip that in Edinburgh. Smaller Jewish communities were also established in Dundee (1874), Aberdeen (1893) and Greenock (1894), and later

Figure 10.1 Portrait of Dr Asher Asher. Courtesy of the Scottish Jewish Archives Centre.

also in Ayr (1904), Falkirk and Inverness (both in 1905) and Dunfermline (1908).[10]

In the first decades of the nineteenth century, Jewish immigrants came mainly from Germany and Holland. However, from the 1860s, as more Jews from eastern Europe passed through Scotland on their way to America, Jews from Russia and Poland gradually came to predominate.[11] Besides the growing merchant element, there were the first Jewish professionals. Two doctors of medicine, Louis Ashenheim (1816–58) in Edinburgh and Asher Asher (1837–89) in Glasgow, were the first Scots-born Jewish university graduates.[12] While some Jews became prosperous, a section of the community, predominantly those who had fled poverty and anti-Semitism in Russia, remained poor.

From the middle of the nineteenth century the growing commercial economy of Glasgow attracted more Jewish traders and merchants. Glasgow became home to the largest Jewish community in Scotland and the burgeoning industries of Victorian Glasgow provided employment for Jewish newcomers. Jewish efforts aided the rapid development of production particularly in the manufacture of clothing, furniture and cigarettes. The concentration of Jews in the tailoring trade from the 1880s led to accusations of sweated labour, charges that community spokesmen were able to rebut.[13]

From early in the nineteenth century many Jewish companies in Hamburg, interested in textiles, set up offices in Dundee and a community was established in 1874 and for a short time supported two synagogues.[14] In 1893, the new Jewish community in Aberdeen was the centre of national attention when it successfully defended a case brought against *shechita* by the local branch of the Society for the Prevention of Cruelty to Animals.[15]

There were about 2,000 Jews in Glasgow in 1891 and this number increased to about 7,000 a decade later. Scottish shipping companies became more involved with transporting emigrants from Europe via Scotland to America from the 1890s. Many travellers decided to end their journey in Scotland, joining family and friends already settled there. Travelling conditions could be difficult with migrants passing through Leith sometimes unfit for onward travel.[16] Some travellers had to remain in Glasgow until they were able to meet American health requirements. As a result, poor and sick Jews became a charge on local Jewish and parish welfare. Although the 1905 Aliens Act limited the number of the poorest immigrants entering Scotland, Jews continued to arrive in the years before World War I. About 10,000 Jews passed through Glasgow in 1908 alone on their way to America and by 1914 there were about 12,000 Jews in Glasgow and about 1,500 in Edinburgh. Between the wars about a thousand Jews arrived in Scotland, mostly refugees from Nazi persecution in Germany.

Some Jews came directly to Glasgow but others were attracted by local employment prospects or by attempts to reduce the numbers settling in London. The London Jewish Board of Guardians returned some 50,000 Jews to eastern Europe between 1880 and 1914 to try to reduce tensions produced by the immigration.[17] In Glasgow repatriation was seen as controversial and was only arranged in a few instances to re-unite families divided during the upheavals of mass migration.[18]

THE EARLY COMMUNITIES

Edinburgh's first synagogue was situated in a lane off Nicolson Street, and a small burial ground was soon purchased in Braid Place, Causewayside. This small burial ground sufficed for about sixty years till the opening of a larger cemetery at Echobank. The synagogue was moved to a hall in Richmond Court in about 1825 where it remained for some decades although for a short period, between 1833 and 1840, there was a breakaway group with its own place of worship nearby.

The first synagogue in Glasgow was situated in the first-floor High Street flat of the community's first minister, and *shochet*, Moses Henry Lisenheim. By 1831, when the first Jewish burial ground at the Glasgow

Figure 10.2 Entrance to the Jewish enclosure, Glasgow Necropolis. Courtesy of SJAC.

Necropolis, big enough only for fifty-one interments, was purchased for 100 guineas, there were forty-seven Jews in Glasgow. Indeed, the first interment at the prestigious new necropolis was of a member of the Jewish community, Joseph Levi, who had died of cholera.

By the middle of the nineteenth century there were about 150 Jews in both Edinburgh and Glasgow. The slow but continuing Jewish immigration

Figure 10.3 Jewish immigrant family, Glasgow, c1910. Courtesy of SJAC.

to the two cities was one strand among the many different groups then settling in Scottish urban centres. The Jews hoped to preserve their distinctive way of life while participating to the full in the commercial, professional and cultural life of their new country. Glasgow Jewry's expansion matched the rise of the city to become the Second City of the British Empire and its attempts to remain as one body were tested in a number of communal rifts and breakaways that plagued the fledgling community all too frequently.

A move by the Glasgow Hebrew Congregation in 1842 to premises leased from the Andersonian University provoked a communal split. The university contained a Medical School with facilities for dissection, which some felt made the premises unsuitable for prayer, especially for *cohenim*. In 1849 the community was re-united with plans to move to new rooms in Howard Street and to obtain new burial grounds at Janefield, bordering on Celtic Park, which were to remain in use for about sixty years. By the 1850s there were about 200 Jews in Glasgow and a larger synagogue was required. In October 1857 a tenement building at the corner of John Street and George Street was purchased, providing a synagogue, classrooms, committee room, accommodation for the minister and a *mikva*. The synagogue was enlarged in 1870 but the community now began to grow more rapidly. The formation and growth during the 1870s of a new community south of the River Clyde in the Gorbals ensured that the days when the Jews of Glasgow could be united in one synagogue were coming to an end.

In Edinburgh, too, the continuing growth of the Jewish community led to the search for more suitable premises. In September 1868 Ross House, in Park Place, was adapted for use as a synagogue. It remained in use for almost thirty years, gradually being surrounded by university buildings, until it was demolished in 1896 as part of a programme of local building improvements. A new synagogue was acquired in Graham Street, and this building, enlarged in 1913, served as the main Edinburgh synagogue until the present synagogue in Salisbury Road opened in 1932.

JEWISH HEALTH AND WELFARE

In response to the arrival of poor and sick newcomers the Jewish community formed a network of charitable and welfare institutions. Philanthropy was seen as an important activity in Victorian times and it was felt that relieving Jewish reliance on parish relief would reduce any potential anti-Semitism. The established community also saw it as their duty to cater for the particular Jewish needs of the poor and sick. Firstly there were the religious obligations, which included the observance of the Sabbath and other holy days and the provision of kosher food. In addition, the Jewish customs, and the close-knit nature of the community, made it easier, and more appropriate, for welfare provision to be provided by the Jews themselves.

The first Jewish welfare charity in Scotland was founded in Edinburgh in 1838[19] and a Glasgow welfare body, the Glasgow Hebrew Philanthropic Society, was in existence soon afterwards, having its own medical officer in 1858.[20] The increasing number of Jews settling led to a considerable increase

Figure 10.4 Odessa Lodge Friendly Society, c1907. Courtesy of SJAC.

in charitable activity, much of it led by the immigrants themselves. Self-help was promoted in Glasgow and Edinburgh by the Jewish friendly societies and by organisations such as the Glasgow Hebrew Benevolent Loan Society (1888), supported by the weekly 1d. subscriptions of their members. Jewish refugees' shelters were opened in Glasgow and Edinburgh in the 1890s to provide temporary accommodation for newcomers and to provide hostel facilities for onward travellers. Thus, the medical and welfare facilities of the immigrant community matched its religious and ethnic needs.

As the number of Jews increased, the local authorities made efforts to welcome the newcomers. In Glasgow, English language classes were provided and the Gorbals Bathhouse, opened in 1886, contained a *mikva*.[21] Local welfare provision also increased. The Edinburgh Jewish Board of Guardians was reconstituted in 1899 to co-ordinate and give support to the various Jewish welfare bodies in the city.

In the first years of the twentieth century, relatives were not allowed to bring in food to patients in hospital. Jews were unfamiliar with the Scottish diet and they not only felt that eating non-kosher food would be a hindrance to their convalescence, but also resented the restriction on their religious observance. The Victoria Infirmary turned down a request in 1910 to provide kosher food as they did not want to establish a precedent that might commit them to offering additional facilities to the Jewish or other communities in the future.[22] Merryflatts, now the Southern General Hospital, run by the Glasgow Corporation, opened a kosher kitchen in 1914 that functioned for many years.[23] The provision of hospital kosher meals remains a Jewish communal priority.

In Glasgow the Jewish Board of Guardians, an austere, remote body, remained based at Garnethill until 1911 despite the need for a welfare centre in the Gorbals, where the Board's clients were now concentrated.[24] The Board's perceived paternalism encouraged the proliferation of many small Gorbals-based welfare charities. In 1891 and 1892 the sums required for local Jewish welfare swamped local resources because of the dramatic increase in immigration. Although the considerable sum of £2,432 had been raised in Glasgow alone, money had to be sought from the Russian Jewish Relief Fund in London.[25]

In Glasgow the traditional small, overcrowded tenement flat was blamed for the city's poor record with tuberculosis (TB).[26] However, in immigrant areas throughout Britain the incidence of TB and other contagious disease was lower in the Jewish population.[27] Even so, the cost of regular grants to TB sufferers was a heavy burden on the Jewish community and the Glasgow Jewish Board of Guardians launched an ambitious Jubilee Fund appeal in 1916.[28] It was hoped to raise funds to send sufferers and their dependants to countries like Australia and South Africa, where the climate might help effect a cure, or to pay for treatment for patients at home or in sanatoria in England.

The slow and relatively small nature of the Jewish immigration and its concentration only in the heavily populated Gorbals area in Glasgow did not provoke much clamour for alien immigration control. The Jewish community was upset by reports that exaggerated their numbers and that gave false information about Jewish housing and health matters. Jewish leaders continually claimed that Jewish housing conditions were better than those of their neighbours.[29] However, there were concerns in the early part of the twentieth century that anti-Semitic attitudes were noticeable in the rented housing market.[30]

Given the sizes of the Scottish communities it was only in Glasgow that residential Jewish institutional care was a possibility. This catered for Jewish children whose parents had died young from diseases such as tuberculosis, providing kosher food and encouraging the children to take part in the life of the community. After World War I the orphanage moved to larger premises to cope with refugee children from Belgium and Hungary as well as local Jewish children. Increasing numbers from Europe necessitated the opening of a children's hostel beside Garnethill Synagogue in 1938 and a Farm School at Whittinghame, on the Balfour Estate, for 160 children in 1939.[31] In 1945 hostels were run in Dunbartonshire and in Midlothian for a group of teenage boys who had survived the horrors of Nazi concentration camps.[32]

A Jewish convalescent home operated during the 1920s at Slamannan, near Falkirk. A small home for the elderly was opened in 1913 in a rented Gorbals apartment but did not survive the war years, and another unsuccessful attempt to set up a Jewish old age home was made in Dixon Avenue in 1929.[33] It was not until 1949 that the Jewish Old Age Home for Scotland, subsequently known as Newark Lodge, was opened in Pollokshields, serving all the Jewish communities of Scotland and providing residential, nursing home and hospital services. With increasing numbers of frail, elderly Jews, the home grew with the addition of hospital wards in the 1960s and additional residential space over the following decades. With the need for more individual facilities and the spread of the Jewish community southwards a new residential development in Newton Mearns was opened in 2000. There are also plans for a new nursing home in Giffnock.

In the 1960s two new Jewish welfare organisations were founded. With poor Jews being relocated from slum clearance in the Gorbals, the Jewish Housing Association, which grew out of the Jewish Welfare Board, provided affordable housing in the Govanhill and Battlefield areas. With changes in care provision for the elderly in the 1990s, it set up facilities for sheltered housing and augmented care for the elderly on two sites within Giffnock with plans for further developments in Newton Mearns and in Glasgow. Cosgrove Care was established in 1960 to look after Jews with learning disabilities. Initially located in Pollokshields, it has now also moved its residential and support services to Giffnock.

During the past few years there has been a remarkable transformation in Jewish welfare in Scotland as agencies have become more professional and institutions have moved to be nearer their client groups. Moving first from its centre in the Gorbals to new facilities in Govanhill, the Glasgow Jewish

Board of Guardians transformed itself during the 1960s into the Glasgow Jewish Welfare Board. Now, as Jewish Care Scotland, it co-ordinates welfare activities from its new centre at the Maccabi Complex in Giffnock operating in conjunction with the Jewish Blind Society and Cosgrove Care. Jewish Care Scotland has developed a unique working partnership with East Renfrewshire Council as a recognised care provider, and delivers care and support to members of the Jewish community matching their religious and ethnic needs.

Jewish welfare provision was always seen as a valuable addition to services provided by the local authorities. There were those in the community who were concerned that the number of Jewish charitable bodies raising money, whether in Scotland or for Jews in Israel or in lands of distress, would affect the income of key communal bodies. History showed that Scotland's Jews were consistently able to meet their charitable obligations at home and abroad.

Paradoxically, one of the stimuli to the better provision of Jewish welfare services was the activity of the Christian missionary groups. Many Jewish organisations, social and welfare, were formed to provide direct competition for missionary facilities targeted at vulnerable newcomers in both Edinburgh and Glasgow. There was much Jewish poverty, misery and disease and the missionaries used financial support and medical aid to entice needy Jews into the mission halls. In Glasgow the Jewish Hospital Fund and Sick Visiting Association was founded in 1899 to counter missionary efforts. It provided money for Jews requiring hospital care and maintained community links with poor, isolated Jews in institutions. The Association continues today as the Glasgow Jewish Sick Visiting Association with a group of voluntary workers who visit Jewish patients in hospitals and residential institutions throughout the west of Scotland.

A Jewish Dispensary opened in the Gorbals in 1910 offering free medicines to counter similar missionary provision. Christian groups devoted considerable sums to Jewish medical relief with well-equipped dispensaries and Yiddish-speaking apostate doctors. In Edinburgh, the apostate Leon Levison ran the Medical Mission to the Jews; however, his attempts to pose as a benefactor to the Jews caused considerable resentment, and his award of a knighthood for services to Russian Jewish relief was widely condemned.[34] Missionaries blamed their lack of success with conversions on the Jewish attitude to Christianity brought over from Russia.[35] However, they remained determined to persist in their work despite the distress and hostility that their activities provoked.

BUSINESS AND THE PROFESSIONS

Much of the early wealth of members of the Jewish communities in Scotland was based on the production and merchandising of a wide range of goods. New Jewish entrepreneurs set up businesses manufacturing a wide range of clothing, including the ubiquitous cloth cap.[36] They were also active in

Figure 10.5 Exterior of Links' shop (with Yiddish lettering on the windows), Main Street, Gorbals, c1907. Courtesy of SJAC.

warehousing, entertainments and whisky. Most of the working members of Scottish Jewry were heavily concentrated in just a few trades and industries such as shop keeping, tailoring, shoe and slipper making and the manufacture and distribution of furniture and picture frames.[37]

In the years before World War I many Jews began their businesses peddling to mining communities. Well into the twentieth century, long after it had ceased to be a Jewish activity in London and Manchester, Jews still travelled from Glasgow to Ayrshire and Lanarkshire, from Edinburgh to Fife, and from Falkirk to Stirlingshire.[38] Eventually, many of these new immigrants set up successful businesses of their own, often aided by loans and credit from the Hebrew Benevolent Loan societies. Despite the creation of individual wealth, much of the Jewish community, especially those in the Gorbals of Glasgow, remained trapped in poverty. There was a continual call on the funds of the Jewish Boards of Guardians for relief. In the 1930s it was not uncommon for over 10 per cent of the community to be receiving such aid.

Despite the financial difficulties, the opportunities offered by the Scottish education system, supported by Carnegie grants to university students, proved increasingly attractive. Encouraged by parents who hoped to see their sons attain the educational standards and financial security that they themselves would never achieve, increasing numbers of Jews entered the universities of Glasgow and Edinburgh, especially in the years after World War I. Higher education was for many the means of escape from the old ghetto trades, and medicine was the most popular career choice as it offered status and income.[39] While many Jews went into teaching, law was less popular at first as it was hard for Jews to break into the conservative

Scottish legal profession. In the post-war period Jewish students began diversifying from the traditional fields of medicine, teaching and the law, although distinguished Jewish physicians and lawyers continued to be among the leaders of these Scottish professions into the twenty-first century.

SYNAGOGUES

When the Delegate Chief Rabbi, Dr Herman Adler, consecrated the Garnethill Synagogue on 9 September 1879 he expressed the hope that the new building would provide a unifying focus for the community. Unfortunately unity was not to be achieved. A *chevra* was founded in Commerce Street to serve the 200 Jews living in the Gorbals who still formed less than a quarter of the whole community.[40] This move had become inevitable given the inconvenience of the distance from Garnethill to the Gorbals. Further small prayer houses opened in the Gorbals during the 1880s but communal unity was preserved by a union, in 1886, between Garnethill and the Gorbals synagogues.

Figure 10.6 Interior of Garnethill Synagogue, Glasgow, opened 1879. Courtesy of SJAC.

The early Gorbals synagogues were small, informal prayer-houses but major places of worship were opened in a hall in Main Street, Gorbals, in 1887 and in a converted Baptist church at the corner of Oxford Street and Buchan Street in 1897. Despite a major extension to the Main Street Synagogue in 1891 the continuing growth of Gorbals Jewry required a larger building, and the Great Synagogue was opened in South Portland Street in 1901. The Great Synagogue was the largest in Glasgow and housed many of the city's religious institutions, including *yeshiva* and *mikva*. It closed in 1974, the last Jewish building in the Gorbals. Other Gorbals congregations came into being in the first years of the twentieth century as the result of schisms in the Oxford Street Synagogue. There were also a couple of Chassidic prayer-houses, while other synagogues reflected the places of origin of the newcomers.

The synagogue union of 1886 and the formation of the United Synagogue of Glasgow in 1898 gave effective leadership of the community to the élite leadership at Garnethill. The United Synagogue broke up in 1906 with disputes about *shechita* and the cost of burials for poor immigrants. Gorbals Jewry then appointed Rabbi Samuel Hillman, from Russia, as community rabbi for Gorbals in 1908 and his traditional stance balanced the more moderate style as exemplified by Garnethill and their minister the Rev. E P Phillips. Phillips' policy of cultivating links with the wider civic and religious society in Glasgow was continued by his successor Rev. Dr I K Cosgrove, minister at Garnethill from 1935 to 1973.

Phillips, minister from 1879 until 1929, had acted as the spiritual leader for all of Glasgow Jewry and was involved in many of the new welfare bodies. He took a high personal profile in the case of Oscar Slater, a foreign Jew who was falsely convicted of murder.[41] Phillips also represented the community on Sunday trading, an important issue for Jewish traders whose religious convictions prevented them opening on the Jewish Sabbath (from Friday afternoon until Saturday night). Rabbi Lurie and Rabbi Atlas provided traditional religious leadership in the inter-war years in the form of a Glasgow *Beth Din*, and this continued in the post-war years under Rabbi Gottlieb, as *Av Beth Din*, and Rabbi Shapiro, *Rosh Yeshiva* from 1932 to 1974.

The provision of Jewish burial grounds continued to be a priority in all the Scottish communities. During the nineteenth century and early twentieth century, Jewish cemeteries were established in Dundee, Greenock, Inverness and Aberdeen. When the first Gorbals synagogue was formed in 1880 a burial ground was obtained in Craigton.[42] In 1895 Garnethill Synagogue purchased burial grounds at Maryhill, which were later shared with some of the Gorbals synagogues. South Portland Street Synagogue opened its own cemetery at Riddrie in 1908 but there were continuing problems with the costs of pauper burials. With adult poverty and the large number of infant and child deaths there were many who could not afford the funeral expenses and the synagogues, the Shechita Board and the Board of Guardians were often asked to contribute.

The Glasgow Hebrew Burial Society was formed in 1908 as a mutual aid society, independent of the synagogues, where members contributed a regular sum to meet the cost of funerals for themselves and their families. At first they obtained a small cemetery at Sandymount, in Shettleston, buying a substantial ten-acre plot at nearby Glenduffhill in 1934 by which time they were the largest Jewish membership organisation in Scotland.

In Edinburgh too the immigrant period was marked by the formation of a number of small synagogues, in Richmond Street, Roxburgh Place and South Clerk Street. Here the newcomers could feel more comfortable than in the Graham Street Synagogue where the more assimilated community members prayed.[43] When Rabbi Salis Daiches came to Edinburgh in 1918 on a call from the various synagogues it was his aim, achieved in 1932, of uniting all of Edinburgh Jewry within the new synagogue in the Newington district. Daiches had a vision of a synthesis of Scottish and Jewish culture replacing the piety and superstition of the immigration generation but it proved to be an elusive dream.[44] The synagogue, now too large for the community, was reconstructed in 1981 to provide not only 500 seats for religious services, but also a community centre. While the Jewish population has dispersed across the city, the synagogue and community centre in Salisbury Road continues to provide a vital centre for Jewish life in Edinburgh.

The movement of Jews to the larger centres of population in Glasgow and Edinburgh meant the end of the small Jewish communities in Greenock,

Figure 10.7 Jewish soldiers outside South Portland Street Synagogue, World War I. Courtesy of SJAC.

Inverness, Falkirk and Dunfermline, which ceased to function in the 1950s and 1960s. The Ayr community survived into the 1970s, while the synagogues in Aberdeen and Dundee continue to be supported by a small but loyal following.

Garnethill Synagogue membership gradually grew as Jews moved from the Gorbals and settled in the Garnethill and Hillhead areas. Garnethill maintained its leading position for many years though new synagogues were being formed in areas to the south of the Gorbals where Jews were settling in much larger numbers. Queens Park Synagogue was founded in 1906 moving to a substantial new building in Falloch Road in 1926. About the same time Langside Synagogue, originally founded in 1915, moved to Niddrie Road. In the inter-war years synagogues were established in Pollokshields (1929), Giffnock and Newlands (1934), and Netherlee and Clarkston (1940) reflecting the continued move of the Jewish community southwards. This movement led to the establishment of the Newton Mearns Synagogue (1954) joined next door by a Reform Synagogue in the 1970s, which had originally opened in Pollokshields in 1931.

The growth of the Jewish community in Giffnock and Newton Mearns led to these communities building substantial synagogues. Giffnock's synagogue (1969) is the largest in Scotland, housing also a *mikva* and community centre within its complex. The community centre provides a wide range of Jewish educational facilities for the youth and adults of the community. At the same time there has been a decline in membership in the synagogues in Garnethill, Langside and Queens Park and the closure of all the Gorbals synagogues as well as those in Crosshill and Pollokshields.

Gradually, the level of religious commitment declined. This was often due to pressures of work and society and the declining influence of the immigrant generation whose religious practices, including *kashrut* and the Sabbath, were widely observed. Some newcomers, however, saw their emigration to Scotland as their escape not just from persecution and economic hardship but also from Judaism.

JEWISH EDUCATION

The first Hebrew classes began in the early years of the nineteenth century within the Glasgow and Edinburgh synagogues. As the communities grew, additional provision had to be made. While the Garnethill Synagogue had its own religion classes meeting two or three times a week, gradually the Gorbals synagogues made their own arrangements.[45] The main provider of Jewish education in Glasgow was the Gorbals-based Talmud Torah, founded in 1895. At first the boys and girls were taught separately, for three hours every day except Friday, with Yiddish as the language of instruction. The language of teaching in the Talmud Torah changed to English in about 1908, indicating that the key to successful integration was fluency in English. During World War I the teaching time was reduced to two hours nightly. With many children attending classes for upwards of twelve hours weekly, a large number of the new generation grew up familiar with the Jewish way of life, Hebrew language and the sacred texts. This was usually reinforced by the traditional pattern of religious home life. As time went by, the number of children committing so much time dropped, and in subsequent decades it was often reckoned that about one-third of Jewish children failed to receive regular Jewish education.

The Talmud Torah, now with over 700 pupils, moved to its own premises in Turriff Street in 1918, and in 1922 a Hebrew College for teenagers was established. Over the next decades it provided its pupils with Hebrew language and Jewish culture and encouraged them to be Hebrew teachers themselves. Despite competition from synagogue Hebrew classes, the Talmud Torah maintained its leading position in Jewish education in Glasgow well after World War II. For some boys the regular Hebrew classes were supplemented with attendance later in the evening at the Glasgow *Yeshiva*. This was founded in 1908 to provide a more intensive learning pattern along the lines found in eastern Europe with an emphasis on teaching of the Talmud.

With the high concentration of children in just a few Gorbals schools it was suggested in 1909 that a Jewish school could be established. Negotiations with the Glasgow School Board proved to be difficult and broke down, as did further discussions in the 1920s and 1930s. A Jewish primary school, Calderwood Lodge, was not established in Glasgow until 1962. Initially an independent school, Calderwood Lodge was taken into the State system in 1982.

In Edinburgh the main classes were held in the basement of the Graham Street Synagogue until 1914, subsequently being held in Sciennes School also for two hours on a daily basis. Similar classes could be found in the small communities around Scotland, from Greenock, Ayr, Falkirk and Dunfermline to Dundee, Inverness and Aberdeen. The Edinburgh, Aberdeen, Newton Mearns and Glasgow New Synagogue still have their own congregational Hebrew classes today.

In the mid-1990s the United Jewish Israel Appeal (UJIA) has developed a Renewal Division that aims to renew Jewish life in Scotland. From its base at the Jewish Community Centre in Giffnock it organises Jewish assemblies in a number of schools where there are a significant number of Jewish pupils. It maintains a resources centre with a library of Jewish books and videos, and provides lectures and discussion groups as well as courses in Hebrew and understanding Judaism.

In the 1970s the Jewish religious facilities in Glasgow were augmented by some new organisations. The Jewish outreach body, the Lubavitch Organisation, established a Scottish base in Glasgow running a nursery school and elementary Hebrew classes. A *Kollel* was set up with some postgraduate rabbinical students who aimed to bring Jewish tradition to children and adults while continuing their own studies. In addition a full-time Jewish student chaplain was appointed to support Jewish students in Scotland and their educational and religious activities. Jewish students in Scotland have their own university Jewish Societies, the first being established in Edinburgh (1909) and Glasgow (1911).

Jewish education was clearly a priority if a new generation was to combat assimilatory trends and grow up sharing the customs and mores of their parents and grandparents. Unfortunately, some of the Hebrew class tuition was unsatisfactory, with teaching conducted by teachers with an imperfect grasp of English. Thus Jewish education was compared unfavourably with the standard of secular teaching in the local schools. Some parents were happy with their children's exposure to traditional eastern European teachers but many children were alienated by the experience.

By the 1970s it was clear that new educational formats would be required to reach a younger generation of Jews. In 1987 the Jewish Teenage Centre replaced the Glasgow Hebrew College and further changes followed the formation of a Jewish Renewal Committee in Glasgow in the 1990s. Living within the secular society the Jewish community showed that they were still determined to provide experience of authentic Jewish life for its members.

COMMUNITY AND POLITICS

Jewish Representative Councils were formed in Glasgow in 1914 and a year later in Edinburgh to act as a unifying force in the community and to provide representation on matters of national and international importance. The Glasgow Council grew out of the committee set up to protest about the false accusations in the Beilis 'blood libel' case in Kiev.[46] There had been no single community-wide structure in Glasgow since the collapse of the United Synagogue in 1906 and it was customary for the Garnethill Synagogue leadership to represent all of Glasgow Jewry.

In Edinburgh the first concerns of the Representative Council were with alien Jews interned under wartime regulations. In Glasgow too, support for alien Jews was a priority and affidavits were provided for all the 2,000 foreign Jews in the city. After the war the Edinburgh community gradually centralised its activities within the synagogue, and the Edinburgh Hebrew Congregation took over all the representative functions.

In Glasgow the Council began by trying to provide a communal structure for *shechita*, supporting moves for a Jewish school and co-ordinating Jewish support for the war effort. During the 1930s the Council continued to protect and represent the Jews of Glasgow, to fight anti-Semitism and to support the economic boycott against Nazi Germany. While there were no clashes with fascists as happened in London, the Council had to steer a cautious role while mounting defensive action against anti-Jewish activities.

The role of the Glasgow Jewish Representative Council was widely seen as the expression of the Jewish community in Scotland. Supported by about fifty affiliated organisations it took the lead in defending Jewish rights and protecting the interests of Jews in Scotland while promoting racial and

Figure 10.8 Sir Myer Galpern, Glasgow's first Jewish Lord Provost (1958). Courtesy of SJAC.

religious harmony. It also provided support for Israel. In 1985 it took the lead in establishing, with the Edinburgh Hebrew Congregation, the Scottish Jewish Standing Committee, which brought together the Jewish communities from Scotland's four main cities. In 1998 the Scottish Council of Jewish Communities was formed to represent the Jews of Scotland in the wake of Scottish devolutionary change.

The leading Jewish political ideology in Scotland was Zionism. Branches of *Chovevei Zion* (Lovers of Zion) were formed in Edinburgh (1890) and Glasgow (1891) to help Jewish settlers in Palestine. There were delegates from Scotland at the Second Zionist Congress in Basel in 1898 and political and fund-raising activities quickly followed in both cities.[47] Zionist social and sporting activities were also organised to improve the physical condition of the local Jewish community and give a place for the community to meet, to learn about Zionism and to study the Hebrew language. In 1909 a co-operative society was founded aiming to settle Glasgow Jews on the land in Palestine showing the importance placed on practical Zionism.[48] The Balfour Declaration of November 1917, indicating British Government support for a Jewish homeland in Palestine, was greeted with much enthusiasm in Scotland. It also gave considerable impetus to Zionist political and fund-raising activities.

The Jews of Scotland traditionally have strong ties with Israel and there has been a steady movement of Jews there. These ties have been fostered by a network of organisations, both cultural and fund-raising, based mainly in Glasgow but also in Edinburgh. Since Israel's independence in 1948, Scottish Jewish support for Israel has been steady and loyal, but not always uncritical.

The community was also active in the political sphere. A number of Jewish immigrants had been active in the Bund, the Jewish trade union in Russia, and many were strongly committed to socialism. There were Jewish trade unions in the Glasgow tailoring and cabinet-making industries and many Jews took part in wider union activities. Jews were also active in local politics. Michael Simons, a leading member of the Jewish community, was elected a baillie in the 1880s, and Manny Shinwell was co-opted as a member of the Fairfield Ward in Govan in 1916.[49]

Jews were to be found in all the political parties but it was the Labour Party which received most of the community's political support in the inter-war years, especially with the rise of fascism in the 1930s. The major political activity, especially after the Nazis seized power in Germany in 1933, was aiding Jewish refugees and organising a boycott of German goods.[50] In the post-war period there were Jews involved in the Labour and Conservative parties and there were Scottish Jewish MPs in both parties. One of these MPs, Malcolm Rifkind, was elected in Edinburgh Pentlands in 1974 and subsequently served in the Cabinet as Secretary of State for Scotland, then as Defence Secretary and, finally, as Foreign Secretary.

SOCIAL AND CULTURAL ACTIVITIES

The newcomers considerably enhanced the cultural life of the Jewish community. Jewish Literary Societies were formed in Edinburgh (1888) and in Glasgow (1893) and these societies kept their members informed on a wide variety of Jewish and general topics. They sponsored English language classes and helped the newcomers to adjust to their new life. The Edinburgh

Society continues to provide a platform for matters of Jewish interest and also attracts members not affiliated to the synagogue.

Choral and dramatic societies were formed in Glasgow and Edinburgh. Zionist groups also had a wide range of cultural and social activities, as did the friendly societies. Masonic Lodges were formed in Glasgow, as Lodge Montefiore (1888), and in Edinburgh, as Lodge Solomon. Jewish drama flowered in Scotland with the arrival in the 1920s of Avrom Greenbaum, and with the formation of the Glasgow Jewish Institute Players in 1936 new standards in local Jewish dramatic art were set.[51] Greenbaum wrote many of the plays, which reflected his preoccupation with man's fight for freedom. Visual arts were represented by Benno Schotz, who arrived in Glasgow from Estonia in 1912 and was later rewarded with the appointment as the Queen's Sculptor-in-Ordinary in Scotland.[52]

Although Yiddish newspapers appeared irregularly in Glasgow around the turn of the century, serious Jewish journalism did not begin until Zevi Golombok started the *Glasgow Jewish Evening Times* in 1914. After World War I, Golombok launched a Yiddish monthly the *Yiddishe Shtimme* (Jewish Voice) but recognising the local decline in Yiddish he founded the English language *Jewish Echo* in 1928, which was published every Friday in Glasgow serving all the Jewish communities of Scotland. The paper closed in June 1992 when the Manchester *Jewish Telegraph* set up a Scottish edition.

As the community numbers fell and old-established Jewish institutions and synagogues closed, the Scottish Jewish Archives Centre, based in Garnethill Synagogue, was set up in 1987. The Centre has the task of preserving, cataloguing and displaying the records of the history of all the Jewish communities of Scotland.

Figure 10.9 Jewish Institute Players performing a play. Courtesy of SJAC.

Figure 10.10 Jewish scouts, c1958. Courtesy of SJAC.

From the first years of the twentieth century youth groups were formed, including youth sections of Zionist and Jewish literary groups, as well as scouts and the Jewish Lads' Brigade. The Brigade, based on the Boys' Brigade, aimed to make loyal British citizens out of the immigrants and its Glasgow branch dates back to 1903.[53] The first Jewish sports club, the Bar Kochba, was set up in the Talmud Torah building in 1933. This formed the precursor of the successful Glasgow Maccabi, which developed its own base in Giffnock in 1969. A Jewish golf club was also established near Eaglesham as Jews still found it difficult, even in the 1950s and 1960s, to gain entry into golf clubs. Jewish Institutes were established in Glasgow and Edinburgh from the turn of the century with regular social and cultural events in the winter and sporting activities in the summer. The Glasgow Jewish Institute achieved its greatest success as a community centre when situated in South Portland Street in the 1930s.

Jewish youth organisations continue to operate in Glasgow and Edinburgh with links between the youth groups in the two cities. In Glasgow there is an informal Jewish Youth Forum supported by a full-time youth worker, based at the Jewish Community Centre. A number of the Jewish youth groups are Zionist in orientation and have a strong educational and cultural influence on the life of their members.

CONCLUSION

While the growth of the Scottish Jewish community has been considerable, at its peak between 1935 and 1960 its number never exceeded 20,000. It was estimated that there were about 15,000 Jews in Glasgow in 1939 and over 2,000 in Edinburgh. Scottish Jewry derives in the main from those Jews who arrived in the country in the two decades leading up to 1914, with a smaller element deriving from the immigration of Jewish refugees from Nazi terror in the 1930s.

Glasgow in the twentieth century was one of the major Jewish communities in Britain, exceeded in size only by London, Manchester and Leeds. Its Gorbals community was still at its peak in 1945 but the movement of Jews southwards was accelerating. While some Jews moved to the west end of Glasgow, and reinforced the membership at Garnethill, most Jews moved further south. The population shift did not just mean the closure of old synagogues and the building of new ones, but was accompanied by a move of some of the institutions out of the Gorbals.

The opening of the synagogue in Salisbury Road gave Edinburgh Jewry a cohesion that made it the envy of many larger communities. Led by a scholarly rabbi, Salis Daiches, they attempted a synthesis of the best of Jewish and Scottish life but had to accept a drift from traditional observance while showing a Jewish vitality that belied its numerical strength. The smaller Jewish communities around Scotland declined in numbers as their members migrated to Glasgow and Edinburgh where Jewish facilities were concentrated. In essence Scottish Jewry became a 'tale of two cities'.

The challenge to Scottish Jewry was not only to survive beyond the immigrant generation, but also to establish new institutions that would foster communal cohesion and maintain Jewish identity without the Yiddish language. Despite much continuing poverty, there was considerable economic and educational advance, and there was little serious anti-Semitism. By the time of World War II, Jews had become an established part of the Scottish scene with achievements in many areas of community, local and national life.

By the start of the twenty-first century new challenges had appeared. Emigration, usually to England or Israel, but also to America, Canada and Australia, marrying out of the faith, and the effects of an ageing community, reduced the numbers of Jews in Scotland to below 10,000. The Jewish community in Glasgow responded by strengthening the infrastructure of its care bodies and by looking for newer and more innovative ways of providing its members with education, religion and culture. Despite smaller numbers, Scottish Jews continued not only to be represented on the leadership of major British Jewish bodies, but also to support a lively and active local community that was involved in wider Scottish issues.

NOTES

1. Barrow, 41–3.
2. Renwick, R and Lindsay J. *History of Glasgow: Vol. 1: The Pre-Reformation Period*, Glasgow, 1921, 80–1.
3. Barrow, 1971, 476.
4. Levy, 1–8.
5. Levy, 1958, 8–10; Boog Watson, C R, ed. *Roll of Edinburgh Burgesses and Guild-Brethren: 1406–1840*, Edinburgh, 1929–1933; Wood, M, ed. *Extracts from the Records of the Burgh of Edinburgh: 1642–1680*, Edinburgh and London, 1938–1950; Wood, M and Armet, H, eds. *Extracts from the Records of the Burgh of Edinburgh: 1681–1691*, Edinburgh and London, 1954. For these references, see Levy, 1958.
6. Collins, 1988, 29–40.
7. The first Jew to graduate in Scotland was Jacob de Castro Sarmento (1691–1762) who was awarded the MD by Marischal College, Aberdeen, in July 1739 on an affidavit from two colleague physicians. Sarmento was a native of Portugal who fled that country from the Inquisition, settling in London where he had a prominent medical career. See Barnett, R. Dr Jacob de Castro Sarmento and Sephardim in medical practice in 18th century London, *Transactions of the Jewish Historical Society of England*, 27 (1982), 84–103; Collins, 1988, 34–5, 174–5.
8. Collins, 1988, 173; Munk, W. *Roll of the Royal College of Physicians of London*, London, 1876, II, 376.
9. Phillips, 1979; Collins, 1990.
10. Collins, 1987, 35–40.
11. Collins, 1990, 34.
12. Collins, 1988, 50–4, 163–4.
13. Collins, 1990, 59–62.
14. Aronsfeld, C C. German Jews in Dundee, *Jewish Chronicle*, 20/11/1953.
15. *Jewish Chronicle*, 20 October 1893, 13–15. See also *Jewish Chronicle*, 6 October 1893, 2 February 1894.
16. Collins, 1990, 57–66.
17. Bermant, 1969, 25–7.
18. *Jewish Chronicle*, 15 December 1905.
19. Roth, 1950, 57–8.
20. *Jewish Chronicle*, 27 December 1861.
21. Records of Glasgow Kosher Baths, in *Glasgow Bathhouse Records: 1885–1891*, Glasgow City Archives, Mitchell Library; *Jewish Chronicle*, 28 October 1892, 31 January 1893.
22. Victoria Infirmary, House Committee Minutes, 24 October 1908, 6 April 1909, Greater Glasgow Health Board Archives, Mitchell Library, Glasgow.

23. Glasgow Infirmaries Consultative Committee, in Glasgow Royal Infirmary Minutes, 17 November 1915, Greater Glasgow Health Board Archives.
24. Collins, 1990, 157–9.
25. *Jewish Chronicle*, 21 October 1892.
26. McFarlane, N. Hospitals, housing and tuberculosis in Glasgow: 1911–1951, *Social History of Medicine*, 2 (1989), 59–86.
27. Feldman, W M. Tuberculosis and the Jew, *The Tuberculosis Year Book: 1913–1914*, 48–54.
28. *Jewish Chronicle*, 29 September 1916.
29. Report of the Royal Commission on Alien Immigration, 1903, evidence of Julius Pinto, 20991–20998; Glasgow Municipal Commission on Housing for the Poor, Glasgow, 1904, 352, 358.
30. Report by T Binnie and Sons to Mitchels, Johnstone and Co., Glasgow University Archives; Fyfe, P. *Backlands and their Inhabitants*, Glasgow, 1901, 15.
31. Kolmel, R. German Jewish Refugees in Scotland. In Collins, 1987, 63.
32. Gilbert, 1996, 326, 343–5.
33. *Jewish Chronicle*, 12 September 1913, 7 November 1913, 22 February 1929.
34. *Jewish Chronicle*, 24 August 1919. The editorial in this issue castigated the Jewish Board of Deputies for their weak protest over the award of a knighthood to Leon Levison. See also, Daiches, D. *Two Worlds*, Sussex, 1957, 93, 100–4; Daiches, D. *Was: A Pastime From Time Past*, London, 1975, 35–6.
35. *Glasgow Herald*, 23 November 1906.
36. Collins, 1990, 150–1.
37. Miss Meiklejohn. Study of Alien Jews working in factories and Workshops in Glasgow, *Jewish Chronicle*, 31 July 1908.
38. Gartner, 1973, 60.
39. Block, 1938, 370–1.
40. For details of the history of the Gorbals' synagogues see Collins, 1990.
41. There is a voluminous literature about the Oscar Slater case. See, for example, Toughill, T. *Oscar Slater: The Mystery Solved*, Edinburgh, 1993.
42. For details of the history of Jewish cemeteries in Glasgow, see Collins, 1990.
43. Collins, 1987, 42–6.
44. Daiches, 1957, 82–6, 121.
45. For details of early developments of Jewish education in Glasgow, see Collins, 1987.
46. *Jewish Chronicle*, 19 September 1913; *Glasgow Herald*, 11 October 1913.
47. For details of the early Zionist movement in Glasgow, see Collins, 1990, and in Edinburgh see Collins, 1987, 45.
48. *Jewish Chronicle*, 5 November 1909.
49. *Jewish Chronicle*, 25 August 1916.
50. Kolmel, 1987, 58–9.
51. Greenberg, A. New Stages, *Hashanah: the Glasgow Jewish Yearbook 1955–1956*, Glasgow, 1955, 117–20, in Scottish Jewish Archives Centre, Garnethill Synagogue, Glasgow.
52. Schotz, B. *Bronze in My Blood*, Edinburgh, 1981.
53. Minutes of the Garnethill Hebrew Congregation, May 1902, in Scottish Jewish Archives Centre.

BIBLIOGRAPHY

Barrow, G W S, et al. *Regesta Regum Scottorum II: The Acts of William I, King of Scots 1165–1214*, Edinburgh, 1971, 476.
Barrow, G W S. A twelfth century Newbattle document, *SHR*, 30 (1951), 41–3.

Bermant, C. *Coming Home*, London, 1976.
Bermant, C. *Troubled Eden*, London, 1969, 25–7.
Block G D M and Schwab, H. A survey of Jewish students at the British universities, *Jewish Year Book: 1938*, 370–1.
Collins, K. *Second City Jewry: The Jews of Glasgow in the Age of Expansion 1790–1919*, Glasgow, 1990.
Collins, K. *Go and Learn: The International Story of Jews and Medicine in Scotland: 1739–1945*, Aberdeen, 1988, 29–40.
Collins, K. The growth and development of Scottish Jewry: 1880–1940. In Collins, K, ed. *Aspects of Scottish Jewry*, Glasgow, 1987, 35–40.
Cowan, E. *Spring Remembered: A Scottish Jewish Childhood*, Glasgow, 1974.
Denton, H and Wilson, J C. *The Happy Land*, Edinburgh, 1991.
Gartner, L P. *The Jewish Immigrant in England: 1870–1914*, London, 1973, 60.
Gilbert, M. *The Boys: Triumph over Adversity*, London, 1996, 326, 342–5.
Glasser, R. *Growing up in the Gorbals*, London, 1986.
Glasser, R. *Gorbals Boy at Oxford*, London, 1988.
Glasser, R. *Gorbals Voices, Siren Songs*, London, 1990.
Kaplan, H and Hutt, C. *The Scottish Shtetl: Jewish Life in the Gorbals 1880–1974*, Glasgow, 1974.
Kay, B, ed. *Odyssey, the Second Collection: Voices from Scotland's Recent Past*, Edinburgh, 1982.
Levy, A. The origins of Scottish Jewry, paper delivered to the Jewish Historical Society of England, 13 January 1958, privately printed.
Phillips, A. *A History of the Origins of the First Jewish Community in Scotland – Edinburgh 1816*, Edinburgh, 1979.
Roth, C. *The Rise of Provincial Jewry*, London, 1950, 57–8.

11 Islam in Scotland After 1945

MONA SIDDIQUI

Islam is one of many religions lived and practised in Scotland. It exists within various communities, is expressed through a number of languages and is visible in a variety of cultures. Those who belong to this faith are considered to form an intrinsic and essential element of multi-faith Britain, the large majority of them being British national citizens with families established and settled all over Britain for decades. Although the faith has been a well-recognised and living part of Scotland for over half a century, a result largely of the post-1945 travel and migration of Muslims from the Commonwealth into the United Kingdom, there is very little evidence of any in-depth research on the influence and spread of Islam in Scotland from an ethnological or sociological perspective. Beyond a few names and statistics dotted around and held principally by local authority networks or government institutions as reference material, information about the faith is gathered and distributed in various forms mainly by the followers of the religion either for self-propagation or for the purpose of self-advertisement. A classic example of this is that, even by the 1970s, over twenty years after the first real entry of Indians and Pakistanis into Scotland, the only documented evidence pertaining to this community was to be found in the *Gorbals View*, a local monthly paper which reflected the news and views of those living in the Gorbals and Glasgow area.[1] Outside these places, information on the Indo-Pakistani community was non-existent. The only area where there is any meaningful analysis of either Islam or Muslims is in the area of academic research on Islam and Europe; again there are a only a few books of this kind and they refer not to Scotland as such but more broadly to the United Kingdom or Britain.

One of the reasons for this has been quite simply the relative newness of academic research regarding religious and ethnic groups in post-war Europe and especially Muslim minorities in Western Europe. This kind of analysis appears from the 1980s onwards and for the most part in the works written by twentieth-century social historians, demographers, political scientists, anthropologists and sociologists, and a few Islamicists. It focuses on Muslim communities and their living religious culture in Europe and the West, the changes brought about by large-scale immigration, the contributions, the tensions, the political discourses and social changes. Neither the religious dogma nor the history of Islam as a world faith are the subjects of these studies. Most of this type of research concerns itself with the challenge of understanding Muslims as a social and ethnic group within post-modern Europe and the faith organisations, both political and social, that have been the gradual consequence of large-scale migration.[2] These works are perhaps one of the most contemporary contributions to studies on Islam/Muslims but they should be viewed against the background of at least two centuries of very different and quite monumental scholarship on Islam.

Although it is not the purpose of this essay to outline a detailed history of the western study of the Islamic world or the Muslim faith, the following quotation is a summary of Princeton historian Bernard Lewis' view of the gradual development of this phenomenon since the Middle Ages:

> The original impetus for Europeans to study Islam sprang from two motives. The first was to learn more about the classical heritage that had been preserved in Arabic translations and commentaries; the second was to mount an informed Christian polemic against Islam. As Christendom pulled ahead of the Islamic world in science and political strength toward the end of the Middle Ages, and as it became apparent that the conversion of Muslims could never be accomplished on a large scale, both motives began to fade. Instead with the Renaissance came new reasons to study Islam. First was the new intellectual curiosity about alien culture ... The Renaissance also brought about a revival of interest in classical philology (which became the paradigm for understanding other cultures) and increased travel to oriental lands, largely in the service of European economic interests. And, the Renaissance saw the rise of biblical and Semitic studies, for which many scholars viewed the study of Arabic language and texts as useful tools.[3]

But it was not till the nineteenth century that Islam became a specifically serious focus for study, a development that was marked by the fact that oriental studies became a separate discipline in European universities. Although the word orientalist is now viewed as a negative criticism of anyone who carries certain prejudices regarding the academic study of Islam or the Middle East, it must be acknowledged that the impetus set by the European orientalists of the nineteenth century ensured a standard of academic excellence that was nothing less than awe-inspiring to those of other disciplines. Despite either the religious or political prejudices that a few had, they bequeathed a legacy of scholarship that was truly rigorous and impressive. The sheer weight of erudition that most of these scholars possessed vis-à-vis knowledge of oriental languages and mastery of huge quantities of texts meant that they achieved in a relatively short time the very basis of what was to gradually emerge as Middle-Eastern Studies and Islamic Studies during the twentieth century. As Richard Martin says:

> What had taken centuries to edit, translate, codify and interpret in the Greco-Roman and Hebrew classics had been tolerably well imitated

by a handful of orientalists in the nineteenth century, the majority of whom trained as classicists or Semiticists.[4]

Discovering the Islamic world involved interest not only in textual studies of the Muslim faith, but also in the travelogues and encounters between Muslims and Europeans from the seventeenth century onwards, in which was to be found a variety of descriptions about the Islamic world and the religious, cultural and political experiences that ensued from Islamic imperialism into Europe and then subsequently the effects of British and European imperialism into many of the Islamic lands. In his excellent research of Islam in Britain, Nabil Matar relates a wide variety of encounters between Muslims and Europeans and the perceptions and legacies left by each in the others' world. He claims:

> Throughout the period roughly extending from the accession of Queen Elizabeth in 1558 until the death of Charles II in 1685, Britons and other Europeans met Muslims from the Atlantic Ocean to the Mediterranean and Arabian Seas. For the peoples of Spain, France, Italy and Germany, the physical proximity with Islam in continental Europe and the Mediterranean made encounters with the Muslims inevitable; as a result, their literature frequently alludes to Islam and the Muslims. In the case of the British Isles, there was a vast difference between London or Edinburgh and Istanbul, but still, the likelihood of an Englishman's or Scotsman's meeting a Muslim was more likely than meeting a native American or a sub-Saharan African.[5]

The current angle of interest in Muslims and their contribution and influence in non-Muslim lands has one particular aspect that distinguishes it from the encounters between Europeans and Muslims of earlier centuries. The post-1945 migration of Muslims to Britain is significant not because this was the first time that Muslims had come to either England or Scotland, but because in the late 1950s and 1960s Muslims from Pakistan formed the other major immigrant community alongside the immigrants from the West Indies. The arrival of these two communities in large numbers led in many ways to the debate on 'coloured' immigration, eventually resulting in the relatively restrictive Commonwealth Immigrants Act of 1962. This was part of a political and social debate on religion and society and led many to view immigrants from the Commonwealth with less sympathy and more scepticism. England and Scotland were part of a wave of migration to Europe as a whole. The three most commonly cited areas are the Turkish presence in Germany, the North-African presence in France and the predominantly Indo-Pakistani element of Britain, which has brought with it Hinduism and Sikhism as well as Islam. Although the large majority of Muslims who came to Britain reside mainly in England, there is evidence that, apart from those migrants who came directly to Scotland, over the years many who initially settled in England subsequently came to Scotland for either economic or family reasons.

The Islamic component in Scottish society is expressed through the diverse ethnic communities in Scotland which have grown enormously over the last fifty to sixty years. The main countries of origin are Pakistan, Bangladesh, Malaysia, Turkey, Iran and the North African states. Different groups came for different reasons and it would be erroneous to think that they all came over simply for socio-economic reasons. It is, however, true to say that the first wave of Asian immigrants from Pakistan and India came primarily to act as a cheap, unskilled labour force for Britain's manufacturing industries. In Scotland, Indians, of whom a sizeable proportion were Muslim, had started arriving mainly in the port towns of Edinburgh, Glasgow and Dundee from the 1920s onwards, and again there were various reasons why an Indian Muslim would be found in Scotland. Nielson writes that the early history of Islam in Britain saw its origins in the expansion of British and colonial involvement in India. In the latter part of the eighteenth century, the East India Company was recruiting heavily in Indian ports and, when their ships were docked in Britain, these men were left to fend for themselves. It was not till 1822, following an investigation by the anti-slavery campaigner Thomas Clarkson, that the East India Company was made to arrange for boarding houses for these people, culminating ultimately in 1857 with the establishment of a home for 'Asiatics, Africans, and South Sea Islanders' in London.[6] But these seamen and servants were also to be found in Scotland, so that by the 1870s Joseph Salter states:

> Even in Scotland Asiatics are to be found especially in the Autumn months. They have been met and spoken to at Dundee, Glasgow, Perth, Greenock and Edinburgh. Passing a short time in Glasgow, Stirling, Leith and Edinburgh I passed onto Sunderland. On this journey I met and spoke to 81 Asiatics.[7]

In the early part of the twentieth century, more and more Indian Muslims were coming over to the cities of Glasgow and Edinburgh either to join their fellow seamen or in the hope of improved economic prosperity. Aside from the few professionals, mainly doctors who came for further qualifications, the large majority of those who came were from rural communities in the north-west province of Punjab and were either totally illiterate or had very little formal education. Many took up peddling to earn a living and almost all of them were Muslim. A number of students, primarily of medicine, had also started studying at Scottish universities, particularly at Edinburgh and Glasgow, and by 1960 there were some 250 students from India and Pakistan attending the various Scottish universities.[8] Studying medicine in Scotland was for a long time more popular than studying medicine in England, one of the reasons being that the Scots themselves had made the most significant contribution in the medical field in India.[9]

One must realise that the majority of these people came as single men and many had established a social life within their own and the indigenous communities over the years. Though many eventually converged in the west of Scotland, particularly in Glasgow, it is reasonable to claim that, despite

Figure 11.1 Laying out prayer mats at Pilrig mosque, Edinburgh, 1993. Photograph by Herman Jai Rodriguez. Courtesy of the Trustees of the National Museums of Scotland, NMS SLA E/9357.

the life they had created for themselves, subconsciously the thought of returning to the subcontinent may have never left them.

By the 1950s, there were about 600 Asians in Scotland.[10] It was around this time that the Muslim men who had initially arrived on their own began to invite their wives and families over, thus laying down the psychological and personal basis for a more permanent stay in Scotland. The economic prosperity in Britain at that time resulted in a far greater number of Muslims from Pakistan arriving in the English cities of Bradford and Birmingham to work in the textile mills and factories. The change in the racial and social makeup of these cities came much more quickly than it did in Scotland. And yet, with the economic decline in the late 1950s, and the subsequent closure of many of the mills, many of the Muslims who had become redundant in England travelled north of the border to fill up vacancies in Scotland, particularly in the transport industry. By the end of 1960 there were about 4,000 Asians in Scotland with the largest number of around 3,000 located in Glasgow.[11]

The question of how Islam manifested itself in these years is a complex one. This is for two reasons. Firstly, when we look at the presence of Islam in Scotland we can only identify any substantial evidence of living Islam when we examine the anecdotal and personal experience of the individuals who had made a name for themselves in some form or other and happened to be Muslim. This then becomes an eclectic mix of different personalities with different professions, where the only thing that binds them is a common faith and yet even having a common faith does not mean that the followers practise or reflect it in the same way. Furthermore, a more comprehensive picture of Islam in Scotland would necessitate an analysis of the developments in the theology and practice of the faith that resulted not just from within but from outside, from external political factors and from social and economic developments both in Britain and throughout the world. Secondly, it is also difficult to extract what could be termed the more Islamic elements from the cultural traditions to which the majority of Muslims belonged and which they endeavoured to recreate for themselves in Scotland. This is not just limited to ritual worship, dress or dietary habits but more importantly to a whole mindset so that it would be fair to say that the version of Islam that had developed within, and defined, rural Punjab over the centuries was in a very short time to be transplanted into a Western, industrial country like Scotland. This was done quite simply because the psychological shift that was required to leave a homeland where Islam was the cultural norm, for a place where the social construct was so very different, had not really taken place, and it was notable particularly within the rural community. Thus, those Muslims who came from the more professional and educated backgrounds, or who lived in the cities rather than the villages of India and Pakistan, found that their perspective on life in Scotland was more open and more receptive to change. Perhaps the change that eventually resulted in Islam becoming a social determinant in Scottish life emerged only in the 1970s with the arrival of the wives and families of those men who had already been working here for some time. The rapid increase in the flow of families saw the formation of communities that grouped together socially and geographically. The Islamic faith became more of an identity and, as Pakistani Muslims prospered into entrepreneurial ventures forming large financial empires, the communities which were either part of these empires or mere onlookers became without any doubt the public face of Islam in Scotland.

MUSLIM COMMUNITIES AND INSTITUTIONAL ISLAM

The Muslim population in Scotland ranges somewhere between 45,000 and 50,000,[12] but they are not a monolithic group. Not only are there ethnic and cultural differences between the Indian, Pakistani, Turkish, Arab, Malaysian and North-African groups, but there is also a fundamental sectarian difference between sunni and shi'ite Islam. Shi'ite Islam is predominant particularly among the Iranian community as Iran is the only Muslim country which has shi'ism as its official religion. It is not within the scope of this chapter to explain in much detail the differences between the two visions of Islam but a brief explanation is necessary. The majority of Muslims in Britain as a whole and also in Scotland are sunni Muslims. Loosely speaking, this term implies that they belong to the orthodox understanding of how events unfolded in the early history of Islam and especially the years immediately after the Prophet's death. Thus, sunni history claims that the Prophet Muhammad had left no successor after his death and the community itself elected Abu Bakr Siddiq, who was the Prophet's father-in-law as well as a dear friend, to succeed him as a the first caliph or leader. Sunni Islam accepts not only this leadership as legitimate, but recognises the literary and religious traditions that bear witness to this succession as well. Thus, the unfolding of religious and political events that followed from this are part of the orthodox understanding of the development and spread of Islam, as is the caliphal line that followed the death of Abu Bakr. The sequence of succession resulted in the Prophet's cousin and son-in-law, Caliph Ali, becoming the fourth Caliph. The shi'ite claim that Ali ibn Abi Talib should have been the first successor as he was related directly to the Prophet and that only a member of the Prophet's family should be leader of the growing Muslim community. After Ali's leadership, the period of the 'rightly guided caliphs' came to an end and leadership of the faithful entered a new political and historical phase. There was by now a deep divide as to who should lead the Muslim community, which had grown astonishingly fast, but still the shi'ite clung to their claim that the authority of Ali had been usurped by the sunnis. As a result the shi'ite do not recognise the earlier caliphs prior to Ali, nor do they recognise the validity of any of the statements by the early sunni hadith transmitters who report on what the Prophet is supposed to have said and done.

Aside from this, there are deep theological differences. Traditionally speaking, sacred authority in sunni Islam comes ultimately from the Divine revelation of the Qur'an and the life and sayings of the Prophet Muhammad. After the Prophet's death, it was textual authority that became the primary force in maintaining and shaping the thought processes of the Islamic community. This was done primarily through the emergence and development of law schools and sunnism has basically four schools of law which became responsible for the transmission of most of the legal thinking down the centuries. They are known as the Maliki, Shafi'I, Hanafi and Hanbali schools of thought and jurisprudence, named after their founders. Today, Muslims in the West and in Islamic lands will be well aware of which school they belong to but, as the differences between the schools amount to endless debates about practical piety rather than creed and dogma, many regard these differences as nothing more than petty abstractions, irrelevant to the spirit of Islam and conflicting with the ideal of a unified Islam transcending all human variances. Textual authority took on its own life force in sunnism quite simply because sunni Islam does not have in its formal structure an equivalent of a priesthood or clerical hierarchy. The nature of worship in Islam is such that Muslims claim formal prayers are a direct link with God; priesthoods are not necessary because one of their fundamental roles is to act as an intermediate between man and God. The concept of imamate or leadership in shi'ism, however, is linked to a different concept of how God reflects his authority in this world. The shi'ite claim that the imams are on this earth to guide Muslims and that their status is not that of a Prophet, as

Muhammad was the last Prophet, but that they are nevertheless imbued with divine light, infallible and chosen by God as the best people to guide the faithful and to interpret God's laws. Thus shi'ism has the notion of a clergy whose ultimate role is to provide the believers with living guidance.

In Scotland, shi'ites make up 6 per cent of the total Muslim population with the majority settled in the west of Scotland.[13] Shi'ism is manifest through a diverse range of communities including the Iranians, Indians and Pakistanis and also the Iraqis. Many came as students and have subsequently settled here, or they came as migrants seeking refuge in Scotland from the poverty or conflict in their homelands. The shi'ite community have their own mosques and there are two such mosques in Glasgow, one in Edinburgh and one in Dundee. Moreover, the Iranians will proudly adhere to the pre-Islamic Iranian celebrations such as the celebration of the Iranian new year. One other group, the Ahmadiyya, not recognised as Muslim by some mainstream Muslim organisations for their messianic beliefs about their leader, are also to be found in small numbers in Scotland. They have various associations in Scotland and official statistics report that there are currently 600 Ahmadiyya in Scotland with the majority residing in Glasgow.

The emergence of Muslims as a community with their own organisations and faith-based institutions was a gradual process. As families settled in Scotland, it became necessary to find suitable religious education for the children and create space for communal activities. During the 1970s and 1980s, when it became common to talk of a 'multicultural' Britain, the most visible sign of Islamic identity became the mosque. What had initially started as simple rooms in terrace buildings or even run-down flats, paid for or rented by small communities, was to manifest itself in far grander buildings and locations. The mosque is probably the most visible symbol of Islam in many of the Scottish cities, binding Muslims from all backgrounds and generations together, as well as providing a focal point for social and educational activity within groups. However, it is a common misconception among non-Muslims that the mosque is the Islamic equivalent of the Christian church. The mosque quite simply provides a public space for congregational prayers and although every *salat* is obligatory on all Muslims, Friday midday prayers are special within the Islamic tradition and therefore these prayers usually attract the largest number of worshippers. As a site, the mosque is functional space, not sacred per se, nor is it compulsory to hold any event with a ritual content, such as a wedding, there. Of course, during certain religious festivals and in the sacred month of Ramadan, the mosque is overflowing with worshippers, many of whom will stay, sleep and worship in the mosque for days. The Eid prayers are run at various times in the morning to accommodate the huge increase in worshippers who will most likely take time off work to be able to pray at a central or even local mosque.

Many of the mosques during the 1980s had their origins in foreign funding since a number of Muslim countries that were flush with money, such as Saudi Arabia, Kuwait and the United Arab Emirates, decided to

Figure 11.2 Edinburgh Central Mosque. Photograph by David Veitch.

become benevolent donors. Apart from being willing to support other Muslims, the governments of these countries were also keen to be regarded as influential within certain segments of the Muslim communities. An example of such a mosque is the Edinburgh mosque which opened in July 1998 (Fig. 11.2).[14] It was financed by King Fahd of Saudi Arabia and opened by his son, Prince Abdul Aziz Bin Fahd. The mosque, which took more than six years to build and cost £3.5million, is unusual for mixing Islamic and Scottish architectural styles. The opening of the building was a big occasion and attracted a number of dignitaries, politicians, journalists and academics. In fact, the then Home Affairs Minister, Henry McLeish, praised the cultural diversity in Scotland and said:

> The government is strongly committed to the concept of community and the inclusion of the rich variety of faith groups, who contribute so much to life in Scotland.[15]

Most Scottish cities, including Aberdeen, Dundee, Edinburgh and Glasgow, now have a proper mosque building. This reflects the growing Muslim community and the increase in the number of Muslim students at Scottish universities who require the facility of an established place for worship and for student activities. Some of the mosques have become landmarks of the city in which they are located and one such example is the Central Mosque in Glasgow, which was opened in 1984 on a site next to the Clyde river. The Central Mosque is larger than the Regent Park Mosque in London and benefits from a recently built Islamic Centre adjacent to it in 2 acres of ground. The Islamic Centre runs various social events, recreational activities and educational workshops for the community.[16] It also houses a small library and works as an advice centre for the community as a whole.

There are currently over a dozen mosques in Glasgow alone, and in Dundee and Edinburgh there are about four or five mosques. The growing number of mosques, combined with recent immigration from a diverse range of countries, has seen a change in mosque culture; ethnic and national divisions have resulted in splintering some of the solidarity within communities and many will go only to certain mosques. It should, however, be borne in mind that Muslims do not need to be seen in mosques to practise Islam. True, the mosque provides space for bringing people together for specific events pertaining to the Muslim community, but it is not imperative to individual worship in Islam. Furthermore, there are many people who are sceptical of mosques' managements and feel that they would rather not get involved in the political and often petty internal wrangling that unfortunately becomes part of running a religious institution dependent on external finances. The imam may also convey political and religious slants, thus affecting a person's choice and preference. One of the reasons why this is particularly manifest in the mosque is that these places become targets for individual loyalties to specific teachings, which ultimately originate from one or another Muslim country, mostly India or Pakistan. These ideologies are seen as an intrinsic aspect of putting forward the ideal of Islam as a society and faith system quite often superior to Western teachings; therefore the quality of Friday sermons among other things is crucial to the image and popularity of any one mosque.

Three specific movements that have their roots in colonial India and which developed fundamentally out of opposition to British rule are the Deobandi, Brelwi and Ahl al-Hadith.[17] The Deobandi movement originated from a town near Delhi called Deoband, which was a major centre of traditional Islamic learning during the Mughal Empire. It began with a particular form of teaching which concentrated on the study of the Qur'an, *hadith* and *shari'a* and logic and philosophy. Teaching was in Urdu and the Deobandis became instrumental in gradually making Urdu the common language of teaching throughout the subcontinent. They have the reputation of being the more puritanical strand of Indian Islam and for acting as sources of authority for legal opinions, the *fatawa*. The ahl-i-hadith group of scholars say that the hadith, i.e. the words of the Prophet, are the ultimate source of explanation for the Qur'an and the understanding of *shari'a*. They rejected all institutional forms of Sufism and devoted most of their learning towards the *hadith* and *sunna*. The Brelwi group were very different for they went in the direction of reasserting the popular traditions within Sufism, the formal celebration of certain events such as the Prophet's birthday and commemoration of the saints' lives, in fact most of the things that had been denounced as alien to the purity and simplicity of Islam by the Deobandis.

The above exist as living Islam in Scotland within many of the communities since these ideologies accompanied the first wave of migrants and have been kept alive by the zeal of their followers. They are also expressed through more active piety through socio-political organisations such as the Tablighi-jama'at who are really reflecting the viewpoint of the Deobandis.[18] Their relentless preaching, which takes them from door to door as well as country to country, has enabled them to achieve huge international success.

Since the mid-1980s there has been a significant growth in different kinds of movement among the variety of Muslim societies that are settled in Scotland. There are of course internal as well as external factors for this, but key events such as the Iranian revolution, the 'Rushdie affair', the rise in militancy and terrorism in many conflict areas of the Middle East, the continued Palestinian–Israeli conflict, the domestic issues of asylum seekers and refugees, have all been instrumental in making many Muslims aware that Western countries often have a fairly negative image of Islam. Furthermore, the increasing challenge that is posed to religions as a rise in secularist and consumerist cultures means that Muslims too face the task of how to make their religion meaningful and practicable in today's changing society. Various Muslim organisations do not ally themselves to the traditional ideological movements originating from the subcontinent or the Middle East, but are trying to achieve the ideal of a new Islam that is born and bred in the West but at the same time challenges some of the cultural patterns to be found in the West. Some, such as the Young Muslims, have grown at a national level, reaching out to Muslims of all backgrounds and cultures. They have been successful in recruiting numbers and trying to mobilise the generations to work in groups for the greater good of the community as well as Islam itself. The Young Muslims, together with the older wing of their organisation, the Islamic Society of Britain, have worked with the ethos that Islam has to work at an educational level, a faith level and be a complete way of life if it is to reassert itself as a positive force. However, most of these organisations achieve only limited success because they are unable to reach out to those who are suspicious of organised movements as well as those who live on the margins of their own communities, isolated and confused.

Contemporary Islam in Scotland comes in many different forms because of differences of generations, cultures and languages. There are third-generation Muslims whose lives have hardly changed from that of their grandparents; traditions have been kept alive either through fear of alternative lifestyles that were viewed as too Western, or quite simply because there was never seen to be a need or desire for an alternative lifestyle. By the same token, there are many Muslim families who are not part of any collective community or any particular trend, but who choose to live both physically and psychologically away from ghetto-like communities. They are Scottish in many of the choices they have made for themselves but in terms of religious law, such as dietary regulations, traditional marriage

Figure 11.3 Four mosques join together to celebrate Ramadan, the Meadows, Edinburgh, 1993. Photograph by Herman Jai Rodriguez. Courtesy of the Trustees of the National Museums of Scotland, NMS SLA E/9353.

systems and the extended family, the importance of Qur'an learning and religious education as a whole, they are Muslim and they do not regard this as in any way a source of conflict. It serves no purpose to refer to the Muslim community in Scotland as if they are a united group in any sense – they are as diverse and as individual as any other group with their own subcultures.

The bigger cities in Scotland have clearly been receptive to the changing social climate. There are more and more Muslims working at professional levels, not solely in the traditional fields of medicine or dentistry, but as lawyers, journalists and lecturers. Local authorities are becoming increasingly aware of local Muslim businesses and their needs. This has broadened far beyond simple awareness of halal meat as the one element that defined Muslims. Scotland has become multi-faith and to some extent this is being reflected at all levels of society. Museums and art galleries, such as St Mungo's Museum of Religion in Glasgow, regularly hold not only inter-faith events but exhibitions that have an Islamic element. In 1995 Glasgow University's Divinity Faculty broadened its commitment to pluralism in the teaching of religion by appointing the first ever woman who was also non-white and Muslim, Dr Mona Siddiqui, to teach a range of subjects in the field of Islamic Studies. Muslim schools such as Iqra Academy have been set up specifically to provide a more religious-based education for Muslim children. The most obvious Islamic symbol, the *hijab*, is now a common sight on the streets, reflecting a growing desire by many Muslims to be seen as Muslims. The United Kingdom's first Muslim Member of Parliament, Mohammed Sarwar, is a Scottish Muslim. However, words such as 'integration' and 'assimilation', which are so commonly used in debates on the nature of contemporary society, mean different things for different people and should be used with caution. If such words denote some sort of 'merging in' with Scottish society, in which traditional culture is being ignored or forgotten, then many people will refuse to say that they have assimilated. The majority of people may well be living as Scottish citizens but their Islamism and cultural allegiances continue to set parameters beyond which they refuse to step and they happily lead parallel lives which are both Islamic and part of a broad European culture.

Perhaps the most significant aspect of Islam in Scotland and in Europe as a whole is that there is a big increase in the numbers of non-Muslims who are converting to Islam. This has arisen out of various factors – interest in Islam, a search for some spiritual solace and the mixing of Muslims and non-Muslims at a social and professional level. It would be fair to say that in a world where more and more people feel an increasing alienation from formal religion, Muslims seem determined to hold on to their individual and collective visions of Islam, and Scotland has been no exception to this phenomenon.

NOTES

1. Maan, 1992, 1. This book, although ten years old, remains the only source of information on the communities from Asia that migrated to Scotland. It deals not just with Muslims and relies heavily on personal and anecdotal experience by the author himself, but is nevertheless a relatively useful tool for an overall picture of Asian ethnic communities living in Scotland.
2. A number of books can be found in the bibliography. An interesting survey in Nielsen, 1995, is a helpful summary of the Muslim presence in Europe as a whole and the different Muslim movements that have emerged over the years.
3. Martin, 1985, 11.
4. Martin, 1985, 9–12.
5. Matar, 1998, 2.
6. Nielsen, 1995, 4.
7. Maan, 1992, 84.
8. Martin, 1985, 75–8.
9. Martin, 1985, 54.
10. Martin, 1985, 160.
11. Martin, 1985, 162.
12. This is the current figure issued by the Glasgow Central Mosque.
13. Having spoken to a number of shi'ites in Glasgow, this was given as an official estimate.
14. The information on the mosque has been taken from the BBC News website, dated Friday 31 July 1998.
15. BBC News website as above.
16. The information on the Central Mosque is taken from the website of the Central Mosque, which has much information about the mosque, its activities and its commitment to the Muslim community.
17. Nielsen, 1995, 131–6.
18. Nielsen, 1995, 135–6.

BIBLIOGRAPHY

Denny, F. *An Introduction to Islam*, New York, 1993.
Maan, B. *The New Scots*, Edinburgh, 1992.
Martin, R C. Islam and religious studies. In Martin, R C, ed. *Approaches to Islam in Religious Studies*, Arizona, 1985.
Matar, N. *Islam in Britain 1558–1685*, Cambridge, 1998.
Nielsen, J. *Muslims in Western Europe*, 2nd edn, Edinburgh, 1995.

12 Reflecting World Faiths: BBC Scotland's Experience

JOHNSTON McKAY

One of the aims of religious broadcasting in the BBC, as approved by the BBC's Board of Governors, is 'to seek to reflect the worship, thought and action of the principal religious traditions in Britain, recognising that these traditions are mainly, but not exclusively Christian'. The inclusion of the reference to traditions other than Christian was the most significant change to the policy towards religious broadcasting, introduced by the Annan Committee of Enquiry, which reported in 1977.

By way of introduction, it may be useful to refer to three different reactions to that stated aim. Pauline Webb, a former Organiser of Religious Broadcasting for the BBC's World Service, and someone who has championed the cause both of multi-faith broadcasting and of a multi-faith society, has described 'the problem of persuading the representatives of the various faiths to present their material in ways that made it comprehensible to those outside their own faith community, for it had to be emphasised that religious broadcasting is not there primarily to provide for the needs of the adherents of the various faiths, but for the public as a whole'.[1]

Speaking at a dinner given in his honour by the Independent Television Commission, the then Moderator of the General Assembly of the Church of Scotland, Rt Rev. Andrew McLellan, quoted the Chief Rabbi as saying 'that the part played by religious broadcasting in smoothing the way to a multicultural society had been enormous and unacknowledged'.[2] However, the writer was present at another meal given in honour of a Moderator several years ago, on this occasion within the BBC, and was as shocked as many of the BBC executives were to hear the Moderator that year insist that faiths other than Christianity 'are to be tolerated in this country, but only tolerated'.

These reactions illustrate the variety in the climate of opinion in which religious broadcasters in Scotland have had to pursue their stated aim of reflecting that the religious scene in the country is not exclusively Christian. It is only within recent years that this aim has been taken seriously.

COMING TO TERMS WITH WORLD FAITHS

The BBC's official policy in 1955 stated that the 'content of religious broadcasting should be what is actually taught and practised by the principal organised expression of the religious life of the country – the Christian Churches'.[3]

Writing in 1968 the former regional director, later controller of BBC Scotland, Rev. Dr Melville Dinwiddie, was able to state unashamedly that 'the unacknowledged aim of the BBC in its religious broadcasting was to make Britain a more Christian country',[4] and that view is reinforced in what Dinwiddie has to say about overseas broadcasting. Its purpose was:

> to project Britain to all who could hear outside this tiny island, and its aim was to transmit the best of our way of life. It was therefore natural that religious items should become a feature of these programmes and their inclusion was welcomed, but raised all the old problems of form and practice as well as new ones. To whom should these items be directed? To Christians living abroad with home church ties, or to any who might be able to hear them – pagan and Christian, Jew and Gentile, Muslim and Hindu? Were the broadcasts to be aimed at the inhabitants of the Empire, the majority of whom were non-Christian though loyal to one earthly monarch? Again, if the Christian religion was to be the radio pattern what form of worship should be used? Would the result be to bring people together or to emphasise their differences?[5]

The assumption that Christianity is the faith which religious broadcasting existed to promote is clear in what Dinwiddie writes, and was just as clear when the man who was in charge of religious broadcasting in Scotland from 1945 to 1971 reflected on its aims the year after his retirement. He quotes with approval the statement of a former Head of Religious Broadcasting in London that 'religious broadcasting is the hand-maid of the churches'.[6]

Shortly after Ronald Falconer retired, the climate of religious broadcasting within BBC Scotland changed, and under one of his successors, Ian Mackenzie, moves were begun both to widen the programme base and to wrest religious broadcasting from its umbilical relationship with the Churches in general and the Church of Scotland in particular.

In March 1977 Mackenzie reported to BBC Scotland's Religious Advisory Committee[7] that he had been having discussions with the Glasgow Inter-Faith Committee, which existed to promote inter-faith discussion and dialogue, and he asked the committee if it would approve his investigating further whether representatives of the various faith communities would be interested in being given access to religious broadcasting.

Six months later the Advisory Committee was told that the Glasgow Inter-Faith Committee felt very strongly about Christianity's virtual monopoly. Significantly the Inter-Faith Committee indicated that it was not interested in programmes *about* other faiths but in expressions of them. This has proved a point of considerable tension between programme makers and the representatives of the various faiths.

In April 1979, one of the leading figures in the Glasgow Inter-Faith

Committee, a deaconess with the Church of Scotland, Stella Reekie, joined BBC Scotland's Religious Advisory Committee. While this was an acknowledgment of her considerable interest in inter-faith dialogue, it still reflected the view of world faiths filtered through a Christian, and indeed a Church of Scotland, perspective.

Four years later, in April 1983, the then Secretary to the Broadcasting Council for Scotland, John McCormick, raised with the Advisory Committee a letter which had been received from the Strathclyde Community Relations Council, asking that contributors to *Thought for the Day* on Radio Scotland should not be drawn solely from adherents of Christianity. After some discussion, the Committee noted that Mr McCormick proposed that a reply be sent, indicating that the matter was under discussion, and the Committee agreed to consider the issue of contributions to *Thought for the Day* further, after there had been consultation between the Head of Religious Programmes and the Head of Radio Scotland.

It was two years before the matter was discussed again at BBC Scotland's Advisory Committee. Then, according to the minute of the meeting, the committee accepted 'that consideration had to be given with some urgency to the involvement of faiths other than Christianity'. At its next meeting, six months later, the Advisory Committee accepted the principle that opportunities to contribute to religious broadcasting should be open to people from all faith communities, and asked that immediate steps be taken to implement this principle.

The minute of the Committee's meeting in April 1986 records that 'some, but not enough progress, had been made in (the matter of) ... a periodic thought by someone from a faith other than Christianity'.

On one of the first of these periodic occasions when a Muslim was invited to record a *Thought for the Day*, the invited contributor arrived at the BBC studio, armed with several books in Arabic from which he proposed to read. That probably said as much about the amount of advance discussion the producer assumed was necessary as it did about the potential contributor's ignorance of broadcasting expectations, and is an example of the sort of in-built, if unconscious, bias towards Christians and the sort of atmosphere where they were comfortable, which had to be overcome if any progress was to be made.

It had been nearly ten years since the Head of Religious Programmes had tentatively raised the issue of contributions from faith communities other than the Christian, and it is clear, both from the minutes of the Advisory Committee and from the writer's recollection as chairman of that Committee for much of the time, that there was a considerable discrepancy between the enthusiasm of the Head of Religious Programmes and some members of the Advisory Committee for the inclusion of contributors who were not Christian and those who were then on the staff of the Religious Broadcasting Department, most of whom had been brought into the Department by Ronald Falconer who, as has been seen, regarded religious broadcasting as 'the hand-maid of the churches'.

Shortly afterwards, a distinguished Jewish consultant surgeon, Henry Tankel, was the first non-Christian to be invited to join the Advisory Committee. Significantly, perhaps, at the first meeting attended by Mr Tankel, the Advisory Committee listened to extracts from a series of programmes made for Radio Scotland, *In Good Faith*. The producer of the programmes explained to the committee that the aim of the programmes had been to include in each programme 'information about a faith other than Christianity, excerpts from the worship and scripture of the particular religion being dealt with, and as much of a flavour of the religious community as possible'.

At the start of the twenty-first century, the membership of the BBC's Religious Advisory Committee in Scotland includes not only a wide representation of Christian denominations (with the Church of Scotland representatives now clearly in a minority on the Committee) but also representatives of the Jewish, Muslim and Hindu faiths. To be fair, the accelerated concern to represent the variety of faiths has reflected the growing awareness in Scottish society of its multi-religious and multicultural nature, and it would be unfair to have expected the BBC to reflect that before it had become a feature of Scottish society's own self-awareness.

Thought for the Day is no longer the exclusive Christian preserve which it had been. There are regular contributions from the faith communities of Islam, Hinduism and Judaism, and in 2001, during Holy Week, following four Christian contributions reflecting on the significance for Christianity of the death of Jesus, on Good Friday, a Muslim reflected from his standpoint on this crucial moment of another faith. Only one letter of complaint was received. A year or so earlier, at a time of tension and violence in the Middle East, the contributors to *Thought for the Day* on two successive days for a month were a Muslim and a Rabbi.

The continuation of *Thought for the Day* in a multi-faith, multicultural society is an important part of the tradition of public service broadcasting. This is not just because it acknowledges the place of the communities of faith within the public life of the country, but because it also reflects an ideal described by the Chief Rabbi, Dr Jonathan Sacks, in his Reith Lectures, as a 'sense of being part of a single moral community in which very different people are brought together under a canopy of shared values'.

WORLD FAITHS: TO DESCRIBE OR TO EXPRESS?

When, in 1977, the first tentative moves towards multi-faith broadcasting were made through contact between the BBC and the Glasgow Inter-Faith Committee, that Committee expressed the view that what was wanted was not programmes *about* the faith held by Muslims, Hindus, Bahais and Jews but *expressions* of the various faiths.

There have been considerable advances in programmes either about the various faiths or contributed to by representatives of these faiths, but there have been very few (if any) which could be said to have met the demand that they express the faith of those who belong to the great world faiths, insofar as that faith is expressed in the *worship* of the faith communities. It would have been somewhat perverse to reflect Hindu or Muslim worship at a time when the trend in religious broadcasting was increasingly to depart from broadcasting Christian worship. However, to offer that as a reason for the worship of ethnic minority faiths being ignored would be to fail to recognise the effect of changes in attitudes towards public service broadcasting on religious broadcasting in general and its reflection of minority faiths in particular.

Since the 1980s there has been increasing pressure on the BBC to justify its reliance on a licence fee by the size of its audiences. This has led those responsible for the radio and television output increasingly to regard the branding of their station as vital to attracting an audience. As the BBC and ITV have faced increasing competition from highly branded, very focused satellite broadcasters, that need for branding and focus has become even more important. In such a culture, the BBC will always have to strike a balance between its avowed intention to celebrate and communicate diversity and creating the impression that its programmes are not aimed at the potential majority audience. The difference in culture between the BBC in the 1970s, when, as we have seen, the debate in Scotland about opening up religious broadcasting to the wider faith communities was just beginning, and today, when the participation of all faiths is taken for granted, can easily be illustrated.

In the 1970s, the writer can recall hearing regularly on Sunday mornings, on BBC Radio Scotland, programmes aimed at the growing Asian community in Scotland and featuring long tracks from the music of the Indian sub-continent. The audience was probably extremely small. There was no attempt to include the majority of the Scottish population. This was niche broadcasting for a tiny ethnic minority. And since the size of the audience was not the dominant issue for a public service broadcaster, the size of the niche was irrelevant. Such a policy would be inconceivable in today's broadcasting culture. That may be a cause for regret, but anyone within the BBC who ignored the competitive culture would very quickly become aware of the need to earn the licence fee. That is not to say that there are not programmes, especially on radio in Scotland, which reflect the life and experience of ethnic minorities, for there are. But, as is the case with religious programmes, they must be made in a way and style which matches the station controller's view of what will maximise the audience for the station.

An illustration of the occasional tension between the expectations of faith communities and the demands of broadcasting, and also of the conflicting signals sent out by faith communities, was given by the then Head of Religious Broadcasting for the BBC, Rev. Ernest Rea, when he spoke to BBC Scotland's Religious Advisory Committee in November 1998. He was reporting on a multi-faith seminar, which had been held in advance of Radio 2's World of Faith Week that year, and expressed the opinion that the BBC had not actually dealt very well with reflecting all the faith communities within the United Kingdom. However, Mr Rea said, the expectations of the faith communities had to be realistic and had to recognise, for example, that Radio 2 was an entertainment channel and that if it reflected world faiths it had to be within the context of the channel's aims and policy. He quoted a Muslim member, attending his first meeting of the BBC's Central Religious Advisory Committee, asking when Muslim worship would be broadcast in Arabic. Other Muslims say that worship need not be in Arabic. Some would regard subtitles as offensive, others would not.

Broadcasters who have no difficulty understanding the divisions within Christianity often assume that other world faiths are more monolithic whereas the experience of those working within religious broadcasting is of the diversity within world faiths. The Scottish experience illustrates this. To invite a Muslim to contribute to a programme does not automatically satisfy the entire Muslim community any more than to invite a Free Presbyterian would satisfy the entire Christian community of Scotland. Those who have converted to Islam are not always regarded as acceptable representatives by those who have been born into that faith. Those who came to Scotland as first-generation immigrants, bringing faith with them from their native country, often regard those whose Islamic convictions have been worked out entirely within their new, more secular homeland as not representing the Islam of their forefathers. And the growing significance of Islam as part of the cultural background, but not necessarily as part of religious conviction (somewhat akin to the difference between secular and religious Jews), is a further complicating factor.

However, the 2001 version of the BBC's producer guidelines instruct producers to take account of the diversity within some of the world faiths:

> People and countries should not be defined by their religions unless it is strictly relevant. Particular religious groups or factions should not be portrayed as speaking for their religion as a whole. Thoughtless portrayal can be offensive, especially if it implies that a particular faith is hostile or alien to all outside it. For example, footage of chanting crowds of Islamic activists should not be used to illustrate the whole Muslim world.[8]

Throughout the 1970s and 1980s there was a temptation among religious broadcasters to regard these conflicting strands, particularly within Islam, as a reason for avoiding Muslim contributors, and so avoiding issues with which they were unfamiliar. Growing and important contacts with the Muslim community, however, have given broadcasters greater sensitivity to the cross-currents within the Islamic community. Inevitably, however, religious broadcasters tended to rely on contributors from ethic communities that had developed an understanding of Western culture best, for they were the most 'natural' broadcasters.

RADIO PROGRAMMES TRANSMITTED

During the 1990s, on Radio Scotland, programmes were planned, produced and developed which involved contributors from the world faiths. In 1996 a series of six programmes was called 'Insights on Offer' in which people belonging to the major world faiths were invited to describe the experience of living their faith in Scotland. Ground-breaking though this series was, it has not been preserved in the BBC's archives.

The following year 'Making a Meal of it' looked at the links between food and faith for Jews, Hindus, Buddhists and Muslims:

Muslim contributor (1)
   Thursday is special because when somebody has died – like my father has died – I cook for my father. He comes to the home.

Interviewer
   Do you actually have it in your head that you're making enough for him to eat as well?

Contributor
   Yes

Interviewer
   What a lovely way of remembering him.

Muslim Contributor (2)
   On a Thursday, they say the souls are more about the house than everywhere else. That's why it's more special on a Thursday. Friday is Jumaah, Prayer Day, that's what we call it, Jumaah, everybody goes to read prayers together. Even if your husband or father is working, they try to go at this time because it's a very holy day.

Interviewer
   You're almost preparing for this holy day, Friday, when you're eating together on Thursday night?

Contributor
   Well you all sit and talk religion and about going to prayers tomorrow.[9]

In the next programme in the series, a Buddhist, who worked in the kitchens at a retreat, spoke about preparing food in the religious context he recognised:

Buddhist Contributor
   Before any food is cooked, we do an offering to the shrine. There's the Guru and the Buddha, so before anything is prepared or cooked, we put offerings on a bowl, and we say a few thousand mantras; so when it is offered to the shrine, it is being offered to every sentient being before it is even cooked. When the food is starting to cook,

> we're actually doing mantras all the time, from five in the morning to twelve thirty when it is cooked. So by the time the food is served, it has been well hyped with mantras and it tastes absolutely wonderful.
>
> *Interviewer*
> When you say 'thousands of mantras being said' how is that done, when you're going through the practicalities of getting the right quantities?
>
> *Contributor*
> Say, for instance, I am doing pastry for five pies. I stop and think about that pastry and I get that done. Then I clock off again. I can potter about the kitchen and do mantras. A lot of it does go into the cooking, into the herbs you put into it. When you add certain herbs to your food to get the benefit of the whole flavour, there's a spiritual part too, and the two of them combined make it taste a lot better.[10]

In the summer of 1998, within a series based on *The Planets* by Holst, one of the programmes was about mysticism and explored the role of mystics in Christianity, Buddhism, Hinduism, Sikhism, Judaism and Islam:

> Mysticism has been part and parcel of the Hindu religion for centuries. You may be aware that Hinduism is actually created out of scriptures written by the sages over the last five thousand years. It's not a religion which was created by any god or prophet who gives certain dictums from time to time and the people follow those religions. These sages have been doing a lot of soul-searching and meditating over a period of time to understand what is right and what is wrong, what is good and what is bad, and trying to evaluate the values of life. Now, when you say 'mysticism', mysticism comes in from the practice of meditation by those yogis and sages who have been able to control their desire so that they can be totally in tune with themselves and away from this material world.[11]

How the major faith traditions use and regard music was the theme of a series of three programmes in November 1999 exploring the place of liturgy and chant, the role of musical instruments in worship and the effect of music and performance in personal devotion and meditation:

> Certain mosques you will go into, you will hear chanting. Different mosques have different traditions. Some mosques even in this area have different traditions of what is done in the mosque. In Glasgow certainly there are some who will chant remembrance, names of God, praises of the Prophet, these kinds of non-Qur'anic vocalisations of a religious nature which you'll hear in local mosques. In other mosques it would be Qur'an and nothing else.[12]

In the next programme in the series, a Hindu woman reflected on the place of worship in worship as she experienced it:

> It's a sense of community when people get together in a Hindu temple and they will sing these songs, and everybody joins them. Its one form of establishing a link between immortals and mortals, and between God and the worshipper. In temples, when you go, you're not just sitting there and having fun by singing certain songs ... There is a purpose by getting everybody to sing together, bringing them into a synergy; and by singing you're got into a mood where you're ready to absorb information and the good things which are there. The typical thing of which we are reminded again and again is be good to each other. If I was given a text to memorise I would find it difficult to memorise but if there is a song, I'd sing it and memorise it far more easily, and that's why those devotional songs have been an important carrier of religion and passing it on.

In the millennium year, a series of four programmes, 'Faith of a Woman', looked at issues particularly affecting women within the major faith traditions, including contributions from Muslims, Hindus, Jews, Buddhists and Christians on the themes of religious leadership, the place of faith in the life of the home and the nurturing of children, links between women's physical and spiritual rituals, associated with, for example, menstruation, and male and female images of the divine:

> By and large the majority of people who are in public religious posts in Islam tend to be men, and that's because of the nature of Muslim societies which are ... you know, women have been the victims of legislation. They've never really talked about women's rights until recently, and so societies are modelled on a very patriarchal system of learning, interpretation, and of gender roles. Really, talking about religious leaders in Islam is difficult because unlike many religions it doesn't have the notion of priesthood in the strict sense of a kind of clergy. The Imam is, strictly speaking, someone who leads a congregation in prayer, and in different communities and in different countries, the Imam may have varying roles but he is in no way a priest or a minister or someone who has sacred authority. As far as women and the status of an Imam is concerned, it's not something that women have debated strongly because it's not seen as a kind of vocation, or a sign of devotion to God in some way, so that Islam doesn't have that sense of vocation, and so women fighting for equality do so, but not through religion.[13]

In the same series a Hindu woman reflected on the place of women from the perspective of her religious faith:

> In Hinduism, women are regarded as sacred members of the family unit. The reason is that God is referred to as 'Mother' and motherhood

is synonymous with women, and woman is synonymous with motherhood, so there is a sacredness built into being the female member of a family; and the changes that develop in a woman's body are interestingly given a status of respect, though it may not be seen as such or understood as such by the onlooker.[14]

But a Jewish woman reflected on the very different attitude of her religion to menstruation:

> Menstruation brings no difference to the practice of one's faith. I still go to synagogue. Life goes on the same at home except that there is no contact with one's husband. The *mikvah* is a ritual bath, used by women, mainly. Men also use the *mikvah* when they want to make a ritual immersion before praying. Women use the *mikvah* as a ritual after menstruation and a period of separation from their husband. The *mikvah* bath is the place women go before having contact with their husband.[15]

What these programmes illustrate is a policy of looking for issues of common humanity which can be examined from the standpoint of the different faith traditions, or areas of experience which the different religions all address, and examining these in a comparative way. While religious broadcasting within the public service tradition must reflect the diversity of the religious traditions within Scotland, it has a commitment to do so in a way which illustrates the richness of diversity without accentuating the reasons for division.

In the early 1990s an experimental worship programme involving a number of faith traditions was made by the writer. A Christian, a Muslim, and a Buddhist, whose religions all recognise the place of pilgrimage, shared a pilgrimage which is significant for Christians in Scotland: a journey to Whithorn in Galloway, associated with Saint Ninian and the bringing of his faith to Scotland. The participants shared conversation, reflection, some of each other's wisdom, tradition and prayer. The result was a series of somewhat polite exchanges, largely because within a meditative and reflective context, the common denominators turned out to be respect for each other and for the earth. These are quite proper ideals but the expression of them can pall over the course of a 45 minute programme! It was largely as a result of this experience that the route taken by the BBC in Scotland since the 1990s has been chosen and proved a much more fruitful way of exploring the beliefs, practices and ideals of the major faith traditions that are now such an important part of Scotland's religious scene.

NOTES

1. Webb, 1991, 129.
2. McLellan, 2001, 101.
3. BBC, 1955, 57.
4. Dinwiddie, 1968, 47.

5. Dinwiddie, 1968, 101.
6. Falconer, 1977, 117.
7. This reference to the proceedings of the Scottish Religious Advisory Committee, and those which follow, are taken from the approved minutes of the committee.
8. Producer Guidelines, 2000, 110.
9. Radio Scotland, April 1995.
10. Radio Scotland, May 1995.
11. Radio Scotland, June 1998.
12. 'Sounds of Glory' Programme 1, November 1999.
13. 'Faith of a Woman', April 2000.
14. 'Faith of a Woman', May 2000.
15. 'Faith of a Woman', May 2000.

BIBLIOGRAPHY

BBC Handbook 1995.
BBC, *Producer's Guidelines, The BBC's Values and Standards*, London, 2000.
Dinwiddie, M. *Religion by Radio*, London, 1968.
Falconer, R H W. *Message Media Mission*, Edinburgh, 1977.
McLellan, A R C. *Gentle and Passionate*, Edinburgh, 2001.
Webb, P. Shall faith speak peace unto faith. In Elvy, P, ed. *Opportunities and Limitations in Religious Broadcasting*, Edinburgh, 1991, 125–37.

13 Reflecting World Faiths: St Mungo Museum of Religious Life and Art

HARRY DUNLOP AND ALISON KELLY

> When it comes to exhibiting the sacred, a fundamental challenge is, quite simply, how do you picture the unpicturable; how do you mount a display about what, at root, is resistant to all forms of expression; how do you convey to visitors that what religions themselves see as of primary importance is something which lies beyond all the carefully assembled material which museums present for their scrutiny?[1]

INTRODUCTION: EXHIBITING RELIGION

The St Mungo Museum of Religious Life and Art stands beside the medieval cathedral of Glasgow, in the heart of the old city, and draws around 180,000 visitors a year. Museums exclusively devoted to exhibiting religions are rare.[2] This lies in the fact that 'religion' is a term that defies easy interpretation – the physical manifestations of religious behaviour in the world are almost infinite. Problems also lie in the conflict between the traditional role of a museum, where interpretations are based on objects, and 'religion', which in very general terms deals holistically with the person's search, on an individual and a communal level, for meaning in the world. If religion is the unity yet also the tension between abstract belief and concrete practices, how can you exhibit it? This question begins a journey that opens up the many and varied ways in which religion has been understood, defined and practised by people across the world and across time. It also leads to an exciting and creative way of understanding museums.

In a multi-faith city, the mission statement of St Mungo Museum is to promote mutual understanding and respect between people of different faiths and of none. This makes people the focus of the museum. This is not to downgrade the importance of an object for a museum, but it realises and is honest about the role people have in making meaning from objects, and, in a museum of religion, the role people have in making objects sacred. Being conscious of the visitor's role in making meaning in the museum does not mean that the museum seeks to dumb down religious, theological and philosophical concepts. It encourages the visitor to explore the complexity of life, and to be able to engage with and respect new and different ways of thinking. Religious objects especially have the power to stimulate deep emotional responses because of the different layers of meaning attached to them. When a statue of St Patrick was displayed in an exhibition charting the story of the Roman Catholic community in Glasgow, the museum staff became increasingly worried that visitors were – sometimes within the same hour – kissing the statue's feet, while others spat at it.

HISTORY OF THE MUSEUM

St Mungo Museum of Religious Life and Art opened on 1 April 1993 having cost six million pounds. The building, designed by Scottish architect Ian Begg in the Scots baronial style, was not originally intended as a museum but as a visitor centre for the cathedral. The Society of Friends of Glasgow Cathedral, the Scottish Tourist Board, Glasgow Development Agency, the European Regional Development Fund, Glasgow City Council and many individual and corporate funders financed the project. In 1990 the building was taken over by Glasgow City Council when the project ran into financial difficulties. It provided money for the completion of the building and paid debts incurred by the Society of Friends. The Director of Museums agreed to take on the management of the building and various ideas were suggested for an appropriate subject in such an historic site. One idea was for a museum of the medieval world drawing on the reserve collections at the Burrell Collection. This gave birth to the suggestion of a museum that not only examined medieval Christianity but also looked at the importance of religion in people's lives across time and across the world.

THE LAYOUT OF THE MUSEUM

Visitors are introduced to the museum by the award-winning welcome video, the first of its kind in Glasgow Museums' venues. This professionally produced video consists of eight people from different religions speaking personally and candidly about their faith. It is intended to orientate visitors within the building and to introduce, right at the beginning of the museum experience, the concept that religion is ultimately about people and the personal experiences individuals have within the structures of their particular traditions.

The museum is divided into four main exhibition areas: The Gallery of Religious Art, the Gallery of Religious Life, the Scottish Gallery, and a temporary exhibition space. In all the galleries the key interpretative text for each display is translated into Gaelic, Mandarin, Punjabi and Urdu for those local people for whom English is not a first language. This also makes a powerful political statement about the multicultural face of Scottish society.

The Gallery of Religious Art is a space that exhibits beautiful artistic works from different religions. Using the principle that sacred art should challenge simple categorical systems, the display encourages the visitor to look beyond the art history of the objects and to focus on their religious and,

Figure 13.1 The Gallery of Religious Art, St Mungo Museum of Religious Art and Life, Glasgow. Copyright: Glasgow City Council (Museums).

in some cases, cultural significance as manifestations of living faiths. Each religion has a general framework that influences artistic expression, whether figurative representation as indicative of Hinduism or the aniconic traditions of Judaism and Islam. Working within particular traditions the artists have created stunning interpretations of religious concepts. For the Jewish artist Dora Holzhandler, a painting of a family around a table preparing for the Sabbath meal encapsulated her understanding of Judaism as, she said, 'God is in the family'. This can be a provocative space for some visitors, with *Christ of St John of the Cross* by the Spanish surrealist artist Salvador Dali surrounded by Buddhist iconography, an Islamic prayer rug and a Kalabari ancestral screen. Rabbi Julia Neuburger commented during a visit before the formal opening that the museum would be a challenge for believers, especially Christians. This prophecy was soon fulfilled, when, a month after opening, the cast bronze image of the Hindu deity, Shiva Nataraja, was seriously damaged by an evangelical Christian enraged by the inclusion of 'pagan idols' in the displays.

The initial brief for the Gallery of Religious Life was directed by a traditional museological interpretative approach with emphasis on taxonomic displays such as ritual equipment, iconography and religious costume. A more radical brief, however, was developed that allowed the displays to become more human and to give people a lever into the subject through empathy. The displays raise questions and explore issues through powerful object juxtapositions, images and oral testimony. Oral testimony plays a large role in this gallery. It allows for contradictory opinions to be expressed and gives ordinary Glaswegians the freedom to talk personally about their faith. For example, one display records the thoughts of Asma Shaikh, a Muslim, on arranged marriages:

> My main concern is that my parents are happy. As long as my parents are happy I'm happy. I trust them completely, absolutely 100%. Some people don't always understand arranged marriages, but it really shows how much we trust our parents.[3]

The Gallery of Religious Life attempts to communicate what people believe and share in common (such as rites of passage) and what is unique and particular in each religious tradition. Thus religion is explored through the human life cycle and also through separate displays centred on each of the six major world religions. This gallery explores religion as a lived experience and the human person as an historical, social and gendered being. The gallery also stresses that while religion has played a very positive role in the lives of many, for others it has resulted in being ostracised or even murdered. Therefore some displays in this gallery focus on the darker side of religion, including religious persecution and violence. A small display on the Holocaust communicates the horror of mass extermination in the name of race and religion in the twentieth century. Given the sensibilities involved, the Jewish Representative Council and the education department were both consulted on how this subject should be approached. The result is a display showing a prayer book used by a local rabbi while imprisoned in a concentration camp, juxtaposed against a photograph of a mass burial pit.

The Scottish Gallery was the most difficult gallery to tackle. The initial brief was chronological, outlining the development of religion in Scotland from earliest times to the present. The difficulty in this approach arose when it was noted that there were so few objects to interpret the medieval period. The Calvinist prohibition on the use of lavish ceremony and ornamentation severely restricted the range of objects available for display. As a way of compensating for the lack of objects, the gallery adopts a thematic approach. The themes finally selected for the six cases in the gallery were headed: Religion in Scotland; Keeping the Faith; Catholics and Protestants; Charity; Missions and Missionaries; and People and Places.

These themes were chosen as they were felt to be crucial in understanding the character of Scottish religious culture both past and present, including the sensitive subject of sectarianism. There was a fine line between underplaying the reality of sectarianism in Scottish culture, and marginalising those Scottish Catholics and Protestants who distance themselves from religious tribalism. In an attempt to solve this problem, objects relating to the major Christian traditions in Scotland were displayed against a background of partially hidden photographic images of Rangers and Celtic football matches.

One thing that shocks many people in this gallery is the inclusion of material from the ethnic minority faith communities in Glasgow. 'Keep them out – they are not Scottish', said some visitors. However, the museum is about the future as much as the past and present. Today the concept of Scottish religion is being expanded as Buddhists, Hindus, Jews, Muslims, Pagans, Sikhs, and many others living in Glasgow and across Scotland, wish to celebrate their Scottishness.

As the religious story of Scotland continues to unfold, the displays will continue to be modified and changed. The talk-back boards in the gallery bear witness to the variety of emotions and passions that abound, feelings that we as professional curators need to be constantly aware of if we are to take our public seriously.

STAYING RELEVANT IN TWENTY-FIRST-CENTURY GLASGOW

In the new millennium, increased globalisation, travel and immigration has resulted in heightened awareness of and interaction between people of different cultures and religions. The advantage that a museum has for the promotion of inter-faith work is that it is a space in and out of which people can drop. Visitors can choose to be part of a formal workshop or lecture or they can take their time and visit the museum in a manner that is preferable to them. This opens up the possibility about learning about each other in a

Figure 13.2 Chinese calligraphy at 'Meet Your Neighbour', St Mungo Museum of Religious Art and Life, Glasgow, 2004. Copyright: Glasgow City Council (Museums).

less organised and less formal manner. The challenge facing the museum team is how do you explore the idea of religion as being something dynamic through static permanent displays? Since 1993 the museum has tried to sustain existing audiences, especially the faith communities, and to attract new audiences through a varied exhibition and events programme. The main event is the now annual 'Meet Your Neighbour' multi-faith event, a festival involving many of the faith communities in the city, with music, dance, workshops and visits to different places of worship. This event is organised in conjunction with a number of inter-faith agencies including the Scottish Interfaith Council, Glasgow Sharing of Faiths Group and the Churches Agency for Interfaith Relations in Scotland.

Using the permanent exhibitions, the education and access team at the museum have organised fascinating and fun workshops on different religions for children and have created new family backpacks, which are full of activities to encourage closer exploration of the museum displays. Currently the education team at the museum is working on a citizenship education programme. Using key objects in the museum and specially created handling boxes, the team aims to work with school children on the issue of sectarianism.

Images in the media of religious fundamentalism and international conflict can polarise communities into constructed caricatures and stereotypes. The museum does not wish to shy away from such thorny issues. Neither does it wish to gloss over differences, nor to give a false sense of unity within and between religious traditions. Instead it hopes to engage with them in a constructive manner and to provide a space within which disagreements can be aired in a respectful manner by discussion and debate, to aim to have what Gadamer calls a 'conversation'[4] within which each participant expands their own horizons. During Islam awareness week in November 2003, the curator of Islamic civilisations and the curator of World Religions led a series of 'conversations' based around the topics of Islam and terrorism, and Islam and women – topics that have created heated debate. The aim was that in a safe space people could voice their opinions and discuss and deconstruct them in a positive manner. Based on the success of this, the museum intends to continue its 'Faith to Faith' seminar series, focusing in future years on contemporary ethical and religious issues current in the media.

Since the visitor's experience is often determined by the potential to be able to 'see themselves' and their story in the museum, the temporary exhibition space is an opportunity for faiths not represented within the permanent displays to have a voice in the museum. It is also a chance for important concepts and diversity within a particular religious tradition to be explored in greater detail. Since 2002, Glasgow Museums photographer, Jim Dunn, has been undertaking a museum project to document religious life in Glasgow in the twenty-first century. The first wave of photographs resulted in the highly successful exhibition 'Faithfully Yours', which opened in May 2003, celebrating the museum's tenth anniversary. By focusing on the ordinary people of Glasgow and the diversity and vibrancy of religious faith represented in the city, a new and fresh perspective can be given to the permanent objects in the display.

St Mungo Museum is becoming aware that while its mission statement is to 'promote respect between people of different faiths and of none', those with no faith have been largely ignored. This omission is especially pertinent, as when the museum first opened, its closest comparison was the former Museum of Atheism in St Petersburg. Therefore the museum has been working with groups such as the Humanist Society of Scotland and this year a workshop on Humanism took place in the museum as part of the seminar series 'Faith to Faith'.

CONCLUSION

The museum will never be finished in a traditional sense. There is, for a museum of religion, a radical potential in constant realisation. By discussing religious differences and developing narratives that provide a new, hopefully less dangerous, way of expressing them, respect between faith communities can be developed. At the core of all our work is the belief that by allowing the museum to be a safe space for discussion, debate and disagreement, we can in some small way work towards the creation of a Scottish society within which diversity of both faith and practice is celebrated rather than feared.

NOTES

1. Arthur, 2000, 2.
2. There are currently at least three other museums in the world devoted exclusively to religion: Marburg, Taiwan and St Petersburg.
3. Extract from an oral history transcription.
4. Gadamer, 1981.

BIBLIOGRAPHY

Arthur, C. Exhibiting the sacred. In Paine, C, ed. *Godly Things: Museums, Objects and Religion*, London and New York, 2000, 1–27.
Gadamer, H G. *Truth and Method*, London, 1981.
Lovelace, A, Carnegie, E, and Dunlop, H. St Mungo Museum of Religious Life and Art: A new development in Glasgow, *Journal of Museum Ethnography*, 7 (1995), 63–78.
O'Neill, M. Serious earth, *Museums Journal*, 2 (1994), 28–31.

14 Alternative Beliefs and Practices

STEVEN J SUTCLIFFE

DISAGGREGATING THE MYTH OF 'PRESBYTERIAN SCOTLAND'

It is a truism that modern representations of the cultural landscape of Scotland have favoured the Presbyterian. The sociologist John Highet inadvertently underscored this point by repeating an old joke about

> the Scots scientist who was 'ranting and raving' against the 'Bishops in Presbytery' proposals. 'Why are *you* so worked up?', asked his friend. 'I thought you were an atheist?'. 'So I am', replied the scientist, 'but I'm a *Presbyterian* atheist'.[1]

But such a witticism would have found little favour with migrant Irish Catholics faced with an often Kirk-sanctioned racism in the 1920s and 1930s: for example, a 1923 General Assembly report noted 'alarm and anxiety' over Catholic 'incursion' into Scotland and in 1935 there was mob violence against Catholics in Edinburgh.[2] Nor, arguably, has sectarian discrimination entirely withered at the turn of the twenty-first century.[3] Despite – or perhaps because of – such grassroots cultural differences and tensions, a rhetoric of consensus and common ground, rather than appreciation of (and legislation for) plurality and diversity, has dominated public debates about Scottish identity in the modern period. The rise of ecumenicism, seen in such initiatives as ACTS (Action of Churches Together in Scotland), in the popular cultural diffusion of a pan-Christian 'Celtic' spirituality, and more recently in the growth of the inter-faith movement (the Scottish Inter Faith Council was set up in 1999), provide indices of how a search for common ground has itself diversified as religion has become increasingly subject to both attritions and opportunities of rapid cultural change – the primary attrition being the 'haemorrhage' in mainstream church attendance figures from the 1960s onwards.[4]

Highet's earlier survey, *The Scottish Churches* (1960), had been at least partially alert to the politics of cultural identity, picking out and clarifying doctrinal and liturgical differences between Presbyterian congregations (and thereby alluding to the moral significance of secession, or principled 'difference', within Presbyterianism itself).[5] But pursuing a post-Reformation ideal of cultural consensus has resulted in a historiography of religion in Scotland dominated by two main (and partly complementary) theories: the 'secular-

Figure 14.1 White Buddha in the garden of the Samye Ling Tibetan Centre, Eskdalemuir, Dumfriesshire, 1999. Offerings of coins, stones and shells have been left around its ledges and base. Photograph by the author.

ization thesis' on the one hand, positing an inevitable decline in religion under the conditions of modern scientific rationality, and a Catholic/Protestant antisyzygy on the other, reducing public religion to a stand-off between two monolithic groups: the Church of Scotland and the Roman Catholic Church. As a result the extent of diversity even within the Christian traditions often fails to register in surveys of Scottish religion. For example, there are not just Presbyterian and Roman Catholic expressions of Christian identity in contemporary Scotland, but Episcopal, Orthodox, Baptist, Methodist and Pentecostal, and these traditions can in turn be differentiated according to internal ethnic-cultural tradition: Orthodox Christianity alone, for example, can be disaggregated into Greek, Bulgarian, Russian, Ukrainian, Syrian, Ancient, British, Celtic and Coptic branches in Scotland.[6]

However, if the term 'Christian', and even 'Presbyterian', can function as a fuzzy collective noun that muffles internal differentiation, at least these traditions carry relatively detailed historiographies.[7] Other genealogies have largely been written out of the history and ethnology of Scottish religion, distorting the fact that a diffuse – and expanding – religio-cultural heterogeneity has been spreading beneath the surface of homogeneous representations of 'Presbyterian Scotland'. Its most obvious representatives are the south Asian minority ethnic religions practised by Muslims in particular (over 42,000 in 2001),[8] but also Hindus and Sikhs, present in differing population densities from at least the early nineteenth century onwards. There has been a Jewish community in Glasgow since at least the early nineteenth century, and the first synagogue opened in Edinburgh in 1816.[9] Buddhism, too, has been overlooked, despite its post-war appeal to Scottish converts. This is attested by the presence of the Glasgow Buddhist Centre, a branch of the Friends of the Western Buddhist Order, in Sauchiehall Street, Glasgow, since 1973; by meditation groups in Theravadan and Zen traditions in Edinburgh; and by Tibetan Buddhists at the Samye Ling centre in Eskdalemuir, Dumfriesshire, since 1967, where one of the largest Buddhist temples in Europe was formally opened in 1988 (Fig. 14.1). Other minorities, particularly Irish, Italian and Polish, have developed a hybrid Scottish Catholicism. The beliefs and practices of Chinese communities in Scotland (with over 12,000 adult members at the 2001 census)[10] also remain largely invisible. And the growth in post-war English migration into Scotland (from 4.5 per cent of the population in 1951 to 7.1 per cent in 1991 and 8.1 per cent in 2001),[11] coupled with an episodic Anglo-sceptic, even anti-English, rhetoric in popular and élite Scottish cultures alike, suggests another elided 'minority ethnic' history in urgent need of recovery and analysis.[12]

In short, representations of Scotland as a homogeneous Presbyterian culture look increasingly difficult to sustain as we enter the twenty-first century. The launch in September 2002 by the Scottish Executive of a high-profile multiculturalist campaign, 'One Scotland, Many Cultures', arguably reflected belated political acknowledgement of the need for civic representation of cultural diversity. It is in this broader context of 'post-Presbyterian' religious and cultural diversity – including 'folk' and 'popular' expressions, alongside official institutional pronouncements – that we must locate, and explain, distinctively 'alternative' beliefs and practices in Scotland.

ALTERNATIVE BELIEFS AND PRACTICES

> Cultic beliefs of all kinds are now closer in cultural distance to the prevailing orthodoxies of society than they were at the turn of the [twentieth] century. What has been traditionally treated as categorically deviant and subject to secular sanctions as well as ecclesiastical wrath is gradually becoming merely variant.[13]

As early as 1975 the directory *Alternative Scotland* was reporting 'a great upsurge of non-Christian religion' in Scotland and 'an increasing number of groups interested in borrowing from any religion or none for the purpose of developing the potential of the individual'.[14] By the early 1990s a report to the Church of Scotland's Board of Social Responsibility advised that 'most unchurched people in Scotland today are more likely to construct their worldview from aspects of the New Age outlook than from elements of mainstream Christianity'.[15]

Figure 14.2 Green Man, May Day Parade, Stirling Castle Esplanade, 2000. In the late 1990s, Stirling Council and local tourist bodies sponsored an annual 'Beltane fire festival' in and around the city centre, including a torchlight procession and a burning wolf sculpture. A promotional postcard from Stirling Initiative noted: 'Beltane was the Celtic festival of Spring. Fire was used as an agent of renewal, driving out malign spirits, protecting the harvest and cattle.' Photograph by the author.

It is instructive to scan the list of religious organisations in *Alternative Scotland* in order to gauge the cultural and demographic affinities of 'alternative' religion. On the face of it there are few 'indigenous' Scottish projects: the list is implicitly weighted towards migrant traditions from the United States of America and the Indian subcontinent.[16] Some of these, like Christian Science and Scientology, are home-grown North American movements of the nineteenth and twentieth centuries; others, such as the International Society for Krishna Consciousness, Sri Chinmoy, and the Rajneesh movement, are rooted in Indian traditions (and *gurus*), mediated to Scotland (and the rest of Europe) via the USA and post-war American popular religion.

I discuss the salience of these cultural-demographic traits further in the conclusion. But one 'alternative' still too deviant to mention in a mid-1970s directory like *Alternative Scotland* is the new Paganism.[17] Its fortunes in the early twenty-first century provide a useful litmus test on two counts: the public confidence of 'alternative' practitioners, and the bounds of Scottish inter-faith inclusivity. Ancient stone monuments such as those at Callanish on Lewis and at Kilmartin in Argyllshire, as well as the revival in the 1990s of a Beltane Fire Festival on Calton Hill in Edinburgh, have attracted interest from self-styled 'Pagans' of whom there were around 2,000 in Scotland at the 2001 Census. There is a Scottish sector of the United Kingdom-wide Pagan Federation (est. 1971); a 2001 issue of its quarterly magazine, *Pagan Dawn*, lists regular 'moots' or meetings in Aberdeen, Ayr, Edinburgh, Galashiels, Glasgow, Perth, Prestwick, St Andrews, Stirling and Stranraer.

On the one hand, such activity is a sign of the increasingly confident public profile of this most explicitly, even theatrically, 'alternative' religious identity. On the other hand, the new Paganism has encountered both terminological ignorance and political exclusion from the Scottish establishment. For example, the June 2002 cover of the Church of Scotland's magazine *Life and Work* carried the banner 'It's Official: Scotland is Pagan', referring to a sociological survey cataloguing the decline of Christian practice. Here 'Pagan' is used to refer pejoratively to a collective 'other' set over against 'Christianity', rather than signifying the collective self-identity adopted by contemporary druids and witches, amongst others. Similarly, while Pagans are represented in some inter-faith activities – for example, Caer Clud, a local Druid group, participated in the annual 'Meet Your Neighbour' inter-faith gathering in Glasgow in 2002 – this has often been a contentious issue: at the time of writing the Pagan Federation has only observer, rather than participant, status on the Scottish Inter Faith Council.

But although a collective 'Pagan' identity still has the power to generate tension, the local vitality of Pagan groups is a sign of a broader flourishing of innovative religion in Scottish culture. At the corporate level alone, a 'delicious chaos' of nearly forty new religions has been identified by Frank Whaling in Scotland in recent decades, including neo-Hindu groups like the Brahma Kumaris, Sahaja Yoga and Transcendental Meditation.[18] This heterogeneous field has been largely overlooked by anthropologists, as have the historical origins of several internationally significant 'alternative' projects in Scotland. The two most obvious candidates are the Camphill Community movement and the Findhorn community, both co-incidentally finding a foothold in the north-east. The first Camphill community was set up in 1940 in Newton Dee, Aberdeen, by Karl König (1902–66), an Austrian paediatrician who came to Scotland as a refugee from Vienna following its annexation by Germany in 1938. Camphill communities provide long-term care for children and adults with learning difficulties, and are linked to the Anthroposophical movement of the Hungarian occultist and activist, Rudolf Steiner (1861–1925); there are currently 12 Camphill communities in Scotland and nearly 50 all-told in the UK and Ireland. On a similar pattern, in 1962 a small colony was set up in a caravan park at Findhorn, Moray, by an English family and some friends. By the late 1970s this had grown into the Findhorn Community, an internationally recognised community of some 200 residents and a platform for 'New Age' religious activism.

These particular projects were in turn preceded by rich loams of 'alternative' urban religion, such as Spiritualism and the Theosophical Society, which found footholds in plebeian and middle-class cultures respectively. For example, in Glasgow by 1880 there were some 250 Spiritualists and 20 practising mediums, and as late as 1951 there were still nearly 2,000 Scottish Spiritualists.[19] In its heyday Scottish Spiritualism produced internationally renowned mediums like D D Home (1833–86) and Helen Duncan (1898–1956) as well as a vigorous convert in Arthur Conan Doyle (1859–1930). Meanwhile, the Theosophical Society in Scotland shared in the international high water mark of Theosophical membership in the 1920s, and as late as 1944 the United Kingdom annual directory listed thirty Theosophical lodges or centres scattered across the Lowlands and the eastern Highlands, from Inverness to Aberdeen and Stranraer to St Andrews, with seven in Glasgow alone.

More recently, by the 1980s and 1990s – and epitomising the dissemination of a pragmatic, populist Anglo-American culture against the grain of Presbyterian rectitude – a thriving local culture of alternative and holistic healing was emerging, from Highland networks to centres, fairs and workshops scattered across central Scotland, including one of the longest-running holistic health venues in the United Kingdom, the Westbank Natural Health Centre in Strathmiglo, Fife (est. 1959). There has also been a renewed, self-conscious interest in 'folklore' and 'lived experiences' as recreational, educational and even 'spiritual' resources: expressions of such interest include local museums, tourist attractions and popular culture icons. Amongst museums we can mention in particular two projects from the 1990s: the St Mungo Museum of Religious Life and Art in Glasgow (est. 1993) and the Carfin Pilgrimage Centre in Lanarkshire, a small Catholic exhibition on cross-cultural pilgrimage (est. 1996); both have complementary functions as foci for inter-faith encounter and expressions of popular piety. At Rosemarkie in the Black Isle, the Groam House museum of Pictish stones and art (est. 1989) provides for some an indigenous identity resource to rival

Figure 14.3 Celtic Christian pilgrimage, Ninian's Cave, Whithorn, Wigtownshire, 2001. A driftwood cross beside an early stone carving. Small icons have also been left in the cave in this contemporary revival of an earlier pilgrimage tradition. Photograph by the author.

the broader territory staked out by 'Celtic spirituality'.[20] The Breadalbane Folklore Centre at Killin should also be mentioned: it holds a collection of healing stones traditionally associated with a seventh-century Christian saint, Fillan, which are occasionally handled for healing purposes by visitors and some locals. A more populist tourist attraction with a supernatural or mystical tint was 'Highland Mysteryworld' at Glencoe, where visitors were enjoined to 'discover the myths, learn the legends and sense the superstitions in a magical world of the Ancient Highlands', but the attraction is now closed. Finally such subjects as the 'monster' of Loch Ness, the ghostly 'big grey man' of Ben MacDhui, UFO activity in the Bonnybridge area and the Templar mysteries of Rosslyn Chapel continue to attract an international following amongst *cognoscenti*.

The foregoing sketch begs the question of the suitability of 'alternative' as a descriptor of these beliefs and practices. If – as seems incontrovertible – 'mainstream' religion, and Presbyterianism in particular, has lost, or is losing, credibility as a primary source of social and cultural consensus, while at the same time 'alternative' beliefs and practices are quite widely diffused in popular and élite cultures, in what sense can these practices continue to be deemed 'alternative' (rather than, say, 'popular', or even 'common')? This is a big question for the sociology of knowledge; for present purposes I must limit my analysis to the implications of practitioners' rhetoric. This is predicated upon a straw man, 'organised religion', which (while less than fair to the nuances of the latter) does serve to erect and maintain a collective focus, in the form of a common foe, among potentially fissiparous groups and networks. 'Religion' in this context is typically represented in terms of 'mundane rules', 'Church control' and 'excessive dogma'. 'The Church dissatisfies me', I was told by the leader of a meditation group (described below), 'I want more, you know?'. Another activist ruminates: 'Why do we need a [priest] to arbitrate between ourselves and God when we have our own personal link?'. One American writer goes so far as to call religion 'one of the greatest forces for evil at work in the world today'.[21]

'Organised religion' is constructed in this idiom, at best, as an honourable but socially constricting set of traditions; at worst, as an unhealthy culture of deference. It is contrasted unfavourably with a vivid, vital, creative 'spirituality'. In a questionnaire I conducted at the Findhorn community in 1995, 'spirituality' was overwhelmingly preferred by respondents to 'religion' as a description of their practices and identity. 'Religion' was associated with such concepts as 'the system', 'dogma', 'organised belief ', and 'narrow' outlooks, whereas spirituality was linked to 'living experience' and to 'open', 'inner' and 'natural' discourse.

This rhetoric has increasingly found expression in popular and mainstream contexts. By the 1990s, for example, significant percentages in one American survey were describing their identities as explicitly spiritual *and* non-religious. These individuals were reported to be 'less likely to evaluate religiousness positively' and 'less likely to engage in traditional forms of worship such as church attendance and prayer'. In contrast, they were

> more likely to be independent from others … to engage in group experiences related to spiritual growth … to characterise religiousness and spirituality as different and nonoverlapping concepts … to hold nontraditional 'new age' beliefs [and] to have had mystical experiences.[22]

Another study from the United States of America emphasises the reflexive qualities of Anglo-American popular piety, in which the word 'spiritual' is 'invoked positively as a basis of self-identity' by a variegated constituency who simultaneously use the word 'religious' as a 'counter-identity for clarifying who they are not'.[23] My own research in Scotland confirms this shift in language: three-quarters of the audience sample at an alternative health fair in Edinburgh readily separated out 'spirituality' from 'religion', positively defining the former as 'deep feelings', 'your own oneness with spirit' and 'existence beyond material life'.[24]

The overall evidence for preferring 'popular' over 'alternative', however, remains inconclusive. Notwithstanding widespread and continuing popular acculturation of hitherto 'alternative' religious practices and values, rhetoric on the hegemony of 'organised religion' is pervasive, and in this limited discursive sense is sufficient evidence to allow us to continue conceptualising the field in 'dissident' terms. I will now describe some significant sites, attitudes and qualities, as well as the particular demographic cohort whose interests these express, through a brief survey of the pre-eminent domain of alternative beliefs and practices: 'New Age'.

'NEW AGE' IN SCOTLAND

> There are many younger generations in Scotland who are already involved in a diffuse search for some kind of spiritual identity. Whether it's bookstore shelves groaning with volumes devoted to 'mind, body and spirit', or lifestyle sections of newspapers proposing all manner of therapies and self-actualisations, the popular appetite for some element of the non-material in the midst of our materialist lives is undeniable.[25]

This brisk newspaper editorial at the millennium had, of course, been anticipated by the aforementioned Church of Scotland report describing 'the active promotion of New Age ideas and practices in Scotland' through 'well organised and business-like communities' supplying 'courses and literature dealing with astrology, alchemy, the occult, tarot, aromatherapy, I Ching, reflexology, the paranormal and reincarnation'.[26] These variegated expressions of spirituality – however diffusive and attenuated – can no longer be confined to 'deviant' subcultures or a sequestered 'private' sphere. For example, in line with trends in Anglo-American business practice, many Scottish firms from small entrepreneurs to major companies are now exploring the benefits for morale (and productivity) of techniques to enhance 'spiritual well-being': to mark the third birthday of the bank First Direct, staff at its Hamilton headquarters were offered 'reflexology, yoga, facials and positive imaging seminars'.[27] And 'spirituality' has an established small business profile, from the successful bookshop 'Body and Soul' (est. 1988) in Edinburgh, which also organises psychic and healing fairs across the central belt, to small businesses like 'Celtic Trails Scotland', which offers residential courses 'exploring Scotland's spiritual landscape', and 'Journey of the Spirit', a 'guided personal mystery tour' organised by the Order of the Ascending Spirit in Argyllshire. Larger projects include courses in holistic education, meditation and eco-consciousness run by the Findhorn Foundation in Moray, and specialist healing centres like the Westbank Centre in Fife, mentioned above, and the Salisbury Centre in Edinburgh (est. 1973). There is now a Highland network of independent healers and therapists,[28] and other cities and towns host dedicated groups or small centres, such as Aberdeen's 'East-West' group, which in 1995 was busy promoting 'the responsibility of the individual for his or her own health based on Shiatsu therapy, macrobiotic dietary philosophy and other healing disciplines'.

Figure 14.4 Wildwood Books, Edinburgh. One of several New Age, Pagan and Celtic shops in the capital in the late 1990s and early 2000s, which also included Crystal Clear and Body and Soul. These shops sell books, music, incense and ritual objects, and also function as contact points. Photograph by the author.

In the late 1980s and early 1990s, body regimes and practices such as T'ai Chi, aromatherapy and Reiki were filtering into hired rooms in church halls and council premises, and into the curricula of local authority adult education classes. At the Maclaren leisure centre in Callander in the summer of 2001, for example, an 'alternative therapist' offered reflexology, aromatherapy and Indian head massage alongside the more standard attractions of swimming pool, climbing wall and badminton court. Like yoga in the 1960s and 1970s, these contemporary 'alternative' and 'holistic' therapies carry at least implicit 'spiritual' meanings and may, as desired, be articulated within a more fully-blown popular theology or cosmology.

Trade turnover is brisk but not necessarily secure. The brief career of 'Healthworks', an avowedly 'New Age' shop which opened in Stirling town centre in 1997, exemplifies the fickleness of 'alternative' practice. The premises had been refurbished and freshly painted in pastel washes; it sold incense, oils, small ritual objects, ambient music CDs and books on healing and spirituality; and hired out rooms for local teachers and therapists. The proprietor told me that before opening for business, the premises had been checked over by a local expert in *feng shui* (the Chinese folk art of creating 'sacred' space) who, she said, had 'given it the thumbs up'. Yet such is the risk in spiritual entrepreneurship that two years later the shop was gone: a sign on the door thanked regular customers and 'blessed' their unfolding 'spiritual journeys'.

How, then, shall we categorise this fluid and ephemeral field of beliefs and practices? Among the chief contenders for an umbrella title is the notoriously slippery term 'New Age'. This term or emblem is primarily expounded in a series of occult texts written by Anglican-turned-Theosophist Alice Bailey (1880–1949) in the 1930s and early 1940s. In several volumes composed in New York, Bailey describes the imminent transformation of the world through the return of 'the Christ' (in indeterminate form). According to Bailey, training in 'new age discipleship' through esoteric instruction in small groups will prepare the way for Christ's return, mitigating the world crises which must precede the new revelation, and ultimately ushering in a 'new age' of global unity and spiritual illumination.

Although Bailey was born in Manchester, England, and spent her most productive years in the United States of America, her first contact with the occult 'Master' inspiring her books took place at an aunt's estate at Gatehouse-of-Fleet, Kirkcudbrightshire, in 1895.[29] A second, less transient, Scottish dimension to the genealogy of 'New Age' is the aforementioned Findhorn colony, discussed further below, which was a key international nexus in the exchange of 'New Age' beliefs, practices and people between the 1960s and 1990s.[30] But despite this distinctive – if somewhat accidental – early Scottish topography, the expression 'New Age' increasingly connoted in popular Anglo-American culture from the late 1960s onwards a bewildering variety of phenomena: humanistic values and positive thinking, 'channelling' and mediumship, alternative communities, the earthly salience of astrological cycles, alternative forms of health and healing and – above all – a popular lay discourse on 'spirituality'. On the one hand, 'New Age' was said to be 'a blend of pagan religions, Eastern philosophies, and occult-psychic phenomena'; on the other it represented a 'highly optimistic, celebratory, utopian and spiritual form of humanism'.[31] Typically it has been represented as a 'movement', even though the emblem has been linked with very different social groups, from urban middle-class 'baby-boomers' to the migrant underclass constituting the so-called 'New Age travellers'. But 'New Age' has few if any of the structures, boundaries and political goals common to recognised social or religious movements. As the American activist David Spangler (b. 1945), resident at Findhorn in the early 1970s, puts it: 'There is no dogma, no orthodoxy, and essentially, no agreement on where the boundaries of the movement are and who is or is not part of it. In this sense, it is not so much a movement as a sprawl.'[32]

For this and other reasons, by the late 1980s and 1990s explicit use of the expression 'New Age' had dwindled. Other umbrella terms in popular use now included 'holistic', 'mind body spirit' and even plain 'spiritual'. In all cases, however, these terms continue, weakly or strongly, to be set rhetorically in opposition to the 'organised religion' of majority and minority ethnic groups. Practitioners constitute a diffuse and amorphous collectivity, typically urban-based or 'down-sizing' metropolitan migrants, linked through loose networks, small groups and a pervasive popular discourse on the need for personal and planetary healing and the cultivation of a personal 'spiritual' consciousness. To illustrate the ethnographical richness of the field I will sketch three examples of activity in Scotland.

The most well-known site in Scotland – indeed, in Europe – is Findhorn.[33] Its official title is the 'Findhorn Foundation and Community', which differentiates the legal functions of the core charitable Foundation (est. 1972) from the wider, looser community of residents and affiliates clustered in the locality. The settlement is said by one activist to be 'the most important New Age centre on the planet', although the principal founders of Findhorn play down 'New Age' in their memoirs.[34] Findhorn advertises itself in these terms:

> While we have no formal doctrine or creed, we believe that humanity is involved in an evolutionary expansion of consciousness which is creating new patterns of civilization and a planetary culture infused with spiritual values.[35]

One resident describes Findhorn as 'a sort of spiritual supermarket, where you can pick and mix and try to find something which suits you'. Another sees Findhorn as 'a training ground for spiritual seekers wishing to understand and express their own unique spirituality'.[36]

The Findhorn colony comprises two geographical sites in the market town of Forres and the peninsulate village of Findhorn on the Moray Firth. The original, and larger, site is called 'The Park', after 'Findhorn Bay Caravan Park', where the pioneers set up home in a family caravan in 1962. These pioneers – an English couple, Eileen and Peter Caddy, and their children, and Dorothy Maclean, a Canadian – had already followed eclectic careers in alternative religion, most recently under the tutelage of Sheena Govan (1912–67), an unorthodox 'spiritual director' and daughter of the founders of a Scots evangelical Christian organisation, the Faith Mission. (From this perspective, the settlement at Findhorn was less an unprecedented innovation than the latest chapter in the unfolding spiritual quests of its founders.) 'The Park' is a small section of land between Findhorn Bay and RAF Kinloss. Caravans, chalets and wooden houses are set in attractive gardens; the latest phase of expansion is a new building plot dubbed 'field of dreams'. Communal buildings include a community centre and an impressive stone hall which seats 300. Services include the Phoenix Shop, a sizable bookshop and foodstore catering to alternative diets, and an apothecary selling homeopathic and herbal medicines. Group meditation takes place in two 'sanctuaries': a large wooden chalet and a small, semi-underground chamber called the Nature Sanctuary (Fig. 14.5). An ecological agenda is expressed in some turf-roofed buildings, a biological sewage treatment plant, a wind-powered turbine, and 'Trees for Life', a charity replanting native forest in the central Highlands. There are a few cars, and some bicycles, but people

Figure 14.5 The Nature Sanctuary, Findhorn Foundation Community, Moray. A semi-underground meditation chamber constructed in 1986, with circular bench seating for around twenty people. The site is said to have been revealed to the architect by a nature spirit. Photograph by the author.

mostly walk around the Park. In 1975 the Foundation acquired its second site: Cluny Hill Hotel, a former Victorian Spa building on the outskirts of Forres. This large, busy building has sloping garden grounds, several floors and around 200 rooms, accommodating the bulk of visitors. These two sites – and a few other satellite buildings, including the Moray Rudolf Steiner school – are linked by a regular Findhorn minibus service, resulting in a fairly self-contained settlement, yet with 'spillover' into the local community, some of which now comprises Findhorn ex-residents, friends and family.

Colony residents number around 200; including the 'spillover' element, around 400. Almost two-thirds of the resident core are women, nearly three-quarters are aged between 30 and 50, and the majority come from England, the United States of America and western Germany[37] – underscoring some effects of 'white' immigration upon 'New Age' demography in Scotland. Findhorn residents reflect a wider constituency of 'alternative' practitioners in other ways: in a small survey in 1995, I found that they typically repudiate the label 'New Age', identify their practices as 'spiritual' rather than 'religious' and ubiquitously practise alternative healing. Revenue largely stems from a rolling programme of residential 'workshops' in spirituality, psychology, healing and ecology. Since the 1970s a shift has been noted in visitor profile, from countercultural to 'mainstream' social sectors: two-thirds of visitors in the 1990s were pursuing professional careers, and half of these again described themselves as 'business people'.[38] This returns us to my earlier comment about the appropriateness of 'alternative' as a descriptor; the overall picture remains ambiguous, although what was previously categorised as 'deviant' is increasingly accepted as merely 'variant'.

In contrast to the sizable regional 'footprint' made by Findhorn is a meditation group in one of Scotland's cities. Around half a dozen people meet monthly in a small flat at the time of the full moon to perform a group meditation based on Alice Bailey's work. Three or four regular, or 'core', meditators have a longstanding commitment to this practice. The founder began a group in the West Highlands in the late 1970s: like one or two others, she is a student in the Arcane School, an esoteric teaching course set up by Bailey. Unlike the playful, sensuous experimentation at Findhorn, these meditation meetings are plain, solemn affairs. There is a minimum of social conversation, and little institutional bulk to support: small donations cover occasional expenses, and copious literature is freely available from the United Kingdom Bailey base in London. Meditators come from within a 40 mile radius, although some attend irregularly. In the two years in the mid-1990s in which I participated in the group meditations, a typical gathering numbered about six; a slightly higher representation of men overall was offset by a 2:1 ratio of women to men in the 'core' group of committed meditators; age tended to be over 30; single people were to the fore; and access to the group – which met in the founder's home – came through personal contact or recommendation.

A brief talk on Bailey's occult cosmology precedes the half-hour, highly structured, meditation session. Meditators sit quietly with closed eyes and hands in laps. The leader announces different sections of the meditation quietly, such as 'Group Fusion', 'Alignment', 'Higher Interlude'. Meditators follow written guidelines for each section – such as 'visualise the energies of Light, Love and the Will-to-Good pouring through the planet and becoming anchored on Earth' – and visualise appropriate esoteric images. The meditation ends with the group speaking in unison a prayer by Bailey called 'The Great Invocation', which has circulated in 'New Age' networks since the 1950s. It calls on Christ to return to earth and ends with the appeal: 'Let Light and Love and Power restore the Plan on Earth'. Three 'om' sounds, adapted via occultism from Asian sources, are then hummed to finish.

The group quietly promotes Alice Bailey's work by renting a stall at alternative health fairs. These illustrate a third domain of alternative beliefs and practices, although fairs are only one of a range of expressions of contemporary healing, which also include private practitioners working from home, specialist centres, small co-operatives and – increasingly – complementary clinics in NHS practices. Nevertheless, alternative health fairs are a convenient index of the range and scope of healing practices available and the demography of participants. There are regular fairs in the cities of Glasgow, Edinburgh, Aberdeen, Inverness and Stirling, as well as in burghs like Dunfermline and Dumfries; even rural villages near conurbations may host occasional events. As with Findhorn and the 'core' full moon meditators, the ratio of women to men is typically 2:1, and participants tend to be white and middle-aged. The fairs function as practitioner showcases and audience testing grounds for an extensive range of alternative models of health and healing, some freshly devised, others with a tradition of folk application (herbalism) or élite use (homeopathy, patronised by the Royal Family). 'Taster' sessions can lead to more substantial engagement with a particular therapy, or simply enhance serial exploration. The techniques and systems on offer also function as portals into an enchanting world of recondite anatomies and cosmologies, involving meridians, chakras, 'subtle bodies', zones, and various holistic and vitalistic axioms and premises. Unorthodox healing has long been bound up with alternative piety: as one commentator observes about the current milieu, 'a *health* movement has become the repository of ... aspirations about individuals achieving a new state of spiritual awareness'.[39] The soteriological question posed, in some form, by many religions – 'what must I do to be saved?' – is reinterpreted within alternative healing circles as the amelioration or resolution of an imbalance, or 'dis-ease' (as the word is often rhetorically rendered), in the person as whole: to be *healed* is effectively to be *saved*. The historical record confirms the interaction between spirituality, piety and healing: spiritual healers and dowsers were among the first visitors to Findhorn in the 1960s, for example, when the Westbank Health Centre in Fife was also in close contact with the colony.[40] Hence a meditation group's presence at an alternative health fair follows the norm rather than exception in alternative and popular religion.

THE SALIENCE OF ALTERNATIVE BELIEFS AND PRACTICES IN SCOTTISH CULTURE

Findhorn, a meditation group, alternative healing: these attitudinally diverse expressions of 'alternative spirituality'[41] – self-important (Findhorn), discreet (meditation group), pragmatic (alternative healing) – offer nothing so solid and coherent as a 'new religion' to replace numerically 'haemorrhaging' Presbyterian (and other) traditions. But they do indicate the kind of popular 'post-Presbyterian' beliefs and practices now at work in Scotland's majority ethnic cultures: in particular, in Scots and Anglo-Scots middle-class and lower-middle-class constituencies in the central belt. This demographic point is significant and deserves elaboration. The evidence is that alternative spirituality constitutes a particular style of popular religion for a specific population cohort: 'white', young to middle-aged, well-educated, of relatively secure financial status and with women in the majority by a factor of 2:1. Notable in this domain is the almost complete absence of minority ethnic representation: the cultural practices in question spring from Christian and secular genealogies. Also noteworthy is the relative marginalisation of economically and socially disadvantaged groups. The gender bias is less remarkable, given the history of women's predominance in Christian congregations (but not amongst ordained leadership). In fact, a similar pattern obtains in both alternative spirituality and Christian Churches, where women predominate in the collective body while men tend to attain positions of leadership. It is also the case that viable participation in alternative spirituality requires the kind of economic and social capital provided by relative financial security and an above-average education. In this sense 'alternative spirituality' marks a terrain of popular religious exploration pursued by the so-called 'middle classes', and this is true in Scotland as elsewhere in Europe, and indeed Anglo-American culture at large.

This leads to a related ethnological observation. The cultural roots of 'New Age' and 'Mind Body and Spirit' are largely Anglophone and Anglo-American through a mixture of genealogy and popular cultural trends. As a Dutch scholar notes, 'New Age' discourse on the European mainland – in Netherlands and Germany, for example – remains 'an English-American affair by any standards',[42] and a similar pattern would seem to hold in Scotland. Indeed, given the rise of cultural-political nationalism in Scotland during the same period in which 'New Age' religion emerged, it might be argued that, until comparatively recently, there has been residual ideological *resistance to*, rather than *uptake of*, alternative practices in Scotland. Debating possible reasons for this are beyond the scope of this chapter, but we could note in passing the potential cultural resistance offered by distinctive Scottish ecclesiastical, educational, legal and financial institutions, in addition to more attenuated but no less real ethnological and biographical resistance at the level of personality and character formation, derived from such sources as Presbyterian behavioural codes (rectitude and rationality), the sceptical legacy of the Scottish Enlightenment and the survival of comparatively vigorous folk traditions (for example, in language, literature, music, song, custom and genealogy), which can still function to demarcate and legitimate significant boundaries, norms and values. Certainly, indices of *non*-Scots involvement in seminal sites and practices are readily found; conversely, there are few 'indigenous' Scots (with the exception of Sheena Govan) who could be classed as essential or even major figures in the international development of 'New Age'. Findhorn, for example, was originally settled by Canadian and English nationals who, before meeting Govan, had never set foot in Scotland. Participation in and residence at Findhorn by native Scots over the last 40 years has been minimal. The full moon meditation group looks for texts and inspiration towards the London-New York axis of Bailey operations; and significantly, Bailey's cosmology and metaphysic – like those underpinning alternative health – celebrates abstract and inclusive categories (humanity, the planet, the soul, the Christ) that play down cultural differentiation.

However, the qualitative and subjective appeal of alternative spiritualities and healing modalities to a well-educated, materially-secure and ageing population is likely to increase in line with detraditionalising tendencies in religion and culture in Scotland, as elsewhere in western Europe. And there is some anecdotal evidence that the voluntaristic and egalitarian ethos of alternative spirituality finds common cause in Scots traditions of self-reliance and religious non-conformity. For example, a proprietor of the 'Body and Soul' bookshop in Edinburgh told me in January 2000 that in his opinion, 'New Age' spirituality was 'more widely accepted' across a broader spectrum of Scottish society than in England, where it remained a more strongly and exclusively middle-class preserve. Despite such affirmations, the evidence suggests that Scotland is on the whole less European trailblazer than tardy conformist in alternative religion, and although Drane warns the Church of Scotland that 'unchurched' Scots will increasingly draw upon 'aspects of the New Age outlook',[43] the public status and cultural impact of alternative religion remains volatile and uncertain. There is so far little evidence that alternative practitioners desire the institutionalisation and bureaucratisation of their beliefs and practices on the scale and complexity necessary to become a real social force in contemporary Scotland. Practitioners remain wedded to popular explorations in piety, enthusiasm, self-expression and 'personal growth'. What attracts individuals to alternative spirituality is precisely what makes it sociologically evanescent: its lay ethos, social networks, egalitarian values, legitimation of subjectivities and biographical narratives, interpersonal encounters and oral transmissions. In this sense the practice of alternative spirituality carries ethnological salience over sociological or political causalities: it shows how religion – even (or perhaps especially) in a period of rapid social and cultural change – persists in articulating the subjectivities and idiosyncracies of particular groups in specific contexts.

NOTES

1. Highet, 1972, 268.
2. Brown, 2000, 255, 257.
3. Devine, 2000.
4. On the 'haemorrhage' and its cultural impact, see Brown, 2001, with good details on the Scottish context. The Scottish church census of 1994, for example, found that only 14 per cent of the total adult population attended churches: Brierley and MacDonald, 1995, 112.
5. Highet, 1960.
6. Whaling, 1999, 18–21.
7. Brown, 1997; Brown and Newlands, 2000.
8. General Register Office for Scotland, 2003, table 9.
9. Maan, 1992; Nye, 1995; Edensor and Kelly, 1989; Kay, 1996.
10. General Register Office for Scotland, 2003, table 19.
11. Kenefick, 1998, 115; General Register Office for Scotland, 2003, table 8.
12. See Watson, 2003.
13. Campbell, 1972, 131.
14. Wright and Worsley, 1975, 114. A more recent survey of 'alternative' culture in Scotland is provided by Fleming and Loose, 1992.
15. Drane, 1993, 57.
16. Wright and Worsley, 1975, 114–15.
17. See the British survey in Harvey, 1997.
18. Whaling, 1999, 22–3.

19. Nelson, 1969, 274, 287.
20. Meek, 2000.
21. In order: St Aubyn, 1990, 84–5; Icke, 1991, 127; Spangler and Thompson, 1991, 176.
22. Zinnbauer, 1997, 561.
23. Roof, 1999, 177.
24. Sutcliffe, 1995.
25. *Sunday Herald*, editorial, 20 February 2000.
26. Church of Scotland Board of Social Responsibility, 1993, 44.
27. Information and quotes in this and the next paragraph from leaflets and cuttings in my collection.
28. Forsyth, 1995.
29. Bailey, 1973, 35.
30. Sutcliffe, 2003.
31. York, 1995, 34; Heelas, 1996, 28.
32. Spangler, 1996, 34.
33. See also Riches, D. Communes. In Storrier, S, ed. *A Compendium of Scottish Ethnology, vol. 6: Scotland's Domestic Life*, Edinburgh, forthcoming.
34. The 'most important New Age centre' claim is Bloom, 1991, 2. See the memoirs of E Caddy, 1988, and P Caddy, 1996.
35. Findhorn Programme, April–October 1996, 2
36. *The Guardian*, 8 July 1992; Walker, 1994, 17.
37. Riddell, 1991, 132.
38. Metcalf, 1993, 11.
39. Coward, 1989, 12.
40. More widely, see Fuller, 1989.
41. Sutcliffe and Bowman, 2000.
42. Hanegraaff, 1996, 13.
43. Drane, 1993.

BIBLIOGRAPHY

Bailey, A. *The Unfinished Autobiography*, New York, 1973.
Bloom, W, ed. *The New Age: An Anthology of Essential Writings*, London, 1991.
Brierley, P and Macdonald, F, eds. *Prospects for Scotland 2000: Trends and Tables from the 1994 Scottish Church Census*, Edinburgh and London, 1995.
Brown, C. *Religion and Society in Scotland since 1707*, Edinburgh, 1997.
Brown, C. *The Death of Christian Britain: Understanding Secularization 1800–2000*, London, 2001.
Brown, S. Presbyterians and Catholics in twentieth-century Scotland. In Brown and Newlands, 2000, 255–81.
Brown, S and Newlands, G, eds. *Scottish Christianity in the Modern World*, Edinburgh, 2000.
Caddy, E. *Flight Into Freedom*, Shaftesbury, 1988.
Caddy, P. *In Perfect Timing: Memoirs of a Man for the New Millennium*, Findhorn, 1996.
Campbell, C. The cult, the cultic milieu and secularization. In Hill, M. *A Sociological Yearbook of Religion in Britain 5*, London, 1972, 119–36.
Church of Scotland Board of Social Responsibility. Young People and the Media. In *Report of the Church of Scotland Board of Social Responsibility*, Edinburgh, 1993, 28–73.
Coward, R. *The Whole Truth: The Myth of Alternative Health*, London, 1989.
Devine, T, ed. *Scotland's Shame? Bigotry and Sectarianism in Modern Scotland*, Edinburgh, 2000.
Drane, J. Coming to terms with the New Age Movement. In *Report of the Church of Scotland Board of Social Responsibility*, Edinburgh, 1993, 54–7.

Edensor, T and Kelly, M, eds. *Moving Worlds*, Edinburgh, n.d. [1989].
Fleming, B and Loose, G, eds. *The Holistic Handbook for Scotland*, Glasgow, 1992.
Forsyth, L. *Directory of Holistic Health Care in the Highlands and Islands of Scotland 1995–1996*, Achnasheen, 1995.
Fuller, R. *Alternative Medicine and American Religious Life*, New York, 1989.
General Register Office for Scotland, *Scotland's Census 2001: The Registrar General's 2001 Census Report to the Scottish Parliament*, Edinburgh, 2003.
Hanegraaff, W. *New Age Religion and Western Culture*, Leiden, 1996.
Harvey, G. *Listening People, Speaking Earth: Contemporary Paganism*, London, 1997.
Heelas, P. *The New Age Movement*, Oxford, 1996.
Highet, J. *The Scottish Churches: A Review of their State 400 years after the Reformation*, London, 1960.
Highet, J. Great Britain: Scotland. In Mol, H. *Western Religion: A Country by Country Sociological Enquiry*, The Hague, 1972, 249–69.
Icke, D. *The Truth Vibrations*, London, 1991.
Kay, B, ed. *The Complete Odyssey: Voices from Scotland's Recent Past*, Edinburgh, 1996.
Kenefick, B. Demography. In Cooke, A, et al. *Modern Scottish History 1707 to the Present, Volume 2: The Modernization of Scotland, 1850 to the Present*, East Linton, 1998, 95–118.
Maan, B. *The New Scots: The Story of Asians in Scotland*, Edinburgh, 1992.
Meek, D. *The Quest for Celtic Christianity*, Boat of Garten, 2000.
Metcalf, W. Findhorn: The routinization of charisma, *Communal Studies*, 13 (1993), 1–21.
Nelson, G. *Spiritualism and Society*, London, 1969.
Nye, M. *A Place for our Gods: The Construction of an Edinburgh Hindu Temple Community*, Richmond, 1995.
Riddell, C. *The Findhorn Community: Creating a Human Identity for the 21st Century*, Findhorn, 1991.
Roof, W. *Spiritual Marketplace: Baby Boomers and the Re-making of American Religion*, Princeton, New Jersey, 1999.
St Aubyn, L. *The New Age in a Nutshell*, Bath, 1990.
Spangler, D. *Pilgrim in Aquarius*, Findhorn, 1996.
Spangler, D and Thompson, W. *Reimagination of the World: A Critique of the New Age, Science and Popular Culture*, Santa Fe, New Mexico, 1991.
Sutcliffe, S. Alternative Health Questionnaire, *Connections*, 26 (1995), 48–9.
Sutcliffe, S. *Children of the New Age: A History of Spiritual Practices*, London, 2003.
Sutcliffe, S and Bowman, M, eds. *Beyond New Age: Exploring Alternative Spirituality*, Edinburgh, 2000.
Walker, A, ed. *The Kingdom Within: A Guide to the Spiritual Work of the Findhorn Community*, Findhorn, 1994.
Watson, M. *Being English in Scotland*, Edinburgh, 2003.
Whaling, F. Religious Diversity in Scotland. In *Soundings II: Proceedings of a Day Conference for the Methodist Church in Scotland*, Edinburgh, 1999, 10–45.
Wright, B and Worsley, C, eds. *Alternative Scotland*, Edinburgh, 1975.
York, M. *The Emerging Network: A Sociology of the New Age and neo-Pagan Movements*, Lanham, Maryland, 1995.
Zinnbauer, B, et al. Religion and spirituality: Unfuzzying the fuzzy, *Journal for the Scientific Study of Religion*, 36 (1997), 549–64.

PART THREE

Language, Literature and Media

15 The English and Scots Languages in Scottish Religious Life

GRAHAM TULLOCH

BEGINNINGS

One of the earliest surviving Christian texts in English was found and is still to be found in Scotland. The English language arrived in south-eastern Scotland with the northern expansion of the Anglo-Saxon kingdom of Northumbria in the early seventh century. About this time the Northumbrians were converted to Christianity and within a hundred years a stone cross at Ruthwell in Dumfriesshire had been engraved with words from the poem we now know as The Dream of the Rood. 'Krist wæs on rodi' (Christ was on the cross) – in these and other words the central event of the Christian story was recorded in the Northumbrian dialect of Old English. From this time on, English has been a language of Christian discourse in Scotland.

The Anglo-Saxons also brought with them Latin, the official language of the Church, but Latin had already reached Scotland as the language of Christians in the late Roman period and was reinforced in the sixth century by St Columba and others from Ireland. Latin and English were to exist side by side as media for Christian life for hundreds of years but were to have rather different roles. Latin was the language of Scriptures, of biblical commentary, of monastic rules, of saints' lives and of the official communication of the Church and had in many ways a superior status. English, on the other hand, was a vernacular of the people and thus the vehicle for popular religious expression, mostly spoken (in the form of prayers and sermons) but on occasions written down, as the Ruthwell Cross shows. Some of the written Latin documents survive to this day but for English we must largely rely on deduction from general circumstances. We know that the Anglo-Saxon Church put a high value on sermons in English and there is no reason to believe that this was any different in Scotland. The Anglo-Saxons also had a well-developed tradition of literary works in English on Christian subjects and some of this may have been known in Scotland as well.

Figure 15.1 The Ruthwell Cross from Dumfriesshire, with The Dream of the Rood inscribed in runes.

THE MIDDLE AGES

As the use of English spread in medieval Scotland so its use in certain religious contexts would have spread, but we need to wait for some centuries before written documents of any length emerge. Meanwhile Latin continued in its established role as the language of the liturgy and the Church. In the later fourteenth century a continuous tradition of writing in Scots begins with Barbour's *Brus* and we get occasional glimpses of the use of Scots in the expression of religious ideas. (The language we call Scots was not at that time called by that name – it was still referred to as 'Inglis' – but because it had by this stage developed as a form distinct from northern Middle English it is appropriate for us to refer to it as Scots). Thus Barbour reports Bruce as offering thanks to God during the battle of Bannockburn:

> Lordingis we aucht to love & luff
> All mychty God yat syttis abuff
> Yat sendis ws sa fayr begynnyng.[1]

While these are almost certainly not the king's actual words and are influenced by the poetic form in which they occur, they nevertheless give us at least some idea of the sort of language that might have been adopted on such occasions.

By the end of the Middle Ages the number of surviving texts in Scots is much larger and we begin to encounter specifically religious texts. While Latin was fully established as the language of the Church and of public worship it did not entirely dominate religious life. Even within the world of formal religious ceremony the Scots language could sometimes find a place. Because the coronation of a king served both a religious and a political purpose, it needed to be expressed in the language of the people. Consequently we find a record of coronation oaths from a manuscript of about 1445 in which, for example, the oath of a bailie ends with these words: 'Sa help me God and myn ain hand, and this halidome and all halidomes, and all that God maid on vi dayis and vii nichtis, under erd and abune'.[2]

Outside such formal occasions, popular forms of Christian material, if they were to be accessible to the ordinary population, had to be made available in the vernacular; an instance of this is a surviving manuscript of a fourteenth-century Scots version of one of the most popular religious texts of the Middle Ages, the *Legenda Aurea* of Jacobus de Voragine, a collection of saints' lives.[3] It is also clear that another great medieval popular Christian genre written in the vernacular, the mystery plays, flourished in Scotland although we have no surviving texts. The mystery plays of Scotland were no doubt similar to those of England, where the great events of the Bible story were presented to the audience in lively contemporary English. We know for example that a passion play called *Haliblude* was performed in Aberdeen in the mid-fifteenth century and in 1531 we find the town council of Aberdeen making arrangements for all the trades to 'furneis thair Pageane conform to the auld statut maid in the yeir of God 1510' in which, for example, the

'Wrichtis, Masonis, Sclateris and Cuparis' performed a play about the resurrection.[4] In part out of this tradition arose the first surviving Scottish play, Sir David Lyndsay's *Ane Satyre of the Thrie Estaitis*. While this is not primarily a religious play, religion figures prominently in its subject matter and it includes, among other things, a short sermon by a 'Doctour'. We need to remember this is a literary text and that the sermon is a set piece within a verse drama, but it provides a window on the kind of preaching that would have been heard in the king's court. Fittingly for its pre-Reformation setting, the Bible is quoted from the Vulgate:

> *Si vis ad vitam ingredi, serva mandata.*
> Devoit peopill, Sanct Paull the preichour sayis,
> The fervent luife and fatherlie pitie,
> Quhilk God almichtie hes schawin mony wayis
> To man in his corrupt fragilitie,
> Exceids all luife in earth, sa far that we
> May never to God make recompence conding.[5]

Private devotions were also a natural place for the use of Scots, although Latin could also be used for this purpose both by the learned and by people who did not understand Latin but knew by heart the *Pater Noster* and *Ave Maria*. For those who could read there were resources of private devotions in the vernacular such as the prayers and poems collected around 1500 in what is now British Library MS Arundel 285. As its editor has remarked, it provides 'a valuable guide to the practices of private devotion observed in Scotland on the eve of the Reformation'.[6] Placing a heavy emphasis on the sufferings of Christ and Mary, it includes four poems (and a long prose meditation) on the Passion and a series of prayers addressed to Mary concluding with a rosary, *The Lang Rosair*. While items such as the prose rendering of the *Stabat Mater* may have been intended for use as a crib in conjunction with the Latin text, most of the prayers are clearly intended for independent use. The language reflects some of the range of diction found in other Scottish writing of the time, extending from the highly aureate address to Mary as 'the schynyng gem of cleyn virginite, the refulgent stern, the flurisand refute of Iesse ... ye flour delice of ye redolent rois, maist delitabill to the Fathir in Trinite'[7] to the more everyday language of the description of hell as 'ane mekle mirk hoill full of fire, reik, and of sa gret merknes yat it may be graipit; with wyrmes yat deis neuer',[8] but in general the language suggests it was produced for a highly educated readership. We are not dealing with the language of the people when we find petitions such as the following addressed to Jesus: 'Grant me till exeme & present all myne entencionis first to ressoun, and all my dedis proceid efter ressoun.'[9] Even in this vernacular text we feel the underlying presence of a Latin Bible and liturgy, for example when we are exhorted, 'Think how ye angell apperit to ye hirdis yat war walkand thair scheipe a myle fra Bathalem, syngand: "Gloria in exelcis Deo".'[10] Similarly in *The Meroure of Wyssdome*, written for James IV by the scholar John Ireland, the biblical version referred to is the

Vulgate although the text is in Scots as in this discussion of part of the Lord's Prayer:

> And finaly, þis wourd *'Noster'*, schawis þat þai þat are full of orguele and prid, hatrent, jnwy, ore cuuatiss has litile ore na part in þis haly Orisoune and prayere. For thire manere of pepil lufis nocht as brethir and scisteris.[11]

These are written texts from an age when books were precious and rare items and are intended for well-off and well-educated audiences, including kings. We do not know how much spoken explication and expression of Christian ideas was available beyond these classes through sermons and other forms of spoken language but we can safely assume that there was some, however inadequate it might be seen to be with the coming of the Reformation.

In this world the combination of Latin and Scots in religious texts was a natural one, as demonstrated by the poems of Dunbar, such as *Ane Ballat of Our Lady*, which begins 'Hale, sterne superne! Hale, in eterne, / In Godis sicht to schyne!' and has throughout the refrain 'Ave Maria, gracia plena'. The language here may strike the modern reader, accustomed to seeing Scots as a largely colloquial language, as English rather than Scots but the aureate diction is just one of the many styles of Scots available to Scottish writers of the time. An even more intimate connection between Latin and Scots can be found in macaronic verse of which the anonymous 'In dulci jubilo' is an example. This fifteenth-century carol, popular throughout Europe, found its way to Scotland and even survived into Protestant times despite its Latin phrases. Here are two lines as printed in *The Gude and Godlie Ballatis* (of which more later):

> *Ubi sunt gaudia*, in ony place bot thair,
> Qhuair that the angellis sing *Nova cantica*.

THE REFORMATION

The coming of the Reformation, first informally and then formally with the actions of the Reformation Parliament of 1560, was to have a profound effect on the language of Scottish religion. As we have seen, by the end of the Middle Ages, Scots had become identifiably different from the English (or Englishes) of England and it had an established place alongside Latin as a language of religious activity in the Lowlands. At the same time the people of Scotland continued to apply the term 'Inglis' to Scots for some time. The noticeably Scottish form of this name does not detract from the implication that the language of England and the language of Scotland were still seen as in some sense identical. This is not to say that Scottish people were unaware of the differences between their language in both its written and spoken forms and the language of the English; this could hardly be the case. Rather, much like Americans and English people today, they were no doubt aware that they spoke and wrote their language differently but they still considered themselves to be speaking the same tongue. This lack of a clear mental distinction between Scots and English was to have significant effects on the language of religion in Scotland.

The Protestant belief in the importance of all people being able to read and study the Scripture in their native language gave a new impetus to the use of the vernacular in religious contexts. However, the printed texts that were available were in English. The easiest way to obtain a copy of the Scriptures in a language that was intelligible to ordinary people was to import the New Testament or the whole Bible in the editions and versions that were being printed in England and in continental Europe. The Church attempted to thwart the importation of vernacular Scriptures but without much success. We learn, for example, that the New Testament in English was being freely imported into Scotland in 1527 from the Low Countries.[12] At first these imported Scriptures would have been copies of Tyndale's version of the New Testament but as other versions were printed they too would have been imported along with other religious texts in English. As a result, English Bibles and other religious texts in English had been circulating in Scotland for some time before the formal move to Protestantism in 1560. Furthermore many key figures of the Reformation, including John Knox, spent time in exile both in England and in English-speaking communities on the Continent, in particular Geneva, where the version of the Bible for long favoured by Scottish Protestants was to appear in 1560.[13] These exiles were familiar with texts in English and were used to writing on religious matters for English audiences. Thus, before the time of the formal acceptance of Protestantism, English had already established itself as a key language for Scottish religion. In these circumstances it is no surprise that the Scottish Reformation failed to produce a printed Scots version of the Bible or even the New Testament. Knox, for instance, used the Geneva Bible after its appearance. When the General Assembly, in order to make a vernacular Bible more readily available, approved the first complete Bible printed in Scotland – the so-called Bassendyne Bible of 1579 – its text was taken unchanged from the 1562 edition of the Geneva Bible.

Thus the most important document of the Christian faith did not appear in a printed Scots translation. Indeed it was only in 1901 that the full New Testament appeared in a Scots translation while there has never been a translation of the whole of the Bible into Scots. The New Testament was translated from Wyclif's version into Scots by Murdoch Nisbet in the earlier sixteenth century but it seems there was only one manuscript copy and the translation was not printed until the first years of the twentieth century. There can be no doubt that the failure to produce a Scots Bible or even a printed Scots New Testament at the time of the Reformation had a significant impact on the status of Scots as a religious language in Scotland: in the long run, English became the dominant language of Scottish religion. The major printed documents of Scottish Christianity were in English: the Bibles, the catechisms, the hymns, the liturgies, the histories were all in English, albeit with occasional scotticisms.

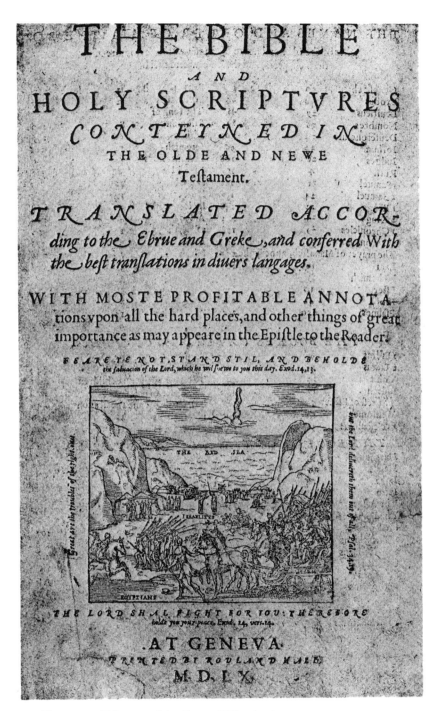

Figure 15.2 Title-page of the Geneva Bible of 1560.

The impact was not only on religious language. The appearance of a Bible in Scots could have given Scots a recognised standard written form and could conceivably have helped to maintain Scots as a written language used for a wide range of purposes in Scotland, the status it had in the earlier sixteenth century. Instead, for a variety of reasons, written Scots went into a decline for a century and a half before a partial revival for strictly limited purposes in the eighteenth century.

As the status of written Scots in general declined so the specific status of Scots as a written religious language also declined. This has long been recognised but it is easy to over-emphasise the effect of the absence of a Scots Bible. In redressing the balance a little, we need firstly to recognise that the Reformation and pre-Reformation period did produce a few Scots documents; secondly we must take into account the degree to which texts, though written as English, could be read as Scots; and thirdly we need to remember the crucial role of Scots as a spoken language. In fact what we see from this point on is not the absence of Scots in religious discourse but its coexistence with English.

Firstly, we need to take note of such prose works in Scots as John Gau's *The Richt Vay to the Kingdome of Heuine*, printed in Malmö in 1533. Despite some orthographical quirks and some lexical influence from his Danish original, Gau's language is recognisably Scots rather than English as the following passage on the first petition of the Lord's Prayer will make clear:

> *Our fader thou quhilk is in the heuine*
>
> Almichtine God sane thow of thy singlar grace and marcie hes noth alanerlie maid wsz / bot alsua commandit and lerit wsz be thy sone Iesus Christ that we suld cal and hald the for our best belowit hewinlie fader / supposz thow may richtuslie be ane scherp iuge apone wsz pwir sinners for our innumerabil sinnis quhilk ve haiff dwne aganis thy commandis and godlie wil in mony vays / and thairthrow hes greitlie offendit thy maiestie.[14]

Apart from the distinctive Scots spelling and forms (*quhilk, marcie, maid, suld, aganis*), there is Scots vocabulary such as *lerit* 'taught' and *supposz* 'even though' and Scots grammar as in the use of *is* after *thou* and *hes* after *ve (we)*, which clearly identifies this as a Scots text. Nor was the use of Scots confined to the reformers: Knox's opponent Ninian Winzet wrote in Scots and famously accused Knox of not doing so: 'Gif ze, throw curiositie of nouationis, hes forzet our auld plane Scottish quhilk zour mother lerit zou, in tymes cuming I sall write to zou my mind in Latin, for I am nocht acquyntit with zour Southeroun'.[15] Apart from this polemical writing there were also official documents of both the Catholic and Protestant Churches: Archbishop Hamilton's catechism of 1552, part of his attempt to reform the failing Church, was in Scots, while the reformers, even though they generally relied on English documents, did produce at least one homegrown document in a form of Scots, the 1560 Confession of Faith. Mairi Robinson, in her detailed examination of the language of this work, describes it thus: 'In vocabulary the authors of the Confession have almost always selected the neutral Scots/English item, and in grammar there is little that is specifically Scots. In phonology and spelling, however, the original draft would appear to have chosen the Scots option rather more frequently' and she concludes that the mixed language of the Confession 'does not support the idea that the reformers had any particular love of English for its own sake, or any positive antagonism to Scots'.[16] But even in the case of a text revealing in its original form no 'positive antagonism to Scots', printed versions showed no consistent practice of maintaining distinctively Scots features, as can be seen by comparing the titles of the two printings of 1561, John Scot's *The Confessioun of faith professit and belevit be the Protestantes within the Realme of Scotland* and Richard Lekpreuik's *The Confessione of the fayth and doctrin beleued and professed by the Protestants of the Realme of Scotland*.

Probably of more significance than these prose texts was the collection of verse that came to be known as *The Gude and Godlie Ballatis*, which has been ascribed to the brothers John and James Wedderburn. Although the first known edition of the *Ballatis* is from 1567, it is likely that some of the songs in it were in circulation before this and indeed John Knox records George Wishart as singing in 1546 the version of Psalm 51 that is found in the *Ballatis*.[17] The popularity of the *Ballatis* can be discerned from its reprinting in 1578, 1600 and 1621. The spelling of the 1567 edition generally follows Scots conventions although the influence of English can be felt. The collection includes both hymns and paraphrases and also polemical songs such as this castigating the Church of Rome:

> Thay brint, and heryit Christin men,
> And flemit thame full far;
> Thay said, thay did bot erre,
> That spak the Commandementis ten,
> Or red the word of Jesus Christ ...
>
> Nobill Lordis of greit renowne,
> That fauoris the treuth,
> On zour Saulis haif reuth,
> And put thir Antechristis downe,
> Quhilk wald suppress the Word of Christ.[18]

Once again spelling (*thame*), vocabulary (*heryit, flemit, thir*) and grammar (the use of an *-is* form after the plural subject *Lordis*) identify this as Scots.

Thus, alongside the large number of English texts circulating in Scotland, there were some Scots ones. However, the printed texts are only part of the story. If we imagine Scots people making use of these we must assume that they would have generally read them aloud with a Scots pronunciation whether they were printed as Scots or as English. The evidence is that Scots continued to be the spoken language of Scotland with all but the aristocracy into the eighteenth century. A very large number of Scots words are shared with English and readers would have had no trouble in identifying such words as the ones they were using in everyday speech. Thus Robinson in her analysis of some biblical passages of the 1560 Confession concludes that some texts, by the way they are spelt, 'seem to prove that Tindale was being pronounced in Scotland as Scots, not English',[19] a conclusion that can be generalised to cover all the English texts of the Bible that were in use in Scotland around this time as well as to the Protestant Church's liturgy as embodied in *The Forme of Prayers ... Approved and Received by the Churche of Scotland*, which came to be known as *The Book of Common Order*.

Given that it was generally only in the course of the eighteenth century that the middle classes began to aim for English speech, we can presumably conclude that this practice of reading English as Scots continued to be followed for some time. Furthermore the reading aloud of texts forms only a part of the use of spoken Scots in religious contexts. The Reformed Church put a strong emphasis on preaching and, even from as far back as Old English, sermons had been in the vernacular. Whereas the Catholic Church under Archbishop Hamilton had been forced in 1549, in a vain attempt to remedy the dearth of adequate preaching, to stipulate that bishops should preach at least four times a year,[20] Knox believed deeply in the importance of teaching 'by tong and livelye voyce'.[21] Among the reformers Knox was not alone in this belief, and even if his own language had been influenced by his sojourn among English-speaking communities, the prominence given to preaching in the reformed Church would immediately have increased the use of Scots in a most important religious context.

Public and private prayer is another area in which Scots is certain to have played a large part. The published liturgies of the Reformed Church included some prayers for use on public occasions but as an increasingly strong value was placed on spontaneous and extempore prayer as opposed to the use of written models then the vernacular language was likely to have come fully into its own. This is not to say that English would have had no effect on the Scots used in religious contexts. The record of the trial of Thomas Forret who was condemned 'to be Burnt, for the vseing of ... the New Testament in Inglis' contains an exchange in which Forret claims, 'Brother, my people are so roode and ignorant they vnderstand no Latin ... The Apostle Paule sayeth in his doctrine to ye Corinthians, that he haid rather speak fyve words to the understanding and edefing of his people, then ten thowsand in a strainge tongue whilk they understand not'. To this the accuser replies, 'Where finds thow that?'.[22] In a situation where the accuser's use of the Scots form *finds* with *thow* suggests a context of Scots speech, it may or may not be significant that Forret, a reader of English Bibles, uses the English form *sayeth*. It is possible that we see here an early case of the disjunction between Scots as the normal language of ordinary discourse and English as one of the options as the language of religion but

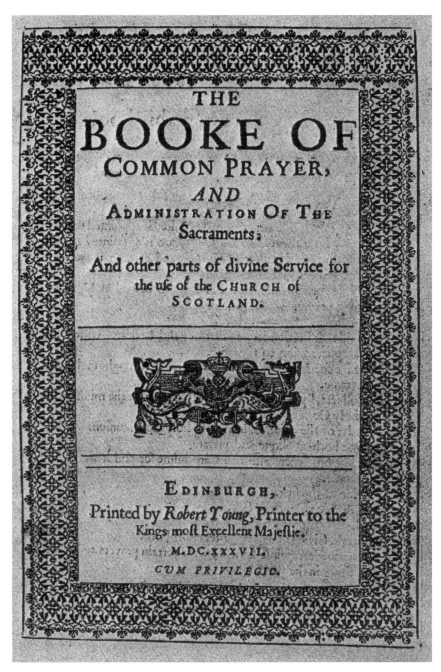

Figure 15.3 Title-page of *The Book of Common Prayer*, 1637.

we could equally be merely seeing the scribe or printer using the elegant variation of the English spelling.

One of the most powerful uses of the spoken vernacular in religious contexts, religious drama, survived the Reformation but not for long. Although in 1574 the St Andrew kirk session gave approval for the staging of 'the comedie mentionit in Sanct Lucas Evangel, of the Forlorn Soul', the next year the General Assembly ruled 'that na clerk playis, comedeis, or tragedeis be maid of the Canonicall Scripturis, alsweill New as Old, on Sabbath nor wark day, in tyme comying'.[23] Despite this particular setback we must nevertheless assume that, by the later part of the sixteenth century, Scots was well established as a language of religion, playing a significant part in all forms of spoken language and particularly in sermons and prayers. In written language English predominated (although there were some Scots texts) but it would have been common practice to read English texts aloud with a Scottish pronunciation and perhaps with some occasional, and not always conscious, scotticisation of the grammar. This situation would have continued into the seventeenth century. However, the removal of the king to London in 1603 deprived the Scottish Church of a resident Scots-speaking king who might serve as a model for other worshippers. Moreover James VI, by sponsoring the Authorised Version of the Bible, gave extra weight to English in Scottish religious life. The authority that he lent to the Authorised Version was enough to ensure that it became the accepted version of the Bible in Scotland and precluded any real chance of the printing of a genuinely Scots Bible. On the other hand his attempts to reform the liturgy were not successful and would probably have contributed little more to the forces of anglicisation since *The Book of Common Order* was already very English in language. Much the same may be said of *The Book of Common Prayer* ('Laud's Liturgy') of 1637, which was promoted by Charles I, in so far as it came to be used in Scotland despite protests against it. However, by extending the areas in which the language of worship was prescribed rather than optional it perhaps gave its English language a slightly more prominent part within Scottish worship.

The pre-eminence of English as the written language of the Church in Scotland was further confirmed later in the same century by the Westminster Assembly of 1643–8, which produced a series of documents that became fundamental to the Church's teaching and worship. The assembly brought together divines from England and Scotland with the aim of producing a uniform and Presbyterian Church throughout the two countries; this aim was not in the end fulfilled and the documents produced by the assembly were ultimately used only by the Church of Scotland but because of the Westminster setting in which they were created they were almost entirely English in language. The Assembly produced a liturgy, *The Directory for Public Worship*, a description of Church government, two catechisms (the Longer and the Shorter) and a Confession of Faith, and also approved Francis Rous' version of the Psalter. All of these works, all in English, were to be central documents for the Church of Scotland for the next three

Figure 15.4 Etching by Wenceslaus Hollar of the riot at St Giles', Edinburgh, 23 July 1637, sparked by the introduction of the Book of Common Prayer. From *Sight of the Transactions of these Latter Yeares*, 1646.

centuries and collectively they reinforced the dominance of English in written religious discourse. It is true that the Psalter, which was partly based on earlier versions including some Scottish ones, exhibited some Scots features but they are not very noticeable in what is overwhelmingly an English text – if they had been they would not have been approved by the Westminster Assembly. What is more, the Shorter Catechism (known, when printed without the scriptural proofs, as the Single Carritch, and full of English forms like *consisteth* and *requireth*) became the standard text for teaching Scottish children to read. This association of literacy with an English text and particularly with a religious English text meant that both learning and religion were irrevocably linked to the English language from the early days of a Scot's life. English as the language of religion also acquired all the secular prestige that rests with the language of learning. The implication was that Scots, if it were to have a differentiated role, must find it in the sphere of the non-religious and non-learned. For Scottish Presbyterians the three texts that were fundamental to religious life – the Bible (the basis of everything else), the Shorter Catechism (the primary statement of belief) and the Metrical Psalter (a crucial part of all public worship) – all came in English.

LATER SEVENTEENTH TO EARLIER NINETEENTH CENTURIES

This enshrining of English as the written language of the Scottish Church coincided with moves towards greater use of English in Scottish society in general. Although eighteenth-century poets like Ramsay, Fergusson and Burns secured a place for Scots in certain kinds of poetry, and in the early nineteenth century Scott, Hogg and others established it as a language for the dialogue of working-class characters in novels and short stories, printed texts were generally in English. While it seems as if many middle-class Scots and most working-class Scots continued to speak Scots on formal and informal occasions for some time, during the course of the eighteenth and earlier nineteenth centuries middle-class people increasingly moved to English in speaking. At first this was for formal speech and they retained the ability to speak Scots on other occasions as is implied by John Galt's comment of 1803: 'In polite companies a Scotsman is prohibited, by the imputation of vulgarity, from using the common language of the country, in which he expresses himself with most ease and vivacity, and, clothed in which, his earliest and most distinctive impressions always arise to his mind'.[24] As time went on, many people lost all knowledge of Scots, prompting Lord Cockburn in 1842 to write that 'Scotch is pretty deeply engrained into the people, but among the gentry it is receding shockingly'.[25] Cockburn was sympathetic to the Scots language but for many the pursuit and rooting out of scotticisms had been a major preoccupation for some time. It is against this background of the declining use and status of Scots in general that we must consider its specific use in religious writing and speech.

The redefinition of Scots as an informal language and then as a purely working-class language had considerable consequences for its status. Galt was not alone in believing that the use of Scots carried an 'imputation of vulgarity' and the notion that Scots was 'vulgar' was to prove a continuing impediment to its use in religious contexts. The major religious writings of the period, the sermons and religious commentaries, were in English and even the few scotticisms in the Metrical Psalter began to attract unfavourable attention. William Tennant, although he was himself a fine writer of Scots verse, felt compelled to write in the *Edinburgh Literary Journal* in 1829 that 'it cannot but fill an Episcopalian stranger with pity ... to hear the ministers of our church, the best educated men of our country, whose sermons are penned and uttered with taste and grammatical accuracy, reading to their people from a psalter where they must of necessity, at every second page, stumble upon and flounder through the most vulgar Scotticisms, obsolete accentuation, and erroneous grammar'.[26] Hoping to expurgate even minor remnants of Scots, Tennant objected to the use of terms like *kythe* (Psalm 18.26), which had by the eighteenth century become confined to Scotland and rhymes that required Scottish pronunciations like *reign/king* (Psalm 136: 19–20). Such attitudes were in keeping with an age where many found Scots vulgar and where vulgarity and religion were seen as worlds apart. At this stage the total ascendancy of English in religious writing must have seemed not only the most desirable outcome but also the most likely one.

Scots in religious speech was another matter. Ministers may be roughly classified as middle class, and all ministers were educated. As the educated middle classes in general moved to speaking English, ministers would increasingly be expected to speak English, but there were pressures the other way as well. They would have spoken English with a Scottish pronunciation that makes for an easy transition between Scottish English and Scots. Some would have been from humble backgrounds and would have grown up speaking Scots; others, though of middle-class origins, would have heard Scots all around them and may have used it for informal occasions. The use of Scots in sermons and public prayers offered certain advantages: it enabled the minister to express his connection with a Scots-speaking flock by adopting their language and it satisfied the Protestant demand for religion to be conveyed in the language of the people. At the same time biblical language offered no barrier to people who heard it regularly in church and encountered it as soon as they learned to read (as so many did). The minister could choose therefore between English with all the solemnity it invoked and Scots with its implication of everyday relevance, and indeed could mix the two as he felt appropriate.

Throughout the eighteenth and nineteenth centuries, strong feelings against prescribed forms of prayer (including even the Lord's Prayer) meant that a minister had scope for choosing his language not just in his sermon but in extempore prayer as well. Since extempore prayer is by its nature unwritten and since *published* sermons conformed to the general expectation that printed texts would be in English, we do not have much direct evidence of the language of public religious speech but any preaching and prayer that combined the solemn and the everyday is likely to have drawn on both English and Scots although not everybody would have been equally happy to make use of Scots. At the end of the eighteenth century, in the first *Statistical Account of Scotland*, the minister for Symington in Lanarkshire offers readers a vignette of the public use of Scots in prayer but in a way that suggests his own disapproval. After recording that the parishioners are 'well affected to the established constitution, in church and state', he records that 'An attempt was made, at the beginning of the Secession [i.e. in the 1730s], to settle a dissenting congregation in Symington, and many violent harrangues [sic] were delivered'. He adds in a footnote an example of mixed Scots and English in public speech, a prayer offered up by Secession preacher for the minister of the Established Church: 'Thou knowest that the silly snivelling body is not worthy even to keep a door in thy house. Cut him down as a cumberer of the ground; tear him up, root and branch, and cast the wild rotten stump out of thy vineyard. Thresh him, Lord, and dinna spare! O thresh him tightly, with the flail of thy wrath, and mak' a strae wisp o' him to stap the mouth of hell!'[27]

Literary representations of ministers also suggest that some used

Figure 15.5 A nineteenth-century French image of Morton finding Balfour of Burley reading the Bible. From Walter Scott, *Les Puritains d'Écosse*, translated by M P Louisy, Paris, 1882, 46; illustration by D Maillart.

Scots. For instance Scott in *The Bride of Lammermoor*, a novel set in the early 1700s but probably following his normal practice of reflecting the sociolinguistic position of his own time, presents Mr Bide-the-Bent as speaking a mixed language in addressing the Scots-speaking John Girder: 'Aweel, John ... and dinna ye see a high judgment in this? — The seed of the righteous are not seen begging their bread — think of the son of a powerful oppressor being brought to the pass of supporting his household from your fulness ... I trust ... ye hae reflected weel on what ye have done, lest you should minister cause of strife, of which it is my duty to say, he who affordeth matter is no manner guiltless'.[28]

When we turn to family prayer there is further evidence of a mixed language. In a chapter on prayers in *The Shepherd's Calendar*, James Hogg claims that 'There is ... no class of men professing the Protestant faith, so truly devout as the shepherds of Scotland. They get all the learning that the parish schools afford; are thoroughly acquainted with the Scriptures of truth [and] deeply read in theological works'.[29] He proceeds to show how they combine their knowledge of religious English with their native Scots as in these examples of the 'family eloquence' of Adam Scott in Upper Dalgleish:

And moreover and aboon, do thou bless us a' wi' thy best warldly blessings – wi' bread for the belly an' theeking for the back, a lang stride an' a clear ee-sight. Keep us from a' proud prossing and upsetting – from foul flaips, and stray steps, and from all unnecessary trouble.

Bring down the tyrant and his lang neb, for he has done muckle ill the year, and gie him a cup o' thy wrath, and gin he winna tak that, gie him kelty.[30]

Yet, while taking pride in recording such language, Hogg shows the unease about any association between religion and colloquial Scots that is typical of his age. Although he praises these passages 'sometimes for rude eloquence and pathos, at other times for a nondescript sort of pomp', he also feels they 'not infrequently' show 'a plain and somewhat unbecoming familiarity'.[31] Burns seems to feel a similar disquiet about a too familiar address to God in 'Holy Willie's Prayer' where hypocrisy and an intimate Scots address to God seem to go hand in hand.

The implication of Hogg's chapter is that he is describing something a little different to the typical religious language of the time while Burns, when praying in his own voice, uses English (e.g. 'A Prayer in the Prospect of Death'). Moreover we can set beside Hogg's shepherds' prayers examples of more thoroughly English diction in private prayer. For instance, in Scott's *Heart of Mid-Lothian*, Davie Deans, a Scots-speaker, at a moment of deeply felt and partly religious emotion when he meets Jeanie after her journey to London uses biblical English rather than Scots:

Figure 15.6 'The priest-like father reads the sacred page'. An illustration of Burns's 'The Cotter's Saturday Night'. From Robert Burns, *The Poetical Works*, London, nd, frontispiece.

Jeanie – my ain Jeanie – my best – my maist dutiful bairn – the Lord of Israel be thy father, for I am hardly worthy of thee! Thou hast redeemed our captivity – brought back the honour of our house – Bless thee, my bairn, with mercies promised and purchased! – But he *has* blessed thee, in the good of which He has made thee the instrument.[32]

Other kinds of speech would clearly have been in English as well such as readings aloud from the Authorised Version. However, scruples about the proper use of the Bible and of other sacred texts such as the Metrical Psalms gave Scots a place on the edges of religious discourse. The Metrical Psalms were for a long time felt to be too sacred to by profaned by use outside formal worship. Consequently the practice arose of using alternative words when practising the singing of the Psalm, like this one, set to the hymn tune 'Bangor':

> O BANGOR'S notes are unco high,
> An' try the lassies sair;
> They pech an' grane an' skirl an' skriech,
> Till they can sing nae mair.

Others were influenced by words of hymns, like the one beginning 'O mither dear, Tod Lowrie's lum, / Whan sweepit will it be?' that draws on the hymn 'O mother dear, Jerusalem'.[33]

LATER NINETEENTH AND EARLIER TWENTIETH CENTURIES

By the mid-nineteenth century the Scots language had regained a place, albeit restricted, in both prose and verse in Scottish literature but English remained unchallenged as the written language of religion. Eventually, the revived fortunes of Scots in literary texts were bound to have an impact on religious texts. However, the first moves towards rendering the Bible into Scots arose from a motive quite different to the emulation of literary Scots and indirectly provided continuing evidence of the way in which the Authorised Version had impressed itself on the minds of Scottish people as the model for religious language.

Prince Louis Lucien Bonaparte, a nephew of Napoleon and a philologist, paid for the printing of a very large number of translations of books of the Bible into European dialects including seven translations into Scots. However, the translations he commissioned were printed in very small runs and many copies remained undistributed at this death.[34] Consequently very few of these books would ever have been used for private or public worship in Scotland. Nevertheless they provide an indication of how strongly the language of the Authorised Version had taken hold on Scottish minds and of how difficult it would therefore be for Scottish Christians to produce modern Scots translations that were not overly dependent on early seventeenth-century English. The prince commissioned Henry Scott Riddell (a poet from the Scottish Borders) and Joseph Philip Robson (a poet from Newcastle in *England*) as well as an unidentified George Henderson, and another unnamed translator to produce translations of Matthew (Riddell and Henderson), the Psalms (Riddell) and the Song of Solomon (all four translators).

Riddell's translations show some contradictory impulses with a move to modernise the language on the one hand and a tendency to retain archaic features or even introduce some of his own on the other. The result is a compromise that is neither fully Scots nor fully English. It is clear that he is reluctant to move any further than absolutely necessary from the revered text of the Authorised Version: while he scotticises some of its vocabulary he also retains shared Scots/English words even where a distinctively Scots term was now the norm in spoken Scots, as with his preference for the Authorised Version's *child* over *bairn*, the term normally used in his own area of the country. Henderson builds on Riddell but produces a more thoroughly Scots text. In remarking in his preface that 'Various vernacular words, such as *lug* for "ear", *crack* for "talk", *cuddy* for "ass" &c. have been discarded as unsuitable in a translation of a portion of the Holy Scriptures',[35] Henderson is the first to articulate a continuing problem for modern Scots translators of the Bible: in the division of roles between Scots and English in Scottish literature English had taken on the role of providing formal language while Scots was very largely confined to informal contexts such as informal speech, love poetry, satire and sentimental verse. Where there was a choice of diction it was often between a formal English term and a colloquial Scots one. Translators did not want their translations to sound too remote from the everyday spoken language that provided their ultimate source of diction but neither did they want to make the language of the Bible sound common and vulgar. In spite of this, Henderson's tone is more colloquial than Riddell's as is that of the anonymous translator while Robson seems at ease with a more fully colloquial tone than any of the others.

These translations give us some insights into the difficulties faced by translators but, because of their very limited circulation, they could, with one exception, play very little part in the religious life of Scottish people. The same is largely true of the work of James A H Murray, later editor of the *Oxford English Dictionary*. Murray's translation of Ruth was only ever printed as an appendix to his 1873 book *The Dialect of the Southern Counties of Scotland*. The translation is of great linguistic interest but is presented in a phonetic spelling that precludes its everyday use. The exception among the Bonaparte-sponsored translations was Henderson's Song of Solomon, which appeared late in the century in revised form as a two-penny pamphlet. It is unfortunate that the Song of Solomon, though attractive to the prince as a short text for multiple translations, is not a key text in Scottish Christian life but, poorly printed as it is, this pamphlet offered the only chance for Prince Louis Lucien's project to contribute to everyday Scots religion.

However, by this time a much more important figure had appeared

Figure 15.7 Title-page of Waddell's translation of the Psalms. From P Hately Waddell, *The Psalms: Frae Hebrew Intil Scots*, Edinburgh, 1891.

on the scene in the person of P Hately Waddell: indeed a quotation from Waddell's Isaiah appeared on the front page of the pamphlet. Waddell was an independent minister in Glasgow with a range of literary interests and the first to translate the Psalms (1871) and Isaiah (1879) into Scots directly from the Hebrew although he also consulted other translations in English and other languages including the Geneva Bible 'in which our own most distinguished Reformers had a share'.[36] Waddell's language is much more thoroughly Scots than that of any of his nineteenth-century predecessors and he produced a powerful and often poetic text, but he ran into some of the same problems as those who preceded him. While claiming to use language that was 'in daily use by all well-educated peasants and country gentlemen of the last generation ... men who represented the true vernacular of the country, from the days of the Reformation and of the Covenant',[37] he recognised that much of the very colloquial language used in the last century or so by writers like Burns 'could not, for very obvious reasons, be admitted in a translation of the Bible'.[38] Waddell's solution was the introduction of a number of new words of his own, either new compounds or words borrowed from cognate languages such as Dutch. As a result his language is often difficult – even for someone well acquainted with an English translation – partly because he freely varies the structure of his sentences from the familiar structures of the Authorised Version. However, Waddell was evidently somewhat of a charismatic figure since he managed to maintain independent congregations for over thirty years and he is likely to have used his translations in his own church. What is more, the translations, which were presented in lined double columns with marginal notes and head notes in a style familiar in English Bibles of the time, sold well: my own copy of the Psalms is from a run which had reached 11,000. Unlike the Song of Solomon, the Psalms have been a central text of Scottish Christianity, as has to a lesser extent Isaiah. By providing these two texts in a vigorous if slightly quirky modern Scots, Waddell gave the written language a chance to play a real role in Scottish religion.

Over the next fifty years Waddell's example was followed by a number of other translators, most notably William Wye Smith, whose complete *New Testament in Braid Scots* was published in Scotland in 1901. Born in Jedburgh, Smith lived in Canada: not the last overseas Scot to produce a biblical translation into Scots. Alert to the generally informal nature of modern Scots, he notes in an address to the reader Smith that:

> Many of us have wished – not as a public version, but for private use – for a familiar and colloquial rendering of the New Testament. It is a thing which will never be satisfactorily done in English ... But, under the cover of 'Scotch' the present Translator humbly intimates that he has done that very thing.[39]

It is refreshing to see what had been considered as a drawback redefined as an advantage and consequently it is disappointing to find that Smith's practice falls somewhat short of his theory. Using the Revised Version as his base he scotticises the vocabulary, spelling and, to some extent, grammar but he fails to produce a thoroughly colloquial text. All the same, that his translation went into three editions by 1924 implies a reasonably large audience for private if not public use. Smith was followed in 1917 by Thomas Whyte Paterson with a free rendition of Proverbs that maintains a consistently colloquial and modern Scots and in 1921 by Henry Paterson Cameron whose *Genesis in Scots*, written in Australia, shows clear influence from Waddell in its language.

These renditions of the Bible into Scots were clearly influential in validating the wider use of Scots in printed religious texts since, in the midst of this period of active biblical translation into Scots, we see for the first time the appearance of Scots sermons in print. In 1910 David Gibb Mitchell published *Sermons in Braid Scots* and followed it in 1917 *The Kirk i' the Clachan*. However, even though some ministers undoubtedly continued to enrich their preaching with the use of Scots, Mitchell's example of printing whole sermons in Scots was not followed, perhaps because the practice of printing sermons was falling out of favour. Nevertheless, the general reinvigoration of literary Scots in the earlier twentieth century occasionally produced poetry on religious topics such as Hugh MacDiarmid's 'I Heard Christ Sing' and 'O Jesu Parvule' in *Sangschaw* (1925). In a conscious effort to *re*establish Scots as a language for a wide range of subjects including religion, MacDiarmid quite deliberately, in the case of 'O Jesu Parvule', drew on the language of the popular Reformation collection, the *Gude and Godlie Ballatis*, written at a time when Scots was still in wide use as a language of religion.

Meantime in areas where the tradition of speaking Scots remained strong, it could still be used in talking about religious matters: the north-east poet Flora Garry (1900–2000) recalled how, in her family, discussion of all matters was carried on in Scots: her mother in the kitchen once asked about the discussion in the parlour and was told, 'O, aye, they're giein Free Trade and Protection a tak throu han,' to which she replied 'Awa, than, and tell them to come ben for a bit supper, afore they get yokit in tull Free Will and Predestination, or they'll be here aa nicht'.[40]

MacDiarmid's use of the Latin phrase reminds us that, for quite different reasons, Latin had made a comeback as a language of religious expression in Scotland. The Roman Catholic Church had not entirely disappeared from Scotland at the time of the Reformation but had been confined to very small numbers of people and the use of Latin in religious worship was necessarily limited to these few people. However, as numbers recovered in the course of the following centuries to the point where in 1878 about 9 per cent of the population of Scotland were Roman Catholics, Latin would have been used in religious contexts by more and more people. Catholics, like Protestants, also made use of English (rather than Scots) translations of the Bible, including the Douai version, for private reading. Later, with the shift to services in the vernacular after the Second Vatican Council, English also became the language of public worship.

LATER TWENTIETH CENTURY TO THE PRESENT

The impulse towards translating the Bible into Scots seems to have faltered for a while after Cameron's Genesis of 1921, but in the meantime Scotland was making a powerful contribution to the translation of the Bible into English. James Moffatt's *Historical New Testament* appeared in 1901, the same year as Smith's Scots New Testament, but Moffatt was destined to make a far more important and more widely known contribution to Bible translation. After the publication of another translation of the New Testament in 1913 and one of the Old Testament in 1924, these were combined into *The Complete Moffatt Bible* in 1926. George Anderson, in offering an assessment of Moffatt's second translation of the New Testament, praises 'what is probably the chief merit of this translation, that it clarifies the meaning of the text at many points ... where the older versions leave it obscure'[41] and this virtue of clarity is characteristic of Moffatt's work in general. Unlike most other translations by individuals, the Moffatt Bible spawned a whole secondary literature including a concordance and a set of biblical commentaries. This, along with the clarity and vigour of the translation itself, ensured that it was widely used, not merely in Scotland but throughout the English-speaking world. Although Moffatt used a few scotticisms, such as *factor* in the parable of the unjust steward (Luke 16: 1–8), the language is English, sometimes rather colloquial but not uniformly so. Unlike many of the nineteenth-century Scots translators, Moffatt was not reluctant to depart from the language of the Authorised Version. He was not content to confine himself to traditionally sanctioned language and was happy to draw on the resources of English as a world language, using words like *khan* (Exodus 4: 24) and *sheikh* (Genesis 25: 16). At times this aroused hostility, so great was the authority of the version that had been so long used in Scotland, but there can be no doubt that Moffatt offered his readers a genuinely new version of the Bible and for the first time an English version wholly produced by a Scot. In the end the popularity of his version has waned and other versions have displaced it, including the Revised Standard Version for which Moffatt served as executive secretary of the translation committee, and the New English Bible to which Scottish translators made a major contribution. By such efforts Scottish biblical scholars have helped to ensure that Bibles in English continue to play a crucial role to Scottish religion.

In the second half of the twentieth century, however, translation into Scots has recovered its impetus with a vengeance. This new movement reached its highest point in W L Lorimer's New Testament but it was preceded by Alex Borrowman's *The Book o Ruth and Ither Wark in Lallans* of 1979. Two significant events separate Borrowman from the earlier translators, the Scots Renascence led by Hugh MacDiarmid and the popularity of the new Bible translations in English. The influence of MacDiarmid and his advocacy of Synthetic Scots can be felt in Borrowman's eclectic vocabulary ('a mixie-maxy' as he calls it),[42] which draws on Scots language from many parts of Scotland and from various periods although not so much as to make it difficult to follow. Borrowman was a member of the Lallans Society (later known as the Scots Language Society) and his book is part of the coherent program of promoting the wider use of Scots which the society espouses. As he writes, 'We hae a wheen makars in Lallans ... but we have a needcessity for mair screevers o prose'[43] and the *Ither Wark* of the title includes sermons (with the dates at which they were given) and a commentary on Ruth alongside translations of other passages from the Bible. All this makes it clear that Borrowman, for many years a Church of Scotland minister, was keen to promote the use of Scots not just for private reading of the Scriptures but also in the public worship of the Church. However, the choice of Ruth perhaps reflects a continuance of the traditional attitude that Scots is particularly appropriate for expressing the familiar and domestic. While he had many forerunners in preaching in Scots, Borrowman was an important innovator in developing orders of service, which, since the Reformation, had been uniformly written in English. Others have followed this lead, notably David Ogston, minister at St John's Kirk, Perth, who has written a communion service, a wedding service and a litany in north-eastern Scots[44] but we are yet to see the Church of Scotland or any other Scottish Church produce a full service book in Scots.

Borrowman's modest and modestly printed booklet was followed by publication of Lorimer's splendid and splendidly printed *New Testament in*

Scots. Largely completed by the time of his death in 1967 the text was edited by his son R L C Lorimer and appeared in 1983. Its appearance was greeted with great enthusiasm and 2,500 copies were sold within a fortnight so that the reprint, which came soon after, was quickly followed by a Penguin paperback. It has recently been reissued in both hardback and paperback by Canongate. Lorimer had spent most of his life teaching Greek in universities and, about a hundred years after Waddell was the first to translate the Psalms direct from Hebrew into Scots, Lorimer was the first to produce a Scots New Testament directly translated from the Greek. While Lorimer aimed for an accurate translation, he allowed himself some freedom in adding extra words or altering the sentence structure to bring out the appropriate tone. As a consequence we have a translation unhampered by regrets for the language of the Authorised Version and one which brings alive the vigour of the original with startling effectiveness. This may be seen most strikingly in passages of abuse, like in Christ's clearing of the Temple, but permeates the whole text, even in more low-key passages. Not himself a Scots-speaker, Lorimer learnt from the speakers around him and was also for many years involved in the work of the *Scottish National Dictionary*. Like Borrowman, Lorimer saw his translation as part of a broader agenda, being, according to his son, 'well aware that ... he would also be setting out to resuscitate and recreate Scots prose'[45] but, unlike Borrowman, he was not sympathetic to the notion of synthetic Scots and abjured the revival of obscure archaisms or the creation of his own terms, relying rather on the huge resources of the Scots language over the last two centuries to provide him with appropriate diction and grammar. He was thus able to bring to his translation all the colloquial vitality of recent Scots, but the breadth of his vocabulary means that much of it is unfamiliar to modern speakers of Scots. Nevertheless, Lorimer's New Testament can provide, and no doubt has provided, a stimulating alternative to English versions for private study and, despite some unfamiliar diction, most passages are suitable for use in public worship.

That we can only speak of Lorimer's version, so powerful and so adroit in using all the resources of modern Scots, as an alternative alongside English versions is testimony to the enduring and inescapable presence of English in Scottish religious life. It was a similar desire to offer an alternative to English that led a more recent translator, Jamie Stuart, into his work. Attending a 'sensational' performance by Alec McCowen of the whole of Mark's Gospel in the English of the Authorised Version at the Edinburgh Festival of 1981, Stuart, who in a varied career had spent some years as an actor, was inspired to do something similar in Scots.[46] After sensibly rejecting Smith's Gospels as a performance text, Stuart decided to make his own version and performed excerpts from the gospels in a fairly standard modern Scots in Scotland and Canada before the text was published as an audio-cassette and then in book form as *A Scots Gospel* in 1985. The book attracted media attention, and a number of ministers in Glasgow encouraged Stuart to work on a version in Glaswegian Scots. Though effective in many ways,

A Scots Gospel presents a rather old-fashioned and nostalgic view of Scots and the ministers apparently wanted to draw on the liveliness of contemporary urban Scots in its best-known manifestation. Stuart was unsure, recognising that a widely stigmatised form of speech would not be seen by everyone as a suitable medium for the Bible story: 'Many people love the pithy, pungent patois of Glaswegians while many others have little regard for it.'[47] In the event the notion was vindicated. After producing the gospel story and tales from the Old Testament in Scots, Stuart combined them with some new Old Testament stories as *A Glasgow Gospel* in 1997. In terms of the number of reprintings this has been one of the most successful of Scots Bible translations, being reprinted three times in 1997, twice in 1998, and once each in 1999 and 2000. This suggests a wide and enthusiastic readership, which has without doubt been enhanced by the attractive format of the books and the production of a video version of *The Glasgow Gospel* starring a number of well-known Scottish actors. Compared to Lorimer, Stuart has very little distinctively Scots diction but the use of a recognisably contemporary Scots spelt in a loosely phonetic style obviously rang a chord with the audience. A few well-placed local phrases are enough to give it a distinct identity as in the following passage:

> The heid priests and pharisees moaned, "Whit are we gauny dae with this heid-banger? If we lee him alane, the hale country'll believe whit he says. There'll be a right stushie an the Romans'll pit the hems on us aw!"[48]

Compare the last sentence with the same verse in *A Scots Gospel* – 'The Romans will come and wrack doon oor Temple an oor hale nation!'[49] – and the more contemporary and Glaswegian feel is very evident. At the same time one can recognise that he has followed his own injunction: 'Every big city has unpleasant slang terms of communication so it was important to avoid colloquial language with unacceptable words and usages.'[50] Modern translators have exploited the colloquial strength of modern Scots but cannot entirely escape the limitations of a speech that is nowadays almost wholly confined to informal and intimate contexts.

Stuart has shown that a combination of the traditional public performance and book with the full paraphernalia of modern media, audio, video, radio and television, can win a place for Scots in Scottish religious activity even in a world dominated by global English. As a popular text at the other extreme to Lorimer's scholarly approach, Stuart's Glasgow Bible can hold its own in a world where such similarly colloquial texts as the Good News Bible vie for attention with the more scholarly and formal Jerusalem Bible and Revised English Bible. It is only by entering the more intensely competitive global context of contemporary religious discourse, where multiple texts and translation abound, and also to some extent by bending to its popularism that Scots can maintain a place for itself in Scottish religion. The price to be paid, however, is a dilution of the distinctively Scots vocabulary that makes a version like Lorimer's so special and interesting.

Scots, despite the many predictions of its imminent demise, is not yet dead. Its continuing spoken presence insures that it is no doubt still used in private prayer and worship on some occasions by some speakers, encouraged and stimulated by the presence of works like Stuart's. In literature contemporary urban Scots has found a new voice in which religion, though less commonly a topic than in the past, can find new and often unconventional forms of expression as in Tom Leonard's 'The Good Thief'.

At the beginning of the twenty-first century, Latin has largely disappeared from the Scottish religious scene but the language of the Anglo-Saxons, in the guise of its two descendants, Scots and English, continues to have a key role. The triumph of the vernacular as the language of religious expression has been complete but the form the vernacular takes remains interestingly complex.

NOTES

1. Book 12, lines 171–3, Barbour, 1980–5, 3.33.
2. Quoted from Mure Mackenzie, 1946, 73.
3. See Metcalfe, 1889–96.
4. Quoted from Mure Mackenzie, 1948, 90.
5. Lines 3444–53, quoting Matthew 19:17; Lindsay, 1931–4, 2.321.
6. Introduction, Bennett, 1955, xxiv.
7. *The Lang Rosair*, lines 65–7, 69–70, Bennett, 1955, 324.
8. *Remembrance of the Passion*, lines 319–21, Bennett, 1955, 224.
9. *Remembrance of the Passion*, lines 74–5, Bennett, 1955, 216.
10. *Remembrance of the Passion*, lines 119–210, Bennett, 1955, 217.
11. Johannes de Irlandia, 1926, 28.
12. Brown, 1993, 694.
13. The Geneva version of the New Testament had already appeared in 1557.
14. Gau, 1888, 86–7; contractions in the text have been expanded.
15. Winzet, 1888–90, 1.138.
16. Robinson, 1983, 72.
17. Knox, 1846–64, 1.139–40.
18. Mitchell, 1897, 182–3.
19. Robinson, 1983, 62.
20. Cowan, 1982, 79.
21. Knox, 1846–64, 6.229.
22. Quoted in Wright, 1988, 162.
23. Mure Mackenzie, 1948, 96.
24. Galt, 1803, 13–14.
25. Cockburn, 1874, 2.88.
26. Quoted in Watson, 1984, 65.
27. Sinclair, 1791–9, 8.589–90.
28. Scott, 1995, 112, 113.
29. Hogg, 1995, 97.
30. Hogg, 1995, 99; *kelty*, as Hogg tells us, means double measure.
31. Hogg, 1995, 98.
32. Scott, 1994, 425.
33. Patrick, 1949, 174–5.
34. Tulloch, 1989, 19.

35. Henderson, 1862, v.
36. Waddell, 1871, 2.
37. Waddell, 1871, 2.
38. Waddell, 1871, 1.
39. Smith, 1898, inside front cover.
40. Milton, 2001, 6.
41. Anderson, 1988, 46.
42. Borrowman, 1979, 10.
43. Borrowman, 1979, 10.
44. Ogston, 1989, 1990, 1991. The wedding service is also available on the World Wide Web at www.monikie.org.uk/wedd.old.htm
45. Lorimer, 1983, xiv.
46. Stuart, 2000, 92.
47. Stuart, 2000, 104.
48. Stuart, 1997, 127.
49. Stuart, 1985, 61.
50. Stuart, 2000, 105.

BIBLIOGRAPHY

Anderson, G J. Moffatt: Bible Translator. In Wright, D, ed. *The Bible in Scottish Life and Literature*, Edinburgh, 1988, 41–51.

Anon. *The Song of Solomon in Lowland Scotch from the Authorised English Version*, London, 1860.

Anon. *The Song of Solomon. Printed in ye olde Scottish dialect*, Glasgow, nd.

Barbour, J. *Barbour's Bruce*, eds. M P McDiarmid and J A C Stevenson, (Scottish Text Society [STS], 4th ser., 12, 13, 15), Edinburgh, 1980–85.

Bennett, J A W, ed. *Devotional Pieces in Verse and Prose, from MS. Arundel 285 and MS Harleian 6919*, (STS 3rd ser., 23), Edinburgh, 1955.

Borrowman, A. *The Buik o Ruth and Ither Wark in Lallans*, Edinburgh, 1979.

Brown, E. Reformation, Scottish. In *DSCHT*, 693–9.

Bruce, F F. *The English Bible*, London, 1961.

Cameron, H P, trans. *Of the Imitation of Christ by Thomas à Kempis frae Latin intil Scots*, Paisley, 1913.

Cameron, H P, trans. *Genesis in Scots*, Paisley, 1921.

Cockburn, H. *Journal*, Edinburgh, 1874.

Cowan, I B. *The Scottish Reformation: Church and society in sixteenth century Scotland*, London, 1982.

Galt, J. Biographical sketch of John Wilson. In Leyden, J, ed. *Scottish Descriptive Poems*, Edinburgh, 1803.

Gau, J. *The richt vay to the Kingdome of Heuine*, ed. T G Law, (STS, 1st ser, 12), Edinburgh, 1888.

Henderson, G, trans. *The Gospel of Matthew, translated into Lowland Scotch*, London, 1862.

Herbert, M. The Bible in early Iona. In Wright, D, ed. *The Bible in Scottish Life and Literature*, Edinburgh, 1988, 131–9.

Hogg, J. *The Shepherd's Calendar*, ed. D S Mack, Edinburgh, 1995.

Ireland, J. (Johannes de Irlandia) *The Meroure of Wissdome*, vol. 1, (STS, 2nd ser., 19), Edinburgh, 1926.

Knox, J. *Works*, ed. D Laing, 6 vols., Edinburgh, 1846–64.

Lindsay, D. *Works*, ed. D Hamer, Edinburgh, 1931–4.

Lorimer, W L, trans. *The New Testament in Scots*, Edinburgh, 1983.

Metcalfe, W M, ed. *Legends of the Saints*, (STS, 1st ser., 18, 23, 35, 37), Edinburgh, 1889–96.

Milton, C. Flora Garry and Buchan Scots, *FRLSU Newsletter* (2001), 4–7.
Mitchell, A F, ed. *A Compendious Book of Godly and Spiritual Songs Commonly Known as 'The Gude and Godlie Ballatis'*, (STS, 1st ser., 39), Edinburgh, 1897.
Mure Mackenzie, A. *Scottish Pageant*, Edinburgh, 1946.
Mure Mackenzie, A. *Scottish Pageant 1513–1625*, Edinburgh, 1948.
Murray, J A H, trans. The Book of Ruth. In his *Dialect of the Southern Counties of Scotland*, London, 1873.
Nisbet, M, trans. *The New Testament in Scots*, (STS, 1st ser., 46, 49, 52), Edinburgh, 1901–5.
Ogston, D. A Scots Communion Order, *Life and Work*, April 1989, 21–3.
Ogston, D. A Scots Wedding Service, *Life and Work*, March 1990, 21–3.
Ogston, D. A Scots Litany, *Scots Glasnost*, 4 (1991), 2ff.
Paterson, T W. *The Wyse–Sayin's o' Solomon: The Proverbs rendered in Scots*, Paisley, 1917.
Patrick, M. *Four Centuries of Scottish Psalmody*, London, 1949.
Riddell, H S, trans. *The Gospel of St Matthew translated into Lowland Scotch*, London, 1856.
Riddell, H S, trans. *The Book of Psalms in Lowland Scotch from the Authorised Version*, London, 1857.
Riddell, H S, trans. *The Song of Solomon in Lowland Scotch from the Authorised English Version*, London, 1860.
Robson, J P, trans. *The Song of Solomon in Lowland Scots*, London, 1861.
Robinson, M. Language Choice in the Reformation: The Scots Confession of 1560. In McClure, J D, ed. *Scotland and the Lowland Tongue: Studies in the language and literature of Lowland Scotland in honour of David D Murison*, Aberdeen, 1983, 59–78.
Scott, Sir W. *The Bride of Lammermoor*, ed. J H Alexander, (Edinburgh Edition of the Waverley Novels), Edinburgh, 1995.
Scott, Sir W. *The Heart of Mid-Lothian*, ed. T Inglis, London, 1994.
Sinclair, Sir J. *The Statistical Account of Scotland; drawn up from the communications of the ministers of the different parishes*, 21 vols., Edinburgh, 1791–9.
Smith, W W, trans. *The Gospel of Matthew in Broad Scotch*, Toronto, 1898.
Smith, W W, trans. *The New Testament in Braid Scots*, Paisley, 1901.
Stuart, J. *A Scots Gospel*, Edinburgh, 1985.
Stuart, J. *A Glasgow Bible*, Edinburgh, 1997.
Stuart, J. *Will I Be Called an Author*, Glasgow, 2000.
Tulloch, G. *A History of the Scots Bible with Selected Texts*, Aberdeen, 1989.
Waddell, P H, trans. *The Psalms frae Hebrew intil Scottis*, Edinburgh, 1871.
Waddell, P H, trans. *Isaiah frae Hebrew intil Scottis*, Edinburgh, 1879.
Watson, H D. William Tennant, the Ettrick Shepherd and the Psalms of David: A linguistic controversy, *Scottish Language*, 3 (1984), 60–70.
Winzet, N. *Works*, ed. J K Hewison, (STS, 1st ser., 15, 22), Edinburgh, 1888–90.
Wright, D F. 'The Commoun Buke of the Kirke': the Bible in the Scottish Reformation. In Wright, D F, ed. *The Bible in Scottish Life and Literature*, Edinburgh, 1988, 155–78.

16 Gaelic and the Churches

DONALD E MEEK

The maintenance and development of Gaelic have been closely intertwined with the religious fabric of Scotland throughout the centuries. Indeed, it could be argued that the Churches have been the most significant of all Scottish institutions in developing and preserving the higher domains of Gaelic in preaching, religious verse and prose. Nevertheless, it is a commonly held notion that the Churches, and especially the post-Reformation bodies, have been broadly hostile to Gaelic, seeking only to consign the language, and especially its secular songs and stories, to oblivion. Such an observation is not, however, borne out by the evidence. The extensive appropriation of Gaelic for spiritual purposes, particularly after the Reformation, strengthened the language, diversified its literature, and imparted a special ethos to Gaelic worship which can still be experienced in Gaelic services. Only comparatively recently has the picture begun to change.

Gaelic is a Celtic language once spoken fairly generally in Scotland, but now used principally by crofting/fishing communities on the north-west Highland mainland and in the Hebrides. Gaelic speakers are, however, found extensively in the Lowlands, especially in Glasgow and Clydeside, as a result of migration. The 1981 Census recorded 82,620 people in Scotland who could speak, read or write Gaelic, representing 1.6 per cent of the total Scottish population. All were bilingual in Gaelic and English. The 2001 Census recorded some 62,000 Gaelic speakers. In 1891 there were 210,677 bilingual speakers of Gaelic in Scotland, and also 43,738 people capable of speaking Gaelic only.

ORIGIN AND DISTRIBUTION OF GAELIC

Gaelic, which is closely related to Irish, was brought to Scotland from Ireland by settlers (Scoti) mainly from the Irish kingdom of Dal Riata in north-east Ulster. Migration from Ireland was under way by the end of the fifth century, with initial settlements in present-day Galloway and Argyll, and was consolidated after AD 500, when a second kingdom called Dal Riata was established within an area corresponding roughly to modern Argyll. The prestige of the Scottish kingdom was considerably enhanced in 563, when Columba established a monastic house in Iona. This led to a powerful alliance between the saint and the royal line, especially during the reign of Aedán mac Gabráin.

The expansionist policies of Aedán and the missionary activity of Iona-based monks were influential in the first phase of the advance of the Gaelic language beyond Dal Riata. The name Atholl (from Ath-Fhódla, 'Second Ireland') bears witness to early Scotic settlement beyond Argyll.

The Scoti of Dal Riata were surrounded by neighbours who spoke other Celtic languages. Chief among them were the Picts, whose principal language was essentially of P-Celtic stock (and thus related to the ancestor-language of modern Welsh). The Picts held sway over territory from the Hebrides to southern Perthshire, where they have left enduring marks in place-names (beginning with the prefix *Pit-*) and in symbol-stones, often beautifully sculptured, and incised with crosses. The Picts appear to have been absorbed gradually during the expansion of the Scots of Dal Riata, and their culture was assimilated to that of their overlords. By 847 the Picts and the Scots were under a single king.

The presence of Gaelic speakers in most of present-day Scotland by 1100 is demonstrated by literary remains and place-names, many of which have ecclesiastical connections. The Book of Deer records, in Gaelic, land grants and other deeds associated with the monastery at Deer in Aberdeenshire, and apparently made in the twelfth century.[1] Place-names with ecclesiastical elements, such as *eaglais* ('church'), *annaid* ('patron saint's church', latterly perhaps 'enclosure'), *crois* ('cross') and *cill* ('monastic cell'), now represented as *Kil-*, bear witness to a pervasive Christian influence on Scottish toponymy. Some church-related Gaelic place-names are now heavily disguised through borrowing into Scots and English. The place-name *Bantaskin*, in Falkirk (itself known in Gaelic as *An Eaglais Bhreac*, 'the speckled church'), derives from *Peit an t-Soisgeil*, 'portion of the Gospel', i.e. a portion of land set apart for the maintenance of church and ministry.[2]

Gaelic had begun its regression from what we now call the Lowlands by 1200. The language was probably weakest and most transient in Lothian; its gradual decay elsewhere in the south was hastened, even in areas of relative strength, by several factors, including English influences on the Scottish crown, the coming of Norman French settlers, and the establishing of burghs. By 1500 Gaelic would have been spoken mainly in areas to the north of the 'Highland Line', but also in Kyle and Carrick, and in Galloway.

GAELIC AND THE MEDIEVAL CHURCH

Gaelic had a prominent place in the Church of the Catholic Middle Ages. It was employed in the composition of hymns in the early medieval Church in Gaelic Scotland and Ireland, in sermons and lectionaries, and in the recording of land grants (as at Deer). Nevertheless, there is a scarcity of Gaelic writing from early Scottish scriptoria. Most writing would have been in Latin, since ecclesiastical documents would have had priority, both in their creation and in their preservation. During the period 1200–1600 we find clear evidence that Gaelic-speaking churchmen were active in the recording of sacred and secular material. The most valuable Scottish collection of Gaelic (including Irish and Scottish) poetry to have survived from the Middle Ages is a manuscript called the Book of the Dean of Lismore, compiled in the Fortingall district of Perthshire between 1512 and 1542, by James MacGregor, titular Dean of Lismore, and his brother, Duncan. The MacGregor brothers wrote their material in a spelling-system based on Middle Scots and not on the 'normal' Gaelic system developed by the bardic schools of Scotland and Ireland.[3]

In the West Highlands, especially in the region under the sway of the Lords of the Isles, the conventions of the bardic schools were more rigidly preserved. The Lords encouraged the development of a Gaelic 'civil service', consisting of clerics, medical men, judges, genealogists and sculptors who were capable of writing in Classical Common Gaelic, the lingua franca of the literary class in both countries. There exists a Gaelic charter of 1408, granting lands in Kilchoman in Islay to a cleric, Brian Vicar MacKay.[4]

The tradition of writing Classical Common Gaelic was transmitted by its earlier Catholic practitioners to the Reformed Church, especially through priests who had been trained in bardic schools, and who subsequently became ministers. The most prominent of these is John Carswell, who translated the Book of Common Order into Classical Gaelic, and published his work in 1567. As the first Gaelic printed book, it ensured that future publications would employ the spelling system of the bardic schools. Vernacular Scottish Gaelic (i.e., Gaelic as spoken by the ordinary people) had begun to show features distinguishing it from vernacular Irish by at least 1300, but the Classical Gaelic tradition, modified to suit vernacular Scottish Gaelic practice, is represented in the language of the Gaelic Bible and the Gaelic Psalter. Another literary form of Gaelic, indebted to English style and particularly to that of the English Puritan divines, was developed after 1600.[5]

SCHOOLMASTERS AND TRANSLATORS: 1600–1800

By 1600 Gaelic had lost much of the high status that it enjoyed under the patronage of the Lords of the Isles; the Lordship had become a threat to the Scottish crown by the fifteenth century, and it was forfeited in 1493. As the crown moved to suppress alleged lawlessness in the west, especially during the reign of James VI, Gaelic was equated with incivility. In the drive against 'these unhallowed people, with that unChristian language' (an official description of 21 December 1615), the equation of Gaelic with insurrection was doubtless (and paradoxically) increased by Highland support for the Royalist and Jacobite causes in the course of the seventeenth century.

In 'civilising' the Highlands the Presbyterian Church played an important role, reinforced by the erection of schools in Highland parishes, from Perthshire to Lewis, during the seventeenth century. Surviving records are meagre, and distribution patterns are uneven, but the number of such schools (established by heritors) appears to have increased towards 1700. Their commitment to English and Latin, especially in grammar schools, would have been very much greater than their commitment to Gaelic, but

Gaelic writing was apparently taught occasionally. In practice, too, the Jacobean Church recognised the importance of supplying Gaelic-speaking ministers for Highland parishes.[6]

Schoolmasters were more potent than ministers in the thrust against Gaelic, but schools supported by heritors were probably less effective than those of a society with a direct policy. In 1709, the Society in Scotland for Propagating Christian Knowledge (SSPCK), a very influential body whose schools initially taught English and aimed to eradicate Gaelic, began to work in the Highlands. SSPCK schools had a significant anglicising effect in the parishes on the eastern and southern edges of the Highlands.[7]

Protestant interest in the Highlands in the eighteenth century, spearheaded by schoolmasters and ministers, established a strong tradition of printed Gaelic prose, but it was prose of a mainly non-indigenous kind. A very high proportion of Gaelic prose texts were, in fact, translations from English religious writings: 86 out of a total of 146 by 1800. Of the 86, 68 were produced in the period 1751–1800, suggesting that Gaelic was more favourably regarded in literary circles after the Jacobite Rebellion of 1745. Important translators included the Rev. Alexander MacFarlane of Kilninver and Arrochar (translator of Richard Baxter's *Call to the Unconverted*, Edinburgh, 1750) and Patrick MacFarlane, a schoolmaster in Appin (translator of Bunyan's *Pilgrim's Progress*, Edinburgh 1812). The Catholic Church also encouraged translation of key texts, though in much smaller quantity. A fine and refreshingly idiomatic translation, by the Rev. Robert Menzies, of Thomas á Kempis's *Imitatio Christi* was published in Edinburgh in 1785. Poetry, especially of a hortatory or didactic nature, was stimulated also, and is represented pre-eminently in the work of Dugald Buchanan (1716–68), SSPCK schoolmaster at Kinloch Rannoch, Perthshire. Volumes of verse topped the list of original religious works appearing in Gaelic in 1741–99. Buchanan's verse was, however, deeply indebted to composers in English, notably Isaac Watts and Edward Young, and in some instances Buchanan translated directly from Watts, without acknowledgement.

Imitation and translation of external models, mostly in English, thus tended to stifle the growth of a properly indigenous Gaelic spiritual literature, particularly in prose. By the 1790s the balance was being redressed to a certain extent by the gradual publication of printed sermons, composed and delivered by Gaelic-speaking ministers. Curiously, the first printed Gaelic sermons were published in 1791 in Fayetteville, North Carolina, where many Gaelic-speaking people had settled from the late 1730s. The author of these sermons was the Rev. Dugald Crawford, a native of Arran, who later served as minister of the parish of Kilmorie.[8] It is remarkable that it should have taken so long to produce Gaelic sermons in printed format. Part of the reason for such 'delay' may be that, in a Gaelic evangelical context, sermons were regarded as essentially 'oral' in their inspiration and delivery, and that printing came about largely in response to changed circumstances, such as emigration, in which it was more difficult to have access to regular Gaelic ministry. It is also noteworthy that the printing of

Gaelic sermons began at much the same time as the Scottish Gaelic Bible was edging its way slowly towards completion.

GAELIC BIBLES

The need for a printed Gaelic Bible for both Ireland and Scotland was first recognised by John Carswell, Superintendent of Argyll, in 1567. However, the Gaelic Bible now used in the Highlands was translated two centuries later, and financed by the SSPCK. This translation, specifically aimed at a Scottish Gaelic readership, was made after the failure of other attempts to provide the Highland people with a version of the Scriptures which they could understand. These early attempts included the translation programme of the Synod of Argyll and the use of Classical Gaelic Bibles translated in Ireland.[9]

Fifty years before the formation of the SSPCK, efforts were made to produce a Gaelic translation of the Bible for Scottish use. In 1657 the Synod of Argyll, which was engaged in the provision of Gaelic catechisms and a Gaelic version of the Metrical Psalter, tried to expedite an existing plan to translate the Old Testament. The work was delayed by the struggle between Presbyterianism and Episcopacy, which led to the ejection of most of the translators from their parishes after 1660. Nevertheless, a complete Gaelic text of the Old Testament, the work of the Rev. Dugald Campbell of Knapdale, is said to have been available in manuscript by 1673, but it never reached print.

Classical Gaelic Bibles
In the late 1680s Classical Gaelic Bibles were imported from Ireland as an interim solution, largely through the initiative of Episcopalians who had links with the (episcopal) Church of Ireland. In Ireland considerable progress had been made with the translation of the Bible into formal Classical Gaelic, the language used by the trained literati of Ireland and Scotland in the period 1200–1600. A translation of the New Testament had been made by William O'Donnell (London, 1602–3), and William Bedell's translation of the Old Testament became available by 1686 (London, 1685–6). O'Donnell's translation of the New Testament may have been in use in Argyll in the earlier seventeenth century. Through the intervention of the Rev. James Kirkwood, 207 copies of Bedell's translation of the Old Testament and also some copies of O'Donnell's New Testament were acquired for Highland parishes. Distribution was slow and hardly effective. In 1688–90 the Rev. Robert Kirk undertook to modify the Classical Gaelic versions for Scottish use. After 1611, the King James Bible was available to those who became literate in English. Some ministers and lay readers made their own *ad hoc* translations from English.

The Scottish Gaelic Bible, 1767–1807
Although the evangelisation of the Highlands was readily perceived as a

priority by the SSPCK, antipathy towards the Gaelic language delayed its involvement in the provision of a specifically Scottish Gaelic Bible. The replacement of Gaelic by English was one of the aims of the Scottish crown and also of the SSPCK prior to 1750, and it was feared that the existence of a Gaelic Bible would prolong the life of the language. Indeed, the need for a Gaelic Bible was accepted only with reluctance by the SSPCK, who hoped that it would be instrumental in promoting literacy, first in Gaelic and subsequently in English.

The role of the SSPCK in producing a Scottish Gaelic Bible reflected two important considerations. First, the society had become aware of the lack of success with which it was pursuing its policy of eradicating Gaelic by enforced use of English in its schools. Thus it began to permit the gradual adoption of the 'comparative method' in the teaching of English in its Highland schools. This allowed some limited use of Gaelic to explain English forms, and was employed as early as the 1720s. By the mid-1750s the society's policy had been modified sufficiently to allow some tentative steps to be taken towards the provision of a Scottish Gaelic Bible. Second, the perceived need for a specifically Scottish Gaelic Bible, rather than a Hiberno-Scottish Classical Gaelic Bible, acknowledged the fact that Scottish Gaelic speakers were having difficulties with the existing Classical Gaelic versions. Indeed, by 1700, if not earlier, Gaelic Scotland had developed its own variant form of Classical Gaelic, in which the verbal and syntactic structures had been modified to accord more closely with those of the spoken language.

In the early 1750s the society signalled its change of linguistic direction by trying to reproduce Kirk's New Testament with Gaelic and English on facing pages, but, finding its way blocked, it commissioned its own translation of the New Testament. The task was originally given to the Rev. Alexander MacFarlane, but it was eventually undertaken (from 1758) by the Rev. James Stuart of Killin, with the help of the Rev. James Fraser, Alness, and was printed (under the care of Dugald Buchanan) in Edinburgh in 1767. Work on the Old Testament (in four volumes) continued thereafter. The first two volumes were translated under the supervision of Rev. John Stuart of Luss, who also translated the third volume (completed last in 1801). The fourth volume was translated by the Rev. John Smith of Campbeltown.

The translators of the Gaelic Bible used the most recent scholarship in their understanding of the original texts. This is particularly evident in their rendering of parts of the Old Testament. Smith made extensive use of the writings of Robert Lowth in his work on the Prophets, and his volume anticipates several of the readings of the English Revised Version of 1885. This was not appreciated at the time, and objections were raised to Smith's 'liberties', which became apparent when set alongside the King James Bible. Smith's volume was extensively revised by the Rev. Alexander Stewart for the 1807 edition. Several further revisions of the Gaelic Bible, mainly affecting style and orthography, were made throughout the nineteenth century.

In its style the Gaelic Bible resembles its Classical Gaelic predecessors, and is to some extent indebted to 'Kirk's Bible' of 1690. It is also influenced by the King James Bible, which it matches in its overall tone. Although morphology and syntax are generally in keeping with the practice of spoken Scottish Gaelic, the language of the Gaelic Bible is formal 'differentiated register', with words and idioms which are today regarded as 'Bible Gaelic', in contrast to everyday, spoken Gaelic. It is best described as a form of 'Classical Scottish Gaelic'.

The translation of the Bible into Scottish Gaelic had great significance for the language and its speakers. It signalled the full emergence of a new phase in the literary life of Gaelic Scotland, in which Scotland was shown to be, to a large degree, independent of Ireland. Further literary development, especially of prose, was stimulated. It also enhanced the status of Scottish Gaelic by providing a major text in 'upper register'. This, in turn, helped to stabilise the language, providing a standard for spelling and establishing a religious lingua franca which, after some initial reservations, came to be used throughout the Highlands. The Scottish Gaelic Bible made a major spiritual impact through the Gaelic School Societies.

Fr MacEachen's Gaelic New Testament
The translation of the Scottish Gaelic Bible was made by a Protestant society, and has been used pre-eminently in the reformed Churches of the Highlands. The Roman Catholic Church did, however, produce a translation of the New Testament in the nineteenth century (Aberdeen, 1875). This was evidently the work of the Rev. Fr Ewen MacEachen of Arisaig (d. 1849), and was based on the Vulgate. The style of this translation is noticeably closer to spoken Scottish Gaelic than that of the 1767-1807 Gaelic Bible.

GAELIC AND METRICAL PSALTERS

Scotland owes the Gaelic Psalter to the labours of the Synod of Argyll, which regarded the translation of the psalms into Gaelic as a major priority. Indeed, it appears to have pursued the project with more vigour than its proposed translation of the Old Testament, mainly because of the importance attached to the psalms within the Reformed tradition and the complete lack of any 'Irish' version. In 1653 the Synod decided to translate the psalms 'so as they may be sung with the common tunes'. Exerting considerable pressure on the initial translators (Revs Dugald Campbell, John Stewart and Alexander MacLaine), it had procured a Gaelic rendering of the first fifty psalms by May 1658, but the translation needed to be revised immediately as it was found to be 'defective in syllabilization'; Common Metre was 'strange and unknown to the Gaelic tongue' and had posed problems. The adjustments were undertaken by Mr David Simpson and John McMarquess, 'ane old man able in the Irish [i.e. Gaelic] song'. An Ceud Chaogad do Shalmaibh Dhaibhidh ('The First Fifty of the Psalms of David') was duly published in 1659.

A programme for translating the remainder of the Psalter was immediately drawn up, and it is possible that some work was carried out. However, the post-1660 troubles intervened, and the project was delayed still further. In the meantime, the Rev. Robert Kirk, Episcopalian incumbent of Aberfoyle, produced a psalter. This work, which supplemented the Synod's 'First Fifty' psalms with Kirk's rendering of the remainder, was published in 1684 with an eleven-year privilege. The Synod of Argyll nevertheless retained its interest, and a manuscript of the remaining 100 psalms was procured in 1691. The entire Psalter, incorporating revisions of the 'First Fifty', was published by the Synod in 1694, and gained great popular esteem, eclipsing Kirk's rendering.

Revision of the Psalter then became an important issue. Rev. Alexander MacFarlane's mainly orthographic revision was published in 1753, and subsequently adjusted by the Rev. Thomas Ross. A revision which affected style and phrasing was produced by the Rev. John Smith in 1787. In 1826 the General Assembly of the Church of Scotland issued an edition which attempted to combine the best features of MacFarlane and Smith.

Given the syntactic structures of the Gaelic language, it is understandable that the first translators had difficulty with Common Metre. However, the use of a Scottish form of Classical Gaelic allowed flexibility, and provided a supply of monosyllabic preverbs and adverbs to satisfy the metre. The overall style is less contrived and contorted than that of the Scottish Metrical Psalter, and subsequent revision has gently removed older morphology and arcane vocabulary, while keeping much of the 1694 texts. Despite their older style and vocabulary, the Gaelic metrical psalms have retained their central position in the Highland Presbyterian Churches. Long-line psalm-singing, with the precentor 'giving out' the line, is still a standard feature of services in the main Gaelic-speaking areas.

GAELIC AND EVANGELICALISM AFTER 1800

Backed at long last by the authority of a vernacular Scottish Gaelic Bible as well as a Metrical Psalter, and empowered by a potent missionary thrust into the Highlands and Islands, Gaelic came to be identified very closely with the evangelical experience of many Highlanders after 1800. The foundation of Gaelic School Societies, the first in Edinburgh in 1810, made the Gaelic Bible accessible to people who were taught to read it by travelling schoolmasters. Revivals resulted in various districts, and there was a strong tradition of powerful evangelical preaching by ministers such as the Rev. John MacDonald of Ferintosh.[10] The link between Gaelic and evangelicalism was reinforced by the Disruption; the Free Church of Scotland became one of the chief bastions of Gaelic in the religious context. Missionary bodies active in the Highlands after 1800 employed Gaelic-speaking missionaries of a similar status to the people themselves. The commitment of the Established Church to Gaelic was also strengthened, largely through the formation of General Assembly Schools and the literary labours of the Rev. Dr Norman MacLeod (1783–1862), who founded and edited the first printed Gaelic journals.

These contained dialogues, sermons and folktales, generally for didactic purposes.[11]

Evangelical attitudes to Gaelic culture are difficult to summarise accurately and fairly. In the absence (to date) of a fully rounded study, sweeping generalisations (usually hostile to the evangelical position) are not uncommon. They ought to be avoided, as the evidence is more subtle, and requires very careful evaluation. Complexities, contradictions and compromises abound at various levels, and it is clear that attitudes were determined as much by individual ministers, catechists and missionaries in the field as by the institutions or the Churches that employed them. Certain aspects of Gaelic culture, such as residual paganism and superstition, drew the wrath of most evangelical bodies, as did (increasingly) intemperance. Secular tales and music could sometimes be dismissed as 'vanity', and as a complete waste of time which could be devoted more constructively to spiritual matters. Yet noted evangelical ministers, such as the Rev. John MacDonald of Ferintosh, and Baptist missionaries, such as Peter Grant (1783–1867) of Grantown on Spey, were exponents of bagpipe and fiddle music, and interacted vigorously with secular Gaelic culture.[12] Archibald Farquharson (1801–78), a native of Strathardle, Perthshire, who became the founder of the Congregational church in Tiree, was noted for his musical talents, and composed hymns which he set to jaunty secular tunes, including pipe tunes. He was also a champion of the Gaelic language. His contemporary in the Baptist Church in Tiree, Duncan MacDougall (d. 1850), likewise resorted to song to convey the message of the Gospel, and, in so doing, regarded himself as casting a net which might attract the attention of the unevangelised community. MacDougall belonged to a talented family in the Ross of Mull, and his sister, Mary MacDonald (1789–1872), is still remembered as the composer of the 'carol' known in English as 'Child in the Manger'. She also composed secular verse.[13]

Evangelical conversion therefore did not always result in renunciation of secular interests. An organised campaign against Gaelic culture is nowhere attested. Nevertheless, it has to be noted that, particularly at times of religious revival and dramatic conversion, individuals and groups were much more likely to renounce their former 'vain' ways and pastimes. Nineteenth-century collectors of folklore, such as Alexander Carmichael, sometimes found that their informants would suddenly refuse to provide material, as happened to Carmichael when he was collecting in Carbost, Skye, in the early 1860s at a time of religious revival in the locality.[14] Some Free Church ministers were also less than supportive of his ventures. It is also apparent that, with the reassertion of the Westminster Standards by the Free Presbyterian Church, which emerged from the Free Church in 1893, and by the post-1900 Free Church, doctrinal rectitude became a dominant issue (which has remained live to the present day). This is bound to have had a bearing on the manner in which these Churches related to Gaelic culture.[15]

On balance, it would be fair to conclude that evangelical activity and experience produced conflicting responses, depending on time, place, people and context. Overall, an alternative Gaelic evangelical culture emerged which, while often rejecting the more 'godless' dimensions of contemporary society, drew selectively on secular models in several fields, such as hymnology, where popular secular tunes were widely used. Traditional forms of poetry, such as panegyric, were refashioned to honour the 'great divines' rather than clan chiefs. Gaelic sermons must have owed something to indigenous rhetorical patterns, and religious cottage-meetings were probably indebted to the concept of the ceilidh-house. Even the great open-air communions, attended by thousands of Highlanders, especially in times of revival, may have drawn on earlier patterns of communal gatherings.

GAELIC SCHOLARSHIP AND CREATIVITY

Ministers and members of all the main Churches active in the Highlands have made noteworthy contributions to Gaelic scholarship, both in the collection and the analysis of data. Within the Established Church, conspicuous collectors of Gaelic songs, traditions and stories include the Revs Alexander Pope (minister of Reay 1734–82) (heroic ballads), James McLagan (1728–1805) (songs and heroic ballads), John Gregorson Campbell (1836–91) (folktales and superstitions), John MacDonald (1779–1849) (Gaelic ballads, at the behest of Sir John Sinclair), Thomas Sinton (1855–1923) (Gaelic poetry from Badenoch), Duncan MacInnes (d. 1903) (folktales), Duncan MacGregor Campbell (1854–1938) (proverbs), and Kenneth MacLeod (1871–1955) (tales and songs), who collaborated with Marjory Kennedy Fraser.

With the exception of MacDonald, these ministerial collectors would probably have identified themselves with the Moderate wing of the Church of Scotland. MacDonald joined the Free Church in 1843. The Free Church had fewer collectors in its ranks, and evangelicals generally tended to avoid fieldwork. Evangelicals, however, were ready to engage in textual analysis and scholarly editing of Gaelic works. The Free Church contained a distinguished group of evangelical scholars who contributed greatly to the analysis and understanding of Gaelic language and poetry in the late nineteenth century and the early twentieth century, notably the Revs Thomas McLauchlan (1815–86), Alexander Cameron (1827–88), John Kennedy and Donald MacLean (1869–1943). Indeed, it could be said that McLauchlan, Cameron and Kennedy (one of the early teachers of the brilliant, pioneering Celtic scholar, Dr Alexander MacBain) laid the foundation of modern Gaelic scholarship in the fifty years before the creation of Scotland's first Chair of Celtic at Edinburgh in 1882.[16] Edinburgh's Chair of Celtic, initially occupied by Professor Donald MacKinnon, had a special responsibility for equipping divinity students who intended to preach in Gaelic.

Ministerial scholars have also contributed to Gaelic lexicography, including the Revs Norman MacLeod and Daniel Dewar (1788–1867) in the nineteenth century, and the Rev. Malcolm MacLennan (1862–1931) in the twentieth. Roman Catholic scholars have placed an emphasis on the production of lexical resources with a strong interest in particular dialects and localities, such as the dictionaries of Frs Ewen MacEachen (Perth, 1842) and Henry Dieckhoff (Edinburgh, 1932), and the word-book of Fr Allan MacDonald of Eriskay (1859–1905).[17] The Roman Catholic islands (Barra, Eriskay, South Uist) have furnished Gaelic with many folk-songs, of which a large number have been collected and made accessible through the outstanding work of Dr John Lorne Campbell of Canna and his wife, Margaret Fay Shaw.[18] It is noticeable that the smaller missionary bodies, such as Baptists and Congregationalists, have contributed much less to Gaelic scholarship than the main-line Churches, although their pastors have produced numerous volumes of Gaelic hymns.

The full orb of ministerial scholarship in a Gaelic context has seldom been better represented than in the work of the Rev. Dr Thomas Moffat Murchison (1907–84). Murchison, whose family roots lay in Kylerhea (Skye) and Glenelg, was best known as minister of St Columba-Copland Road Church in Glasgow, and he gained a degree of national recognition when he became Moderator of the General Assembly of the Church of Scotland in 1967. He was a quiet, but powerfully influential, contributor to Highland economic development from the mid-1930s, when he and Dr Lachlan Grant of Ballachulish founded the Highland Development League. From the 1940s to the 1970s he was a well-known voice on Gaelic radio, delivering explanatory talks on crofting law and history, and on many other subjects. Besides his regular contributions to *Life and Work*, the Gaelic Supplement of which he edited from 1951 to 1980, Murchison wrote extensively in Gaelic and English for several Highland newspapers, and edited three significant volumes of Gaelic prose by different authors. He was also a reviser of the Gaelic Bible, and one of a number of highly respected ministers who were staunch supporters of, and office-bearers in, An Comunn Gaidhealach, among them the Revs Neil Shaw (1881–1961), Malcolm MacLeod (1881–1946) and Malcolm MacLean (1895–1960). Murchison's initiative, alongside the Rev. Malcolm MacLean, in founding the short-lived periodical *Alba* in 1948, encouraged the establishment of the all-Gaelic periodical *Gairm* (1952–2002). His poetic talent was evident when he won the Bardic Crown at the 1958 Mod. In the late 1940s Murchison was also involved in the creation of the Folklore Institute of Scotland.[19] Such a combination of interests is, of course, rare among Highland Presbyterian clergy. Even in the 'broad' Church of Scotland, Murchison was probably exceptional in his range of scholarly gifts, and in his holistic concept of ministry, which extended to the rebuilding of the Highland economy. However, it is not unusual to find other Highland ministers, including evangelicals, who, while operating on a smaller scale, have made important contributions to Highland politics, education and local government, or who have enhanced Gaelic literature. Roman Catholic priests from Barra and South Uist have also provided strong leadership in political and cultural contexts which have benefited Gaelic.

Although Gaelic literature since 1950 has owed more to the laity than to the clergy, and specifically to the founding of *Gairm* by Derick Thomson and Finlay J MacDonald (himself a former candidate for the ministry of the

Church of Scotland), the clergy have made major contributions to literary development. The Rev. Donald Lamont (1874–1958), T M Murchison's predecessor as editor of the Gaelic Supplement of *Life and Work*, helped to pioneer the modern Gaelic short story, chiefly through his humorous sketches of 'Cille Sgumain', an imaginary Highland parish, with its range of 'typical' characters. In addition to Lamont and Murchison, clerical writers regularly featured in *Gairm* have included the Revs Donald N MacDonald (short stories), Colin N MacKenzie (short stories), Roderick MacDonald (poetry) and Girvan MacKay (essays on language and culture, sometimes with a South American flavour). MacKay, born in Bolivia and for a period minister of the parish of Strath, Skye, has continued to write engaging and important articles for *Gath* (2003–), the successor to *Gairm*. Localised Gaelic publishing has occasionally been developed by parish ministers such as the Rev. Dr Roderick MacLeod, formerly of Bernera, North Uist, but now in Cumlodden, Argyll, who is the current editor of *Life and Work*'s Gaelic Supplement. Since 1900, however, and more noticeably since 1945, the Gaelic scholarly activity of the clergy of the larger Churches has diminished markedly.

DENOMINATIONAL ATTITUDES TO GAELIC

The main denominations in the Highlands have generally always been aware of their obligation to provide appropriate means for the Gaelic-speaking people. The fulfilment of that obligation has, however, faced many practical difficulties over the years, and it is unfair to accuse the Churches of neglecting the language and its people. Unquestionably, there is evidence of an institutional lack of will to promote the language at particular periods; the delay in providing a Scottish Gaelic Bible represents that attitude at its worst. Nevertheless, the individual Churches have striven to find appropriate candidates for parishes and mission stations, and several ministers and priests have learned Gaelic for their parish duties. In 1831 the Episcopal Church of Scotland formed the Gaelic Episcopal Society for training Gaelic-speaking clergy, and retained a commitment to Gaelic well into the twentieth century.[20] The Roman Catholic Church in the Outer Hebrides has been more successful than the Protestant bodies in maintaining a small but effective supply of Gaelic-speaking priests, who are usually natives of Barra and South Uist.

Failure to provide Gaelic-speaking clergy can be caused by factors other than ill-will towards the language. Most obviously, the supply of candidates can decline or terminate, for reasons which reflect a wider malaise in the life of Churches and their communities. The loss of Gaelic preaching in the Highland Baptist churches after 1965, for example, was determined by failure of supply, matched by the decline of congregations and their increasing inability to evangelise their localities. The Baptists' evangelical emphasis on salvation, through the medium of the most expedient language, may also have hastened the eventual dominance of English.[21] This model of language shift is already operating within the larger bodies, which face similar difficulties of recruitment; maintenance of regular services becomes more urgent than language loyalty. Hitherto the principal denominations have maintained Gaelic-speaking ministries most effectively in Skye and the Outer Hebrides, but this has begun to change in the course of the last decade. In 1989 the appointment of a non-Gaelic-speaking parish minister to the charge of Carloway was recognised as 'a new chapter in the ecclesiastical history of rural Lewis' (*Life and Work*, February 1989, 27). The minister concerned then learned Gaelic, and persevered until he was able to preach in the language.

Institutional commitment to Gaelic can be measured beyond the parish in the space given to the language in the main organs of the Churches, such as the Established Church's *Life and Work*, which has had an influential Gaelic Supplement for many years, and the Gaelic page of the *Monthly Record* of the Free Church of Scotland. It is noticeable, however, that these Gaelic pages are parts of larger national organs, thus testifying that a fully self-sustaining and solely Gaelic denomination has never emerged in the Highlands, although the main Scottish Churches have developed strong Highland congregations.

Overall, it can be said that denominational commitment to Gaelic is broadly in proportion to the size of the denomination itself. The larger churches, namely the Church of Scotland and the Free Church of Scotland, sustain the greater numbers of Gaelic ministers. The smaller Churches, such as the Free Presbyterian Church (1893) and the Free Church (Continuing) (2000), contain only small numbers of Gaelic ministers. Recent disturbances in the Free Church (productive of the Free Church (Continuing)) and in the Free Presbyterian Church (productive of the Associated Presbyterian Churches) have acted against the wider Gaelic interest by reducing the number of Gaelic ministers available to the original bodies. Baptists and Congregationalists have no clerical Gaelic-preaching capacity at the present time. Levels of toleration of Gaelic culture also tend to follow this pattern, with greater receptivity to language and culture in the Church of Scotland and the Free Church. The Free Church draws a high proportion of its support from Gaelic speakers in the Western Isles, and many of these are active in Gaelic education and local history. The Free Church also holds an annual Gaelic Camp for young people, and it received financial support for this venture from Bòrd na Gàidhlig in 2003. From time to time, however, it becomes clear that the Free Church is suspicious of Gaelic language activism by 'the Gaelic lobby', which it tends to regard as non-Christian and anti-Church. This perception stems in part from the loss of broad-minded, scholarly, Gaelic-speaking ministers within the Church, who would be able to cross the various chasms which are now appearing between the Churches and the secular world.[22]

CONCLUSION

The Churches active in the Highlands, and especially the Protestant bodies, became, with time, a powerful bastion for maintaining the Gaelic language in the prestigious upper domains of preaching and exposition. Ministers also had a key role in Gaelic arts and scholarship. This remained the case until c1950. Unfortunately, in the subsequent half-century, it has become evident that the Protestant bodies in particular have lost much of their interest in, and commitment to, Gaelic, while the number of Gaelic preachers and 'Gaelic essential' parishes has declined sharply in the same period. Gaelic scholarship and creativity formerly undertaken by ministers, such as the revision of successive editions of the Gaelic Bible, have now passed in large measure to a very few scholarly laymen who have a sense of duty towards both the Highland Churches and the Gaelic language. In the clerical ranks, linguistic pragmatism prevails. English is seen increasingly as the language through which the largest number of souls can be reached most effectively, and, although Gaelic preaching is maintained in the Gaelic-speaking areas and in Lowland cities such as Edinburgh and Glasgow, it is hard to be optimistic about its future in those very Churches which once espoused it as a principal medium for their message of salvation.[23]

NOTES

1. Jackson, 1972.
2. Watson, 1973, 244–69.
3. Meek, 2002b, 86–8.
4. Munro and Munro, 1986, 21–7.
5. Meek, 2002b, 91–3.
6. Kirk, 1986.
7. Withers, 1984, 120–33.
8. Meek, 2002b, 96–106.
9. Meek, 1988b.
10. Meek, 1988b.
11. Kidd, 2000.
12. Meek, 2002a.
13. Meek, 1988a, 24–9.
14. Adv. 50.1.12, fos. 123, 333a.
15. Meek, 2002b, 109–11.
16. Meek, 2001a.
17. MacDonald, A. *Gaelic Words and Expressions from South Uist and Eriskay*, Dublin, 1958.
18. MacCormick, D. *Hebridean Folksongs*, 3 vols, ed. J L Campbell, Oxford, 1969–81.
19. Meek, (forthcoming).
20. Strong, 2002, 70–137.
21. Meek, 1989.
22. Meek, 2001b.
23. This article is an updated and expanded version of three entries which first appeared in *DSCHT*. For recent and more detailed evaluation of most of the themes discussed above, the reader should refer to the present writer's work, and especially that published since 1993, as cited in the Bibliography.

BIBLIOGRAPHY

Historical and cultural perspectives

Jackson, K. *The Gaelic Notes in the Book of Deer*, Cambridge, 1972.
Meek, D E. 'Beachdan Ura à Inbhir Nis: New Opinions from Inverness'; Alexander MacBain and the foundation of Celtic Studies in Scotland, *PSAS*, 131 (2001a), 23–39.
Munro, J and R W, eds. *Acts of the Lords of the Isles 1336–1493*, Edinburgh, 1986.
Steer, K and Bannerman, J W M. *Late Medieval Monumental Sculpture in the West Highlands*, Edinburgh, 1977.
Thomson, D S, ed. *The Companion to Gaelic Scotland*, Oxford, 1983.
Watson, W J. *The History of the Celtic Place-Names of Scotland*, 2nd edn, Shannon, 1973.

Churches and Gaelic culture

Bebbington, D, ed. *The Baptists in Scotland*, Glasgow, 1988, 280–308.
Durkacz, V E. *The Decline of the Celtic Languages*, Edinburgh, 1983.
Kidd, S M. Social Control and Social Criticism: The Nineteenth–century Còmhradh, *Scottish Gaelic Studies*, 20 (2000), 67–87.
Kirk, J. The Jacobean Church in the Highlands, 1567–1625. In MacLean of Dochgarroch, L, ed. *The Seventeenth Century in the Highlands*, Inverness, 1986, 24–51.
MacDonald, K D. Prose, religious (eighteenth century). In Thomson, 1983, 240–2.
MacDonald, K D. Verse, religious. In Thomson, 1983, 298–301.
MacInnes, J. *The Evangelical Movement in the Highlands of Scotland*, Aberdeen, 1951.
Meek, D E. *Island Harvest: A History of Tiree Baptist Church*, Edinburgh, 1988a.
Meek, D E. Baptists and Highland Culture, *Baptist Quarterly*, 33 no 4 (Autumn 1989), 155–73.
Meek, D E. *The Scottish Highlands: The Churches and Gaelic Culture*, Geneva, 1996, reprinted 2000.
Meek, D E. The Reformation and Gaelic culture: perspectives on patronage, language and literature in John Carswell's translation of the Book of Common Order. In Kirk, J, ed. *Church in Highlands*, Edinburgh, 1998, 37–62.
Meek, D E. Gaelic Bible, Revival and Mission: The Spiritual Rebirth of the Nineteenth-century Highlands. In Kirk, 1998, 114–45.
Meek, D E. God and Gaelic: The Highland churches and Gaelic cultural identity. In McCoy, G and Scott, M, eds. *Aithne na nGael: Gaelic Identities*, Belfast, 2000, 28–47.
Meek, D E. The language of heaven? The Highland Churches, culture shift and the erosion of Gaelic identity in the twentieth century. In Pope, R, ed. *Religion and National Identity: Wales and Scotland c. 1700–2000*, Cardiff, 2001b, 307–37.
Meek, D E. 'The glory of the lamb': The Gaelic hymns of Peter Grant. In Bebbington, D, ed. *The Gospel in the World*, Carlisle, 2002a, 129–64.
Meek, D E. The pulpit and the pen: Clergy, orality and print in the Scottish Gaelic world. In Fox, A and Woolf, D, eds. *The Spoken Word: Oral Culture in Britain, 1500–1850*, Manchester, 2002b, 84–118.
Meek, D E, ed. *Os Cionn Gleadhraich nan Sràidean* [Gaelic Prose Writings of T M Murchison], forthcoming.
Murchison, T M. Prose, religious (post-1800). In Thomson, 1983, 242–3.
Strong, R. *Episcopalianism in Nineteenth–century Scotland*, Oxford, 2002, 70–137.
Thomson, R L, ed. *Adtimchiol an Chreidimh: The Gaelic Version of John Calvin's Catechismus Ecclesiae Genevensis*, Edinburgh, 1962. This also contains (231–50) an edition of the Gaelic version of the Westminster Shorter Catechism.
Thomson, R L, ed. *Foirm na n-Urrnuidheadh*, Edinburgh, 1970.
Withrington, D. Education in the 17th Century Highlands. In MacLaine, 1986, 60–9.

Gaelic Bibles and Psalters
Durkacz, V E. *The Decline of the Celtic Languages*, Edinburgh, 1983, 52–72.
MacKinnon, D. *The Gaelic Bible and Psalter*, Dingwall, 1930.
MacTavish, D C, ed. *The Gaelic Psalms 1694*, Lochgilphead, 1934.
Meek, D E. The Gaelic Bible, and The Bible and social change in the nineteenth-century Highlands. In Wright, D F, ed. *The Bible in Scottish Life and Literature*, Edinburgh, 1988b, 9–23, 179–91.
Meek, D E. Language and Style in the Scottish Gaelic Bible (1767–1807), *Scottish Language*, 9 (Winter 1990), 2–16.
Withers, C W. *Gaelic in Scotland 1689–1981*, Edinburgh, 1984, 116–37.

17 Words of Power: Literature, Drama and Religion

DONALD SMITH

The relationship between literature and religion in Scotland has been intimate and at times combustible. For most of the historical period the Christian Church has been the main institutional player in religious terms, though the influence of primal religion and the druidic faith can still be traced. The late twentieth-century landscape was transformed as other world religions took root in Scotland through migration and cultural transmission.

The pre-Christian traditions of the people loosely characterised as Celtic were handed on orally and the emergence of written literature is closely associated with Christianity. Adomnán's *Life of St Columba*,[1] which was completed about a century after Columba's death in 597, is a *locus classicus* for tracing this process and is in itself a defining work of Scottish literature. Adomnán's immediate sources are the traditions, oral and written, of the Columban *familia*, but his literary models are the Bible, the Church Fathers and earlier Mediterranean hagiography, such as Athanasius' *Life of St Antony*.

Throughout the *Life*, Columba is portrayed as a passionate and devoted scribe. Miracles retailing the quasi-magical influence of his manuscripts, the extraordinary preservation of his books and his visionary powers of scriptural interpretation are recounted. On the eve of his death Columba is transcribing the psalms:

> When he had come down from the hill and returned to the monastery, he sat in his hut writing out a copy of the psalms. As he reached that verse of the thirty-fourth psalm where it is written, 'They that seek the Lord shall not want for anything that is good', he said: 'Here at the end of the page I must stop. Let Baithéne write what follows.'
>
> The last verse he wrote was very appropriate for our holy predecessor, who will never lack the good things of eternal life. The verse that follows is 'Come, my sons, hear me; I shall teach you the fear of the Lord.' This is appropriate for Baithéne his successor, a father and teacher of spiritual sons, who, as his predecessor enjoined, followed him not only as a teacher but also as a scribe.[2]

Irish traditions gathered by Manus O'Donnell in his *Life of Columba*[3] go even further, ascribing the battle of Cul Drebene and Columba's exile in 561 to a dispute over the copying of a psalter. The *Cathach* or Battler, supposedly a psalter copied by Columba, was carried as a sacred relic into battle by the O'Donnells. Irish tradition, however, is also careful to characterise Columba as a poet and defender of the oral tradition. Words spoken and written have power and Columba's voice in prayer, blessing or curse is a formidable weapon. One unfortunate Skye boar according to Adomnán was 'killed by the power of his terrible word'.[4]

The spoken voice and dramatic action of the Eucharistic ritual or of ecclesiastical imprecation carry immediate effect, but before electronic recording it was the written record that provided lasting influence. Hence Adomnán's own work in establishing a 'superior' written culture, not just through the *Life* but in written laws and in political jurisdictions. The book reflects an eternal order, an idea which is vividly embodied in Adomnán's account of the 'ordination' of Áedán as king of Dal Riata, which in turn draws on Old Testament kingship and the imagery of biblical apocalypse. Text, or at least an ideal text, governs the ritual and the political succession.

> Once, when the praiseworthy man was living in the island of Hinba, he saw one night in a mental trance an angel of the Lord sent to him. He had in his hand a glass book of the ordination of kings, which St Columba received from him and which at the angel's bidding he began to read. In the book the command was given him that he should ordain Áedán as king, which St Columba refused to do because he held Áedán's brother Éoganán in higher regard. Whereupon the angel reached out and struck the saint with a whip, the scar from which remained with him for the rest of his life. Then the angel addressed him sternly:
>
> 'Know then as a certain truth, I am sent to you by God with the glass book in order that you should ordain Áedán to the kingship according to the words you have read in it. But if you refuse to obey this command I shall strike you again.'
>
> In this way the angel of the Lord appeared to St Columba on three successive nights, each time having the same glass book, and each time making the same demand that he should ordain Áedán as king. The holy man obeyed the word of the Lord and sailed from Hinba to Iona, where Áedán had arrived at this time, and he ordained him king in accordance with the Lord's command. As he was performing the ordination, St Columba also prophesied the future of Áedán's sons and grandsons and great-grandsons, then he laid his hand on Áedán's head in ordination and blessed him.[5]

The significance of this formative passage is its depiction of Christendom as the primary spiritual, social and cultural reality expressed in written words and ritual, a concept which displayed remarkable stamina in Scotland.

The cultural ambitions of the Celtic missionaries were played out over several centuries but it was not until the later medieval period that

Christendom in Scotland became securely established, and then it was on mainstream European rather than distinctively Celtic lines. The literary outworkings of this development were also decisive for Scottish culture. Until the fifteenth century, Scotland had no universities, and the Scottish Church sent its best and brightest to Europe, where individual intellectual stars such as the philosopher Duns Scotus shone radiantly. But churchmen were also instrumental in forming a national literature.

As a small European nation, Scotland had to account for its independent origins, and clerical chroniclers turned to their Celtic predecessors to provide an origin legend that linked the foundation of Scotland via Irish migrations to the biblical and sacred history of the world. In the 1420s Andrew of Wyntoun combined the sacred universal history with what little was known of the early Scottish kingdom and with the saga of the wars of independence in a verse narrative in Scots.[6] Walter Bower, the abbot of Inchcolm, then absorbed the origin legend and Wyntoun's chronicle as well as the more source-critical history of another cleric, John of Fordun, to create his chronicle of the Scottish nation – the *Scotichronicon* – which is written like Adomnán's *Life* in the language of cultural authority, Latin.[7]

It is hard to evaluate objectively Bower's master narrative since its conception of a small nation struggling justly against over-mighty neighbours for its legitimate freedom is still a primary constituent of Scotland's self-understanding. In addition, Bower's vivid storytelling gifts have lived on in the popular imagination. To him we owe transmission of the legend of the stone of destiny, the translation of the bones of St Andrew from Greece to Scotland, the supposed intervention of St Andrew at the battle of Athelstaneford, the story of Macbeth as later embroidered by Hector Boece and Shakespeare, and the combined saga of William Wallace and Robert the Bruce. At all times Bower is a churchman relating the fortunes of the nation to those of an independent Scottish Church, and his greatest heroes are those such as Saint and Queen Margaret, who upheld both.

Bower's narrative of the wars of independence against England is influenced by another literary cleric, John Barbour of St Machar's Cathedral in Aberdeen. Barbour's *The Bruce* is a vernacular epic in the style and manner of the *Chanson de Roland* linking the geste of chivalry – in the shape of Bruce and the Good Sir James Douglas – to the realistic crisis of Scottish nationhood in Bruce's time and at Barbour's time of writing in the 1370s. Barbour is conscious of uniting the Romance tradition with historical epic, and the oral tradition of recitation and declamation with written literature:

> Storys to rede ar delitabill
> Suppos that thai be nocht bot fabill,
> Than suld storys that suthfast wer
> And thai wer said in gud maner
> Have doubill pleasance in herying[8]

The Romance tradition was not, of course, an entirely clerical one, and a century later Blind Hary was to produce a heroic biography of William

Wallace based on many earlier legends. Hary may have been a professional poet or even a soldier but was certainly no cleric. Nonetheless in *The Wallace* the national epic remains saturated with religious imagery and paradigms. Wallace himself is a Christ figure sacrificed for the people, and he receives a sacred commission as defender of Scotland in a dream vision of the Virgin Mary herself – an episode conveniently re-edited in later Protestant translations from Hary's rich Scots.[9]

Clerical influence on popular culture was also mediated through drama. In addition to the liturgical or 'clerks' plays' which supplemented Church teaching, the Feast of Corpus Christi in June provided the occasion for a major annual civic festival sponsored and performed by the Craft Guilds to texts devised by clerics. Unfortunately no substantial Scottish texts survive but the evidence for this dramatic festival traverses Scotland.[10] The purpose of the Corpus Christi cycle was to reinforce the salvation history in all its universal and cosmic dimensions as a drama with immediate relevance to Scottish character and society. The folk plays of Robin Hude in May, though tolerated in the medieval centuries, remained outwith Church influence, ruled in topsy-turvy fashion by the Abbot of Unreason. The official festival with appropriate finance and cultural prestige was the Corpus Christi play cycle.

The mainstream liturgy of Catholic Christendom was also a profound literary influence. The main point of the liturgy was its universality embodied in the universal language of Latin, but the late medieval period brought an increasing official emphasis on the Scottish saints. This trend is encapsulated in Bishop Elphinstone's *Aberdeen Breviary* (published as one of Scotland's earliest printed books from 1509),[11] which constitutes one of the great unsung masterpieces of Scottish literature. Crafted for recitation, the *Breviary* brings together the feasts of the universal saints with distinctively Scottish figures, all united within the calendar of the Christian year. In November, following All Hallows, the faithful heard about St Martin of Tours (a probable influence on the earliest mission to Whithorn) and St Machar of Aberdeen. In July, lessons were drawn from the lives of Mary Magdalene and St Thenew or St Enoch, the mother of St Mungo. In February, Candlemas embraced St Brigid and the Purification of the Virgin Mary. Vernacular sermons – where these occurred – would utilise this liturgical material while drawing on folk traditions and stories of the supernatural to drive home the Christian message to a potentially credulous populace. Abbot Bower's *Scotichronicon* undoubtedly provided the matter for many medieval Scottish sermons, the aural and dramatic impact of which we are now sadly denied.

The influence of these literary forms was all the wider because their transmission was not textbound, but the Church's contribution to literacy and literature was also notably channelled through the burgh grammar schools. Robert Henryson, who moved from the staff of Glasgow University to become schoolmaster at Dunfermline (c1468), is an exemplar in this regard. His work evidences wide learning drawn from scholastic and humanistic sources and English poetry, but it is also in the best sense of the term didactic.[12] The classical tales of Troilus and Cresseid and Orpheus and Eurydice are made warmly human, while his retelling of Aesop's Fables is crafted for a Scottish audience. Henryson seems often to have his pupils, their families and their community before his eyes. But not literally, for Henryson was a scholar–poet, writing in his study, as he himself tells us at the start of his masterpiece *The Testament of Cresseid*, which was undertaken in the 'doolie sessoun' of Lent when showers of hail descend from the north:

> I mend the fyre and beikit me about,
> Than tuik ane drink my spreitis to comfort
> And armit me weill fra the cauld thairout:
> To cut the winter nicht and mak it schort
> I tuik ane Quair [book] and left all uther sport,
> Writtin be worthie Chaucer glorious,
> Of fair Creisseid, and worthie Troylus.[13]

It was the spread of literacy among the laity, along with the growth of towns and the emergence of a prosperous, professional and trading middle class, that reshaped the cultural and religious landscape of Europe in the fifteenth and sixteenth centuries. Humanism in education and literature fuelled Catholic and later Protestant reform movements, reconnecting with source texts in their original languages and provoking a furore of competing biblical interpretation. Once again religion and power politics could not be separated and the future of the Scottish kingdom was contested as theological justification was sought for the respective roles of the monarchy, parliament and the Church. Four writers (Robert Wedderburn, Sir David Lyndsay, George Buchanan and James VI) are selected here to show the range of literary responses within this ongoing arena of debate.

Robert Wedderburn was a priest who was probably acting as prior of the preceptory of the Knights of St John at Torphichen when he wrote *The Complaynt of Scotland* (c1550).[14] Wedderburn's service in this capacity enabled the prior, Sir James Sandilands, to travel as an emissary for Marie de Guise to France during this critical period of Mary Queen of Scots' minority. Despite the reforming passions of his two Dundonian brothers, John and James Wedderburn, Robert's *Complaynt* draws on European humanism to argue in vernacular Scots for the joint reform of Church and State within a traditional moral and theological framework. In a vision sequence Dame Scotia attacks the three estates of the realm for their failure to provide unity, meritocratic leadership and social justice. The Complayner employs a range of rhetorical devices, such as dialogue and lists or catalogues, to pen a portrait of the nation, which covers everything from its herbal remedies to its favourite folktales. The influences of humanism and nationalism are harnessed to renew and popularise the traditional institutions. The older order was not necessarily exhausted or totally corrupt as Wedderburn's literary achievement demonstrates.

Something of the same paradox is evident in Sir David Lyndsay's monumental drama *Ane Satyre of the Thrie Estaitis*,[15] which was first produced in its present form in 1552. Lyndsay goes much further in his demand for reform and in his fierce denunciation of ecclesiastical tyranny and corruption. In a 'suthfast' stroke of realism Lyndsay allows John Commonweal to speak on behalf of the Scottish people and voice their demand for justice. Nonetheless the whole framework of the play is institutionally traditional, promoting moral monarchy as the main instrument of reform in Church and State. Lyndsay wrote and produced the play in his capacity as the royal herald, an inherently aristocratic role.

Though the format and context of the *Satyre* may be conservative and traditional, three sources add a radical dimension. Firstly, there is Lyndsay's use of the folk tradition and its pungent, earthy Scots vernacular to present a genuinely inclusive social portrait. Secondly, Lyndsay shows clear support for the moral and social criticisms, though not necessarily the theology, of John Knox and the Protestant reformers. Thirdly, Lyndsay's ambassadorial travels on behalf of the Scottish court seemed to have acquainted him with European drama. In the *coup de théâtre* of the three estates of Parliament 'ganging backwards' and in the satirical or farcical 'Praise of Folly' with which the play ends, he adds a new artistic agenda to his social purpose. The audience dynamic created by the play's public performances in the presence of the court takes literature beyond the humanist study and beyond the scholastic canons of the Church into a new and dangerous environment of change and conflict. This is confirmed by the fact that, from the 1550s, Catholic ecclesiastics and Protestant reformers clearly considered theatre to be a powerful and highly dangerous weapon.[16]

A third important writer, George Buchanan, also took literature beyond the study.[17] Imprisoned for his Catholic reforming sympathies in Europe and in particular his poetic lampoon of the Franciscan friars (compare Lyndsay's Pardoner in the *Satyre*), Buchanan returned to Scotland and eventually threw in his lot with the reformed Church of Scotland. Although Knox is the prime voice of the Reformation in Scotland and the master of vernacular preaching and propaganda, Buchanan was the more culturally authoritative figure in his time. He reformed university education, tutored the monarch, bolstered the Protestant regime with a new interpretation of Scottish history, and articulated a radical theory of limited kingship which was more republican than royalist. Along the way, Buchanan had founded French neo-classical drama and established a reputation as the finest Latin poet of his time. For present purposes, however, the key point is that Buchanan was neither priest nor aristocrat. His career was a product of new cultural conditions – the power of the book and of a learned professional class that did not require the sanction of court or Church to guarantee their influence. Buchanan's fame has waned along with the notion of a pan-European, Latin culture but he was the literary arm of the Reformation and as influential in the shaping of modern Scotland as the oft remembered but little understood Knox.

The effect of an enlarged public culture fostered by printed books and urban institutions is also seen in the literary career of James VI of Scotland and I of Great Britain and Ireland. Tutored as a child by George Buchanan, James never forgave or forgot his harsh task master and formed a political agenda in direct opposition to Protestant radicalism.[18] But James employed not just the levers of royal power and the cultural patronage of his court but the power of his written words in the battle of ideas. His *Basilicon Doron* of 1599, written ostensibly for his heir Prince Henry, and his more extended *The True Lawe of Free Monarchies* became bulwarks of the case for royal absolutism in Great Britain. Sir David Lyndsay's *Rex Humanum* was of course expected to discourse upon 'true governance' but James also promulgated his views on tobacco, witchcraft and poetic metre in treatises and tracts. Perhaps in the end this talented, intellectual monarch was a credit to his early education at Buchanan's hands. He certainly epitomises a culture in which the Church's monopoly of ideas and their dissemination had long since evaporated.

This did not mean that religion had become less socially, politically or culturally significant. In fact it had become the touchstone for conflict and, throughout the seventeenth century, lay and clerical authors alike joined battle in a metaphorical war of ideas matched only by the political warfare in Scottish government and actual bloodshed. The literature produced in this century is not necessarily inferior to anything that went before or came after, but it is troubled by the lack of social stability or consensus and by a linguistic uneasiness as Scots and English competed for authorial loyalties. It is this century in which a communication gap opened up between the oral and popular literature of Scotland on the one hand and the learned or élite traditions on the other, which had previously interacted so fruitfully with each other. Theatre was limited to rhetorical exercises in the burgh grammar schools, while Scottish ballads and folktales became bearers of an alternative dramatic tradition in Scots.

The Presbyterian Settlement of 1690 did not in itself put an end to conflict but it did provide a context of relative stability and ultimately a basis for economic growth. The different strands of Christianity that had warred in the seventeenth century all maintained or increased their cultural productivity in the eighteenth and early nineteenth centuries, but this productivity depended on a broad interaction between Church and society.

The dominant cultural tradition that grew out of the Presbyterian establishment of 1690 became known as Moderatism. Far from provoking radical dissent, Calvinist orthodoxy provided a secure background for rational enquiry in many branches of learning. In the absence of a Scottish Parliament after 1707, a powerful alliance developed between the Church of Scotland, the universities, the legal establishment, the lairds and the burghs to promote a godly, ordered and educated society. At the centre of this enterprise was an educated ministry which produced luminaries such as William Robertson the historian, Hugh Blair the first professor of literature in a Scottish university, Adam Fergusson the pioneer sociologist, and notable philosophers such as Francis Hutcheson and Thomas Reid, the principal exponent of the Common Sense School. When in 1791 Sir John Sinclair began the first *Statistical Account of Scotland* – a defining Enlightenment project – it was to the parish ministers of the Church of Scotland that he looked to describe and report on the condition of Scotland.[19]

The Scottish Enlightenment was not a wholly clerical affair as David Hume's career clearly demonstrates but, unlike the French Enlightenment, it is unimaginable without the active participation of the Church. At points, such as the literati's backing of a new national drama in the shape of the Rev. John Home's *Douglas* in 1756, Church sensibilities were offended by cultural progressiveness. But more often the Church of Scotland was shrewdly managed by Moderates, such as William Robertson and Alexander Carlyle, to enable an agenda of humane rationalism – liberal thinking, cultured living and common sense – based on a conservative hierarchy of wealth and property in which the ministers enjoyed a modest but nonetheless influential position.

Throughout this period, however, evangelical piety remained a powerful theme in Scottish religious and Church life. Culturally the evangelical mindset promoted introspection and a strong sense of psychological theatre – awareness of sin, repentance, forgiveness and sanctification – all played out in relation to a stern but loving Jehovah. This spiritual drama is evident in a rather grim variant of the works of the seventeenth-century Calvinist Thomas Boston,[20] and in a more winning and passionate form in the letters of Samuel Rutherford.[21] What is less understood is the link between this introspective psycho-drama and the later confessional outpourings of James Boswell, the first-person narrations of John Galt, and the novels of Walter Scott and James Hogg. The Romantic movement in painting, literature, drama and politics drew heavily in Scotland on the evangelical consciousness, even though the Evangelicals were a minority tradition of opposition and dissent in ecclesiastical affairs throughout the eighteenth century.

Episcopalianism provided a third strand in Scottish Church life. Although Episcopalians became politically 'tainted' in official eyes through their association with the Jacobites, they provided a backbone of intellectual life in the seventeenth century, and continued to make their own distinctive contribution to scholarship, literature and science. The Episcopalian tradition, represented by figures such as Archibald Pitcairne (author of *The Assembly*, a dramatic satire on Presbyterianism) and Thomas Ruddiman, maintained the ethos of sixteenth-century Christian Humanism. Their attitude was founded on a more optimistic assessment of human potential than Calvinism and a more positive evaluation of emotion as well as thought. Ruddiman the scholar–printer is a bridge between the latinate literature of the earlier humanists and the eighteenth-century vernacular revival associated with Allan Ramsay and Robert Fergusson. Ramsay's popular drama *The Gentle Shepherd* (1725) is a showpiece of vernacular humanism which celebrates royalist restoration and draws on folk tradition. Episcopalianism, though, like Roman Catholicism, had to await a nineteenth-century revival to renew its contribution to national life on a larger scale.

The vernacular revival also reconnected with the oral traditions of ballad and folktale in the work of Robert Burns. Burns could be classed in matters religious as a moderate Presbyterian, but he also has clear affinities with more radical secular strains of Enlightenment thought, especially in politics, as well as with the older culture of folk festivals. In 'The Holy Fair' and 'The Ordination', Burns puts contemporary religious life in the context of communal festivity, reaching back to medieval poems such as 'Peblis to the Play' and 'Christis Kirk on the Grene'.[22] However, in a more complex narrative work, 'Tam O Shanter', Burns evokes the pagan side of folk festivity as a lost world mediated through a mock heroic classical literary mode. Despite his immense subsequent influence, Burns' work marks the end of a Scots rural culture that had survived religious and political conflict only to succumb to economic improvement.

The same tale of change is rehearsed by John Galt through the eyes of his clerical narrator in *Annals of the Parish* (1821), Rev. Micah Balwhidder. This is the world of the *Statistical Account* in which the minister is the measure of society and culture, but Galt's focus is ironic. Balwhidder's sincere but complacent piety and his limited world view only serve to highlight the dramatic, political, economic, social and religious changes affecting his parish with the onset of both the French and industrial revolutions. What would Balwhidder have made of Robert Burns, given his advice to one young parishioner that 'poem-making was a profane and unprofitable trade, and he would do well to turn his talents to something of more solidity'? This potential to 'do nae gude' goes on to publish a successful book, so Balwhidder has to change tack: 'thus has our parish walked sidy for sidy with all the national improvements having an author of its own, and getting a literary character in the ancient and famous republic of letters'.[23] Creativity is frowned on but cultural achievement when secured is a creditable asset – an attitude which has lingered long in Scottish society. Galt himself is to be commended for capturing this pivotal period so sharply; the nineteenth century was to transform Scotland beyond Micah Balwhidder's wildest imaginings.

Urbanisation brought a huge increase in institutional religious activity in Scotland, despite churchmen's rhetoric of the heathen masses who had been displaced from their rooted parish allegiance. But denominationalism replaced the ideal of Christendom. Thomas Chalmers, an heir of both the Enlightenment and Evangelical movement, led a heroic attempt to renew a Scottish Christendom in the disruption of 1843, but the end result was a major new denomination – the Free Church of Scotland – commanding its own sectional loyalties.

While sociologically religious belief and practice were on the increase in the cities, there was a cultural and intellectual perception that religion was being marginalised by the all-compelling forces of industrialisation and the free market. The most significant responses to this underlying spiritual crisis arguably came, not from clerics or theologians, but from imaginative writers who grappled with the expression of cultural values which could counter-balance economic liberalism and transcend denominational dispute.

Sir Walter Scott is the crucial figure in this context. By rediscovering Scotland's history, Scott was trying to restore the community to a more unified sense of itself. His intention was not to deny the Enlightenment science of universal human nature or the importance of religion, but to encompass these in a form that was relevant and appealing to Scots. Perhaps unintentionally, therefore, Scott emphasised the truth of actual human experience – collective and personal – as the most important reference point for self-understanding. Scott's Scotland is a cultural invention but it is one which helped reconnect a rapidly changing nation with its past, and provided a framework of cultural values with which the majority of Scots came to identify through literature, painting, architecture and theatre.

Reference has already been made to the parallel importance of John Galt, whose reading of the Scottish past is more sympathetic to Presbyterianism than that of Scott. Yet, whatever the particular perspectives on key episodes in Scottish history, the overall authorial understanding is in both cases humane and historical. James Hogg, the third of these early nineteenth-century novelists, is a more complex case since his multi-layered narratives affirm a supernatural and spiritual dimension to experience that potentially transcends science and history. This complex world view is memorably embodied in *The Private Memoirs and Confessions of a Justified Sinner* (1824), Hogg's acknowledged masterpiece. The *Confessions*, however, are not an exposition or critique of Calvinism and Scottish Protestantism, but a subtle spiritual fable grounded in Scottish social and historical reality.[24] In this regard Hogg anticipates R L Stevenson's *The Strange Case of Dr Jekyll and Mr Hyde* (1886) and *The Master of Ballantrae* (1888). Hogg's is a spiritual philosophy but he is a cultural interpreter and a literary artist.

The trend by which educated Scots looked beyond ecclesiastical influence was defined in the work of Thomas Carlyle. Brought up in a pious Presbyterian household as a classic 'lad o pairts', family logic dictated a glorious clerical and scholarly career for the Sage of Ecclefechan. Instead Carlyle had a troubled formation as an imaginative writer, historian and cultural commentator. The crucial text is *Sartor Resartus* (1833–4), which in the guise of a fictionally edited, fictional first-person narrative traces Carlyle's own spiritual development. This progresses through Romanticism, through mechanistic despair, through indifference, to 'The Everlasting Yea', a personal affirmation of spiritual life and its dynamic expression in creative work and social duty:

> May we not say, however, that the hour of Spiritual Enfranchisement is even this: When your Ideal World, wherein the whole man has been dimly struggling and inexpressibly languishing to work, becomes revealed, and thrown open; and you discover, with amazement enough, like the Lothario in *Wilhelm Meister*, that your 'America is here or nowhere?' The Situation that has not its Duty, its Ideal, was never yet occupied by man. Yes here, in this poor miserable hampered despicable Actual, wherein thou even now standest, here or nowhere

is the Ideal; work it out therefrom; and working, believe, live, be free. Fool! The ideal is in thyself, the Impediment too is in thyself: thy Condition is but the stuff thou art to shape that same ideal out of: what matters whether such stuff be of this sort of or that, so the Form thou give it be heroic, be poetic? O thou that pinest in the imprisonment of the Actual, and criest bitterly to the gods for a kingdom wherein to rule and create, know this of a truth: the thing thou seekest is already with thee, 'here or nowhere', couldst thou only see![25]

'The felt indubitable certainty of Experience' gives Carlyle an existential centre and the growth of the person in time and society through this spiritual ideal becomes the *raison d'être* of biological existence. Carlyle goes beyond Scott's emphasis on history and experience to figure a new form of spirituality which, acknowledging his debt to Goethe, he labels 'natural supernaturalism'.[26]

Carlyle famously translated to Chelsea, going on to influence English-speaking culture across the world, but his effect on Scotland should not be underestimated. There is a direct line from Carlyle to R L Stevenson's late nineteenth-century existential faith and Hugh MacDiarmid's early twentieth-century poetic and spiritual eruptions. Carlyle changed religion in Scotland by expounding a cultural alternative or at least complement to doctrinal paradigms, so shifting the relationship between Church and culture as well as the longer-term ethos of mainstream Scottish theology.

Carlyle's lay influence on religion was not an isolated phenomenon as access to education spread across Scottish society. The brothers William and Robert Chambers spearheaded the provision of scientific, social, literary and cultural knowledge to all classes without denominational bias through *Chambers Journal*. Robert was also the anonymous author of the 1844 *Vestiges of the Natural History of Creation*, which anticipated Darwin's theory of evolution and provoked widespread controversy. However, it was Carlyle's distinction to interpret the spiritual mood or experience of his time from outwith the structures of organised religion, and in this at least Carlyle was prophetic since the cultural influence and authority of the Christian Church in Scotland was set for a long period of decline.

This process is mirrored in the changing perceptions of ministers. Although the clergy of all denominations continued to play an important role in Scottish society, their status as intellectual leaders steadily diminished. From the nineteenth century, fictional portraits of ministers tend to emphasise their shortcomings as definers or exemplars of society's values. An interesting example of this is J M Barrie's *The Little Minister* (1891) since it is set in the world of his Kirriemuir childhood – the fictional Thrums – in which matters ecclesiastical rule supreme. Nonetheless, Barrie's narrative, which was popularised on stage and screen, shows how Gavin Dishart's humanity, with its passions and imaginings, cannot be accommodated within his ministerial role. Some twentieth-century portrayals of ministers, such as Fionn MacColla's *The Ministers* (1979) or Robin Jenkins' *The*

Awakening of George Darroch (1985), are metaphors for the flaws, oppressions and hypocrisies of Scottish society, though others, such as James Bridie's play *Mr Bolfry* (1943), retain the minister as a focus of moral and spiritual integrity.

The twentieth century brought secularisation to Scotland, accelerating from the 1960s to produce a largely de-churched community.[27] Writers meanwhile, haunted by a sense of moral and spiritual emptiness, dreamed of a political or social 'kingdom of heaven' realised on Scottish earth, and imagined a renewed Body of Scotland rather than an ecclesiastical Body of Christ.

Poetic voices from Hugh MacDiarmid, Edwin Muir and Sorley MacLean articulated a sense of Scottishness burdened by the tragedies of the past and the moral pressures of the present, yet resilient with new possibilities. Lewis Grassic Gibbon's trilogy *A Scots Quair* (1932–4), Neil Gunn's Highland novels published in the 1930s and 1940s, and Naomi Mitchison's *The Bull Calves* (1947) all preached their own visions of communal renewal. In *A Scots Quair* Chris Guthrie marries successively the land, religion and socialism. From the 1950s the tone became more sceptical and ironically humane in the writing of Robin Jenkins and Ian Crichton Smith, though Muriel Spark continued to probe a metaphysical subtext from a vantage point outwith Scottish society. By the end of the century James Kelman was pushing the Enlightenment project of human understanding to the limits of optimism and beyond, in compelling portraits of felt, but apparently disposable, lives in Scottish society.

In the twentieth century the theatre again came into its own culturally and politically, registering change with a sharp immediacy. J M Barrie disguised profound pessimism and despair with brilliant theatrical games and myth-making. James Bridie questioned the agnosticism and moral apathy of modern Scotland while playing the fool. After the political fireworks of John MacGrath's 7:84, theatre settled for a more literary and less threatening existence as one of 'the arts', but the live event continued to exercise a latent potency within Scottish society.

The Presbyterian Church of Scotland itself conformed to the new positive evaluation of creativity and the arts by opening the Gateway Theatre in 1946 and The Netherbow Arts Centre in 1972. The aim was to participate in and draw inspiration from the renaissance of Scottish culture. In 1995 a Church and Nation Committee report, 'Faith in the Arts', affirmed a creative shift in Scottish society and depicted theology, spirituality and the creative imagination as intimately related activities.[28]

This change in theological climate seemed to mirror a more positive late-century cultural evaluation of religion and spirituality within the context of a multi-faith society. Writers as diverse as Alan Spence and George Mackay Brown had continued to make their religious faith – Buddhism and Christianity respectively – central to their creative work, while poets ranging from Kathleen Jamie to Don Paterson and Angus Peter Campbell sustained Scottish poetry's close relationship with metaphysical concerns. A revival of oral tradition in song and story seemed to be partly driven by reaction against economic materialism and impersonal technological change.[29] Not long before his death, George Mackay Brown placed his own vocation as a storyteller within a wider sense of spiritual enlightenment:

> Going over tales I've written during the last decade or so, I was not too surprised to see that many of them are calendar tales, that yield their best treasure in midwinter when the barns are full.
>
> The mystery of light out of darkness has been with us since the builders of Maeshowe five thousand years ago. The Celtic missionaries gave the mystery breadth and depth.
>
> I like to think I am part of that tradition.[30]

Any trace of millennium optimism, however, still had to come to terms with a century in which humanity had torn itself apart in two gruelling world wars as well as a host of bloody ethnic and religious conflicts, created and implemented the means of mass destruction, and brought the natural basis of life to the brink of destruction. At the start of the twenty-first century these issues evoke the moral and spiritual prescriptions of the major world religions in a stark form, yet Scots seem more inclined to personal lifestyle choices or to a renewed political process. The roles of written literature, drama and ritual are themselves under scrutiny in a media-driven culture that remains in search of common symbols or even a common language of human value. Perhaps Scottish culture will reaffirm its distinctiveness by rediscovering the power of words.

NOTES

1. See Adomnán of Iona. *Life of St Columba*, translated with an introduction by Richard Sharpe, London, 1995.
2. *Life of St Columba*, 228.
3. See O'Donnell, Manus. *The Life of Colum Cille*, ed. Brian Lacey, Dublin, 1998, 97–108, 172–90.
4. *Life of St Columba*, 175.
5. *Life of St Columba*, 208–9.
6. See Amours, F J, ed. *The Original Chronicle of Andrew of Wyntoun*, Edinburgh, 1902–1914.
7. For a general introduction to Bower's monumental work see Bower, W. *A History Book for Scots: Selections from the Scotichronicon*, ed. D E R Watt, Edinburgh, 1998.
8. Barbour, J. *The Bruce*, ed. A A M Duncan, Edinburgh, 1997, 47.
9. See McDiarmid, M P, ed. *Hary's Wallace*, 2 vols., Edinburgh, 1968, and Hamilton of Gilbertfield, W. *Blind Hary's Wallace*, Edinburgh, 1998.
10. See Findlay, B, ed. *A History of Scottish Theatre*, Edinburgh, 1998, 9–15.
11. See Blew, W J, ed. *Breviarum Aberdonense*, 2 vols., Edinburgh and Glasgow, 1854–55.
12. See MacQueen, J. *Robert Henryson: A Study of the Major Narrative Poems*, Oxford, 1967; McDiarmid, M P. *Robert Henryson*, Edinburgh, 1981.
13. Henryson, R. *Poems and Fables*, ed. H Harvey Wood, Edinburgh, 1978, 106.
14. Wedderburn, R. *The Complaynt of Scotland*, ed. A M Stewart, Edinburgh, 1979.
15. Lindsay, Sir David. *Ane Satyr of the Thrie Estaitis*, ed. Roderick Lyall, Edinburgh, 1989.

16. The first statutes against religious theatre were passed by the pre-Reformation Scottish Church in 1555.
17. See McFarlane, I D. *Buchanan*, London, 1981.
18. See Mason, R A. *Kingship and the Commonweal: Political thought in Renaissance Scotland*, East Linton, 1998.
19. For a convenient survey of the Enlightenment, including the role of churchmen, see Broadie, A, ed. *The Scottish Enlightenment: An Anthology*, Edinburgh, 1997.
20. See for example Boston, T. *Human Nature in its Fourfold State of Primitive Integrity entire Depravity Begun Recovery and Consummate Happiness or Misery*, London, 1964.
21. See Bonar, A A, ed. *Letters of Samuel Rutherford*, London and Edinburgh, n.d.
22. See MacLaine, A H, ed. *The Christis Kirk Tradition: Scots Poems of Folk Festivity*, Glasgow, 1996.
23. See Galt, J. *Annals of the Parish*, Year 1801 'An Account of Colin Travis, who becomes a poet'. The most recent scholarly edition is edited by James Kinsley, Oxford, 1967.
24. For a detailed discussion see MacQueen, 1989, 234–65.
25. Carlyle, T. *Sartor Resartus*, ed. Kerry McSweeney and Peter Sabor, Oxford, 1987, 148–9.
26. For a good general introduction to these themes, see Carlyle, 1987, vii–xxxiii.
27. See Brown, C G. *The Death of Christian Britain: Understanding Secularization 1800–2000*, London and New York, 2001.
28. See Church and Nation Committee report, Faith in the Arts, Edinburgh, 1995.
29. See Smith, D. *Storytelling Scotland: A Nation in Narrative*, Edinburgh, 2001, for an extended discussion of this theme.
30. Brown, G M. *Winter Tales*, London, 1995, foreword.

BIBLIOGRAPHY

Craig, C, ed. *The History of Scottish Literature*, 4 vols., Aberdeen, 1987–8.
Daiches, D, ed. *The New Companion to Scottish Culture*, Edinburgh, 1993.
Findlay, B, ed. *A History of Scottish Theatre*, Edinburgh, 1998.
MacQueen, J. *Progress and Poetry*, Edinburgh, 1982.
MacQueen, J. *The Rise of the Historical Novel*, Edinburgh, 1989.
Scott, P H, ed. *Scotland: A Concise Cultural History*, Edinburgh, 1993.
Sher, R. *Church and the University in the Scottish Enlightenment*, Edinburgh, 1985.

18 Medium for Godly Messages

COLIN MACLEAN

> It's thirty year, said ye, it's forty an' mair,
>   Sin' last we were licket at squeel;
> The Dominie's deid, an' forgotten for lang,
>   An a' oor buik learnin' as weel.
> The size o' a park – wi' the gushets left oot –
>   We'll guess geyan near, I daur say;
> Or the wecht o' a stot, but we wouldna gyang far
>   Gin we tried noo the coontin' in 'Gray'.
> 'Effectual Callin'' we canna rin throu'
>   Wha kent it aince clear as the text,
> We can say 'Man's Chief En' an' the shorter 'Commands',
>   But fat was the 'Reasons Annexed'?
> Oor heads micht be riddels for a' they haud in
>   O Catechis, coontin' or date,
> Yet I'll wauger we min' on the mornin's lang syne
>   When it wasna oor wyte we were late.[1]

In the final lines of his well-known poem Charles Murray sums up the uncertain long-term results of school education, not least of religious education. Noteworthy for our immediate purpose are Murray's references to the Shorter Catechism, product of the Assembly of Divines at Westminster and completed in 1647. Charles Murray's education, at Gallowhill School, Alford, was in the 1870s. A century earlier, Robert Burns had written:

> An' ay on Sundays duly, nightly,
> I on the Questions tairge them tightly;
> Till, faith! wee Davock's grown sae gleg,
> Tho' scarcely langer than your leg
> He'll screed ye aff Effectual Calling
> As fast as ony in the dwalling.[2]

When Murray was young, the evangelist D L Moody visited Scotland: addressing a gathering of children in Edinburgh, he happened to ask the rhetorical question, 'What is prayer?' and was astonished when hundreds of hands were raised and a chorus of voices shouted the answer to Question 98:

> Answer: Prayer is an offering up of our desires unto God for things agreeable to His will, in the name of Christ, with confession of our sins, and thankful acknowledgement of his mercies.

Writing about her early years in the 1890s and 1900s, Isabella G MacLean reports:

> Having tried Welfare of Youth exams at Sunday School and Donaldson exams as a pupil-teacher, the Shorter Catechism was as familiar to me as the alphabet.[3]

In the Shorter Catechism there were 107 Questions and Answers (Charles Murray refers to Q&As 1, 31 and 32, 52 and 56; Burns to 31 and 32). The Multiplication Table was printed on the back cover of Catechisms published by Collins. Constantly reprinted, as learning matter for many generations of children at home, school or church (sometimes all three), the Catechism was impressed upon folk thought and memory in Scotland for at least two and a half centuries: which presumably makes it rank next only to the King James Bible in impact and durability; one part of the Old Testament being secured in many memories by the prolonged choral monopoly which was assigned to the Metrical Psalms. The Bible was widely used in schools for learning to read as well as for religious instruction.

What other school textbooks have ever survived for so long? Bearing what fruits? 'Wha kent it aince clear ... Oor heads micht be riddels for a' they haud in'. Then what of hand and heart? An Aberdeen Unitarian minister, the Rev. Alexander Webster, wrote in 1915 that the language of the Catechism 'haunted the mind with a weird authority'. But he regretted the use and effect of Catechism and Bible. The primary influence of Calvinism, he said, was traceable to the Shorter Catechism; also, in casting off one tyranny the reformers had not seen that they had taken on another. 'The Infallible Bible involved a bondage as close and sore as the Infallible Pope had proved. The movement for liberation of the mind from bondage to autocratic authority was arrested from within'.[4]

Webster argued at the same time that George MacDonald, along with Carlyle, Stevenson and others, was 'a representative of a type of mind which though born and bred in strict orthodoxy departed from it ... his departure from the sphere of spiritual bondage was caused partly by a constitutional disposition towards freedom, and by irresistible outdrawing influences operating in the atmosphere of his time'.[5]

The formation of the mind, young or old, the individual's reactions and powers of retention, all are conjectural and unpredictable. As there is endless speculation about cause and effect in all communication, there must be about the impact of literature, not least about religious literature, by no means least about the Book itself, and then about the countless communications with which the Book and the Catechism, the long and the short of religious education, have had to compete. In *The Printed Word* Christopher Small points out that among works printed in the first hundred years after

Gutenberg were 'not only Theses against Indulgences but the Indulgences themselves, not only Erasmus's In Praise of Folly but the master-handbook of witch-hunters, the Malleus Maleficarum ...' and so on. With reference to the Bible, Small concludes:

> In this process the printed and, especially, the vernacular Bible had a unique and central place, but one that also is ambiguous in implication; profoundly seditious, yet used as a seal of authority; the common treasure of all, yet the most fruitful source of division and strife; vehicle of the Word made Flesh, yet tending more than any of the former vagaries of religious teaching to make flesh word, to isolate words from feelings and to appropriate all value to itself; instrument for the destruction of idolatry, replacing all others with the idolatry of the book.[6]

The reformers decreed that every church was to be provided with a Bible in English: both Old and New Testaments were to be read through systematically at church services. Church funds should support not only the churches but also education in school and university, this in addition to the relief of the poor. Such a programme – introducing 'a decidedly democratic tendency', says Professor G D Henderson[7] – laid emphasis on respect for, but especially on the responsibility of the individual – an individual who heard, understood and, it was hoped, could read the Word of God. By the middle of the seventeenth century the King James Bible was displacing the Geneva Bible, an English version produced on the Continent in the middle of the sixteenth century by exiles from England and Scotland. Tyndale's English translation of the New Testament had been printed on the Continent in the 1520s. In 1543 the Scottish Parliament was persuaded to permit the reading of English and Gaelic translations of the Bible. John Knox wrote:

> Then might have been seen the Bible lying almost upon every gentleman's table. The New Testament was borne about in many men's hands.[8]

No survey is offered here of all the printers, publishers or booksellers (sometimes three in one) who over the centuries brought the printed word to the reader. In many cases the producer was far distant from the consumer. Few barriers of a nationalist kind have operated (say, between Britain and the Continent, or between Scotland and England), all of which may be seen as appropriate in relation to a religion that may have had some local characteristics but was nevertheless seen to be a faith for the world. Printers were at times subject to local regulations and pressures, some were licensed or granted monopoly, some were even imprisoned. But the printed word, 'profoundly seditious', kept breaking free.

In *The Dictionary of Scottish Church History and Theology* Dr John A H Dempster says there has been no systematic survey of Scottish religious publishing. In fact, neither moderate nor evangelical (if the terms may be used in their widest application) authors, nor publishers, have submitted to national constraints, and in recent times publishers have often employed distant printers. Dempster lists the many printers of the sixteenth and seventeenth centuries, from Walter Chepman and Andrew Myllar to Evan Tyler, who first issued the Scottish Metrical Psalms, and Andrew Anderson and his wife who enjoyed the title of King's Printer for forty-one years. By the end of the eighteenth century, religious works were being issued in fifteen Scottish towns. In the nineteenth century, local printers proliferated, and major Scottish religious imprints came into being and flourished – Blackwood, A & C Black, Nelson, Blackie, Collins, T & T Clark. Works published in England have often been printed in Scotland: the Third Edition of The Church Hymnary, published by Oxford University Press, was set by OUP but printed by Robert MacLehose and Co Ltd, The University Press, Glasgow. From the mid-nineteenth century, large numbers of hymnaries were being produced. In his history of *The Oxford English Dictionary* Simon Winchester records that Oxford University Press (this by the end of the nineteenth century) 'had long made the bulk of its money from the publishing of Bibles, hymnals and prayer books'; also that the profitability of what was called the Bible Side of the Press subsidised 'the indulgent obscurities' of the Learned Side.[9]

The printing and sale of Bibles, and generous promotional activities by Bible societies, had made it possible for a steadily increasing number of people – not only gentlemen – to read the Scriptures. No parents were 'suffered to neglect keeping their children at school, till they can read the Scriptures distinctly' (Synod of Glasgow and Ayr 1700). By the eighteenth century many Scottish households possessed their own family Bible; in The Cotter's Saturday Night, Burns wrote:

> The sire turns o'er wi patriarchal grace
> The big haa Bible, aince his father's pride.

Little is known about how, how well, or in what numbers people learned to read in the years following the invention of printing. Who taught whom, with what teaching skills, using what reading matter? For more than 200 years, until literacy – and the availability of books etc – was widespread, much of what was written in Scotland was by ministers, and for comparatively few readers, many of them ministers. What Arthur Herman in *The Scottish Enlightenment* calls 'the old order of Calvinist ayatollahs'[10] began to lose their power early in the eighteenth century and 'that same fundamentalist Calvinist kirk had actually laid the foundations for modern Scotland in surprising and striking ways'. Those foundations were literate, mostly Moderate and scholarly. It is of or from scholars that history has often spoken. At the same time, those Evangelicals, who in the late eighteenth century began to produce and distribute tracts, were assuming literacy in an appreciable number of people, and at some level of competence.

Tracts, pamphlets issued with a religious or moral purpose, were produced from – indeed before – the earliest days of printing. In England of the sixteenth and seventeenth centuries the production of tracts was stimulated by controversies. The Edinburgh Religious Tract and Book Society was founded in 1793 by an Edinburgh ironmonger John Campbell who later became minister of a London chapel. In the mid-nineteenth century the society was to be renamed the Religious Tract and Book Society of Scotland, then in 1908 the Tract and Colportage Society of Scotland. For various Scottish tract societies, colporteurs took tracts and Bibles, sometimes in Gaelic, to the towns and villages of Scotland. Their activities and their wares were often subsidised by well-wishers and societies, some of whom were especially eager to counteract the influence of chapbooks and other material which hawkers were selling to the impressionable masses. Some booksellers made appreciable profits from a wide range of religious matter. One society, the Aberdeen Book Agent and Colportage Society, by 1859 employed five agents, selling in one year 28,491 periodicals. By 1876 the Glasgow Committee for Colportage had 31 colporteurs who sold 138,821 periodicals and 185,000 tracts.[11]

While many religious tracts were sold, some were freely distributed – at times, it is recorded, to people who could not read. The religious message they carried came to depend very much upon the denominational commitment of the society or well-wisher who financed the project. The message was sometimes promoted, indeed at times determined, by the colporteur who distributed the tracts: in some cases he was responsible for both printing and content. One such was Duncan Matheson who was born in Huntly in 1824, son of a mail-runner between Huntly and Banff. In 1846, a builder's labourer, Matheson was 'born again' and then for the remaining twenty-plus years of his life 'strove with all his might to win souls'. The Duchess of Gordon, hearing of his 'zealous and successful labours' offered to employ him as a missionary at a salary of £40 a year: he accepted. His biographer tells us that Matheson spent a large proportion of his small salary in the relief of the poor and in the purchase of tracts. The Duchess, Lady Aberdeen and other titled and landed personages encouraged and often funded the spreading of the Word by such as Matheson. For his later efforts in the east, taking in Gibraltar, Malta and Italy on the way, he was 'liberally aided by the Countess of Effingham and others'.[12] 'Not satisfied with the efforts of his voice', the penniless Matheson was determined to devise means for:

> the circulation of tracts on the widest scale ... He began to cry to God for aid. One night in prayer, the thought came into his mind, 'If I could get a printing press I could make as many tracts as I could use'. On this he began to pray for a printing press, and for several months continued to supplicate this gift from his God. The prayer was unexpectedly answered. Accidentally discovering that an old printing press was for sale, he made inquiries as to the terms, although he did not possess the means of purchase. Much to his astonishment, the person whose property it was let him have it, with a set of old worn types, at a merely nominal price. Never did warrior bear away the trophies of victory with deeper joy than he felt in carrying the old printing

machinery to his father's house. On reaching home, he wrote upon it FOR GOD AND ETERNITY and then, hastening to his closet, fell upon his knees, and asked the needed skill to work it.

To ensure that Matheson acquired the needed skill 'the two great teachers of all heroic souls and successful workers, to wit, Failure and Perseverance' were, one assumes, appointed by the Almighty who then answered further prayers by securing supplies of paper, as well as sufficient customers and distributors. For his travels abroad, Matheson's stores of Christian literature 'for gratuitous distribution' were immense and varied, including Bibles, the Shorter Catechism, tracts and books 'in the several languages of the East'. In Constantinople he distributed tracts and Bibles among the French and the Turks. 'Since my arrival here 6,600 copies of the Scriptures, in all languages, have been distributed.'

By 1857 Matheson, in Scotland, was producing *The Herald of Mercy*, which gains interesting mention in the advertisement (see Fig. 18.1) at the end of the biography by the Rev. John Macpherson, *Life and Labours of Duncan Matheson, the Scottish Evangelist* ('Fifty Eighth Thousand'). A greater element of calculated commerce was apparent in 1907 than there had been in 1857. It may be noted that the Magazine – and much else – was not always to be for 'gratuitous distribution'. 'Localization' was aimed for; illustrations were selling features; Gospel-filled Volumes – such as *Happy Hearts* and *Tom Dawson's Deliverance* – available at a shilling each, plus 1d. postage. And there were 'lots of good anecdotes' useful for sermons.

Early in the nineteenth century a number of Bible societies had been established in Scotland; the first of these formed in Glasgow in 1805, followed by the Edinburgh Bible Society, the Scottish Bible Society and others in half a dozen centres, all of which worked along with the British and Foreign Bible Society based in London.[13] Thomas Chalmers (1780–1847), in a pamphlet *The Influence of Bible Societies as the Necessities of the Poor*, argued for the widespread distribution of 'the pure Christianity of the original record ... to the hearts of the ignorant' – of whom there were still many, in spite of two centuries of Presbyterian zealous preaching and teaching. The societies supplied Bibles not only to poor people in Scotland but also in large numbers to the rest of Britain and to missionary endeavours abroad.

Chalmers' pamphlet was but one of a range of publications bearing his name: *The Collected Works of Thomas Chalmers* filled 25 volumes, *Posthumous Works* by Chalmers ten volumes. Hazlitt reported how some of Chalmers' sermons 'ran like wildfire through the country, were the darlings of watering places, were laid in the windows of inns, and were to be met with in all places of public resort'. Sermons were to be published widely in the nineteenth century: in the late eighteenth, John Gillies (1712–96) Evangelical minister of the College (Blackfriars) Church in Glasgow for over 50 years, had published many of his sermons, some in the 1750s in the weekly *Exhortations*.[14] John McLeod Campbell (1800–72) provoked, at first, a response less favourable than that accorded to Chalmers. In 1832 Campbell's

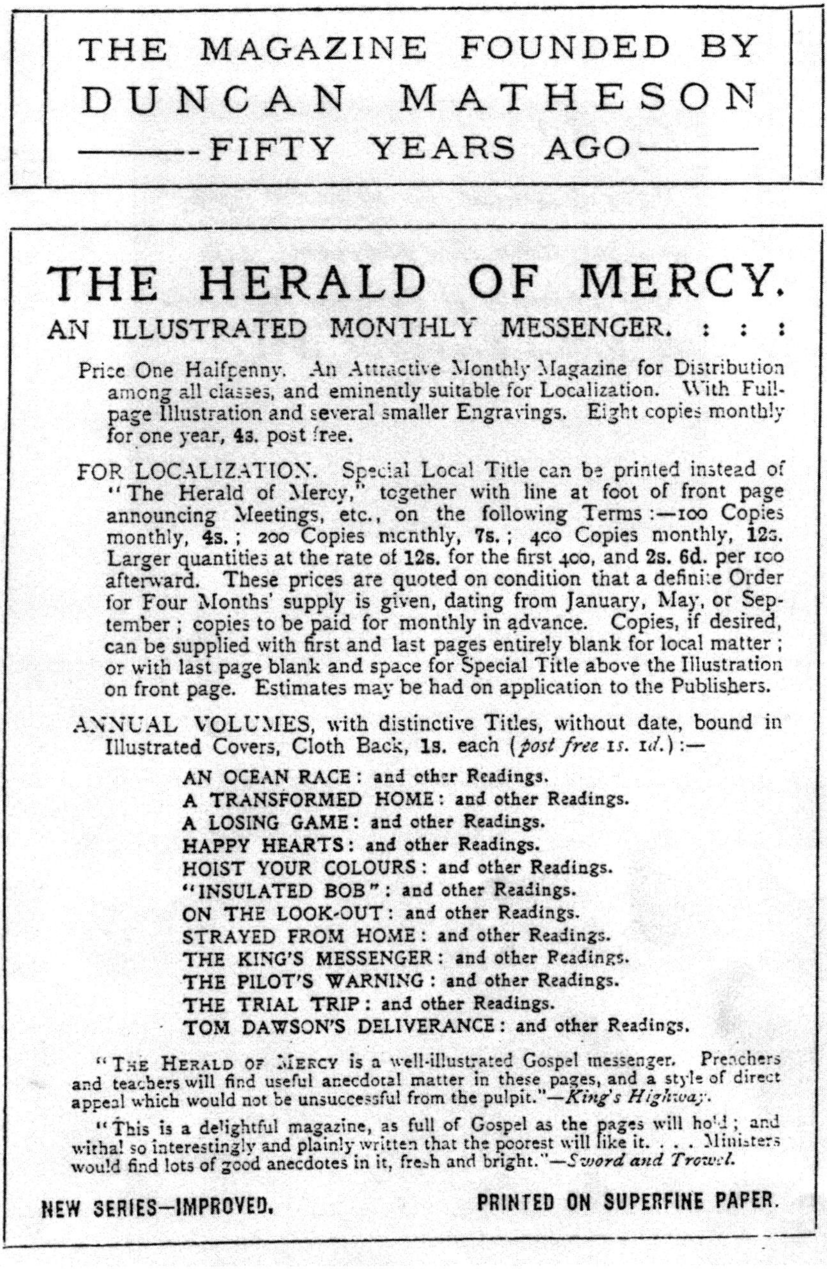

> THE MAGAZINE FOUNDED BY
> DUNCAN MATHESON
> ——FIFTY YEARS AGO——

THE HERALD OF MERCY.
AN ILLUSTRATED MONTHLY MESSENGER. : : :

Price One Halfpenny. An Attractive Monthly Magazine for Distribution among all classes, and eminently suitable for Localization. With Full-page Illustration and several smaller Engravings. Eight copies monthly for one year, 4s. post free.

FOR LOCALIZATION. Special Local Title can be printed instead of "The Herald of Mercy," together with line at foot of front page announcing Meetings, etc., on the following Terms :—100 Copies monthly, 4s.; 200 Copies monthly, 7s.; 400 Copies monthly, 12s. Larger quantities at the rate of 12s. for the first 400, and 2s. 6d. per 100 afterward. These prices are quoted on condition that a definite Order for Four Months' supply is given, dating from January, May, or September; copies to be paid for monthly in advance. Copies, if desired, can be supplied with first and last pages entirely blank for local matter; or with last page blank and space for Special Title above the Illustration on front page. Estimates may be had on application to the Publishers.

ANNUAL VOLUMES, with distinctive Titles, without date, bound in Illustrated Covers, Cloth Back, 1s. each (*post free* 1s. 1d.) :—

- AN OCEAN RACE : and other Readings.
- A TRANSFORMED HOME : and other Readings.
- A LOSING GAME : and other Readings.
- HAPPY HEARTS : and other Readings.
- HOIST YOUR COLOURS : and other Readings.
- "INSULATED BOB" : and other Readings.
- ON THE LOOK-OUT : and other Readings.
- STRAYED FROM HOME : and other Readings.
- THE KING'S MESSENGER : and other Readings.
- THE PILOT'S WARNING : and other Readings.
- THE TRIAL TRIP : and other Readings.
- TOM DAWSON'S DELIVERANCE : and other Readings.

"THE HERALD OF MERCY is a well-illustrated Gospel messenger. Preachers and teachers will find useful anecdotal matter in these pages, and a style of direct appeal which would not be unsuccessful from the pulpit."—*King's Highway*.

"This is a delightful magazine, as full of Gospel as the pages will hold; and withal so interestingly and plainly written that the poorest will like it. . . . Ministers would find lots of good anecdotes in it, fresh and bright."—*Sword and Trowel*.

NEW SERIES—IMPROVED.   **PRINTED ON SUPERFINE PAPER.**

Figure 18.1 Advertisement for *Herald of Mercy*, in Rev. John Macpherson, *Life and Labours of Duncan Matheson, the Scottish Evangelist*, 1871.

Figure 18.2 Thomas Chalmers. From Thomas Brown, *Annals of the Disruption*, 1890.

Sermons and Lectures was published in two volumes, the year after he had been deposed by the General Assembly of the Church of Scotland because he argued for a doctrine of universal pardon. In subsequent years he gained a widespread following for his statement of the doctrine of atonement and his emphasis on the fatherhood of God. When the Rev. Professor John Caird (1820–98) preached before Queen Victoria at Crathie Church a sermon on 'Religion in Common Life', the Queen demanded its publication. The sermon went through many editions and was widely translated. Arthur Stanley, Dean of Westminster, said it was the greatest single sermon of the century.[15] However, Caird's 'radical and revisionist books of *Scotch Sermons*' which 'demanded that the churches' first duty was to this world and in this world' enraged conservative ministers of the time but gained support from such as Donald MacLeod (1831–1916), editor of *Good Words* (1872–1905), who claimed that 'the spirit of Christianity is essentially socialistic' and argued for State intervention in matters of 'general social welfare'.[16]

It has been claimed that sermons were at that time the most popular form of reading material in Scotland, a claim perhaps difficult to sustain in view of all the reading matter by then available, but clearly there was a good market for some categories of sermons, as for many popular religious writers, in the nineteenth century and beyond. Professor James Iverach (1839–1922), who wrote widely on the challenge of science to religion, published two collections of sermons and addresses, and his *Life of Moses*, published in 1881, eventually sold 20,000 copies. Some sermons or addresses by the theologian Henry Drummond (1851–97) are still published, including his *The Greatest Thing in the World* (on 1 Corinthians 13). His *Natural Law in the Spiritual World* (1883) was a best-seller in its time, and *The Ascent of Man* from a lecture delivered in Boston had to be brought quickly into print because pirate copies were already being produced.[17]

Reviews of *The Herald of Mercy* suggest that preachers and teachers will find it useful, then we read that it is 'so interestingly and plausibly written that the poorest will like it'. Appeal to 'the poorest' was important, indeed a well-considered art, or craft. Thanks to the 1870 Education Act most servants could read. Between 1890 and 1904 no fewer than forty-eight new weekly or monthly magazines appeared specially aimed at the servant classes. The editor of *The Servants' Magazine* wrote:

> Servants are fond of reading, but it is of vast importance that what they read should be adapted to suit their real welfare, to render them useful in their station, more contented with the arrangements of a kind Providence which has placed them in it.[18]

Onward and Upward was the title of the magazine of the Onward and Upward Association founded at the end of the century by Lady Aberdeen 'for the material, mental and moral elevation of women'. Henry Drummond, whose sermon *The Greatest Thing in the World* was first preached in the chapel of Haddo House to the Aberdeen family and their servants, provided (in an article in *The Spectator*) guidelines for the writing of stories of an elevating nature for servant girls:

> Cultivated ladies of leisure will be delighted to find out the marvellous power exercised by a writer who likes girls and who likes books. Working girls want kind, cultured friends and simple tales in which they can find their own lives mirrored in a pure and friendly light. It is worth stooping if lives can be brightened and improved by showing little kindnesses and trivial tales. Small things lead to great: our fires of coal will burn solidly and brightly enough if only they can be lit with light rubbish.[19]

The serial story *Her Day of Service*, by Mrs Isabella Fyvie Mayo, author also of *Equal to the Occasion*, *At Any Cost* etc., is – and who would argue? – categorised as 'light rubbish' by James Drummond in *Onward and Upward*, his selection of material (1891–6), including the Mayo tale, from the said magazine. Mrs Mayo's heroine Margaret Ede is an embodiment of the belief that in a girl's innocence lies her safety:

> So wise in all she ought to know
> So ignorant of all besides.

The journalist W T Stead offered guidance to Lady Aberdeen not only on the production and distribution of *Onward and Upward* but also on content – avoid the topic of women's rights, he said, and items on sexual morality should be muted.

Figure 18.3 'There came an evening when he followed Margaret into the garden where she had chosen to sit at work. Mr Fraser stood beside her and his head was high among the blossom branches of the little green tree, so that his slightest motion sent the petals flying down over Margaret's dark grey dress'. Illustration in *Her Day of Service* by Mrs Isabella Fyvie Mayo in *Onward and Upward* magazine, 1891.

In his innovative and provocative study *Popular Literature in Victorian Scotland: Language, Fiction and the Press*,[20] William Donaldson devotes a chapter to Didactic Fiction in the Scottish Press. The serialisation of original fiction in newspapers, he says, appears to have been introduced into Scotland during the 1850s. Among early examples of such fiction were stories by David Pae, 'probably the most widely read author of fiction in Victorian Scotland', but little known as such because he invariably published anonymously. He was caught up in the liberal evangelicalism of his time and chose 'to become a novelist as the most effective way of expressing his ideas, just as he chose newspaper rather than book publication as the most effective means of spreading them'. His earliest tale, *George Sandford; or the Draper's Assistant* (1853), 'might strike the reader as a novel struggling to grow out of a tract and not quite managing'.

Pae's first full-length serial novel, *Jessie Melville; or the Double Sacrifice*, was a story about contemporary city life, a novel 'considerably at variance with the idea that fiction in Victorian Scotland concerned itself exclusively with an idealised rural past'. The novel, Donaldson says, 'at every step ... subverts Pae's attempt to transform it into a vehicle of popular morality'. And later, 'David Pae's career as a novelist spans that period during which a genuinely popular fiction came into being in Scotland free from external religious constraints. He had begun using fiction merely to subvert reader resistance to the message he wished to convey, but by the end of his career fictional values predominated ... Although precipitated, ironically, by evangelical liberals struggling to perpetuate a basically theocentric world view, this marked a decisive step in the secularisation of Scottish society'. In a later chapter, 'Swallowed Up in London: A view of the Kailyard with Some Conclusions and a Postscript', Donaldson writes:

> Within a few years of the popular novel in Scotland shaking itself free from external religious constraints, there was a determined attempt by a group of expatriates in England to recover the form for the purposes of evangelical propaganda.[21]

It may be argued that the Victorian novel in general was much concerned with moral and religious constraints – none more so than novels by George MacDonald (1824–1905) – which may suggest that the expatriates in question were not alone (we shall return to them). But there were magazines other than *Onward and Upward* for which the serialised novel or short story continued to be employed as a channel for enlightenment. Fiction was included in the monthly *Edinburgh Christian Magazine* (1849–59) edited by Norman MacLeod (1812–72), who then edited *Good Words* (1860–72) and was succeeded in that post until 1885 by his brother Donald, by which time fiction was less frowned upon than it had been by some churchfolk.

A noteworthy account, from a Free Church household, of reading matter for school and leisure purposes, from Shorter Catechism to magazines and Victorian novels, including appropriate Sunday reading, is to be found in *Children of the Manse, Growing Up in Victorian Aberdeenshire*. The account is by Alice Thiele Smith (1858–1943), one of eleven children of the studious minister and teacher William Pirie Smith, and sister of the theologian William Robertson Smith. Her account was first published in 2004.

Isabella G MacLean recalls family reading in a shoemaker's household in the last decade of the century:

> We never seemed to be ill-off for something to read. My father had a fairish collection of heavily bound books – Scott, Dickens, church history, bound volumes of *Good Words*, *The Quiver* and the *Sunday at Home* ... We got the *Christian Herald* and read it from cover to cover. There was always a good serial story, a sermon by C H Spurgeon or Dr Talmage.[22]

Of her first visit, in 1910, to her husband's home in Contin (his father was a mason) Isabella MacLean writes:

> In common with most Highland students of his day (he) had never read a novel ... There were very few books in Contin – William Law, Thomas Boston, Life and Sermons of Dr Kennedy of Dingwall, Sunday School prizes and the like ... The Contin folk maybe didn't read but they were in no doubt about what other people should not read. Novels were dangerous lies and no conscientious person would read one.[23]

There were learned periodicals – theological and philosophical – also those which served the purposes of worthy causes (such as temperance) or of denominations, such as the variously titled groups of Brethren in north-east Scotland. Some books served denominational causes, not least the bulky *Annals of the Disruption 1843*[24] for which in 1876 the Free Church authorised publication, which took place in 1884, then in a new edition in 1893:

> It is with no desire to provoke controversy that these Annals have been prepared, but rather in the full belief that if the facts as they actually took place were better known, the hostility of adversaries would be to a great extent disarmed, and the attachment of friends confirmed and strengthened. It is true that when men are describing sacrifices and suffering – their own or others' – there is a difficulty in avoiding a certain amount of feeling ... If, however, we are to have a truthful view of Disruption times, such circumstances cannot be wholly suppressed; and surely, they may now be spoken of all the more calmly and frankly when the keen feelings of former days have to so great an extent passed away.

Surely the most succinct record of Church divisions was produced by T & T Clark, Edinburgh publishers: *The Scottish Church 1590–1920, A Graphic Chart*,[25] compiled by Robert Adams, assistant city librarian at the Mitchell Library, Glasgow. Prominent within this vast chart (30in x 42in), alongside listed chronological Prominent Events in the Churches, we read:

> Dedicated to all true lovers of the Scottish Church in all its branches whose first desire is that a Christian Via-Media be earnestly sought for and found whereby all these divisions herein charted become One under the Great Head of the Church. The Compiler.

On the front of the card envelope for the 3s. 6d. paper version of the Chart, T & T Clark proclaim with noteworthy optimism:

> Enables anyone to understand at a glance the puzzling ramifications of Scottish Religious history. For use in the Minister's Study, the College Class-Room, the Sabbath-School and the Home. May also be had cloth-mounted, varnished, on roller for hanging price 12s. net.

A much smaller chart was to be provided in J H S Burleigh's *A Church History of Scotland* (1960) and then reproduced in *The Concise Scots Dictionary* (1985). In these the 1929 Union is recorded.

Many periodicals and books were identified with, or reported, or indeed provoked religious controversies. The last *Encyclopaedia Britannica* to be produced in Scotland (1875) carried the article 'Bible' by William Robertson Smith which eventually led to his being removed in 1878 from his professorial chair at the Free Church College in Aberdeen. Smith, a scholar of repute, eventually a Cambridge professor, dared in the article to suggest that the Bible, though a revelation from God to man, was not sound as history.

In the 1870s, other dominant issues in religious thinking were being explored by ordinary readers. The works of Hugh Miller (1802–56) were being re-issued – 'New Cheap Re-Issue. In Monthly Volumes, price five shillings each'. These included *The Testimony of the Rocks*, *The Old Red Sandstone*, *Footprints of the Creator* and others in which:

> Mr Miller has brought his subject to the point at which science in its onward progress now stands – Agassiz. From Preface to American Edition of the *Footprints*.

Donald J Withrington writes that Miller had:

> aimed to prove that there was no necessary opposition between the implications of recent advances in man's geological understanding and the Biblical account of the creation. For Miller what recent scientific research demonstrated was nothing more nor less than a hitherto unrevealed design in geological evolution, but that still depended, of course, on there being a Great Designer. *The Vestiges of Creation* provided in Scotland a kind of geological prelude, with very extensive public discussion, to the similar arguments based on evolutionary biology which were set off by Darwin's *Origin of Species* (1859).[26]

It has been said that Robert Chambers's *Vestiges* (1844) sold many more copies than did Darwin's *Origin of Species*. Miller had, from 1840 to 1856, edited the pro-Free Church newspaper *The Witness* whose circulation

Figure 18.4 Hugh Miller. From Thomas Brown, *Annals of the Disruption*, 1890.

at one point was greater than that of the rival *Scotsman*. Miller's writings, widely read long after his death, were greatly admired by William Morrice, the shoemaker father of Isabella G MacLean: he 'had read everything he (Miller) ever wrote and remembered it'.[27] Readers such as William Morrice were apparently well served by the introduction to geology which Miller, the lay scientist, offered. Discussion at a Miller bicentenary conference in 2002 concluded that:

> His objectivity and commitment to fact and truth were felt to be without doubt, so there was no question of his having bent his observations to fit religious beliefs.[28]

Hugh Miller's widow, Lydia, had oversight of most of the editing and reprinting of his works. At the same time she was building her own reputation as a writer (mostly under the name of Harriet Myrtle) of books for children. These included *Always Do Your Best*, and *Lizzie Lindsay, Home and its Pleasures: Simple Stories for Young People*, and *Cats and Dogs, Nature's Warriors and God's Workers: or Mrs Myrtle's Lessons in Natural History*. Of Lydia Miller's books, her biographer writes:

> they show Lydia's undoubted talent for teaching small children, and an understanding of how to keep their interest, as well as her eagerness to instil in them all the Christian virtues.

Of cats and dogs, Lydia herself wrote:

> All creatures manifest the eternal thought and the eternal will. Their physical conformity displays the unfailing harmony which is demonstrated in all the works of the Creator.[29]

Then there were the stories, mostly of a sterner sort, from the mission fields, none more popular than the works by and about David Livingstone (1830–73). In *David Livingstone, Mission and Empire*, Andrew Ross notes that John Murray, one of the most successful publishers in Britain, had written 'before it was known whether Livingstone would emerge safely out of the bush at Quilemane' offering 'to bear all the costs of publication himself, including the engraving of maps and drawings, and to give Livingstone two-thirds of the profits made by the book'. *Missionary Travels and Researches in South Africa* was to be 'an outstanding publishing success':

> Twelve thousand people paid pre-publication subscriptions for the first edition. A reprint of the same number of copies followed almost immediately and was sold out very quickly. Other reprints were to follow. As early as the spring of 1858, he was able to deposit with Coutts something over £9000 produced by royalties ... Livingstone gained an enviable degree of economic security.[30]

The diversity and profusion – and to a great extent the profitability – of religious writing in nineteenth-century Scotland, especially in the later decades, have no equal in times before or since. Norman MacLeod, addressing the letterpress printers of Glasgow in 1860, declared:

> I cannot imagine anything on the face of the earth so wonderful as printing or so wonderful as books ... I can state assuredly that the press is one of the greatest powers under God for the advancement of the civilisation of man.[31]

So it may have seemed, at the time. Churches and churchmen kept a lot of printers in work, and publishers and booksellers in profit, and much of what was printed must have brought assurance and inspiration and enlightenment to many. The Word, preached and printed, held sway. In print it would continue well into the twentieth century, in no small part thanks to one man.

In 1885 a Free Churchman resigned his charge in Kelso, because of ill health, and went to London to become editor of *The Expositor*. Thus began a career which may be seen as marking a climax, and in one sense a conclusion, to the era of the press as a significant power under God. William Robertson Nicoll (1851–1923) was also editor of *The British Weekly* from 1886 until he died, likewise editor of *The Bookman* from 1891, and *The Woman at Home* from 1893. He edited *The Expositor's Bible* and *The Expositor's Greek Testament*. He was the author of a large number of books and in *The British Weekly* he wrote a regular column under the name Claudius Clear. 'He made popular journalism literary and he made religious journalism interesting'. He was knighted in 1909 and made a Companion of Honour in 1921. On his

Figure 18.5 W R Nicoll. From *Under the Bay Tree*, 1934.

death *The Times* wrote, 'The mark that he has left on his time is undoubtedly real and deep'. In 1925 Sir George Adam Smith wrote of Nicoll:

> Poet and publisher, mystic and manager, trustful yet shrewd, affectionate yet unsparing, theologian, politician, and one of the foremost of our literary critics, he stands alone among his contemporaries.[32]

Twenty-six years later, on the centenary of Nicoll's birth, Professor G D Henderson wrote:

> He was a very obvious factor in the religious, political and literary controversies of the early twentieth century, and a far from negligible force during the anxieties of the First World War; but he belonged essentially to a period that is gone and to a serious-minded nonconformity, a passionate liberalism, an enthusiastic admiration for Tennyson, Emerson and Dickens of which little remains.[33]

Nicoll was the son of the Rev. Harry Nicoll, a founding member of the

Free Presbytery of Alford whose overriding passion was reading: he amassed a library of 17,000 books. The young Nicoll inherited his father's 'idolatry of the book'. His own writing output was prodigious. In his day and through all his writings Nicoll was relied upon by many as a provider of theological and literary advice, intellectual stimulus and 'good reading'. He was guide and often comforter to churchfolk as they endured the storms of Higher Criticism controversy or entered the maze surrounding the Historical Jesus. Those who had no real width of literary knowledge could gain from him an introduction both to the classics and to contemporary writers. The period may indeed be gone. Possibly his last published work was the reprint of *The Doctrine of the Holy Assembly*, a slight leaflet published privately in 1961 by his daughter Mildred in his memory (and in memory of his wife Catherine who died in 1960) in the year of the third Assembly of the World Council of Churches: *The Holy Assembly* was from a lecture on Christian Mysticism delivered by Nicoll in 1905 at the Glasgow Summer School of Theology.

It is unlikely that Nicoll's name will feature in scholarly histories as a literary critic of significance or indeed as a theologian – or mystic – of note. But his name has kept appearing and is most likely to survive because in 1895 J H Millar, in the *New Review*, coined the derisive nickname Kailyard for some of the fiction then being produced by what William Donaldson has called an expatriate group of writers, their milieu 'noticeably bourgeois', with Nicoll as their 'Godfather'.[34] Actually Millar blamed J M Barrie for inventing the Kailyard,[35] but the term stuck to the group, to the style, and eventually to Nicoll who, mainly through *The British Weekly* and Hodder and Stoughton, did so much to promote these writers, who included J M Barrie, Ian Maclaren (Dr John Watson), Samuel Rutherford Crockett and Annie S Swan. Ian Maclaren used as a motto to the first collection of Scottish rural tales, *Beside the Bonnie Brier Bush*, the lines from an old Scottish song:

> There grows a bonnie brier bush in our kailyard,
> And white are the blossoms on't in our kailyard.

Our concern here is not with literary criticism – or with theological evaluations. The Kailyard school belongs here partly because it is thought to have served 'the purposes of evangelical propaganda', partly because Nicoll was undoubtedly involved in it and also because the works of those writers were very much part of the package bought and read and enjoyed by the religious community of the day, a community large enough to contribute appreciably to the profits of authors and publishers. The books were without doubt immensely popular but have been denied the variably-used label of Popular Culture. The stories were popular with readers outside Scotland, many of them expatriate, and this has been seen as proof that the stories fed little more than a sentimental nostalgia. (Compare Charles Murray, like so many other Scots expatriate for nearly all his working life: his 'we min' on the mornin's lang syne' – nostalgic like so much else he wrote – did little to prejudice, though his dialect limited, his popularity.) The home readership for Kailyard was unquestionably large:

It was commonly supposed that every Friday morning all the housewives, when the family had departed for work and school, sat with their feet on the fender and read the new instalment of an Annie Swan story in the "People's Friend". I know it was the case with a great many and if they met afterwards on the stair or in the backgreen they would say, "Well what do you think of so and so today?"[36]

Many serials were the 'soaps' of their day. But some of the condemned products of the Kailyard were greatly superior to today's soaps. Crockett and Barrie have benefited in recent years from critical rehabilitation, Crockett by Islay Murray Donaldson,[37] and Barrie by R D S Jack, who notes that Thomas Hardy 'of all people' was able to call *A Window in Thrums* 'a faithful representation of reality'.[38]

Nicoll's judgement was of course fallible: he thought Barrie a genius but George Eliot something less. The bookishness of his childhood and background had not been much concerned with the appreciation of fiction. But Nicoll knew what would sell. No penniless Duncan Matheson he! He had no need of funding from the landed gentry; rather, he mixed easily with the highest in the land. In *Under the Bay Tree*, reminiscences published privately in 1934 by Nicoll's widow,[39] it is recorded that in 1921, suffering from failing health, he enjoyed a long holiday in Lumsden (at what had been his father's manse, and which became a family holiday home), then returned to a celebration of his 70th birthday. At a private dinner given by Sir Ernest and Lady Hodder-Williams, the guests included the Prime Minister (Lloyd George), John Buchan and Sir James Barrie. Many messages of congratulations were received, including letters from Thomas Hardy and Edmund Gosse.

While Nicoll capitalised on his own command of contemporary theology and Church affairs, he also knew how to cater for people whose childhood had been informed by Sunday school prizes in the category of Amy Le Feuvre's *Odd* ('a graceful and touching story, full of gospel teaching') or *Christie's Old Organ* by Mrs O F Walton, a product of the Religious Tract Society. Some of his readers had been brought up on the *Herald of Mercy*, some had been members of the Onward and Upward Association. Nicoll guided many adults to contemporary writing of quality: such adults, even if they enjoyed Kailyard stories, usually read more substantial works: some graduated tentatively, some eagerly, upwards and onwards from the Kailyard and suchlike. In *Blood and Fire* Roy Hattersley writes of William Booth:

> For years he believed that fiction only aroused temptation and he constantly warned his children against Charles Dickens. However, some time during the late 1860s he read and enjoyed *Jane Eyre*, *Les Miserables* and the collected works of James Fenimore Cooper. His interests expanded from Red Indians to the French Revolution.[40]

Like Booth, Nicoll had many things other than fiction on his mind and in his diary. His own writing was varied and extensive. In addition he administered and advised on the writing of many authors, some of them eminent in their day. Decades after Nicoll died, historians of communication would be writing about the medium and the message: one of them, Walter J Ong, explored the development of communication from one mode to another, from pre-literate oral thought, 'primary orality', to and through the written and printed traditions to the 'secondary orality' of manufactured sound in the twentieth century.[41] Nicoll died as secondary orality was being born. He communicated extensively through the medium of print (he was not much of a preacher; his large congregation – of readers – was not bound by the limits of orality). He is especially noteworthy and should be remembered as a manager of his medium, the printed word, and as a manager of innumerable messages transmitted via that medium. He had inherited from his father a passion for books, to which he added a talent for promoting books. He operated when the printed word still had a virtual monopoly so far as mass communication was concerned. The printed religious word, however, was already losing some of its grip. The Shorter Catechism was disappearing from classroom, Sunday school and home. The King James Bible was no longer sacrosanct: James Moffat's translations of the New (1913) and then the Old (1924) Testaments were under way, and the Word as folk had known it was seen to be fallible.

Nicoll died in 1923, the year when the first BBC station was opened in Scotland: in charge of this new, secondary orality was another son of a Free Church manse, John Reith.

NOTES

1. Murray, 1979, It wasna his wyte.
2. Burns, Robert. The Inventory.
3. MacLean, 1998, 75.
4. Webster, 1915, 15.
5. Webster, 1915, 75.
6. Small, 1982, 13.
7. Henderson, 1939, 53.
8. Henderson, 1939, 43.
9. Winchester, Simon. *The Meaning of Everything*, Oxford, 2003, 136.
10. Herman, 2001, 10.
11. *DSCHT*, 825–6.
12. Macpherson, 1898, 89.
13. *DSCHT*, 71–2.
14. *DSCHT*, 361–2.
15. Drummond, 1975, 297.
16. Withrington, D. A ferment of change: Aspirations, ideas and ideals in nineteenth–century Scotland. In Gifford, 1988, 54.
17. *DSCHT*, 257–8.
18. Drummond, 1983, 2.
19. Drummond, 1983, 5.
20. Donaldson, 1986, 72.
21. Donaldson, 1986, 147.
22. MacLean, 1998, 36.

23. MacLean, 1998, 88.
24. Brown, 1893, vi.
25. Adams, 1923.
26. Withrington, 1988, 55.
27. MacLean, 1998, 23.
28. Borley, 2002, 135.
29. Sutherland, 2002, 92.
30. Ross, 2002, 115.
31. Donaldson, 1986, 34.
32. *British Weekly*, 17 September 1925.
33. *Press & Journal*, 10 October 1951.
34. Donaldson, 1986, 145.
35. Donaldson, 1989, 72.
36. MacLean, 1998, 89.
37. Donaldson, 1989.
38. Jack, 1991, 23.
39. Nicoll, 1934, 289.
40. Hattersley, 1999, 172.
41. Ong, W J. *Orality and Literacy: The Technology of the Word*, 1982; *The Times* obituary, 26 August 2003.

BIBLIOGRAPHY

Adams, R. *The Scottish Church 1500–1920. A Graphic Chart*, Edinburgh, 1923.
Borley, L, ed. *Celebrating the Life and Times of Hugh Miller*, Aberdeen, 2002.
Brown, T. *Annals of the Disruption*, Edinburgh, 1893.
Burleigh, J H S. *A Church History of Scotland*, Oxford, 1960.
Cameron, N de S, ed. *Dictionary of Scottish Church History and Theology*, Edinburgh, 1993.
Darlow, T H. *William Robertson Nicoll, Life and Letters*, London, 1925.
Donaldson, I M. *The Life and Work of Samuel Rutherford Crockett*, Aberdeen, 1989.
Donaldson, W. *Popular Literature in Victorian Scotland*, Aberdeen, 1986.
Drummond, A L and Bulloch, James. *The Church in Victorian Scotland*, Edinburgh, 1975.
Drummond, J, ed. *Onward and Upward*, Aberdeen, 1983.
Gifford, D, ed. *The History of Scottish Literature, vol. 3: Nineteenth Century*, Aberdeen, 1988.
Hattersley, R. *Blood and Fire*, London, 1999.
Henderson, G D. *The Church of Scotland*, Edinburgh, 1939.
Herman, A. *The Scottish Enlightenment*, London, 2001.
Jack, R D S. *The Road to the Never Land*, Aberdeen, 1991.
MacLean, I G. *Your Father and I*, East Linton, 1998.
Macpherson, J. *Duncan Matheson, the Scottish Evangelist*, London, 1898.
Murray, C. *Hamewith, the Complete Poems*, Aberdeen, 1979.
Nicoll, C R. *Under the Bay Tree*, London, 1934.
Ross, A. *David Livingstone, Mission and Empire*, London, 2002.
Small, C. *The Printed Word, An Instrument of Popularity*, Aberdeen, 1982.
Smith, A T. *Children of the Manse, Growing Up in Victorian Aberdeenshire*, Edinburgh, 2004.
Sutherland, E. *Lydia, Wife of Hugh Miller of Cromarty*, East Linton, 2002.
Webster, A. *Theology in Scotland, Reviewed by a Heretic*, London, 1915.

19 Marvellous New Trumpets: The Media 1920s–2001

COLIN MACLEAN

In the year 2000 a thousand American journalists voted Gutenberg Man of the Millennium. Commenting on this, Richard Morrison argued in *The Times* that when Gutenberg printed 200 copies of his 42-line Bible in 1452 'he was unwittingly throwing a bomb into the heart of the medieval world'; that by being able to print his 95 Theses, Luther blew Christianity apart; further, that by 1500 the work of 2,000 printers in Europe ignited the torch of modern science, education and reason.[1]

Looking back on media progress in the twentieth century one may claim that in and around the 1920s a number of remarkable media developments took place which would impact forcefully upon all sectors of society. Did those developments, like Gutenberg, have explosive results? In some respects, they were more explosive because more immediate and more general. With Gutenberg there was an appreciable delay mechanism: there was still a need for people to learn, not merely to read in the vernacular but to read at all. Five centuries later there were still many millions of people in the world, quite a number in the western world, who could not read. In contrast, the new twentieth-century media of sound and vision offered instant access for all who could hear or see. By the 1930s access to the new media was available or being promised to virtually all of the western world. At the end of the century the unschooled illiterate of the third world who turned the handle of a mechanically powered radio could hear and appreciate song or story from a local radio station.

The terms *medium* and *mass media* – used first in relation to advertising – originated in the USA in the 1920s. The use of *broadcasting* to mean transmitting by radio or television also began in the 1920s.[2] By mid-century *medium* and *media* were being given wider frames of reference: in 1946 J S Huxley described 'the media of mass communication' as 'the somewhat cumbrous title (commonly abbreviated to Mass Media) proposed for agencies, such as the radio, the cinema and the popular press, which are capable of the mass dissemination of word or image ... Regarded from this angle, the mass media fall into the same general category as the libraries and museums ... that of servicing agencies for man's higher activities.'[3]

The medium of the printed word was to lose little of its impact in the twentieth century; indeed, with extended education, it gained much. In the 1920s, however, printed words lost much of their usefulness in the film industry when sub-titles were no longer required: 'talkies' and a fast-growing film industry arrived, and along with them electrical recording, the gramophone and the record industry. Film, record and fast-developing radio would all help to promote one another. The 1920s also brought the first demonstrations of television.

Up to the 1920s, music and talk in the home had, for the most part, been home-made. To what extent, then, did the noisy newcomers, the cluster of invasive media, merit the title of 'servicing agencies for man's higher activities'? No one would have been readier to make such a claim for broadcasting than John Reith, who introduced 'the wireless' to the British public. In his eyes no activity was higher than the religious.

PROTECTION

John Reith, son of a Scottish Free Church minister, in 1922 became the first general manager of the British Broadcasting Company (later Corporation, of which he was director-general from 1927 till 1938). The first BBC station in Scotland was opened in 1923. From the beginning, Reith wrote, 'the Christian religion and the Sabbath were given positions of privilege and protection in the broadcasting service'.[4] In 1933 Reith appointed the Rev. Melville Dinwiddie, minister of St Machar Cathedral, Aberdeen, as the first BBC regional director for Scotland: Dinwiddie continued in command in Scotland till 1957. In 1945 the Rev. Ronald Falconer, a minister at Coatdyke, was appointed to the first full-time religious broadcasting post in Scotland: till 1971 he was responsible for the direction of radio and then television religious programmes in Scotland.

The introduction of religion to British radio in the 1920s was tentative, partly because of denominational tensions, partly because the original Broadcasting Company had the commercial aim of attracting purchasers of receivers by providing pleasant home entertainment, 'and religion might have been regarded as an unnecessary extra'.[5] On Christmas Eve 1923 J A Mayo, rector of Whitechapel, gave an address and was followed at midnight on New Year by Dr Archibald Fleming, minister of St Columba's (Church of Scotland), London. Doubts were expressed about whether a full religious service would be suitable for radio. An advisory committee was formed: its members wrestled with questions of form, purpose, teaching, preaching, Bible versions, choice of hymns.

On 6 January 1924 a service led by Dick Sheppard was relayed from St Martin's in the Field, where he was vicar. Prominent preachers from all over Britain were invited to provide religious services. Some clergymen were critical of worship being listened to on earphones: later, with the use of loudspeakers, programmes could be listened to by groups 'with the family as the normal unit'.[6] Some thought that complete acts of worship on radio might be blasphemous. The suspicion was voiced that services would be received in an irreverent manner and might even be heard by persons in public houses with their hats on.[7]

For the first ten years of broadcasting in Britain there was no special staff for the planning and presentation of religious items. In 1922, 15 minutes each week were given to religion: ten years later there were four hours each week,[8] with an audience then of ten million licence-holders and their households. The initial opposition of Churches had turned into an eagerness 'to use this medium for the proclamation of the Gospel'. In the 1930s the broadcasting of secular items on a Sunday provoked criticism from the Lord's Day Observance Society and others. Dinwiddie wrote:

> As was to be expected, war with its national and personal tensions aroused greater interest in religion and gave radio a unique opportunity to meet this need ... One of the first innovations in wartime was an early morning item, a kind of spiritual exercise, coupled with a series of physical jerks, at 7.30 on weekdays.[9]

Memorable in the wartime story is the arrival on the religious broadcasting scene of its first celebrity, though the term was not then in vogue. On the new Forces Programme the radio padre was to be heard: he was Ronald Selby Wright (from the Canongate Kirk in Edinburgh), an army chaplain. He spoke to a sizable audience of serving personnel, their families and friends.

Figure 19.1 Reverend Dr Ronald Selby Wright. Courtesy of the Church of Scotland.

His talks continued after the war. The Forces Programme also introduced a Sunday half-hour of community hymn singing. Dinwiddie, however, argues that the most significant development of the war years was the presentation of a cycle of twelve plays by Dorothy L Sayers, *The Man Born to be King*, in which the voice of Jesus had to be impersonated – a radio 'first' (the actor was Robert Speaight). Blasphemy, said some critics.[10]

Dinwiddie recalls that towards the end of the war 'a more positive and vigorous proclamation of the Gospel was noticeable' in religious broadcasting. 'More emphasis was placed on evangelism and it was found that the united witness of the churches, so much a feature of these tragic years of conflict, made a more profound impression on the listening public'.[11]

By 1947 'the most difficult and fundamental policy decision was taken': the service of Holy Communion was broadcast – and widely welcomed. In 1949 the Report of the Broadcasting Committee under Lord Beveridge restated the BBC policy on religion: 'a positive attitude towards Christian values, to safeguard them and foster acceptance of them'. The acknowledged aim of the BBC in religious broadcasting, Dinwiddie wrote, 'was to make Britain a more Christian country'.[12] The concern was expressed that people might settle for listening at home and not going to church. But church attendance in the 1950s seemed to counter this.

In the first twenty years of religious broadcasting a marked control was exerted on religious views expressed and on those expressing them: broadcast services were to be provided only by 'churches and denominations in the main stream of the Christian tradition'. Rationalists, freethinkers, humanists and sceptics had minimal access to the microphone, even these (by the late 1930s) always with the Christian view also being expressed.[13] In wartime, pacificism was a problem – solved simply by the rule 'Pacifist views may not be broadcast',[14] which debarred some noted preachers, including George MacLeod, and philosophers, including John MacMurray.

Media attitudes changed with time and experience. At first, churchmen opposed or suspected religious broadcasts: eventually they competed for exposure, personal and general. The book trade and the press began by resenting broadcasting, often refusing coverage or publication: then they were to feed richly off many aspects of broadcasting. Dinwiddie reports a troublesome interview with the Scottish Football Association in the 1930s when the BBC sought SFA permission 'to relay even 30 minutes of an unnamed soccer match each Saturday'.[15]

Dinwiddie and Falconer worked in BBC partnership for twelve years (1945–57), critical years of recovery from World War II, years in which some broadcasters, along with some churchmen, identified a need and grasped the opportunity to engage broadcasting in evangelism. Falconer had strong backing from Dinwiddie, not least for the first Radio Mission which was launched in 1950, with the support of a number of denominations though not the Roman Catholics.

An impressive participant in the 1950 Mission was the Rev. Tom Allan. Dinwiddie and Falconer decided to promote another Mission in 1952 with

Allan as 'chief missioner'.[16] A conference was mounted from which grew the Tell Scotland Movement of Evangelism, with Allan as its national leader. Falconer claims that the Movement 'was to dominate the Churches' thinking and action for a decade',[17] the Churches being Scottish and still non-Catholic. Falconer's English colleagues watched 'with interest and some envy', though one member of the English Religious Advisory Committee, Canon (later Cardinal) Heenan, thought the idea of a radio mission preposterous. The Tell Scotland steering committee met in the BBC's Edinburgh quarters.

> Like a bombshell into the middle of our small-groups technique was thrown the explosion of Mass Evangelism through the person of Billy Graham and his All Scotland Crusade ... Most of us were against having anything to do with the All Scotland Crusade. But Tom Allan, our leader, had come to believe deeply in its efficacy ... in the end we closed our ranks and went with him.[18]

Tell Scotland invited Billy Graham. He agreed. When Graham had run a Wembley Crusade in 1954 BBC London's religious department had avoided him, leaving most coverage to News Division. Graham was back, now in Scotland in 1955, to be televised and broadcast on radio from the Kelvin Hall on Good Friday, with an audience in the hall of 20,000.

Graham's 1955 visit to Scotland drew vast crowds and lively coverage. He returned six years later, this time to muted reaction. There had been general agreement that Tell Scotland never recovered from the part played in it by Graham. Falconer had achieved much for ecumenism by bringing Churches together, only to find that Graham's highly publicised intrusion on the Scottish religious scene contributed to well-publicised division of opinion among churchmen.

This crisis coincided with a drop in Scottish Church membership, also with the period when more and more people in Scotland were acquiring television sets, years in which Falconer was still enthusiastically applying his conviction and skill to television, first with the strong support of Dinwiddie, then with that of Andrew Stewart, the next BBC Controller in Scotland. Stewart drew on the guidance of Sir Thomas Taylor, lawyer and principal of Aberdeen University, to assert that the Church of Scotland, the National Church by law established, was allowed parity in broadcasting alongside the Church of England.[19] In the early 1960s BBC Scotland was providing half of all BBC religious television on Scottish screens.

Falconer ensured a noteworthy provision on television from Scotland of broadcast church services, of General Assembly coverage, religious documentaries, films, discussions and talks. He brought to Scottish screens churchmen of world repute such as Martin Niemoller from Germany and Josef Hromadka from Czechoslovakia. The sports department – quite a change in twenty years – began to grumble when Falconer was given priority in use of outside broadcasting facilities.

Under Falconer a number of Scottish churchmen appeared regularly on television, among them Hugh Douglas, Campbell Maclean and Colin Day.

A striking impact was made by the series of radio and television lectures and also sermons by the Rev. Professor William Barclay, a man of gruff voice and direct approach whose broadcasts helped promote his numerous books. The New Testament volumes of his Daily Study Bible series, published in a number of languages, sold six million copies and over the years ensured the financial security of the Church of Scotland's St Andrew Press.[20]

Some churchmen did not need broadcasting to bring them fame, for instance George MacLeod, who broadcast often. Some, though preachers of national and international repute, did not translate well to television: one such was J S Stewart. Falconer kept driving hard but his self-assurance was not blinkered: BBC Scotland hosted the 1966 conference of the World Association for Christian Broadcasting, at the end of which the Very Rev. A C Craig was asked to comment. Falconer quotes Craig:

> I've been shown a marvellous new trumpet, the tones of which can penetrate to the ends of the earth. I've seen men handling its keys with great dexterity. Yet the sounds I've heard coming out of the trumpeters' mouths have been more like the incoherent twitterings before the conductor ascends his rostrum, than any Clarion-Call.[21]

Falconer reached the conclusion:

> that the programmes which had gone furthest and deepest with *the viewers* were those which had been unashamedly Christian in content, provided their truths had been presented in direct, understandable ways, through personalities of integrity and sincerity.[22]

This writer knew and sometimes worked with Falconer. He was a man of his time, a time in which concepts of confident authority still prevailed in some quarters, but were eventually challenged. Falconer responded to and stimulated the religious mood of the time, as did Professor Barclay. The Rev. James Dey, who succeeded Falconer, soon went south to head Religious Broadcasting in London: he was followed in 1974 by the Rev. Ian Mackenzie who was to work under and establish rapport with Alastair Hetherington, Controller of BBC Scotland from 1975 to 1980. These two believed there should be a sharp reduction of 'worship' programmes.[23] *Songs of Praise*, 'the great, incredible survivor of religious broadcasting', continued, and remains a regular Sunday evening feature.

Mackenzie devised new styles of presenting religion, from *We've Got a Hymn* (from a new housing area), and *Spirit of Scotland*, a series of lectures, to *Eighth Day*, which discussed Christian ethics and the implications for society, and a programme mostly shot at Inverness railway station:

> By meeting people coming off or getting on to trains, Paul Streather made what Mackenzie speaks of as a 'brilliant' film. He had simply got the travellers to talk about values, with freedom to say what they wanted. It was not 'clouded by theologising'.[24]

Kenneth Roy presented *The Yes, No, Don't Know Show*. Roy proved to have a pleasant way of moving up and down a studio audience, encouraging people to speak their minds as they wished, but not for too long.[25]

Mackenzie also produced *Coast to Coast* which linked areas of Scotland, apparently talking to each other: 'metaphor, debate, issues in communities and, not to be forgotten, "worship"'. Hetherington wrote:

> Ian Mackenzie was doing his utmost to bring a new and wider understanding of a changing world – changes that some of the more senior Church of Scotland people did not always recognise. He felt it his duty to promote debate and discussion, while at the same time Mackenzie himself was fully committed to his Church.[26]

Mackenzie admired the way Hetherington responded to complaints about some contributions to *Thought for the Day*, the short mid-news morning chat. He writes of Hetherington:

> … he took an intense interest in the content of our programmes, and every significant boat we pushed out seemed to receive his vigorous approval.[27]

Of *The Yes, No, Don't Know Show* Mackenzie wrote in 1998:

> The format is now *de rigueur*, but I believe we were the first in Britain to attempt it. Kenneth Roy leapt around the audience as Kilroy does, and an audience voted on ethical and theological questions. What would now be regarded as merely lively audience participation was then seen by theologians, church dignitaries, and some BBC traditionalists as unstructured chaos likely to lead to the end of civilisation as we know it.[28]

To critics, Hetherington and Mackenzie were able to show 'rocketing' audience figures. 'Are you saying all these people are wrong?', Hetherington asked.

The commercial station, Scottish Television, was founded in 1957. In 1962 Robert McPherson, a schoolteacher, became head of its educational and religious programmes. For twenty years from 1967 the head of religious programmes on STV was the Rev. Dr Nelson Gray, a Congregational minister. In those years STV made more religious programmes than any other independent television company. Dr Gray brought a vigorously evangelical talent to bear on the broadcast services, the documentaries and the discussions he produced, always encouraging new television talent, ministerial and otherwise. At Grampian Television, founded in 1961, James Buchan, head of programmes, was responsible for some impressive documentaries which were religious in tone.[29] Grampian, however, could claim some kind of record: its *Reflections*, a short epilogue chat, lasted nearly forty years: it followed the *News at Ten*, was then 'floated' to later slots, then floated out of existence.

ITV in England had the resources to tackle religious drama, notably

Jesus of Nazareth (1970s), also religious documentary, such as Melvyn Bragg's £1 million twenty-part *Two Thousand Years* in 1999 – some parts of which Scottish and Grampian displaced to 1.15am from its slot elsewhere of 10.45pm.

Independent Radio for Scotland began in 1974 with Radio Clyde. Radio Forth followed in 1975, and other regional companies in later years. Radio Clyde won a prize at a New York broadcasting festival in 1999 for its Sunday evening programme *Down to Earth*, presented by the Rev. Alan Sorensen, who said he tried not to sound like a minister. Dr Nelson Gray believed that radio and television had 'probably been the most effective agents of ecumenical development in Scotland'.[30]

By 1968 Reith had a gloomy view of the effectiveness of religious broadcasting. In a foreword to Dinwiddie's *Religion by Radio* he wrote that religion by radio had been ineffectual, not because of failure or inadequacy in what was broadcast – 'millions of pounds worth of advertising had been done for them (the churches) free' – but because of the churches' failure to follow up and capitalise on what was so freely and generously promoted in religious broadcasting. If only, he said, the Churches had moved in to encourage those whose interest had been revived by broadcast religion 'there might have been a national revival on a scale hitherto unimagined'.[31]

Dinwiddie had a calmer grasp of realities:

> Broadcasting is a scattering abroad over the whole field. It is there for the taking or leaving, and the individual at the receiving end can accept or reject what is heard or seen, can react passively or with appreciation or criticism, whether the speaker is a member of the Royal Family, a Prime Minister or a bishop, a shop steward or a soccer idol.[32]

Those at the receiving end have reacted more favourably to the soccer idol than to the shop steward. Royal family, prime minister and bishop (or minister or priest) have all been given some degree of privilege in British broadcasting. Only a head of religious broadcasting has been encouraged to go ahead and run a Radio Mission as Falconer did in the 1950s, but the next half-century put an end to that privilege.

The effectiveness of prime ministers or politicians in the media is not considered here. As for the royal family, they have given television the opportunity to present some most impressive occasions, none more impressive than the coronation ceremony in 1953, which was the first television programme seen by many Scots. Great national occasions cannot be conjured up frequently; the rarity value may count for much, as does the impact made possible by a magnificent religious service in a historic setting, complete with dignified procession and impressive music. The impact of such broadcasts on the viewer may relate to a sense of being part of a large community, infinitely larger than, but represented by, the company shown on the television screen.

Over the years, producers of religious television programmes, especially church services, have experimented with countless techniques of presentation. To begin with, many of them were overawed merely at the thought of pioneering the presentation of religious observance in any of its forms – prayer, preaching, communion etc. By 2001 ITV companies seemed to have settled on the simple, undemanding practice of watching all of a Sunday morning church service, the location changing each week.

The formulae now used in presenting the BBC's *Songs of Praise* vary considerably, often depending on the location and on how many interruptions to the hymns are created for purposes of solo performance or interview, usually with someone speaking of committed belief or action. Persistent close-ups of people who have been required to sing enthusiastically may for some viewers prevent, rather than enable, a sense of worship. Sometimes a cleric makes a brief appearance pronouncing a benediction, but the deity thus invoked is rendered remote by the obligatory presence of a – revered if not worshipped – celebrity. 'Very exciting to meet you!', said the lady of the house welcoming the celebrity interviewer (Sally Magnusson) to Abbotsford in a *Songs of Praise* in 2001.

Songs of Praise may survive because no one can think of an alternative, or possibly because the programme provides at least a token and unprovocative foothold for religion. It is said that people do not now whistle in the street as they used to: at least *Songs of Praise* keeps some whistleable tunes in circulation.

MEDIA ATTENTION

Even when the Christian Church enjoyed privilege and protection, it was also, along with other sectors of society, the object of media attention. By the end of the century such attention was determined by news departments, current affairs producers, chat show hosts, satirists, dramatists, all looking for and using whatever suited their varied purposes. The full force of what became known as the media, usually with reference to press and broadcasting, was not brought to bear until mid-century. BBC radio news had mattered a lot during World War II, but television news did not begin to have importance until the late 1950s.

At the end of the nineteenth century an admiring son wrote of his father's ministry in Aberdeenshire:

> The story of a parish minister's life is and should be rarely told in print. No profession is more likely to be devoid of incident than the ministry of a country parish ... I take it that he sought to make no mark upon his time save the stamp that the doing of duty leaves. His lifework is written on the characters of the people whom he influenced.[33]

A century later, perhaps we should say 'in the media' instead of 'in print'. During the century a number of Scottish churchmen were to make well-publicised marks upon their time. Some did so determinedly, some dramatically. More promising than the parish ministry as a springboard to media fame was a position of the kind provided by the Church of Scotland's Church and Nation Committee, established in 1919 with John White as its first convenor. Problems of Church unity and ecumenism would persist through the century. The challenge of relationship with the Roman Catholics was the most sensitive and serious, though it changed materially in character and context. Irish immigration alarmed White: like John Buchan he felt a need to protect the Protestantism of Scottishness. In a book prepared as a memorial of the great event of 1929, Buchan wrote:

> ... the haste to get rich had led masters to import and exploit cheap alien labour, Irish and Continental, and today one-eighth of the people of Scotland are Irish in race and Roman Catholic in creed – a proportion yearly increasing. From such a fermenting vat strange vapours might be looked for.

and

> There are many hard problems yet to be solved before we are in sight even of Protestant union, many more before there is hope of a united Christendom; but the impulse is there, and it is difficult to believe that it will weaken ...[34]

In the year 2000 the last sentence would have been uttered without media reaction: the one about a fermenting vat would have appealed to the media, who made much of the same topic in August 1999 when the composer James MacMillan gave a lecture on Scotland's Shame, arguing that Scotland was a land of 'sleepwalking bigotry' where 'visceral anti-Catholicism' disfigured the professions, academe, politics and the media, a view which some people may have thought to have been more relevant in 1929 than 1999. The media excitement over MacMillan resulted in a book *Scotland's Shame?*, noteworthy for the fact that nearly all of its twenty-two contributors were academics.[35] In the later decades of the century, academics devoted an appreciable amount of attention to religion and the Church in Scotland.

Writers of books are glad to be quoted, and publicised. Sometimes the person who has gained media attention is then better placed to win support for book publication, then to have further media attention for the book. This may be said to have happened in the 1990s in the case of Bishop Holloway, Episcopal bishop of Edinburgh, who voiced opinions on such controversial topics as gay rights, cannabis, promiscuity, abortion, the Virgin Birth, church establishment and the appointment of women within the Episcopal Church. He was the subject of a series of television programmes, *Holloway's Road*.

The case of the late Cardinal Thomas Winning was different. As head of the Catholic Church in Scotland he enunciated what were unarguably Catholic attitudes, and his opinions were usually sought and published, on such subjects as homosexuality, abortion, euthanasia and sex education. The last was the focus of prolonged and acrimonious argument in the media and the new Scottish Parliament, when the Scottish Executive sought to repeal the Section 28 rulings on the syllabus for sex education in schools.

Holding office for some years, also holding views which they expressed forcefully, these two men gained media fame of a kind not normally sustained by ministers of the Church of Scotland, where moderators and committee convenors have brief tenure of office and are reminded, though some forget, that they are primus only inter pares.

John White and George MacLeod achieved renown before television added to the volume of fame. MacLeod was identified not only with Iona – the abbey which he helped renovate, and the community which he founded – but also with pacifism and with debates on Africa in the 1950s. His books sold well. Other Church of Scotland ministers have gained occasional prominence over the years, and in Assembly debates several themes have gained recurring prominence – Church unity (Beaverbrook's then powerful *Scottish Daily Express* waged war on the Bishops Report in the late 1950s), nuclear weapons, decolonisation, the ordination of women, gambling, immigration and asylum seekers, homosexuality, extra-marital sex, and the government of Scotland. Through succeeding decades the claim was made that the General Assembly and especially the Church and Nation Day were the nearest Scotland came to having a parliament. For many years the committee argued and the Assembly agreed that Scotland should have its own parliament: some of the committee's convenors gave the impression of enjoying their political role – and the media limelight.

In 1929 Buchan had written with confidence that Scotland required a leader if she was to keep what she cherished in the older Scotland and adapt herself courageously to the demands of the new, 'and her natural leader, now as in the past, is that Church which is her most idiomatic possession'.[36] The devolution which had been sought by Church and Nation helped put an end to that, as did the growing media indifference to the Churches. Except when something was said that allowed scope for sensational treatment the Scottish media came to have little interest in Church affairs and certainly lost any belief in the Established Church as Scotland's natural leader. While the media expanded steadily in the course of the century, less prominence, in time or space, was devoted to Church matters. Coverage of the General Assembly decreased markedly. Technology made the old breed of expert sub-editor redundant: those in office at 121 George Street, Church of Scotland's headquarters, despaired of the Scottish media's ignorance of Church terminology or fact.

Undiminished media interest, however, was ensured for any hint of scandal within any Church group, large or small. Roderick Wright, Catholic bishop of Argyll and the Isles, disappeared in 1996, having eloped with a local married auxiliary nurse. News of this led a woman in Eastbourne to reveal that she was the mother of Wright's son. 'Runaway Roddie' may be said to hold a position of singular pre-eminence in the century, partly because by the time of his offence the media had grown in size and appetite.[37]

It may be argued that the media merely mirror, though indeed they mould, public appetite and opinion and – by omission – ignorance, an ignorance nowadays not only of Church affairs but also of the Bible. In the 1930s and 1940s poems by the popular Glasgow poet W D Cocker appeared first in the Glasgow-based newspapers *Daily Record* and *Evening News*. Cocker and the papers' editors were able to assume Bible knowledge in the readership. Adam and Eve, Elisha, Joseph, Pharaoh, Moses, Naaman, Ruth, Balaam, Abraham, Cain and Abel, Daniel, Job – Cocker retold all their stories in comic verse.[38] By December 1995 religious sensitivities could be set aside: the *Record* ran a £50,000 Christmas Advent Scratchcard competition which required no biblical knowledge on the part of contestants.

On 21 April 2000 the *Scottish Sun* on page 1 had an:

Easter Eggs-clusive
The greatest story ever told

Easter's here, folks – but a recent opinion poll discovered that half of us don't have a clue what it is all about.

So today we begin a three-part special on the trial, crucifixion and resurrection of Jesus Christ, told as the *Sun* might have reported it 2,000 years ago.

It's the greatest tale in history. So forget those chocolate eggs for a while – and turn to Page 46 for a story to Saviour.

At Christmas-time 2000 the newspapers, tabloid and broadsheet, gave front-page treatment to the christening in Dornoch Cathedral of Rocco, son of the pop singer Madonna: she married the father of her baby the following day. *The Daily Telegraph*'s first leader began:

A child in time.

Yesterday's front pages were dominated by pictures of a mother, a father and a babe in arms. We may be repelled by the global glitz that surrounds this celebrity christening and wedding. Yet the universal, primeval appeal of the image calls to mind another family, in another time and another place.

In the adjacent correspondence columns were letters arguing for and against a Christmas-free Christmas which had been recommended in a feature article.

So far as the printed word in the bookshop is concerned, the most striking development is related to the Bible itself. Pride of place in many bookshops early in the century was given to Bibles, predominantly Authorised Version, and the more prominently they were displayed the more likely it was that they were locked behind glass, not as an act of reverence but because they were said to be the books most often stolen. At the end of the century the general bookseller had difficulty in determining how many translations to stock – AV, RSV, Moffat, Phillips, Barclay, New English, Good News, the People's Bible and then the vernacular, from Lorimer's New Testament to other regional versions. An English teacher in a London school translated St Mark's Gospel into rhyming slang – 'Oi Mary, you are going to give birth to a currant bun (son)'. Surveys in 1998 indicated that 30 million people had a Bible in their home; more than 60 per cent had not read it in the previous year; 3 per cent read something from the Bible every day.[39]

The Scottish publisher Canongate had some success with a series of small paperbacks, Pocket Canons, first published in 1998, each with a well-known writer – P D James, Ruth Rendell, Alasdair Gray, A N Wilson etc. – introducing one book from the Bible (AV). The tone of the publicity surrounding the launch of the series gave, as did some of the introductions, the impression that the potential readership would be coming new to the Bible rather than sustaining an established interest or knowledge. At the beginning of the century, for most people in the UK, the Bible was the Authorised Version and for many the AV was the Word of God. It was as dominant as Shakespeare in the vocabulary and imagery and allusion of the written and the spoken word. By 2000, that part of Christian culture had disintegrated.

One is likely to find the term Spirituality rather than religion being given prominence in books on those shelves now marked 'Mind, Body and Spirit', a novel trinity beloved of many bookshops, with specifically biblical titles sometimes within, sometimes separate from, the category. No attempt is made here to survey the wide range of the century's published theological works. The popularity of William Barclay has been mentioned. Probably no theological work drew greater media attention than Bishop Robinson's *Honest to God* (1963), which scandalised religious conservatives and became a best-seller. The media are intrigued by controversy, as by scandal, in the Church, sometimes even bringing new life to controversy which some churchmen have been engaged in for many years. It was, after all, a Scottish academic, William Robertson Smith, who in the *Encyclopaedia Britannica* at the end of the nineteenth century challenged the historical authenticity of parts of the Bible; some tabloids might make a meal of such a story even today.

At the turn of the century, two categories of publication have been noteworthy for their contrasting presentation and content. The writing of some historians now stimulates, or is published to accompany, television programmes. Inevitably, it seems, the two media join to highlight or sharpen whatever conflict is to be found in the story of Scotland. Some aspects of Church history lend themselves to such a treatment. In contrast, the colourfully illustrated brochures and books provided specifically for consumption by tourists tend to soften or even ignore much of the inherent but complex and sensitive drama of Church history. Controversy is avoided in favour of the easily assimilable; emphasis is given to the picturesque, not least to the older cathedrals.

DISPLACEMENT

Religious broadcasting within and without a regime of privilege, and also the media attention granted to churchmen, as considered above, are important aspects of twentieth-century Church history which merit and reward careful study. These, however, have much less significance than the combined impact of the new twentieth-century media upon Churches and Church life in general. Here we return to the comparison with the Gutenberg bomb in the heart of the medieval world. The media disturbances in the twentieth century may be described as a powerful cluster of bombs, and whereas the Gutenberg explosion is said to have led to an enlivening of the Christian Church, twentieth-century media developments may well be seen as damaging. Their impact is one of attention distracted, of diversion and displacement.

If we accept J S Huxley's definition of mass media as agencies capable of the mass dissemination of word or image, then we must allow that for many centuries the Christian Church was itself a mass medium. In Scotland for many years it was given support and authentication by the education provided in schools and universities. In 1900 many Scottish schoolchildren were required to show some familiarity with the Shorter Catechism; for how many children or adults today is the question 'What is man's chief end?' remotely meaningful? Where the State provides compulsory education for a wide age range such education may properly be described as the medium with greatest mass. The Christian Church is no longer the mass medium it was: it no longer has back-up from the medium of education. Nor today is it supported by mainstream broadcasting, though serious, informed discussion about religion may sometimes be heard on BBC's Radio Four in the morning and late at night.

The increase in immigration from Africa and Asia, then the gradual disappearance from most schools of Christian religious education, or of Christian-oriented assemblies, or of Church-affiliated teachers, all of these have weakened links between school and Church. The award-winning television schools series *Stop, Look and Listen* presented in the late 1990s by Channel 4 for primary school children, introduced 'key RE strands': five world religions (Hinduism, Islam, Judaism, Buddhism and Sikhism) were described: 'support materials will ensure good coverage of basic issues in RE'. On television in 2001 those young viewers could see in the TV news bulletins 30 million people gathering in the waters of India 'to cleanse their sins ... It is faith alone that brings these people here', said the BBC's reporter.

Broadcasting has, of course, done much more than present world religions to viewers, young and old. It has in many respects been a servicing agency for man's higher activities. Along with other media it has helped to widen horizons – of the world, of space and time. Holiday travel, a new careers mobility and the rehousing of millions have done much to dislodge churches as recognised centres for family and community life. In 1990 Gordon Donaldson wrote that 'in the complexities of modern life churchgoing has far more competitors than it used to have ... Far from church attendance being in any way a matter of social conformity, it is now considered eccentric'.[40] Other agencies now provide what once the Church provided.

Radio and television, both available eventually on many channels for

24 hours a day, were together the major media innovation of the twentieth century; broadcasting, too, operated in alliance with other significant mass media. 'Talkies' have been described as the first modern mass media.[41] In the 1930s Aberdeen and Glasgow had remarkably high ratios of cinema seats per head of population: Aberdeen spent more on screen entertainments than any other city of comparable size in Britain.[42] The challenge to Churches created by the cinema was not fully recognised until cinemas were open on Sundays: by which time religion and the Christian story were no longer treated with the reverence which film producers had at first observed. By the 1970s the Monty Python team were producing *The Life of Brian*, followed by *The Meaning of Life*, films which adopted a comic, indeed mocking, approach to the Christian Church, an approach less controversial than that of Martin Scorsese in *The Last Temptation of Christ* in which the sexuality of Christ was explored. At a conference in Edinburgh in 1999 Dr Lynn Schofield Clark, of the University of Colorado's centre for mass media research, said that research showed that films such as *Terminator 2*, *Titanic* and *Star Wars* offered to teenagers deeper religious experiences than conventional Churches.[43] Films, along with broadcasting and recorded music, as developed in the vast record industry, have raised the eyes and the adulation of millions in worship of many idols – stars, personalities, celebrities. 'The Beatles are bigger than Jesus Christ', said John Lennon.

Christ, however, was allowed into the pop picture by some: the 1999 Royal Command Variety Performance, shown on television, ended with Cliff Richard and the entire company singing his top-selling rendering of The Lord's Prayer to the tune of Auld Lang Syne. The 2000 Royal Variety ended with the company singing Christmas Carols. A pop group, Slade, presented in the 1970s a song 'Merry Christmas' which is repeated year after year though not quite attaining the all-time popularity of 'I'm Dreaming of a White Christmas' as sung by Bing Crosby. On BBC's *Heaven and Earth Show* the comedian Jimmy Tarbuck described God as someone always 'perched on my shoulder, bringing me good luck'. Footballer Ally McCoist was reported in the *Scotsman*: 'I am not heavy into going to church or heavy into God. But it's great to know He's there.'[44] These may be seen merely as modern manifestations of a long-established though intellectually loose religiosity not easily distinguishable from the themes of remembered children's sermons or of mission-hall witness.

In some cases Vox Pop (or Vox Celebrity) has displaced – it has certainly clouded – the Word of God and of religious vocabulary, but it has done so in the company of many other agencies, including academics and royalty. We have been told that football has become Scotland's religion; also, we are told, the 'crucial insight' of a historian, Michael Burleigh, that Hitler's movement was not just a political party but a religion 'makes his book authentically new'.[45] What does that mean? Does *Star Wars* really offer a 'religious' experience? When Prince Charles suggests that he become Defender of Faith he loses much of what was a comparatively precise meaning, without saying what he is to defend Faith against. The question

'Do you believe in God?' is asked – and readily answered, and the answers respected – for surveys, academic projects, etc., this with no definition given to either belief or God. Where there is belief, or spirituality – this is the writer's impression – the focus is more on God than on Christ: it is seldom Christ-centred, or Bible(any Bible)-centred. All this the result, one may suggest, partly of political correctness, but mainly of widespread detachment from the 'eccentricity' of Church association, certainly from the stimulating demands imposed by a preaching–teaching ministry.

Today it is believed that Gutenberg helped to bring about a much-needed reformation in the Church. It may be too early to assume that the twentieth-century cluster of media bombs have been wholly damaging. And it may be something of a miracle that Churches have survived as they have in the face of media and other challenges. Some church services – dull in preaching, dreary in music – are still acceptable to their congregations.

Should Reith, Dinwiddie, Falconer and the rest ever have expected the new media to serve as convenient support rather than as disturbing challenge? That is what, in effect, Gutenberg was to the then protected and privileged Church. Falconer produced – latterly he fought for – worship programmes, assuming that radio and television could sustain a congregation of the air, this in a context of limited channel choice and of protection (a strong compound of both mirror and mould). But today's media in the UK are not mission media: truly healthy media should make mission difficult. The marvellous trumpets, as Dr Craig called them, may not after all be especially appropriate instruments for Church use. At the beginning of a new millennium we can only speculate about the potential usefulness or effect of new media technologies now being developed: 'virtual' religious experience may yet be welcomed – or deplored. Meantime Church people may best be advised simply to maintain what can best be mirrored. The example of the Aberdeenshire parish minister (above) is probably as good as any: it was, we should note, eventually recorded in print – the medium of his time.

A final note on new media technology, which has further extended application of the term *broadcasting* and has also made life easier for some church treasurers: over the past two centuries, large sums of money were expended on the sometimes competitive building of church spires, pointing dramatically and symbolically to God's Heaven. Now the mounting cost of maintaining scores of church spires across Scotland is substantially assisted by the rentals paid for the mobile-aerial masts discreetly constructed within the spires.

NOTES

1. *The Times*, 8 April 2000.
2. Ayto, John, ed. *Twentieth Century Words*, Oxford, 1999, 153.
3. *Oxford English Dictionary*, 2nd edn.
4. Dinwiddie, 1968, 8.
5. Dinwiddie, 1968, 18.
6. Dinwiddie, 1968, 22.

7. Dinwiddie, 1968, 20.
8. Dinwiddie, 1968, 24.
9. Dinwiddie, 1968, 26–7.
10. Dinwiddie, 1968, 28.
11. Dinwiddie, 1968, 28.
12. Dinwiddie, 1968, 47.
13. Dinwiddie, 1968, 84.
14. Dinwiddie, 1968, 42.
15. Dinwiddie, 1968, 131.
16. Falconer, 1978, 75.
17. Falconer, 1978, 75.
18. Falconer, 1978, 77–8.
19. Falconer, 1978, 125.
20. *DSCHT*, 63.
21. Falconer, 1978, 147.
22. Falconer, 1978, 147.
23. Hetherington, 1992, 57.
24. Hetherington, 1992, 59.
25. Hetherington, 1992, 59.
26. Hetherington, 1992, 60.
27. Roy, 1998, 96.
28. Roy, 1998, 96.
29. *DSCHT*, 703–4.
30. *DSCHT*, 704.
31. Dinwiddie, 1968, 9.
32. Dinwiddie, 1968, 111.
33. Davidson, J. *Old Aberdeenshire Ministers*, Aberdeen, 1895, 5.
34. Buchan and Smith, 1930, 235 and 242.
35. Devine, T M, ed. *Scotland's Shame? Bigotry and Sectarianism in Modern Scotland*, Edinburgh, 2000.
36. Buchan and Smith, 1930, 239.
37. Parris, M. *The Great Unfrocked: 2000 Years of Church Scandal*, London, 1999, 182–9.
38. Cocker, W D. *New Poems*, Glasgow, 1949, and *Random Rhymes and Ballads*, Glasgow, 1955.
39. *The Times*, 15 December 1998.
40. Donaldson, G. *The Faith of the Scots*, 1990, 139.
41. Sklar, R. *Movie-made America*, New York, 1994.
42. Thomson, M. *Silver Screen in the Silver City*, Aberdeen, 1988.
43. *Independent*, 24 July 1999.
44. *Scotsman*, 9 June 2001.
45. *The Times*, 13 June 2001, Niall Ferguson, historian, one of the judging panel for Samuel Johnson prize, commenting on Michael Burleigh's *The Third Reich. A New History*, London, 2000.

BIBLIOGRAPHY

Buchan, J and Smith, G A. *The Kirk in Scotland 1560–1929*, London, 1930.
Dinwiddie, M. *Religion by Radio*, London, 1968.
Falconer, R. *The Kilt Beneath My Cassock*, Edinburgh, 1978.
Hetherington, A. *Inside BBC Scotland 1975–1980: A Personal View*, Aberdeen, 1992.
Roy, K, ed. *Alastair Hetherington: A Man of His Word*, Irvine, 1998.

PART FOUR

•

The Religious Community

20 Ministers and Society in Scotland 1560–c1800

IAN WHYTE

In 1821 John Galt's *Annals of the Parish* was published. A 'kind of theoretical history'[1] rather than a novel, it traced the life and times of a rural parish in the western Lowlands from 1760 to 1810 through the eyes of its minister, Mr Micah Balwhidder, from his installation until his retirement. Mr Balwhidder is portrayed in gently comic but sympathetic terms as a simple, homely and modest man, strongly traditional and conservative, suspicious of change and increasingly out of step with the social and economic transformations which affected his flock. Educated at Glasgow University and three times married, he spent his entire career in the parish of Dalmailing. He was on fairly equal terms with the local gentry though always conscious that, as the heritors of the parish, they were responsible for paying his stipend, and maintaining the fabric of the church and manse. He was suitably deferential to the nobility but developed a good relationship with Lord Eglesham based on a mutual liking and respect which allowed Balwhidder to approach him for patronage to support improvements in the parish, such as better roads, and for helping to finance the education of poor scholars. The minister was not, however, above gently chiding Lord Eglesham for keeping a mistress. Balwhidder also seems to have been on good terms with the farmers and tradesmen in his parish, some of whom served as kirk session elders. He regularly visited the sick, helped the poor, and gave advice in cases of moral dilemmas. If not 'gifted with the power of a kirk-filling eloquence'[2] he was assiduous in his duties. When, after nearly 20 years in his parish, he was summoned to attend the General Assembly in Edinburgh, his period of absence, three weeks and five days, was longer than all his time out of the parish in the previous two decades.

Mr Balwhidder tells us little about his social origins though he was clearly not a wealthy man when he began his ministry in Dalmailing. However, he ended up quite comfortably off due largely to the industry of his second wife who managed the glebe as well as the household, making butter and cheese for sale and spinning wool. He and his family were able to live on the profits of the glebe and put his stipend into the bank so that he was able to set his son and daughter up with substantial portions.

Like many of Galt's characters in the *Annals*, Mr Balwhidder was based on people whom he had known in his youth in and around the burgh of Irvine. But to what extent did he epitomise the parish ministers of his day, the men who, with only one exception, contributed to Sir John Sinclair's *Statistical Account of Scotland*, and in what ways did the ministers of Balwhidder's day differ from those of earlier times?

Despite their central importance in Scottish society during the period from the Reformation to the late eighteenth century, and despite their being a particularly well-chronicled group, the social dimensions of the lives of Scottish ministers have received remarkably little attention. Church of Scotland ministers formed an organised, tight-knit professional élite with a common ethos.[3] However, while a tremendous amount has been written about fluctuations and factions within the Church, we know far less about how ministers interacted on a day-to-day basis with the inhabitants of the communities which they served. Biographies of prominent Scottish religious figures have tended to concentrate on their political activities and their theology rather than their social lives.[4] The social origins of ministers in the mid-seventeenth century have been studied by Makey[5] but there is less published information for earlier or later periods. Even the nature of the ministry as a career is far from clear. To what extent did ministers tend to stay in the parishes to which they were first called? How frequent was it for them to move through a succession of parishes and what influences governed their mobility? Was mobility a positive feature representing promotion within a definite career structure or was it sometimes a negative process separating difficult ministers from hostile congregations? Some research on these themes has been undertaken relating to the clergy of the Church of England: in what ways were Church of Scotland ministers comparable or different?

Ministers represented a new element in the structure of early modern Scottish society; they enjoyed good middle-rank incomes, which were boosted in the latter half of the reign of James VI and under Charles I. Due to the absence of pluralism there were far fewer inequalities in incomes among Church of Scotland ministers than occurred in the Church of England. As a new professional group in late sixteenth-century Scotland, ministers were supported by mutual discipline, and were better and more uniformly educated than the pre-Reformation priesthood, to an even greater degree than the clergy of the Church of England.[6] The common ethos and maintenance of shared standards and practices through the presbyteries and the General Assembly turned the reformed clergy into a tight professional élite; there were about 900 parishes in the early eighteenth century, compared with over 11,000 in England, and some 62 presbyteries.[7] Ministers could not get away with failing to perform their jobs adequately due to the tight control over their activities exerted by the presbyteries and they were not allowed to hire poorly-paid substitutes to act for them.[8] A further element of solidarity among ministers as a social group was the tendency of clerical families to intermarry and for son to follow father, often in the same parish.

THE DEVELOPMENT OF THE REFORMED KIRK

The new class of Protestant ministers did not, however, appear overnight following the Reformation of 1560. The Reformed Kirk made slow progress in its earliest years due partly to a crippling lack of finance.[9] The aim was to create a 'godly commonwealth', a partnership between the Church and the State in which the State governed in accordance with the Kirk, which in turn had the backing of the secular authorities in imposing moral discipline. The reformers focused on the preaching of the word, prayers in the vernacular and education in religion for everyone. The need was also seen for the new system to operate effectively by the inspection and monitoring of ministers. Education was also an important element in the Reformed Kirk's programme, with the aim of installing a schoolmaster in every parish to provide education for even the poorest. Provision of assistance for the poor was a further target.[10] This ambitious programme was, however, limited at first by lack of money. In practice the Reformation succeeded best where it changed least; in a conservative society, conscious of rank, privilege and tradition, it only made progress where it did not disturb the fabric of society too much. The early reformers succeeded by being flexible in adapting the organisation of the Church to circumstances and merging with society as they found it.[11] There was a good deal of continuity in personnel with the old Church: at least a quarter, in some areas a half, of the Catholic clergy came over to the Protestants.[12] Even so the Kirk remained desperately short of trained ministers in its early years.

Following the ideas of Andrew Melville, a more strict Calvinist who returned to Scotland from Geneva in 1574, the leaders of the Kirk determined to create a Church of dedicated professionals who, rather than reflecting society, would seek to transform it.[13] To this end it was essential that ministers should be graduates. This emphasised the important role of the universities and meant that for many years after 1560 a substantial proportion of parishes in many presbyteries lacked fully trained ministers and had to make do with exhorters and readers as stopgaps. By the early seventeenth century the ministry had become a graduate one, though a similar process was occurring at about the same time in the Church of England.[14]

MINISTERS' SOCIAL ORIGINS AND WEALTH

Ministers were relatively well off financially and were often upwardly mobile socially. In the mid-seventeenth century ministers whose social origins are known were drawn mainly from the ranks of landed proprietors, burgesses and, in particular, from the families of ministers.[15] Many ministers at this time were the younger sons of landowners with modest estates though often having connections with more elevated families; the younger sons of the nobility rarely became ministers. In 1648 27 per cent of ministers came from a clerical background; 6 per cent were the sons of burgesses; 13 per cent came from the families of landed proprietors; and 49 per cent came from unidentified origins. Makey has suggested that a significant proportion of these were probably drawn from the ranks of small feuars, tenant farmers and tradesmen.[16] The proportion of ministers drawn from the lower levels of society tended to fall over time though some ministers from humble backgrounds continued to be recorded, while the percentage of ministers whose fathers had themselves been ministers rose, with sons often following fathers not only in the same career but even in the same parish.[17] Thus, in the eighteenth century, the ministry became more of a closed profession and it must have been more difficult for men of modest origins to break into it.

From the start the aim of the Reformed Church was to pay ministers a reasonable stipend that would allow them to carry out their duties properly. Although less money was available than had been hoped, by the early seventeenth century ministers were sufficiently well off, in general, to form a significant new element of middle class wealth. The relative value of ministers' stipends fluctuated over time. Not particularly generous in the later sixteenth century, they nevertheless seem to have been an improvement on the incomes of most pre-Reformation parish clergy. The incomes of ministers came from two main sources; stipends payable by the landowners of the parish and the produce of their glebe lands. Glebes were seen as essential for the maintenance of ministers and their size and significance have received little attention. The First Book of Discipline wanted stipends to be from 100 to 300 merks (£66.13.4d – £200 Scots) a year, a level of remuneration which was impractical at first. John Knox, as Edinburgh's first Protestant minister, was promised £400 Scots but actually received less than half this amount. The authorities in the larger towns competed to attract notable preachers, offering stipends that in the event they often could not manage to pay. In 1617 a sufficient stipend was defined as five chalders of victual or 500 merks (£333 Scots).[18] By the 1620s the increase in stipends was running behind inflation. Under Charles I ministers did well with rises ahead of inflation. Minimum stipends were raised to eight chalders of victual or 800 merks (£533 Scots). At this time most stipends included a substantial proportion of grain which helped to shield them from the effects of inflation.[19] As a result many ministers by the mid-seventeenth century had incomes comparable to, or greater than, small estate owners.[20] They were some 12–15 times as wealthy as the tenants and cottars who made up the bulk of their congregations.[21] It has been suggested that the real value of stipends fell during the eighteenth century with rising costs of living,[22] though when stipends were still being paid partly in kind this would have acted as a hedge against inflation. Careful cultivation of glebes, giving produce not merely for direct consumption by the minister's household but also for sale, would have benefited many ministers in the later eighteenth century with rising food prices. Ministers could also augment their incomes by charging annual rents, or interest, on loans. By the later seventeenth century they were becoming a significant source of credit in Scottish society.[23] In addition some ministers enjoyed incomes from landed property, either inherited through their families and their wives, or bought as an investment. Many wives who came from landed backgrounds were provided with substantial dowries.

An important point is that average levels of stipends do not seem to have varied too greatly between different regions. In the mid-seventeenth century the average stipend for Scotland as a whole was around £674 Scots. Stipends were slightly higher in some Lowland synods while north of the Tay the average was £593, but while there was a north/south gradient it does not seem to have been an extreme one. In the Lowlands south of the Tay stipends in more remote upland parishes – such as the presbytery of Biggar – were often as high as in rural parishes close to major towns.[24] Despite their substantial incomes, ministers, in the seventeenth century at any rate, often seem to have lived modestly, more on a par with tenants than with the lairds. As a result they were often able to build up substantial reserves of cash.[25]

THE MINISTRY AS A CAREER

In the seventeenth and early eighteenth centuries the career path for entry to the ministry was fairly clear for a young man with a university education although success was far from being guaranteed. After graduating with an MA from a Scottish university, postgraduate study of divinity led to an examination by a presbytery assessing a candidate's suitability for the ministry. The system of patronage which operated for much of the period must inevitably have favoured the local recruitment of ministers as did the growing tendency for son to follow father on the same parish but the relationship between ministers' origins, the university that they attended and the location of their first parish requires further research. It is clear, however, that links through family and patronage were often critical in the selection of candidates. When Alexander Carlyle was put forward for a vacancy in the parish of Inveresk, his father, himself a minister, used his friendship with a law lord to have his son recommended to the Duke of Buccleuch, the patron of the living.[26] The practice in some synods of requiring the ministers in each presbytery to make regular donations towards maintaining divinity students would also have tended to favour local men.[27] Once a would-be minister had been granted a licence he might then serve as a temporary stand-in for various ministers within a presbytery, or he could act as an assistant on a more permanent basis to the minister of a particular parish, particularly if he was becoming old or infirm. Ideally he would serve as apprentice to his father, though such successions did not operate without the approval of the congregations concerned.[28]

Alternatively, would-be ministers waiting for vacant parishes might take up posts as regents in universities, serve as chaplains-cum-tutors in landed families, or work as humble parish schoolmasters. Private chaplains to landowners were sometimes older probationers who had failed to get a parish.[29] With the expansion of the army in the eighteenth century, posts as regimental chaplains provided further opportunities to gain experience. Even after a man was licensed it might take some time for a suitable parish to fall vacant. Until patronage was abolished in 1649 the process of admitting a minister to his charge began with an agreement between the patron and a suitable candidate. The agreement of the presbytery and kirk session, and the consent of the congregation, was also required. It was normal for a minister to preach a trial sermon to his intended congregation so that they could assess his ability and in the Highlands a test of a minister's ability to speak and preach in Gaelic would also be required.[30] Doubtless there were variations in the mechanisms used between parishes and over time.[31] Following 1649, presbyteries were given more of a say in the selection of ministers but elders and congregations were also consulted. Between 1662 and 1688 the induction of ministers was by presbytery and bishop, after 1690 by presbytery alone.[32] The whole process of training a minister might span six or seven years following graduation.[33] As a result ministers commonly entered their first parish in their late twenties or early thirties, and most were unlikely to consider marriage until they had this security. There was, however, no guarantee of a parish at the end of all the training. An unknown proportion of hopeful young men must have ended up as 'stickit ministers', continuing as schoolmasters, sometimes too poor to marry, like Scott's prodigious caricature Dominie Sampson in *Guy Mannering*. Once installed in a parish, however, the job was usually one for life. Ministers sometimes retired if they were incapacitated by old age or infirmity but it was much more common for them to continue their duties, perhaps helped by an assistant, until they died.

What were the personal qualities that made a successful minister? The ability to preach effectively was probably the one most often cited in the brief character sketches of ministers in Hew Scott's nineteenth-century compendium of data on Church of Scotland personnel, *Fasti Ecclesiae Scoticanae*.[34] In the seventeenth and early eighteenth centuries, when many ordinary people were extremely pious, the standards by which they judged ministers were probably very high. The duties of ministers were seen as primarily preaching and the administration of the sacraments. The qualities that were most valued in preaching were fluency, audibility, and the ability to compose well-structured sermons on a wide variety of subjects. Showy preachers who aimed for superficial popularity were not favoured.[35] Some ministers, otherwise acceptable, were deemed unsatisfactory because of their weak voices which, in some cases, might have been adequate for a small parish kirk but not for a large burgh one.[36] The Sunday services were something of a test of stamina as well as intellectual ability. The morning service usually began at 10am with the singing of a psalm, followed by a lecture from the minister on a passage from the Scriptures, prayers and then a sermon, after which came prayers and more psalms. The same order was followed in the afternoon only without the lecture. It was also usual to hold a midweek service, often on market day. The composition of three sermons and a lecture each week must have occupied a good deal of time for most ministers as the sermons were very long, and could not be read directly from notes, a practice which only began to become acceptable during the later eighteenth century.[37] Ministers' ingenuity must have been further stretched by the limited range of conventional themes used in sermons and by the custom of preparing discourses from the same passage of Scripture for several consecutive weeks. To preach sermons of this length, often highly charged with emotion, could be a physically draining task; John Forbes of Abercorn, an able and zealous preacher, 'delivered himself with so much vehemency that he was obliged to change his linen after every sermon from perspiration'.[38] Services focused on the reading and expounding of the Bible because even by 1700 the bulk of ordinary households did not have Bibles and this was the main way in which knowledge of its contents was imparted. Communion was held only once a year in many parishes, sometime only once every two or three years, with several parishes joining together.[39] Sermons preached at such times by popular preachers might attract visitors from a wider area

In terms of their parochial duties ministers were required to visit each family a certain number of times a year in order to catechise all the 'examinable persons' of 12 years of age and upwards, including servants. In large upland parishes this might involve a considerable amount of local travelling. Ministers also presided over meetings of their kirk sessions, normally weekly.[40] Presbytery meetings, which ministers were expected to attend unless they had a good excuse, were often held once a week during the summer months and once or twice a month during the winter[41] though less frequently in more remote parts of the Highlands.[42]

Once installed in their first parishes, many ministers, like Mr Balwhidder, remained there for their entire careers. Because stipends, though not completely equal, did not vary very much from one parish to another there was relatively little financial advantage in moving unless it was to one of the larger burghs where higher stipends were offered to notable preachers.[43] Other avenues for promotion included posts as university regents and chairs in divinity, which were often held in conjunction with the ministry of an urban parish. Promotion to a bishopric was also possible during periods of episcopacy. In the early seventeenth century many of the men presented to bishoprics under James VI were from the families of lairds or even more modest origins.[44] In other cases ministers were moved for less positive reasons; because they had fallen out with their congregations for instance. On the other hand, Alexander Hamilton, minister of Dalmeny, was moved by his bishop in 1677 to a parish more distant from Edinburgh because his sermons were too popular, attracting many people out from the city to hear him.[45] When moves were made they usually seem to have been within a particular synod; moves from one end of the country to another seem to have been exceptional.[46]

Some indication of career patterns within the ministry can be gained from a study of the average length of time spent by them in particular parishes. Table 1 shows the contrasting experience, and variations over time, for urban parishes in Edinburgh, and rural parishes in presbyteries in the Lothians, Biggar and Peebles, extracted from Scott's *Fasti*. The high figures for the later sixteenth century can be discounted because data are incomplete and ministers who stayed in parishes for only short periods are more likely to have been missed from the records. The shorter length of stay in the early seventeenth century compared with the eighteenth may have been due in part to a lower average life expectancy in the earlier period, but was also due to the need to move ministers to where they were most urgently needed at a time when many parishes still lacked incumbents. The marked drop in average lengths of stay in the later seventeenth century was due overwhelmingly to expulsions during the civil wars, after 1662, during the early 1680s when a number of ministers were ejected from their parishes for refusing to take the Test, and following the Glorious Revolution of 1688. From the 1690s the Church settled down and it became much more common for ministers to spend thirty, forty or even fifty years in a single parish. Throughout the period the shorter average length of time spent in urban parishes may reflect in part less healthy urban mortality rates but is also likely to have been influenced by the attraction of older, more experienced preachers to city parishes.

Table 1. Average length of time spent by ministers in particular parishes

	Edinburgh Parishes	*Rural Parishes in Lothians & Presbyteries of Biggar & Peebles*
1560–79	20.7 years	34.5 years
1580–99	8.2	17.4
1600–19	10.7	18.0
1629–39	6.6	17.1
1649–59	8.8	14.2
1660–79	8.5	9.5
1680–99	6.2	10.6
1700–19	19.6	15.9
1720–39	13.8	29.9
1740–59	16.2	21.3
1760–79	16.1	23.1
1780–99	14.9	23.8

THE POLITICAL CONTEXT

The continuity of ministers within their parishes, and their relations with their congregations, was disturbed before the eighteenth century by a political situation which was often difficult and which led to periods of major disruption within the Kirk. Such upheavals, in particular the alternation of Episcopal and Presbyterian rule in the seventeenth century, meant that there cannot have been many ministers whose careers were not affected at some point by difficult ethical and political choices. Many ministers who made the wrong choice were driven out and ended their days in poverty. As a result there must have been less continuity of personnel and more upset within the Church of Scotland than in the Church of England over the same period.

During the latter part of the reign of James VI a number of ministers were imprisoned, placed under house arrest or banished for challenging royal authority on ecclesiastical matters. The outbreak of the Covenanting wars and the Great Rebellion in 1638 saw some ministers driven out for adhering to the new Prayer Book. With the restoration of episcopacy in 1662 there was widespread removal of Presbyterian ministers, while at the Revolution of 1688, and in the years immediately following, some 600 Episcopalian ministers, or about two thirds of the total, were 'outed' with varying degrees of violence, while others, foreseeing what was likely to happen to them, fled before they were ejected. Of the Presbyterian ministers who had been ejected from their parishes in 1662 when Episcopacy was established, about sixty old men, the 'diluvians' or survivors of those who had been ejected after 1662, were restored.[47] The General Assembly which met in 1690 was a thin gathering and it took many years to fill all the vacant parishes though some 300 Episcopalian incumbents were left undisturbed in their parishes, especially in the north east, where attempts to install Presbyterian ministers were met in some parishes by violent mobs.[48] Concessionary legislation in the 1690s allowed some Episcopalian ministers to return to the Church.[49] The major upheavals of 1662 and 1688 especially meant that many men were drafted into the ministry without adequate training and of a poorer quality than would otherwise have been accepted, though Mitchison suggests that in 1662 the vacancies were easily filled with young divinity graduates already in training.[50] Ferguson on the other hand believes that it took some thirty years for the Kirk to recover from the self-inflicted wounds of the Revolution.[51] Such a situation can hardly have instilled confidence in their congregations. Problems continued during the eighteenth century, though on a reduced scale. The re-introduction of patronage in 1712 led to a number of violent scenes when landowners attempted to install ministers without the approval of congregations. By the third quarter of the eighteenth century secession from the Church of Scotland was beginning to occur on a significant scale.

MINISTERS' FAMILIES AND HOUSEHOLDS

For most Scottish men in the seventeenth and eighteenth centuries choice of a marriage partner depended not solely on affection but also on the need to choose a wife who would be a competent household manager and helpmate. For a minister it was presumably of considerable importance to choose a woman of good moral character who would support him in his parish activities by visiting the sick and doing charitable work. Contemporary sources emphasise piety, meekness and frugality as the qualities ministers valued most in their wives. Bequests to the poor made by ministers' widows suggest their active involvement in helping the needy. In the larger towns it was easier for ministers' wives to maintain some independence as dressmakers, milliners and shopkeepers.[52]

A sample of 2,471 marriages in 25 presbyteries[53] shows that the percentage of ministers' wives whose origins are known and who came from the families of ministers rose from 25 per cent in the late sixteenth century to 40 per cent in the late eighteenth century. The proportion from landed backgrounds stood as high as 47 per cent in the mid-seventeenth century when the Church, during the Covenanting era, was especially influential, but was more usually between 30 per cent and 40 per cent. However, the social origins of many ministers' wives is not given in Scott's *Fasti*, and it is likely that many of them came from among the tenantry, a suggestion reinforced by cases where ministers were charged with antenuptial fornication with their household servants.[54]

Ministers, as we have seen, tended to marry relatively late. In a sample studied by Whyte and Whyte[55] the average age of first marriage for ministers during the period 1560–1649 was 31.0, rising to 31.7 for the last quarter of the seventeenth century and levelling out at 34.5 for the eighteenth century. The median figure was more stable at 32 for the seventeenth century and 33 for the eighteenth. This relatively late age at marriage can be explained by the long and rigorous training which candidates for the ministry had to undergo. Ministers generally delayed marriage until they had been entered into their first parishes. 16 per cent of a sample of ministers married in the same year that they entered their first parish, 70 per cent of them within five years but only 3 per cent before they obtained a parish. It is likely that many ministers must have been engaged to their brides-to-be for a number of years, postponing marriage until financial security was confirmed. The average age of ministers' wives at marriage during the eighteenth century was 26–27.

Courtship was initiated by family links and by proximity in the same community. The General Assembly in Edinburgh also acted as an important marriage market where ministers could meet the daughters of their more senior colleagues, and girls from landowning backgrounds – for Edinburgh in the eighteenth century was a major marriage market for Scotland's landed and professional classes generally.[56]

The work of running a ministers' household and cultivating a glebe was accomplished with the aid of servants. One male and two female servants was a common establishment in the households of ministers though they sometimes had even more.[57] At home ministers were expected to conduct morning and evening prayers and reading of scripture, and to catechise family members and servants.[58]

MINISTERS AND THEIR CONGREGATIONS

It was the relationship of a minister to his flock, and his acceptance by the Church as a whole, which gave him his status,[59] but the nature of this relationship has received little attention. The Scottish clergy had a high concept of the importance of a minister's office, as the man through whom God communicated with the congregation. Despite the swings between Episcopacy and Presbyterianism there was remarkably little variation in dogma and little tendency towards schism in the Church of Scotland before the mid-eighteenth century. Ministers presided over kirk sessions, groups of elders chosen from among the more respectable men of the parish, often lairds, feuars and larger tenants.[60] Kirk sessions ensured moral conformity within their parishes and regular meetings helped to ensure that ministers and their more prominent supporters were in accord.[61] Of course in the event of any unsuitable behaviour by a minister his session could send an unfavourable report to the presbytery.

Pastoral work was probably a more important function of ministers than their pre-Reformation equivalents, as was the case in England.[62] Such activities, including visiting the sick, was seen as an area in which ministers might prove to be dedicated but which was of lesser importance than the ability to preach. John Erskine (Greyfriars 1767–1804) took great delight in his parish ministry and visiting the humblest people in their homes.[63]

Most ministers are likely to have been deferential to the nobility and other major landowners within their parishes, especially where they were patrons whose influence had led to their installation. They were probably on a more equal footing with the lairds and feuars, a group with which they often had close connections through birth and marriage. Smaller tenants, tradesmen and cottars would have been distinctly lower in status than their ministers. To what extent ministers were able to act as mediators between landlords and tenant, master and servant, bridging some of the gulfs that existed within Scottish society, is not clear but would repay further research. Another important part of parish ministry was care for the deserving poor, the aged, infirm, sick and disabled who were unable to work but not able-bodied beggars and vagrants who were seen, in the sixteenth and seventeenth centuries, as a social menace.[64]

Presbyteries were important institutions in ensuring uniformity of practice by ministers at a regional level; in the seventeenth century presbyteries might meet monthly or even weekly in summer, less often in winter, though in remote Highland areas ministers might be excused from attending presbytery meetings unless in summer due to the difficulties of travel.[65] Attendance was compulsory and ministers were fined for failing to appear if their excuses were unconvincing. Presbyteries regulated the work of their ministers by regularly examining the records of their sessions.[66] Kirk sessions also ensured conformity, for aggrieved elders could report to the presbytery any shortcomings of conduct by their ministers.

Ministers also had some responsibility for overseeing education in their parishes. The Reformed Kirk laid great stress on education and the period between 1633 and 1650, when the clergy were especially influential socially and politically, was the most important one for the establishment of a network of parish schools. Nevertheless, ministers as a group did little to secure adequate incomes for parish schoolmasters who remained poorly paid throughout the seventeenth and eighteenth centuries.[67] Schoolmasters were, in fact, recruited mainly from the ranks of those who had failed to enter the ministry. Ministers might take on the role of schoolmaster,[68] but this tended to be frowned on by their presbyteries.

The attitudes of ministers towards their flocks are sometimes displayed in the legacies which they left to them. It was relatively rare for ministers to leave large sums of money for the poor, or for the education of poor scholars. The ties of kinship tended to keep the bulk of the assets of most ministers within their families. However, in some cases substantial legacies were left, sometimes simply to help the poor in their parishes but also frequently more specifically to fund the education of poor scholars. In some cases at least this may reflect the early struggles of ministers from less privileged backgrounds.[69] William Struther, minister at the High Kirk in Edinburgh, who died in 1633, left 6,000 merks (£4,000 Scots) to the universities of Edinburgh and Glasgow to finance two bursaries at each institution to enable poor students to study theology. He also made bequests to the poor and to Edinburgh's Trinity Hospital.[70] His contemporary, Walter Balanqual, left 1,000 merks (£666.13s. 4d. Scots) to maintain a chair in divinity at the University of Edinburgh.[71]

With regular attendance at presbytery meetings, and occasional trips further afield to the General Assembly, ministers probably travelled more widely outside as well as within their parishes than many of their congregation, though they were not expected to absent themselves from their parishes for significant periods without permission from their presbyteries and the provision of a temporary substitute. Travel, even locally, could be arduous and not only in remote upland parishes. Several ministers were drowned as a result of crossing swollen streams, like John Spark of Currie, who drowned in the Water of Leith in 1739 returning from a presbytery meeting,[72] while James Anderson, minister of Falkirk, witnessed his son drown in the River Carron and narrowly escaped the same fate himself, when on the way to the ordination of a fellow minister.[73]

Although as 'God's elect', ministers' status should have ensured their personal safety this was not always the case. The moral and social standing of ministers in relation to their congregations did not automatically protect them from censure or even physical assault. The various upheavals of the mid- and late seventeenth century are likely to have undermined the position of ministers as a group in the eyes of their congregations. In the later seventeenth century in southern Scotland Covenanters caused problems for many ministers. Not only did members of their congregations absent themselves to attend conventicles but zealous Covenanters also disrupted the work of ministers directly. In 1682 Mr David Spence, minister of Kirkurd in Peeblesshire, was prevented from preaching by a rabble of strangers. Four months later a group of men stole the parish poor box, mortcloth and other articles belonging to the church.[74] Spence nevertheless fared better than the minister of Manor, David Thomsone, who, in 1680, was attacked and left for dead by an armed gang who plundered his house and stole his horses. He was eventually forced to retire due to deafness caused by the wounds he had received.[75] The most widespread violence occurred at the Revolution of 1688 when large numbers of Episcopalian ministers were 'rabbled' and driven from their parishes. Some of them were marked men, having been involved in the apprehending of Covenanters, but many were not; some who had been popular with their flocks were ejected by wandering bands of zealots.[76] However, unpopular or unlucky ministers were not infrequently assaulted on other occasions for reasons which may have ranged from the doctrinal to the purely personal. A woman was condemned to death for beating Mr Alexander Ramsay, minister of St Giles' in Edinburgh, in 1682.[77] One of his predecessors, David Fletcher, had been assaulted and maltreated by several women in 1638, presumably for being insufficiently enthusiastic in support of the Covenant. The relative wealth of ministers sometimes made them the target for housebreakers or highway robbers. Some seventeenth-century ministers at least kept weapons at home and even wore swords in public.

In the early eighteenth century patronage disputes could also lead to violence. Thomas Lawrie, minister of Bathgate 1717–31, was supported by one faction of landowners and elders when he was put up for the parish but not by another, who put forward a rival candidate. The synod and presbytery required Lawrie to be entered as minister but it took a troop of dragoons as a guard to accomplish this. Most of his parishioners then refused to accept him so that he had a congregation of only about 50 in a parish with a population of some 1,200.[78] In circumstances where a patron forced through the installation of a particular minister against the wishes of the congregation, violent opposition was virtually guaranteed. Although the number of disputed presentations is unclear, and more research is needed, their significance was so great that they have been described as the most persistent and widespread cause of popular unrest in Scotland between 1730 and 1843.[79] A minister who started his career in this way might have an uphill struggle for years before he became accepted. Later in the eighteenth century there was less deference towards both ministers and elders, shown by the growing refusal to accept the discipline of kirk sessions and in the biting satires of Robert Burns.[80]

THE DARK SIDE

Ministers were only human and, despite the high standards which they set themselves as a professional group, they sometimes fell short of these ideals. In some respects it is invidious to highlight a few extreme and bizarre cases where ministers transgressed moral and legal codes, but such instances nevertheless serve to highlight some of the problems and temptations which they faced. Ministers can be found committing a range of sins and crimes for which they might be reprimanded, deposed or, in rare cases, imprisoned and even executed.

Not surprisingly, at a period when a substantial proportion of the average man's calorie intake came in liquid form, drunkenness was the cause of a number of ministers being suspended or deposed. Debt was the cause of some ministers' downfall. The minister of Aberlady fled his charge in 1711 for the sanctuary at Holyrood.[81] Also recorded were cases of fornication and adultery. Robert Menteith of Duddingston 1630–3 had an illicit affair with Dame Anna Hepburn, the lady of Sir James Hamilton of Priestfield. When this was discovered he fled to France, became a Catholic and, under the patronage of Cardinal Richelieu, ended up as a canon of Nôtre Dame in Paris.[82] Cases of dismissal for theft were rare but are not unknown. Duncan MacCaig of Edinburgh's Gaelic Chapel from 1823 to 1831 was tried by the High Court of Justiciary for stealing books and was sentenced to fourteen years transportation to Van Diemen's Land, where he ended up as a schoolmaster.[83]

Cases of violence by, rather than to, ministers also occurred. Thomas Lamb, minister of Kirkurd, Peeblesshire, belied his name and clearly had the temper of an ill-natured lion. After striking a man and causing his death he was suspended from his ministry but in 1641 was re-instated by the General Assembly. The day after this decision had been taken he was walking from Leith to Edinburgh. He stepped into a rig of corn to relieve himself. The owner of the crop came up and objected to his conduct. An argument developed during which Lamb drew his sword and killed the man with a single stroke. This time he was deposed and executed.[84] At least he committed murder in hot blood. John Kello, minister of Spott, strangled his wife with a towel and then hung her from a roof beam in the manse to make it look as if she had committed suicide. He then went to church to preach, pretending to discover her body when he returned to the manse. Suspicions were nevertheless aroused and he eventually confessed, being sentenced to be hanged and burnt.[85]

MINISTERS AS AN EDUCATED ÉLITE

Ministers must often, with the exception of the laird, have been the only graduates in their parishes. To what extent this set them apart from the bulk of their congregations is not clear but it must often have been a lonely life.[86] The nature of their education during the late sixteenth and seventeenth centuries must often have been quite limited and narrow. This problem was aggravated by the necessity of drafting in new, not always well-trained, ministers after the purges of 1662 and 1688. When episcopacy was re-established in 1662 and when Presbyterianism triumphed after the revolution of 1688, it was widely acknowledged that a number of the candidates hastily examined and installed in vacant parishes, were sub-standard in ability.[87] The limited size and value of ministers' libraries in the first half of the seventeenth century, noted in their testaments and listed by Scott, also indicates the limitations of their education.

By the later seventeenth century, however, increasing wealth among ministers and the wider availability of books in Scotland was reflected by many ministers owning larger libraries. Their education may in many cases have been put to no more than local use but a number of ministers gained reputations as scholars within a wider world. In the seventeenth century this was generally purely within the realms of theology, but in the eighteenth century also in a range of other fields including history, agricultural improvement and the natural sciences. Ministers also provided the parish descriptions for the *Old* and *New Statistical Accounts* of the 1790s and 1830s. In such activities, as well as in their theology, they are likely to have benefited from regular meetings with fellow ministers at presbytery and synod and, of course, the interchange of ideas at the General Assembly.

In terms of the pastimes and interests they could pursue, ministers had to watch their behaviour before the later eighteenth century; James Law of Kirkliston in the late sixteenth century was rebuked by his synod for playing football on the Sabbath,[88] though field sports seem to have been more acceptable.[89] By the late eighteenth century the interests and activities of ministers were more wide-ranging or perhaps better recorded. John Thomson, minister of Duddingston from 1805 to 1840, was an ingenious mechanic, an accomplished musician and a distinguished landscape painter.[90] From the mid-eighteenth century there was more scope for ministers to indulge in a wider range of activities, though as late as 1757 Thomas Whyte, minister of Liberton, was called before his presbytery and was suspended for three weeks for going to see John Home's play *Douglas* despite his plea that he had only gone once and had taken care to conceal himself in a corner to avoid giving offence![91] The congregation of Haddington were doubtless not appreciative when their scientifically minded minister, Mr David Wark, absented himself from the pulpit in order to see an eclipse.[92]

In the mid- and late eighteenth centuries many ministers belonging to the more liberal and broad-minded Moderate Party had wider interests. From the first production of Home's *Douglas*, written by a minister, and the effective defeat of the outcry it caused, it became increasingly possible for ministers to publish polite as well as sacred material, and a number of them did so.[93] Some ministers made notable contributions to the Scottish Enlightenment.[94] Among historians, William Robertson, minister of Greyfriars 1761–94 and Principal of Edinburgh University from 1762, had an international reputation with his histories of Scotland, America and the Emperor Charles V. Hugh Blair had few equals as a writer of sermons and was awarded a pension of £200 from George III as a result, though he also wrote a critical dissertation on the poems of Ossian.[95] Some ministers had considerable mathematical abilities; the talents of Alexander Webster, minister of the Tolbooth Kirk in Edinburgh 1737–54,[96] and Robert Wallace, minister at the New North Kirk, Edinburgh 1738–71,[97] were put to good use in calculations for the Kirk's pension fund.

A number of ministers became noted as agricultural improvers though at local rather than national levels as they were rarely landowners on a large scale. Typical, perhaps, was William Wilkie who, while still a divinity student at Edinburgh, inherited the unexpired lease of a farm near the city, gaining a reputation as an expert farmer and the nickname of 'Potato' Wilkie.[98] John Fleming, minister of Colinton 1804–23, inherited a small property when still young and, improving it to great advantage, acquired the reputation of a capable improver so that his services as a valuer of estates were in demand even after he had entered the ministry.[99] John Fleming, minister of Cockpen at the same period, had also attracted the attention of major landowners for his ideas on agricultural improvement before he went into the church. He became factor to the Earl of Rosebery who eventually presented him with a vacant parish.[100] A more unusual man was David Ure, minister of Uphall at the end of the eighteenth century. Trained initially as a weaver, he went into the ministry and became a correspondent of Sir John Sinclair, contributing to three of the Board of Agriculture county reports.[101]

CONCLUSION

This chapter began with Mr Balwhidder, rather bewildered and ill at ease with the social and economic changes that were affecting late eighteenth-century Scotland. It is only fitting that it should end with the work of Sir John Sinclair as the man who mobilised Scottish ministers into contributing to the 21-volume *Statistical Account of Scotland* in the 1790s. This provides an impressive monument to the range of interests and talents of ministers at the close of our period. Although the individual parish descriptions vary in length and detail, taken together they are an impressive showcase for the range of knowledge and enthusiasm of ministers, spanning the natural and physical sciences, history and archaeology, demography and an almost ubiquitous concern for social reform and economic progress that places them in the mainstream of contemporary intellectual thought, still with a central role in the guidance of Scottish society. This chapter has covered a very broad area of Scottish social history, many aspects of which remain little explored. While emphasising the central role of ministers in Scottish society from the seventeenth to the early nineteenth century it also shows that there is much that we do not know about ministers and their social context during this period. A good deal of information on many of the themes which have been touched on here should be recoverable from church records such as kirk session material, testaments and other official documents, as well as from private correspondence and other material. The author hopes that the manifest gaps and shortcoming of this chapter will encourage further research into this neglected yet fascinating field.

NOTES

1. Galt, 1986,vii.
2. Galt, 1986, 95.
3. Mitchison, 1982, 2.
4. Buckroyd, 1987.
5. Makey, 1979.
6. Mitchison, 1982, 1.
7. Mitchison, 1982, 4; Makey, 1979, 85.
8. Mitchison, 1982, 1.

9. Smout, 1972, 58.
10. Smout, 1972, 84.
11. Makey, 1979, 10.
12. Smout, 1972, 58.
13. Makey, 1979, 10.
14. O'Day, 1976, 61–8.
15. Makey, 1979, 94–105.
16. Makey, 1979, 94–105.
17. Makey, 1979, 96–7.
18. Makey, 1979, 108.
19. Makey, 1979, 108.
20. Makey, 1979, 13.
21. Makey, 1979, 116–7.
22. Graham, 1950.
23. Whyte and Whyte, 1984; Makey, 1979, 106.
24. Makey, 1979, 112.
25. Makey, 1979, 116.
26. Sher, 1985, 33.
27. MacTavish, 1943, 6.
28. MacTavish, 1943, 27.
29. Graham, 1950, 25.
30. MacTavish, 1943, 17.
31. Makey, 1979, 79.
32. Mitchison and Leneman, 1989.
33. Whyte and Whyte, 1999.
34. Scott, 1866–71.
35. Scott, 1866–71, I, 48.
36. Scott, 1866–71, I, 20.
37. Graham, 1950, 293–4.
38. Scott, 1866–71, I, 164.
39. Graham, 1950, 303.
40. Makey, 1979, 85.
41. Graham, 1950, 283–5; Makey, 1979, 85.
42. MacTavish, 1943, 5.
43. Makey, 1979, 118.
44. Mitchison, 1983, 181.
45. Scott, 1866–71, I, 181.
46. Mitchison and Leneman, 1989, 25.
47. Graham, 1950, 269; Ferguson, 1968, 107.
48. Graham, 1950, 271–2.
49. Donnachie and Hewitt, 1989, 168.
50. Mitchison, 1983, 72–3.
51. Ferguson, 1968, 105.
52. Whyte and Whyte, 1999, 229.
53. Whyte and Whyte, 1999.
54. Whyte and Whyte, 1999, 223–2.
55. Whyte and Whyte, 1999.
56. Goudie, 1889, 27.
57. Scott, 1866–71, I, 173–4.
58. MacTavish, 1943, 8.
59. Mitchison, 1982, 1.
60. Makey, 1979, 20.
61. Whatley, 2000, 146–9.

62. O'Day, 1976, 55.
63. Scott, 1866–71, I, 32.
64. Makey, 1979, 10.
65. MacTavish, 1943, 5.
66. MacTavish, 1943, 6.
67. Mitchison, 1982, 10.
68. Scott, 1866–71, I, 311.
69. Makey, 1979, 102.
70. Scott, 1866–71, I, 19.
71. Scott, 1866–71, I, 31.
72. Scott, 1866–71, I, 145.
73. Scott, 1866–71, I, 187.
74. Scott, 1866–71, I, 244.
75. Scott, 1866–71, I, 250.
76. Ferguson, 1968, 54.
77. Scott, 1866–71, I, 11.
78. Scott, 1866–71, I, 168.
79. Whatley, 2000, 168.
80. Whatley, 2000, 165.
81. Scott, 1866–71, I, 318.
82. Scott, 1866–71, I, 110.
83. Scott, 1866–71, I, 76.
84. Scott, 1866–71, I, 243.
85. Scott, 1866–71, I, 380.
86. Makey, 1979, 103.
87. Graham, 1950, 275.
88. Scott, 1866–71, I, 189.
89. Scott, 1866–71, I, 103.
90. Scott, 1866–71, I, 113.
91. Scott, 1866–71, I, 115.
92. Scott, 1866–71, I, 315.
93. Sher, 1985, 77–85.
94. Sher, 1985, *passim*; Chitnis, 1976, *passim*; Rendall, 1978, *passim*.
95. Scott, 1866–71, I, 27.
96. Scott, 1866–71, I, 51.
97. Scott, 1866–71, I, 51.
98. Scott, 1866–71, I, 149.
99. Scott, 1866–71, I, 271.
100. Scott, 1866–71, I, 206.
101. Scott, 1866–71, I, 206.

BIBLIOGRAPHY

Buckroyd, J. *The Life of James Sharp, Archbishop of St. Andrews 1618–69. A Political Biography*, Edinburgh, 1987.

Chitnis, A C. *The Scottish 'Enlightenment': A Social History*, London, 1976.

Donnachie, I and Hewitt, G. *A Companion to Scottish History From the Reformation to the Present*, London, 1989.

Ferguson, W. *Scotland 1689 to the Present*, Edinburgh, 1968.

Galt, J. *Annals of the Parish*, Oxford, 1986.

Goudie, G, ed. *The Diary of the Rev John Mill, Minister of the Parishes of Dunrossness, Sandwick and Cunningsburgh in Shetland 1740–1803*, Edinburgh, 1889.

Graham, H G. *The Social Life of Scotland in the Eighteenth Century*, London, 1950.
Handley, J E. *Scottish Farming in the Eighteenth Century*, London, 1953.
Mactavish, D C. *Minutes of the Synod of Argyll 1639–1651*, Edinburgh, 1943.
Makey, W. *The Church of the Covenant*, Edinburgh, 1979.
Mitchison, R. *Life In Scotland*, London, 1978.
Mitchison, R. The social impact of the clergy of the reformed kirk of Scotland, *Scotia*, 6 (1982), 1–13.
Mitchison, R. *Lordship to Patronage: Scotland 1603–1745*, London, 1983.
Mitchison, R and Leneman, L. *Sexuality and Social Control. Scotland 1660–1780*, London, 1989.
O'Day, R. The reformation of the ministry 1558–1642. In O'Day, R and Heal, F, eds. *Continuity and Change. Personnel and Administration of the Church of England 1500–1642*, Leicester, 1976, 55–75.
Rendall, J. *The Origins of the Scottish Enlightenment*, London, 1978.
Scott, H. *Fasti Ecclesiae Scoticanae*, 1st edn, London, 1866–71.
Sher, R. *Church and University in the Scottish Enlightenment*, Edinburgh, 1985.
Smout, T C. *A History of the Scottish People 1560–1630*, London, 1972.
Whatley, C. *Scottish Society 1707–1830*, Manchester, 2000.
Whyte, I and Whyte, K. Geographical mobility in a seventeenth-century Scottish rural community, *Local Population Studies*, 32 (1984), 45–53.
Whyte, I D and Whyte, K A. Debt and credit, poverty and prosperity in a seventeenth-century Scottish rural community. In Mitchison, R and Roebuck, P, eds. *Economy and Society in Scotland and Ireland 1500–1939*, Edinburgh, 1988, 70–80.
Whyte, I D and Whyte, K A. Wed to the manse: The wives of Scottish ministers c1560–c1800. In Ewan, E and Meikle, M, eds. *Women in Scotland c1100–c1750*, East Linton, 1999, 221–32.

21 The Church Social

JOHNSTON McKAY

When he arrived in Scotland in February 1947 to deliver a series of lectures, the American theologian Reinhold Niebuhr was quoted in *The Glasgow Herald* as saying that Americans were more 'church-minded' than any other nation 'with the possible exception of Scotland'.[1] At the start of the second half of the twentieth century, nearly two million Scots, or around 55 per cent of the population, claimed membership of one or other of the Churches in Scotland, and 75 per cent of those belonged to the Church of Scotland,[2] with whose social activities this study is primarily concerned. The early 1950s represented the peak of Scottish church attendance and affiliation. Whatever spiritual benefits attendance at church brought, the affiliation with the church provided the context within which, for a large proportion of the population, social contact took place. It is not an exaggeration that, for many, life 'revolved around the church'. Within it, organisations met and events took place which catered for a very wide variety of social needs, and a popular song of the time, made famous by the entertainer Harry Gordon, 'the Kirk Soiree', captured a social occasion which had resonance for many and also traced one aspect of the changing mores through the decades under review:

> My greatest ambition, it was, you see
> To tell a little story at the Kirk Soiree.

The story clearly was slightly risqué, but as time went on its content became more acceptable in 'polite company', and the teller of the story less 'bold' for telling it, until:

> The minister's youngest daughter at her father's jubilee
> Told the story that I started at the Kirk Soiree

There were several different Scotlands in the 1950s, and in each of them the church fulfilled a very important social role. When the Rev. Fraser McLuskey went to be a minister in Bearsden in 1955, he found that 'young couples poured in as soon as there were houses ready for them or their firms to buy. In many cases they already had church connections or were very ready to make them. It would have been surprising if the congregation had not grown steadily or failed to be representative of the business and professional life of Scotland's largest city'.[3] In Edinburgh's Cramond 'The kirk was the social hub of the community. Its organisational life was strong and the welter of activities ensured that there was something for everyone'.[4]

The pattern of church life in Scotland's suburbs in the 1950s was one which had been set by the Free Church and the United Presbyterian Church in the nineteenth century and the United Free Church in the first half of the twentieth century, where, as the late Professor Ian Henderson used to like to say to his students, the clergy were no longer ministers to parishes but became chaplains to congregations. These congregations had halls adjacent to the church which illustrated that the significance of the social activities which were housed in the church halls was at least as great as the spiritual significance of the worship held in the sanctuary. Congregations which belonged to the 'auld kirk', the Church of Scotland 'by law established', seldom boasted such halls. They understood the parish church to be a reflection of the whole community at worship and not a group gathered for worship.

In many parts of rural Scotland in the 1950s that understanding of the church still prevailed. The Very Rev. Gilleasbuig Macmillan of St Giles', Edinburgh, described in a radio broadcast how his father, when minister of Appin in the 1950s, saw no distinction between church and parish.[5] And in rural Perthshire, the Rev. Kenneth MacVicar of Kenmore found kirk and community two different aspects of the same reality. The entire village was involved in the social life of each.

The 1950s saw the growth of the sprawling housing schemes which were such a feature of post-war Scotland, and there especially the role of the church in providing a place for social contact and the opportunity for social activity was vital. Tom Cromar, an elder in St Ninians in the Larkfield housing scheme on the outskirts of Greenock, where a hall-church was opened in 1955, recalls how the hall provided a focal point for a population which had been uprooted from long-established communities and was 'converted' into a church for Sunday worship. In his view the greatest mistake made was to build a separate church, against the wishes of many in the congregation, who saw the connection between worship and social activity being broken.

Pollok was a large housing estate on the edge of Glasgow's south side. James Currie went to be a minister there in 1955, and in what had been the hall-church before a church building from the Pollokshields area of the city was moved stone by stone into the housing scheme, almost Currie's first move was to start a youth club because amenities were few and far between, and in the era of gangs, the club provided somewhere for young people who lived a costly journey from the city's social scene to meet and socialise. Currie's biographer claims that though the club's methods and organisation were unconventional 'they prevented many a battle (and) were influential in many a life'.[6]

In middle-class suburb, rural community, or huge housing scheme, the social role of the church was clear and unquestioned. There were, however, places where it was less clear what the church's social role should be. These were where churches were set either in areas where members had once lived nearby but had now joined the emigration outwards from what were formerly the city's inner suburbs but in the 1950s began to change character, or else were in parts of the city where middle-class and working-class housing existed close to each other within a single geographical area. One such area was Partick in Glasgow, where the Rev. Harry Whitley was called to be minister in 1950.

> With great enthusiasm we launched into a mission to the parish to discover that it consisted of the Western Infirmary, two flourmills, a small bit of the busy Dumbarton Road, one school, and a few tenements of reasonable housing and half a dozen streets of thoroughly bad slum property. Incidentally it was soon quite clear that our congregation was drawn from widely scattered areas of Glasgow.
>
> As if to mark out this division, this segregation, there were two Sunday schools: one in the gloomy church halls, for the children of members of the congregation; the other in a dirty, decrepit hall on the other side of Dumbarton Road, for the children of the parish. It was done innocently enough, and represented the desire of the congregation to do mission work. It was the hang-over of the era of 'doing good for the poor'. Partick was not unique in this regard.[7]

It is worth noting that many congregations in the 1950s still maintained the practice of 'seat rents' by which, for the payment of a sum of money, families 'bought' pews in which they were entitled to sit. Some congregations at the time, such as St Giles' and St George's West in Edinburgh operated a system whereby only those who paid seat rents were entitled to enter the church up until a set time before the service started, and all others had to queue.

At the start of the second half of the twentieth century, the social function which the church fulfilled varied because there were differing levels of social cohesion and a variety of social roles requiring to be played. The vehicles through which the church fulfilled its differing social functions were very similar. In the buoyant church of the 1950s it was assumed that patterns of church life and activity which were appropriate for a pre-war society still had value in a changing Scotland, and that the sort of activities which met the needs of the church in relatively stable communities would continue to attract people in a more geographically mobile and socially fluid Scotland. The assumption was wrong, but it does allow a degree of generalisation in the description of the social side of mainly Church of Scotland life during the period under review.

THE SUNDAY SCHOOL

The primary purpose of Sunday schools was to provide for the Christian nurture of children. In 1950, just over 281,000 children attended Church of Scotland Sunday schools, representing 31.7 per cent of children of school age, and Bible classes were attended by just under 50,000 young people. The numbers attending Sunday school continued to rise until 1956, when 325,200 children are recorded as having attended. Figures for Bible class attendance show more fluctuation. For example, after a peak of just over 65,500 in 1955, the subsequent two years show a drop. However in 1958 the total rises again dramatically to 67,888.[8]

When the numbers attending began to drop in 1958 the Church of Scotland's Committee on the Instruction of Youth agreed to support a survey into possible reasons for the decline, and the conclusions of the survey provide interesting evidence of what Sunday schools were like. The survey revealed that a large proportion of those attending Sunday schools came from families where there was little or no contact with the Church. The report concluded:

> Young people whose parents attended church regularly remained in attendance (at Sunday School and Bible Class) longer than the other young people. In addition, many of the children who were interviewed after leaving Sunday School or Bible Class came from homes where there was little or no interest in the Church, or where the attitude was openly antagonistic.[9]

It is interesting to note in this connection that until World War II, the percentage of children of school age attending Sunday schools was higher than the percentage of adults in the population who were members of the church. The report comments:

> It may be that as regards adult church membership the Church is living on the capital gains of previous generations, and that if an analysis by age of our present-day congregations were to be carried out, it would be found out that the average age was much higher than that of twenty years ago and that it was the continued membership of these older people that was giving to church membership a greater degree of stability than could be found in Sunday School attendance.[10]

Since 1956, attendance both at church and Sunday school has continued to decline. In 1967, 30,000 fewer children attended Sunday school and 20,000 fewer were members of Bible classes. By 1977 the number attending Sunday schools had halved again, and by 1987, just over 90,000 attended Sunday schools, and Bible classes attracted 20,000.[11] At the beginning of the twenty-first century, the historian of religion, Dr Callum Brown, described the Sunday school as 'in terminal decline'.[12]

It is clear that although church and Sunday school attendance peaked in 1957, the seeds of the subsequent decline were already germinating. The report into the decline in Sunday school attendance quotes two reasons given for children stopping going to Sunday school,[13] both of which hint at considerations which are important from the point of view of this study. The fact was that children clearly wanted to distance themselves from anything associated with childhood and there is considerable evidence that the Sunday schools of the 1950s and 1960s failed to recognise earlier maturity and the tensions of adolescence. The writer recalls the Sunday school he attended in

Glasgow during the 1950s, and with which he continued to be associated during the early 1960s, where the Sunday school met on a Sunday afternoon and the format took the shape of the normal morning church service of hymns, prayers, Bible readings, offering. Instead of the sermon, however, the Sunday school broke into class groups where well-intentioned Sunday school teachers taught lessons culled from a Sunday school teachers' magazine. The language was consistently less mature than at each stage of the child's development, and the atmosphere was very like school. There was an annual Bible examination and prize giving. At a time when considerable advances were being made in the understanding of children's psychological, emotional and intellectual development, Sunday schools were, by and large, places where such development was ignored.

The other reason given for the decline in Sunday school attendance which is significant in this context is that:

> the majority of the teachers in Sunday School are very young and lacking in experience. In many cases also they may just be lacking in training. The Sunday Schools, in other words, are being staffed by young, immature, partially trained girls, each giving a few years of service.[14]

It is important to note that throughout the 1950s and 1960s Sunday schools provided a safe social atmosphere for 'young, immature' girls, many of whose parents were church members and who encouraged their daughters to volunteer to teach in Sunday school because it was an acceptable way of their spare time being occupied. On the other hand, these girls were introduced to adult society in a context which often encouraged them to find adult support at a difficult stage in their development outside their own families.

There was an important social side to many Sunday schools. The annual Sunday school trip of a town or city congregation was an important social event right up until the 1970s when car ownership became widespread and foreign travel offered more glamour than a day in the countryside. Its heyday was in the 1950s and 1960s, when fleets of hired buses took sometimes hundreds of children and parents often to a field owned by a farmer who was known to someone in the church, or to a park in the country, or to a seaside venue, where an advance party of adults, often making use of a lorry made available by a church member or parent, had set up benches brought from the church hall, boiled large urns of hot water to make gallons of tea, marked out an area where 'the races' were to take place, and had prepared a vast number of paper bags containing perhaps two sandwiches, a cake and a biscuit. Sometimes meat pies were ordered from a nearby bakery and delivered hot to the field. In town and city churches the Sunday school trip was often held on the day before the Sunday when the Sunday school closed for the summer and a Sunday school prize-giving was held.

Two other aspects of the social side of the Sunday school should be mentioned. Many Sunday schools had what was called a 'junior choir'. The writer can recall in the Sunday school he attended the junior choir sitting on the platform along with the Sunday school superintendent (in much the same way as in church the chancel choir sat beside the pulpit), and occasionally singing 'an anthem'. This pattern of Sunday school meeting was by no means unusual, and many congregations enjoyed each year a junior choir concert.

Often allied to the junior choir was an annual children's performance. Some churches specialised in operettas, specially written for children. Others preferred a children's pantomime. Others performed plays or shows written for the children by members of the Sunday school staff or the congregation. For many children these were formative experiences in developing confidence and communication skills.

So important was the Sunday school thought to be that the BBC Scottish Home Service broadcast a *Fireside Sunday School* with hymns sung by a children's choir and prayers and a talk given by Rev. Donald MacFarlan, a Church of Scotland minister who taught religious education at Jordanhill College of Education.

From the mid 1960s, the Church of Scotland attempted to reshape its Sunday school teaching. A much more child-centred programme was developed which doubtless was more effective in Christian nurture but could do little to resist the decline of the Sunday school, and as the influence of television grew and children and young people developed tastes which commercial interests both created and met, the Sunday school's role as a social as well as a religious agency declined dramatically.

SEASIDE/SUMMER MISSION

Allied to the work of the Sunday school was the Church of Scotland's organising of Seaside (later to be called Summer) Mission. This was begun in 1934 by the Church's evangelist, D P Thomson, with himself and one divinity student conducting a mission in Millport. It grew in the years after World War II to become a very big exercise throughout most of the second half of the twentieth century, involving over a thousand young people in a wide variety of locations over a summer.

The young people were usually drawn from a congregation where the minister had a particular interest in Summer Mission, and for some of them the two-week mission was their summer holiday. Sometimes two ministers and members of their congregations combined together to run a mission. Whatever the composition of the mission team, and in whichever of the twenty to thirty locations from Aberdeen to Ballantrae, the format of the mission was similar. After breakfast each morning in the hall where the mission team was housed, usually a local church hall, the team held an hour's Bible study, then took to the beach and organised games for children: treasure hunts, rounders, races, sandcastle building, tide fights. In the afternoon there was a children's meeting. Before it started, the team (with the help of the children) built a large sand pulpit, from which the team leader would speak. Simple choruses were sung, and often the meetings were focused on a well-tried theme which involved the children in some preparation work beforehand. A 'museum' service involved children bringing objects which illustrated texts from the Bible. 'Who am I?' services involved children acting out a small scene illustrative of a biblical character. Parable plays were staged. In the evening the teams held adult services in the open air where members of the team gave their own testimony to faith or were questioned about their beliefs by another member of the team. In some centres a youth café was held. Very often the fortnight's mission ended with a huge beach barbeque.

It is tempting to assume that Summer Mission was at its height in the 1950s (before the age of the cheap package holiday) when on one evening the two putting greens at Girvan were closed to allow over 2,000 people to listen to the legendary minister Tom Allan who inspired a whole generation of those involved in Summer Mission. However, the Very Rev. Sandy Macdonald, who organised summer missions in the 1970s, describes a barbeque for over 500 people at Millport in the 1980s and, in the same decade, addressing over 1,000 people at an adult service in Aberdeen.

Both Sandy Macdonald and Rev. Bill Shannon, who was responsible for Summer Mission when he was a Church of Scotland evangelist in the 1970s, admit that summer missions probably had more impact on the mission teams themselves than on those they were aimed at. When asked what summer mission did for those who took part in it, Sandy Macdonald said 'The traditional church offered traditional church activities. Summer Mission was very free but gave young people the opportunity to explore what it means to be the church. Young people were put into the position where they had to explore their own faith'.[15]

A considerable number of people who were involved in Summer Mission made the decision to become ministers of the Church of Scotland.

THE BOYS' BRIGADE

Although churches provided homes for many groups of cubs and scouts, brownies and guides, these were not Church organisations. The kirk session of the church did not appoint the leaders, as it did in the case of the Boys' Brigade and the Girls' Brigade, though it is questionable whether this connection was as beneficial to the Church as Boys' Brigade apologists liked to claim. One minister who had himself been in the Boys' Brigade described the movement as the greatest inoculation against church membership that had been invented!

In 1950 those in the Boys' Brigade (including its younger group, then called 'the Life Boys', now called, more prosaically, 'the Junior Section') totalled over 32,000 in Scotland. By the end of the century that number had dropped to below 9,000.[16] Towards the end of the twentieth century, changes were made in the organisation and style of the Boys' Brigade, to some extent to meet and combat the decline in numbers. Boys' Brigade companies were given the choice as to whether to wear the traditional uniform of cap, haversack and belt. If a company decided, officers were no longer obliged to wear the traditional uniform of black suit and Glengarry. There was greater emphasis on outdoor activities. For example, the captain of one company, Brian Fraser, who co-authored the definitive history of the Boys' Brigade, said in a recording for a radio broadcast in 2001, 'The Boys' Brigade in recent years has changed its uniform, discarded its titles and there is a move away from saluting. One of our Scottish leaders said that the organisation had to get rid of the perception of military baggage. It is possible that the BB will discard completely its military activities and terminology for they play such a small part now'.[17]

However for most of the 1950s and 1960s the military activities and terminology played an extremely important part in the ethos of the Boys' Brigade, and there was then no variety in the styles of Boys' Brigade company. The movement was organised very much as it had been organised by the founder, Sir William Smith:

> The company was and is the basic unit, being a detachment of boys connected with a particular church or religious meeting place which has a local designation numbered according to order of formation. The company was placed under the command of a captain, assisted by lieutenants ... The uniform adopted, however, was simple and inexpensive. The most junior rank of private had a forage cap (with a chin strap added later) having two rows of eight inch white braid and the company number at the front, a white linen haversack, and a leather waist belt with the BB crest on the buckle. Lance corporals and corporals wore the same uniform as a private, with the addition of a one barred chevron for the former and a two barred chevron for the latter, worn on the right arm. The sergeant's uniform consisted of a cap with a straight peak, two rows of quarter inch white braid, and a three barred chevron ... The forage or 'pillbox' cap remained in use until 1970–71 when it was replaced by a field service style of hat in blue terylene and cotton[18]

The two principal meetings of each Boys' Brigade Company were the Friday evening parade and the Sunday morning Bible class. The Friday evening meeting usually began with a formal parade of the whole company, divided in two squads, followed by a period of marching and drill as a full company and in individual squads. Companies took part in Battalion drill competitions and squads in Company drill competitions. The opening parade and drill was followed by the company dividing into groups studying for one of the variety of badges a boy could earn: for first aid, physical training, playing in a band, playing drums, and a whole range of other activities. The Company would gather on parade at the end of each evening and the meeting would conclude with a short act of worship.

The Boys' Brigade Bible class usually met prior to a church's Sunday morning service. Callum Brown makes a good deal, perhaps too much, of the significance of this:

> In the 'BBs' the curriculum may be regarded as perceptively gendered, with compulsory attendance at the weekly prayer meeting qualifying the boys to participate in the Saturday football ... An act of (feminine) piety was required before admission to the (masculine) sport.[19]

Certainly in the Boys' Brigade company best known to the writer, the Bible class was compulsory. It was conducted by the company captain, and, like Sunday schools of the time, followed the rough outline of a church service, the address being given each week by a different officer on a theme dealt with each week in the *Boys' Brigade Handbook*. Many of these handbooks were written by the Rev. (later Professor) William Barclay and were reissued as part of his prolific output in the 1960s and 1970s.

One feature of the Bible class which the writer attended was the weekly 'essay' written by one of the NCOs (lance-corporal, corporal or sergeant). This 'essay' was read, lasted three or four minutes and consisted of the writer's reflections on the theme of the previous week's Bible class.

Two other features of the Boys' Brigade deserve mention. Most battalions ran at least one football league, and matches between companies caused considerable rivalry. Matches were usually played on Saturday afternoons on local authority football pitches but some companies, such as the 1st Glasgow Company, had their own football pitch and changing facilities.

Each company provided a summer camp for boys, and at least in the 1950s and 1960s it was a common sight on a July Saturday morning to see Boys' Brigade companies marching through the streets of Glasgow to the appropriate railway station or quay to meet the transport which would take them to a fortnight spent under canvas. Some companies, such as the 1st Glasgow Company, had a permanent camp site (at Portavadie on Loch Fyne). Others went to the same site each year. King's Cross and Brodick on Arran or Ettrick Bay on Bute were favourite places for companies' annual camps.

THE YOUTH FELLOWSHIP

While Bible classes were provided for the teaching of adolescents, many churches in cities and towns provided an organisation for, and usually run by, young people in their final years at school or early years at university. Running through the winter months, the Youth Fellowship or Young Peoples' Society prepared a syllabus of Sunday evening meetings, which were sometimes addressed by speakers who described their charity or voluntary work. At other meetings members themselves provided a panel for a quiz, debate or balloon night.

These organisations were particularly successful in middle-class areas such as Netherlee and Kelvinside in Glasgow, and Greenbank and Cramond in Edinburgh. In his biography of the minister/politician Geoff Shaw, who became the first Convener of Strathclyde Regional Council when it was established in the 1970s, Ron Ferguson writes of Shaw's involvement with Cramond Kirk: 'Geoff felt at home in the church. He joined the Young People's Fellowship. Here, with other young people, he learned to socialise, stretch his wings, test ideas, flirt, learn to speak in public and find room to grow emotionally and intellectually'.[20]

Many of these organisations, especially in the middle-class areas of cities and towns, held an annual concert or show. This was an important part of the calendar into the early 1970s when commercial and other pressures meant that the church was no longer a natural focus for social activity. Some held out against the trend much longer than others. 'Quest', which was the youth society attached to Greenbank Church in Edinburgh, still flourished well into the 1980s and derived its strong membership and social cohesion from the fact that many of those involved shared schooling at George Watson's in the city. Fifty or sixty members of Quest still held a weekend retreat/break at Wiston Lodge near Biggar in the 1980s.

THE WOMAN'S GUILD

In 1962, the then minister of St Giles' in Edinburgh, Harry Whitley, wrote of the Woman's Guild:

> This organisation has become a Church within the Church – and while they have done much good work in raising funds and keeping the plant of the Church running, they have assumed such authority at local and assembly level that they now constitute a challenge to Presbyterian government. I am assured on all sides that without them the Church would die. I think the Church of Scotland ought to risk dying.[21]

That view may owe more than a little to Whitley's dislike of women playing any real part in the running of the Church and to his tendency to see church life in terms of conspiracies, but it certainly reflects how significant the Woman's Guild was in many congregations.

In her history of the Woman's Guild, Mamie Magnusson describes the Guild's ethos at the beginning of the second half of the twentieth century and what women were looking for:

> What they wanted after years of war and shortages and austerity was, above all, space to breathe again, time to lick wounds and heal broken hearts, a return to safe, familiar surroundings and simple, familiar pleasures.

For them, the Guild represented comfortable continuity and security. Its membership was at a record post-war peak of 150,000. It was an institution now, firmly settled into the fabric of the Church, second only to the kirk session in importance in the parish, with its own set of procedures and ways of doing things. It was not exactly a closed society, but like all long-established institutions it had a mystique and an aura of exclusiveness about it which could frighten off outsiders.[22]

However, lest that be allowed to paint too cosy and a comfortable picture of the Woman's Guild in the post-war years, it should be noted that in the immediate post-war years the Guild made contact with women in Germany and began project work to help in the work of German reconstruction.

In 2001 the Guild (now no longer a shortened form of 'The Woman's Guild' but the name chosen for the organisation after a painful restructuring exercise during the 1990s) numbered between 44,000 and 45,000. In addition to factors common to the decline in church membership generally, the Guild was profoundly affected by the most significant change within the Church of Scotland during the second half of the twentieth century: the opening up first of the eldership in 1961 and then, in 1966, of the ministry to women. In the years of which Mamie Magnusson wrote, the Woman's Guild was at its height when there were few other options for women to take part in the life of the Church, and when women took their place in the courts of the Church, the Woman's Guild inevitably suffered a loss of the leadership which would have been theirs. However the present General Secretary of the Guild, Alison Twaddle, sees a benefit to the Guild in that it 'opened up opportunities for other women, who might be terribly reticent, and might take steps, for example, to lead prayer in the Guild where they would die doing that in church. They have been allowed to uncover gifts they didn't know they had'.[23]

In another of the most significant developments in the past few years, leadership of Guild branches is no longer assumed to be the responsibility of the minister's wife. This is partly because a growing number of ministers have a husband rather than a wife, but more importantly because many ministers' wives now are in paid employment and contribute more than significantly to the manse economy. Indeed one of most important changes in the life of the Church of Scotland has been the reluctance of most ministers to move from one parish to another because the minister's wife has a job which cannot transfer as easily as the minister's. The well-off, so-called 'desirable' congregations of the 1960s now find that when they are vacant they can count on the fingers of two hands the number of applications they receive, when, forty years ago, they might have received applications running into hundreds.

Dr Archibald Charteris, a leading nineteenth-century minister who founded the Woman's Guild, did so because he wanted to harness the practical energies and spiritual gifts of women, and, as Harry Whitley of St Giles' acknowledged, the practical energies were very significant in raising funds for congregations. Most branches of the Woman's Guild held an annual Sale of Work during the 1950s, 1960s and 1970s which often played a very important part either in balancing the congregational finances or in providing a good proportion of the contributions which each congregation was expected to make towards the central organised activities and mission of the church.

The sort of project work which was undertaken in Germany in the years after World War II is still an important part of the Guild's role: providing resources, for example, to combat Aids in Malawi, or funds to furbish a nursing school in India or an Asian Advice Centre in Glasgow.

While the painful restructuring of the 1990s, which brought about changes in name, constitution and style which were far from universally welcomed, has resulted in a more contemporary image for the Guild, it is still true that many Guild branches are made up of older members and meetings are structured in a very traditional way, with opening devotions, a speaker, and a time for social interaction over tea. According to Alison Twaddle, a recent survey reported that most people joined the Guild because a friend or neighbour invited them to come along, and when radio producer Anna Magnusson visited Guild meetings in connection with her radio series *A Church without Walls* she was told that 'the fellowship' of Guild meetings was what members found most appealing.

However for the same radio programme, Alison Twaddle told Anna Magnusson:

> A fortnight ago I was in Croatia for the opening of a church built with Guild money. When I got up to speak, I didn't talk about the £80,000 raised. I talked about Margaret the fisherman's wife, and Mary the doctor, and Jean who cleans toilets and Peggy the farmer's wife and said 'These are the people who have heard the name of your village. Women in cathedrals and tiny churches have learned about the bombed out buildings and the land mines and your poems. And they have responded'.[24]

In the 1960s and 1970s, the Guild encouraged the formation of Young Women's Groups in each congregation, whose organisation, activities and meetings mirrored the parent organisation. The Young Women's Groups provided a meeting place for women with young families but perhaps for that reason found leadership difficult to sustain and continuity of membership difficult to maintain. By the 1990s many of them had ceased to meet and/or had been assimilated in the branch of the Guild.

THE CHOIR

There was a choir in most churches, leading the praise either from the gallery of the church or from a chancel or quasi-chancel area in front of the congregation. In the 1950s and 1960s, many suburban congregations had a quartet or double quartet of professional singers who were under an obligation to turn out morning and evening, and who usually undertook solo or ensemble parts in the anthem. Other members of the choir were volunteers. In some congregations it was the tradition for the choir to be robed. For some reason, Thursday evening tended to be the popular evening for choir practices, and these were as much social occasions as musical ones, with tea being served either at a break in the middle of the practice or at the conclusion of the rehearsal.

Church choirs were of three sorts. Those in places like St Giles' in Edinburgh, Glasgow Cathedral and Paisley Abbey provided church music of a very high quality, and their repertoire drew on the tradition of English church music. These were untypical of both Scotland's church music and liturgical tradition. In the major cities of Scotland there were, however, suburban congregations, or congregations which drew on the suburban population, which had very competent choirs which reached a high musical standard. Places like Netherlee and Newlands South in the south side of Glasgow, Wellington and New Kilpatrick in the north of the city, St George's West, St Cuthberts and Greyfriars in Edinburgh, and, in Aberdeen, Queen's Cross and St Nicholas, all reflected a quality of musical performance which was common and expected in the early decades of the second half of the twentieth century. Their choirs were able to attract applicants of high calibre for the post of organist.

However, many ordinary parishes and congregations could provide a choir of enthusiastic and able singers and in those and the suburban congregations (which tended to be the ones employing a number of professional singers) annual performances of works from the repertoire of sacred oratorios and cantatas made important contributions to the amateur musical life of Scotland. Handel's 'Messiah' was frequently performed, often by choirs with relatively meagre resources in comparison to some. At Easter, performances of Stainer's 'Crucifixion' and Maunder's 'From Olivet to Calvary' were a regular feature of the musical life of many congregations, and the rehearsals for these through the winter months an important focus of social contact.

The writer can recall long queues outside a Glasgow west-end church in advance of a performance of Handel's 'Messiah', which involved the church choir, augmented by friends of the organist, performing along with an amateur orchestra of some ability, all conducted by the congregational organist who was not a professional musician in the sense of making his living from playing the organ. Rows of floral arrangements decorated a temporary stage from which the choir sang, and from below which the orchestra played. And the church, which seated around one thousand, was packed. That was not an uncommon feature of the church calendar in the 1950s and early 1960s.

The Billy Graham campaign in Glasgow's Kelvin Hall had, for some years before and afterwards, an effect on church choirs because each night during the campaign, the massed choir, drawn from congregations mainly in the west of Scotland, had a pivotal role in the event. However the effect of the Billy Graham campaign was much wider than the area from which the choir was drawn because the style of music used became popular in some congregations where a sort of mission praise took root. This had the result, in some communities, of polarising the sort of music and style of choir between different congregations.

In Glasgow and Edinburgh especially, church choirs competed in the annual festivals which were held for amateur musicians, but there were also 'festivals of praise' in which church choirs competed against each other in performances of sacred music. For many years there was a festival of junior praise held in Glasgow which attracted Sunday school choirs and child soloists drawn from them.

This lively tradition of church choirs started to die out in the decades towards the end of the twentieth century for a number of reasons. Fewer people chose to learn to play the organ and many congregations found it difficult to attract applicants for the post of organist. The teaching of music throughout schools became less important, and the ability to read music was an accomplishment fewer people possessed.

Mention should be made of the immense contribution to the musical life not just of churches but of Scotland as a whole by St Giles' in Edinburgh and the organist and master of the music there for almost the whole of the second half of the twentieth century, Herrick Bunney. Each Sunday evening between 1955 and 1970, during the ministry of Harry Whitley, 'Music in St Giles' preceded a short evening service, and mainly choral but also instrumental groups from churches, schools, colleges as well as amateur choirs were given the opportunity to perform for a very appreciative audience in surroundings which created an atmosphere of musical excellence. This is, in a way, what all the churches, whatever their musical tradition and the quality of their choirs, provided across Scotland: opportunities for those who had exceptional talent to begin their careers. The now world-famous tenor Neil Mackie started singing in the choir of Paisley Abbey, the acclaimed ensemble, Capella Nova, drew its membership from those who sang in parish choirs, while countless amateur singers were given the opportunity to do what they most enjoyed.

OTHER ACTIVITIES

There were a number of other interests and activities catered for through the agency of the churches which were perhaps not as widespread as the provision of youth work and the Woman's Guild but nevertheless contributed considerably to the social fabric of Scotland, especially in the early decades of the period under review.

Most churches, certainly in the cities and towns, hosted a Badminton Club. Church halls almost invariably were marked out as a badminton court and once or twice a week those keen on badminton met to play. Marjorie Forrester, who played badminton for Scotland, played as a member of a church badminton team in Glasgow, and Billy Gilliland, Scottish champion and international, was introduced to the game through a team linked to a church in Greenock. Most competitive badminton in Scotland in the 1950s and early 1960s was played in leagues of teams run in and through churches.

There were churches' football leagues, though those who remember playing in them have a clear recollection that most participants were noted neither for their subtlety of play nor for their religious devotion. In the most popular evening newspaper of the 1950s in Glasgow, the *Glasgow Evening Citizen*, a considerable amount of space was devoted to the results in the churches' badminton and football leagues.

Many churches in the post-war years, especially in the towns and cities, supported a dramatic club which performed three-act plays. Agatha Christie's 'Murder at the Vicarage' and the comedies of Joe Corrie and William Douglas Home were favourites of church dramatic clubs, and a number of actors who became household names, like Stanley Baxter and Willie Joss, owed their first introduction to the stage to a local church drama club. It was quite common for churches to have stages built in sections which could extend platforms in church halls or be placed over and beyond them to make a reasonably sized stage.

THE WIDER EFFECTS

In addition to the activities which were organised by, and specifically related to, the work of the Church, the second half of the twentieth century saw a huge increase in the use of church premises by organisations not formally attached to the Church. The financial pressure put on local authorities in the 1970s and 1980s resulted in those community groups without relatively large funds, which had previously been accommodated in schools and community halls, being deprived of a place to meet. Churches provided meeting places for such organisations as Alcoholics Anonymous, flower clubs, community councils, tenants' associations, drug awareness groups, keep-fit classes and many more.

Churches also made a considerable contribution towards the social capital of Scotland. They provided places where those who would go on to take part in various levels of the democratic process were nurtured in the sort of skills such as chairing meetings and attending committees necessary for involvement in democratic and political activity. Some social commentators have noted that the decline in church membership could result in a decline in the stock of Scotland's social capital.

FALKIRK: A CASE STUDY

In the late 1960s, the sociologist Peter Sissons, on behalf of the Church and Ministry Department of the Church of Scotland, conducted a survey of church membership in the Falkirk area. His conclusions could well have been drawn from many other areas of Scotland:

> Church attendance in Falkirk was strongly weighted in favour of the professional, intermediate and skilled worker groups of the Registrar General's classification. This is not to say that partly-skilled and unskilled workers and their families did not attend church, it is to say that they were greatly unrepresented in Falkirk congregations. The membership of the churches is heavily weighted in favour of the middle classes, but there are important differences between the denominational groups. The Protestant congregations have a monopoly of the professionals whilst the Catholic Church has the

largest representation of partly skilled and unskilled workers ... Class and church membership were equally prominent in the development of the church members' social networks. A large majority of the Church of Scotland members made their friends within the context of their own denomination. Although the Catholics were more likely than the Church of Scotland members to make their friends within the context of their own parish church a majority of them made their friends within the same socio-economic class. This was particularly the case with the members who belonged to the higher social classifications. This was much less the case with those who belonged to the lower classifications, and the members of the lower classifications were much more likely to count the middle classes amongst their friends than were the middle classes to count members of the lower classes amongst their friends.[25]

The social significance of the churches in Falkirk would also be likely to reflect the social significance of churches in Scotland as a whole:

> The churches in Falkirk touch the life of the population in many ways, not least through church-based organisations for women and young people. The thirty-nine churches in the burgh sustain more than fifty women's organisations and about the same number of youth organisations excluding Sunday schools and Bible classes. Some two thousand women attend the women's organisations and about two and a half thousand children and adolescents are involved in the Sunday schools and youth organisations.

At that time, the total population of Falkirk was 37,500. Then (in 1967) it was possible to conclude:

> In a particular part of central Scotland membership of the churches is an important social phenomenon which, in diverse ways, influences the attitudes and judgements which prevail in the life of a small industrial town. The factors which contribute to the formation of distinctive styles of church membership are not always the apparently obvious ones of religious commitment, belief and doctrine, indeed in many instances these play but a small part in the development of the characteristics of religious belonging when compared with the influence of the social worlds which focus upon the denominations and upon the congregations and factions within the congregations.[26]

NOTES

1. *Glasgow Herald*, 5 February 1947.
2. Highet, 1950, 74, 75.
3. McLuskey, 1993, 113.
4. Ferguson, 1979, 21.
5. *A Sense of Place*, BBC Radio 4, September 1996.

6. Coffey, 1988, 104.
7. Whitley, 1962, 146.
8. Sutherland, 1960, 18.
9. Sutherland, 1960, 123.
10. Sutherland, 1960, 22.
11. Statistics contained in the annual *Reports to the General Assembly of the Church of Scotland*.
12. *Assembly Lines*, BBC 2 Scotland, May 2001.
13. Sutherland, 1960, 123.
14. Sutherland, 1960, 122.
15. *The Church without Walls*, Radio Scotland, May 2001.
16. Statistics contained in the annual *Reports to the General Assembly of the Church of Scotland*.
17. *The Church without Walls*, Radio Scotland, May 2001.
18. Springall, Fraser and Hoare, 1991, 43.
19. Brown, 2000, 110.
20. Ferguson, 1979, 22.
21. Whitley, 1962, 154.
22. Magnusson, 1987, 110.
23. *The Church without Walls*, Radio Scotland, May 2001.
24. *The Church without Walls*, Radio Scotland, May 2001.
25. Sissons, 1973, 284.
26. Sissons, 1973, 298.

BIBLIOGRAPHY

Brown, C. *The Death of Christian Britain*, London, 2000.
Coffey, W. *God's Conman*, Moffat, 1988.
Highet, J. *The Churches in Scotland Today*, Glasgow, 1950.
Ferguson, R. *Geoff*, Gartocharn, 1979.
Magnusson, M. *Out of Silence*, Edinburgh, 1987.
McLuskey, F. *The Cloud and the Fire*, Durham, 1993.
Sissons, P L. *The Social Significance of Church Membership in the Burgh of Falkirk* (The Church of Scotland), Edinburgh, 1973.
Springall, J, Fraser B and Hoare, M. *Sure and Stedfast. A History of the Boys' Brigade*, London, 1991.
Sutherland, J. *A Survey of Sunday Schools and Bible Classes in the Church of Scotland* (Church of Scotland Youth Committee), Edinburgh, 1960.
Whitley, H. *Laughter in Heaven*, London, 1962.

22 Occasions in the Reformed Church

HENRY R SEFTON

COMMUNION

'The Occasion' in Scottish ecclesiastical parlance was the Communion season. In the Reformed Church it has been an infrequent event but the very infrequency has heightened the sense of occasion. At its best it has been a very carefully prepared occasion; at its worst it has been either the Holy Fair satirised by Robert Burns or a means employed by fringe members of the Church to keep their names on the roll of communicants of their local church. It has often been asserted that infrequent Communion is a distinctive characteristic of Presbyterianism, but during the periods of Episcopal rule in Scotland Communion was also very infrequently celebrated. It is, however, odd that a sacramental occasion, which should be the norm of Christian worship, should be so infrequent in Scotland.

In pre-Reformation Scotland, Mass was frequently offered on Sundays and week-days at high altars and at side altars. Most people received communion very infrequently, indeed as seldom as once a year in most cases. Normally only the officiating clergy received the sacrament. The laity had an obligation only to be present at the offering of Mass and most were content to witness the elevation of the Host (or sacramental bread) rather than receive the sacrament. The reluctance to receive Communion was in part because of the requirement to make confession and receive absolution before partaking. In other words, due preparation was necessary before receiving the sacrament.

The reformers in Scotland were particularly critical of the Mass. John Knox declared that he would rather that foreign troops invade Scotland than one Mass be offered in the kingdom. This of course put him on a collision course with Mary, Queen of Scots, who claimed her right to have Mass in her private chapel. The central objection to the Mass was the obscuring of the Lord's Supper in the manner of its celebration. The laity, when they did receive Communion, received only the bread as they knelt at the altar. Knox and the other reformers insisted that the cup be restored to all the people of God. It was also felt that the proper posture was not kneeling but sitting at the Lord's table alongside the minister. The reformers were also critical of infrequent Communion and wished to have at least a monthly observance of the Sacrament. This was not achieved until the twentieth century, partly because of the instinctive reluctance to receive Communion often and partly

Figure 22.1 Silver Communion cups from Leslie church, Aberdeenshire. Engraved with branches of vines and grapes along with the inscription 'I am the true vine and ye are the branches'. Fashioned by James Penman of Edinburgh, c1688. Courtesy of Aberdeen Art Gallery and Museums.

because the reformers insisted on careful preparation before one could dare to approach the Lord's table.

The period of preparation for the Communion season could be over several months. The minister would visit throughout the parish, meeting with groups of parishioners and catechising them in the style recommended by the *Shorter Catechism*, which had been approved by the General Assembly of 1648 as 'a directory for catechising such as are of weaker capacity'. This would be followed by a series of sermons on sacramental doctrine. During the week immediately before the observance of the Sacrament there would be a fast day on which at least two sermons would be preached and two other days of services. On the Saturday metal tokens would be issued to those deemed worthy and these tokens gave admission to the Communion table. The tokens were handed in and re-used at subsequent seasons. Although prior to 1560 the reformers had celebrated the Lord's Supper in private houses, such observances were forbidden after the Reformation, the parish church being regarded as the proper place for the celebration of the Sacrament. Communicants would come in relays to sit at the table until all had received. This might last most of the Sunday. On Monday there would be one or more services of thanksgiving.

Preparation on such a scale meant a considerable strain on the parish minister and it became customary for him to be assisted by two or more other ministers. It also meant that it was impracticable to observe the

Sacrament more frequently than once or at most twice a year. Because of the rarity of Communion seasons it became the custom for people to attend Communion seasons in parishes other than their own. Gatherings for Communion were often quite large and imposed considerable burdens of hospitality on the host parish. Despite this, hospitality was usually on a generous scale and much merriment ensued, in marked contrast to the solemnity of the services. This was the Holy Fair described by Burns. It has been suggested that Burns' satire was more effective than the Acts of the General Assembly in bringing this style of Communion to an end.

The nineteenth century saw considerable changes in Communion occasions. They became more regular on fixed Sundays and were held twice a year or quarterly. Preparation was limited to a service on the preceding Friday, which was the occasion on which first communicants were formally admitted on profession of faith. This was the culmination of several weekly meetings for instruction in Christian doctrine and practice and replaced the former catechisings. In larger parishes there would be more than one service of Communion on the Sunday and a service of thanksgiving on the Sunday evening. In order to shorten the length of the Communion service, it gradually became the custom to serve the bread and wine to communicants in their pews and to discontinue the practice of going forward to the Communion table. The book boards of the pews were covered with white cloths to symbolise that each was an extension of the Communion table. The innovation was condemned by the General Assemblies of 1825 and 1826, but popular opinion favoured the practice and it became general. It has since been felt that something of value was lost by the change and it is now the practice in some churches, such as St Giles' in Edinburgh, to go forward to the table to receive the sacramental elements. In others the opportunity is taken whenever possible to have communicants sit round a table.

Figure 22.2 Free Church Communion, Plockton, early twentieth century. Courtesy of Reverend Norman Macrae.

Communion tokens were discontinued and replaced by cards bearing the name and address of the communicant. These were distributed by elders to the communicants residing in their district but without catechising. The purpose of the card was to record attendance at the Communion, not to regulate admission like the former token. Failure to be present at Communion over a period of three years was held to be a possible reason for the kirk session to remove that person's name from the roll of communicants. This gave rise to the impression that attendance once every three years was sufficient to retain Church membership. For this reason communion cards are being discontinued in many parishes today.

Early in the twentieth century a further innovation was introduced, possibly following American practice and certainly because of concerns about hygiene and infection. This was the introduction of the individual cup, or rather small glass, instead of the common cup. It has become widespread in Scotland and very few parishes offer the Sacrament in the traditional manner to communicants despite the act of the General Assembly of 1909, which established their right to do so. It is also seldom possible to break bread for in most parishes the bread is served in small cubes.

Communion occasions are now more frequent in many parishes. Monthly Communions are fairly common and Easter, Pentecost and Christmas Communions are not unusual. Very few parishes have weekly Communions.

BAPTISM

The Lord's Supper was clearly intended by Jesus to be a frequent occasion. By contrast baptism was to be a unique occasion for each individual. That much would be conceded by all branches of the Church. Controversy has surrounded these questions: to whom and by whom should baptism be given? Where and when should it be administered?

The Directory for the Publick Worship of God, which an Act of the General Assembly of 1645 'established and put in execution', says that while baptism is not to be unnecessarily delayed, it is not to be administered in any case by any private person but by a minister of Christ, called to be a steward of the mysteries of God. This excludes emergency lay baptism in the case of an infant in danger of death. The *Directory* forbids baptism in private places or privately and orders that it should be given in the place of public worship in the face of the congregation where the people may most conveniently see and hear. Fonts at the door at the back of the church are explicitly forbidden.

The *Directory* speaks only of the baptism of children. After notice given to the minister the day before, the child is to be presented by the father (or in the case of his necessary absence) by some Christian friend in his place, professing his earnest desire that the child may be baptised. The *Directory* is not a Book of Common Prayer or even a Book of Common Order, but does give fairly detailed guidance to the minister in this and other services. It is suggested that the minister, in outlining the doctrine of baptism, should point

Figure 22.3 John Philip, *Baptism in Scotland* (1850). Courtesy of Aberdeen Art Gallery and Museum.

Figure 22.4 Baptismal ewer, Greyfriars, Edinburgh, 1707–8. Courtesy of the Trustees of the National Museums of Scotland, NMS Q.L.1950.24.

Figure 22.5 Baptismal certificate, United Free Church of Scotland, 1925. Courtesy of the Trustees of the National Museums of Scotland, NMS SLA C29980.

out that it is not so necessary to salvation that an unbaptised child is in danger of damnation. There is much of exhortation to congregation and parents but no explicit vows or promises by the parents are detailed, apart from a warning of the consequences of failing to bring up the child 'in the knowledge of the grounds of the Christian religion, and in the nurture and admonition of the Lord'. The manner of baptism is to be by pouring or sprinkling water on the face of the child, without adding any other ceremony.

Since the promulgation of the *Directory*, practice has differed considerably except that lay baptism is unknown in the Church of Scotland and other Reformed Churches. Baptism of children is frequently delayed and in many cases not requested. Baptism in the child's home rather than in the church was almost the norm until the latter part of the twentieth century. Sometimes it was given in the manse or in the vestry of the church. Emergency baptism of sickly infants is quite common but is usually given by the hospital chaplain if the parish minister is not available. The writer's first baptism after his ordination was of a sickly infant in a maternity hospital. His second was of an adult on profession of faith. Both were in private.

MARRIAGE

The Directory for the Publick Worship of God has a good deal to say about marriage. It recognises that marriage is not a sacrament nor peculiar to the church of God but common to all mankind, but because those who marry 'have special need of instruction, direction, and exhortation from the word of God' it is judged expedient that marriage be solemnised by a lawful minister of the word that he may counsel them and pray for a blessing upon them. Act I of the General Assembly of 1977 re-affirmed that only an ordained minister may solemnise marriage in the Church of Scotland.

The *Directory* states that marriage is to be betwixt one man and one woman only, but probably had polygamy rather than same sex marriage in mind! The minister is to solemnise the marriage in the 'place appointed by authority for publick worship', that is, in church. This requirement has been frequently ignored. When the writer's parents were married in 1928 in Clepington Church, Dundee, this was regarded as unusual. During World War II marriages were frequently held in the manse before two witnesses. In the latter part of the twentieth century it became fashionable to have weddings in church. There is a contemporary trend to have marriages in hotels or even public houses. The Chancellor of the Exchequer, Gordon Brown, wedded his wife in his own home. The lack in Scotland of any legal requirement for a religious marriage to be held in a place of worship or a building registered for the purpose has led to weddings being held in eccentric locations such as hill tops or under water.

The *Directory* makes no mention of irregular marriages but these were common prior to 1940 when new legal enactments came into force. The most popular irregular marriage was a simple declaration that the parties took

Figure 22.6 Wedding group, Davidson's Mains parish church, Edinburgh, 1965. Courtesy of the Trustees of the National Museums of Scotland, NMS SLA C29980.

each other as man and wife, sometimes, but not necessarily, before witnesses. Another form was a simple promise followed by intercourse. These two forms were abolished by the Marriage (Scotland) Act 1939, but it is still possible for a marriage to be recognised by habit and repute. This is the case of a couple free to marry who have lived together as man and wife and have been reputed to be so. The act of 1939 introduced civil marriages by a duly authorised registrar in the registrar's office. This latter requirement made the unrestricted location of religious ceremonies attractive. Civil weddings are now possible in designated locations other than the registry office.

FUNERALS

The Book of Discipline (1560) has this to say about 'Buriall': 'We think it most expedient that the dead be conveyed to the place of buriall with some honest company of the kirk, without either singing or reading; yea without all kind of ceremony heretofore used'. The *Directory* is of the same mind, but adds: 'This shall not extend to deny any civil respects or deferences at the buriall, suitable to the rank and condition of the deceased, while he was living'. The *Book of Discipline* discusses the desirability of sermons at funerals and points out that the preparation of them take a great deal of the minister's time and may tempt him to preach only at the burial of the rich, a discrimination to be condemned. Despite this dissuasive sermons were preached at funerals. John Knox preached at the funeral of the Regent Moray in 1570 and offered a prayer of thanksgiving. The Regent was buried in the south aisle of St Giles', Edinburgh, again a practice condemned by the *Book of Discipline*.

Figure 22.7 George Rankin (1864–1937), *The Crofters Funeral*. Courtesy of the Trustees of the National Museums of Scotland, NMS SLA C29976.

Figure 22.8 James Guthrie, *A Highland Funeral* (1881). Courtesy of Glasgow Museums: Art Gallery and Museum, Kelvingrove.

The Forme and Manner of Buriall usit in the Kirk of Montrois, preserved in Volume I of the Wodrow Miscellany, shows that even in their own day the views of the compilers of the *Book of Discipline* were not everywhere adopted or enforced. The Montrose order of service (dated prior to 1581) provides for an address followed by a prayer and a hymn. The prayer is almost identical with a prayer in the 1552 *Book of Common Prayer of the Church of England*. The hymn is taken from the *Gude and Godlie Ballates* compiled by Robert Wedderburn prior to 1578.

The Glasgow General Assembly of 1638 passed an act forbidding funeral services, but the *Directory* says that the minister 'as upon other occasions, so at this time, if he be present, may put them in remembrance of their duty'. In some parts of Scotland however it was not customary for ministers to be present at burials until well into the nineteenth century. The lack of religious services at burials created a vacuum which was filled with excessive drinking and even merriment.

The re-entry of prayers at burials was by way of an extended grace at the meal which followed. This was offered by the minister if present but also by elders if he were not. *The Scotch Minister's Assistant* of 1802 makes no provision for funeral services but includes a prayer when 'Death visits a family'. In 1867 the Church Service Society issued *Euchologion or Book of Prayers* which included funeral services at the house and in public. The services are preceded by this comment: 'The gradual and general resumption of prayer at funerals has long since proclaimed the universal conviction that no good reason for omitting on such occasions the devotional observances solemnizing and comforting to the living can be found in the fact of their having once been regarded as beneficial to the dead'.

It was not until 1897 that the Church of Scotland gave official sanction to services at the burial of the dead. These were included in *Prayers for Social and Family Worship,* a hand book for those who for any reason were 'deprived of the ordinary services of a Christian ministry'. In 1923 *Prayers for Divine Service* was published by authority of the General Assembly of the Church of Scotland for optional use by ministers. This book includes a service to be conducted at the house or in church and a committal and prayer to be used at the grave.

When the United Free Church of Scotland issued a *Book of Common Order* in 1928 provision was made for services at the cremation as well as the burial of the dead. In cities such as Aberdeen cremation is now the norm and interments are infrequent. Recent *Books of Common Order* (1979 and 1994) have 'funeral services' rather than services for the burial of the dead. They also suggest that an 'address' may be given. Such addresses are usually eulogies or celebrations of the life of the departed and only rarely could be described as sermons. Although it is usually expected that a minister will conduct the service, the address is quite often given by a close friend or a member of the family.

BIBLIOGRAPHY

Burnet, G B. *The Holy Communion in the Reformed Church of Scotland*, Edinburgh and London, 1960.
Cameron, J K, ed. *The First Book of Discipline*, Edinburgh, 1972.
Forrester, D and Murray, D, eds. *Studies in the History of Worship in Scotland*, 2nd edn, Edinburgh, 1995.
Kerr, J. *The Renascence of Worship*, Edinburgh, 1909.
MacLean, C. *Going to Church*, Edinburgh, 1997.
Maxwell, W D. *A History of Worship in the Church of Scotland*, Oxford, 1955.
Schmidt, L E. *Holy Fairs – Scottish Communions and American Revivals in the Early Modern Period*, Princeton, New Jersey, 1989.
Sprott, G W. *The Worship and Offices of the Church of Scotland*, Edinburgh and London, 1882.

Service Books
The Book of Common Order 1564.
The Directory for the Publick Worship of God 1645.
The Scotch Minister's Assistant 1802.
Prayers for Social and Family Worship 1858.
Euchologion or Book of Prayers 1867.
Prayers for Social and Family Worship 1898.
Prayers for Divine Service in Church and Home 1923.
Book of Common Order for use in Services and Offices of the Church 1928.
Prayers for Divine Service 1929.
Book of Common Order of the Church of Scotland 1940.
The Book of Common Order 1979.
Book of Common Order of the Church of Scotland 1994.

23 Missions and Missionaries: Home

FRANK BARDGETT

ORIGINS – EVANGELICAL[1] CONCERNS AND SOCIETIES: 1790S–1820S

The tasks of the Christian Church have been described as being fivefold: evangelistic (summoning of men and women into the fellowship and service of Jesus Christ); educational (instruction in the meaning and significance of Christian faith); pastoral (assisting people with moral, spiritual and personal needs); prophetic (as the conscience of the world challenging things that are and should not be, or promoting things that are not and should be); and social (offering service in the name of Christ without thought of return).[2] To this list, others would add a sixth: maintaining and encouraging the worship and praise of God through the ordinances of religion. The sum of all these tasks may be said to be the mission of the Christian Church: her reason for being, her proper service to Jesus Christ. In Scotland, home mission has, since around 1800, comprised a large variety of movements each seeking to concentrate on one or another or some combination of all these tasks, by means additional to the normal parish ministry or congregational life. Throughout, evangelism has been at the heart of home mission.

In the late eighteenth century increasing numbers throughout Britain came to a new realisation of the urgency of the mission of the Church. For example 'Puritanicus', writing in July 1797, asserted: 'The state of this country [Scotland] at present, from the progress of infidelity and luxury, and from the total relaxation of domestic discipline, is very alarming. It is certain, that common methods are not sufficient to remedy the dreadful evil'.[3] In their desire to go beyond 'common methods', the home and overseas missionary movements came from the same stock, shared (at least at first) something of the same structures, and began largely outside the existing denominational institutions. Between 1796 and 1797, following the example of evangelicals in England, a monthly prayer meeting was begun in Edinburgh 'for the revival of religion at home, and for the success of the Gospel abroad'.[4] The vocabulary used, the bi-polar focus on 'home' in conjunction with and opposition to 'abroad', had long roots in English language and British thought. With the increasingly global reach during the seventeenth century of British trade and territories, and of the Royal Navy, the division of the world into concerns 'home' and 'abroad' became commonplace and the new missionary constituency easily adopted an existing terminology. 'Puritanicus', indeed, argued:

> I am one of those who still think that something more effectual should be attempted for the advance of religion *at home*; and that, while Societies are honourably exerting themselves to convey the knowledge of the Gospel to distant countries, they should, at the same time, look a little into the state of their own. *Here* is an ample field for the display of Missionary ardour and activity; and I know not but it may require a greater degree of Christian zeal, and self-denial, to meet opposition from the ignorant, the self righteous, the careless, and the abandoned, where Christianity is long established, and is little known in its power and spirit, than even to instruct the superstitious natives of Hindustan.[5]

One society, formed in London in 1819 to promote preaching tours to the countryside, in fact took the name of 'The Home Missionary Society'.[6]

The prime means adopted for these twin spheres for mission was the voluntary society or association. Indeed, the Society in Scotland for the Propagation of Christian Knowledge (SSPCK), founded by a number of 'private gentlemen' in Edinburgh in 1701, already sought voluntary giving to fund schools supplementing the parish schools in the Highlands and Islands, and to appoint its own missionaries and catechists for both the Highlands and Canada. By the end of the century, the growth of what has been described as a 'maritime consciousness', especially in relation to India and the South Seas, completed the preconditions needed for the birth of the modern British missionary movement and a host of societies came into existence, some with a British reach, some with local reach.[7] This generation of evangelicals, finding their establishments unwilling or opposed to taking any further measures to proclaim the gospel, determined nevertheless to go forward in mission, using every lawful method with or without denominational sanction.[8] Their monthly magazines (*The Evangelical Magazine*, published in London from 1793, and *The Missionary Magazine*, published in Edinburgh from 1796), by which the enthusiasts of the 1790s shared news ('intelligence'), hence invited attention to mission whether at home or abroad. So Rev. Grenville Ewing, at the time a Church of Scotland minister, secretary of the Edinburgh Missionary Society and editor of the *Missionary Magazine*, included in its first issue both news about Moravian overseas missions and a review of a book *Genuine Religion the Best Friend of the People* and commented: 'At a time when the great and animating design of spreading the knowledge of the Gospel among the heathen engages so much of the attention of the friends of Christianity in this country, we are happy to find that the no less interesting object of diffusing its influence at home, is by no means overlooked or neglected'.[9]

Together with the brothers James and Robert Haldane, Grenville Ewing shares in the origins of home mission in Scotland as a movement at the least supplementing the normal parish ministry, and often using 'all lawful means' independently of it, by means of specially funded denominational or interdenominational committees, societies or associations.

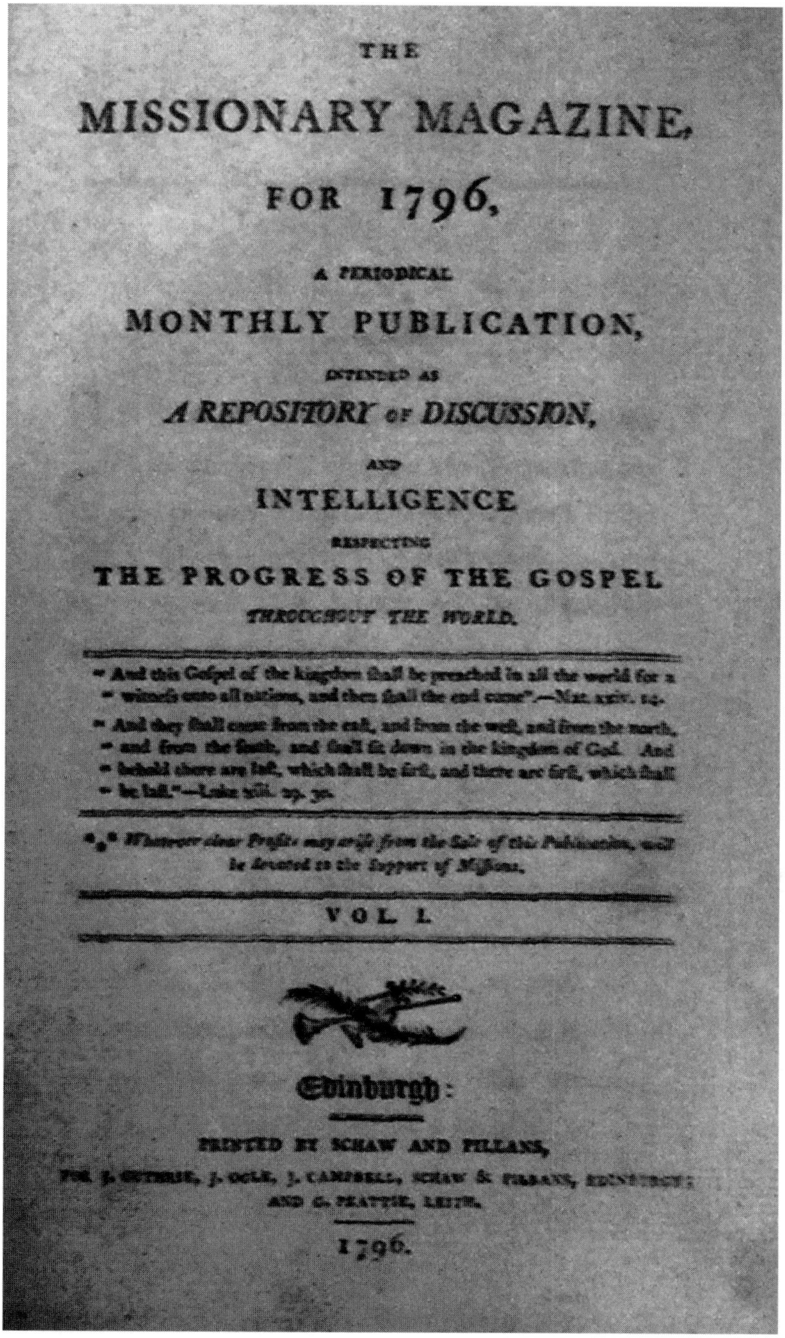

THE

MISSIONARY MAGAZINE,

FOR 1796,

A PERIODICAL

MONTHLY PUBLICATION,

INTENDED AS

A REPOSITORY OF DISCUSSION,

AND

INTELLIGENCE

RESPECTING

THE PROGRESS OF THE GOSPEL

THROUGHOUT THE WORLD.

"And this Gospel of the kingdom shall be preached in all the world for a witness unto all nations, and then shall the end come".—Mat. xxiv. 14.

"And they shall come from the east, and from the west, and from the north, and from the south, and shall sit down in the kingdom of God. And behold there are last, which shall be first, and there are first, which shall be last."—Luke xiii. 29, 30.

*** *Whatever clear Profits may arise from the Sale of this Publication, will be devoted to the Support of Missions.*

VOL. I.

Edinburgh:

PRINTED BY SCHAW AND PILLANS,

FOR J. GUTHRIE, J. OGLE, J. CAMPBELL, SCHAW & PILLANS, EDINBURGH;

AND G. PEATTIE, LEITH.

1796.

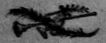

Figure 23.1 Cover page of *The Missionary Magazine* for 1796. From a private collection.

Frustrated in his desire to go himself as a missionary to Bengal, through the *Missionary Magazine* Ewing encouraged the introduction to Scotland of itinerating preaching, a missionary scheme already well used by dissenters in England.[10] With the Haldanes he helped to found in December 1797 the Society for Propagating the Gospel at Home [SPGH], which recruited preachers and catechists, and offered training and funds for shorter or longer preaching tours of the counties of Scotland, especially to areas where they believed existing ministries did not effectively communicate the gospel. Having resigned from the ministry of the Church of Scotland, Ewing superintended theological training for the SPGH's recruits, and himself undertook preaching tours. By 1810, his practice on arrival at a small town was to pay a fee to the local bellman to have a sermon publicly announced. Thus he gathered the people, choosing a central location, himself only equipped with a Bible, chair and desk, for reading a psalm and preaching. This method, he urged, had 'all the simple elegancy ... of ancient times'.[11] In 1812, Ewing was instrumental in the creation of the Congregational Union in Scotland as an association for the mutual support and fellowship of the new independent congregations. Wealthier churches and individuals donated funds to the Union for the financial support of new causes: 'The institution now formed was, essentially, a Home Missionary Society' concluded Ewing's biographer, his daughter.[12]

Faced with vehement opposition from the Church of Scotland, and considered seditious by the government during the French Revolutionary and Napoleonic wars, by the early years of the nineteenth century these itinerant missions had developed into the deliberate creation of new congregations by what were to become the Scottish Congregational and Baptist denominations.[13] The brothers Haldane had adopted Baptist principles by 1808. After the initial period of the first itinerancies, the practice of the preachers was to 'separate the brethren', calling on those who had come to faith or who professed an evangelical faith to found a new, separate congregation of the committed, for fellowship, prayer and teaching. Those converted by SPGH missionaries thus often faced opposition. An account by Robert Little, pastor of a fellowship in Perth, of John Farquharson's mission in Breadalbane (1800–4) mentions that those who responded to Farquharson's ministry were:

> ... traduced for leaving the church of their forefathers – for casting cruel and unjust reflections on those who lived and died in that church. [Nevertheless] ... many of them walk ten or twelve miles or more, through bad roads, and even heavy rain, to sit on the side of a mountain, (for they have no house to meet in) and listen to the joyful sound. On the Sabbath I was with them, it rained most of the day, yet they continued without dispersing for five hours, without any shelter.[14]

Other such Highland congregations besides discomfort accepted eviction and emigration rather than return to the Church of Scotland.[15]

Apart from those of the SPGH, Methodist missionaries were also active in Scotland. Their most successful sphere of operations was Shetland where John Nicolson, John Raby, Samuel Dunn and others built up congregations after 1819. Nicolson was himself a Shetlander who returned to his home islands on his own initiative to undertake preaching, while Raby and Dunn were sent north by the Methodist Conference and supported by funds raised in England through the agency of the President of the 1822 Conference, Dr Adam Clarke. Starting from scratch (as far as Methodism in Shetland was concerned), the denomination had grown by 1828, the year of Nicolson's death, to some three thousand in congregations across the islands, supported by six ministers, possessing seven or eight chapels and numerous other preaching stations.[16] As with the SPGH preachers, the Methodists initially intimated a sermon by whatever means of public announcement were practical; met in borrowed parish or independent churches, halls or schools, or sail-lofts, or in the open air; gathered the converted into their denomination's distinctive classes for discipline and fellowship, and then built chapels.

From its beginning, evangelical preaching and church planting were thus high priorities for those engaged in mission at home. Education was also part of the core vision of the founders of the movement. Missionaries and catechists of the SSPCK routinely organised Sabbath schools in the Highlands. Grenville Ewing also strongly advocated the value of Sabbath schools: the praying societies in Edinburgh with which he was connected founded the Edinburgh Gratis Sabbath School Society in 1797. As in England, the Sunday school movement had begun in Scotland as an adjunct to the normal parish schools. Sunday schools catered particularly for children employed during the week, perhaps in factories, and hence unable to attend day schools; or else for those unable to afford day school fees. A minimal basic education in reading was offered, together with the same instruction in the catechism and the psalms as was integral to the fuller parish school syllabus. Such Sunday schools had by the 1780s gained wide support, and could be funded by Burgh Councils as well as by private subscription. A society was thus established in Paisley in January 1798 for supporting Sabbath and weekday evening schools: its funds came via occasional special collections at evening worship at Paisley High Kirk (the Abbey). The Gratis school movement of the 1790s, however, aroused the same opposition as the itinerant preachers with whom it was associated. Often attended by adults as well as children, with a strong element of Christian doctrine, and with teaching that transformed into preaching, these schools were suspected of subverting the established constitutional order and hence, with the missionaries, were condemned by the General Assembly in its Pastoral Admonition of 1799.[17] Nevertheless by 1809 the Edinburgh Gratis Society was supporting some 40 schools, attended by more than 1,500 children: its twelfth annual meeting heard that there was more 'to be done for the instruction of the children of the poor, the ignorant and the careless in this city and its adjoining villages'.[18] Similarly in 1813 the Paisley Society, connected with the Established Church, was reporting that it supported 37 Sabbath and two weekday evening schools, with 1,365 children, 110 young men and 272 young women enrolled.[19] By 1819, a total of 452 Sabbath schools could be identified across Scotland, largely without official countenance from the Established Church, but often maintained by district societies.[20] Sunday schools rapidly became and still remain a normal and not an extraordinary part of congregational life.

Besides Sabbath schools, the educational wing to the home mission movement also comprised societies promoting the wider circulation, reading and knowledge of the Bible, and of other Christian literature and tracts. Since the Reformation one of the tasks of the Church had been to encourage personal knowledge of the Bible and of the doctrines founded on it and taught in the catechism – and this knowledge in turn presupposed schooling to a level of education sufficient to read the Scriptures, and preferably personal possession of a Bible (or at least a New Testament) to feed personal faith. As Sabbath school societies supplemented the work of the parish day schools, the 'common means' of the regular ministry of the Church was supplemented after the 1790s with a multitude of Bible and Religious Tract societies. Widely supported in Scotland, the British and Foreign Bible Society (founded London 1804) was often the model followed. The Paisley and Eastern Renfrewshire Bible Society, for example, had a constitution that required it to supply Bibles and Testaments and to aid the funds of the British and Foreign Bible Society and others; to follow the principles of the British and Foreign in that the Bibles it promoted would be in the languages of the United Kingdom and of the Authorised version only, 'without note or comment'; and that they should be offered for sale at a low price rather than generally supplied free of charge. At the Society's annual meeting for 1813, with the Provost of Paisley in the chair, office-bearers included the Rt Hon. the Earl of Glasgow, Lord Lieutenant of Renfrewshire, as President, his Vice-Lieutenant, the county MP, the Sheriff Depute, the Sheriff Substitute, the Provosts of Paisley and Renfrew and other notables.[21]

While the British and Foreign Bible Society attracted support from the higher ranges of society and across the theological spectrum, the evangelical wing of the Churches, including the non-established but presbyterian Relief and Secession denominations, often gathered in more general associations aimed at sharing missionary news and offering gospel-focused financial support whether for Bible Society, mission, schools or tracts work – or for combinations of these causes. Many of these supported the prestigious London Missionary Society, and its accounts thus offer information on a range of Scottish evangelical activity. In 1827, for example, the LMS acknowledged receipt of financial contributions from: in <u>February</u>: the Brechin Society for Missions, Schools and Tracts (Rev. D Blackadder), the LMS Glasgow Auxiliary Society, Kilmory (collection at parish church), Galston Bible and Missionary Society, Peterhead Missionary Society; in <u>March</u>: Dundee Juvenile Bible and Missionary Society, Dunfermline Missionary and School Society, Nairnshire Society for the Propagation of the Gospel; in <u>April</u>:

Dove Hill Relief Congregation in Glasgow, Cumnock Society for Religious Purposes, Carmunnock Bible and Missionary Society; in May: both the Glasgow and Greenock LMS Auxiliary Societies, Dumbarton Bible and Missionary Society, Dunkeld Missionary Society, the Cumbraes Missionary and Bible Association, Keith Missionary Association and the Leith LMS Auxiliary Society.[22] Behind these donations lie gatherings of serious Christians, meeting to exchange 'intelligence', to pray, to raise funds and to support their chosen mission causes, whether at home or abroad. Action, too, was sponsored by the societies. The *Evangelical Magazine* for September 1810 reported the first annual meeting of the Edinburgh Bible Society. More than £1,700 had been raised to finance the distribution of Bibles in Danish, Dutch and French translations to prisoners of war held in Scotland, and of English Bibles to the army, the navy and the poor. Money had also been given to the British and Foreign Bible Society.[23]

Appealing to the more evangelical, the Tract Societies went beyond the simple placing of Scripture in a neighbour's hands; through the medium of story, sermon or an appeal in letter form, tracts pressed home to the reader the urgency of his or her need to respond to the gospel. Edinburgh's praying evangelicals founded the Edinburgh Religious Tract and Book Society in 1793, and many other such societies followed. The Glasgow Religious Tract Society was formed in May 1803, with Grenville Ewing prominent among its promoters. Members subscribed one shilling at entry and a minimum of six pence a quarter, and were entitled to tracts equal to half their contribution – these they then distributed free to neighbours, chance acquaintances or passing strangers. By 1805, it was calculated that some 13,834 tracts had been put into circulation by the Glasgow Society, gratuitously 'to diffuse the saving knowledge, and promote the practical influence of divine truth'.[24] In Haddington, the Circulating Tract Society required a quarterly subscription of either one shilling or six pence, but members were entitled to choose tracts to the full value of their subscription: 'Meetings to be held upon the first Thursday of February, May, August and November for receiving tracts, conversing upon the state and appearance of Religion, and the various ways of advancing its interest, and recommending it to the serious consideration of those around us; every meeting shall begin and be concluded in Prayer'.[25]

THE 'AGGRESSIVE SYSTEM': THE 1820S AND BEYOND

After the end of the Napoleonic wars, a new generation of evangelical ministers came to hold increasing sway in the established Church of Scotland. Their concern focused on the cities, with populations swelled by industrialisation. In addition, they sought to defend the principle of establishment, of State support for an established Church, as the best means by which the gospel could be taken to the poor, to those unable to afford their own ministry. Through the example and leadership of Thomas Chalmers, these two causes came together. Chalmers described the work of the non-established Churches, whether presbyterian or independent, as 'attractive', the gathering together of committed, self-supporting congregations by a preaching ministry that literally attracted. He argued that such ministries were inadequate to solve the urgent destitution, both physical and spiritual, of the urban masses. For this, a new home missionary system was needed: the 'aggressive' system, based on the territorially-defined parishes of an established Church. During his ministry at Glasgow's St John's Parish (1819–23), Chalmers sought to create a model of this new method of working, which called for the parish to be divided into districts, each with its own allocated visitors: elders, deacons and Sabbath school teachers, whose task would be personally to visit all in their area. This visitation of homes, the pressing home of the claims of the gospel in a systematic way that ensured the coverage of the entire population of a parish, together with church-based poor relief, constituted Chalmers' aggressive system. He sought to support the preaching of the gospel with city-wide personal contacts, and hence to bring the energies and subscriptions of the evangelical supporters of mission to complement the work of the ordained minister.[26]

The application and development of the aggressive, territorial system became an evangelical orthodoxy within the Established Church after Chalmers' ministry at St John's. A network of societies was created in Glasgow with the dual aims of supporting and extending the Established Church: 'the parochial institutions of Scotland'. Hence St David's parish created a Parish Society in 1836, bringing together the minister, missionary and elders with all other like-minded members, to promote Christian fellowship and visitation by districts and, through the visiting, the provision of subsidised Bibles, recruitment to Sabbath schools, the distribution of tracts and the purchase (from Glasgow Burgh Council) of pew sittings in the church for sub-renting to impoverished families.[27] Such parish associations or societies provided a means of raising additional funds for new work and as autonomous institutions under their own directors by-passed the courts of the Church. The object was by new means to bring the new, industrial centres of population back within a traditional Scottish religious and social discipline.

The 1830s saw the development of the aggressive system by the engagement of missionaries as paid agents to support parish visitation within towns and cities: adopted by independent and interdenominational home missionary societies, the system also outgrew the parish structure envisaged by Chalmers. Missionaries for home stations were not unknown, however, within the Established Church. Indeed, they predated by several decades those of the itinerating movement of the 1790s. Among the Committees of the General Assembly of the Established Church was the Royal Bounty Committee, founded in 1725 to manage a royal grant given for the employment of additional ministers and catechists in large Highland parishes: 'for the reformation of the Highlands and Islands of Scotland, for promoting the knowledge of true religion, suppressing popery and profaneness ...'. Previously, itinerant ministers had been sent on mission to the Highlands on short-term commissions for the same purposes. By 1795 the official Scheme of the Royal Bounty Committee described its ministers as 'missionaries', or 'missionary ministers', as were those appointed by the SSPCK to its stations. Such ministers were appointed to Highland stations serving districts remote from the parish church and exercised a normal ministry there, though, like colleagues appointed to SSPCK stations, they were denied membership of the presbytery for the area.[28] By the 1830s similar missionary appointments were being made for urban ministry by voluntary societies on the basis of locally-raised funds.

Hence a Town Mission was established in Greenock, on an interdenominational basis, specifically to appoint home missionaries. In 1830 the second report of the directors emphasised the remit given to their agents: '1st, In visiting the lower classes in their own houses; 2nd, In collecting into one house individuals living in the same neighbourhood, for the purpose of reading and expounding the Scriptures; and 3rd, In an investigation into the state of the community generally'. The purposes of the third heading were to ensure an adequate provision of parish day schools and congregational Sabbath schools, and to assess the proportion of the population for whom accommodation in existing church buildings was available.[29] While the clergy of the town attended the AGM of the Mission, its office-bearers and directors were laymen: their missionaries, however, were clergy – ministers in training (probationers) or those who though qualified were yet to be called to a charge (licentiates). Similarly the 1835 *Annual Report of the Society of Probationers of the Church of Scotland* (instituted 1832) recorded the work of the probationers in mission stations throughout Edinburgh, distributing tracts and Bibles, opening parish libraries and visiting the poor. Recognising that probationers held only temporary appointments, they urged that permanent employment should be offered:

> During the last year, paid agency has been introduced into several parishes of the city, and, while in all, it cannot fail to do good, the spiritual destitution of many districts aggravates the urgency of its immediate and universal adoption. The benefit which one paid agent in each parish, aided and encouraged by its Minister and Elders, would soon effect is incalculable. He would go in and out among the people, bearing with him the influence and authority of an accredited messenger, and would give his Sabbath ministrations an interest and power quite irresistible ... Had each of the parishes in which the districts inhabited by the poor lie, a paid Missionary, pious, zealous, prudent, well educated, and receiving the cordial co-operation of the Clergyman and Kirk Session, we are convinced that the face of the society in them would present a very different aspect from what it does at present.[30]

Already in May 1832, a committee of Edinburgh gentlemen had created the Edinburgh City Mission in order to operate just such a scheme as the Probationers' Society recommended. In their 1834 Report they explained the origins of the Mission: 'Persuaded that the mass of ignorance, vice, and superstition, would continue untouched, or nearly so, by ordinary means, – that if an impression was to be made on their wretched subjects, they must have spiritual instruction brought to them,' its members proposed 'engaging the services of individuals willing to penetrate the abodes of misery, to go from door to door, inviting, compelling attention to the soul and eternity'. More particularly, they divided the city and suburbs into thirty districts:

> ... resolving to locate in each of these, not occupied by any similar labourer already, an Agent, who, aided and encouraged by visitors, or the voluntary services of private Christians, should visit the inhabitants with a view to religious conversation, reading the Scriptures, – form meetings for prayer and Christian instruction, – promote the circulation of the Bible and religious tracts, – stimulate to regular attendance on the preaching of the gospel, – advance Scriptural education, by Sabbath or Infant schools, or otherwise, etc etc.[31]

Members of the Society were required to be persons 'of decided piety and evangelical principles', 'irrespective of denominational distinction'. They believed that their mission reached 'a class of characters which seldom or never comes under the ordinary means of grace, and which can by aggressive operations alone ever be brought to hear the sound of the gospel of grace'.

Edinburgh, however, was not the first to found a city mission. The Glasgow City Mission, under the title 'The Society for Promoting the Religious Interests of the Poor of Glasgow and its Vicinity, or the Glasgow City Mission', dates from 1826 and is held to be the first of its type in the world. The constitution of the interdenominational society stated: 'That the object of the Society shall be to promote the spiritual welfare of the poor of this city and its neighbourhood, by employing persons of decided piety, and otherwise properly qualified, to visit the poor in their own homes' for the overall purpose of increasing 'a knowledge of Evangelical truth'. Approved agents were expected to devote at least four hours daily between 11am and 9pm, except only Saturdays, to the task, and were organised on a district basis. Preaching stations were allocated to each district, at which ordained ministers and preachers might be invited to support the work. Agents might be selected 'from all Evangelical denominations of professing Christians', and detailed printed instructions were provided for them. Visits, for example, would not normally exceed 15 minutes per home. All missionaries were to gather on Monday evenings for prayer, and to keep a daily journal. By 1833, the Glasgow City Mission employed 21 agents who each month contacted 5,374 destitute families. A sermon was supplied for the Seaman's Chapel and a religious address at the Glasgow Police Office. Besides religious instruction classes for 'young men and adult females' and a women's general education class, the Committee was seeking admission to factories to gather employees to 'a practical discourse' given by a missionary. Bible and tract distributions were, of course, also part of the work.[32]

Aggressive visiting, whether by volunteers or paid agency, especially to poorer areas of cities and linked with mission worship and an offer of pastoral support, remained a hallmark of home missionary strategy well into the twentieth century. Edinburgh's Carrubbers Close Mission, for example, begun in 1858, still functions as an independent evangelical congregation, as does Arbroath Town Mission (founded 1849), in 2005 offering 'Christian fellowship / Companionship / Leisure / Café facilities / Lunch Club for Senior Citizens / Activities for Children/Youths'.[33] *District Work: or The Reclaiming of Waste Land* was a pamphlet published anonymously in 1854 in aid of the Glasgow City Mission, claiming to reflect the experience of one 'who speaks from personal experience of what district work really is.', 'Having visited for some time in a missionary district in one of the worst localities in the city, I have seen much to humble and grieve me. The scenes of wretchedness and vice, the filthy and unwholesome dens, in which men and women are crowded together'. The mission work began with the inauguration by the minister of a Sabbath evening sermon ('opposed by Papists and heathens') in a large but miserable apartment in a lodging house, with tin sconces for candles suspended from the ceiling by wires. These services were attended by 'half a dozen pious young men who act as Sabbath school teachers and visitors, two or three persons of respectability – the rest composed of outcasts'. The first service was 'for most of the audience an exercise in learning to keep still'. By the time of publication, it was claimed that some 300 were regularly attending the mission services, that more had obtained membership of the parish church, and that the other features of aggressive work had taken root:

> Sabbath schools, visiting agents, and other means of doing good. Tracts and good books left in houses at the end of the week are, I think, really useful, and supplant in many cases pernicious reading. Reading the Bible to those who cannot do for themselves, and holding prayer meetings are also useful. We have a parish missionary who gives his time and his heart to the work, and is likewise doing great good. Many cases of real improvement are met with by those visiting the district.

Both the paid agent and the volunteer had a place in the active service of the mission. 'Infidelity and socialism are also giant evils that are slaying their thousands; therefore we must have missionaries who are men of God, mighty in the Scriptures, to cope with them'. However, the 'main stay' of the mission was considered to be visiting of the working classes by the working classes: 'They can go amongst their neighbours without fear of being met with the prejudices which unhappily often exist between the poor and those above them in society'. Nor did working men take extended holidays down the Clyde: 'those men do not go to the coast, they are always at hand, and may be depended on, not only in winter but also in summer'.[34]

MISSIONS TO THE HIGHLANDS

The British and Foreign Bible Society authorised an edition of the Bible in Gaelic in 1801. This edition built on an earlier Gaelic translation of the New Testament undertaken for the SSPCK in 1767, and was of use also to ministers in the Lowlands whose parishes hosted a transitory influx of Highlanders, earning cash assisting with the harvest. That Gaelic was now recognised as a proper mode of Christian instruction was in part a result of the evangelical movement, with its emphasis on personal faith; enthusiasts for mission both home and abroad advocated the work of translation to place a readable Bible in the hands of those to whom they preached. Originally the SSPCK had sought to replace Gaelic with English: their task in the early eighteenth century was understood as the bringing of the British, protestant religion to those they considered ignorant, popish, superstitious and probably Jacobite Highlanders. English was seen as the medium of the one culture, and Gaelic as that of the other.[35] From the recognition of a Gaelic-medium Bible, however, came societies to promote Gaelic-medium schools. After an appeal conveyed throughout Britain in the *Evangelical Magazine*, a Gaelic Schools Society was formed in Edinburgh in January 1811. By 1827, the Society managed eighty-one schools with some four to five thousand enrolled. Although Gaelic was their medium, the Society's attitude remained patronising: 'The economy of the Gaelic school is exemplary; masters being only allowed £25 per annum'; 'The extracts of correspondence

Figure 23.2 Mission Hall at Halsary, Caithness, built for Gaelic-speaking Highlanders in Caithness after the Sutherland clearances. Inset reads 'Built by Subscription, Anno Domine 1842'. Photograph by the author.

are particularly interesting and are written in a very primitive style'; 'Its simple design is to bring the benighted mind of the Highlander into contact with the oracles of truth. It places in his hand the record which God has given us in his Son; it teaches him to read it in the language with which he has been familiar since his childhood; and it leaves the issue to Him'.[36] 16,332 Gaelic Bibles, Testaments and Scripture Extracts were reported to have been circulated between 1811 and 1827.

Both the SSPCK and the Church of Scotland's Royal Bounty Committee continued to place missionaries – ministers and catechists – in Highland stations. The former deployed nine missionaries and twenty-nine catechists in 1834, and the latter supported almost ninety missionaries, Scripture readers and catechists by 1853.[37] These established societies were joined by others. Robert Haldane headed the committee of the 'Baptist Home Missionary Society for Scotland, chiefly for the Highlands and Islands'. By 1833 this mission employed nineteen preachers based in such locations as Grantown, Inverness, Kingussie, Skye, Thurso, Tobermory and Westray. These men itinerated between small gatherings of believers; funds were gathered from friends in both Scotland and England. There was also an interdenominational Highland Missionary Society [HMS] under the patronage of the Marquis of Bute in existence by the 1820s, like the others seeking to send preachers 'into those districts of the country where they are most needed'. Unlike the Baptist Society, the HMS sought to send out 'pious, prudent, and active Missionaries, who, instead of propagating the tenets of a party, shall devote their time and talents to the sole object of instructing the people in the great truths of the common salvation'. Nevertheless from the grants given, it is apparent that the mission worked closely with the Committee for Gaelic Missions of the United Secession Church, a denomination that took a keen interest in missions. The HMS also gave funds to the work of the Paisley Society for Gaelic Missions, which in turn helped to support itinerants commissioned by the Congregational Union.[38]

In its fourth annual report, 1824, the HMS spelt out why it believed Highlanders should be the object of evangelical attentions. There was the argument that parishes were too large for any regular ministry to have an impact: 'Does any one need to be reminded of the deep glens and rocky steeps, lakes, firths and arms of the sea debarring the inhabitants from religious instruction?'. The people themselves were considered 'most interesting', with reference to the patriotic debt owed to Highland regiments, and to traditional Highland hospitality. The report pleaded that poets, novelists and romantic travellers were all interested in the Highlands – so why not Christians? 'Were men but as concerned about the Highlander's moral state as they are about his heroic qualities – were there but half as great a taste for saving his soul as there is for exploring his scenery'; 'We should be meeting them at their own doors'. The Highlands, distinct in culture, language and geography, were perhaps viewed as a form of *foreign*, or at least overseas, mission.

Not that all parts of the Highlands needed to be so evangelised.[39] In 1824, 'a lay member of the Established Church' contrasted the state of religion in Argyll with that prevailing in the Church of Scotland presbyteries of Ross and Inverness. He asked what the 'Moderate' clergy of Argyll had done 'in support of Education, Missionary or Bible Societies?'. In the east, however, he offered an impressive list of evangelical activity: the Inverness Society for Educating the Poor in the Highlands (and a number of other Inverness-based societies), the Northern Missionary Society (meeting in Tain), the Cromarty Ladies Missionary and Bible Society, the Cromarty Religious Association, the Black Isle Bible Society, the Dingwall Ladies' Association for aiding Bible and Missionary Societies (and another of the same in Easter Ross), the Rosskeen Religious Association, and the Kilmuir Easter Bible Society.[40] Outside the Gaelic Highlands, in the county of Orkney, the most effective church planting of this period was achieved not by visiting itinerants (who certainly mounted several missions across the Pentland Firth: Quaker preachers even reached Shetland) but by the efforts of the Kirkwall United Secession congregation, founded c1793. Between 1797 and 1870, a further thirteen United Secession congregations were formed across Orkney Mainland and the North and South Isles, so that the Secession became the largest denomination in the county.[41]

SOCIAL POLICY AND CHURCH EXTENSION: THE MID-NINETEENTH CENTURY

By mid-century urban home missionary work was developing into distinctive streams or specialities. On the one hand, aggressive visiting was used in most Scottish denominations by wealthier congregations to found mission stations for the working classes. Some of these stations developed into congregations and eventually became parish or self-standing churches in their own right: this was mainstream home mission work. Arising from this work and in time developing a life of its own was an evangelically-motivated prophetic concern for social policy. Aggressive evangelical work was also carried forward by ad hoc societies seeking to bring the gospel not just to territorial districts but also to discrete sections of the population: canal workers, seamen, railway men, women, children, the travelling people, navvies. The home mission movement therefore offered employment to increasing numbers, and not just of ministers. A sizable publications industry had developed out of the tract societies, so that the Religious Tract and Book Society of Scotland claimed in 1871 an aggregated annual total of 'nearly a million and a half' items sold, including 100,000 copies of the journal *The British Workman*, 19,866 Testaments, 15,843 Bibles, 7,000 Bogatsky's *Treasury*, 3,100 Reid's *Blood of Jesus*, 2,585 Dean Law's *Works*, 2,426 Bonar's *God's Way of Peace* and 2,100 McCheyne's *Memoir*. The Society recruited pious working men to sell tracts, magazines and books from door to door within set districts and paid them £60 a year from donations raised locally and from the profits of their sales. It was claimed that these colporteurs: 'also act as catechists and missionaries in no small degree. They know the truth and love it, and therefore spread it with lips as well as with the printed page'.

> Neither the efforts of the physician, nor of the magistrate, nor of the city-missionary, nor of the minister, nor of the schoolmaster, nor of the temperance agent, nor of the lady-visitor, nor any one else, can ordinarily avail to reclaim them [ie the working classes] to sobriety, or to elevate their condition. It is all, or nearly all, good labour wasted and thrown away; whereas, if you can get them into decent, healthy, and cheerful abodes, you may work all these agencies with delightful encouragement, and with the best hopes, through the blessing of God, of rearing a sober, happy and pious population.[42]

So suggested a leading Free Church minister, the Rev. W G Blaikie, in 1863, advocating the work of housing reform.[43] Having by visiting the working classes discovered something of the conditions in which they lived, Scotland's supporters of mission realised that there were physical or social as well as spiritual obstacles to the reception of the gospel. The historian Callum Brown has identified these constraints as they were tackled through the century: illiteracy, drunkenness, poverty and overcrowded homes. Those engaged in home mission were, by the mid-nineteenth century, thus also likely to be concerned about education (whether day or Sabbath school), temperance, the relief of poverty, better sanitation, housing reform and urban renewal.[44] The Scottish Temperance movement developed out of these evangelical concerns into a mass movement in its own right, with a host of its own voluntary societies and associations.[45] Rev. Robert Buchanan DD, minister of the Free Tron Church in Glasgow, who started the Wynd Mission around 1845, was also involved in advocating slum clearance in the 1860s. The evangelicals defined such constituents of social policy in moral terms: they were part of the great work of the 'moral improvement' of the people. The whole missionary enterprise might also be advocated as helping to prevent revolution. Buchanan, speaking to the Free Church's General Assembly in 1851 on a motion concerning the spiritual destitution of the cities, contrasted the poverty of the poor of Glasgow with their city of 'merchant princes' and their 'wealth and refinement and splendour'. In the light of the European revolutions of 1848, he spoke of the disorder coming from 'those demon-like figures that were seen mounting the barricades of Paris, and shedding blood like water in the late revolution'. He warned that in Glasgow, 'there are elements of mischief gathering deep in those dark and dismal recesses'.[46]

In emphasising the spiritual aspects of destitution of Glasgow and the cities, however, Robert Buchanan was calling primarily for a new programme of church building – of church extension. Recalling the work of Thomas Chalmers, he had earlier argued at the Free Church Presbytery of Glasgow for 'the necessity of a great Home Mission movement in this city'. The population of Glasgow was rapidly increasing, from 150,000 in

Chalmers' time to near 400,000. Buchanan projected this forward another thirty years; his city was heading towards equality with Paris: 'We may have the numbers of the French capital, but we shall have their infidelity, their Popery, their licentiousness, and their lawlessness too'.[47] Buchanan sought a new phase of aggressive work that would create not just city missions or mission stations but churches, led not by missionaries but by able ministers in full status. He looked back to the pre-Disruption phase of Chalmers' work as Convener of the Church of Scotland's Church Accommodation Committee, that had successfully raised funds to build churches for over 200 new congregations across Scotland between 1834 and 1841. Renamed the Home Mission Committee in 1842, its work had been inherited by the Home Mission Committees of both the Free and the Established Churches after 1843. Between 1846 and 1899, the Established Church found funds to endow 405 new parishes, serving a combined population of almost one and a half million. Promoting the extension work of the Free Church, and unlike W G Blaikie, Buchanan believed that social issues would only find their solution if priority was given to the spiritual, because only thus would a moral climate prevail: '... considered simply as a question of money, it is infinitely cheaper to govern society by the Bible than by the sword. Churches cost far less than jails; and schools than poor-law workhouses', he argued.

Engaging in home mission was thus a routine part of the work of wealthier urban congregations of all the main protestant denominations. In the Church of Scotland, the work of Rev. Norman MacLeod at Glasgow's Barony Parish was thought exemplary. In addition to the regular Sunday morning services for his congregation, mission-style evening services were also held, for which working-class clothing was the norm. Because not to have appropriate church-going clothes was considered a major deterrent to attending public worship, the well-dressed might even be excluded. Besides the main parish church, the Barony also managed a mission chapel of its own and up to twelve Sabbath schools, a congregational Penny Savings Bank and a refreshment room offering cheap, well-cooked food. By 1869, the staff included a minister-missionary for the chapel, four lay missionaries, four female workers and 98 Sabbath school volunteers. During the course of MacLeod's ministry (1851–72), six new daughter churches were created within the bounds of the former parish.[48] The key agents in this type of mission work were not, of course, the ministers who supplied the vision, but those who actually walked the streets and visited the homes: volunteer lady visitors and men like Robert McEwen at Glasgow's Bridgeton Village Mission or James Goodfellow of the United Presbyterian Mission to Edinburgh's Newington district.[49] The latter, appointed in 1855, when in retirement wrote a book of memoirs of his forty-year task; for the former, the tribute at his funeral was published. When Robert McEwen died in 1861, he had been engaged in home mission work in the same area for over fifty years: for thirty years as a cotton-spinner, leading a Sabbath school and engaged in visiting before and after work; and then as a paid agent (though with a reduced salary). An elder of the United Presbyterian Church of

Greenhead, of McEwen his minister said: 'Before city missions existed or were heard of, Mr McEwen had embodied the idea in his own history ...'. Women also shared in this grass-roots mission work: the Dundee East Church Female Domestic Mission, for example, sought 'to reach, through means of a female missionary, a class of destitute and degraded females'.[50] This was mainstream home mission: evangelistic, educational and pastoral.

The decades around 1850 may be seen, following Thomas Chalmers' vocabulary of aggression, as a battle for the cities. It was certainly perceived as such by those engaged in home mission. Besides countering sheer ignorance of the Christian faith, those presbyterians and independents involved saw themselves seeking to retain the urban populations within a protestant, sober and moral Scottish culture: the battle could be presented in Buchanan's terms as a battle of civilisations. The massive growth of the population, together with an equally massive movement of people from the country to the towns, threatened to run ahead of the resources of the Churches. Simultaneous large-scale migration from Ireland greatly added to the Roman Catholic population of Scotland, at a time when evangelicals of all Protestant denominations held as a matter of faith that the pope was antichrist. The Roman Catholic Church in Scotland had its own particular vocabulary of 'home mission': this was because, until the re-establishment of the Scottish hierarchy in 1878, the entire Church operated as an outreach of the Vatican under the authority of the Office of Propaganda, with all its priests considered as missionaries serving particular missions. It is clear from documents of the time that considerable rivalry, even hostility, existed between the priests and the presbyterian agencies. With their denomination lacking local funds, and their client population among the poorest in west central Scotland, the Vicars–Apostolic 1835 report to the Holy See emphasised the poverty of their priests, sometimes needing even homes of their own. As the century progressed, the Roman Catholic Church created her own organisations, using the vocabulary of her own tradition, to tackle the challenges of the cities: infidelity, drunkenness, lack of education, ignorance and poverty. The League of the Cross, for example, was the catholic temperance organisation.[51] Parish missions were led by members of the preaching orders – Redemptorists, Passionists or Jesuits, and as with the protestant denominations, good numbers of new churches and congregations opened.[52] That by the end of the century probably the majority of Christian congregations (protestant or catholic) were composed of the skilled working classes, and that Christian attitudes and values were predominant throughout Scottish society (whether church-attending or not) show that the Churches' missions were not without fruit in the cities.[53]

MISSION SOCIETIES AND MISSION CAMPAIGNS: THE LATER NINETEENTH CENTURY

Home mission in Scotland began as a dissident movement within the Established Church and retained a potential for going its own way. The independent voluntary society as a vehicle for mission remained attractive, particularly for those seeking to focus on distinct sectors of the population, and to those with a particular call to evangelism. Sailors and seamen were identified by the early Bible Societies as requiring special agencies if they were to be reached with the gospel. Missions to seamen on a British basis followed, and Seaman's Friend Societies based in the major Scottish ports were established in the 1820s. To apply the principles of Thomas Chalmers to the fishermen's and other coastal communities, Thomas Rosie was particularly instrumental in setting up the Scottish Coast Mission, the West Coast (later Glasgow and West Coast) Mission, and the North-East Coast Mission in the 1850s. Rosie was to die in 1860 while serving as Missionary for Bombay Harbour Mission. By 1862, some thirty missionaries were employed in coastal work in Scotland under principles he had laid down: 'thorough catholicity [ie interdenominational], strictly home missionary, aggression'.[54] The coast missions tended to concentrate on homes in harbours and ports and played a part in the creation of networks of Christian Brethren congregations around the Scottish coast, while the Hebridean stations of the

Figure 23.3 Reverend Thomas Rosie (d. 1862), a Coastal Missionary, associated with the Scottish Coast Mission, the West Coast Mission and the North-East Coast Mission. From a biography published in 1862.

West Coast Mission were eventually assimilated (in the main) by the Church of Scotland.[55] The Royal National Mission to Deep Sea Fishermen (1881) continued the offer of Christian pastoral care, fellowship and worship through its stations around the British coast, as did the Seaman's Bethel and the Seaman's Chapel on the south and north sides of the Clyde.

Glasgow, named 'Gospel City' as early as the 1830s, was home to a great number of sector-based missions, perhaps having been a particular focus of evangelical concern since the time of Chalmers.[56] By the 1950s, the mission halls still played a large part in popular culture. In his 1999 autobiography *Comfy Glasgow*, George Mitchell recalls his boyhood in the city during the 1940s and 50s, and lists the Gorbals Medical Mission (founded 1867), the railway missions at Maryhill, Springburn, Townhead and Oatlands, Artisans' Hall (1891), the Foundry Boys' Mission, the Canal Boatman's Institute and others. As a young man, Mitchell joined the Lambhill Gospel Band, a brass band that accompanied music with Christian testimony as an agency of the Lambhill Evangelistic Mission (1895). He recalls, besides the performances, the 'brightness and warmth' of the Christmas programmes offered by the halls.[57] By Mitchell's boyhood, the original trade-based emphasis had tended to recede often leaving in place independent evangelical congregations preserving their own working-class ethos and free from wider denominational structures, compromise and control. Sometimes, perhaps even regularly, this protestantism in the aggressive tradition of mission could focus on sectarian anti-catholicism – said to be the 'most prominent characteristic of proletarian evangelisation in Victorian Scotland'. The Glasgow Working Men's Evangelistic Association (1870) and the United Working Men's Christian Mission in particular campaigned against catholicism.[58]

Of all the Glasgow mission halls, the Glasgow United Evangelical Association's Tent Hall was considered by George Mitchell to be 'the flagship of the fleet'.[59] Named in order to show its origins in the marquee on Glasgow Green in which the Moody and Sankey Mission of 1873–4 was held, the Tent Hall had an auditorium to hold 2,200 – in its time the largest in Britain. From this base was run an extensive programme of both evangelism and basic care for the poor: free breakfasts, an annual Hogmanay supper, a Fresh Air Fund to finance breaks 'doun the watter', the Cripple League.

> The Free Breakfast presented even to the casual visitor an appalling sight. Here there would gather 1,800 to 2,300 men and women in varying degrees of prodigality. From four or five o'clock on Sunday morning the queues would begin to form until they circled round the whole block where the Hall is situated. Some were weary and bleary-eyed from lying out all night on the Green; some had wandered about the streets, harried from pillar to post, with no friendly stairhead on which to rest peacefully the whole night long; some were from the 'models' (lodging houses) low, lower, lowest, with which Glasgow

Figure 23.4 Tent Hall postcard c1920, 'A Salvation March'. Original owned by Mrs Marsaili Mackinnon.

abounded: others were from respectable homes, careful, thrifty and Godly, glad of the meal and the Gospel message. That great gathering had no equal anywhere in Christendom, either in its unique character, or in its unique attendants.[60]

The Glasgow Bible College (now the International Christian College) also sprang from the vision of the GUEA and its association with Dwight Moody: at the College were trained many generations of Scottish missionaries in the evangelical tradition of the Tent Hall.

Dwight Moody was an American mass evangelist; with his musical colleague Ira D Sankey, he preached to huge crowds during his Scottish tour of 1873–4. Further tours took place in 1881–2 and 1891–2. The rallies focused on a sermon calling for personal commitment to Jesus Christ. Moody's rallies also featured mass hymn-singing, accompanied by an organ, and so did much to popularise hymn-singing and church music in Scotland where the presbyterian tradition allowed the use of the psalms and paraphrases, unaccompanied. The mass attendances seem also to have helped to validate among the denominations a new pattern of one-off missions. In 1874, for example, the Episcopal Church of Scotland's congregations in Edinburgh held a series of mission services, led by a delegation of preachers from the Cowley Fathers of Oxford – perhaps ironically this mission was criticised by the conservative evangelical rector of St Thomas (a Church of England congregation in Edinburgh under the authority of the bishop of Carlisle) for mixing evangelical teaching with High Church dogma and ritualism.[61] The

Free Church of Scotland also described as a 'mission' a preaching tour of the Highlands by Rev. A N Somerville DD during his year as Moderator of the Free General Assembly, 1886–7. At the invitation of twenty Highland Presbyteries, Somerville's remit was to 'pass from one region to another, preaching the old and loving Gospel'. Between July and the next May he delivered 250 addresses in 189 locations, accompanied on occasions by former moderators, his minister son, and teams of medical and divinity students to assist with local follow-up. For such missions to have a lasting impact in conversions and changed lives, Somerville wrote that they had to begin with prayer and expectation, and had to be followed up with prayer and active personal evangelism. He also suggested that Free Church congregations, committed to using only the metrical psalms in public worship, might usefully deploy a wider range of tunes than the basic three that he had encountered; in this comment, as well as in the essential structure of the mission, the influence of Moody can be seen.[62]

Beyond the traditional denominations, crusading evangelism proliferated. The Faith Mission was founded in Scotland in 1886 as an inter-denominational society for the evangelisation of the rural areas of Great Britain. While returning to the 1790s style of itinerant preaching by 'pilgrims', the Mission was and is based on the 'faith principle' of never appealing for money but relying on God to give direction and to supply its needs in response to prayer. A residential centre/Bible college of its own followed in 1912, after a series of training missions to Rothesay. As with the earlier period of itinerancy, the 1880s pilgrims also met opposition. A Rothesay minister, Rev. C A Salmon, objected both to the theology of the Mission and to their style: while appreciating the warmth of the Christian love of the young men and women involved, he objected to serious hymns about judgment being sung to such popular tunes as 'Jock o' Hazeldean'. Neither did he appreciate the pilgrims' description of Rothesay as a place 'where there has been little or no real aggressive or spiritual work for years'.[63] In taking their evangelism down the Clyde to the island of Bute's summer holiday resort-town, however, the Faith Mission was in the long and still continuing tradition of summer missions. One particularly fruitful branch of evangelism targeting holiday-makers is that of children's beach missions: the origins of the Children's Special Service Mission (now Scripture Union) lay in the experience of its founder, Payson Hammond, assisting at a Free Church children's mission at Musselburgh in 1859.[64] Children's holiday clubs still occupy the energies of congregations and volunteers of all denominations across Scotland.

Local Evangelical Associations also grew up later in the nineteenth century, deploying their own evangelists in shorter or longer campaigns.[65] Typically under the leadership of businessmen, the associations gathered together supporters from across the presbyterian denominational spectrum and went wider with strong support from Baptists, the Open Brethren and the other smaller or independent evangelistic Churches. Founded following the example of the Ayrshire Christian Union, the Lanarkshire Christian

Union [LCU] gained much from the support and leadership of John Colville MP, of a family of iron-masters.[66] The Union organised monthly conferences and rallies 'promoting fraternal acquaintances and Christian fellowship amongst those interested in aggressive work'. Time-limited local mission campaigns were held in tents or halls. In its first season, LCU evangelist Richard Murphy and his tent visited Bellshill, Cambuslang, Coatbridge, Hamilton, Shotts, Stonehouse, Strathaven and Wishaw. The Union also sponsored Sunday evening meetings or evangelistic services, whether in churches or, if necessary, in the upper hall of a Carluke pub: 'The audacity of the venture, the idea of a Gospel mission in a pub, the people singing and praying in one room, and men drinking beer next door, so upset all ideas of the fitness of things that out of curiosity the people came to see, and presently some of them remained to pray'. As the numbers gathered, an approach was made to transfer the meeting to the parish church, which had no evening service: but the elders set the impossible condition that Ira Sankey's songs were not to be used. The nearby Free Church had no such reservations. Eventually, permanent rooms in a tenement were secured and a Christian Union mission congregation formed, independent of both the presbyterian kirks. 'After the mission was over it was felt that some city of refuge would have to be found for the young life that had sprung up'. Such mission halls drew upon a pool of evangelists and preachers often without formal training or qualifications – the Carluke pub campaign was led by one of Moody's disciples, John Scroggie, an ex-blacksmith.[67] These congregations could continue for years after the initial campaigns as small, independent churches; the Gordon Evangelistic Mission in Aberdeen (1856), for example, celebrated its centenary in 1956.[68]

The evangelism of the SPGH itinerants of the 1790s had quickly broken away from the existing denominations. Though Thomas Chalmers and his heirs had established their aggressive methods at the heart of presbyterian church life, the home mission movement's success meant that it overflowed denominational structures and created its own. The interdenominational evangelism of the societies and the associations, though it claimed the honoured label of 'aggressive', owed no allegiance at all to the 'territorial' or parish principle so strongly advocated by Chalmers, nor to the academic theology of the presbyterian Churches. Though still frowned upon by the ministerially-dominated presbyterians, lay preaching came to be an essential aspect of mission hall life. This evangelical tradition generated its own sub-culture focused on 'the blood of the Lamb': a way of life that as time passed was not always appreciated. Hymns with the chorus 'Death-beds are coming' are now probably rarely sung, but were once popular. Colin MacLean remembers his mother's aversion to her uncles, leaders of Aberdeen's Gordon Mission: 'I wouldn't willingly have been left alone with one of my uncles in case they would ask me if I was saved'.[69]

THE LONG TWENTIETH CENTURY: A CENTURY OF TURMOIL

The various strands of home mission already established were extended in the twentieth century, not so much by way of principle as via technology: the car, the radio, modern management techniques. The technological and social changes that had so transformed Scotland during the nineteenth century continued, with an ever-increasing rate of change. The coming of modern sanitation, piped water, motor transport, electricity and, latterly, the computer and the information/communication technology revolution vastly improved ordinary living conditions for the entire population of the United Kingdom. As homes became more private and society less collective, the essential strategy of visiting was undermined.[70] The century also saw mass unemployment, mass emigration and immigration, the near economic collapse of the Great Depression, and unprecedented social strife – besides two World Wars, the end of Empire and the long Cold War. All of these presented their own challenges to the Christian Church, at a time when doubt was increasing about the doctrines of the faith. Home mission was conducted in the twentieth century against a rising tide of indifference to organised religion, a tide swelled by such factors as those identified by T C Smout: 'the death of hell, the rise of class, and the spread of other entertainment'.[71] Moreover, while heavy industry departed from the country and the centres of population moved and moved again, the Churches had to struggle to adapt what they had built during the nineteenth century: the network of congregations and church buildings suited for the Scotland of 1880.

At the beginning of the twentieth century the Church of Scotland held firmly to its calling to take the gospel to all parts of the country as a (or as *the*) national Church. In response to this calling, home mission stations had been created during the nineteenth century so that the ordinances of religion might be offered in locations inconveniently far from existing parish churches. By 1914 the budget of the Established Church's Home Mission Committee supported 182 such posts, some controlled by particular congregations while to others the Committee itself appointed ministers or lay missionaries, a position first brought under central control in 1902. Twelve stations were in Shetland alone, and many of Scotland's smallest islands had their own home mission missionaries – Burra Isle, Skerries, Papa Stour in Shetland; Graemsay and Papa Westray in Orkney; the island of Luing in the presbytery of Lorn. Two-thirds of the 1914 Mission Stations of the Established Church, however, were in industrial central Scotland: twenty-six in the presbyteries of Glasgow and Hamilton. The parish of Bothwell maintained two missions: Palace Collieries and Thornwood. In Glasgow, parish missions staffed by lay missionaries included Cadder's Mavisvalley, Chryston's Auchengeich and Townhead's Cobden Street. Elsewhere, Fife's expanding coal mining industry had attracted missions: Glencraig and Lochore in the presbytery of Dunfermline and Kinross, and St Fothad's and

Figure 23.5 Mr John Murchison, Lay Missionary at the Glasgow and West Coast Mission's station at Kylerhea (1921–54), in his motor boat. Courtesy of the Board of National Mission of the Church of Scotland.

Coaltown (of Wemyss) in the presbytery of Kirkcaldy. In the presbytery of Dalkeith, similar mining areas supported Inveresk's missions of Smeaton and Wallyford. The United Free Church inherited most of the mission stations of the United Presbyterian and Free Churches, with the result that after the 1929 union of the UF Church with the Church of Scotland there was massive duplication of effort. In Shetland, for example, there was roughly one full-time presbyterian worker to every 500 of the population. Rationalising these competing agencies was the work of the Union and Readjustment Committee: it took the remainder of the century to establish the mission stations as self-supporting congregations, or to link or unite them with others, or to close them.[72]

In the Gaelic-speaking Highlands and Islands the role of the lay missionary was especially valued, for such men were usually themselves called from the scattered communities they served. Rev. Dr Roderick Macleod, Superintendent of Missions for the UF Church and then for the Church of Scotland, was an especially effective advocate of his stations and missionaries, as his 1927 Report to the UF General Assembly makes clear:

> The work of the Committee's Missionaries does not shine in a Report, for it is usually a long, patient sowing and a piecemeal harvest. In little sequestered townships and thinly-peopled areas striking increases in membership are impossible, 'shock-tactics' are impracticable, and the Guilds and Societies of a town Congregation are unknown. Yet the

Highlands and Islands Missionary has little time hanging on his hands, his work has compensations, and his 'uneventful annals' are rich for himself and his cure. He has few people, but he knows every one of them, and time and need conspire to make him father-confessor to them all. His five or ten miles' walk to visit outlying families is an event to them, and each renewal of it opens new windows in their welcome and, to him, new interests. Then with the people nearer his Station he shares an unavoidable intimacy, and they and he see one another in all weathers and in all moods. They rarely lock their doors in the Outer Islands and along the West Coast, and with a man whom they trust they make sincerity the greater part of their courtesy. So whether he likes it or no, the Missionary, in his witness and preaching and neighbourliness, submits to the acid test of intimacy. The Committee is thankful for Agents who have stood the test again and again, and have come out approved.[73]

Besides the denominations, two home missionary societies maintained stations in the Western Isles in the twentieth century: the Glasgow and West Coast Mission, and the Ladies' Highland Association [LHA]. The LHA began as an independent society in November 1850, associated with the then Free Church of Scotland. Its enduring purpose being to support the work of the gospel in the Highlands, the LHA's actual work altered with circumstances. Originally its staff had run schools, remembered in each community as the 'sgoil na ladies'; with the coming of state education, they turned to nursing.

Figure 23.6 Mission house and church at Craigton, Strath Halladale, Sutherland. As built by the United Free Church, c1920: corrugated iron on wood frame. Photograph by the author.

With the coming of the Health Service, the lady missionaries continued running their own Sabbath schools, providing personal pastoral care, and holding handicraft classes for boys and girls:

> Where parishes are enormous and villages are scattered and weather is often bad, the minister and the missionary cannot reach all the children who are eager to be taught. So the ladies of the LHA have their Sunday Schools from the Butt of Lewis to the eastern border of Skye, and several hundred children gather every week to recite the Psalms in Gaelic, to unite in prayer and praise, and to hear of the progress of the Church through all the world. There is no sectarian teaching in these Sunday Schools and all the children of the community are made welcome. This in itself is a great contribution to the happier development of life in the Islands.[74]

Another sphere of service of the LHA was its support for other missions. The Glasgow and West Coast Mission agents' reports of the 1930s invariably thanked the LHA for its donations towards their annual or Christmas children's treats: Free Church areas enjoyed annual treats, while Church of Scotland areas celebrated Christmas.

Though the LHA was unusual in describing its staff as lady missionaries, much mission work was undertaken by women.[75] A whole transient community of 'herring lassies' followed the herring fleets as they followed the shoals up and down the length of the British coast, so a whole enterprise of mission moved with them, also largely staffed by women. Similarly, the seasonal gatherings of berry-pickers in Angus and Tayside were supported largely by the Church of Scotland's Home Board female staff at missions whose aim was stated by Rev. Arthur Dunnett, joint Secretary to the Board: 'The church is endeavouring by every means in its power to give the workers the sense that at least some one cares for their well-being. Beyond the ordinary services and Sunday Schools held in the marquees and huts, concerts are arranged, literature and games are provided, old clothing is distributed, and a savings bank is instituted'.[76] The women engaged in full-time home mission work were known at different times and in different Churches as 'Bible Women', 'Church Sisters' and 'Deaconesses'. Their pastoral, social and educational calling led to the birth of the modern diaconate.[77] Before and after World War II, individual deaconesses underpinned mission work in inner city areas or new housing schemes.

Miss Helen L Hardy held a unique position as the key-worker for the ecumenical mission to Scotland's travelling people, known in the inter-war period as the tinkers. The mission had its origins in discussions before World War I, and Miss Hardy was first appointed by the Established Church in 1917 to visit Perthshire's travellers. The mission gained ecumenical support in 1921. She retired in 1944, having seen the work grow to include a Church-backed school near Pitlochry, and Church-supported camping sites. Largely spending her time on the roads, and alone, Miss Hardy appears to have won the trust of a group usually excluded from Scotland's static communities.

Figure 23.7 Scottish Colportage Society's Bible and Gospel Book Van, 1947.

In his book, *The Church in Changing Scotland*, Arthur Dunnett insisted both that 'The church is always ready to interest itself in any gathering of people for seasonal work'[78] and also that 'The church cannot rest satisfied to leave any section of the population uncared for spiritually'.[79] Before and after World War I, Scotland saw major engineering schemes requiring large mobile workforces for manual labour: road building and forestry; and during and after World War II, for the erection of hydro-electric dams. For these men and their camps, the Navvy Missions were created. There was thus a lay missionary appointed for the gangs of men who drove a road through Glencoe. In 1933 he was based at Tyndrum, provided with a car, and set to look after the worship and pastoral needs of four camps of men towards Loch Awe and another in Glencoe. The road having been built, the missionary was transferred to Glenbranter where a Ministry of Labour Training Camp for 200 men was being erected in co-operation with the Forestry Commission. The task of serving as pastor to these temporary camps of working men, gaining the trust of people in such transient situations and creating conditions for regular worship, must have been no mean challenge to those appointed, especially as many of the navvies were likely to be Irish or of Irish descent, and were often Roman Catholic by faith.

National Church Extension was the response of the Church of Scotland to the new housing schemes developed throughout central Scotland in the 1920s and 1930s. Under the leadership of the Very Rev. John White, the Church committed itself to raising some £180,000 to fund the creation of between thirty and forty new parishes complete with church buildings and suites of halls, so that the ordinances of religion could be offered in such places as Glasgow's Balornock, Dundee's Mid-Craigie and Aberdeen's Middlefield. Where previously individual congregations had established their own missions, which had gradually grown towards independence, now the central committee identified sites, built both church and manse, and appointed and paid the minister. Thus kick-started, the congregation in time hopefully came to full self-supporting status as a parish church. Similar programmes were funded by the Church of Scotland after 1945 for new towns such as Glenrothes and Cumbernauld and, under the title of New Charge Development, were still proceeding at the start of the twenty-first century. The Episcopal Church, too, engaged in home mission by church extension. A Million Shilling Fund was opened in 1914 to build six churches around Glasgow. The Home Mission Appeal built ten new churches across Scotland in the late 1920s, and the Home Mission Crusade programme, begun 1944, created a further five.[80] The Baptist Union of Scotland, the Free Church and others also promoted Church Extension.

While Rev. George Macleod supported the Church Extension movement, his concern after World War I was not so much the continuation in new buildings of older patterns of church life and worship, as how to bridge the gap between the church and the working people of Scotland. The 'Mission of Friendship' (1933–4) he inspired during his ministry at Govan caught the attention of many. Organised like a military campaign, the mission saw every household in Govan visited twice by teams of church members, supported by a continuous chain of prayer in the church. Meetings for the discussion of social issues were held. The unemployed were recruited (and fed) to renovate the church garden; the church choir took to the streets, supporting open-air preaching.[81] From such experiences came the Iona Community (1938). Macleod took care that the Community, though reporting to the General Assembly via the Home Board, should be independent and not subject to the control of the Board. His concept of mission included work for peace and social justice as well as the renewal of worship, and did not fit comfortably with existing orthodoxy.

'We think of missions as an extra, whether it be foreign missions or a "mission" at home. We think of them as extras added on to the ordinary work of the Church. They make us feel uncomfortable because they speak a language unrelated to our ordinary lives. We think of their aims as excessive and regard those who go in for them as rather peculiar'. So began an Iona Youth Trust publication around 1943.[82] While sympathetic to those who 'are alive to social and economic needs' but 'are suspicious of anything that commits them to a definite faith' (reflecting perhaps the mutual lack of sympathy then common between those of the Iona Community and Scotland's conservative evangelicals), Ralph Morton's booklet urged that youth work within the churches be understood as mission in its best sense.

He argued that girls and boys leaving school in the mid-twentieth century were entering a 'jungle', a 'secular society', 'a new continent, a continent that is strange and pathless'. A new language had to be learnt by those who wished to support them, and a fellowship meaningful in their own terms had to be created for those who wished to join. Service was the new watchword, rather than aggression, for the boys' and youth clubs pioneered by churches in this period. For example Rev. R Selby Wright's Canongate Church operated the Canongate Boys' Club and owned Skateraw, a hut camp on the east coast beyond Dunbar. Glasgow's Bridgeton's Saint Francis-in-the-East opened Church House in 1942 as a place where young people might feel they belonged and were welcome.

Inherent in the home missionary movement in Scotland has been an implicit criticism of the 'common methods' of the mainstream Church – Ralph Morton was correct that home mission has often been seen as 'extras added on to the ordinary work of the Church'. The ecumenical 'Tell Scotland' movement of the 1950s sought therefore to place mission at the heart of normal congregational life, to see it as the 'continuing responsibility of the whole body of believers'. Rev. Tom Allan stated the three key principles of 'Tell Scotland':

- Effective evangelism is not a sporadic or occasional encounter with the world, but a continuing engagement at every level.

- The agent of effective mission is the Church itself, the redeemed Community, representing the Saving Life of the Risen Lord in the context of the whole life of man.

- The place of the layman is decisive.[83]

Tom Allan's emphasis on the importance of the entire congregation came from his experience of a visitation mission in his parish of North Kelvinside in 1947, led by the Home Board's evangelist, Rev. D P Thomson, with at its core a visiting team of young students and volunteers, supported by a number from the congregation – a mission similar to many conducted by 'D P'. Very substantial numbers of people were contacted and attracted to the church, their faith [re]kindled – but then the problems arose. The vast majority of the congregation had played no part in the mission; some were hostile, 'suspicious of the very word "evangelism"'. For many of the new Christians, the congregation had nothing to offer but this suspicion and they could not be assimilated. Indeed, they were frozen out. Tom Allan's realistic, perceptive and thoughtful account of the 1947 mission and its outcome in North Kelvinside, *The Face of My Parish*,[84] helped to set the theological agenda for 'Tell Scotland'. He was advocating a rethinking of the nature of the Church. He envisioned a committed Christian community that emphasised not the professional ministry but God's calling to all Christians to share their faith: a change perhaps requiring the 'destruction of the institutional church as we know it'. Perhaps ironically, in retrospect it can be seen that church membership in Scotland during the 1950s had enjoyed a modest

Figure 23.8 Cover page of the Song Book of the 1955 Billy Graham All-Scotland Crusade. © Billy Graham Evangelistic Association, used by permission.

recovery: thanks to a century and more of home mission, by mid-century still 'a vibrant Christian identity remained central to British popular culture'.[85]

Similar ideas as those considered by Tom Allan were to resurface in the Church of Scotland in 2001, with the 'Church without Walls' report. In 1955, however, whatever impetus 'Tell Scotland' had had towards radical rethinking was channelled into the Billy Graham All-Scotland Crusade. Although the Crusade had not been an original part of 'Tell Scotland', the movement's secretary, Tom Allan, fully participated in the invitation to the American crusade evangelist to visit Scotland. The Crusade eventually reached some 1,185,360 people – the total that attended meetings of one kind or another, whether in Glasgow's Kelvin Hall, the massive open-air rallies, or the radio Relay Meetings across Scotland.[86] Supported by the experience of the Billy Graham Organisation, counsellors for those who came forward in response to a pulpit call referred them to congregations of their choice, so that (in Tom Allan's words) 'The All-Scotland Crusade was no tip-and-run raid by a visiting evangelist. It was essentially part of a great national movement'.[87] The Billy Graham Crusade, supported by eight of the protestant Churches of Scotland, was remembered by many as a turning point in their following of Christ: nevertheless, there were also criticisms. In retrospect, one supporter wrote, '"Tell Scotland" never recovered from the trauma of 1955'.[88] Perhaps Tom Allan had in any case underestimated the institutional conservatism of the Church, in allowing simply a year for the second phase of 'Tell Scotland': one year to recruit and train and mobilise the laity for mission. The theological and cultural suspicions regarding overt crusade evangelism were also strong: no such event has followed since 1955 to unite the Scottish Churches. The 1960s saw a decisive down-turn in church membership, and a growing loss of connection between the Christian faith and the wider national culture. The year 2000 saw a marked disjunction between the Churches' widespread celebration of a Christian millennium and popular (and even some official) determination to see little or no religious significance to the party.

OVERVIEW

Driven by evangelical zeal (though of differing strands of theology), home mission in Scotland sought to extend and make more effective the normal ministry of the Churches. Indeed, it raised a key theological question: what ought to be normal for the Christian Church, if not mission? Some of its innovations were in time adopted as common practice – the Sunday school movement, for example. Radical in his day, Thomas Chalmers' principles of sustained aggressive visiting within a defined territorial area or parish became the presbyterian norm for almost two centuries. 'Visitation Evangelism' was at the heart of D P Thomson's work as an evangelist and Organiser for Evangelism for the Home Board in the 1950s and 1960s. At St Mary's, Motherwell, he extended the concept beyond homes to places of work and leisure, so that teams also visited offices, warehouses, hospitals, the cinema, schools and the greyhound track.[89] Arising in part as a theologically-driven protest against the Established Church, home mission evangelism retained a two-sided cutting edge: the newly-enthusiastic may (as did the Faith Mission at Rothesay) justify their campaign by criticism of existing Christian agencies and churches. Home mission could thus be divisive: sincere Christians criticised both the Dwight Moody and the Billy Graham campaigns, the one for departing from strict Calvinist orthodoxy and practice, the other for its potential to lose sight of the social aspects of the gospel. George Macleod's Iona Community and his political radicalism were by no means universally welcomed. Home mission has in its time fostered sectarian divisions between Protestant and Roman Catholic; tensions have also arisen between the denominations and the interdenominational networks, societies and associations.

Based in a country that had been officially Christian for centuries, the home missionaries tended to assume that, provided the gospel was properly and clearly set before the people, they would respond with faith. Preaching continued to draw the crowds, from the itinerants of the 1790s to the mid-twentieth century. As the twentieth century progressed, however, the assumption that people would come if invited became dubious. Ron Ferguson, George Macleod's biographer, thus asks whether the principle behind the Govan mission was correct, that the people of Govan had no link with the Church because they were unasked: 'assumptions that were by now highly questionable. Was it really the case that the men of Govan were not coming to church simply because they were "unshepherded"?'.[90] In an age of mass communication, moreover, the prevalent attitudes in the British media became distanced from the Churches or even hostile: the Radio Missions (1950–2) would now be unthinkable via public service channels. The historical leading emphases of home mission, preaching and visiting, cope with difficulty with the current 'famine of hearing' and a shrinking and ageing church membership supplies fewer ministers and activists.

Often both a rebuke and a challenge to Church orthodoxy, at its best home mission led the Christian response to social change: to the early stages of industrialisation, to nineteenth-century urban destitution, to the class divisions of the early twentieth century and to soulless housing schemes and estates. Even within the slowly-changing presbyterian Churches, the

Figure 23.9 Congregation of the Bilston Mission, Midlothian, at their jubilee service, 1988. Photograph by E W Hutchison DCS.

Figure 23.10 New Charge church built by the Church of Scotland, 2002–03, for Inverness Inches between a shopping mall and new housing. Photograph by the author.

movement created new avenues of service: both for volunteers and via the 'paid agency' that allowed non-university educated, un-ordained working men and women to dedicate their lives in full-time service, so that the male lay missionaries and the female deaconesses eventually were recognised with their distinctive diaconal calling. That all the Churches coped as well as they did with tremendous movements of population as Scotland was re-housed again and again and again within two centuries was a result of the movement's vigour and repeated Church Extension campaigns. Home mission insisted that those on the fringes of society – whether in remote islands, mining communities, the travellers or navvies, or the urban poor – deserved the attention of Christian concern, prayer, generosity and service. Though often the vision came from the ordained ministry (even if from mavericks or radicals), the strength of home mission finally came from the people of the Churches: the members of the early Bible and Tract Societies, the lady visitors, the volunteers for beach and parish missions; people rising to the challenge to 'zeal and self-denial' set by 'Puritanicus' in 1797.

By the beginning of the twenty-first century home mission had reached the end of its road as a concept generally understood within the Scottish Churches and interdenominational movements. Its origin as a movement to supplement (or even subvert) *the common methods* of the Churches meant that some home mission activities were seen as additional to core congregational life. As accumulated capital reserves were spent, finances became tight and all work thought of *as an extra* was liable to be pruned or terminated. By the end of the twentieth century much of the Victorian era home mission industry had been wound up: the numerous Bible and Tract Societies, the subordinate mission congregations for the poor, the Sunday evening evangelical mission services, the preaching vocation of lay missionary. Sustained by the generosity of a still lively Scottish evangelical constituency, a few of the societies survived, perhaps even revitalised: the children's and youth work of Scripture Union Scotland, for example, or the relief work of the Glasgow City Mission, together with such younger organisations as the Abernethy Trust that engaged in Christian outreach via outdoor activities. While congregational 'missions' were still held, their scope was increasingly confined to children's work, children being the last major slice of society for whom organised entertainment linked to Christian teaching still held some appeal. Meanwhile the denominations – faced with falling membership – sought to enhance their outreach by emphasising that an understanding of mission, more broadly defined as comprising outreach and service, should underpin all their core activities. Paradoxically, identifying the congregation as the true home of mission meant that the older self-standing home missionary structures were no longer required. The 2004 General Assembly of the Church of Scotland was thus able warmly to affirm a new commitment to mission as essential to a revitalised Church while simultaneously agreeing to dissolve its Board of National Mission, merging its remaining activities with other departments. Certainly congregations that sustain vigorous outreach can be identified across denominations. As an identifiable movement reaching across Scotland, however, home mission is no more; whether its spirit will live on remains to be seen.

NOTES

1. The 'evangel' is the good news, or gospel, of salvation through Jesus Christ. 'Evangelistic' and 'evangelism' thus relate to the process or activity of summoning of men and women into the fellowship and service of Jesus Christ, by offering them the evangel to believe. 'Evangelical', when used of Christians, describes those people who make it their overriding priority to promote the evangelistic task of the church. Being focused on the evangel, on occasions the loyalty of evangelicals to, say, their denominational structures or traditions may take second place or be overridden. Only a few evangelicals are full-time 'evangelists': one who calls others personally to commitment.
2. Thomson, 1968, 7–8.
3. *The Missionary Magazine* (July 1797), 315–17.
4. Matheson, 1843, 138.
5. Thomson, 1968, 7–8.
6. *The Evangelical Magazine* (September 1828), 399–401, reported that the Home Mission Society by that date supported 30–40 agents involved in itinerating village preaching, and some 60 Sunday schools; places of worship had been erected and Christian churches founded.
7. Walls, A F. The missionary movement: a lay fiefdom? In Lovegrove, D W, ed. *Rise of the Laity in Evangelical Protestantism*, London and New York, 2002, 167–87. Walls lists these preconditions as: (a) 'a body of people with the degree of commitment needed to live on someone else's terms, together with the mental equipment for coping'; (b) 'a form of organisation which could mobilise committed people'; (c) 'sustained access to overseas locations, with the capacity to maintain communication over long periods. This implies what might be called maritime consciousness, with maritime capability and logistical support.': page 172.

8. Carey, W. *Enquiry into the Obligations of Christians to use Means for the Conversion of the Heathens*, Leicester, 1792, 3. The Baptist missionary summarised the vision of the generation: 'As our blessed Lord has required us to pray that his kingdom may come, and his will be done on earth as it is in heaven, it becomes us not only to express our desires of that event by words, but to use *every lawful method* to spread the knowledge of his name'. [*Emphasis* added.]
9. *The Missionary Magazine* (July 1796), 43.
10. Lovegrove, 1988.
11. Matheson, 1843, 139, 372–3.
12. Matheson, 1843, 392.
13. McNaughton, 2003; McNaughton, 1993; Escott, 1960; Bebbington, 1988; Meek, D E. Evangelical missionaries in the early nineteenth century Highlands, *Scottish Studies*, 28 (1987), 2–34.
14. *The Missionary Magazine* (July 1803), 310–12.
15. See, for example, McNaughton, W D. *The Congregational Witness in Kintyre 1800–c. 1878*, Glasgow, 1994, 4.
16. Bowes, H R. The Launching of Methodism in Shetland, 1822, *Proceedings of the Wesley Historical Society*, (August 1972), 2–10.
17. Brown, C G. The Sunday-School Movement in Scotland, 1780–1914, *RSCHS*, 21 (1983), 3–19.
18. *The Missionary Magazine* (May 1809), 203.
19. *The Missionary Magazine* (August 1813), 353.
20. Brown, 1983, 12.
21. *The Missionary Magazine* (August 1813), 351.
22. *The Missionary Chronicle of the London Missionary Society*, editions for 1827: February page 87; March page 132; April page 175; May page 233.
23. *The Evangelical Magazine* (September 1810), 373.
24. *The Missionary Magazine* (March 1805), 238–9.
25. *The Missionary Magazine* (March 1803), 141–2.
26. Brown, 1982, 101–4; Cheyne, A C. Introduction: Thomas Chalmers, then and now. In Cheyne, 1985, 16–17; Brown, C G. 'To be aglow with civic ardours': the 'Godly Commonwealth', *RSCHS*, 26 (1996), 170: 'The home-mission industry was a vast legacy to his [i.e. Thomas Chalmers'] ideals'.
27. *First Report of the Society for Promoting the Interests of the Church of Scotland and of Religion Generally in the Parish and Congregation of St David's*, Glasgow, 1836. See also: *Address to the Central Committee for Supporting and Extending the Parochial Institutions of Scotland, June 17, 1835, by John C. Colquhoun Esq. of Killermont* with *Hints towards the Organisation of District Societies in Glasgow and suburbs*, both printed Glasgow by W. Collins & Co., no date, presumably 1835.
28. Bardgett, 2002, 21–35.
29. *Second Annual Report of the Greenock Town Mission*, Greenock, 1830.
30. *The Second Annual Report of the Society of Probationers of the Church of Scotland*, Edinburgh, 1835, 11.
31. *First Report of the Edinburgh City Mission*, Edinburgh, 1834; *Second Biennial Report of the Edinburgh City Mission, with a list of subscriptions and donations, for 1834–1835*, Edinburgh, 1835.
32. *Seventh Annual Report of the Society for Promoting the Religious Interests of the Poor of Glasgow and its Vicinity; or, The Glasgow City Mission*, Glasgow, 1833.
33. See the Arbroath Town Mission website (http://www.atmc.freeuk.com at October 2005): 'The ministry of the Arbroath Town Mission dates back to 1849. It is therefore a long established interdenominational church fellowship which offers the gospel message, Bible teaching, warm Christian fellowship and friendship for all who come within our walls'.

34. *District Work: or The Reclaiming of Waste Land*, Glasgow, 1854.
35. Bardgett, 2002, 12.
36. *The Evangelical Magazine* (May 1827), 199. The 1810 July edition, 294, had summarised a letter reporting that among those Highlanders coming south for the harvest, 'a minister found that not one in eight could read … More is needed to provide a means to instruct the people first in elementary books in their own vernacular tongue – a sum of money raised among ourselves to support a few teachers – and so fit them to read the Gaelic Bible'. The writer hoped his statement of the case would 'affect our English friends and countrymen, both in Scotland and England and especially in the metropolis'.
37. Bardgett, 2002, 24, 73.
38. *Fourth annual report of the Highland Missionary Society, 1824 with a list of subscribers and benefactors, state of accounts etc*, Edinburgh, 1824.
39. Macinnes, 1951.
40. 'A lay member of the established church', *An account of the present state of religion throughout the Highlands of Scotland in which the comparison instituted between the clergy of the 'Western Districts' and those of 'Ross-shire' in a speech before the General Assembly in 1824 is examined, and the true merits of both parties exhibited*, Edinburgh, 1827.
41. Bardgett, F D. *Two Millennia of Church and Community in Orkney*, Durham, 2000, 101–43.
42. Boyd, W. *The Colportage Operations of the Religious Tract and Book Society of Scotland: Paper read at a Conference on Mission Work, held at Glasgow on 31st Aug. 1871*, no publication details but presumably Glasgow, 1871.
43. Blaikie, W G. *Better Days for Working People*, London, 1863, 68–9; Brown, 1981, II, 191.
44. Brown, 1981, I, 18.
45. King, E. *Scotland Sober and Free: The Temperance Movement 1829–1979*, Glasgow Museums and Art Galleries, Glasgow, 1979.
46. *Speeches of Dr Buchanan, Rev. A. Gray, and Sanderson Kirkwood Esq delivered in the General Assembly of the Free Church of Scotland on the Spiritual Destitution of the Masses in Glasgow with the decision of the Assembly thereon*, Edinburgh, 1851, 9.
47. Buchanan, R. *The Spiritual Destitution of the Masses in Glasgow: Its Alarming Increase, Its Fearful Amount, and the only Effectual Cure. A speech delivered to the Free Church Presbytery of Glasgow on Wednesday Jan. 8, 1851 by R. Buchanan DD*, Glasgow, Edinburgh and London, 1851.
48. Bardgett, 2002, 53.
49. Edwards, J. *Living like Jesus: or, Going About Doing Good: a tribute to the memory of Robert McEwen, agent of the Bridgeton Village Mission*, Glasgow, 1861; Goodfellow, 1906, cited and summarised in Bardgett, 2002, 79–86.
50. Macdonald, L O. Women in Presbyterian Churches. In *DSCHT*, 884.
51. Brown, 1987, 160–4; Bellesheim, A. *History of the Catholic Church in Scotland 1625–1878*, Edinburgh and London, 1898, IV, 283–4.
52. McCaffrey, J. Roman Catholics in Scotland in the nineteenth and twentieth centuries, *RSCHS*, 21 (1983), 275–300.
53. Brown, 1987, 165–7; Brown, 2001, 154–6.
54. Bardgett, 2002, 60–4.
55. Meek, D E. Fisherfolk, missions to. In *DSCHT*, 323–3; for the Glasgow and West Coast Mission see Bardgett, 2002, 191–9 and Appendix 8, 310–26. One of the remaining GWCM stations, that at Lemreway, joined the Free Church.
56. Hillis, 1987, 46–62, concludes that the 'take-off' period for home mission work in Glasgow was in the 1830s and 1840s as the growing prosperity of the city allowed those with wealth to channel it towards mission.

57. Mitchell, G. *Comfy Glasgow – An Expression of Thanks*, Fearn, 2000, 68–79.
58. Brown, 1981, II, 319–23.
59. Mitchell, 2000, 72.
60. Mrs McRosie, widow of a former Superintendent of the Tent Hall, quoted in the pamphlet *Christ in the City 1874–1974*, Glasgow United Evangelical Association, 1974.
61. Drummond, D T K. *Letter to the Very Rev. The Dean of Carlisle on some Matters Affecting the Interests of English Episcopalians in Scotland*, Edinburgh and London, 1876.
62. *Speech delivered by the Rev. A. N. Somerville DD in the Free Church General Assembly on May 26, 1887 in Connection with his Mission to the Highlands and Islands during the Year of his Moderatorship, 1886–87*, Glasgow, 1887.
63. Peckham, C N. *The New Faith Mission Bible College*, The Faith Mission, Edinburgh, 1994; *Perfectionism: the False and the True. A lecture delivered by the Rev. C. A. Salmon MA, Rothesay, on 30 December 1888 with special reference to the teaching of the Faith Mission Pilgrims*, Glasgow, 1889.
64. Butler, J M F. Scripture Union. In *DSCHT*, 763; Findlay, 1988.
65. Bardgett, 2002, 65–6.
66. Bryson, N W, ed. *History of the Lanarkshire Christian Union instituted 1882*, Strathaven, 1937, 14, 70.
67. Bryson, 1937, 19, 116–19.
68. MacLean, C. *Your Father and I: A Family's Story*, East Linton, 1998, 57–66.
69. MacLean, C. *Monkeys, Bears and Gutta Percha*, East Linton, 2001, 42.
70. Dunnett, 1953, 61–4. Considering the wealthier, west-end parishes of Glasgow, Rev. Arthur Dunnett wrote: 'The method of approach is hard indeed to find. Will door-to-door visiting be welcome or in any degree effective? If it were so, during what hours of day or night could it usefully be done?'.
71. Smout, 1986, 208. By 'the death of hell', Smout refers to a growing loss of the certainties about heaven and hell which had driven the energies, zeal and generosity of earlier generations of evangelicals. The whole of Chapter VIII, 'Churchgoing', builds to the conclusion cited here.
72. Bardgett, 2002, covers this history in detail.
73. *Report of the Committee for the Highlands and Islands*, Reports to General Assembly of the United Free Church of Scotland, 1927, 8; cited in Bardgett, 2002, 132.
74. *Report of the Home Board*, Reports to the General Assembly of the Church of Scotland, 1947, 245.
75. Macdonald, 2000, 41–104.
76. Dunnett, 1953, 171.
77. Bloesch, D G, ed. *Servants of Christ – Deaconesses in Renewal*, Minnesota, 1971.
78. Dunnett, 1953, 173.
79. Dunnett, 1953, 147.
80. Stranraer–Mull, G. *A Church for Scotland* as published on the historical pages of the website of the Episcopal Church of Scotland, autumn 2003.
81. Ferguson, R. *George Macleod: Founder of the Iona Community*, London, 1990, 116–17.
82. Morton, T R. *Can You Get On Without It? 3: Missionary principles for the home front*, Iona Youth Trust publications, no date given: c1943.
83. Allan, T. *The Agent of Mission: the Lay Group in Evangelism, its significance and task*, 'Tell Scotland', no date or place given: c1954.
84. Allan, 1954.
85. Brown, 2001, 169.
86. Allan, 1955, 8.
87. Allan, 1955, 21.

88. Bisset, P T. Tell Scotland. In *DSCHT*, 815.
89. Thomson, 1968; also *The Sutherland Adventure*, Crieff, c1956, 31–2; and Thomson, 1954.
90. Ferguson, 1990, 116.

BIBLIOGRAPHY

Allan, T. *The Face of My Parish*, London, 1954.
Allan, T, ed. *Crusade in Scotland: Billy Graham*, London, 1955.
Bardgett, F D. *Devoted Service Rendered: the Lay Missionaries of the Church of Scotland*, Edinburgh, 2002.
Bebbington, D W, ed. *The Baptists in Scotland: a History*, Glasgow, 1988.
Brown, C G. *The Death of Christian Britain: Understanding secularisation 1800–2000*, London and New York, 2001.
Brown, C G. To be aglow with civic ardours: the Godly Commonwealth, *RSCHS*, 26 (1996), 169–96.
Brown, C G. *The Social History of Religion in Scotland since 1730*, London and New York, 1987.
Brown, C G. 'Religion and the development of an urban society: Glasgow 1780–1914', PhD thesis, Faculty of Social Sciences, University of Glasgow, 2 vols., 1981.
Brown, S J. *Thomas Chalmers and the Godly Commonwealth in Scotland*, Oxford, 1982.
Cheyne, A C, ed. *The Practical and the Pious*, Edinburgh, 1985.
Dunnett, A. *The Church in Changing Scotland*, London, 1953.
Escott, H. *A History of Scottish Congregationalism*, Glasgow, 1960.
Findlay, J, et al, eds. *A Place to Pitch Your Tents: The Story of Scripture Union Camps in Scotland*, Scripture Union Scotland, Glasgow, 1988.
Goodfellow, J. *The Print of his Shoe: Forty Years' Missionary Experience in the Southside of Edinburgh*, Edinburgh, 1906.
Hillis, P. Education and evangelisation: Presbyterian missions in mid-nineteenth century Glasgow, *SHR*, 66 (1987), 46–62.
Lovegrove, D W. *Established Church, Sectarian People: Itinerancy and the Transformation of English Dissent 1780–1830*, Cambridge, 1988.
Macdonald, L A O. *A Unique and Glorious Mission: Women and Presbyterianism in Scotland 1830–1930*, Edinburgh, 2000.
Macinnes, J. *The Evangelical Movement in the Highlands of Scotland 1688 to 1880*, Aberdeen, 1951.
McNaughton, W D. Revival and reality: Congregationalists and religious revival in nineteenth-century Scotland, *RSCHS*, 33 (2003), 165–216.
McNaughton, W D. *Early Congregational Independency in the Highlands and Islands and the North-east of Scotland*, Tiree, 2003.
McNaughton, W D. *The Scottish Congregational Ministry 1794–1993*, The Congregational Union of Scotland, Glasgow, 1993.
Matheson, J J. *A Memoir of Grenville Ewing, Minister of the Gospel*, London, 1843.
Meek, D E. Evangelical missionaries in the early nineteenth century Highlands, *Scottish Studies*, 28 (1987), 1–34.
Smout, T C. *A Century of the Scottish People 1830–1950*, London, 1986.
Thomson, D P. *Aspects of Evangelism*, Crieff, 1968.
Thomson, D P, ed. *Two by Two! The Rank and File of the Church Report on What Happened When They Went Out Visiting*, Crieff, 1954.

24 Missions and Missionaries: Foreign

ANDREW C ROSS

Although a small country, Scotland has played a disproportionately large part in the initiation and expansion of Protestant missionary activity outside Europe in what K S Latourette called 'The Great Century', i.e. 1790–1914, and since. From the birth of Protestant foreign missions, Scots were involved as organisers of missionary societies. They were also some of the first missionaries sent furth of the United Kingdom by these societies. At a time when the institutional Churches were indifferent, when not hostile, to the call to missionary action abroad, the voluntary society was the characteristic organisation which carried out this new development in the Protestant world. This was the case in the Protestant world generally, not only in Scotland. The vast majority of these Scottish pioneers were men and women outside or on the margins of Scottish Presbyterianism, whether in the form of the Established Kirk or the Relief and Secession Kirks.

In 1795 The Missionary Society, later re-named the London Missionary Society (LMS), was founded. Scottish ministers working in England played a significant part in the creation of the Society. Three were of particular importance. The first of these, Alec Waugh (1754–1827), minister of Wall Street church in London, drafted the Society's famous 'fundamental principle', which insisted that it was the aim of the society to send out the Gospel rather than any particular denomination of the Christian Church. The second was David Bogue (1750–1825), minister in Gosport, who set up an academy there which trained many of the early LMS missionaries. James Love (1766–1840), the third of this distinguished trio, was the Society's secretary until 1800. In 1796, there were founded the Glasgow Missionary Society (GMS), with which Love was closely associated, and the Scottish Missionary Society (SMS). These two societies supported the LMS but also sent missionaries abroad themselves. The most successful of these efforts were the GMS mission to the Cape Colony in 1821 and the SMS mission to Bombay in 1822. A number of other 'missionary societies' were founded shortly after in Stirling, Paisley, Greenock, Perth and Dundee. However, these societies acted simply as auxiliary organisations for the LMS, as did the 'praying societies' formed in a number of the towns in the north east, including Aberdeen, Elgin and Nairn.

The Evangelical Revival of the mid-eighteenth century had a significant impact on the development of these societies. It was also influenced by the increasing confidence of newly prosperous groups in Scotland, the expanding commercial middle class and the skilled artisans, notably the weavers.[1] It was from these skilled artisans that the overwhelming majority of both Scottish and English missionaries was drawn until the late nineteenth century.[2] Men from these two groups also supplied the leadership in the Relief Kirk and in the congregations of the various branches into which the Seceders had divided. They and their families also provided the majority of the members of the new Congregational and Baptist congregations that appeared as a result of the missions led by the Haldane brothers in the 1790s. These new congregations should be distinguished from the older groups, however, as they were not at all concerned about the Covenanting and Calvinist disputes which resulted in the eighteenth-century secessions from the Kirk. The new groups were explicitly concerned about preaching the Good News to humankind who were all deemed free to accept or reject it. They also provided many of the Scots who supported parliamentary reform, the abolition of slavery and other reforming causes of the day.

These associations with reform partly explain why it was that, when the issue of the support of foreign missions was raised at the General Assembly of 1796, not only was the motion rejected but also a leading elder had the support of the majority of commissioners when he vehemently asserted that the missionary and other voluntary religious societies were seditious. This attitude was deeply entrenched in the Moderate-dominated Assemblies of the 1790s. The almost paranoid fear created by the French Revolution was another element in this opposition. A third element is the fact that, although the new evangelicalism had a strong presence in the Kirk and in the Relief and Secession Kirks, the main evangelical excitement in the 1790s lay among people outside the main structure of Church and State, which had been built up so carefully in Scotland during the Moderate ascendancy. Thus in 1799, the Assembly issued a pastoral letter to all the parishes of Scotland, which condemned the lay preaching and Sunday School movements developed by the Society for the Propagation of the Gospel at Home, an organisation founded by the Haldanes. The letter also explicitly linked support for the LMS with this seditious movement. That these activities were going on 'without the consent of presbytery, minister or Heritors' appeared to be proof enough of their guilt. It has also to be noted, however, that the Relief and Secession Kirks adopted a similar stance towards missionary work abroad. As early as 1796 the Cameronians excommunicated some members for attending a service at which the minister had called for support for the new LMS, while in 1798 the General Associate Synod condemned the societies at its meeting. The Burgher Synod proved to be the exception, formally supporting the LMS from the beginning.

The activity of Claudius Buchanan (1766–1815) was separate from these developments but of importance for the cause of foreign missions because of the fame he achieved. Born in Cambuslang, he was converted while working in London, by John Newton (1725–1807), one of the outstanding evangelical Anglican clergyman of the period. Buchanan became an Anglican and studied theology at Cambridge. He was ordained in 1796 and was appointed to an East India Company chaplaincy in Calcutta. He co-operated there with William Carey, leader of the Baptist mission at nearby Serampore, in linguistic and biblical translation work. The Serampore mission was the first of the new movement sent abroad from Britain. At the Company's request Buchanan founded Fort William College and went on to be an outstanding scholar in the field of Indian and biblical languages, and a prominent publicist for the cause of mission throughout the British Isles when he returned there. He was the first in a long line of Scots who contributed to the understanding in the West of the various cultures of India and to the introduction and dissemination of western academic education in the sub-continent.

It was in connection with India that the first signs of change in the attitude of the pre-Disruption Kirk towards mission took place. When the charter of the East India Company (EIC) came up for renewal in 1813, the General Assembly of that year appealed to the government for India to be opened up for missionary work. The same Assembly also appealed to the EIC to make financial provision for Church of Scotland chaplains in their territories as they had already done for the Church of England. James Bryce was the first such chaplain and was appointed to Calcutta. In 1824 he wrote to the General Assembly appealing for the Kirk to send missionaries to work with him in approaching the upper caste Hindus, the Brahmins, in their own language. As we shall see, the approach taken by Alexander Duff (1806–1878), sent by the General Assembly to Calcutta in 1828, was precisely the opposite. Duff sought to reach the Brahmins through western higher education in English. This contrast in approach was a matter of dispute among missionaries for the rest of the century, a subject to which we will return.

The General Assembly of 1824, which had received Bryce's appeal, also received petitions from the synods of Aberdeen and Moray and the presbyteries of Edinburgh and Linlithgow for the establishment of foreign missions by the Church of Scotland. This time the Assembly voted in favour of missionary action, which was to be supported by a general collection in all parish churches. Much has been made by Scottish churchmen in the last two hundred years of this decision by a denomination to take up foreign mission as part of its essential task. At a time when it was the norm in Europe and North America for overseas missions to be conducted through voluntary societies, this was undoubtedly of symbolic importance. However, when the Assembly's decision is studied closely it is clear that in reality the Kirk had given its approval to a *de facto* denominational voluntary society. (The situation was no different after 1843 with the Free Kirk, though the base of support was somewhat wider than in the Auld Kirk.) Of the 900 parishes and 54 chapels of ease in existence in 1824 only 59 parish churches and 16 chapels made a collection. This pattern was still the same in the period 1875–1914, often seen as a 'Golden Age' of foreign missions in Scotland.[3] A minority in a minority of congregations *voluntarily* supported the work of the Foreign Mission Committee of the Kirk and that of the Free Kirk.

The decision to initiate foreign missions was taken in 1824 but it was not until 1829 that Alexander Duff was ordained as a missionary to India and arrived at Calcutta, where his work led to the creation of the Assembly's College. Meanwhile in 1828, John (1804–1875) and Margaret Wilson (1798–1835) went to Bombay (now Mumbai) to expand the mainly educational work there of the Scottish Missionary Society. Margaret was unusual in that she had studied unofficially at Aberdeen University and was a *de facto* missionary. She started western-style education for girls in western India. In 1832 the Wilsons successfully petitioned the General Assembly for the SMS mission in Bombay to be received as part of the work of the Foreign Mission Committee of the Kirk. These two missions in Calcutta and Bombay played important roles in the growth and development of the universities of these two cities. Wilson was Vice-Chancellor of Bombay University from 1868 to 1870.

The Disruption saw all of the missionaries and most of the mission supporters joining the Free Kirk. A sign, however, that there was firm, if small, support for mission in the Established Church was that its mission funds had returned to their 1842 level by 1845. The 1840s saw another significant change when the missions and missionaries of the GMS in Jamaica, the Cape Colony and Nigeria were incorporated into the new United Presbyterian Church when it came into existence in 1847.

Meanwhile three of the best-known Scottish missionaries, and arguably the most important missionaries in the history of Christianity in southern Africa, were already at work in Africa. All three had been sent there by the LMS. If one were to ask a passer-by in the street anywhere in southern Africa to name a missionary most would reply with the name of one of the three: John Philip (1775–1851); Robert Moffat (1795–1883); David Livingstone (1812–1873). These were men produced not by the Kirk in any of its forms but by the independent chapel movement that led to both Scottish Congregationalism and the Scottish Baptist Church. All three began as skilled artisans, Philip a weaver, Moffat a gardener, and Livingstone, who studied at the Andersonian, now Strathclyde University, a cotton-spinner.

Philip was the Resident Director of the LMS in southern Africa from 1819 till the year before his death. As such he not only oversaw all the society's work inside and outside the Cape Colony, but also campaigned incessantly for what he called 'equal civil rights for all His Majesty's subjects'.[4] In 1828 his influence on the British government through the leaders of the anti-slavery movement led the government, by an Order-in-Council, to make it impossible for the Cape Colony to enact any legislation that discriminated between people on the basis of race other than by the specific permission of the British government. It was at the time of Philip's death that the final triumph of his persistent campaigning was achieved and the new Cape parliament came into being with a colour blind, low qualification franchise. This tradition of racial egalitarianism was only reversed when Britain united the Cape Colony with the Transvaal, the Orange Free State and Natal in the Union of South Africa in 1910. The Liberal government of Britain agreed to a white male adult suffrage for the new Union Parliament. They made one gesture to the more liberal past and left the 'Coloured' and African voters of the Cape on the new voters' role. In 1924, however, the Africans were removed, to be followed by the 'Coloureds' in 1948.

Robert Moffat who, with his wife Mary, created what was often held up as the model missionary station at Kuruman, in what is now the northern Cape, was seen by many as the ideal missionary until surpassed by his son-in-law David Livingstone, though only after the latter's death. Perhaps Moffat made his greatest contribution to the cause of Christianity in southern Africa by his translation of the whole Bible into seTswana, a language widely understood across southern Africa. Moffat also helped to arouse public interest in missions by writing *Missionary Labours and Scenes in Southern Africa* (1842), one of the first missionary 'best-sellers'. In terms of success in publishing he was, yet again, to be passed by his son-in-law David Livingstone, a man of Highland stock and one of the first generation to be English- rather than Gaelic-speaking.

There are ninety-one entries relating to David Livingstone in the catalogue of Edinburgh University library, and over eighty in the library of Yale University. This is a measure of the impact he made on the consciousness of the English-speaking world in the hundred years after his death. He died on 1 May 1873 at Chitambo's village in what is now north-eastern Zambia. The astonishing story of his followers burying his heart and then carrying his mummified body to the coast to be returned to Britain, where he was accorded a *de facto* State funeral in Westminster Abbey, made headlines in the world's press. There then followed a period when Livingstone was 'written up' as a Protestant missionary saint. From the middle of the 1880s until well into the twentieth century another school of writers, some of whom were also part of the previous group, chose to portray him, in books, journals and newspapers, as the hero of liberal or paternalist imperialism. It was, they insisted, his call for Europe to intervene to end the east African slave-trade, that elicited the response by the European powers which led, between 1885 and 1895, to their conquest of all Africa, save Ethiopia, 'for the good of its inhabitants'. Even Leopold of Belgium insisted that it was this philanthropic motive which inspired his creation of the Congo Free State, the most brutal regime of a brutal era. Livingstone's role as the icon of liberal imperialism as opposed to the brutality of the Congo, lasted a long time, and he was still thus described by Cecil Northcott in his *David Livingstone* (1973).

A number of other 'Livingstones' were presented to the public. As John Mackenzie has pointed out, the image of Livingstone as the quintessential Scottish 'lad o'pairts' played an important part in a renewal of Scottish self-consciousness and assertiveness in the first decades of the twentieth century. Mackenzie also interestingly notes that Livingstone became part of the ideology of African nationalism in the 1960s and 1970s.[5] At the dedication of a new monument to Livingstone at Chitambo's in 1973, President Kaunda of Zambia called him 'the first freedom-fighter'. It is notable that in Africa south of the Sahara the colonial name of every town or district has been changed to an African name save that of Livingstone in Zambia and Livingstonia and Blantyre in Malawi.

Kaunda's encomium of Livingstone was nearer the mark than the other iconographic Livingstones.[6] Biographers of Livingstone, until the last twenty years of the twentieth century, have too often been unable to see vital elements in Livingstone's life because of the way their vision was limited by their preconceived notions. In some cases these elements were deliberately ignored by men who, despite their admiration for the man, simply did not understand Livingstone. This process began straightaway in 1874, the year of his funeral, with Waller's edition of Livingstone's last journals. In this two-volume work Waller edited out Livingstone's warm admiration for the earlier missions of the Jesuits in Africa, something also omitted in Blaikie's classic biography.[7] These two authors also removed from the record the two women, Ntaoeka and Halima, who were important members of the group of Africans faithful to Livingstone to the last. Most biographers have omitted Livingstone's support for the Xhosa in their fight against the British in the Cape frontier war of 1851–2. Again almost universally omitted from the record was Livingstone's insistence that Africans had the right to take up arms against attempts by Europeans to conquer them. Indeed, Livingstone repeatedly told his correspondents that obtaining firearms was the only hope Africans had to maintain their independence. All biographers of Livingstone before Timothy Holmes have omitted any reference to this often-repeated assertion in the private letters of Livingstone.[8] Perhaps the most astonishing misrepresentation of all is that Livingstone's famous march from the centre of southern Africa (what is now western Zambia) to the west coast, then back to the Makololo capital at Linyanti, before marching on to the east coast, has been consistently represented as an example of European achievement. Livingstone himself pointed out unambiguously in his *Missionary Travels in South Africa* that it was an expedition of the Makololo chieftaincy, which he headed. Livingstone had exhausted all his resources by the time he reached the Makololo capital. The new expedition was entirely financed and staffed by Sekeletu of the Makololo. It was an expedition in which Livingstone played the role of the chief 's *nduna* or representative. This was recognised by the Portuguese authorities in Angola who received Livingstone formally as an ambassador of Sekeletu. So many protagonists of one cause or another have shaped Livingstone as a hero of their cause that it has been only at the very end of the twentieth century that attempts have been made to get back through the iconography and the critics of the iconography to try to see who Livingstone was.

India, meanwhile, was the focus of an intellectual argument within the Church of Scotland that continued to influence thinking in both the Auld and Free Kirks after the Disruption. The debate began in the 1770s, perhaps earlier, and involved what Ian Maxwell has designated as the 'rational Calvinists' on one side and the 'evangelical Calvinists' on the other.[9] The rational Calvinist approach was based on a providential understanding of the growth of the Church. They argued that once the appropriate level of education had been reached and what they called 'civil society' had developed sufficiently, then the conditions were ready for the appearance of Christianity. Professor Hill of St Andrews, their leading thinker, perceived 'the rise of Christianity as a necessary historical moment to be repeated in every civilized society, given the correct conditions'. The evangelical Calvinists insisted that straightforward biblical preaching, which always took priority in mission over any concern about civilisation and Christianity, could be propagated effectively in 'primitive' societies.

Thomas Chalmers, in his writing and lecturing, tried to show that the apparently permanent division between those who insisted that 'civilisation' must precede the attempt to spread Christianity and those who insisted that the preaching of the Gospel was all that was necessary, was in fact bridgeable. This point of view was shared by John Philip in South Africa, who insisted that civilisation inevitably accompanied the growth of Christianity but it was not a precondition. Philip was one of the many Scots who got on with mission while the Kirk did nothing as the arguments between the two understandings of mission raged on. It was because certain rational Calvinists became convinced that Bengal was approaching that necessary level of civilisation that they joined with evangelicals in 1824. This alliance gained the necessary majority in the General Assembly and foreign mission became part of the life of the Kirk, a tradition later to be adopted by the Free Kirk. When Alexander Duff returned to Scotland after his first tour of service as the head of the General Assembly's Institution in Calcutta, he was convinced that the way ahead was the conversion of the youth of the Brahmin leadership of Bengali society through western higher education. He believed he had to persuade the overwhelmingly evangelical majority among the supporters of missions in Scotland to accept this view. If he failed, the voluntary giving of mission supporters would go to missions that conformed to the evangelical Calvinist image rather than to his Institute and its supremely rational Calvinist approach. So, at the General Assembly of 1835, in a brilliant speech, followed up by a widely distributed pamphlet, he persuaded the Kirk to adopt what was essentially the rational Calvinist approach to mission. Brahmin students were to be converted to Christianity through western education in English. This would then enable them to understand logical argument and so comprehend the truth of the classic evidential arguments based on history, miracles and prophecy. They had to be freed from Hindu irrationality. Duff was so successful that this rational Calvinist approach continued as part of the thinking of the missionary movement in the Free Kirk when it came into being in 1843.

However, this argument, so persuasively described by Ian Maxwell, is not the whole story. We have seen how John Wilson (who befriended David Livingstone on the latter's two visits to Bombay) developed educational work which was part of the development of the university there. However, Wilson also became a scholar in Marathi and Gujerati and a supporter of village and bazaar preaching, which Duff had dismissed as fruitless in his famous address to the General Assembly. Indeed, although Wilson's college work led to the conversion of some leading Parsi and Brahmin young men, the real growth of the Christian Church occurred among the low and outcaste groups (known today as Dalits).

This combination of concern for higher education in English with concern for the local culture and for the vernacular and vernacular preaching, was to be characteristic of Scottish missionaries whether they served the Auld, Free or UP Kirks or the LMS. Perhaps the classic example of this combination of focus on English higher education with concern for indigenous culture was the Lovedale Institution in the Tyume valley in the eastern Cape Colony. In 1841 the Institution was opened under the leadership of William Govan. It was in an area where work had been begun by the Glasgow Missionary Society, which had then been handed over to the Kirk in 1838. (In 1843 the mission and Institution became Free Kirk.) The Institution, which had no denominational test for entry and whose language of instruction was always English, started as a primary and high school. Later its campus grew to contain a primary and high school, teacher training college, technical college, theological seminary, and a hospital and nurses' training school. A unique feature of the Institution was that in the multi-racial Cape it was a multi-racial school where white, coloured and black met together in classroom and dining hall. (This situation continued until 1896 when the British authorities were seeking more and more to placate white opinion and forbade the enrolment of white children and young people in the Institution.) The Institution could be seen as the supreme example of the rational Calvinist model of mission, the more so as Govan was succeeded as principal in 1870 by James Stewart (1831–1905), a self-confessed admirer of Duff. The Lovedale Institution was also involved in the creation of literature in the local Xhosa language and, through the activity of some of its most distinguished alumni, it became a centre for the study of the culture and traditions of the Xhosa and other African peoples. The Rev. Tiyo Soga, one of the first pupils at the Institution, later graduated from Glasgow University and the UP Kirk seminary in Edinburgh. When he returned to South Africa it was as an equal with the other Scottish missionaries, but he did not cease to be a Xhosa and was an outstanding poet in his own language. He had become a Christian before his education and his conversion was not dependent on gaining western rationality as Duff had argued in his Assembly speech.

Soga was a classic example of what the other root of the Scottish missionary concern for higher education sought to achieve. This root was much older than the rational Calvinism of the second half of the eighteenth century. It was the tradition in classical Calvinism which held that the production of cultivated and well-educated Christians was not only necessary for the propagation of the faith but also for the creation of a godly and just nation. This was the tradition to which John Philip, with his political activism, belonged. This old tradition is as much part of the explanation of the Scottish concern for education in missions as the dominance of the rational Calvinist model among leaders in Scotland. The tradition was one deeply rooted in Scottish Protestant culture.

The leaders of the LMS at its inception were closely related to the movement for the abolition of slavery. At the annual general meeting of the LMS what nowadays would be called a 'keynote address' was preached. Again and again in those special sermons the preacher insisted that British Christians owed the peoples of Africa a debt. African people had been enslaved and transported to the Americas and their societies profoundly disrupted, all to bring financial profit to British people. It was an obligation laid on British Christians, then, not only to preach the Gospel in Africa but also to try to do something to repay African people through the development of legitimate commerce. This, it was hoped, would bring about what at the end of the twentieth century was referred to as development. This vision of the transforming power of Christianity and commerce was at the root of the idea so widely publicised by Fowell Buxton, successor to Wilberforce as leader of the anti-slavery movement. It has to be noted that Buxton began to propound this idea after he had spent well over a year in regular discussions with John Philip when the latter was in Britain during 1827 and 1828. Philip had certainly worked with this principle in mind in South Africa before this visit, even though he had not coined the phrase. He just as often wrote of Christianity and civilisation as Christianity and commerce, but meant the same thing. For Philip and Buxton, as for Livingstone who made the phrase famous, Christianity and commerce were going to bring about not only the Christianisation of Africa but also its economic transformation.

Livingstone is one of those many Scots who prized higher education and civilisation yet had deep concern for and sympathy with the traditional society in which they were placed. For example, most missionaries of the time condemned the *bogwera* and *bojale*, the male and female initiation rites among the Tswana people, as irredeemably evil. Livingstone, who had a better scientific training than most nineteenth-century missionaries, on the contrary insisted they were social customs essential to the community and something to which the Church should not object. Again he knew the difference between the *nganga* or healer and the witchfinder in African society, lumped together as 'witchdoctors' by most European observers until late in the twentieth century, and Livingstone always treated the *nganga* as a medical colleague. He was followed in this tradition by Duff Macdonald (1850–1929) the first ordained missionary at Blantyre, Malawi, the mission founded to fulfil Livingstone's hopes for that land. In his two-volume *Africana* (1885), Macdonald produced one of the first detailed and careful examinations of African traditional life and beliefs. Macdonald's successor in Blantyre, David Clement Scott, passionately committed to offering Africans the best possible education in English, followed up *Africana* with his *Cyclopaedic Dictionary of the Cimang'anja Language* in 1892. This book was a deeply sympathetic guide to Malawi culture as well as the basis for all subsequent dictionaries of the language known as Nyanja in Zambia and Mozambique and as Chewa in Malawi.

This combination of high academic qualifications and concern about 'civilisation' combined with understanding of and sympathy for the local culture was extraordinarily widespread among many leading Scottish missionaries. In India we have noticed that at the beginning of the Scottish enterprise James Wilson was not only deeply involved in the development of western higher education but also in Marathi and Gujerati language and culture. A number of outstanding Scottish missionaries subsequently followed Wilson's example. The first was William Miller (1838–1923), appointed to Madras in 1862 and principal of the Christian College from 1870. He was a clear successor to Duff in the tradition of rational Calvinism. Higher education was for him a means whereby Hindu thought and Hindu society would be transformed and Christianity would then spread. The difference between Miller and Duff was that Miller admired Hinduism and sought to transform it, not destroy it. John Nicol Farquhar (1861–1929) was another of Wilson's heirs. After a brilliant career at Aberdeen and Oxford Universities he went to teach at Bhowanipur College under the auspices of the LMS, but then transferred his activities to student evangelism with the YMCA. He became a leading scholar in both Sanskrit and Bengali. Strangely, although an LMS missionary, he held a view of the future of Christianity in India which was close to the evolutionary development ideas expressed by Professor Hill in his articulation of rational Calvinism. The difference was that, as with Millar, Farqhuar saw Christianity as fulfilling the deepest needs of Hinduism.[10] His near contemporary, Nicol MacNicol (1870–1952) developed this concept further. What was outstanding about MacNicol was that he was a scholar of Hinduism, with many Brahmin intellectuals as friends, but he also had great knowledge of and sympathy for village Hinduism. His *Hindu Theism* published in 1915 is still a useful guide to the elements of theistic thought in Hinduism. The last of these Scottish missionaries was A G Hogg (1875–1954). Like Miller, Hogg served as principal of Madras Christian College. In the 1930s and 1940s he was an able and articulate defender of the continuity between Hindu faith and Christian faith against the then very popular discontinuity theory of the relations between Christian faith and other forms of faith. This position had been forcibly asserted by Hendrik Kraemer at the International Mission Conference of 1938 held on the campus of the Madras Christian College, where Hogg was both host and delegate.

Although Scottish missionaries only made up a small percentage of the total number of Protestant missionaries who served in India in the two centuries after the arrival of the Serampore trio in 1796, their contribution to the intellectual and spiritual understanding of Hinduism was vastly disproportionate to their small number. Though to a lesser extent, the same can be said of Scottish women missionaries as pioneers of western education for Indian women.

This was begun in the 1820s, as we have seen, by Margaret Wilson in Bombay. After her early death, her two sisters, Anna and Hay Bayne, continued her pioneering work in Bombay and Pune. Her work was rapidly followed up in Madras, Calcutta and elsewhere in India through the initiative of women in Scotland. There was a significant body of Scottish women committed to the work of mission from early in the nineteenth century, from before the time when 'women's work' came to be accepted in the Kirk at home. This activity drew its supporters from the same circles that also produced the very active women's committees in Glasgow and Edinburgh in support of the abolition of slavery, first in the Empire but also in the USA. Again the key institution in all of this was the voluntary society. The organised work began in 1837 when the Edinburgh Association of Ladies for the Advancement of Female Education in India was formed. After the Disruption this movement flourished in the Free Kirk; the Auld Kirk, however, had much greater difficulty in finding staff until the last two decades of the nineteenth century. The UP Kirk inherited its women's work from the old GMS where women had been active from the beginning and from that tradition there emerged three working-class young women who went on to become outstanding missionaries: Christina Thompson, Euphemia Millar and Mary Slessor. The enterprise of Christian women in Scotland dedicated to overseas mission was astonishing. Both the Free and Auld Kirks, for example, sent out women doctors as missionaries before the Scottish universities had admitted women for medical training. The first two pioneers were Dr Letitia Bernard of the Church of Scotland, who went to Bombay in 1884, and Dr Matilda MacPhail of the Free Kirk, who was appointed to Madras in 1888. Dr Bernard went to join her two sisters, Emily and Amy, who were already stalwarts of the Church of Scotland work in Calcutta and Bombay. The sending of these two women marked the beginning of an era, 1880s–1914, which has been called elsewhere that of 'the feminisation of foreign missions'. Women's role at this period is emphasised by the fact that by 1896 the UP Kirk alone had five women candidates for overseas service studying medicine. At the same time, when the number of male volunteers was static, the number of Scottish women teachers serving all three Scots Kirks and with societies like the China Inland Mission and the LMS was growing dramatically.[11]

When we look at Scottish missionaries in China, one thing stands out. This is that a surprising number of them were sympathetic to Chinese culture, particularly to Confucian philosophy. The majority of the Protestant missionaries to China, when told of the attempts by the Jesuits of the seventeenth century to develop Christian thought in China in relation to Confucian philosophy, as St Thomas had done in Europe with Aristotelian philosophy, dismissed the whole enterprise. As one of the most prominent American Presbyterian leaders said 'it matters not what Confucius said. We have come to preach the Gospel'. A number of Scots nevertheless stood out against this attitude, not least Huntly-born James Legge (1815–97). A graduate of Aberdeen University, Legge never strayed from a straightforward evangelical faith, but in the Chinese College at Malacca, where his career began, and then in Hong Kong, he learned Mandarin and translated the Confucian classics. He came to conclusions similar to those of the Jesuit missionaries of the past, that Christianity and Confucian thought were compatible and that 'original' Confucianism had been monotheistic. This opinion, which provoked a great deal of suspicion and opposition from many Protestant missionaries in China, was shared by another Scot, John Ross (1842–1915) of the UP mission to China. A Gaelic speaker and another brilliant linguist, Ross also revered the classics and insisted that the Jesuits had been correct in their general approach to Confucian thought and culture. His contact with Korean visitors to his station at Mukden took him off on a new initiative. He learned Korean, producing a primer in the language in 1877 and a grammar later. More importantly, he supervised the first translation of the New Testament into Korean. His work was the foundation upon which the first Korean Protestant Churches were built. These two Scots stand out from the general Protestant approach to China, though it is another Scot who is lodged in the Scottish imagination with regard to missions in China, the Olympic 400 metres champion, Eric Liddell (1902–45) of the LMS.

Foreign missions have had an impact not only on societies abroad, but also on Scotland. The clearest example of this influence is the close relationship which has existed since the last decades of the nineteenth century between the peoples of Scotland and Malawi. The Livingstonia Mission (1875) of the Free Kirk and the Blantyre Mission (1876) of the Auld Kirk were sent to Malawi to honour Livingstone's memory. Despite the still bitter division between the two Churches in Scotland at that time, the missions co-operated from the beginning. What was even more dramatic was the fact that, in response to the pleas of the missionaries in the field, the two Churches back in Scotland co-operated closely for over a decade to try to influence the British Government. Thus in 1888 to 1891, after many public meetings and a mass petition, they successfully persuaded the British to keep the Zanzibaris and the Portuguese out of what is now Malawi. Even more striking was the alliance of Auld and Free Kirk leaders which brought in the leading Scots Liberal and Tory politicians to a massively supported campaign in Scotland. This campaign persuaded Lord Salisbury to prevent Rhodes and his British South Africa Company from taking over Malawi first in 1893, then again in 1896. This took place at a time when probably nothing else would have brought those same people together, Free and Auld Kirk, Tory and Liberal.

The story does not end there. In 1927 and again in 1938 there were attempts to amalgamate into one white-dominated, self-governing Dominion, Northern and Southern Rhodesia with Nyasaland, as Malawi was then called. On each occasion the missions, Malawi Christians and the Church at home lobbied the government against it, pointing out the opposition of the African people to such a move. On each occasion they were heeded. In 1949–50 the same groups protested but were not listened to and the Federation of the three territories came into being. The Nyasaland African Congress and the African Church in Malawi looked to Scotland for support in their opposition to Federation. They were not disappointed. Thus it was in 1959, when the future of the Federation was in crisis, that the General Assembly of the Kirk debated the Church's stance on this issue.

Kanyama Chiume, the only leader of the Nyasaland African Congress not in prison at the time, cabled the Moderator thus:

> In this perilous time in the history of my country when the civil liberties and freedom of my people are shamefully scorned and suppressed, may I, as a member of the Church, assure you of the unfailing friendship and brotherhood between the peoples of Scotland and Nyasaland.

Despite further trials and tribulations in Malawi, this bond between the two peoples has endured. Thus, when Orton Chirwa, Malawi's first Minister of Justice, died in prison under the Banda dictatorship in 1989, the High Kirk of St Giles was packed full for his memorial service. That same year, international pressure led by the Scottish branch of Amnesty International obtained the freedom of his imprisoned wife, Dr Vera Chirwa. That Dr Bakili Muluzi, the president of the newly democratic Malawi, was the first foreign dignitary to address the new Scottish Parliament is but the latest symbol of this enduring relationship.

NOTES

1. For these changes in Scottish society see Smout, T C. *A History of the Scottish People, 1560–1830*, Glasgow, 1972, 338–400.
2. See Warren, 1965; Warren, 1967.
3. Ross, 1972, 52–72.
4. Ross, 1986.
5. Mackenzie, J M. David Livingstone: The construction of the myth. In Walker, G and Gallagher, T. *Sermons and Battle Hymns*, Edinburgh, 1990.
6. See Ross, 2002.
7. Waller, H, ed. *The Last Journals of David Livingstone*, 2 vols., London, 1874; Blaikie, W G. *The Personal Life of David Livingstone*, London, 1880.
8. Holmes, 1993.
9. Maxwell, 2001.
10. See his *Crown of Hinduism*, London, 1913.
11. See the excellent entry by Macdonald. In *DSCHT*.

BIBLIOGRAPHY

Holmes, T. *Journey to Livingstone*, Edinburgh, 1993.
Macdonald, L. Women in Presbyterian Missions. In *DSCHT*, 888–91.
Maxwell, I D. Civilisation or Christianity? The Scottish Debate on Mission Methods. In Stanley, B, ed. *Christian Missions and the Enlightenment*, London, 2001.
Ross, A C. Scottish missionary concern 1874–1914: A golden age? *SHR*, 51 (1972), 52–72.
Ross, A C. *John Philip: Mission, Race and Politics in South Africa 1775–1851*, Aberdeen, 1986.
Ross, A C. *David Livingstone: Mission and Empire*, London, 2002.
Warren, M. *The Missionary Movement from Britain in Modern History*, London, 1965.
Warren, M. *Social History and Christian Mission*, London, 1967.

25 Church, Law and the Individual in the Twentieth Century

FRANCIS LYALL

INTRODUCTION

Questions of Church and State, and of religion and the law, continue to cause discussion in the twenty-first century, but it has to be said that such matters appear not to be as important to the generality of the population as they once were.[1] However, this development is not confined to matters religious. All forms of authority, social structure and tradition have been increasingly questioned, not to say, disparaged. While criticism is always healthy, when directed to the improvement of obvious defects, the urge to pull down when unaccompanied by constructive proposals for building, may be dangerous for society as a whole.

That the Scottish Parliament, established under the Scotland Act 1998, c. 46, sat, albeit temporarily, in the remodelled Assembly Hall of the Church of Scotland on The Mound in Edinburgh, reflected the apparent decline of the role of matters religious in the life of Scotland from that indicated in my previous chapter.[2] Matters religious appear to carry less weight among the populace in the twentieth and twenty-first centuries than they did in the sixteenth and seventeenth centuries. As a member of the Church of Scotland I regret that decline, but can see why it has happened. Secularisation has both occurred and appears to be continuing. Simultaneously there has been a growth of other religions and religious beliefs and practices, both ancient and modern. Not only has there been a natural change or development in matters cultural, there has also been an alteration in the basic premises or the shared religious geology of Scottish society. Matters which would formerly have been taken as settled by theological perceptions have been opened for discussion and deliberation. Most modern law is sought to be justified on a basis other than theology or deduction from theological propositions. On occasion the result is the same as the reformers would have deemed right. In other matters their views, for example as to sexual morality, are challenged on the ground of 'human rights'. This has had considerable effects in law, in practice, and in what we may term the broad cultural environment.

In his Preface to *The General Assemblies of Scotland 1560–1600*, Duncan Shaw says that his book 'traces the rise of what was once the most influential national church council in Europe ...'.³ That was written nigh on forty years ago: its elegiac element becomes almost painful when one contemplates the effect of those decades. Successive statistical analyses show the decline of all the former major institutional Churches, although it may also be noted that certain of the smaller denominations such as the Baptists, are less affected, and that within the major denominations congregations of traditional orthodox belief tend to resist the general trend.⁴ On the other hand, the diminution of the position of the Churches in general, and particularly that of the Church of Scotland from its former primary position, has been accompanied by a relatively peaceful transition within Scotland to a pluralist society. That peace is to be welcomed, provided that it does permit the denominations to retain their individualities, and their ability to disseminate their views as to true theology, even if (or when) that conflicts with the views of other theologies.

In addition, questions of the Church, the individual and the law in Scotland, have been affected by the constitutional arrangements of the United Kingdom. Usually this has meant that Scottish affairs have ranked low on the agenda of central government and the Westminster legislature. Subsequent to the Union of the Parliaments, while, as noted below, the Union arrangements protected the Church of Scotland and the Scottish legal system, Scotland did not again have its own legislature until the setting up of the Scottish Parliament in 1999. Further, although a Secretary of State for Scotland did exist after the 1707 Union of the Parliaments until 1746, thereafter Scottish responsibilities were allocated among various ministers, notably the Home Secretary. But though that was the constitutional position, in practice Scottish affairs were largely run informally by the Lord Advocate, normally the leader of the government's party in Scotland,⁵ and Scottish affairs were not of great interest to the rest of Britain although the controversy that eventuated in the Disruption (see below) is an exception to that. In 1885 the Scottish Office was set up with a Secretary for Scotland, and to him over the years were transferred many responsibilities. The office of Secretary of State for Scotland was created in 1926, with Cabinet status. Until devolution in 1999 the Scottish Secretary and the Scottish Office did function to some extent as if it were a Scottish government, having within the Office responsibilities which in England were committed to separate Departments. While that meant that on the one hand there could be a greater sensitivity to Scottish opinion, on the other hand in practice legislative action could be long delayed. To these elements of political organisation may be added the suspicion that at least in the past those accustomed to the English scene, moulded as it is by the episcopacy of its established Church, have had difficulty in comprehending Presbyterian traditions and viewpoints. It remains to be seen what effect the creation of the Scottish Parliament with lawmaking powers (subject to a list of matters 'reserved' to Westminster) and a formal Scottish Executive may have. Finally, as affecting the law other ways, I would note that although Scotland has a legal system and administration separate from that in England, decisions as to prosecutions in Scotland can be affected by events and decisions south of the Border, and civil law may also be altered for Scotland in the wake of English agitation or discussion. For this reason I will occasionally cite English material as authoritative for Scottish practice.[6]

CHURCH INSTITUTIONS

We must begin before the twentieth century in order to understand its developments. In 1690 the Scottish Parliament had fully confirmed the Church of Scotland as Presbyterian, and its Confession of Faith, the Westminster Confession.[7] Protected by the terms of the negotiations and agreement of the Treaty of Union of 1707,[8] the Church of Scotland remained the dominant denomination for roughly a century or more.

The Church of Scotland
The Church of Scotland as now constituted operates under a constitution which is not wholly reduced to writing, but in which important documents include the Basis and Plan of Union of 1929 and the Articles Declaratory of the Constitution of the Church of Scotland in Matters Spiritual, which were recognised as lawful for the Church to hold by the Church of Scotland Act 1921, c. 29. As their dates indicate, these two documents were major developments of the twentieth century, but they build on history.

In the seventeenth and eighteenth centuries, protected by law as indicated above, the Church of Scotland had few institutional problems.[9] Its parochial system and supervision were important in education, poor relief and other matters,[10] being in effect the major territorial division for local government as that was developed. However, during these centuries seeds were sown which resulted in the Church's major crisis, the Disruption of 1843. Further, changing views as to the proper responsibilities of the State led to the Church's loss of position in education, poor relief and other matters of local government.

'The Disruption' occurred over the relationships between State and Church authority, which manifested itself in two different ways. The first was whether the Church had power of itself to create new parishes with full ecclesiastical authority for their ministers. The second arose over debate and dispute over patronage in the election and settlement of ministers.[11] Broadly the opposing views within the Church reflected a division between a group which came to be known as the Moderates, and the Evangelicals. The Moderates were reputed to sit easy as to their ecclesiastical obligations, but were very interested in literature, philosophy and science, while the Evangelicals were firmer on 'the fundamentals of the faith' and inculcating these among the people. In 1834 the Evangelical group became a majority in the General Assembly, which passed two crucial Acts, Act IX, the 'Chapel Act', and Act XII, the 'Veto Act'.

By the nineteenth century the boundaries of many parishes of the Kirk had become obsolescent, not to say obsolete. The growth of towns and cities which accompanied a rise in population, a move from country to town, and the start of the Industrial Revolution, meant that the old parishes often were not well situated. At that time the parish was still the unit of local government, and had responsibilities for education and welfare, met by its kirk session through endowments and collections. The financial responsibility for the building and maintenance of ecclesiastical buildings within each parish was that of the heritors – those who owned heritable property (i.e. land or buildings) above a certain value in each parish. To meet the new situation the obvious solution was to create within the boundaries of an older parish a new parish with full standing – a parish *quoad omnia*. However, because such a step affected the rights and duties of the heritors of the existing parish, as well as affecting responsibilities with regard to education and social welfare – all matters involving civil right – the creation of a new parish could be accomplished only with the authority of the Court of Session, sitting as the Court of Teinds, and only with the consent of three quarters of the heritors of the original parish. An alternative was for a presbytery to establish a chapel of ease, a preaching station, created *quoad sacra* – for 'sacred matters' only. As a *quoad sacra* charge had no civil responsibilities, it could be created by the relevant presbytery itself and it existed for religious purposes only – the preaching of the gospel and the bringing of the other ordinances of religion to those within its boundaries. The Evangelicals were anxious to establish these new preaching stations to meet the social and ecclesiastical needs of areas to which the population had moved and which were remote from the older parish churches.[12] However, to comply with the civil rules administered by the Court of Teinds as to the creation of 'real' new parishes, until 1834 the ministers of parishes *quoad sacra* had no right to sit in the courts of the Church. The General Assembly's Chapel Act, Act IX of 1834, purported to give them that right, but, in a series of cases culminating in 1843, this piece of Church legislation was held to be unlawful, and decisions made by a court of the Church improperly composed because of the presence of ministers of *quoad sacra* charges were set aside.[13] The civil courts held that the Church could not of itself alone adopt legislation which had the effect of altering the membership of presbyteries by adding persons not inducted to a charge *quoad omnia*.

The Veto Act, Act XII also of 1834, sought to alter the rules as to Patronage. Patronage, the right to present a minister to a charge, is a pre-Reformation concept, and in many legal systems is in law a separate right of property. Usually it gives the presentee right to stipend and the other emoluments attaching to the charge to which he is presented. After the Reformation patronage was a matter of contention in Scotland. As a matter of law it was re-established by the Church Patronage (Scotland) Act, 1711, (10 Anne c. 10 (c. 9 in some editions)) in breach of the Union arrangements, and became one root of important secessions in 1733 and 1752. In the next hundred years opinion polarised between those who sought to defend the right of property, and those who considered that the congregation involved should have the say as to who should be their minister. The Veto Act of 1834 declared it a fundamental law of the Kirk that no minister should be intruded on a congregation contrary to the will of the people as expressed by the male heads of families on the congregational roll. The 'Patronage cases' ensued, by the end of which in 1843 the Court of Session and the House of Lords had held that the Kirk was not at liberty to make law contrary to the civil law by removing or restricting patronage rights, and that attempts to implement, comply with or enforce the Veto Act would be struck down by the civil courts.[14]

Of course, during the period 1834 to 1843, and particularly as the court cases began and proceeded, there was intensive discussion, including with the government of the day.[15] However, no progress satisfactory to the proponents of the Chapel and Veto Acts was made. On Thursday, 18 March 1843, the General Assembly met, and opened with prayer. The Moderator, the Rev. David Welsh, then, departing from normal procedure, read out a Protest to the effect that the liberties of the Kirk had been infringed without redress and that the Assembly should proceed no further. Thereafter he, and a large section of the Assembly withdrew, and, passing to the Canongate Hall, constituted themselves as the General Assembly of the Free Church of Scotland, a body with legal status as a voluntary association. About one third of the ministers and elders Church of Scotland followed them into the new Free Church.

Thereafter, however, matters were eased for the Kirk. The New Parishes (Scotland) Act 1844 made the creation of new parishes much simpler and easier, and gave all ministers in a charge the right to sit in the courts of the Church. Patronage was abolished by the Church Patronage (Scotland) Act 1874, the right to elect a minister being given to the congregation, with the Church having the right to determine whether the electee was suitably qualified. Finally, within the Churches (Scotland) Act 1905, c. 12 (passed mainly for other purposes), opportunity was taken to depart from the rigours of the Ministers Act 1693 (c. 38) and by s. 5 to pass to the Church power to prescribe the formula of subscription to the Westminster Confession.

However, implicit in that series of Acts of Parliament is the proposition that it was for the civil legislature to make law for the Church. 'The Church by law established' was a clarion call, but implied the power of the law to alter matters. This, of course, was contrary both to the Scots Confession of 1560, and to the Westminster Confession of 1647, though both of these had been legislatively recognised in their time. The precise relationship between Church and State was therefore a matter of great concern both in the Kirk, and outside it, particularly among those whose roots were a disagreement with the power of the State over the Church.

Starting in the nineteenth century, moves began to bring separate denominations together. A major step was the coming together of the Relief Church, the New Light elements of the Burgher and Anti-Burgher Seceders (formed in 1733 and 1752) and some others to form the United Presbyterian Church in 1847. That Church and most of the Free Church of Scotland united in 1900 to form the United Free Church, thereby triggering a very important case as indicated below. Thereafter the bulk of the United Free Church and the Church of Scotland united as from 1929, on the Basis and Plan of Union agreed between these Churches, and the 'Articles Declaratory of the Constitution of the Church of Scotland in Matters Spiritual' negotiated between the Churches, and declared lawful for the Church of Scotland to hold by the Church of Scotland Act 1921, the Church of Scotland (Property and Endowments) Act 1925 dealing with most matters of property.[16]

The first of the 'Articles Declaratory' scheduled to the Church of Scotland Act 1921 affirm the Church to be part of the Holy Catholic or Universal Church, and ground the Church firmly in the traditional orthodox faith, and the Scottish Reformation.[17] Article II sets the Westminster Confession of Faith as the principal subordinate standard of the Church, and as containing the sum and substance of the Faith of the Reformed Church. It also provides for the Presbyterian government of the Church. Article III affirms historical continuity with the Kirk of the Scottish Reformation, acknowledging a duty to provide the ordinances of religion throughout Scotland by a territorial ministry. Article IV claims the right of the Church to self-government in doctrine, worship and discipline, including membership, and the election of office-bearers. The Article also affirms that, while the State may recognise the claimed right of self-government, this does not imply that the right is derived from any other authority than the Divine Head of the Church, nor does it give any right to the secular authority to interfere in such matters. Article V expands on the right of self-government in Art. IV, allowing *inter alia* the interpretation of the Confession of Faith, and the relationship of office-bearers to statements of the Faith, while allowing liberty of opinion on matters not entering into the substance of the Faith. Article VI acknowledges that the secular authority is divinely appointed, and has its role to play. Indeed, Church and State owe mutual duties and may 'signally promote each other's welfare'. Each has the right to determine questions as to their rights and duties. Article VII affirms the duty to unite with other Churches 'in which the Word is purely preached, and that such a union will not result in a loss of identity'. Article VII contains provision as to the interpretation, modification of the Articles by the Church or addition to them but always consistent with the First Article, which Art. VIII states to be unalterable. Article IX ratifies and confirms anew the Constitution of the Church of Scotland, subject to the Articles and any future amendment to them.

In terms of the settlement indicated in the 1921 Church of Scotland Act, the Church's judicatories are independent of those of the State,[18] though powers of those of the State may be called in to assist those of the Church.[19] However, we should note that the 1921 provisions have deep roots, and in many ways are not to be taken as innovative.[20] The Church of Scotland remains the 'established Church' in Scotland, but establishment means much less than it does for the Church of England. There is statutory recognition of it as the National Church, and it is looked to as the Church to dignify solemn occasions, but otherwise 'establishment' in Scotland means no more.[21]

The relationship between Church and State is seen in other ways. On accession the Sovereign takes an oath to preserve the settlement of the True Protestant Religion and the Government, Worship, Discipline, Rights and Privileges of the Church of Scotland. The Sovereign of the day appoints a Lord High Commissioner to each General Assembly, but in the General Assembly Hall on the Mound, this individual sits in a gallery, open to the Hall, but not accessible from it.[22] The Lord High Commissioner is not a member of the Assembly, and, though as a matter of courtesy invited to address the Assembly, has no power to initiate or participate in its debates or to control its proceedings. In the days of the Stuarts, who called an Assembly and specified where it would meet were matters of contention. Since 1927 the Assembly sets the date and place for its successor, and the Lord High Commissioner undertakes to inform the Crown. When in Scotland, the Sovereign attends Church of Scotland services[23] as well as those of other denominations.

But finally the question must be asked whether the Church of Scotland exercises any significant influence or authority in Scottish affairs. On one level one cannot disregard the effect of individuals of strong religious conviction active either in politics or otherwise, but more doubt must be cast on the effect of the Church of Scotland as an institution. Certainly many of its members would point to the political and social influence of the Deliverances of the General Assembly (i.e. its formal resolutions), many stemming from the work of committees, such as 'Church and Nation' and 'Social Welfare'. Until the establishment of the Scottish Parliament in 1999, the Assembly was the major gathering of Scots, and as such its debates and decisions were reputed to carry weight, but many would consider that weight overstated. Unlike the episcopally organised Churches, no one really 'speaks for the Church' in its relations with the media, and committee conveners and the Moderator (whose official function really expires with the end of each Assembly) have always to indicate that what they say is not 'the Church' speaking.

Other Churches
As indicated above, the Church of Scotland was afforded protection in terms of the negotiations for and the term of the Treaty of Union of 1707. It remained dominant for many decades thereafter, other Churches and denominations (particularly the non-Presbyterian) operating under various disabilities.[24] Thus although toleration was extended to the Episcopalians in 1711 and extended in 1790, the Episcopalian Church as such was not recognised as anything other than a sect.[25] Disabilities attaching to Roman Catholicism from the Reformation times started to be reduced by the Roman Catholic Emancipation Act 1829, 10 Geo. 4, c. 7, but when the Roman Catholic hierarchy was restored by the Bull *Ex Apostolatus Apice* of Leo VIII of 4 March 1878, the old statutes against Roman Catholicism were still on the statute book.[26]

The Church of Scotland Act 1921, 11 & 12 Geo. 5, c. 29, is concerned mainly with the constitution of that Church. However, s. 2 of the Act specifically states that nothing in that Act 'shall prejudice the recognition of any other Church in Scotland as a Christian Church protected by law in the exercise of its spiritual functions'. The Church Act 1579, c. 6, which affirmed that those who professed the Scots Confession of 1560 were 'the only trew and haly Kirk of Jesus Christ within this realme ...' has therefore been restricted.

That said, the other Christian Churches do not have the same status in law as the Church of Scotland. Technically all Churches other than the Church of Scotland are voluntary associations. Their 'jurisdiction' over their members is, as a matter of civil law at least, simply based upon the ordinary law as to the authority which any club or association may have over a member. While the civil courts will not ordinarily intervene, abuse of procedures or a failure to observe the requirements of natural justice can result in action before the civil courts.[27] Crucially, their property is held by trustees, and the terms of the trust will be enforced by the courts. This can mean that, absent any provision to the contrary, if there is a split within a denomination or a congregation, those who adhere to the 'original principles' of the trust are entitled to the property, even though they may be in a minority. This point was clear in earlier disputes, but the 'Free Church Cases' of 1904, which arose from the union in 1900 between the Free Church of Scotland and the United Free Church of Scotland, is the authority now usually referred to.[28] In that instance some 10 per cent of the Free Church which refused to enter union with the United Presbyterian Church to form the United Free Church were held entitled to the property of the Free Church, since they adhered to the 'original principles' of the trusts on which that property was held. However, it was clear that this minority could not administer and did not need all they had gained. After negotiation and agreement, under the Church (Scotland) Act 1905, 5 Edw. 7, c. 38, the property was divided by Commissioners, the property remaining with the continuing Free Church where one third of the members remained with that Church, and otherwise being transferred to the new United Free Church.

The message of the 1904 Free Church cases is clear, and when the bulk of the United Free Church united with the Church of Scotland in 1929, steps were taken to agree a division of property with those remaining outside that union. But not all denominations have taken the point. Court disputes over property seem likely to arise in respect of recent developments in the argument between the Free Church of Scotland and the Free Church of Scotland (Continuing), which split on 20 January 2000 over matters debated throughout the late 1990s,[29] as also between the Free Presbyterian Church of Scotland and those who removed themselves from it in 1989 to form the Synod of the Associated Presbyterian Churches.[30] The difficulty for the courts in such cases will be that there seems no doctrinal difference between the contending parties.

Finally, as far as 'influence' is concerned, the smaller Presbyterian denominations share the problems of the Church of Scotland. Episcopal constitutions produce bishops, archbishops, and even cardinals, whose media presence may help since their official position as spokesmen for their constituencies is defined. Some have made use of this.

Other Religions
A major change in the last century or so has been in the development and embedding within the culture of Scotland of religions other than the Christian family of denominations. Much of this occurred after World War II with immigration particularly from the Indian sub-continent resulting in defined Hindu and Islamic minorities.[31] This has had effects in education, but otherwise such new religions operate under the normal rules as to the law of charities, as well as that of trusts and of natural justice which bind the non-Church of Scotland Christian denominations. At present their influence in Scotland is weak, relative to the Christian denominations. However, their English-based communities and bodies have a greater effect, and sometimes this spreads into Scotland through United Kingdom legislation.

THE INDIVIDUAL

As noted in the introduction, matters formerly founded upon theological considerations are now sought to be justified on other grounds. That was a development foreshadowed in the so-called 'Institutional Writers', whose work was formative of Scots Law as we know it today during the seventeenth century. Reason was given a place in determining what might be required in a given situation.[32] But, of course, when reason itself turned to examine the bases from which it was supposed to operate, debate arose. In the eighteenth century the Scottish Enlightenment flourished.[33] David Hume, Adam Smith and others were active. Nor should it be thought that Scotland was insular in the intellectual ferment of the times. Deism, scepticism and even atheism became respectable.

In the nineteenth century, social problems, better communications, a growing population and industrialisation resulted in a willingness on the part of government to undertake duties formerly thought not to lie within its responsibilities. The authority of the Church of Scotland itself was diminished, not to say undermined, by the Disruption. Although in general terms its theology was the same as the Free Church, it came to be seen as one option among many. Intellectual change continued, and notions of the place of mankind within the known world altered. The Church began, it would seem, to retreat, allowing 'Education' to take over ground which it had previously held.[34] Darwin's *Origin of Species* (1859) and the popularisation of his work by T H Huxley, as well as predecessors such as Sir Charles Lyell in geology, not to mention the semi-hysterical reactions of some ecclesiastics to the 'new' ideas, contributed to debate. Nietzsche proclaimed 'God is dead', and Karl Marx observed that 'religion is the opiate of the people', and he and his friend Engels questioned the foundations of society. Biblical criticism, in particular the Germanic 'Higher Criticism' began to erode traditional theology and ideas of 'revelation'. The spectacle in 1881 of the trial for heresy of William Robertson Smith, then a professor at the Free Church College in Aberdeen, did much to diminish general respect for theologians and ecclesiastical politicians.[35]

In the twentieth century these processes continued. The impact of a general system of compulsory education (outlined below) meant that more could read, and absorb a variety of beliefs and ideas, albeit usually in the form of impressions rather than well-worked out theories. World War I disturbed previous social patterns, and produced cynicism about civic leaders, which affected the Churches as well. In Scotland the casualties of that war led to an upsurge in Spiritualism, which the Church did little to counter. World War II produced other social change, including an altered balance between the sexes. In the latter half of the century, popular understanding of nineteenth-century biblical criticism and 'modern theology' rendered orthodox belief questionable. To the non-churched, religion became unnecessary. 'Doing good' did not require the employment of ecclesiastics, who appeared not to believe in what remained of the popular understanding of the traditional faith. The 'Tell Scotland' movement with Tom Allan and the Billy Graham crusades in the 1950s and 1960s did not reverse the decline of the Churches' influence. Finally, so-called Post-Modernism has produced an impatience with authority and an antagonism to ideas of 'objective truth' whether it be in religion, or as to the 'meaning' of a piece of literature. One result of three and a half centuries from the Reformation is, therefore, that many areas of law and behaviour, as well as of our common culture, have either ceased to have a religious basis, or have become at least semi-detached from such.

Moral Governance
Before we turn to particular areas of behaviour, it may be noted that as indicated above, the various Christian denominations, in their several forms, as well as the equivalents in non-Christian religions, retain disciplinary powers over their members, subject, in the case of non-Church of Scotland bodies, to review before the civil courts in appropriate instances. The former general control over moral matters exercised by the Church of Scotland has disappeared, in that only its own members are now subject to its processes. Indeed, the Church does not appear properly now to exercise much control or discipline in morals or doctrine over its own members. The law of defamation now curbs former general powers of rebuke that might be exercised by ministers from the pulpit.[36]

Education
The general education of the people has been a major factor in the diminution of the role of any particular religion, and notably that of the Presbyterian Churches, in the twentieth century. People have wanted to make up their own minds, or have been unwilling to pay attention to any one religious system within a plethora of beliefs.

The bulk of primary and secondary educational provision is now a responsibility of the State, although provision is made both for denominational schools and for religious representation within the education authorities.[37] Tertiary education, whether universities or colleges, has no such requirement, except in the case of appointment to certain theological chairs.

Schools
The *First Book of Discipline* of 1560[38] called for the creation of a school for every parish, and a system of bursaries for the 'clever poor', and in practice many parishes did set up such schools, as did the burghs. Under a series of statutes in the seventeenth and eighteenth centuries, schools were made subject to inspection by the local presbytery, teachers being required to subscribe the Westminster Confession of Faith. However, this control weakened. State intervention began with the Parochial Schools (Scotland) Act 1803, 43 Geo. III, c. 160, setting salaries and conditions including accommodation, State subsidies for education, and a system of inspection parallel to that of the Church.[39] In practice, though not in law, Church control over establishments subject to it became erratic. After the Disruption the headmaster of Campbeltown School was deposed by presbytery as he had 'gone out' to the Free Church.[40] An ultimately successful attempt by the local presbytery to enforce through the courts its right of inspection over Elgin Academy was the last straw.[41] The Parochial and Burgh Schoolmasters (Scotland) Act 1861, 24 & 25 Vict. c. 107, abolished the rights of presbyteries in the matter and the requirement that teachers subscribe the Westminster Confession substituting for that an undertaking that the teacher would not subvert its doctrines. A decade later, and in the wake of a similar English legislation, the Education (Scotland) Act 1872, 35 & 36 Vict. c. 62, transformed school education in Scotland, set up a State system of education controlled through local school boards, permitted the transfer of private schools, including denominational schools, to the State system, and made education compulsory from five to thirteen, and, in the State system 'undenominational'. The undertaking anent the Confession disappeared. There should be no inspection in religious knowledge in the State schools, nor should there be public monies for religious instruction although such instruction might continue as in past practice. State schools would be open to children of all denominations, and parents might withdraw their children from religious instruction or religious observances in accordance with their own views. In short, there were so many variants of Christianity that it was considered a primary role could not and should not be given to a particular denomination or its beliefs.

Denominational and other schools had, as indicated, existed for many decades, but on a private basis.[42] Many were in financial trouble and in the years after 1872 took the opportunity to join the State system, albeit that the transfer of the property had to be by way of gift – no compensation was paid. Roman Catholic schools, however, remained outside the State system though with increasing financial difficulties,[43] until by the Education (Scotland) Act 1918, 8 & 9 Geo. 5, c. 48, local education authorities were laid under a duty to accept the transfer of voluntary schools, with powers to pay compensation to their former managers.[44] Schools so taken over were to remain of the same character and status as at the transfer, and retain staff in post. Teachers appointed thereafter were to be approved by the Church or denomination which used to manage the school, religious observances were to continue, the time allocated for religious instruction was not to be diminished, and a supervisor of religious instruction appointed by the Church or denomination was to be given access to the premises.

Broadly the 1918 arrangements remain the position for denominational schools in Scotland. The Roman Catholic Church is the main beneficiary of these arrangements. However, where the 'demand' for places reduces, it is possible, subject to safeguards, for the local authority to discontinue such a denominational school, amalgamate it with another, change its site or arrangements for admission to the school, or, in the ultimate, disapply the religious conditions applicable to a school. Needless to say this is a matter of some contention. Self-governing schools remain competent, although they must comply with general standards of educational requirements. There is a small, but growing, interest in 'home-schooling' by parents who consider that the State or other system may inculcate unacceptable religious views. 'Home schooling' may be made subject to supervision as to quality in general knowledge and understanding, on the model used in certain states of the United States.

As far as education in religion is concerned, where a local authority decides to have an education committee, it is required to appoint to that committee at least three persons, not members of the Authority, but interested in the promotion of religious education. For Scotland other than the Orkney, Shetland and Western Isles authorities, these will include a nominee from the Church of Scotland, the Roman Catholic Church and someone to represent other religious groups. In the three excluded authorities a Roman Catholic nominee is not required, and two persons are nominated to represent non-Church of Scotland interests.[45] 'Religious Studies' is an examinable subject in the school system.

Universities

By the Act 1567, c. 11, and 1697, c. 25, made part of the Union Settlement by the Union with England Act 1706, c. 7, the universities of Scotland were made subject to Church control, all professors being required to subscribe the *Westminster Confession of Faith* and subjected to presbytery supervision in their execution of their duties. In 1839 Professor Blackie on induction as Professor of Humanity (Latin) at Marischal College, Aberdeen, signed the Confession with express reservation as to whether it represented his faith. The effect of his reservation required a court case for its determination.[46] In 1853 the Universities (Scotland) Act, 16 & 17 Vict. c. 89, required subscription of the Confession only by holders of theological chairs, defined as Divinity, Church History, Biblical Criticism and Hebrew. Other professors were required only to undertake not to subvert the Confession, a requirement that vanished in 1889. Two religious elements remain in the university system. First, since a revision in 1950–1, recommendations as to appointments to divinity-related chairs in the four ancient Scottish universities are made for each by separate boards of nomination for theological chairs on which the Church of Scotland appoints one half. There is no requirement as to the appointee. Thus in 1979 Professor Mackey, a Roman Catholic, was appointed to the Thomas Chalmers Chair of Theology in the University of Edinburgh. Second, each board of nomination for theological chairs has a scrutiny and consultative function in relation to divinity courses and curricula, not unlike the equivalents which other professional bodies exercise where university courses are accepted as qualifying for professional status.

Social Welfare and Charity
Outside the burghs, the Church of Scotland parish was an important unit of local government well into Victorian times, and beyond.[47] For almost three centuries after the Reformation throughout Scotland the heritors of parishes *quoad omnia* (and church collections) were the major source of poor relief, administered by kirk sessions, until the Poor Law (Scotland) Act 1845, 8 & 9 Vict. c. 83, established parochial boards for the purpose and authorised assessments on occupants of lands and properties to finance the matter.[48] Now obligations of charitable relief are still recognised by Churches, but social welfare in all its many developed forms is generally seen as the responsibility of the State. Unemployment and disability benefits, old age pensions, and the National Health Service are all areas where in the past the Church of Scotland had a major role as a matter of law, and where other denominations came to play some part. Not many see a religious element in the taxation system, although arguably our social welfare system is based on religious foundations, albeit well-buried ones. That said, various denominations and religions still provide social welfare, including the running of hospices, detoxification units and the like. Their relationship (particularly financial) with both national and local government is causing difficulties.

Charitable activity does seem to have religious roots, although not always clearly articulated. Certainly such activity is to the benefit of the public. Many bodies do good work, most being constituted as trusts in order to provide a proper and clear framework for their activities, and in order to own or lease property, run bank accounts and so on. The establishment of a trust for public benefit is a relatively simple matter in Scotland. However, one of the purposes or desirable outcomes of a trust may be to gain charitable status for tax purposes, with a consequent increase in revenue. Common to all forms of trusts eligible for charitable status is the requirement that the trust should be for the public benefit. Most charities, intent on 'good works' have no difficulty in meeting that requirement. However, attaining charitable status for a 'trust for the advancement of religion' took time in Scotland. In 1888 a Scottish case held that religious purposes were not charitable under the then taxation statutes.[49] However, in a Scottish appeal to the

House of Lords in 1953 it was held that taxation statutes are UK statutes, but they are framed on the basis of English law and its definitions.[50] That meant that 'trusts for the advancement of religion' could have charitable status, as that category had been and continues to be elaborated by many English cases.[51] But it should be noted that under present English law (2002) some 'religions' may not qualify for charitable status since they do not contain a concept of God or of a monotheistic deism.[52] In 1990 the Law Reform (Miscellaneous Provision)(Scotland) Act, c. 40, put charities in Scotland on a new basis, and established mechanisms for their regulation. Trusts for religious purposes still may exist, but by s. 3 of the Act a category of 'designated religious bodies' was established, which exempts from the full requirements of central control. Such bodies must have as their principal purpose the promotion of religious objectives, and as their principal activity the regular holding of acts of public worship. In addition such a body must have been established in Scotland for not less than ten years, have a proven membership of not less than three thousand members of 16 years or older, and have an internal organisation such that one or more authorities in Scotland exercise supervisory and disciplinary functions in respect of its component elements, and that such elements have imposed on them requirements as to the keeping and auditing of accounts, as will meet the other requirements of the 1990 Act in such matters. As of 2001 there were ten religious bodies so designated, covering five Christian denominations. Charity law in Scotland is, however, under review and there may be change.[53]

Religious bodies and organisations are important for their ordinary charitable activities, and for which there is clear justification in granting them charitable status in law. They are also of considerable importance in their cultural function as social groups. Many denominations have come to rely financially on the State being willing to give them charitable status for their religious activities. In the twentieth century therefore we can see an inversion between Church and State in such matters. Churches used to perform their religious functions, and their charitable activities, out of their own income. Now many are dependent on what amounts to State subvention in order to carry out their fundamental function – the dissemination of their particular religious beliefs.

Family Life
Until the twentieth century, family life and sexual morals were largely determined by beliefs and attitudes deriving from the Bible.[54] However, Scottish culture and Scots law, even at the time of the Reformation, did take a more relaxed view of some of these matters than is commonly believed. Marriage by the three 'irregular' forms, which were abolished by the Council of Trent in 1560, remained competent in Scotland until 1940, and one of these, marriage 'by habit and repute', which requires a court decree, still remains possible. In normal form, marriage is entered into either in a church ceremony conducted by a minister of the Church of Scotland, or another Christian denomination recognised for the purpose by the State, or in a

secular ceremony conducted by a Registrar. Marriage cannot be entered into between persons within the 'prohibited degrees' relationship.[55] Non-Christian marriages are in law constituted by the Registrar's civil ceremony, the religious ceremonies of non-Christian religions conducted in Scotland are being as yet without legal effect. Scots law does, however, recognise the validity of non-Christian marriages conducted in a country where that religion is recognised. Certain arranged marriages have caused problems, and cases have been annulled by Scottish courts on grounds of lack of consent.[56]

Divorce for adultery and for desertion were early permitted, and in the former instance the penalty of inability to marry the paramour under the Marriage of Adulterers Act 1600, c. 29, was readily bypassed by failing to name the paramour in the actual decree of divorce. In the twentieth century other grounds of divorce, namely cruelty, incurable insanity, and sodomy and bestiality, were added to those of adultery and desertion by the Divorce (Scotland) Act 1938, 1 & 2 Geo. 6, c. 50. Apart from insanity, these four grounds of divorce implied fault on the part of the party sued, and the divorce was seen as punishment for an offence. This led to acrimonious disputes in some cases and to perjury in others. The interest of the State was seen not as defending marriage by making divorce difficult, but rather as affording a method by which a dead marriage could be quickly buried. 'No-fault' divorce became the call. Now under the Divorce (Scotland) Act 1976, c. 39, irretrievable breakdown of marriage is the ground of divorce, although proof of the case may require establishment of many of the older forms. In the case of desertion, living apart for two years will suffice if the parties so agree, while five years will allow one party alone to proceed. In other words, the concept of 'irretrievable breakdown' is artificial, and in practice the notion of 'fault' or 'penalty' remains in many instances. That said, most divorce actions are now undefended, and divorce is now available through the Sheriff Court system in appropriate cases. The annual publication of the 'Scottish Judicial Statistics' shows that the rate of divorce in Scotland continues to rise. Judicial separation remains a competent though seldom used alternative to divorce, and, anecdotally, is resorted to mostly by Roman Catholic couples.

In family matters husband and wife have equal status,[57] in particular as to the treatment of children. Where there is no marriage the Law Reform (Parent and Child)(Scotland) Act 1986, c. 9, has swept away the disabilities of bastardy. 'Parental rights' or their equivalent, are available under the Children (Scotland) Act 1995, c. 36, to virtually anyone who can show a proper interest, and now even a sperm donor may be given rights of access to his child. The criterion is the 'best interest of the child', not a patriarchal-derived concept of marriage such as the reformers would recognise. Aliment may still be sued for.

In bringing up a child the wishes of the parents are obviously important. This will include whether the child is brought up in a particular religion or in none. Older cases indicating that a child should be brought up as a

Christian are thought to be obsolete. Where parents differ in religion or denomination the courts are neutral as to choice. When a child is sixteen, a young person can choose for him or her self. A parent cannot veto a medical procedure on a child on religious or moral grounds. If necessary the local authority can go to court to ask for parental rights in order to consent to such treatment.[58] Although there seems to be no reported case, and the matter involves medical confidentiality, it is likely that the Scots courts would follow the English in not requiring a parent to be informed if a girl under sixteen seeks contraceptives, or contraceptive advice from a doctor.[59]

Sexual Matters
The end of the previous section itself indicates that in sexual matters the twentieth century saw a major movement away from the law and customs of a society founded upon the traditional, Protestant views as to morality. Contraception by medical means has radically transformed the attitudes of many to sexual morality. Elements of the law have, however, been strengthened, but for reasons other than religious precept. The Incest Act 1567, c. 15, based its provisions on Leviticus 18, as found in the Geneva Bible version. Always strictly construed, that Act remained the law until replaced by the Incest and Related Offences (Scotland) Act 1986, c. 36. The previous law prohibited sexual relationships between blood-relatives and also relationships of affinity.[60] The new legislation adds adoptive relationships and step-relatives.[61] While therefore the religious taboo is still present, the law also clearly is concerned with the protection of children from exploitation by 'family members', widely construed. Short of incest, which involves penetration, there is still the criminal offence of shameless and indecent conduct. Bestiality remains an offence as well, although there seems to have been no recent prosecution. That offence is the only one clearly still based on the biblical prohibition.[62] Homosexual behaviour between consenting adults in private was de-criminalised by s. 80 of the Criminal Justice (Scotland) Act 1980, c. 62, but homosexual partnerships do not have legal effects.[63] There have been suggestions that some sort of 'civil partnership' might be invented and given legal effects, but the idea that such could be termed 'marriage' is contested on the ground that marriage by definition implies the parties are male and female – a ground traceable to religious doctrine.[64]

Other Matters
In various other areas it is clear that religious considerations have declined in their importance in shaping law and behaviour, and sometimes law has been brought in to produce behaviour which religion appears not to have accomplished. Various kinds of discrimination have been made unlawful,[65] but not discrimination on grounds of religion. In other areas an old form persists, but its religious base is attenuated. 'The oath' remains an important element in court trials,[66] as well as in the acceptance of certain offices.[67] However, for most oath-takers the sanction of perjury may be more important than any possibly divine sanction.

Within employment law there is no entitlement to time off for religious observances, no matter what the faith involved, but some employers permit this as a matter of grace, and there is argument as to whether such a right should be introduced. Now holidays are an entitlement within a contract of employment, and specific public holidays have secular, not religious roots. To the reformers, the Church Year was not something important. That attitude persisted well into the twentieth century. Thus Christmas used not to be a matter of serious concern to working people. I can recall the cold-water tank in my mother's house being replaced one Christmas in the 1940s – to the plumber it was an ordinary working day. Despite their pagan connotations, Hogmanay and the New Year were the time for festivity. Now, both Christmas and the New Year are exploited by commerce, and the religious significance of the former is minimal.

Just as the Church Year was unimportant to the reformers, so they and their successors made little of burial,[68] but now a 'good send-off' with secular music is a regular occurrence. Again, culture has changed. Most cemeteries are now established and run by private companies, or local authorities, not the Church. Cremation used to be unthinkable, but in Victorian times pressure for its introduction, particularly in England, increased. Too much land was being swallowed by graveyards, and parish cemeteries were filling.[69] Now we have facilities still regulated under the Cremation Act 1902, 2 Edw. 7, c. 8.[70] In neither burial, nor cremation, is it required that a minister of religion should conduct the ceremonial, although the Church of Scotland considers that the relevant parish minister should act, as a matter of last resort, as part of its self-declared duty to bring the ordinances of religion to the whole of Scotland. Tangentially, we may note that various legislation permits the use of human tissue for transplantation, or educational purposes.[71]

The Scottish Sunday used to be famous. Now, throughout the bulk of the Scottish mainland, Sunday observance has declined. In fact the penal clauses of the old Scots legislation were in abeyance or repealed for centuries before the general shift in civil behaviour began in the late 1980s.[72] The Western Highlands and the Islands remain the only parts of Scotland where Sunday observance remains staunch, but that as a matter of local culture and behaviour, not of law.[73]

Apart from the sexual matters outlined above, the criminal law has few provisions which are religiously based or oriented.[74] Witchcraft was punishable by death in Reformation times and for a century thereafter, but the Witchcraft Act 1735 repealed the former legislation and prohibited prosecutions for witchcraft, sorcery, enchantment or conjuration. However, the pretended use of witchcraft and similar acts for fortune-telling or the recovery of goods remained criminal. Such matters are now covered by the Fraudulent Mediums Act 1951, 14 & 15 Geo. 6, c. 33. The effect of such legislation is that the practice of witchcraft is not a criminal offence *per se*. It remains possible that certain of its rituals and practices may be prosecuted for breach of other ordinary criminal laws of the land. The same applies to other 'new' religions, or to imports from abroad, for example, voodoo, ritual sacrifice of infants, etc.

The Christian religion is no longer protected in the way that it was. As indicated above, various varieties of Christianity have come to be accepted over the centuries, and the Doctrine of the Trinity Act 1812, 53 Geo. 3, c. 160, gave relief throughout the United Kingdom to those who impugned various Christian doctrines. In Scotland spoken blasphemy would now be treated in appropriate circumstances as breach of the peace, and written blasphemy is unlikely to be prosecuted by Crown Office.[75] Other religions are not protected either, and suggestions that Islam should be somehow protected in the wake of the Salman Rushdie affair, were turned down.[76]

Per contra, however, it may be noted that there are voices which would oppose traditional Christian beliefs, arguing that, for example, the assertion that the idea that 'Christ is the sole avenue to salvation', disparages other beliefs, and offends those of a different religion or who reject such ideas. Further, it is said that the view that marriage is 'instituted by God' implies that other 'arrangements' are inferior. Those who are in or are the product of single-parent families or informal 'partnerships' therefore are made to feel inferior. On such a line of argument, the expression of traditional Christian precepts and beliefs should therefore be either unlawful or at least unstated in the public media, or in schools, and be permitted only in churches. The legal position is that newspapers and books are free to publish what they will, subject to the controls of defamation, incitement to race hatred and the like, but not subject to any controls about religion. Broadcasting and television, however, are controlled. Under the BBC Charter and Agreement (for the BBC) and the Broadcasting Act 1990, c. 42, for other licensed broadcasting and television (i.e. commercial broadcasting) religious programming is required for national broadcasters, but the susceptibilities of viewers or abusive treatment of the views of a particular religion or denomination is prohibited.[77]

Turning to the criminal law, in terms of the concepts of punishment, retribution is something which the intellectuals tend to down-play although clearly the general population considers it, and deterrence, as important elements of the criminal process. The reformation of the criminal and the protection of the community at large figure larger with legislators and commentators. Murder no longer attracts the death-penalty,[78] and, subject to safeguards, abortion has been legal since 1967.[79] In general British law, religious belief does not exempt from compliance with the general law, although some statutory exceptions exist.[80] Persons opposed to war on religious or moral grounds may not lawfully seek to dissuade troops from their duty,[81] nor refuse to pay taxes that will finance military expenditures[82] or some other matter of which the opposer disapproves.[83]

There is room, however, for conscience to be recognised within the law, a recognition that need not be based on religious grounds, but can be grounded on morality. Conscientious objection to military service was introduced along with compulsory military service in the Military Service Act

1916, 5 & 6 Geo. 5, c. 104, by its s. 2 and regulated latterly by s. 17 of the National Service Act 1948, 11 & 12 Geo. 6, c. 64.[84] Military personnel are now allowed to leave service should their convictions as to the morality of their service alter. In the case of medical practice, a doctor cannot refuse to perform a treatment to which he has conscientious objection,[85] save in the case of abortion unless the abortion is necessary to save the mother's life.[86] A parent's objection to medical treatment for a child can be set aside and the 'best interests of the child' be used to justify an assumption of parental rights by a local authority in order that consent may be given for appropriate medical treatment where a parent refuses consent or objects to treatment.[87] Since the enactment of the Age of Legal Capacity (Scotland) Act 1991, c. 50, the age at which the consent of the child becomes effective, and therefore when its choice over-rides that of a parent, seems to depend on his or her maturity.

Other twentieth-century advances in medicine have produced other problems in the inter-action of law and religion. The previously impossible has become possible and questions, not previously thinkable, have arisen. As to matters of life and death, in Scotland medical treatment has always been subject to the consent of a patient, allowing a person to refuse treatment if so wished. Where the patient is incapable of choice, treatment may usually be authorised by a relative or competent authority. But medical procedures have now produced the possibility of persistent vegetative state. In Scotland someone in such a state may be permitted to die by the withdrawal of nutrition, hydration and medication, provided the Court of Session has considered and approved in the particular case.[88] The position as to 'assisted suicide' is unclear. It may be that some instances of assisted suicide are not prosecuted in Scotland: Scottish criminal law is distinct, and on occasion different, from that in England.[89]

There are few legal restrictions upon individuals based upon their religion. As noted above, teachers in Roman Catholic public schools must be approved by the authorities of that Church. The previous bar on various ministers of religion sitting in the House of Commons has been repealed by the House of Commons (Removal of Clergy Disqualification) Act 2001, c. 13.[90] Most of the remaining restrictions on Roman Catholicism were finally removed by the Roman Catholic Relief Act 1926, 16 & 17 Geo. 5, c. 55. However, restrictions still apply in the case of the monarch, who may not be nor marry a Roman Catholic.[91] A Roman Catholic cannot act as Regent to an under-age monarch,[92] or be appointed Lord High Commissioner to the General Assembly of the Church of Scotland.[93] If a Roman Catholic were to be appointed as Lord Chancellor (a largely English office) provision exists for those duties which have religious connotations to be exercised by someone else.[94] These legal provisions are under question, not to say attack.[95] They are rooted in history, but do appear curious in modern times, especially as they mean that only Roman Catholicism is a cause of difficulty, and the possibility of a Buddhist or atheist monarch unconsidered. A major problem is, of course, the role of the monarch as the Head of the Church of England,

and the requirement under English law that the monarch shall 'join in communion with' that Church.[96] As well discussed elsewhere, the Church of England is the established Church south of the Border, with all that implies.[97] Whether the removal of these disabilities of the monarch and others would have much effect in Scotland is moot. Presumably the monarch could still undertake to protect the Church of Scotland and the establishment of religion as recognised in law in Scotland.

PROSPECT

Looking forward is difficult. Within Christianity, progress will have to be made combating sectarianism.[98] The General Assembly of the Church of Scotland meeting in May 2002 was told that if present trends continue that Church will cease to exist by 2050.[99] That might be. Certainly the age profile of the Church is skewed, and many members will die off in the next decades, without being replaced. Church membership, or office as elder, no longer carries the social advantages that it did. But whatever happens to that Church, it is certain that religion will not cease to be an important element in Scottish society. Its manifestations may well alter greatly, but diversity will continue. Those who believe will continue to associate together, and will require organisations which the law will recognise and protect.

There will be an increasing Islamic element to be accommodated within Scottish society, and culture. According to some varieties of Islam there should be no distinction between religion and the State – a view with which the Scottish reformers might well have sympathised. Were such an attitude to become prevalent in Scotland or the United Kingdom, difficulties may be foreseen. As it is, conversion from Islam is a capital offence in some foreign states, and is said to be difficult in Scotland, although this is not paralleled in Scottish conversion to Islam. Muslim expectations may have to change if Scottish society is to become fully inclusive and properly multi-faith. The example of Spain pre-Ferdinand and Isabella could provide a model.

Other tenets of the law, formerly based upon religion, but now not upon a generally held or common belief, may come to be questioned. In particular the full effects of the enactment of the Human Rights Act 1998, c. 42, incorporating into domestic law much of the European Convention on Human Rights of 1951 and certain of its Protocols remain to be discovered. By s. 13 of that Act, freedom of thought, conscience and religion for a religious organisation or its members are to be particularly regarded by a court or tribunal if the matter comes before it.[100] Matters of doctrine therefore should be outside attack under the 1998 Act, and a balance held between other freedoms and freedom of religion. There were fears that homosexuals might demand to be 'married' in a church, or that women should demand to become priests.[101] But it may be that we will need cases to see exactly what s. 13 means. For example can a religious body running a hospice insist that its employees should share the tenets which it espouses, or is that discriminatory under other provisions of the Convention? Or could internal Church rules prohibiting office-bearers from certain activities or occupations, for example from being an MP, be challenged under the 1998 Act?

Finally, we may wonder 'who will speak for' religion in its public aspect in the future. Individuals of strong religious conviction will certainly play a role, but religious institutions may decline in their influence. The non-Christian religions such as Islam and Judaism have structures through which views can be articulated and made known to government and the rest of the community. As noted above, given an appropriate individual, the episcopally organised Christian denominations are better placed to solve the 'spokesman' problem. More generally, Roman Catholicism will continue, though the much weaker Scottish Episcopal Church may not. The problem for the Presbyterians will continue. The General Assembly of the Church of Scotland will doubtless have a voice.

Church union will continue to be debated, though I have grave doubts whether any union between a Presbyterian and an episcopally organised Church would succeed.[102] The result might well be three Churches instead of two, with consequent property questions to be dealt with. Whether all the present Presbyterian denominations and groupings will survive, or survive in their present forms, is moot. Notwithstanding, and despite the warnings given to the General Assembly of 2002, I doubt whether Presbyterians, as opposed to the Church of Scotland, will cease to exist. It is more likely that some of the existing varieties of Presbyterianism will dissolve and re-group with a much smaller membership (perhaps including a large slice of the Church of Scotland), but with a clearer commitment to it doctrinal basis. If these can learn not to treat that doctrinal basis as their god, but rather use it as an avenue to God, there is hope.

NOTES

1. I write as an academic lawyer. See generally, Lyall, 1980; *Stair Memorial Encyclopaedia of the Laws of Scotland*, Edinburgh, vol. 3, Churches and Other Religious Bodies, 701–818, paras 1701–87, and vol. 5, Constitutional Law: Church and State, 357–72, paras 679–705. For ecclesiastical perspectives see: Burleigh, 1960 and 1970; Fleming, 1927, and 1933; Henderson, 1957; Drummond and Bulloch, 1973, and 1975, and 1978; Cheyne, 1983.
2. See Chapter 4.
3. Shaw, D. *The General Assemblies of the Church of Scotland 1560–1600*, Edinburgh, 1964, v. The general tenor of Buchan, J and Adam Smith, G. *The Kirk in Scotland, 1560–1929*, London, 1930, written to mark the union of 1929 also has a dark hue as it notes the Church's lessening status.
4. Highet, 1950; Highet, 1960; Wolfe, and Pickford, 1980. See also Brierley and Macdonald, 1985, and 1995; Church of Scotland Board of Social Responsibility, *Lifestyle Survey*, Edinburgh, 1987. Cf. for the United Kingdom, Currie, 1977; Bruce and Glendinning, 2002, 12–15; Glendinning, 2003.
5. Henry Dundas, First Viscount Melville, (1741–1811) and his son Robert (1771–1851) were significant figures: Fry, 1992.
6. Decisions as to criminal prosecution are virtually always a matter for the civil

authorities in Scotland. An English decision not to prosecute in a particular type of case could be persuasive without further debate. Cf. n. 88. See for English law, Robilliard, 1984.
7. The Westminster Confession of Faith, formulated by the Westminster Assembly as part of a projected uniformity of religion in England and Scotland, was adopted in 1647 by the General Assembly of the Church of Scotland, with minor modifications, explanations and reservations: Acts of Assembly 1647, August 27, Sess. 23: see *The Subordinate Standards and other Authoritative Documents of the Free Church*, Edinburgh, 1955, 13–14. Important modifications made by the Assembly related to the claim of the Crown to call Assemblies of the Kirk, and stressed the right of the Kirk to meet of its own volition irrespective of the will of the Civil Magistrate. Other points also relate to the interaction of Church and State.
8. The Act 1705, c. 50 (c. 4, 12mo), appointing the Scottish Commissioners to negotiate with England excluded the 'worship, discipline and government of the Church of this kingdom as now established' from the remit of the Commissioners. See also the Protestant Religion and Presbyterian Church Act 1706, c. 6 (the Act of Security), which *inter alia* confirmed and ratified anew the Confession of Faith Ratification Act of 1690. These Acts were, as required, inserted into the legislation implementing the Treaty of Union, the Union with England Act, 1706, 1707, c. 7 (Scotland), and the Union with Scotland Act, 1706, 6 Anne c. 11 (England). See generally, Lyall, 1980, 21–2.
9. Walker, 1996, 266–315, and 1998, 259–94, and both volumes for their sources.
10. See Chapter 4 above.
11. Lyall, 1980, 23–53. These linked questions caused great debate, many participants playing a variety of roles. Contemporary accounts from either side are Bryce, 1850, and Buchanan, 1854. See also Burleigh, 1960, 334–69; Taylor Innes, 1867, and new edn 1904.
12. The Evangelicals were, for example, very active in the setting up of new churches in Glasgow in the early nineteenth century, and took the lead in discussing social as well as ecclesiastical reform: see, for example, the bibliography of the writings of Thomas Chalmers (1780–1847), a leader of the Evangelicals.
13. Lyall, 1980, 42–6. The Chapel Act cases are: *Livingstone* (1841) 3 D. 1278; *Wilson v. Presbytery of Stranraer* (1842) 4 D. 1294; the Stewarton Case – *Cuninghame v. Presbytery of Irvine* (1843) 5 D. 427, *Cuninghame W. Mutual Cases, record and appendix in cause W. Cuninghame and others, heritors of Stewarton v. Presbytery of Irvine*, 3 parts (1842). Bell, J M. *The Stewarton Case* (Edinburgh, 1843), which contains the judgments. Cf. *Campbell v. Presbytery of Kintyre* (1843) 5 D. 657, where the Court of Session did not intervene on a complaint as to the deposition of a minister for drunkenness as the Presbytery did not contain *quoad sacra* ministers and therefore was properly constituted and acting on a matter within its jurisdiction.
14. Lyall, 1980, 30–42. The Auchterarder cases – *Earl of Kinnoull and Rev. R. Young v. Presbytery of Auchterarder* (1838) 16 S. 661; affirmed by the House of Lords (1839) McL. & Rob. 220: Robertson, C G. *Report of Court of Session proceedings in the Auchterarder Case*, Edinburgh, 1838; *Earl of Kinnoull and Rev. R. Young v. Rev. John Ferguson and others (a majority of the Presbytery of Auchterarder)* (1841) 3 D. 778; (1842) 1 Bell 662; (1843) 3 D. 1010; the Lethendy case – *Clark v. Stirling* (1839) 1 D. 955; Robertson, C G. *Report of the Proceedings in the Court of Session in the Lethendy Case* (Edinburgh, 1839); the Daviot case – *Mackintosh v. Rose* (1839) 2 D. 253; the Strathbogie (Marnoch) cases – *Presbytery of Strathbogie and Rev. J. Cruickshank and Others, Suspenders* (1839) 2 D. 258, (1839) 2 D. 585, (1840 2 D. 1047 and 1380; *Rev. J. Edwards v. Rev. J. Cruickshank and others (a majority of the Presbytery of Strathbogie) and Rev. J. Robertson and others (a minority of said Presbytery)* (1840) 4 D. 1298; *Rev.*

J. Cruickshank v Gordon (1843) 5 D. 909; Rev. J. Edwards v. Rev. H. Leith (1843) 15 Jurist 375: the Culsamond case – Middleton v. Anderson (1842) 4 D. 957.

15. See the Claim of Right, 1842, in which the then majority in the General Assembly of the Church of Scotland stated its view of the relationship of church and state. The Claim is printed in the *Free Church Case* (1904) AC 515 at 737, *The Subordinate Standards and other Authoritative Documents of the Free Church of Scotland*, Edinburgh, 1955, 235–55, Buchanan, 1854, II, 471–5; Taylor Innes, 1904, 162–7.

16. Lyall, 1980, 66–84; Sjolinder, R, trans. Sharpe, E J. *Presbyterian Reunion in Scotland, 1907–1921*, Acta Universitatis Upsalensis Studia Historico-Ecclesiastica Upsalenie No. 4, 1962.

17. What the 'Scottish Reformation' was may be important in future debate. It certainly relates to the Scots Confession of 1560 and the abolition of Papal jurisdiction. In 1963 the Confession of Faith Ratification Act c. 1 and the Papal Jurisdiction Act c. 2, both of 1560, were among the 'spent, obsolete and unnecessary' enactments to be swept away by a Statute Law Revision (Scotland) Bill. Both were dropped from the final form of the legislation. Cf. Lyall, F. The Westminster Confession: The Legal Position. In Heron, A I C, ed. *The Westminster Confession in the Church Today*, Edinburgh, 1982.

18. Thus in *Logan v. Presbytery of Dumbarton* 1995 SLT 1228, a minister sought to have the civil courts consider the refusal by the Presbytery to permit him to accept civil appointment additional to being a minister. The Court of Session held it had no jurisdiction, the matter being one governed by the powers of the Kirk. The Employment Appeal Tribunal has no jurisdiction to review allegations of unfair dismissal or sexual discrimination in relation to a minister: *Percy v. Church of Scotland Board of National Mission*, Employment Appeal Tribunal, Edinburgh (EAT/1415/98) 22 March 1999 (unreported: available through LEXIS). However, it remains possible that in a properly argued case involving property rights, the civil courts may have jurisdiction: See Lyall, 1980, 54–61. The Human Rights Act 1998 applies to procedures of the courts of the Church of Scotland, an attempt to have them excluded as dealing with 'matters spiritual' having failed. See n. 100.

19. In *Presbytery of Lews v Fraser* (1874) 1 R. 888, it was held that the civil courts could compel a witness to appear in a court of the Church of Scotland. One judge would have allowed this in the case of other denominations as well.

20. Thus in the nineteenth century in *Lockhart v Presbytery of Deer* (1851) 13 D. 1296, the courts had said that they would not review a decision by a Church of Scotland court on a matter of discipline. Cf. *Lang v Presbytery of Irvine* (1864) 2 M. 823. Church of Scotland courts are courts of the realm into which the civil courts will not inquire: *Paterson v Presbytery of Dunbar* (1861) 23 D. 720, *Wight v. Presbytery of Dunkeld* (1870) 8 M. 921, *Presbytery of Lews v Fraser* (1874) 1 R. 888. The Human Rights Act 1998 may alter this if the Church of Scotland court behaves quite outrageously, and the appellate structures within the Church fail to correct this. In the Church of Scotland Act 1921 (11 & 12 Geo. 5, c. 29) s. 3 preserves the jurisdiction of the civil courts 'in relation to any matter of a civil nature.'

21. Lyall, 1980, 148–9; Munro, C R. Does Scotland have an Established Church?, *Ecclesiastical Law Journal*, 4 (1996–7), 639–45.

22. Queen Elizabeth has twice attended the Assembly, 1977 and 2002. The High Commissioner may not be a Roman Catholic: Roman Catholic Emancipation Act 1829 s. 12. On the Office, see Mechie, S. *The Office of Lord High Commissioner*, Edinburgh, 1957.

23. Queen Victoria, Supreme Governor of the Church of England, caused debate which closed when, having attended numerous ordinary services at Crathie Church of Scotland, situated just outside the Balmoral Castle estate, she eventually took communion there. See Chadwick, 2000.

24. The Church Jurisdiction Act 1567 c. 12 (confirmed by the Act 1570 c. 7) stated expressly that there was 'na other face of Kirk nor uther face of Religion than is presentlie estabblischeit within this Realme And that thair be na uther jurisdiction ecclesiastical acknawlegeit within this Realme ...'. The Church Act 1579, c. 6, affirmed that those who professed the Scots Confession of 1560 were 'the only trew and haly Kirk of Jesus Christ within this realme ...' and that those of other opinion were not of that Kirk.
25. Scottish Episcopalians Act 1711, 10 Anne c. 10 (c. 7 in some edns); Scottish Episcopalians Relief Act 1790, 32 Geo. 3, c. 63. See also Grub, 1861; Goldie, 1976.
26. Legal Opinion was that while the action of the Pope was unlawful under Scots law, the new prelates would not incur any penalty: Taylor Innes, 1904, 288. The penal clauses of the statutes at issue have all been repealed.
27. *MacMillan v. The General Assembly of the Free Church of Scotland,* (1859) 22 D. 290; (1861) 23 D. 1314; (1862) 24 D. 1282; (1864) 2 M. 1444: *Brentnall v. Free Presbyterian Church of Scotland* 1986 SLT 471.
28. *The General Assembly of the Free Church, Bannatyne and Others v. Lord Overtoun and Others: MacAlister and Others v. Young and Another,* (1904) 7 F. (HL) 1; [1904] AC 515; Orr, R L. *The Free Church of Scotland Appeals,* Edinburgh, 1905. On this and other cases, see Lyall, 1980, 101–12.
29. The cause of this split was the dissatisfaction of a minority who, for a variety of reasons, opposed Professor Macleod in the Free Church College, Edinburgh. Action in the Sheriff Court in 1996 was unsuccessful. The Free Church General Assembly refused to carry on further investigations and trial, in 1999 prohibited further discussion, and in 2000 disciplined some who sought to raise the matter again. On 20th January, the minority withdrew to form the Free Church of Scotland (Continuing). Action commenced in the Court of Session in June 2002 to determine questions as to property: the process of argument and probable appeal may well be protracted.
30. Although the Free Presbyterian Church was showing fissiparous signs earlier, (*Brentnall v. Free Presbyterian Church of Scotland* 1986 SLT 471) the final straw was action taken in 1989 by its Synod against Lord Mackay, then Lord Chancellor, for his attendance at the Roman Catholic funeral Mass for a former judicial colleague. On the origin of the Free Presbyterian Church see, MacLeod, J L. *The Second Disruption,* East Linton, 2000.
31. For the Jews in Scotland, see chapter 10.
32. For example, James Dalrymple, earl of Stair, one of the founders of Scots Law in *The Institutions of the Law of Scotland,* 1681 and 1693, in sec. 3 of Book I Title 1 'Common Principles of Law', acknowledges 'Divine Law written in men's hearts'. This divine law allows men by reason to arrive at principles of natural law. However, Stair earlier defines Law itself as 'the dictate of reason, determining every rational being to that which is congruous and convenient for the nature and condition thereof ', God himself unchangeably determining himself by his goodness, righteousness and truth: Book I Title 1, sec. 1. In so doing Stair, whose last book was theological (*A Vindication of the Divine Perfections,* Edinburgh, 1695), followed, though more hesitantly, in the train of such as Hugo Grotius, who in his *De Iure Belli et Pacis* of 1652 (ET by Kelsey, F W. London 1923–8, rep. New York 1964) had indicated that principles of natural law might be based as much on reason and social order as on revelation, and be valid even if there were no God. See also the Hon. Lord Wark. Law of Nature. In *An Introduction to the Sources and Literature of Scots Law,* Stair Society, vol. 1, Edinburgh, 1936, 249–56.
33. Rendall, 1978; Sher, 1985; Davie, 1994; Wood, 2000; Broadie, 2001; Herman, 2002.
34. Davie, 1961. Cf. Walker, 1994.

35. Smith (who had earlier in his career been a scientist and worked on the physics of electricity) was surrounded by controversy after his articles on matters biblical in the Ninth Edition of the Encyclopaedia Britannica. At the trial he defended himself eloquently, and the indictment was dropped. However, he was removed from his chair. Black and Chrystal, 1912; Johnstone, 1995; MacLeod, 2000, 38–124.
36. *Adam v. Allan* (1841) 3 D. 1058–78, 1147–8.
37. Scotland, 1969; Grant, 1876; Lyall, 1980, 115–23.
38. Cameron, 1972.
39. The obligation of the Kirk (i.e. the kirk session, in practice usually the heritors) to provide for the schoolmaster when there was a school, was not always properly honoured: cf. Steven, 1995, 74–80.
40. There is no official report of the associated court case, save as a note in the Elgin Academy case cited in the next sentence.
41. *The Presbytery of Elgin v. The Magistrates and Town Council of Elgin*, (1861) 23 D. 287.
42. These included parliamentary schools, established by private statute, Roman Catholic schools, 'side schools' provided by kirk sessions *quoad omnia* in additional to the main parish school, 'sessional schools' also provided by the Church of Scotland in *quoad sacra* parishes, 'Assembly schools' provided by the Church of Scotland General Assembly in the Highlands and Islands, schools established by the Free Church after its creation in 1843 and schools maintained in the Highlands and Islands by the Scottish Society for Propagating Christian Knowledge. There were over 150 other schools provided by religious bodies and also private venture schools established by groups of teachers and parents. See Watson, 1956, 105–17, át 106–7.
43. Immigration from Ireland into the Glasgow area in the later nineteenth century exacerbated the problems of providing Roman Catholic schools for Roman Catholic children in that area.
44. Brother Kenneth. The Education (Scotland) Act, 1918, in the Making, IR, 19 (1968), 91–128.
45. See s. 124 of the Local Government (Scotland) Act 1973, c. 65, as substituted by s. 31 of the Local Government (Scotland) Act 1994, c. 39.
46. Taylor Innes, 1904, 122–3 and 158–61; Drummond and Bulloch, 1975, II, 300–1; Bryce, 1850, II, 172–3.
47. Black, 1893.
48. Black, 1893, 74–153; Ferguson, 1956, 177–93. Parish boards also dealt with health matters until replaced by County Councils; Black, 1893, 154–187. Cf. Steven, 1995, 81–90.
49. *Baird's Trustee v. Lord Advocate* (1888) 15 R. 682.
50. *Inland Revenue Commissioners v. Glasgow Police Athletic Association* 1963 SC. (HL) 13.
51. *Commissioners for Special Purposes of Income Tax v. Pemsel* 1891 AC 531, esp. per Lord Macnaughten at 583. English Charity law goes back to the Charitable Uses Act, 1601, and many cases thereafter. See Barker, C R, ed. *Charity Law in Scotland*, Edinburgh, 1996.
52. Thus in *Re South Bank Ethical Society* [1980] 1 WLR 1565 the position of Buddhism was left undecided, and Scientology was turned down as a religion in *R. v. Registrar General ex parte Segerdal* [1970] 2 QB 697.
53. The particular head of charitable activity 'advancement of religion' may disappear, although 'designated religious bodies' may continue to be accorded charitable status. See McFadden, J (Chair). Charity Scotland; the report of the Scottish Charity Law Review Commission, Scottish Executive, 2001, and Spicker, P and others. *Consultation on the Review of Scottish Charity Law: Analysis of the Responses*, Scottish Executive, 2002. See now the Charities and Trustees Investment (Scotland) Act 2005, asp 10.

54. Cf. the Westminster Confession of Faith, 1647, c. XXIV, 'Of marriage and divorce'. Stair, *Institutions*, I.4.1–6.
55. Stair, *Institutions*, 1.4.4: 'The degrees in which marriage is allowed or forbidden are by divine institution ...' and were to be found in Leviticus 20. In the period 1907–31 a series of Acts restricted the prohibitions of marriage within certain degrees of relationship. The 'prohibited degrees' are now contained in Sched. 1 to the Marriage (Scotland) Act 1977, c. 15.
56. *Mahmood* v *Mahmood*, 1993 SLT 589; *Mahmud* v *Mahmud* 1994 SLT 599; *Sohrab* v *Khan*, 2002 FamLR 78 (Greens, Edinburgh); *The Herald*, 24 April 2002.
57. Stair, *Institutions*, I.4.11–17, indicates that in practice the husband controlled his wife, and her property. In I.4.9 Stair quotes Gen. 3: 16 as showing it to be the express ordinance of God that 'the husband is lord, head and ruler over the wife'.
58. Children (Scotland) Act 1995, c. 36, s. 11(1). See Thomson, 1996, 205–10.
59. Cf. *Gillick* v. *West Norfolk and Wisbech Area Health Authority* [1985] 3 All ER 402.
60. The restrictions on the 'prohibited degrees' for marriage purposes made in 1907–1931 were applied to incest from 1938.
61. A table of prohibited relationships for the purpose of defining incest is contained in s. 2A of the Sexual Offences (Scotland) Act 1976, c. 67, added by s. 1 of the Incest and Related Offences (Scotland) Act 1986, c. 36.
62. *Stair Memorial Encyclopaedia of the Laws of Scotland*, Edinburgh, 1995, vol. 7, 'Criminal Law' para. 320, referring to Sir George Mackenzie of Rosehaugh, *The Laws and Customs of Scotland in Matters Criminal* (Edinburgh, 1678; 2nd ed. 1699), I.15.4. A criminal charge used solely to narrate Leviticus 20.
63. It is possible under tenancy law for a same-sex partner of a deceased to establish that she or he comes within the definition of 'family member' for the purpose of transmission of the tenancy, but meeting the required standard of proof is difficult.
64. Thus, Westminster Confession of Faith, 1647, c. XXIV, 'Of Marriage and Divorce' para 1: 'Marriage is to be between one man and one woman: ...'. See now the Civil Partnerships Act 2004, c. 33.
65. Race Relations Act, 1974 c. 74, building on and amending Acts of 1965 and 1968; Sex Discrimination Act, 1975, c. 65, all as amended. Cf. the Disability Discrimination Act 1995, c. 50.
66. Oaths Act 1978, c. 19. Any form acknowledged by a witness as binding him or her is accepted. Affirmation is possible.
67. Cf. the discussion as to the taking of the oath by certain members of the new Scottish Parliament in 1999. Cf. also the refusal of Sinn Fein members to take the oath in order to take their seats as MPs.
68. See 'Concerning Burial of the Dead' in 'The Directory for the Publick Worship of God', adopted by the Westminster Assembly 1645, approved by the General Assembly (Edinburgh 3 February 1645, Sess. 10) and the Scottish Parliament (Charles 1, Parl. 3, Sess. 5), 1645.
69. Parishioners used to have burial rights within the parish graveyard. The right was not, however, in perpetuity, and amounted to a 'right of decomposition', after which the lair might be re-used.
70. Executors or next of kin are not bound by the wishes of a deceased person, as to the disposal of the body. However, a curiosity is that where a local authority is dealing with a corpse, as occasionally it will have to, it may not cremate the body if there is evidence that cremation would have been against the beliefs of the deceased: Social Work (Scotland) Act, 1968, c. 49, s. 28(1).
71. e.g. Human Organs Transplants Act 1989, c. 31; Human Tissue Act 1961, c. 54. Cf. Human Reproductive Cloning Act 2001, c. 23.
72. It may be cynical (but realistic) to note that often the Sunday opening of shops, and

73. Sunday working, was latterly often opposed, unless double-time was paid. Religious considerations were not the prime motive.
73. There are still difficulties in running ferries and opening airports on the Sabbath in the Western Isles.
74. Lyall, 1980, 138–9.
75. Criminal prosecution in Scotland is generally a matter for the Crown Office. Private prosecutions require the consent of the Lord Advocate or the High Court of Justiciary. Thus *R. v Lemon, R. v. Gay News* [1978] 3 All ER 175; [1979] 1 All ER 898 (HL) where Mrs M Whitehouse successfully prosecuted for blasphemy, would not occur in Scotland. The European Court of Human Rights in *Wingrove v United Kingdom*, Case 19/1995, held that the refusal by the British Board of Film Classification to license a video film on the ground that it would infringe the law of blasphemy in England was not a contravention of the right of freedom of expression protected by Art. 10 of the European Convention on Human Rights.
76. *Regina v Chief Metropolitan Stipendiary Magistrate ex parte Choudhury* [1991] 1 QB 429.
77. BBC Charter and Agreement: Broadcasting Act 1990, c. 42, ss. 6(1)(d) and 16(2)(e). The BBC and the Independent Television Commission (which licenses commercial television and radio services) are advised by the Central Religious Advisory Committee, which has representation from all main religions and is appointed by the BBC Governors and the Commission. Sched. 2 Part II(2) of the 1990 Act disqualifies religious bodies from owning a terrestrial broadcasting service, but satellite service licences can be awarded. (The draft Communications Bill 2002 removes this prohibition but leaves the other restrictions in place.) While religious advertising is permitted on commercial television to publicise events, describe activities, offer publications, or sell merchandise, religious belief may not be expounded or it must be made clear that it is only the advertiser's belief. The denigration of other faiths, claims to be the 'only' or 'true' faith, or the use of fear or the consequences of not being religious or holding (or not) a particular belief, are unacceptable. See Independent Television Commission. *Code of Advertising Standards and Practices*, App. 5 – Religious Advertising, London, 1998.
78. Murder (Abolition of Death Penalty), Act 1965, c. 71. As this section was being written the 'tariff' for murder in England was reduced to twelve years.
79. Abortion Act 1967, c. 87. How strictly the safeguards are applied in practice is an occasional question.
80. *R. v. John* [1874] 2 All ER 561. Under the Motor-cycle Crash-helmets (Religious Exemption) Act 1976 Sikhs were first permitted to wear their turbans and were exempt from the requirement laid on all other motorcycle users to wear crash-helmets. See now reg. 4.2 of the Motor Cycles (Protective Helmets) Regulations 1980 (1980 SI 1279) and s. 16.2 of the Road Traffic Act 1988, c. 52. The exemption has spread. Young Sikhs are exempt under reg. 2 of the Horses (Protective Headgear for Young Riders) Regulations 1992 (1992 SI 1201) and all Sikhs are exempt from the general requirements as to safety helmets on construction sites by ss. 11 and 12 of the Employment Act 1989, c. 38. This concession is not required under the European Convention of Human Rights: see *X v UK*, (1978) 14 DR 234. Special rules exist for slaughterhouses operated in accordance with Islamic or Jewish law. These may lawfully use techniques prohibited in other slaughterhouses. Christian Science Nursing Homes are required to draw their affiliation to the attention of their possible clients: Nursing Homes Registration (Scotland) Act 1938, 1 & 2 Geo. 6, c. 73, ss. 6 and 7, as amended.
81. *R. v Arrowsmith* [1975] QB 678, [1975] 1 All ER 463; cf. *Arrowsmith v the United Kingdom*, (1975) 19 DR 5, Application no. 7050/75.
82. *Cheney v Conn* [1968] 1 All ER 779, where nuclear weapons were argued to be contrary to the Geneva Conventions of 1949.

83. *Mohr* v. *Henry* 1992 SLT (Notes) 285 an attempt to argue for exemption from the Poll Tax on grounds of both morality and religion.
84. The 1948 Act was repealed by the Reserve Forces (Safeguard of Employment) Act 1985, c. 17, but should compulsory military service be re-introduced it is likely that similar provision would be made for 'conscientious objectors'.
85. *R.* v. *Bourne* [1939] 1 KB 687.
86. Abortion Act 1967, c. 87, s. 4, which applies to nurses and doctors only, not to others (e.g. secretaries) who may be indirectly involved in abortions through employment in hospital administration.
87. See n.58. Thus the child of Jehovah's Witnesses may be given blood albeit the parents object on religious grounds.
88. *Law Hospital Trust* v. *Lord Advocate* 1996 SLT 848. Each case must be the object of a special petition to the Court of Session in the exercise of its *parens patriae* jurisdiction. A statement of policy by the Lord Advocate, printed as an Appendix to the case report, indicates he will not authorise the prosecution of a qualified medical practitioner acting in good faith and on the authority of the Court of Session. This policy statement deals only with cases involving a persistent vegetative state. Cf. for England: *Re B (adult: refusal of medical treatment)* [2002] EWHL 429 (Fam); [2002] 2 All ER 449; [2002] 1 FLR 1090; [2002] 2 FCR 1; 65 BMLR 149.
89. Assisted suicide is a crime under English law by s. 2(1) of the Suicide Act 1961, 9 & 10 Eliz. 2, c. 60, which applies only to England. The refusal of the authorities to undertake not to prosecute one who assists a suicide is not a breach of the Human Rights of the person wanting to commit suicide. See *R.* v *Director of Public Prosecutions*, [2001] UKHL 61, [2002] 1 All ER 1, [2001] 3 WLR 1598, [2002] 1 FCR 1, [2002] Fam Law 170, 63 BMLR 1, 29 November 2001: *Pretty* v *United Kingdom* (App no 2346/02), European Court Of Human Rights (Fourth Section), [2002] 2 FLR 45, [2002] 2 FCR 97, 29 April 2002.
90. The internal rules of a denomination may still prohibit a minister from becoming an MP or a MSP. Thus Vatican rules prohibit a priest becoming an MP. Such a rule might be challengeable under the Human Rights Act 1998, to which we are coming, but in the case of the Church of Scotland, such a challenge may be barred by the Church of Scotland Act 1921. (See n. 18 and *Logan* v. *Presbytery of Dumbarton* 1995 SLT 1228).
91. Act of Settlement, 1700, 12 & 13. Will. 3, c. 2, ss. 1 and 2. The Crown, succession to the Crown and Regency are matters reserved to the Westminster Parliament by s. 30 and Sched. 5, Part I, para 1(a) of the Scotland Act 1998, c. 46.
92. Regency Act 1937, 1 Edw. 8 & 1 Geo. 6, c. 16, s. 3(2). See also previous note.
93. Roman Catholic Emancipation Act 1829, 10 Geo. 4, c. 7, s. 12.
94. Lord Chancellor (Tenure and Discharge of Ecclesiastical Functions) Act, 1974, c. 25.
95. A resolution of the new Scottish Parliament in 1999 indicated its wish that the discriminatory elements of the Act of Settlement of 1701 (next note) should be repealed: 16 Dec. 1999, Vol. 3 Scot. Parl. Off. Rep. 1774, with debate at 1623–1670.
96. Act of Settlement, 1700, 12 & 13. Will. 3, c. 2, s. 3. The first two Georges were not members of the Church of England, but remained Lutheran throughout their lives. The Lutheran Churches are in communion with the Church of England.
97. Doe, 1996. Most recently see the debate on 'Church and State' in the House of Lords, 22 May 2002, 635 HL Deb. (HL Hansard), 769–816, at the conclusion of which Lord Chancellor Irvine indicated that the government had no plans to change the existing position.
98. 'Sectarianism' in the Church and Nation Report, *Reports to the General Assembly of the Church of Scotland*, Edinburgh 2002, 12/11–19. In this section the Committee apologised for anti-Catholic attacks in its reports of the 1920s and 1930s.

99. National Mission, *Reports to the General Assembly of the Church of Scotland*, Edinburgh 2002, 20/6 para 1.1.3.
100. During the passage of the Human Rights Bill the Government overrode objections by the Church of Scotland to its courts falling into the category of 'public authority' and therefore subject to the new legislation. Cf. the House of Commons Committee Stage debate, 20 May 1998, 312 HC Deb. 977–1076, esp. at 1056–58 and 1063–69. Act III of the General Assembly of 2001 'anent Discipline of Ministers, Licentiates, Graduate Candidates and Deacons' revised the Church's disciplinary procedures to ensure they are not challengeable under the Human Rights Act.
101. The Sex Discrimination Act 1975, c. 65, s. 19, protects from the operation of that Act employment for the purpose of 'an organised religion' where 'the employment is limited to one sex so as to comply with the doctrines of the religion or avoid offending the religious susceptibilities of a significant number of its followers.'
102. Henderson, 1967.

BIBLIOGRAPHY

Barker, C R, ed. *Charity Law in Scotland*, Edinburgh, 1996.
Black, J S and Chrystal, G W. *The Life of William Robertson Smith*, London, 1912.
Black, W G. *Scottish Parochial Law other than Ecclesiastical*, Edinburgh, 1893.
Brierley, P and Macdonald, F. *Prospects for the Eighties*, London, 1980.
Brierley, P and Macdonald, F. *Prospects for Scotland*, Edinburgh, 1985.
Brierley, P and Macdonald, F. *Prospects for Scotland 2000*, Edinburgh, 1995.
Broadie, A. *The Scottish Enlightenment*, Edinburgh, 2001.
Bruce, C and Glendinning, A. Shock report! Scotland is no longer a Christian country, *Life & Work*, June 2002, 12–15 (from Scottish Social Attitudes survey, 2001).
Bryce, J. *Ten Years of the Church of Scotland from 1833 to 1843 with a Historical retrospect from 1560*, 2 vols, Edinburgh, 1850.
Buchanan, R. *The Ten Years' Conflict: Being a history of the Disruption of the Church of Scotland*, 2 vols, Glasgow, 1854.
Burleigh, J H S. *A Church History of Scotland*, Oxford, 1960 and 1970.
Cameron, J K, ed. *The First Book of Discipline*, Edinburgh, 1972.
Chadwick, O. The Sacrament at Crathie, 1873. In Brown, S J and Newlands, G. *Scottish Christianity in the Modern World*, Edinburgh, 2000.
Cheyne, A C. *The Transforming of the Kirk*, Edinburgh, 1983.
Currie, R, et al. *Churches and Church-Goers: Patterns of Church Growth in the British Isles since 1700*, Oxford, 1977.
Davie, G E. *The Democratic Intellect*, Edinburgh, 1961.
Davie, G E. *Essays on the Scottish Enlightenment*, Edinburgh, 1994.
Doe, N. *The Legal Structure of the Church of England*, Oxford, 1996.
Drummond, A L and Bulloch, J. *The Scottish Church 1688–1843*, Edinburgh, 1973.
Drummond, A L and Bulloch, J. *The Church in Victorian Scotland, 1843–1874*, Edinburgh, 1975.
Drummond, A L and Bulloch, J. *The Church in Late Victorian Scotland, 1874–1900*, Edinburgh, 1978.
Ferguson, T. The poor; welfare and social services. In McLarty, 1956, 177–93.
Fleming, J R. *A History of the Church in Scotland 1843–1874*, Edinburgh, 1927.
Fleming, J R. *A History of the Church in Scotland 1875–1929*, Edinburgh, 1933.
Fry, M. *The Dundas Despotism*, Edinburgh, 1992.
Glendinning, A. Religion and the supernatural. In Bromley, C, Hinds, K and Park, A, eds. *Devolution: Scottish Answers to Scottish Questions*, Edinburgh, 2003.
Goldie, F. *A Short History of the Episcopal Church in Scotland*, Edinburgh, 1976.

Grant J. *A History of the Burgh Schools in Scotland*, Edinburgh, 1876.
Grub, G. *An Ecclesiastical History of Scotland*, 4 vols, Edinburgh, 1861.
Henderson, G D. *The Burning Bush*, Edinburgh, 1957.
Henderson, I. *Power without Glory: A Study in Ecumenical Politics*, London, 1967.
Herman, A. *The Scottish Enlightenment*, London, 2002.
Highet, J. *The Churches in Scotland Today*, Glasgow, 1950.
Highet, J. *The Scottish Churches*, London, 1960.
Johnstone, W, ed. *William Robertson Smith: Essays in Reassessment*, Sheffield, 1995.
Lyall, F. *Of Presbyters and Kings: Church and State in the Law of Scotland*, Aberdeen, 1980.
Rendall, J. *The Origin of the Scottish Enlightenment*, London, 1978.
Robilliard, St J A. *Religion and the Law*, Manchester, 1984.
Scotland J. *The History of Scottish Education*, 2 vols., London, 1969.
Sher, R B. *Church and University in the Scottish Enlightenment*, Edinburgh, 1985.
Steven, M. *Parish Life in Eighteenth-Century Scotland*, Dalkeith, 1995, 74–80.
Taylor Innes, A. *The Law of Creeds in Scotland*, Edinburgh, 1867, and new edn 1904.
Thomson, J M. *Family Law in Scotland*, 3rd edn, Edinburgh, 1996, 205–10.
Walker, A L. *The Revival of the Democratic Intellect*, Edinburgh, 1994.
Walker, D M. *A Legal History of Scotland, vol. IV, The Seventeenth Century*, Edinburgh, 1996.
Walker, D M. *A Legal History of Scotland, vol. V, The Eighteenth Century*, Edinburgh, 1998.
Watson, G. Education. In McLarty, M R, ed. *A Source Book and History of Administrative Law in Scotland*, Edinburgh, 1956, 105–17, at 106–7.
Wolfe, J N and Pickford, M. *The Church of Scotland: An Economic Survey*, London, 1980.
Wood, P, ed. *The Scottish Enlightenment*, Rochester, 2000.

26 From Monochrome to Colour

ELIZABETH HENDERSON

ARRIVAL AT GRANTON PARISH CHURCH

In 1986, when I arrived at my first charge of Granton in Edinburgh, I was presented with a copy of the roll, some back editions of the church magazine and a bundle of black and white photographs. Some of these photographs were of the Boys' Brigade from some thirty or forty years before; black and white rows of boys in uniform. If you looked very carefully you could chart the progress of just one boy from section to section through the different pictures. Others contained happy looking pictures of the Woman's Guild. The present-day smaller Guild was composed almost entirely of the younger looking women in the group aged again by some thirty or forty years. There were assorted pictures of the previous ministers of the church and several of the large crowds who came to see the Moderator lay the first stone of the new church in 1934. There were no colour pictures, just black and white. Over the ten years that I was minister of Granton church I would revisit these pictures and reflect on how much had changed in the time since they were taken.

It was said of Archie Charteris when he went to his first charge that 'He was 22 years old, and very frightened.'[1] Charteris only stayed at St Quivox church in Ayrshire for thirteen months but by the end of that time we are told that 'he had 60 married miners attending his class for first communicants.'[2] The year was 1858. By the time I arrived in Granton, at twenty-six years old, a changing church and society meant that for most new ordinands there was no expectation of exponential growth in the congregation. Similarly, if there was a fear, it was only of some day leaving things in a worse state than you had found them.

Apart from wondering whether or not single miners and women joined the church in Charteris' day, I had cause to reflect on the comparative absence of youth in the church in the latter part of the twentieth century. The photographs that I had inherited clearly reflected the fact that people under the age of forty had been well represented in Granton a mere thirty years before my arrival. The handful that remained were but a remnant of the picture of the hordes that went on the train that had taken them on a Sunday School outing. Granton's office-bearers were nearly all in the same posts as they had been at that time. Young men with families were now grandfathers, with a young minister. The kirk session had not all been young but the younger people, who were contemporaries of the minister, held all the posts except for that of Session Clerk. The photographs showed a past energy in Granton church which manifests itself in youth organisations, drama groups, choirs and large church socials. Hope for the future of the church was rooted in this image of an extension charge which was full of such activity. 'Successful' ministers like Charteris had left me an unfortunate legacy; an expectation that I might just be able to provide the key that would unlock the door to a resurgence in the fortunes of the church.

A CONFLICT OF EXPECTATIONS

On a day-to-day basis ministry in my first charge was not a matter of choice. The report that I presented to the Annual General Meeting of the congregation each year was often structured around the statistical information that evidenced the work that I was engaged in. Ninety funerals a year, fifty housebound, ten weddings, fifteen baptisms and fifteen new members a year were the average figures. Beyond these tasks was the responsibility of ensuring the smooth workings of the inherited machinery of a local parish church. While this challenge is shared with the kirk session and congregational board, the minister is still often the last resort for most problems. Granton was just out of the bracket where aid had to be provided from central Church funds and it was proud not to be in this category. Much of the congregation's work revolved around raising the money to keep its head above water. When the church was not raising money it was trying to save money. The property committee spent days in the building, as did the voluntary cleaners. Ministry occasionally involved assuming the role of mediator and peacemaker between these hard-working but human volunteers. The harder task was trying to find willing people to make up their numbers and thus ensure that there were people to lead the organisations, to fill the elders' districts, to tend to the property, to deal with administration, to distribute the flowers and so on. As resources of people and money shrink these tasks become more onerous. The concept of simply not having a choir or flowers or a property committee would have left the church feeling that it was falling down some invisible league table of churches. Schools and community organisations, church and presbytery meetings mean that the minister is never at a loss for something to do. This, however, is where the tension begins because although the minister is busy it can feel like they are simply maintaining the legacy of the past.

It is not realising the worth of the past that is difficult, it is rather seeing its relevance to the present that is often problematic. The boys who wear trainers and casual clothes to school invite ridicule if they don a uniform and attend the Boys' Brigade. Rightly or wrongly no one thinks it is a 'cool' place to be on a Friday night. The younger women are frightened of the stereotype of being a member of the Guild. The choir carefully practise anthems that are not of the kind of music listened to on most car radios. The concepts of joining and duty and commitment that are strong in the

Figure 26.1 Women's Guild, Granton parish church, early 1960s. Photograph courtesy of the author.

church are not universally valued in the community around the building. The struggle between past and present is deeper, however. The people of the church did not live in a vacuum; they were aware of the need to change. They were not reluctant to try new things, but they were keen to hold onto their involvement and the things that were important to them. In the organisations, for example, people understood the criticisms but they knew the merit of their activities and the genuine enjoyment that those who came found in them. The search was for the key that would change the organisation into one as attractive as its photographed past.

It was also true that, while the young and the people of the community wanted something different, there was no single articulation of the nature of that something. Vague notions were the only universal. The church should be bright, lively, relevant, caring and understanding of frailty. The church listened, cared and tried both reformation and invention but there were few people available to set new ideas in motion. The truth was that people still came to the church looking for baptisms and weddings and funerals albeit in smaller numbers than they had in the past. In one five-year period I counted that we had 135 new members. Around 80 per cent of these were under forty years of age. Those who came through communicant's classes were enthusiastic. Many had a very rudimentary knowledge of Christianity. In one class, for example, no one knew the word 'parable', had heard of Moses or knew any of the names of the disciples. These communicants brought the freshness of people reading the gospel for the first time.

Most, however, became paper members, the kind of people who are accused of only joining the church to get the baby baptised. I saw them rather as people who did not become part of the community of the church. They came for a while but despite effort did not become known or involved. Jürgen Moltmann suggests that many people experience the church today as a 'church without community'.[3] He quotes a couple who said of their time worshipping in a new church, 'our hopes for new genuine relationships with our fellow Christians with whom we sat in the pew were not fulfilled. We entered the church as isolated individuals and we left the same way'.[4] The response of this unknown couple is similar to the ones that most people gave to me for not staying. Some who came to worship week by week claimed to feel just as isolated, but were not enthused by the idea of a church social, the politics of a church fair, the organisations or any other activity in a church hall. The church started new activities like an all-age lunch once a week and a keep-fit group and although these proved popular they strained the already over-committed. New things, like old, still lived in the shadow of the past: my youth group of eight teenagers could never be the Youth Fellowship of more than forty who had tended the orchard in the garden of the old manse.

In his book *Ministry Burnout*, John Sanford suggests that the minister is caught up between trying to make people feel valued in traditional roles and find new ways of being the church. The generation of which the younger minister is part is often more enthusiastic about the new, but this is the area that is often squeezed into the time that is left after everything else. Already stretched volunteers find it hard to think about taking on something else, and new people find it hard to begin something new in the midst of an organisation which has well-established patterns of working. It is, moreover, possible that the minister will find some people who are deliberately obstructive. Sanford suggests that the presence of such people may well be a consequence of the decline that the churches have faced when he speaks of 'parishes that have exhausted themselves and are now made up of determined but narrow people who partly keep the shell of the parish alive and partly keep it from growing'.[5] It was never, however, my experience that the numbers of such people exceeded a handful. The negative current generated by a few did, however, have the potential to exert power over innovation whether the direction was worship, education or social activities.

TRANSITION

Charteris only stayed for thirteen months in his charge of St Quivox in Ayr. I was in Granton for ten years before I felt called to move to my present charge. The paper membership of Granton had fallen by about 10 per cent while the worshipping congregation had increased. Most of those who had been there when I arrived were still there. Like me they were ten years older, so the increase in the numbers coming to church has to be offset against the knowledge that the average age of the congregation was still older than when I had arrived. In May of 1997 I was called to Richmond Craigmillar church, which is set in an area of multiple deprivation on the other side of the city from Granton. The move was variously described as inexplicable, courageous, career suicide or just plain daft by my colleagues in the presbytery. I did not feel that it was any of these things. The expectation in the Church of Scotland is still that a minister will always move to a 'better' charge from their first charge. A 'better' charge is thought to be one with a better income and a bigger congregation. I think even the people of Granton felt that I might have had the decency to leave them for an upwards rather than a downwards move. For me it seemed right that people with experience should consider the call to work among marginalised people as a privilege.

ARRIVAL AT RICHMOND CRAIGMILLAR

Richmond Craigmillar church could certainly not be described as prestigious. On the day I arrived there was a large hole in the roof. The congregation was heavily aid-receiving and numbered around twenty-five on a Sunday morning. The guild had dwindled to six people and the Boys' Brigade had only four boys in its company section. A pensioner's lunch club that used to flourish had only ten people coming and the heating system in the hall was so inadequate that mostly it was freezing for those who did come. A 'nearly new' on a Thursday, which again had run for many many years, had as many visitors as it had volunteers running it. The area of the community in which the church is situated is awaiting regeneration and is marred by boarded-up and burnt-out houses. The level of vandalism is high and the outside of the church has suffered as a consequence. The community had suffered from years of receiving the most troubled people in the city. An earlier clearance of the alcoholics from the Grassmarket area of Edinburgh had been followed by years of problematic and vulnerable people being housed as priority tenants. The area has staggeringly high numbers of people who need support of one kind or another. It seemed to me, in the early days, that Community Care for the whole of Edinburgh was taking place in Niddrie. It, additionally, seemed strange that the people of Niddrie, who are often put down by those who live elsewhere in the city, were left to care for such large numbers of people with mental illnesses, physical disabilities and so on. The bad reputation of the area meant that people did not want to move to Niddrie with the result that those who did come were mostly those who could not get a house anywhere else. The community, consequently, contains higher than average numbers of young single parents, homeless people and those suffering from drug addiction. These factors are still influential in the community although it is currently undergoing regeneration.

When I went to Richmond Craigmillar I was handed a copy of the roll, a bundle of familiar and yet unfamiliar black and white photographs and a plan. The only thing that was unexpected was the plan. It was for a grand redevelopment of the church and hall to make it suitable to be used as a community theatre and café as well as a church. The proposal had first been mooted by a community organisation but the people of the church were seemingly enthusiastic. As time went by the community organisation lost interest and the people of the church applied for money from grants and trusts to build a café instead. The café took away a third of the hall but it opened the building to the community every weekday. Employing part-time staff solved the problem of volunteers (partially through grant money). Within months the café was attracting over one hundred visitors a day. Two years after its opening it is still as busy. While the money was being raised to build the café the people of the church opened a food co-operative selling fruit and vegetables at near market prices. Very few people have cars and the nearest supermarket is a bus journey away so the food co-operative was set up to make it easier for people to eat healthy food. Other projects came along in response to requests from people in the community or other organisations. During the week there is now a furniture store, pensioner's lunches, a milk token scheme, a pick-up point for the credit union, the food co-operative, the 'nearly new' and a dental health project.

PEOPLE'S STORIES

The church is a place of activity all week long and it feels like a place that has turned itself inside out to help the community around it. The café is driven by the needs of people in the community. The church is made aware, on a daily basis, that the people who live around its building are often lonely, addicted, bereaved or mentally ill. The people amaze me with their non-judgemental attitude to other people who come in. Only diverse stories can give a flavour of what happens in the building. One day, four of the pensioners were talking when the subject of suicide came up. The first to speak said that no one could really understand it unless they had attempted suicide as she had on two occasions. The next revealed that she had come home to find that her ex-husband and her daughter had hanged themselves. The third had also had a suicide in her family and the fourth listened quietly to the others. On another day, one of the social workers telephoned the church to ask if one of the pensioners that they dealt with was eating in the café. After they received a positive answer they explained that he had refused meals on wheels on the basis that the café was a place where he found company. Some time later the old gentleman died and the social worker contacted the church to say that he asked her to give all his money to the café if anything happened to him. She came to the church with a plastic bag containing all of his money. We counted £17.64.

Gordon is another regular in the café. He comes every week as part of a group of young adults with special needs. His great love is cars and every week the staff of the café give him a toy car. This young man then goes round all of the tables in the café showing his car to everyone and indicating that he wants to know what kind of model it is. Everyone talks to him and

Figure 26.2 Community café, Richmond Craigmillar parish church. Photograph by the author.

some even bring a car to the café in case they see him. Margaret is one of the many young people who are addicted to drugs. She lost her baby to a cot death. She came to the café to ask me to help. Someone went with her to the funeral director and helped her to make all the arrangements. When she returned she told me that she wanted a 'little Catholic service' for the baby but she was adamant that I should conduct the service. In the end I shared a service in the Catholic church across the road with the priest, and café staff arranged a funeral tea for the mourners in the café. It is not unusual for our stories to be tied up with other agencies in the community. The health visitors and the midwives openly admit that when they get stuck they expect the Richmond church to solve the problem. One day they phoned to ask if we could provide some baby clothes for a pregnant mother who had nothing ready for the birth of her child. When we asked when the baby was due we received the reply, 'Anytime now, she is in labour as we speak.' This led us to have layettes ready for mothers who have nothing. On another occasion the same health visitor said 'We don't know what we would do if you weren't here.'

The merging of church and community in the café is intriguing. Out for a meal, the waitress came up to me and asked if I would go and have a word with the rest of the staff. It transpired that she had been telling them that she went to church and they wouldn't believe her. She said, 'Tell them I go to church, I'm there every Monday!' I explained to her that church would be more thought of as a Sunday event and she replied that she was close. The café has blurred the distinction and there is a risk in doing that, but the benefit is that it is also a place where discussions about faith arise naturally and can jump across tables and end up involving everyone. One such memorable discussion started with a group of drug addicts trying to remember the ten commandments. With a lot of help from other people, they went through them one by one with each person putting them into their own words. After some discussion, these young people pronounced that they thought each commandment had worth and should be the way that people conduct their lives.

The traffic is not only one way. When the wall that surrounds the church was knocked out of alignment when a stolen car crashed into it, my project worker's husband recruited one of the men from the café to come and help him push it back into place. As they were standing beside it agreeing that it would be impossible for the two of them a white van drew up and then another and another. At the end sixteen men from the local community, some from notorious local families, pushed the wall back into place, saving the church £1,800.

There have been problems as well. These have ranged from drug addicts trying to use the café as a place to sell drugs to a schizophrenic who came in shouting abuse regularly and finally smashed the window on the door. Solutions to these kind of problems are found but not without thought and concern. There is also the sadness of watching and worrying over the effects of the class A drugs which are readily available in the community. A four-year old came in by himself first thing in the morning to get his breakfast as soon as the café opened. He is still coming on his way to school. He is typical of the many children whose home life is chaotic because of drug misuse. One third of the parents of the local school are seriously addicted to either drugs or alcohol. The associated problems are part of the life of the church and café. Deaths of drug addicts, children coming to the café needing to be fed, parents who sell their cooker to buy drugs, grandparents who worry over their addict children and their vulnerable grandchildren are always somewhere in the minds of both church and community. It is also true that while the activity of the café has grown and the worshipping community has doubled there has not been much effect on the more traditional parts of the church. The church organisations are as low in numbers as they ever were. We have a choir of one and we only have enough money for real flowers on high days and holidays. The cradle roll and the membership roll of the church are both problematic and the elders have more or less been forced to abandon the visiting of districts. While the café is almost financially self-sufficient, the church is still heavily aid-receiving.

One place where the life of the parish and the worshipping community have met in a creative way has been in the construction of a memorial chapel within the church. Not long after the café opened people began to ask if the church could provide a place where they could remember family members and friends who had died. Many churches have books of remembrance, but the desire was for something more than names on paper. In the end people who were bereaved both from the church and from the parish met and the result was a tree made of copper where each name could be engraved on a leaf. A book was provided so that people could have a page for a photograph and a poem or whatever they wanted. The people of the church gathered poems and readings that they had found helpful and put them in a transept of the church along with books on bereavement. The chapel is now a place where people come and remember anniversaries in the quietness of the church. The moving part of the dedication service was hearing that many of the people of the church had sat and quietly wept throughout the service when they thought about how some of the people in the community had suffered.

WORSHIP

Services on a Sunday in Richmond are not very different from the ones that I presided over in Granton except that the numbers of people who come are very much smaller. When I first came it was on a Sunday morning that I noticed the lack of resources more than anything else. No choir, no musician, no church officer and sometimes no heating placed immediate boundaries on what was possible. Later I realized that a paucity of resources had forced the church to be creative and experimental in relation to worship. People are greeted when they come into Richmond by a group of people who sit in the vestibule on a Sunday morning, but the concept of people being on duty had disappeared long before I came because of the lack of people on the congregational board. In a community where many people have difficulty in reading I was also surprised to find that there was more than one person who was willing to preach and many more able to conduct prayers and read lessons. There was another difference that struck me not long after I came. One Sunday I asked a rhetorical question during the sermon only to find that someone shouted out an answer. This was not an isolated event. On another occasion I was interrupted by a voice from the congregation saying that she did not agree with what I was saying. It feels right in the environment here to allow such dialogue to take place and even to encourage it. Often the answers that come from the congregation are both simple and profound. There are advantages in being small. In Granton I had wanted to pray for the bereaved and people who were in need during the service, but I was always stayed from doing this by people's concern that we would miss someone out. In Richmond from the very beginning it was easy and natural to remember the families of the parish. So much so that on the first anniversary of the death of my project worker's son, we all took time to remember the grief that she and her family were experiencing on that day. It had been planned beforehand and friends and family came knowing that the congregation were willing to share their remembrance.

Granton, however, had strengths that Richmond lacked. Five years ago Richmond relied on people from other churches to come and play the old Hammond organ that was used to accompany worship. The choir in Granton had been willing to learn and lead new songs and the organist had been a skilled musician. In Richmond, having different people, of varying ability, playing each week dictated the kind of music that was possible. A boy of thirteen eventually emerged who was able to play by ear but for a long time we had to sing within his repertoire. It seems sad to me that now he is sixteen and a huge asset to the congregation it is only a matter of time before we lose him to a church which is able to pay him much more than we ever could. From time to time he gets such offers from churches whose income exceeds ours. The lack of a choir led to an ad hoc music group being formed who help to introduce new songs and lead the worship. There is a strong leaning in this group, and in the congregation in general, towards music that has arisen from peoples' struggles for justice and freedom. People like to be able to understand the words that they are singing and they make that clear when I try to introduce them to some of the hymns that I love and they do not know. People ask regularly for more choruses and 'upbeat' songs where in Granton there was a leaning towards hymns.

The other place where I think that there is a distinct difference between the two churches is in the way in which they celebrated communion. In Granton we sent out invitations and then on a Sunday many of the elders still wore black ties and there was a pride in the organisation of the distribution of the elements. Like so many ministers in other places I found that this was a service that was hard to alter in any way at all. Communion was served with order and precision. Elders marched slowly to the table, were nervous about 'getting it right' and had a great sense of pride in being an elder on those days. The silences and the dignity were important. On other occasions we would have informal communions but the quarterly communions remained a constant. In Granton's seventy-year history the only real change that I could perceive in this ritual was the elders who served. Imagine for me the difference when I came to Richmond and I looked around during the last hymns to see my small group of elders dancing to the music! The Sacrament of Holy Communion is celebrated with some structure but a lot less formality. There is, as a consequence, a greater desire to experiment even with the main communion services. Smaller numbers make it easier to try different styles such as all moving the seats in order that everyone can sit in a circle or move into the chancel area. History, culture and context have shaped the worship of the two congregations, but the decline of the church has been felt in the smaller one to the extent that it has meant that there is a need to be creative in response to limitations on what is generally regarded as the norm.

POST-MODERNISM

While acknowledging that the terms modern and post modern are often difficult to define accurately both in terms of meaning as well as chronology, Granton and Richmond Craigmillar can loosely be considered a modern and a post-modern church respectively. The assumption here is that the modern era can be characterised by such words as respectability, order, institution, rationality, cohesion and nationalism and that the postmodern era is a place where these concepts have been devalued. The postmodern era (from the 1960s–1970s onwards) is a time where the emphasis has shifted towards fragmentation and globalisation and where there are a lack of agreed responses. Granton, as a church, was a place of structure and order where people valued membership and commitment. Richmond Craigmillar, on the other hand, is poorly structured. The worshipping community struggle to adequately complete the tasks that came easily to Granton. Property schedules, statistical returns, even the organisation of duties on a Sunday morning are often poorly done. Granton organised these and communion with almost military precision. The numbers of those attending worship in Richmond have increased but in line with the theories of postmodernism these are people who are not particularly interested in membership and would rather profess faith as and when they feel that it is appropriate for them. The result is that more than one-third of regular worshippers are not members of the church.

Richmond Craigmillar is a church which seems to fit more easily within a postmodern world. Will Storrar describes a 'sustainable' church as follows, 'it must inhabit its faith as the good news of the suffering Messiah to be shared in vulnerability and, increasingly, from the margins', and again 'It must do so through the gracious witness and service within the common life it shares with its fellow-Christians and its diverse neighbours, and through a shared concern for the common good amid postmodern fragmentation and insecurity'.[6] The vulnerability of Richmond Craigmillar as a church and the service that it has embarked upon suggest that it could be thought of in these terms as a sustainable church. It exists within a community where few people live traditional lives and where it gains respect through its weakness rather than its strength. Paradoxically, as an aid-receiving church it is only able to be such because churches like Granton struggle so hard to pay for the aid which enables it to continue in existence.

The description of Richmond Craigmillar's ministry in its community was described above by the narration of a whole series of unconnected stories. One of the marks of modernism has been said to be that people were all able to subscribe to big stories which defined who they were. Postmodernism is defined as a time when people have rejected these meta-narratives. Those who work in the field of pastoral theology have noted this change. David Lyall wrote this from his experiences as a hospital chaplain: 'Every client or parishioner has a story to tell ... people tell their stories in the attempt to make sense of their experience'.[7] Our experience in the café has been one of listening to peoples' stories and then quietly reflecting where the stories of the Bible and these meet. Speaking only for myself, I know that my faith has been unsettled and enriched by these encounters.

The world is not so simple that it is possible to say that these encounters did not also take place in Granton. The difference in Niddrie is that the whole church is involved on a daily basis in the ordinariness of people's lives. This is partly because it is a smaller church but it is also because there is nothing between us and the community. There is nothing to protect us and nothing to prevent the community from bringing everything to the door of the church. The existence of new kinds of activity such as the café have obviously encouraged this and the lack of other forms of activity mean that everyone is involved in this meeting. Richmond has been able to embrace a model of risk and innovation suggested by writers on postmodernism such as Toulmin who says, 'the task of defining realistic "futurables" is open only to those who are ready to adopt imaginative attitudes, think about the directions in which we might be moving, and recognize that the future will reward those who anticipate the institutions and procedures we shall need'.[8] Innovation is made much easier because the time has been available and the alternative is extinction. It has also been possible because of the support of the modern institution of the church and the aid from other churches. Whether or not Richmond Craigmillar is anticipating the 'institutions and procedures' that are needed for the future will only be able to be judged by those who inhabit that time. We are well aware that experimentation requires failures as well as successes before new patterns emerge. My hope for both churches is that they will be there in some form in the future for a new minister to come and receive a bundle of colour photographs which bear witness to a particular generation of faith.

NOTES

1. Magnusson, 1987, 33.
2. Magnusson, 1987, 34.
3. Moltmann, 1978, 113.
4. Moltmann, 1978, 113.
5. Sanford, 1984, 80.
6. Storrar, 1999, 83.
7. Lyall, 1995, 83.
8. Toulmin, 1990, 203.

BIBLIOGRAPHY

Lyall, D. *Counselling in the Pastoral and Spiritual Context*, Oxford, 1995.
Magnusson, M. *Out of Silence*, Edinburgh, 1987.
Moltmann, J. *The Open Church*, London, 1978.
Sanford, J A. *Ministry Burnout*, London, 1984.
Storrar, W. From *Braveheart* to faint-heart: Worship and culture in postmodern Scotland. In Spink, B D and Torrance, I, eds. *To Glorify God: Essays in Modern Reformed Liturgy*, Edinburgh, 1999, 69–84.
Toulmin, S E. *Cosmopolis: The Hidden Agenda of Modernity*, Chicago, 1990.

PART FIVE

●

Architecture and the Arts

27 Cathedrals and Churches

JAMES A WHYTE

It has been said that the most basic form of architecture is a post in the ground. If that be so, our earliest architectural remains are the standing stones and stone circles, some of which seem to have had a religious purpose. Circles such as those at Callanish in Lewis, Stenness in Orkney and Templewood in Mid-Argyll, have attracted much study and even more speculation, but little can be known of their religious use.

Our more recent ancestors, the Picts, left many symbol stones, some to commemorate battles, some perhaps to mark boundaries. The meaning of this sophisticated language of symbol is open to debate. The earliest stones are incised, the later sculpted. Many of the latter contain Christian symbols, showing the acceptance and importance of Christianity from the seventh century on. The stones show skilful and developed art, some of it influencing or influenced by Irish carving and manuscript illumination. But the Picts have left us the merest fragments of their language and little or nothing of their building. The Scots, the earliest Irish immigrants, have left a little more. It has traditionally been thought that the Church introduced by the Scots was monastic in character, based on abbeys set within large compounds and controlled by abbots. In this scheme, bishops were simply monks who had the power to ordain. There also existed many smaller churches and chapels. They were often located at sites associated with local saints, and may have been served by a priest who administered to the surrounding populace. Chapels were also established within the strongholds and palaces of rulers, as at Forteviot.

These early churches are often identifiable today only through earthworks, carved monuments and place-names. Only rarely have buildings survived, perhaps the most evocative example being the (partly restored) bee-hive cells on Eileach an Naoimh. For the most part, early church sites have either been allowed to decay or been eradicated by frequent rebuilding. Tourists admiring the restored abbey church in Iona, for example, often imagine that they are seeing the church built by Columba, when what they see is a restored medieval Benedictine abbey (Fig. 27.2). Recent excavations at Iona and Whithorn have nevertheless shown that archaeology can reveal much about the life and character of the early Church even when no visible remains survive.

Early church building was probably not very durable. Much may have been of turf, wattle, timber or dry-stane, vulnerable to the rigours of climate

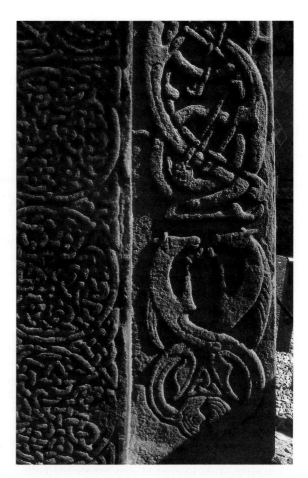

Figure 27.1 Detail of Pictish stone at Aberlemno. Photograph by the author.

and of history. Some early churches, especially in the Western Isles, but also in eastern Scotland, were destroyed by marauding Scandinavians. But many more were demolished and replaced in the late eleventh and early twelfth centuries after the marriage of Malcolm III to Margaret. Before Malcolm came to the throne in 1058, he had spent a long time in exile in England, and when he married (c1069), his queen was deeply attached to the religious ways she had known in Anglo-Saxon England. It was perhaps inevitable that Celtic usages in Scotland should in time conform to those of England and the continent. Margaret initiated the process. She introduced the Benedictine order to Scotland at Dunfermline. Her innovations appear to have been restricted to the royal court, though she extended her patronage to some existing communities, including St Andrews.

Under Margaret's sons, particularly Alexander I and David I, the transformation of the Church and the elimination of Celtic ways continued. Influenced by their own experiences while in exile in Anglo-Norman England, and possibly also by a biography of their mother that portrayed

Figure 27.2 Iona Abbey. Photograph by the author.

her as a great reformer, they brought the Scottish Church into line with that elsewhere in Europe. Abbeys and priories were established for religious orders brought from the continent, the diocesan system was reformed, and a network of parishes instituted. This meant an unprecedented church-building programme, ranging from magnificent abbeys to more humble parish churches. The new churches were built in stone, in the latest Anglo-Norman style, mostly by masons introduced from England. Frequently they were built over or beside existing churches, which were either demolished or left to decay.

Churchmen also took a cavalier attitude to the building work of their predecessors. There is a record of successive bishops in Aberdeen destroying the buildings of their predecessors. Not till modern times do we find a reverential attitude to the buildings of the past. Our ancestors were, each in their own generation, modernists, not hesitating to alter, enlarge, improve, or demolish the work of their predecessors. In the same way, churchmen in the eighteenth century despised Gothic as barbaric, and replaced Gothic features, where they could, with classical styles, and the nineteenth century, under the guise of 'restoration' or improvement, destroyed much medieval and eighteenth-century building. Our church buildings have undergone an almost continuous process of change, from their beginnings to this day.

So the built evidence of the Celtic way largely disappeared, though the Irish-inspired round towers at Abernethy and Brechin (built around the time of Margaret's death in 1093), the multiplicity of churches and places dedicated to Celtic saints, and the medieval pilgrimages to shrines such as St Duthac's in Tain, testify to the continuing vigour of the Celtic tradition in the hearts of the people.

In St Andrews, the first Anglo-Norman bishop, Robert, brought in Augustinian canons to replace the native clergy. The building now known as

Figure 27.3 St Andrews Cathedral. Photograph by the author.

St Rule's, which was to be served by them, was built by masons trained in England. Its lofty tower still dominates the landscape, and must have made a powerful statement in stone to the people living around. However, as early as the 1160s St Rule's began to be superseded by the much grander cathedral, whose building started then and continued through almost two centuries until its consecration in 1318 (Fig. 27.3).

English influence is evident in the abbey church built for David I at Dunfermline. The lower parts of the nave, which included decorated piers to mark the position of the high altar, suggest that these masons had worked on Durham Cathedral. Echoes of Durham are also evident at St Magnus' Cathedral, Kirkwall. Begun in 1137 by Earl Rognvald, it was the first cathedral in what is now Scotland to be built on the grand scale. The church was still not completed by the time of the Reformation, but its nave is the greatest surviving example of Romanesque architecture north of Durham – and is still in use.

A considerable number of twelfth-century parish churches still stand. Though many of them have been altered and extended or 'restored' through the years, one can still see original work. Dalmeny and Leuchars (Fig. 27.4) are fine examples. Early features also survive at Abercorn, Birnie, Kirkliston, Legerwood and Stobo. Most of these churches are rural, and quite small, and were often built at the behest of the local laird. The development of the parish system is illustrated by the labours of Bishop de Bernham of St Andrews, who, in the years between 1240 and 1249, consecrated no fewer than 140 churches in his vast diocese. He was not initiating a great period of church extension. Few, perhaps, of these churches were new. Most had been in use for some time, even before the advent of Margaret and her sons. According to Rome, they had not been properly consecrated. By consecrating them, the bishop was bringing them firmly under his authority and affirming their place in the new parish system.

The transition from the rounded Romanesque arch to the pointed Gothic begins in the late twelfth century. This was a matter not of taste, but of technique, since the latter allowed wider openings at doors and windows still to support the weight of the roof, enabling more light to enter the building. News of the new technique was brought to Scotland by builders from England. The builders might introduce it in mid-construction, as one can see in the remaining south wall of the nave of St Andrews Cathedral. The three easterly bays are Romanesque, the rest Gothic. One can imagine news of this technical advance arriving, being accepted by the builders, and tried out immediately. One can see this transition also at Coldingham Priory. The north and east walls of the chancel remain, and show delicate late Romanesque work at ground level, with a fine pointed arcade above. The building that most clearly illustrates the transition from Romanesque to Gothic is Dundrennan Abbey. Here in the late twelfth century the transepts of the church were remodelled according to the latest Gothic style, probably by a mason from northern England.

During the times of English occupation and the Wars of Independence there was little building, and much destruction caused by English raiders. The return of peace saw restoration and rebuilding. The renewal of trade with the Baltic and the Continent brought prosperity which was marked by the construction of some fine burgh churches. The architecture of these was

Figure 27.4 Leuchars church. Photograph by the author.

strongly influenced by a particular development in religious faith and practice. The belief that a mass, said with the right intention, had the power to benefit souls living and dead meant that those who could afford it paid for masses to be said for themselves and their dead relatives. A multiplicity of masses required a multiplicity of altars, which were constructed mostly in the nave, between the pillars. Thus St Giles', Edinburgh, had 46 altars, St John's, Perth, 33, St Mary's, Dundee, 35, St Nicholas', Aberdeen, 33, and Holy Trinity, St Andrews, 32. The multiplicity of altars required a multiplicity of priests. Holy Trinity had eighteen chaplains. The result was the creation of collegiate churches, the primary function of which was the saying of votive masses for the dead. Some of the great burgh churches, such as St Giles', Edinburgh, gained collegiate status. The majority of the collegiate churches, however, were founded by nobles or barons. Roslin Chapel is perhaps the most remarkable of Scotland's collegiate churches, but other noteworthy examples survive at Dunglass, Lincluden and Seton. Fate was not so kind to what was perhaps the finest collegiate church in Scotland, Trinity College in Edinburgh. It was taken down stone-by-stone to make way for the Waverley Station, with the intention that it would be rebuilt on another site. By the time this was arranged, most of the stones had been lost or stolen. As a result, all that remains of this once great church is the apse, which was rebuilt in Chalmers Close nearby, and serves today as a brass-rubbing centre.

The rise and fall in the popularity of pilgrimages also influenced church building. Glasgow Cathedral is perhaps the best surviving example, not just of Gothic architecture in Scotland, but also of a church designed for pilgrims. The great cathedral in St Andrews housed the relics of the patron saint, as a place of pilgrimage. Since many pilgrims were in search of healing, a hospice, dedicated to St Leonard, was established to care for them. By the end of the fifteenth century pilgrimages had so declined that it was no longer needed, and after a brief period as a home for old women, it was in 1512 transformed into the second college of the university.

Changes in architectural techniques and tastes interacting with changes in religious practice account for the history of many of our buildings. Dunblane Cathedral (Fig. 27.5) has a Norman tower begun in 1150, but the Norman church (probably in poor repair) was demolished by Bishop Clement who began to build the present church in 1237. It has a pointed, arcaded nave and a chancel with a Lady Chapel on the north side, but no transepts. After the Reformation the chancel alone was used for public worship, the unused nave being in time unroofed also. It was restored in 1889–93 to give the church as it looks today, and as it never looked before. The pre-Reformation church was divided by the rood screen into two rooms – a chancel and a nave. The post-restoration church attempts to treat these as one.

The cathedral at Dornoch shows how vulnerable medieval buildings were to the forces of fire, storm, sword and well-meaning restoration. Built by Bishop Gilbert de Moravia in the thirteenth century, it was ruined by fire in 1570 in the course of a clan feud. The choir and transepts were repaired

Figure 27.5 Dunblane Cathedral. Photograph by the author.

and used as the parish church, and the nave left unroofed. The nave was further damaged in a storm in 1605. An attempt to repair it in 1616 came to nothing, but, through the generosity of the duchess of Sutherland, in 1835 it was demolished and rebuilt by William Burn. Dissatisfaction with this led to a further restoration of the cathedral in 1924.

A similar story can be told of Brechin. Apart from the early round tower, the medieval cathedral was substantially of the thirteenth century. It was remodelled in 1806 to make the nave more suitable for preaching, which included raising the aisle walls and roofs and enclosing the clerestory within the building. It was restored in 1900–2 by John Honeyman who, among other things, re-exposed the clerestory and re-built the choir. The round tower has remained inviolate.

Burgh churches, similarly, have undergone extensive change since they were first built. St Mary's, Dundee, suffered by fire in the seventeenth century and only the tower remains of the original building. The nave of St Nicholas', Aberdeen, was rebuilt in 1755 (James Gibbs) and the choir in 1837 (Archibald Simpson), so that only the transepts and the crypt remain of the original. The transepts now contain the chapel of the Oil Industry – a modern version of the ancient guild chapels.

The Reformation in Scotland was more radical than in England, and so also was its attitude to the buildings it inherited from the medieval Church. Two reasons can be advanced for this radical approach. The first is that the

Church in Scotland was more deeply unpopular and corrupt than in England, and attempts at reform within the Church were too little and too late. The second is that, while the Reformation in England was initiated and managed by the Crown with the help of the bishops, in Scotland it was a popular movement, with the support of some of the barons, in defiance of Crown and hierarchy. Three bishops did sign the Scots Confession of Faith, but took no very active part in the Reformed Church thereafter. Some prominent churchmen, however, notably John Winram, sub-prior of St Andrews, and John Douglas, rector of the university, became leaders of the Reformed cause. The success of the movement was a measure of the deep and widely shared religious conviction, and its adaptation of its buildings was seriously theological and severely practical.

The first thing the reformers did was to 'cleanse' the churches. Statues, images of the Virgin and the saints, and their relics, all of which people had been encouraged to venerate, were removed and destroyed. These were objects of superstition, and had no place in a church where the grace of God in Christ was being preached. We today might bemoan their loss as works of art, but that thought had no place in the sixteenth-century mind, Protestant or Catholic. It is, however, charming to go today to St Michael's, Linlithgow, and see on the corner of the building a statue of St Michael, which, by accident or design, escaped the otherwise wholesale cleansing of the building.

With parish churches, the urgent need was repair. The munificence which furnished the religious houses all over Scotland as well as the great collegiate churches, and the three universities of St Andrews, Glasgow and Aberdeen, had been at the expense of the parishes. Revenues were diverted to the order or institution, which would often install a poorly paid vicar and neglect the fabric. In the years before the Reformation, the reformers were much concerned about this. It was a first concern of the reformers themselves, noted in the First Book of Discipline, 1560:

> Least that the word of God and ministration of the Sacraments by unseemliness of the place come in contempt, of necessity it is that the Kirk and place where the people ought publickly to convene be with expedition repaired with dores, windows, thack and with such preparation within as appertaineth as weel to the Magestie of God, as unto the ease and commoditie of the people.

The buildings themselves were often simple rectangles, with the altar at the east end, and a stone font at the west door. It was not difficult to rearrange them internally, with the pulpit on the long south wall, the baptismal basin bracketed to the pulpit, and the long tables for communion, placed, when required, down the length of the church.

The provision for the sacraments is peculiarly Scottish. Communion in the Reformed Church is a corporate action. The Scottish reformers insisted that communion should be received sitting at a table, which was never regarded as an altar. Such 'long tables' might be, and often were, movable features. Baptism also was to be in face of the people, at public worship, informed by the preaching of the Word. So the basin bracketed to the pulpit replaced the stone font at the door.

To adapt the great churches for Reformed worship was more difficult. The large medieval church, with its chancel and nave separated by a rood-screen, its acoustics suited to chant and music but not to the spoken voice, presented problems. Reformed worship was not something chanted or recited in Latin at a distant altar, with the people summoned by the bell to adore the transubstantiated host. It was a gospel proclaimed, prayer offered and psalms sung in the language of the people. It had to be audible and intelligible. Nor did it need a multiplicity of chapels and altars, far less a division between chancel and nave. It required a single space for public worship.

The cathedrals of Aberdeen, Brechin, Dornoch, Dunblane, Dunkeld, Glasgow, Kirkwall and Whithorn continued to be used, in whole or in part, after the Reformation, as did large town churches such as Holy Trinity, St Andrews, St Giles', Edinburgh, and St Nicholas', Aberdeen. In St Giles', the solution was to divide the church into three, the High Kirk, the Laigh Kirk and the West Kirk so that three congregations could worship under the same roof. Edinburgh was a growing city. In St Nicholas', Aberdeen, St John's, Perth, St Mary's, Dundee, and in Glasgow Cathedral, a similar method was followed. But in smaller towns, large cathedrals were bigger than the local congregation required. The solution then was to adapt and use either the chancel (as in Dunkeld and Dunblane), or the nave (as in St Machar's, Old Aberdeen, Holyrood Abbey, Edinburgh, and Dunfermline Abbey), allowing the rest of the building in time to fall into ruin. Galleries were in time inserted, so that a large congregation could be accommodated within earshot of the preacher.

The largest cathedral in Scotland, St Andrews, had been built for prestige and as a pilgrimage church. But as Edinburgh grew, St Andrews declined. There was no need for such a large building especially since there was an adequate parish church in the centre of the town. So the parish church of the Holy Trinity, built in 1411, became the Town Kirk, and was adapted for Reformed use, while the cathedral (which had already been 'cleansed') was simply abandoned. Its ruination began when the government stripped the lead off its roof to make cannonballs. In the eighteenth century, the thrifty burghers of St Andrews began to find in it a fine quarry of hewn stone with which to build their new town houses in South Street. In a wynd off Market Street one can see in a wall a quarter of a consecration cross, clearly once part of the cathedral. Similar plundering must have happened in many towns where an old church or part of it fell into ruin.

In general, the reformers preserved what was useful to them, adapting it to Reformed use. The abbeys and priories were a different matter. It is largely to the piety and generosity of David I ('a sair sanct for the croon') that we owe the large number of abbeys and priories that existed over Scotland. All are now ruins, save Pluscarden, which has been restored. The four border abbeys – Kelso (Tironensian, c1128), Melrose (Cistercian, 1136), Jedburgh (Augustinian, 1138), and Dryburgh (Premonstratensian, 1150) – were badly damaged in Henry VIII's 'rough wooing' of 1544–5. Though the remoteness of some made it impossible to use them, the Kirk was able to use parts of the abbey buildings in Jedburgh, Kelso and Melrose as parish churches until they were deemed unsafe. We have noted above that the naves of Holyrood Abbey and Dunfermline Abbey were used as parish churches. The Cluniac abbey at Paisley (twelfth-fifteenth century), was in the centre of Paisley and the nave became the parish church.

The monastic houses did fare worse at the Reformation than the parish churches and cathedrals. The resentment of the people was keen against the monks and especially the friars, though we may be sceptical about how much demolition work can be done on stone buildings by a mob with bare hands. The nobles also felt that the generosity of their ancestors, in land and gifts, had been gained under false pretences, since they no longer believed the claims of the pre-Reformation church.

When it came to church building the Reformed Church could be both conservative and innovative. When a rural church could not be repaired it was replaced by one not dissimilar to the old – a simple rectangle, or the characteristic T-shape with an aisle opposite the pulpit. One can see this in the ruined church at Kemback in Fife (1582). But when the parish church of Burntisland needed to be rebuilt in 1592, a new site was chosen above the harbour, and a magnificent church was built for this growing and flourishing community (Fig. 27.6). The building stands foursquare, with a central tower, whose weight is carried by four massive pillars with connecting arches and buttresses on the four corners. The windows are square sash windows. There

Figure 27.6 St Columba's church, Burntisland. Photograph by the author.

Figure 27.7 Lauder parish church. Photograph by the author.

are galleries on all four sides, which belonged to the guilds and crafts of the town, not the least significant being the mariners. Some of the eighteenth-century decoration of these lofts has been recovered. The farms had seats in the body of the kirk. The pulpit was against one of the pillars, and on the pillar opposite is a canopied pew (1611), originally for the laird of Rossend Castle, but later for the magistrates of the town. The tables for the sacrament were set out in the central space, which on other Sundays provided accommodation for the poor. Here was a church designed for a whole community to bring its secular life and its family life under the Word of God and round the table of the Lord.

The seventeenth century was marked by conflict between the Crown, which imposed episcopacy, and a large part of the Kirk and people, who opposed it. Church building reflects these conflicts. When Greyfriars Kirk was built in Edinburgh in 1620, it was a simple Gothic nave, with pillars and aisles. The charming church built by Archbishop Spottiswoode at Dairsie, Fife, in 1621, shows Gothic features ('survival Gothic') in a rectangular plan, with a little spire on one corner. It may have been built as a model, to show how a church could accommodate the Anglican practices required by James VI without offending the Presbyterians too much. Eirenical gestures were too late, however, and when Charles I and his archbishop of Canterbury, William Laud, imposed a high Anglican liturgy on the Scots, there was open rebellion. The National Covenant was signed in Greyfriars Kirk in 1638. From then on, till the Revolution Settlement in 1690, times were turbulent.

Nevertheless some fine buildings were erected. The fifteenth-century tower at Cramond, for example, got a fine new cruciform kirk in 1656. When the Duke of Lauderdale was having his castle at Thirlestane rebuilt by Sir William Bruce he asked his architect to design a new kirk for Lauder (1673) (Fig. 27.7). This is in the form of a Greek cross, with a central tower and spire, and four identical arms with pointed windows. Each contains a gallery, though the east gallery was not added till over a century later, and it may originally have been a communion aisle. The pulpit stands against one corner of the central space, with the baptismal basin bracketed to it. The minister is therefore audible and visible from anywhere in the church, though some of the congregation are almost at his back. That this remarkable church was built during the second period of episcopacy may suggest that the worship of the Kirk during this time was not all that different from that in Presbyterian times.

More remarkable is the kirk of the Canongate, Edinburgh, completed when Presbyterianism was being restored in 1690, and replacing Holyrood Abbey, which James VII took as a chapel for the Order of the Thistle. The design by James Smith is a Latin cross, with transepts and a small chancel and apse, and a nave with pillars and three bays. There is a very pleasing façade. Clearly its architect did not anticipate the return of Presbyterianism.

The eighteenth and early nineteenth centuries saw some fine buildings in a classical style. In the New Town of Edinburgh, St Stephen's, St Andrew's (now St Andrew's and St George's), and St Mary's, Bellevue, are fine examples – the two latter with magnificent interiors. Bellie Church, Fochabers, has a fine façade and spire. St George's, Tron, Glasgow (William Stark, 1808), is also a fine example of a burgh church of that period.

One also finds octagonal churches. Kelso Old (James Nisbet, 1773,

Figure 27.8 Kelso Old parish church. Photograph by the author.

altered by William Elliott, 1823) (Fig. 27.8) is one, built when the abbey was considered unsafe. The Glasite church, Dundee (1777, now part of St Andrew's parish church halls), Glenorchy (James Elliot, 1811) and Kilmorich (1816) are attractive examples. The church of Kilarrow (Bowmore) in Islay is round (Daniel Campbell, 1767). Many were simply square, or rectangular, often with galleries on the east, west and sometimes north sides, and the pulpit in the middle of the south wall. But the commonest shape of church is the T-church. Some were built in that form, while others achieved it by building an aisle opposite the pulpit in a rectangular church. In Cromarty, a rectangular church was built in 1700, but by 1770 it was too small. The kirk session invested its poor fund in the building of an aisle, with gallery, opposite the pulpit, and repaid the Fund out of the seat-rents it was able to charge. The aisle came to be known as The Poors Aisle. Kilmodan, Glendaruel (1783, restored 1983) and Yester, Gifford (James Smith, 1710), built when the marquis of Tweeddale moved both church and village out of sight of his grand house, are fine examples of the T-shaped church.

There is a reason for much rebuilding at this time. The provision and upkeep of parish churches was by law the responsibility of the heritors (landowners). Decisions of the Court of Session from 1787 on reinforced the interpretation that the church must accommodate 'two-thirds of the examinable persons in the parish', i.e. those not under twelve years of age. As a result, many medieval churches were too small, and were extended or a new church built. Cupar Old (1785) is one example of the latter. Holy Trinity, St Andrews, was extensively altered (Robert Balfour, 1798–1800) to provide galleries and enable it to accommodate many more people.

In 1817 the Church of Scotland became concerned about the state of the Highland parishes, many of which were without adequate church buildings. Memorials to the Government eventually led to an Act of 1823, to be followed the next year by a more generous Amended Act. This provided for forty churches and manses, and, in some cases, the stipend of ministers. The implementation was entrusted to the commissioners for Highland Roads and Bridges, and Thomas Telford produced the final designs. His Parliamentary Churches can easily be recognised in the Highlands today, with their broad arched windows and bell tower. The design was criticised by the Gothic revival architect William Burn, but it is very attractive. The churches were either rectangular or T-shaped in the traditional arrangement, with the long table in the centre. Good examples can be seen at Plockton (Fig. 27.9) and Portnahaven in Islay. The church on Iona was altered internally in 1939 to suit high Church tastes.

From the time of the Reformation until the 1820s, every Presbyterian in Scotland when communion was celebrated received the sacrament seated at a long table. Sometimes the tables were trestles, erected for the communion season. Sometimes they were permanent features (examples can be seen at Ardchattan, Lochbroom and elsewhere) and sometimes by an ingenious adjustment, pews could be converted into tables (examples can be found at Ceres, Falkland and Kilmany). The congregation, usually too great

Figure 27.9 Pulpit and long table, Plockton parish church. Photograph by the author.

to be seated thus at one time, came in relays to receive the elements. By the end of the eighteenth century the Scots passion for preaching demanded a table address for each sitting, and services became inordinately long. When the great Dr Thomas Chalmers moved to St John's, Glasgow, he found this practice very inconvenient and began to distribute communion to the people sitting in the pews, the pews being covered with white cloths, to show that they are the table. Not till 1824 did the General Assembly reluctantly sanction this, but Chalmers' innovation soon became standard practice. It involved the introduction of a table from which the elements might be distributed by the elders to the people in the pews. This is now usually termed 'the communion table', and even, by some ministers and servicebooks, 'the holy table', although the communion table proper is where the people sit to receive communion.

The nineteenth century saw the Romantic Movement and, in architecture, the Gothic revival. It was also a period of great church building in Scotland since after 1843 the Free Church was building in every parish. The Roman Catholic and Episcopal Churches were also very active. Nineteenth-century taste demanded something a little more plush and comfortable, more in line with the Victorian home. Churches continued to be built in the classical style, Wellington, Glasgow (1884), and Hillhead Baptist (1883), both by T L Watson, being examples. Alexander 'Greek' Thomson, going right against the fashions of the time, produced masterpieces in a classical, almost Egyptian style, of which St Vincent Street, Glasgow, has been preserved. But Gothic became the recognised and recognisable style for church buildings. The Free Presbyterian Church in Oban is a simple building with pointed Gothic windows, while the Free Church there, on an impressive site overlooking the bay, was designed by the English Catholic architect, A W Pugin. Charles Rennie Mackintosh used a form of Gothic in his distinctive Queen's Cross Church, North Kelvinside, Glasgow (1896–9). James Gillespie Graham designed many churches in a Gothic style (e.g. Liberton, 1815, St Mungo's, Alloa, 1819, Clackmannan, 1815). Some of these have been reordered internally. Greenside, Edinburgh (1839) is T-shaped. Hippolyte Blanc designed some fine buildings in a characteristic Gothic revival style (Mayfield, Edinburgh, 1875, St James', 1884, and Coats Memorial Baptist, 1894, in Paisley, St Luke's, Queen Street, Broughty Ferry) as well as the opulent St Cuthbert's, Edinburgh (1894). F T Pilkington designed the exuberant Barclay Church, Edinburgh (1864), and Kelso North (1866) (Fig. 27.11).

The development of gaslight made possible the evening service, and the popular preacher in the towns and cities. Lighting the church with gas was more effective than oil lamps, and gas lamps made the streets safer. The Scots had always been addicted to preaching. The hour-glass on the pulpit was to measure the length of the sermon, not the service, and anything less than an hour was short change. A new style of preaching became popular,

Figure 27.10 Plockton parish church. Photograph by the author.

Figure 27.11 Kelso North church. Photograph by the author.

with the emphasis less on intellectual content and more on emotional appeal. The nineteenth-century church might be square or rectangular, with the pulpit on the narrower end wall, and horseshoe galleries, and the 'platform pulpit' became a platform for the great orator, rather than a throne for the Word of God. In St George's Free Church, Edinburgh (David Bryce, 1869; campanile R Rowand Anderson, 1881), the pulpit was the size of a large room, in which the preacher was free to roam.

The second half of the nineteenth century saw also the rise of the liturgical and high church movements in the Church of Scotland. The gentry and the moneyed people began sending their children to English public-schools, from which they emerged as Anglicans. The high church movement was in part a response to this. G W Sprott maintained that the Church Service Society had stemmed the exodus from the Church of Scotland to the Scottish Episcopal Church. At the end of the century the Scoto-Catholic movement was repeating in Scotland what the Anglo-Catholic movement had done in England more than fifty years before. Its effects in Scotland were not so extreme or divisive, but in church building they were considerable. The Aberdeen (later Scottish) Ecclesiological Society was formed in 1886, and drew members from the Free (soon to be United Free) Church as well as the Parish Church. The great influence in its inception was the Rev. James Cooper, later Professor Cooper, leader of the Scoto-Catholic movement. The Society's transactions show its strong archaeological interest in the medieval, and also its approval of the kind of 'restoration', which could turn a traditional Presbyterian church into an east-facing church with chancel or apse.

Hillhead, Glasgow (now Kelvinside Hillhead), by James Sellars, 1876, has a high-apsed nave, reminiscent of La Sainte Chapelle in Paris. Peter MacGregor Chalmers restored the interior in 1921. MacGregor Chalmers' own churches are almost all long-axis buildings. His preference (following Cooper) was for Romanesque (e.g. Carriden, Bo'ness, 1909; Colvend, 1911; St Leonard's, Dunfermline, 1904) but Cardonald (1889) is an example of his Gothic style.

As had happened with the English cathedrals, the great churches of Scotland began to be 'restored' according to the tastes of the time. In Edinburgh, on the initiative of William Chambers, who had been lord provost, the high kirk was restored (1865–9), galleries taken down and impressive royal and magistrates pews introduced. By 1883 the other congregations vacated the building and the high kirk took over the whole of St Giles. It has been suggested that it was intended as a Scottish equivalent of Westminster Abbey, for great civic occasions. Glasgow Cathedral, which alone had never been roofless, and had retained its medieval rood screen, also became the home of a single congregation, though the choir alone is used on normal occasions. The west end of the church lost its two spires to Victorian taste in the 1840s. St Michael's Linlithgow, which at one time housed two congregations, was restored and the gallery removed, in the 1890s. The nave of Dunblane Cathedral, left unused after the Reformation, was restored (Rowand Anderson, 1889–93). In 1909 P MacGregor Chalmers rebuilt Holy Trinity, St Andrews, in a Norman style he preferred. The ruined Norman choir of Paisley Abbey was restored by MacGregor Chalmers and Robert Lorimer in the early twentieth century. The whole could now be used as a single space, for the first time in its history. St John's, Perth, was restored by Robert Lorimer in 1923. St Mary's, Haddington, whose nave had been the parish church since the Reformation, had its choir and transepts rebuilt (Ian G Lindsay & Partners) as recently as 1973.

These rebuildings and restorations have produced some beautiful buildings for the visitor to admire. Whether they are suitable for Presbyterian worship is another question, but the worship tends to change to suit the building. Moreover, public address systems (introduced in many quite small churches so that the deaf may benefit from the loop) have changed the character of preaching, encouraging a more conversational style, and in many cases (St Giles' being a notable exception) have overcome the problem of bad acoustics in a large medieval building.

It is rare now in the Lowlands to find a country church which has not been 'east-ended', with pulpit, communion table, chairs, lectern and font at the east end and all the pews facing east. One can see the space between the large windows in the long south wall where the pulpit used to be, and where the galleries were in the east and west ends. Sometimes a little chancel or apse has been built on to the east end to house the table etc., while the pulpit is left in the corner of the 'nave'. Many of these (e.g. Crailing, 1907; Bowden, 1909; Linton, 1912; Symington, 1919) are by MacGregor Chalmers. Mostly this has been done with good materials, though there is one dreadful example (Garvald) as late as 1962.

But all new churches before 1960 followed this arrangement also. It had become official in the Church of Scotland. When, in the late 1930s, the Advisory Committee on Artistic Questions published *The Principles of Church Design*, by J Arnott Hamilton, they gave official approval to a design for a rectangular, preferably aisled building, with shallow transepts, and a chancel or apse containing a communion table. Hamilton allowed architects to choose Norman or Gothic styles, though some held that the Norman style was more appropriate to a Presbyterian church. Hamilton's book was succeeded, or supplemented, in 1963 by Esme Gordon's *Principles of Church Building*. Gordon had some useful advice and, unlike Hamilton, he eschewed plans and illustrations. But there is no doubt that his ideal church was very like his predecessor's. Neither had any place for the T-church or the central plan.

One of the features of many country churches was the laird's loft, a gallery, with fine woodwork and chairs, for the laird and his family. This often had a retiring room behind, where he could have lunch between the morning and afternoon services. The laird's loft would be directly opposite the pulpit in a T-shaped church. One effect of the east-ending of such churches was to leave the laird's loft in limbo, at right-angles to the rest of the congregation. This can be seen in Eckford, Bowden and many other places. But the laird is now probably an Episcopalian, or an absentee, or a faceless company.

From 1560 praise in Presbyterian churches was led by a precentor. During the nineteenth century, congregations began to desire organs, and permission to use the organ was gained in the Church of Scotland (1866), the United Presbyterian Church (1872) and the Free Church (1883). Later, Andrew Carnegie gifted pipe-organs to many churches. The organ was not, however, easily accommodated in most Presbyterian buildings. In many the only wall high enough to take the pipes was the wall behind the pulpit, and it became not only the most audible, but also the most visible object in the church. The need to bring the pulpit forward dislocated the entire interior arrangement of the church. Some recently reordered churches have managed to get the organ into the back gallery – a better place for organ and choir to lead the praise, but with the effect of blocking the light from a large window.

In the Church of Scotland a great effort of Church Extension took place in the years before and after World War II, as the church sought to keep pace

Figure 27.12 Pulpit and hour glass at St Salvator's, St Andrews. Photograph by the author.

with new housing developments. Since the local congregation was just coming into being this was centrally controlled and financed. The assumption was that all of these buildings would be rectangular with a small apse or chancel at the east end. Some were dual-purpose buildings, with a stage at the other end, so that social activities of various kinds could take place facing west during the week, and worship, facing east, on a Sunday. What was surprising was that though the explosion of new materials and new techniques had freed the architects from the limitations of the past, the modern church was so often an echo of old styles in modern materials. The brief given to architects at the time, specified the costs, the seating, the stage and the requirements of badminton, but said nothing about theology or worship or religious education. In 1959, Wheeler and Sproson did their own research, and at St Columba's, Glenrothes, were the first to abandon the long-axis design in favour of a central plan – a Greek cross. As a result the church began to send out a theological brief, and the dreary succession of predictable buildings gave way to some original and exciting churches. Alas, the buildings of the mid-twentieth century have not proved as enduring as those of earlier ages.

The nineteenth and twentieth centuries saw much building by the Scottish Episcopal Church and also by the Roman Catholic Church. The latter experienced new freedom after the Relief Act of 1829, and the re-establishment of the hierarchy in 1878, as well as a great increase in membership due to Irish immigration.

The Episcopal Church has some fine Gothic revival buildings, with the internal arrangement on the principles of the Oxford Movement. The greatest of these is St Mary's Cathedral, Edinburgh (Gilbert Scott, 1829), but

Figure 27.13 Bo'ness Roman Catholic church. Photograph by the author.

Figure 27.14 The altar in Bo'ness Roman Catholic church. Photograph by the author.

there are many other examples, among them William Butterfield's Cathedral of the Isles on Bute (1851), and St Ninian's Cathedral, Perth (1850), Gilbert Scotts' St Paul's, Dundee (1855), and St Mary's, Glasgow (1871), and G F Bodley's St Salvador's, Dundee (1868). St Margaret's, Newlands, Glasgow (P MacGregor Chalmers, 1910), is Romanesque.

The Roman Catholic Church has been less thirled to the Gothic. St Mary's Cathedral, Edinburgh, still has the Gothic façade, but that alone, of James Gillespie Graham's 1814 building. St Margaret's, Dunfermline (R Rowand Anderson, 1889), is in a transitional style. St Thomas, Keith (William Robertson 1831), shows Italian influence, with a classical façade and a dome. St Aloysius, Glasgow (Charles Menard, 1910), is a classical church, with high campanile. In the twentieth century, Jack Coia's eccentric genius produced some fine and unusual buildings for the Roman Catholic Church. His St Paul's, Glenrothes (1956–7), gained international acclaim. His budget for church and presbytery was £20,000. He designed the church in the shape of a trapezium, narrow at the back and broad at the front, with unbroken side walls and light coming from the back, which is completely glass. His dramatic east end, lit from a lantern tower above the altar, lost some of its point when a table was set forward so that the priest could celebrate facing the congregation, and the power of his unbroken side walls was lost when the south wall was pierced with doors and windows, to give access and visual access between the church and a new hall beyond. In this otherwise unadorned church, Coia had the Jewish sculptor, Benno Schotz, design a 12ft high metal crucifix, abstract and eloquent, to sit upon the altar. St Mary's,

Figure 27.15 Chapel at Scottish Churches' House, Dunblane. Photograph by the author.

Bo'ness (1962) (Figs. 27.13, 27.14), is in many ways an advance on Glenrothes, with a lovely baptistery. St Bride's, East Kilbride (1964), looked like a fortress, with no windows visible at all, and was known locally as Fort Apache, but the quality of light within the building could be entrancing.

The Presbyterian church has often been regarded as dour and forbidding. Certainly, there was little colour, and representational or symbolic art was frowned on as encouraging superstition. But the woodwork in pulpit and panelling was often beautifully proportioned and exquisitely carved, a proof that art and craftsmanship were not despised. Much of that can be seen still, even in simple country churches. The communion vessels, of silver or pewter, were and still are objects of fine beauty, though with the introduction of the individual cup and the uniting of parishes, many of these have now been disused or sold. The nineteenth century saw the acceptance of stained-glass, and in the early and mid-twentieth century, Scotland could boast of some of the finest stained-glass artists, such as Douglas Strachan, William Wilson, Gordon Webster and Herbert Hendrie, whose work adorns many churches. There was a danger that the enthusiasm for stained glass could make a bright church into a gloomy one. The cross (though never the crucifix) began to find its way into Presbyterian churches, often behind the pulpit, or on the pulpit fall. Gable crosses began to appear on buildings. The danger is that with overuse this becomes an empty sign rather than a powerful symbol. In the ecumenical chapel of Scottish Churches House, Dunblane (1960), Eric Stevenson set a large cross in the middle of the little chapel, so that the people do not look at it but sit under its arms – a totally different experience (Fig. 27.15). The nineteenth-century desire for comfort introduced cushioning in pulpit and pews. It also led to the introduction of the pulpit fall, which became, apart from the organ, the visual focus in the church. Victorian falls were usually of velvet or velour, with the burning bush, the emblem of the Reformed Church, or the monogram IHS, embroidered on them. Few knew what these three letters meant, and many remarkable explanations were believed. In the twentieth century, however, the art of embroidery has come into its own in churches, and many striking examples of pulpit falls and book markers are now to be found.

The church continues to change and so it will, as long as it continues to live.

BIBLIOGRAPHY

Burleigh, J H S. *A Church History of Scotland*, London, 1960.
Cameron, J K, ed. *The First Book of Discipline*, Edinburgh, 1972.
Coltart, J S. *Scottish Church Architecture*, London, 1936.
Fawcett, R. *Scottish Medieval Churches*, HMSO, 1985.
Forrester, D and Murray, D, eds. *Studies in the History of Worship in Scotland*, 2nd edn, Edinburgh, 1996.
Gordon, E. *A Handbook on the Principles of Church Building*, Church of Scotland Advisory Committee on Artistic Questions, Edinburgh, 1963.
Hamilton, J A. *The Principles of Church Design*, Church of Scotland Advisory Committee on Artistic Questions, Edinburgh, npd.
Hay, G. *The Architecture of Scottish Post-Reformation Churches, 1560–1843*, Oxford, 1957.
Lindsay, I G. *The Cathedrals of Scotland*, Edinburgh, 1926.
MacGibbon, D and Ross, T. *The Ecclesiastical Architecture of Scotland*, 3 vols., Edinburgh, 1906.
Rogerson, R W K C. *Jack Coia: His Life and Work*, Glasgow, 1986.
Scotland's Churches Scheme, *Churches to Visit in Scotland*, Edinburgh, 2002.

28 New Uses for Old Churches and Manses

DAVID MAXWELL

CHURCHES

'What is to become of our lovely church' must be the thought passing through the minds of many members of a congregation when the decision has been taken to unite with a neighbouring congregation 'in the other building'. Some members, of course, will take a more prosaic view and will rejoice that the other church offers more car parking spaces and better kitchen facilities. So, as the question arises – do we sell our church to the highest bidder or do we seek out a worthy future use for the building we love?

The highest bidder (perhaps the only bidder) may, of course, be a developer whose aim is to demolish the building and raise a block of flats on the site, as has happened in Edinburgh to the John Kerr Memorial church at Polwarth and the Baptist church in Dublin Street. It is a sobering thought that in the last thirty years over seventy churches in Glasgow have been demolished.

We must not forget, however, that our forebears were never afraid to demolish when they wanted to build something bigger and better (at least in their own estimation) on the site of an existing church. Many of our present-day rural churches now stand on the foundations of one or more earlier buildings, and often incorporate within their structure the stones from the churches they have replaced.

In the Borders alone there are many examples: Hopkirk (1862) is built on a previous Christian site and stones from earlier churches are to be seen in the font; Eskford (1771) stands on an ancient place of worship; Bowden sits astride the old pilgrim route from Melrose to Lindisfarne and embodies in its walls parts of earlier churches of the fifteenth and sixteenth centuries; Oxnam (1738) was altered to a T-plan in 1874 and occupies a medieval site dating from before 1153; Yetholm (1837) stands on a site in use since David I's apportionment of parishes; Hownam was completely remodelled in 1752 as a rectangle from an earlier cruciform church and contains a late fifteenth-century round-headed doorway; Crailing (1775) is built on an ancient place of worship and its bell shows the date 1702.

In Angus, the abandoned church of Tealing proudly displayed on its interior walls two fine mementoes rescued from earlier religious buildings. One, a small stone plaque recording the ministry of a priest in the pre-Reformation church, the other, an imposing relief of a minister brought from the second church on that site.

Such demolition and rebuilding also happened in the cities. Early in the 1970s the Church of Scotland faced the choice of either selling the former Moray-Knox church in the High Street of Edinburgh, or demolishing it to build a centre for the arts on the site. The former church was not adaptable for the projected use and was of no outstanding architectural or historic merit. The decision was taken to replace the building with a purpose-built centre including a theatre, exhibition galleries and restaurant surrounding an open-air courtyard – traditional for the Old Town of Edinburgh. Named the Netherbow after the medieval gateway into the city, which stood at this point, the centre has invigorated the artistic and community life of the Old Town. Recently integrated with the adjacent building, the refurbished John Knox House museum, it provides a base for extending drama and story-telling to parishes and schools all over Scotland.

The quest to find a responsible purchaser may not be as difficult as at first thought. Provided the church is soundly built, has been well cared for and is wind and water tight, it can offer the discerning buyer a solid base for a wide variety of uses. The existence of tall spires and bell towers was once regarded as a decided liability due to the difficulty and cost of maintenance. However, the great Pugin spire on the former Highland Tolbooth, dominating the top of the Royal Mile, did not deter the Edinburgh Festival Society from acquiring and converting the church into its permanent headquarters, opened as The Hub, in 1999. The tall spire of William Leiper's Dowanhill church in Glasgow now presides over the Cottier Community Centre, named after the artist whose stained glass adorns the building, and in Dalkeith, Buccleuch church, whose spire soars over the town, now operates as a spacious workshop for architectural carpentry. In Edinburgh, the former Belford church, with a high spire rising above the Dean village, is now a hostel for back-packers (Fig. 28.1), while the Tron church, with its elaborate stone spire (a replacement for the wooden Dutch steeple destroyed in the fire of 1824), now functions as the Old Town Information Centre (Fig. 28.2).

There are indeed occasions where a steeple is so much a feature of the local skyline that it survives the demolition of its adjoining church. The spires of Park church in Glasgow and Stockbridge church in Edinburgh are fine examples, which now fulfil new roles in their locality (Fig. 28.3).

The arts – music, drama, film, painting, dance, architecture and glass-work – have all found suitable homes in former churches. Glasgow's Trinity Congregational church is now the Henry Wood Hall, home of the Royal Scottish National Orchestra, while Newington St Leonard's in Edinburgh is a successful concert venue, The Queen's Hall. Also in Edinburgh, Lothian Road church has become the Filmhouse, headquarters of The Scottish Film Council. Salisbury church is a dance studio, and the Glasite meeting house (1836) has been restored as the headquarters of the Architectural Heritage

Figure 28.1 Belford Youth Hostel, Edinburgh. Photograph by David Veitch.

Figure 28.2 The Tron church, Edinburgh, now houses the Old Town Information Centre, shops and exhibitions. EERC F.1305.

Society of Scotland. In Glasgow, the Tron church re-opened in 1982 as the Tron Theatre, while in Edinburgh, Morningside High became the Church Hill Theatre and St Bride's a theatre and community centre. Lammington in Lanarkshire and Kilmaurs in Ayrshire have both been converted into studios for stained glass artists, while Old Kilmorach, by Beauly, is a gallery exhibiting local contemporary arts and crafts. In Edinburgh, the former Norwegian church in Leith (1868), the oldest Norwegian church outside Norway, has been transformed into the Leith School of Art. In addition, in

Figure 28.3 Stockbridge church, Edinburgh, was demolished in 1980 and replaced by sheltered housing, but the tower was preserved. The tower had been moved in the 1860s from Lothian Road after its original church was demolished. Photograph by David Veitch.

Glasgow, Cowcaddens parish church now operates as The Piping Centre, and St Andrew's parish church is now The Scottish Music Centre. In Aberdeen, the former North church (1831) is an arts centre, while a former Congregational church functions as a maritime museum (Fig. 28.4).

The architectural style of a former church does not appear to feature

Figure 28.4 Aberdeen Maritime Museum, opened in 1997, is partly housed in a former Congregational church. Courtesy of Aberdeen Art Gallery and Museums.

largely in its conversion to domestic accommodation. Temple in East Lothian, Doune in Perthshire and Newburn in Fife, all typical rural churches, have been altered to form delightful single residences. Saline church in Fife, a listed building (1844) with a distinctive bell-cote, is a luxury home with a heated indoor swimming pool. Large city churches, like St George's in the Field, Glasgow, with its magnificent classical Greek façade, now contains several self-contained flats. Rutherford Free church in Gatehouse-of-Fleet now provides four apartments and two shops, while Crawford church in Lanarkshire offers two separate dwellings. In Angus, the former Kirk of Craig has been converted into a ground floor studio/workshop with a house above; a circular staircase winds its way to the top of the church tower and a penthouse room with panoramic views over the Montrose Basin.

A graveyard surrounding a rural church is a much greater disincentive to its ever becoming a private residence. However, a notable exception can be found at Kilchoman on Islay. Here the conversion of the nineteenth-century church, on the site of a medieval building, stands in the middle of an ancient graveyard with two cross-slabs only yards from the front door.

Local museums are worthwhile uses for former churches and good examples are found throughout Scotland: Farr in Sutherland; Auchinleck in Ayrshire; Fochabers in Moray; Tarbet Old (St Colman's) at Portmahomack in the Black Isle; Crawfordjohn in Lanarkshire; and Broughton in Peebleshire, which is now the John Buchan Centre. The Auld Kirk of Eyemouth now functions as a museum featuring the history of Berwickshire's fishing industry, while the former Church of Scotland church in Ullapool, a fine

Telford building, has been converted with great sensitivity into the Ullapool Museum. The pulpit, traditionally sited on the long south wall, has been carefully retained, as has the laird's loft and several pews. The remainder of the pews have been stored against possible future use.

Some disused churches have retained their religious connections after conversion. Dalmuir Old at Clydebank was purchased by the Roman Catholic Church and renamed Our Lady of Loretto. Wester Coates in Edinburgh is now the headquarters of the National Bible Society of Scotland, while Kelvinside (Botanic Gardens) in Glasgow became the Bible Training Institute. Dull church in Perthshire is used by the Order of Knight's Templar for their services, and St Ninian's in Creiff is an ecumenical conference centre. The True Jesus Church now owns St Luke's in Edinburgh, and Invertiel in Kirkcaldy is home to a Coptic congregation.

Sometimes a church that has become unworkable or unnecessary on its original site can fulfil its purpose at another location. Transportation is a realistic possibility and has been successfully accomplished on at least five well-documented occasions. An Episcopal church in Nairn (established in 1845) was closed in the 1880s and, after lying derelict for many years, was taken down stone by stone and shipped around the Pentland Firth to Lochinver in Sutherland where it was re-erected in 1903 as the Church of Scotland parish church. The parish church of St James' (Pollock) was originally Titwood church in Pollockshields, Glasgow. Pollockshields was an over-churched area with six churches all within five minutes walk of each other, whereas Pollock, then the largest post-war housing scheme in Scotland, was in urgent need of a place of worship with hall accommodation. When Titwood congregation united with its mother church, it was decided to move its buildings to the Pollock estate. Work on moving the hall began in November 1948 and within two years it had been rebuilt and occupied. The main church was then dismantled stone by stone and transported four-and-a-half miles to its new site by one lorry doing up to three journeys per day. Three years later it was reconstructed and its doors opened on 3 September 1953 (Fig. 28.5). The cost of moving these two stone buildings at £65,000 was certainly greater than the cost of building a church and hall in the modern materials available at that time, but whereas St James' (Pollock) is flourishing and expanding its activities, many of its less substantial contemporaries are either long since demolished or face serious fabric problems.

The Scottish Episcopal church of St Michael and All Angels in Inverness was founded in 1877 as a Sunday School in a thatched cottage on the banks of the River Ness. As the congregation grew, a new church was built in 1886. By the turn of the century however, local flooding and population movement prompted the congregation to move to the opposite bank of the River Ness. The church was dismantled and floated across the river in sections on barges to be re-assembled on its present site. It was re-dedicated in 1904.

Orwell parish church (1729) was moved by horse and cart about four miles up the hill from the side of Loch Leven to be re-erected on a site where

Figure 28.5 The church of St James' was moved to Pollock from Pollockshields stone by stone. Photograph by the author.

it continues to fulfil its original function. Perhaps most spectacularly, in 1999 Perceton and Dreghorn parish church was removed from the Ayrshire village and re-assembled with its pews, pulpit and organ as the centre piece of a wedding complex in Tokyo, Japan.

The best laid schemes of town planners gang aft agley, as we well know, and in one particular case the results were irretrievable. Trinity College church, established by papal bull in 1460 as successor to the church and hospice of Holy Trinity at Soutra (founded by King Malcolm IV in 1164), stood on what is now Platform 1 of Edinburgh's Waverley station. The church was taken down in 1848 and the stones were individually numbered and stored on the Calton Hill. In 1871, by the time a new site had been identified in Chalmers Close off the Royal Mile, only one third of the stones remained available, and these were used to construct an apse at one end of a new building. It now serves the community as the city's brass rubbing centre (Fig. 28.6), and the numbered stones can still be seen in the walls.

Education establishments have been worthy of former churches. The University of Strathclyde acquired the Ramshorn church as its drama theatre and the Barony as its assembly and graduation hall. The University of Glasgow uses the former Anderson Free church as its department of theatre, film and television studies. In Edinburgh, Napier University took over

Figure 28.6 Trinity College Apse, Edinburgh, houses a brass rubbing centre.
Photograph by David Veitch.

Morningside Parish church (1838), one of Dr Thomas Chalmers' earliest extension charges, as a lecture hall. Moray House Institute has absorbed the church in St John's Street, New North church has become The Bedlam Theatre of the University of Edinburgh Student's Association (Fig. 28.7), and St Oswald's is the drama department of Boroughmuir High School. In Glasgow, Belmont church is the assembly hall for Laurelpark School.

It is important that churches which stand out on prominent urban sites and form an integral part of a townscape should not only retain their outward appearance but also remain, if possible, in the church domain. Three Edinburgh cases illustrate this point. The former St George's parish church occupies the centre of the west side of Charlotte Square in the New Town, and its townscape value lies in the way its classical façade and unique dome close the vista from George Street across the gardens of the square. Happily it became West Register House, the repository for the archives of Scotland's social history. The Episcopal church of Holy Trinity, whose distinctive tower stands guard over the north end of the Dean Bridge, was converted into an electricity substation when it first ceased to be a place of worship. Although its exterior remained virtually unimpaired, its general appearance was that of a closed and empty building. Now, it has been reopened as a church by Christian Centre Ministries (Fig. 28.8). The Catholic Apostolic church in Mansfield Place, which contains the important murals by Phoebe Anna Traquair, had an uncertain future for many years and functioned for a while as a café and a Festival Fringe venue. It has been acquired by a trust which intends to conserve and restore the murals, and to provide public access.

Nation-wide and local community trusts are stalwart guardians of historic churches. The Scottish Redundant Churches Trust (www.srct.org.uk),

Figure 28.7 The Bedlam Theatre, Edinburgh, formerly New North church. Photograph by David Veitch.

Figure 28.8 The Episcopal church of the Holy Trinity, Dean Bridge, Edinburgh, after a period as an electricity substation, has been re-opened as a church by Christian Centres Ministries. Photograph by David Veitch.

set up in 1996, has recently taken over for preservation three former churches: St Peter's at Sandwick in Orkney; Tibbermore in Perthshire; and Cromarty in the Black Isle. Independent community trusts now care for Nigg with its seventeenth-century sculptured stone, Edderton in Ross and Cromarty, St Duthus in Tain, and Westerkirk in Dumfriesshire.

A fine example of local initiative and enthusiasm is currently being shown by the good folk of Moulin in Perthshire. The community has set up a private enterprise company to secure and develop the Kirk of Moulin, which was rebuilt in 1875 on an historic site going back to the seventh century, but was closed for worship in 1988. Listed as 'of architectural merit' and the focal point of the conservation village, the kirk could provide a variety of facilities for villagers and visitors alike. Suggested uses include a base for the creative arts, an interpretative centre for the history and culture of the area and its people, 'hands on' educational activities for children and young families, and a sanctuary for meditation and private prayer.

Miscellaneous examples of alternative uses for churches in Edinburgh include the following: Broughton, with its imposing classical façade, is now the office and sale-room of a well-known firm of auctioneers; North Morningside is the Eric Liddell Centre for community use; The Albany Trust has acquired the Lockhart Memorial to provide office accommodation and conference facilities for charitable organisations; Newhaven is a thriving indoor rock-climbing centre; the former Queen Street church, after serving for some years as a law firm's office, has reopened as a health and fitness club; while Holyrood Free, built by Queen Victoria in 1850 to serve the citizens of the Canongate, was renovated as an art gallery for exhibiting works from the Royal Collection to mark the present Queen's Golden Jubilee in 2002 (Fig. 28.9). On the island of Canna, the National Trust for Scotland now runs St Edward's Roman Catholic church as a visitor and study centre providing accommodation for naturalists and Gaelic scholars. Two former churches in Inverness now function as funeral parlours and the old Gaelic church (1649), which later served as Greyfriars Free, is now the largest second-hand bookshop in Scotland. In Auchterarder the Barony is a warehouse for antique furniture, and in Comrie the parish church is a lively community centre.

The former Carrington church in its tiny village has the characteristic East Lothian tower and spire projecting from the ecclesiastical south wall. In the building's new guise as a design centre, its exterior appearance has been carefully preserved and is excellently maintained (Fig. 28.10). In Glasgow, the Adelaide Place Baptist church was imaginatively redeveloped in 1995 into a multi-functional centre embracing a sanctuary, guesthouse, café and nursery.

We must remember too, that alternative uses for churches are not nineteenth- and twentieth-century phenomena. St Giles', the High Kirk of Edinburgh, during its long and chequered history, has been divided by internal walls into smaller churches, and has provided space for a police station, a place for settling commercial transactions and a store for the burgh's guillotine, known as 'The Maiden'. St Giles' weathered these storms and is once again solely and wholly a parish church. Similar resurrection may happen to other former churches presently used for new purposes, as we have seen in the case of the church on Edinburgh's Dean Bridge.

These, then, are a few of the more acceptable uses to which disused and redundant churches have been put. Readers will know of many other examples in their localities, and will also be aware of those less desirable solutions such as casinos, night clubs, garages and an ice-cream factory. More instances of both the good and the bad are arising as we write and will continue to do so for years to come.

In conclusion, let us not forget the unusual story of Ayr Baptist church. It was originally built as a theatre in 1850, but closed its doors in 1886. Thereupon, the Baptists bought the building and converted it into a church – a role it still fulfils today.

Figure 28.9 The Queen's Gallery, Edinburgh, formerly Holyrood Free church. Photograph by David Veitch.

Figure 28.10 Carrington church is now a design centre. Photograph by the author.

MANSES

In the eighteenth and nineteenth centuries and early into the twentieth century, parish ministers, certainly in rural areas, were, along with the dominie and the doctor, prominent members of the local community, and the manses they occupied were large imposing houses. However, his standing in the community was not the main reason for the minister requiring a substantial house. Space for a large and growing family and the need for domestic help due to the minister's wife playing an active role in the affairs of the parish were other reasons. There was also the need to provide

Figure 28.11 St Ninian's manse, Leith, has been converted into offices. Photograph by the author.

overnight accommodation for visiting preachers and their horses, so ample stabling was also a necessity. But, above all, absence of ancillary buildings at the church meant that the manse usually had to provide not only a study for the minister, but also space for meetings of the kirk session, the Sunday School and probably the Women's Guild.

The second half of the twentieth century saw great changes taking place. The building of church halls, the move towards smaller families, the unavailability of domestic help and the desire (often the need) for manse wives to seek full-time employment or pursue a career, meant that smaller, more easily run houses were now desirable. This move was further encouraged by the expense of maintaining and upgrading old manses, which was often beyond the resources of the minister and the congregation. Thus, when the opportunity arose, on the calling of a new minister or the union of adjoining parishes, old manses were put up for sale.

Although many of these manses required substantial renovation, they offered prospective purchasers sturdy, handsome properties, often in attractive surroundings. Rural manses had further attractions, including stables, where the minister had kept his horse and, in later times, his car, and a glebe, a small piece of arable land attached to the manse for the minister to cultivate himself.

While many manses have found new life as family homes, others have been converted for commercial use. The manse of Marnock, for example, is now a small country house hotel, while manses in Helmsdale in Sutherland, Newmills in Fife, and Perth have all been transformed into bed and breakfast establishments. One of the most striking examples of a manse now in commercial use is to be found in Leith, where the sixteenth-century St Ninian's manse, a rare pre-Reformation survival with an equally rare seventeenth-century ogee steeple attached, has been converted into offices (Fig. 28.11).

The uses to which churches and manses have been put over the centuries reflect changes in Scottish society as a whole, and the part played in it by religion. Most obviously, the conversion of religious buildings to secular uses in the past thirty years or so mirrors the secularisation of Scottish society. As the preceding discussion suggests, however, changing uses also reflect more subtle changes. The use of former Church of Scotland churches by other denominations, for example, indicates how Scottish religious life has become increasingly diverse since the mid-nineteenth century. The abandonment of large, prominent manses, moreover, reflects that the Church of Scotland recognises not only the financial burdens of such buildings, but also the benefits of its ministers moving towards a simpler lifestyle, with homes similar in size and style to those of their congregation. This is exemplified by one minister, later nominated as Moderator Designate to the General Assembly of the Church of Scotland, who chose to stay with his family in a council flat in Glasgow to be in the midst of his flock.

The age-old practice of converting and adapting church buildings to take account of the changing aspirations and values of society will continue to be one of the most visible and informative signs of the evolving role of religion in Scottish society for many years to come.

APPENDIX: THE PROTECTION OF PLACES OF WORSHIP AND CHURCH RUINS

Robin D A Evetts and Deborah Mays

Places of worship make a significant, and sometimes dramatic, contribution to the physical character of urban and rural Scotland. They are markers in the history of the nation and help define the historic and communal identity of a settlement. A legislative framework exists to identify and protect church buildings or ruins of special architectural or historic interest to ensure that as far as possible, either by statutory scheduling or listing, their character is respected in any changes to the fabric or setting. Demographic change and declining congregations are producing an increasing number of redundant churches and a need to protect them as artefacts, or to convert them to other uses, if they are to continue making their contribution to the historic built environment.

Churches and church ruins are scheduled under the Ancient Monuments and Archaeological Areas Act of 1979 according to a set of established criteria that are applied carefully to determine if the subject possesses the required national importance. These criteria include that the monument belongs to a group or subject which is archaeologically, historically or architecturally important, can be recognised as part of the national consciousness, or that it retains structural, decorative or field characteristics of its kind to a marked degree, or that it offers a significant archaeological resource within a group or subject of acknowledged importance. Buildings still in (or restored to) ecclesiastical use or converted to secular use may not be scheduled, though scheduled status may still be applied to part of a site not in ecclesiastical or secular use, including the archaeological layers beneath the church itself. Scheduled monument consent is required for works that would change the character of scheduled churches, and this is obtained from Scottish Ministers through the Ancient Monuments Division of Historic Scotland. Scheduled monuments are visited on a regular basis by Historic Scotland's Monument Wardens, who report on the state of the monument and any unauthorised works that may have taken place, providing a valuable monitoring service.

Listing and listed building consent is conducted under the Planning (Listed Buildings and Conservation Areas) (Scotland) Act 1997, and follows similar parameters of selection for inclusion to that of scheduled monuments, the criteria weighing up a number of factors ranging from age, technological virtuosity, historic association and significance in the theological development and worship practices of the church, to material and design quality. Listed churches will normally be in ecclesiastical or secular use, but are sometimes ruinous. Listing brings with it the mantle of curtilage, in that any object or structure that is fixed to the listed subject or which falls within its curtilage, and although not fixed to it has formed part of the land since before 1 July 1948, may be treated as part of the listing.

Listed church buildings that are in ecclesiastical use are exempt from the normal requirement for listed building consent for works that would affect their character. A Pilot Scheme of non-statutory listed building control was introduced in 1999 in association with the Scottish Churches Committee. The Scheme applies to external alterations only and is operated through planning authorities and Historic Scotland. Alterations to interiors remain the responsibility of each denomination, but it is a requirement that they set up a Decision Making Body that is the final arbiter in the Scheme.

A wider and less specific form of protection for churches is that arising from Conservation Area legislation. These areas are designated to ensure that any development within them respects or enhances their character. This is achieved through control over demolition, the application of Article 4 Directions that suspend certain classes of permitted development, and the implementation of environmental enhancement schemes. There are currently over 600 Conservation Areas in Scotland.

The prime responsibility for the upkeep of any scheduled or listed building whether ruined or intact rests with its owner. For a series of historical reasons, ownership of pre-Reformation and Church of Scotland structures can sometimes be difficult to ascertain, although ownership of post-Reformation and non-Church of Scotland structures is usually less complex. A number of ecclesiastical ruins and some intact cathedrals and churches are in the care of Scottish Ministers and currently maintained by Historic Scotland, while other ruinous churches are cared for by local authorities, or other bodies and individuals. In 1996 the Scottish Redundant Churches Trust was established to care for some of the finest examples of churches of all denominations that have ceased to be used for worship on a regular basis. In other cases, redundant churches have been converted to a wide variety of new uses. Grant aid may be available from Historic Scotland, the Heritage Lottery Fund, local authorities, the Scottish Churches Architectural Heritage Fund and other sources for ruined, redundant or re-used churches as well as those in ecclesiastical use. In addition, from 1 April 2004 (to be reviewed in 2008) the full amount of VAT paid on eligible repair, maintenace and structural replacement works to listed places of worship may be claimed through the Listed Places of Worship Grant Scheme promoted by the Department of Culture, Media and Sport. This and the various protection methods reflects a recognition of the importance of this class of building to the social, political and spiritual development of Scotland and its built environment.

29 Furnishings in the Reformed Church

HENRY R SEFTON

Church furnishings tell much about those who design and use them. This is illustrated by the furnishings of two places of worship in Aberdeen – King's College Chapel and the West Church of St Nicholas. The former dates from 1500 and was designed for the offering of Mass and the singing of the Daily Offices.[1] The members of the College would spend many hours of each day in the Chapel and so seats were provided for them. These canopied stalls are one of the great treasures of Scotland. A screen separates the college community from the general public and no seating was provided in the ante-chapel. A small balcony projected from the screen on the outer face and from this the visitors could be addressed. The West Church of St Nicholas dates from 1755 and replaces the nave of the mediaeval kirk of St Nicholas (fig. 29.1). It

Figure 29.1 The West Church of St Nicholas, Aberdeen. Photograph by Margaret Dundas.

was designed for the preaching of the Word of God, and the interior is dominated by a huge pulpit with a sounding board. Every available space was filled with box pews with doors so that the maximum number of people could hear and understand what was being said from the pulpit. The West Church is separated from the surviving mediaeval transepts by a permanent wall to provide a space in which the preacher could be seen and heard. The quire of the mediaeval kirk was similarly separated from the transepts for the same reason. Aberdeen thus has the finest mediaeval seating and the most complete eighteenth-century seating in Scotland.

SEATING

In an article contributed to *History*, Andrew Spicer concludes: 'Perhaps it is appropriate to see the provision of some form of seating after 1560 as being as essential as the pulpit for Reformed worship'.[2] He takes issue with the view expressed by the architectural historian George Hay that 'For some time after the Reformation Scottish kirks remained as devoid of seating as they had been earlier'.[3] Both refer to an order made in 1560 by the town council of Edinburgh for 'saittis, furmes and stullis ... for the people to syt upoun the tyme of the sermoun and prayarris within the Kirk'. Hay claims that this was not accomplished until much later but Spicer questions the accuracy of this assessment as it depends on the surviving physical evidence of church furnishings. It is therefore not clear whether Jenny Geddes threw her own stool at Dean Hannay or a movable one provided by the town council! The seating was obviously of varying quality and those who could afford to do so set up their own pews and corporate bodies installed private seated galleries.

There were two reasons why seating was necessary after the Reformation. John Knox and the other reformers insisted that communicants should receive the sacrament seated at a table, so seating had to be available. The transition from mere presence at Mass, lasting usually less than an hour, to lengthy preaching services during which the congregation were expected to hear and understand what was being said from the pulpit also made some form of seating necessary.

It became the obligation of heritors (landowners) and town councils to provide in each church sufficient seating for two thirds of the 'examinable persons' in the parish, i.e. those of twelve years of age and over, as well as an elders' pew and manse pew. The dimensions were to be 29 inches wide with seats of 18 inches in breadth. The extent of each heritor's responsibility was determined by the value of his property and the area of the church was divided up on this basis. The heritor then erected pews on his allocated space for himself, his family and tenants and he remained entitled to this seating as long as he held land in the parish.[4] When the church of Kinneff was rebuilt in 1738 and space allocated, one heritor considered that his space had been encroached upon and he took an axe to his neighbour's pew![5] In the towns the seats were not associated with lands, and allocation was made by the town council with preference for magistrates. In addition to the general seating there were often a baptismal pew, a marriage pew and a 'cutty stool' or place of public repentance, which was usually on a platform in the middle of the church. There is some evidence that men and women were separately seated for some time after the Reformation. In some places the practice died out in the seventeenth century, in others in the eighteenth, but as late as 1827 the women sat by themselves in the church of St Giles in Elgin.[6]

Baptism, marriage and penitential seats have disappeared but the manse pew and the elders' pew survive in the old church at Newbattle. Magistrates' pews are to be seen at Burntisland and in the West Church of St Nicholas in Aberdeen. Elaborate laird's lofts are to be found at the old churches at Pitsligo and Kilbirnie. Pews consisting of long seats and bookboards are still the most common form of church seating but in many churches pews have been replaced with chairs. In a defence of such a replacement in Queen's Cross church, Aberdeen, the minister, Rev. R F Brown, has written 'Pews have no theological or ecclesiastical significance. They were simply a cheap and efficient, and often uncomfortable, way of seating people in a previous age ... they have no special place in churches.'[7] When the north transept of the kirk of St Nicholas was re-furnished in 1990 chairs were designed by Tim Stead which fit the curvature of the human

Figure 29.2 A folding stool of the type used in churches in Scotland before and after the Reformation. Courtesy of the Trustees of the National Museums of Scotland, NMS H.K.L3.

Figure 29.3 Examples of old and new seating, St Giles', Edinburgh. Photograph by David Veitch.

back. By using the initial letters of different kinds of wood the message 'We remember you' was incorporated in each piece of furniture in memory of the victims of the Piper Alpha oil platform disaster.

BELLS

The first *Book of Discipline* does not specifically mention seating as indispensable in every parish church but it does list the following requirements:

> Every Kirk must have dores, close windowes of glasse, thack able to withhold raine, a bell to convocate the people together, a pulpet, a basen for baptizing, and tables for ministration of the Lord's Supper.

A judicial decision of 1777 established that a steeple is not a necessary part of a church but that a bell is. A later decision in 1835 maintained that only the Established Church was entitled 'to assemble her members for public worship by the sound of a bell'. The town council or heritors were obliged to provide a bell but the kirk session had the right to decide when it was to be rung.[8] The normal practice was for the bell to be rung three times on a Sunday morning. The first bell was rung at an early hour to prepare the people for setting out, the second at the commencement of the reader's service, the third to mark the beginning of the minister's service.[9] The writer can recall the ringing of the early bell at Careston, Angus, in the 1930s. The

Figure 29.4 Etching of the Kilmarnock funeral bell, 1639. Courtesy of the Trustees of the National Museums of Scotland, NMS SLA C18652.

bell was usually hung in a belfry but hand bells were also used, especially at funerals. This was known as the 'mort' bell. Many of the bells were of continental origin but late in the seventeenth century bells were made in Scotland following continental patterns. Bells have been imported from England from the eighteenth century until the present day. Change ringing is rare in Scotland but in Aberdeen a carillon in the tower of St Nicholas' is played before the service on Sunday morning and so can be said to 'convocate' the people.

PULPITS

The emphasis on preaching the Word of God in Reformed worship made the provision of a pulpit necessary. This was not an innovation. There is a fine pulpit made for St Machar's Cathedral, but now in King's College Chapel, Aberdeen, which dates from 1531x45. A pulpit said to have been used by John Knox stands in St Salvator's Chapel in St Andrews but was originally in the parish church. It is a two-decker pulpit, the upper stage for the minister and the lower for the reader. This was the normal style of pulpit until the middle of the nineteenth century. Although common in England, only one or two examples of three-decker pulpits occur in Scotland. To aid audibility, a canopy was often attached to a pulpit to act as a sounding board. Few canopies have survived. Another feature which has almost entirely disappeared is an enclosure round the pulpit to be the place of

Figure 29.5 Pulpit (1912), Greyfriars church, Edinburgh. Photograph by David Veitch.

baptism. The baptismal basin was held in a bracket on the side of the pulpit. Several of these brackets survive but the only baptismal enclosure is at the Auld Kirk of Ayr. Some pulpits also have a holder for a sand-glass which measured sermons by the hour. Clocks are often placed opposite the pulpit in view of the preacher but not of most of the congregation.

The pulpit was intended to be a fixed and permanent feature of the furnishing of the church. Movable pulpits are to be found in some modern re-arrangements, as at Sherbrooke St Gilbert's church, Glasgow. In some cases the pulpit has been removed altogether, as at Queen's Cross church, Aberdeen. Theological justification for this can be found in the writings of Karl Barth. For him the ideal solution is a simple wooden table to serve at one and the same time for pulpit, communion table and baptismal font. He comments: 'No matter how it is done, the separation of pulpit, communion table and baptismal font can serve only to dissipate attention and create confusion; such separation could not be justified theologically'.[10]

BAPTISMAL BASINS AND FONTS

The first *Book of Discipline* prescribes 'a basen for baptizing' as a requisite for every church and this took the place of the font which had been hitherto used for baptism. It seems that most of the medieval fonts were destroyed for very few have survived and these are often to be found in Episcopal churches. Medieval fonts were positioned near the main door of the church to symbolise entrance into the church by baptism and this meant that they were usually at the back of the church. As well as objecting to 'superstitious figures' on fonts the reformers considered that the position prevented baptism in the face of the congregation after the preaching of the Word. For this reason the baptismal basin was placed near the pulpit or in a bracket attached to it.

When the writer was baptised in 1931 his parents presented a baptismal basin for use in the church of Pitsligo and in homes when baptism was given there. When a font was presented to the church the baptismal basin disappeared and only recently was discovered being used as a rose bowl. There are still churches, such as West St Nicholas', Aberdeen, which do not have a font, and the Communion table serves as the stand for the baptismal basin. Fonts in many cases have been privately given and are in some cases movable but quite often fixed at the front of the church. In St Leonard's church, St Andrews, the font is located in its own apse. St Margaret's church, Knightswood, Glasgow, has a font located at the front of the church but also near a door.

COMMUNION TABLES

The first *Book of Discipline* prescribes 'tables for ministration of the Lord's Supper' as necessary in each parish church. These were not permanent fixtures. For many years it was customary to erect tables when they were

Figure 29.6 Nineteenth-century, eagle-style lectern, St Giles', Edinburgh. Photograph by David Veitch.

required and take them to pieces after the observance of the Sacrament. In some places this continued up to the eighteenth century. The table was flanked with forms on which the communicants sat and the elements were passed along from one person to another. Sometimes there was only one long table, sometimes more than one. There was usually a 'head board' or cross table at which the minister and his assistants sat and on this table the elements were placed before distribution to the communicants. Frequently there was a temporary open fence enclosing the tables. This enabled elders to restrict admission to the tables to those deemed worthy to approach. Worthiness was determined by the possession of a Communion token which was presented on admission to the enclosure.

The Scots maintained the practice of sitting at the Communion table against the wishes of King James VI, who commanded communicants to kneel, and the English Puritans, who took the elements to communicants in their pews. During the eighteenth century, when fixed seating was generally adopted, long Communion table pews became a central feature of the church building. Pews in the centre of the church were constructed in such a way that they could be converted into a long table with seats round it.

In 1824 an innovation in the manner of communicating was made in St John's, Glasgow. Instead of the people coming forward in relays to the

Figure 29.7 Organ (Peter Collins, 1990), Greyfriars church, Edinburgh. Photograph by David Veitch.

Communion table, the elements were taken by elders to communicants in their pews. This practice was condemned by the General Assembly of 1825 but has become the norm. The cross table became a permanent furnishing and at the time of Communion the table and pews were covered with white cloths. Pew cloths are now gradually being discontinued. Communion tables are usually made of wood and movable but the communion table in Crathie Church is made of marble and is a fixture.

LECTERNS

During the nineteenth century, two-decker pulpits largely went out of use and the 'lattron' or reader's desk came to be replaced by a lectern. Most churches now have a lectern or reading desk apart from the pulpit. They are often made of brass and consist of an eagle bearing a book board for the Bible. The East Church of St Nicholas in Aberdeen is probably unique in having a pelican instead of an eagle.

Dr Stewart Todd has questioned whether the lectern should be a permanent item of church furniture: 'In many churches in Scotland it is scarcely ever used and simply adds to the general clutter ... Why should the place for the reading of the word be two or three or more feet lower than the place for the preaching of the word?'.[11] In St Machar's Cathedral the word is now read and preached from the pulpit.

PRAYER DESKS

Prayer desks are a comparative novelty in Scottish parish churches and were introduced as a result of the concerns of members of the ecclesiological societies that were founded in the nineteenth century. They arose out of the feeling that the pulpit was not a suitable place for leading prayer as it was above the level of the people.

ORGANS

It was not until 1865 that the Church of Scotland General Assembly permitted the introduction of organs subject to the supervision of the presbytery concerned. By the following Assembly, eighteen organs had been installed, and by the end of 1867 at least thirty-one organs were in use. The United Presbyterian Church removed the ban on organs in 1872. The Free Church gave permission to congregations to install organs in 1883. From being absolutely prohibited the organ soon became the most prominent feature of many Scottish Presbyterian churches and still is today.

COLLECTION LADLES

Collection ladles are still in use at Guthrie in Angus and at Balquhidder in Perthshire. These are boxes with a half lid fixed to a pole and are pushed along the pew for the offerings of the congregation. At Newbattle the practice was to have offering plates on stools outside the entrance to the church. This had disadvantages on a windy day as bank notes were apt to be blown away, and it has been discontinued. Many churches have plates for the offerings as the worshippers enter, but inside the vestibule. In a former day, kirk sessions were their own bankers and the offerings were locked in the poor's box. These boxes had two or three different locks so that no one person could have access.[12]

Figure 29.8 Collection ladles from Auchterless parish church, Aberdeenshire. Photograph by Alexander Fenton.

HEARSES

The hearse was originally a triangular frame for holding candles in a church but the term is also used of a hanging chandelier. There is particularly fine example in Brechin Cathedral, which was presented by Bishop Lamb in 1615. It was later converted to take gas and later electricity but has been restored to take candles.[13] The West Church of St Nicholas, Aberdeen, has three fine hearses dating from 1755 and fitted to take electric light in 1939.

BENEFACTION BOARDS

Benefaction boards were a common feature in Scottish post-Reformation churches. It was hoped that the recording of gifts and mortifications (or legacies) on a wooden board in the church would encourage others to follow the good example. Few of these boards survive but in the north transept of St Nicholas', Aberdeen, there are boards recording mortifications between 1629 and 1799. In the vestibule of the West Church of St Nicholas there is a curious monument in the shape of a sail and anchor, which records the benefactions of a shipmaster to individual charities and totals the amount.

MURAL FABRICS

In a former day, fabrics were used to decorate the church on great occasions. In 1688 a set of embroidered panels was sold to the town council of Aberdeen 'for the decorement of the King's loft in Nicholas kirk in dayes of Solemnitie'. Four of them are preserved in the vestibule of the West Church. They are thought to have been the work of Mary Jamesone, the daughter of the painter George Jamesone, and depict the Finding of Moses, Jephthah and his daughter, Esther and Ahasuerus, and Susanna and the Elders.[14]

NOTES

1. For an in-depth analysis of furnishings in the medieval Church, see Fawcett, R. *Scottish Medieval Churches: Architecture and Furnishings*, Stroud, 2002.
2. Spicer, 2003, 422.
3. Hay, G. Scottish post Reformation church furniture, *PSAS*, 88 (1954–56), 53.
4. Black, W G. *A Handbook of the Parochial Ecclesiastical Law of Scotland*, Edinburgh, 1888, 62–3.
5. *Minutes of the Kirk Session of Kinneff 1733–48*, NAS CH 2 / 218.
6. McMillan, 1931, 156.
7. Brown, 2001, 4.
8. Black, 1888, 63–4, 146–7.
9. McCrie, 1892, 63n.
10. Quoted in Bieler, 1965, 92.
11. Todd, 2001, 7.
12. Burns, 1905, 171.
13. Thoms, 1972, 67, 99.
14. Hay, 1957, 222.

BIBLIOGRAPHY

Black, W G. *A Handbook of the Parochial Ecclesiastical Law of Scotland*, Edinburgh, 1888.
Bieler, A. *Architecture in Worship*, Edinburgh and London, 1965.
Brown, R F. Queen's Cross Parish Church, Aberdeen – Sanctuary Alterations, *Record of the Church Service Society*, 37 (2001).
Burns, T. *Church Property*, Edinburgh, 1905.
Cameron, J K, ed. *The First Book of Discipline*, Edinburgh, 1972.
Hay, G. *The Architecture of Scottish post-Reformation Churches 1560–1843*, Oxford, 1957.
Lindsay, I G. *The Scottish Parish Kirk*, Edinburgh, 1960.
McCrie, C G. *The Public Worship of Presbyterian Scotland*, Edinburgh and London, 1892.
McMillan, W. *The Worship of the Scottish Reformed Church 1550–1638*, Dunfermline, 1931.
Maxwell, W D. *Concerning Worship*, Oxford, 1948.
Maxwell, W D. *A History of Worship in the Church of Scotland*, Oxford, 1955.
Sefton, H R. Thomas Chalmers and the Lord's Table, *Record of the Church Service Society*, 28 (1995).
Spicer, A. Preaching, pews and reformed worship in Scotland 1560–1638, *History*, 88 (2003), 405–22.
Thoms, D B. *The Kirk of Brechin in the Seventeenth Century*, The Society of Friends of Brechin Cathedral, Brechin, 1972.
Todd, A S. Church furnishings, *Record of the Church Service Society*, 37 (2001).

30 Music, Church and People

DOUGLAS GALBRAITH

More than any other form, it is the psalm which is identified with the music of the Church in Scotland. Psalms have also punctuated the progress of the history of the nation. Legend claims that a turning point in the conversion of the Picts to Christianity was when King Brude heard 'in amazement and fear' St Columba singing one of the 'royal' psalms, Psalm 45. It was with psalms and instruments that Mary Queen of Scots was greeted in 1561 when she arrived from France. John Durie, outspoken critic of James VI, returning from exile to his pulpit at St Giles' in 1582, was famously greeted at the Netherbow Port with the singing of Psalm 124 by a multitude who accompanied him up Edinburgh's Royal Mile, singing (in four parts, it is said) 'till heavin and erthe resonndit'. Covenanters at Drumclog (1679) defeated Claverhouse with voices upraised in Psalm 76 sung to the thunderous *Martyrs* ('There arrows of the bow He brake'). Victim of the same conflict, the 20 year old Margaret Wilson, tied to a stake in the Solway, defied the rising waves with Psalm 25 ('My sins and faults of youth/ Do Thou, O Lord, forget'). Much later, all feuding over, when the two great processions of the United Free and the Established Churches converged on the Royal Mile on their way to consummating the reunion of 1929 the crowd broke into a spontaneous rendering of Psalm 133 to *Eastgate* – 'Behold, how good a thing it is,/ and how becoming well,/ together such as brethren are/ in unity to dwell'.[1]

Such events were often as much to do with politics as religion but the psalms have spoken for the Scottish people in other more obviously 'secular' contexts. It is recorded that victorious Labour MPs, leaving to take up their seats at Westminster after the election of 1922, led their followers at Glasgow's St Enoch's station in the 'fervent singing of Covenanting psalms',[2] while at the opening of the Scottish Parliament in 1999 'A man's a man for a' that' was partnered by the singing of the 'Old Hundredth' (Psalm 100). Except in the case of Columba, these psalms were 'metrical', made into measured lines so that congregations used to ballads could sing them with ease. If the demands of metre sometimes challenged sense, nevertheless the language, in Sir Walter Scott's words, 'though homely, is plain, forcible, and intelligible, and very often possesses a rude sort of majesty'.[3]

THE EARLIER CENTURIES

The place of the psalms in national life far predated the Scottish Reformation of 1560. Psalms were the staple of both the devotional and the liturgical life of the Columban (Celtic) Church, in whose abbeys it would not be uncommon for two to three dozen psalms to be sung at the night offices on weekdays and up to seventy-four at weekends.[4] An outstanding ability to sing was attributed to numerous saints; the voice of St Columba was said by his biographer Adomnán to be capable of being heard a mile away. That they should relish the singing of psalms is not surprising given the strong artistic traditions of the Irish people from which these Scottish missionaries came. Their art works attest to an intermingling of druidic/bardic and biblical traditions. On the shaft of the Martin Cross outside Iona Abbey the figure of David singing to Saul is carved, while on the Dupplin Cross[5] David is portrayed on a throne like a bard. Monks were said to carry at their belts a small harp (*cruit*) as they tramped the glens, and it is possible that these also accompanied the office psalms in their communities.[6] Rock gongs, bull horns and quadrangular handbells (of which 19 survive), used for assembling the people and in ceremonies of healing, complete the register of sounds surrounding monastic communities and the life of the people in general.[7]

Very little information about church musical practice in this early period survives, and what information there is tends to be in Irish or Welsh sources. However, the medieval *Inchcolm Antiphoner* contains music which may have originated in those earlier centuries. As well as examples of the beautiful and restrained 'Gregorian chant', known throughout the Western Church, there are also chants associated with Columba to whom this Augustinian abbey on an island in the Firth of Forth was dedicated. These settings for an office for his feast day leap and soar with intensity and expressiveness, shaping the music to the rhymes of the text, calling for a wide vocal range, and more akin to the 'showy' Ambrosian chant than to the more disciplined Gregorian tones. A feature of the music is the repetition of individual phrases, suggesting a delight in music for its own sake.[8] Here is music to match the kind of sophisticated designs found on Celtic parchment and stone.

With the marriage of Malcolm III to the Anglo-Saxon princess Margaret, Irish influence on the Church gave way to English and then to French. A greater consistency of practice throughout the country followed from the establishment of monastic houses and the building of cathedrals and churches, Anglo-French in style and skill. The majority of the cathedrals adopted 'Salisbury (or Sarum) Use', the liturgical pattern local to that cathedral, but heavily indigenised through the incorporation of Scottish saints (no less than fifty-five in an Aberdeen source).[9] The Sarum use was a variant of the great Gallican family of liturgies which differed from the Roman use in that their practices were more colourful, more active, more eloquent in gesture, more elaborate in dress and more magnificent in procession.

Scotland was far from being on the receiving end only in these developments. Just as the *Inchcolm Antiphoner* married indigenous composition with settings found in manuscripts furth of Scotland, so also did the manuscript now known as W_1 (from the continental library of Wolfenbüttel, to which it was removed in 1553). The Wolfenbüttel manuscript, possibly compiled and written in St Andrews in the thirteenth century, providing as it does early examples of part music, is of international significance. The development of polyphony, which resulted from the ability to write and sing in independent parts, has been described as 'the greatest evolutionary shift in the history of music'.[10] At the core of the collection are examples of the revolutionary music of the French Nôtre Dame composers. Of the settings likely to have been written in Scotland, one, *Vir Perfecte*, is composed in honour of Scotland's patron saint. It is like the other examples of 'organum' in the manuscript but has a more elaborate, quickly moving and rhythmic top part. Some believe that the compiler of W_1 was the Scottish Dominican monk Simon Tailler, whose four treatises on musical theory were in circulation up until the seventeenth century.[11]

Behind this lively collection was a healthy, country-wide, church musical establishment. The provision for the collegiate church of St Salvator at St Andrews (still in use as the university chapel) was typical. As well as

Figure 30.1 Panel of the Dupplin Cross showing David the Psalmist as a Celtic bard with a harp. © Crown copyright: RCAHMS.

Figure 30.2 Inchcolm Abbey, Firth of Forth, in whose repertoire examples of early Celtic chant were preserved. © Crown copyright: RCAHMS.

the founding 'college' of poor clerks/students, priests and three other teachers/clerics, there were some thirty 'chaplains' to service the altars but who also had to be 'trained in plainsong' so that they could maintain the daily offices and provide music for the Sunday masses.[12] These chaplains might have had their early training in the local 'sang schuil', a feature of all towns of a certain size, whose curriculum was broader than music and from which many were able to pass to the universities. An engaging contemporary account of the choir at Dunkeld Cathedral is found in Alexander Myln's *Vitae Dunkeldensis Ecclesiae Episcoporum* (c1517) from which it is clear that musical establishments were by no means made up only of clerical members. 'Pillars of the choir' (*sic*) here included advocates and notaries as well as trained musicians, chaplains and vicars. Their human and personal skills are also noted, in counselling, gardening, the mending of furniture, the repair of houses (one of the vicars), a command of Gaelic, generosity towards the poor – and Myln notes with approval more than one member as giving an honoured place to his mother in his household.[13]

This lively activity in the churches was paralleled by the musical life of the court. The Stewart dynasty from James I onwards produced monarchs of some musical skill and discernment (including James V, reports of his 'rawky voice' notwithstanding[14]). Music surrounded the life of the court and travelled with it. When in 1538, for example, James V married Marie de Lorraine, after the nuptials when there was a mass and playing on the organs, there were pageants devised by Sir David Lyndsay, with jousting and hunting, banquets and 'gritt mirriness', and it is recorded that at dinner there was 'great mirth of shalmes, trumpettis and diverss utheris instrumentis' for the rest of the day.[15]

CALLS FOR REFORM

Nevertheless, there was a growing discomfort about some developments in the Church. Bishop Kennedy had founded St Salvator's College in 1450 out of a concern about the spiritual decline, as he saw it, among the secular clergy.[16] Sir David Lyndsay's *Ane Satyre of the Thrie Estaitis*, with one of its themes the need for reform in the Church, offers a portrayal of a parson boasting of his prowess at the card table and on the football field. The Church Reform Act of 1541 spoke of 'unhonestie and misreule of kirkmen baith in witt, knawlege and maneris'.[17] The criticism was also made that the music had become too interesting, that it appealed to the carnal rather than the spiritual side of our nature. Perhaps the critics may have had in mind the popular observances which nourished the piety of the population – like the 'Holy Blood' or the 'Five Wounds', or local versions of the liturgical dramas that were known in continental cathedrals where biblical scenes were enacted with costume and special music.[18] Certainly, the inventory of the collegiate church of St Salvator contained 'ane mytyr for sant innocentis beschop', the mitre for the boy bishop who would lead the revels of the choir boys on Holy Innocents' Day (28 December),[19] while Fowlis Easter had a procession of children to the altar the day before. More likely than this much needed turning of authority on its head, however, was the growing elaboration of liturgical music. Among its critics was Robert Richardson, who writing about 1530 attacked the 'frivolity, vanity and delight' of contemporary settings. He was concerned not just by the use of instruments, but by composers' obscuring the text by complex music,[20] charges that were later to be echoed by the reformers in Scotland and by the Council of Trent and others in Europe generally.

One of these composers would undoubtedly have been Robert Carvor,[21] the full glory of whose music (known to contemporaries as 'musick fyne') has only relatively recently been revealed. Carvor was possibly a canon of Scone and of the Chapel Royal (then at Stirling Castle), and lived c1487 until c1566. It was a time when there were strong connections between Scotland and Flanders – as evidenced, for example, in the Edinburgh Burgh Records, which laid down that processions are to be mustered 'lyk as thai haf in the towne of Bruges and siclyk gud townes' – and it was there that Scots went to receive the final polish on their musical skills. Carvor's sumptuous style had echoes of such as Dufay as well as of the English composers of the Eton Choirbook, but some have heard in his music also characteristic features of the popular music of his native country. Yet, complex as it was, he could not be accused of mere display. His immense nineteen-part motet *O bone Jesu* may use many voices, but the way the music is made to flower

Figure 30.3 A Crear McCartney window in St Michael's, Linlithgow, showing the panel (inset) that incorporates the title of the Carvor composition which was an inspiration for the design. Photograph by Iain Walker.

on each invocation of the name 'Jesu' reaches deep into the soul. His great ten-part Mass, *Dum sacrum mysterium*, first performed either on the Feast of St Michael (29 September) or at the coronation of the infant James V after his father's death at Flodden in 1513, symbolises the nine orders of angels but also, by the tenth, a restored humankind. This Mass was the constant companion of the Scottish stained-glass artist, Crear McCartney, as he designed and made the 750th anniversary window (1992) in the medieval St Michael's Church, Linlithgow, an inspiration acknowledged by the inclusion of the Latin *incipit* round the central St Andrew's Cross (Fig 30.3). Another mass written in honour of St Michael, and in nine parts, was by Patrick Hamilton, martyred in 1528, but it does not survive. It appears that in burning a heretic, the religious authorities of the day were also depriving the nation of a promising composer. Others, such as the gifted Robert Johnson, chose voluntary exile.

Carvor's music is of interest in another way also. The music the Church has used for worship, in spite of official pronouncements, has always found ways of assimilating the music of the common life. Carvor was typical of the times in employing a fragment of a plainsong melody or popular song as the principal building block of a work. One of his masses uses the ubiquitous *L'homme armé* in this way. English composers might use the love song *Westron Wynde*[22] – 'Westron wynde, when wilt thou blow, the small raine down can raine. Christ, if my love were in my armes and I in

Figure 30.4 Initials outside the University Chapel at St Andrews marking the site of the martyrdom of Patrick Hamilton, a Reformer who was also a composer of a nine-part Mass setting. Photograph by James A Whyte.

my bedde again!' – and one would not be surprised, even given that the melody was fairly well buried, if these references did not somehow stir the subconscious and allow the mass to 'speak' to the human situation.

THE REFORMATION

Another use of popular song is found in the collection known as the Gude and Godlie Ballatis, an extraordinary anthology which challenges all the stereotypes of dour reformers stifling the natural impulses for music and dance.[23] The anthology, whose full title was *Ane Compendius Buik of Godly and Spirituall Sangis Collectit out of Sundry Partes of the Scripture*, contained, as well as psalms and Lutheran hymns translated into Scots, 'sundry uther Ballatis changeit out of prophaine sangis'. This practice, known as *contrafacta*, rendered the same music serviceable for different contexts or enabled the enjoyment of melodies by those who had renounced the world, without the need to utter their secular and often bawdy words. The purpose here, however, is different.

This collection seems to have been in circulation in the years leading up to the Reformation and had offered one means of familiarising people with the teachings of the continental reformers. No edition survives with melodies, but it is often clear which were intended, whether from the repertoire of Scottish courtly love songs or popular ballad. Thus the new/old doctrine of justification by faith was given currency through the popular love song, 'Johne cum kis me now (and mak nay mair adow)', which had also provided the 'ground' for many a variation by English keyboard composers. Here the parodist derives additional mileage for his message from the original song narrative, leaving the chorus unaltered: the exasperated maiden becomes a passionate God appealing to humankind; 'Johne representit man', as the song is careful to explain; and the 'kis' is the kiss of faith (rather than 'works' or the purchase of indulgences) by which we are restored to a relationship with God. A considerable number of verses is required to put across this teaching. The same doctrine is found expressed in a version of the many 'night visit' songs current at the time, 'Go from my windo', where again the maiden in her attic is God in heaven, calling on humankind to depart from unscriptural manipulative practices and to come round instead to the open door of faith. The book, which continued in publication until 1621, also included 'a table to show when the moon ryseth and when she shineth', essential information for the planning of evening meetings or services in those unlit times.

The Reformation in Scotland, unlike in England, was achieved with very little violence. In terms of music, more final than any deliberate destruction of artefacts and manuscripts through excess of zeal was the critique, and subsequent dismantling, of the structures which had supported music in the cathedrals and collegiate churches, which were condemned as channelling resources away from the local parish and its needs. At the same time, the fashion for simplified, more tightly constructed music was already spreading from the continent. Most important of all, however, were the Reformation principles that worship be in the language of the people rather than in Latin, that congregations should not be spectators but participate in the liturgy through singing, and that the content of worship should be thoroughly biblical. This recipe led to the establishment of the metrical psalm which, after centuries of dominance, provided the model for the hymn.

Very early in the Reformation period (1564) a service book-cum-psalter was published, containing all 150 psalms in several metres and with their own tunes, the melody only being printed. Soon after, parliament passed an act which required households of substance to possess a 'bible and psalme buke in vulgare language'. The contents of this psalter and its successors were to become part of the national consciousness and in the centuries to follow few would not know *'Dundee's* wild warbling rise' or 'plaintive *Martyrs,* worthy of the name' or count *Elgin* as 'the sweetest far of Scotia's holy lays' (Robert Burns, *The Cottar's Saturday Night*). The dignity and solemnity of these melodies were ensured by the long 'gathering note' at the beginning and ending of each line. Some of the tunes are not found elsewhere and were probably Scottish compositions. When parts were sung, the 'church part' was in the tenor, a practice found also in secular song. Many melodies derived their appeal from their being written in the old ecclesiastical 'modes', like the Dorian or Phrygian. However, the irrepressibility of musicians and their determination to employ their art despite what the authorities of the day might say issued in another type of setting, the 'psalm in reports', where voices entered imitatively and in sequence, creating a pleasingly complex texture with a certain rough dignity.[24]

These changes naturally caused distress to those who loved music and who saw this withdrawal of the Church as presaging the demise of music as a living art in Scotland. It was on behalf of these that Thomas Wode spoke when he wrote: 'I cannot understand, bot musike sall pereishe in this land alutterlye'. Nevertheless a well-known composer, David Peebles, was independently commissioned to provide parts for other voices, not elaborate or 'curious' but plain and 'dulce'. Wode collected these in the *St Andrews Psalter*, which appeared only two years after the first melody-only 'official' book. More settings, together with canticles, motets, anthems, and some instrumental compositions, were added over the next thirty years. Andro Blackhall, John Angus and Andro Kemp were also contributors.[25] Perhaps the fact that the crowd welcoming Durie in 1582 was able to sing in four parts was a result of this initiative.

One of the striking differences between the new Scottish psalters and their English counterparts was the greater variety of metre used in the former – no less than thirty in number.[26] However, by the time of the influential 1650 psalter (which was published without tunes), most of the versions were in the ballad 'common metre' with only two other metres included. It had been too much to expect congregations to learn the 105 tunes of the first psalter, and in so many metres, and by 1615 a body of twelve 'common tunes' was recognised, each of which could do duty for a number

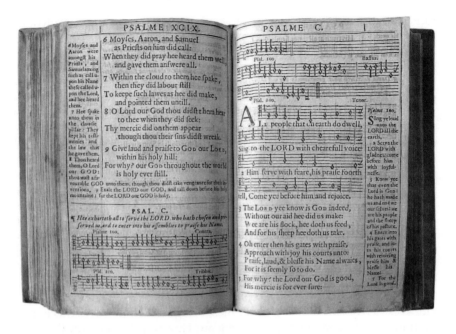

Figure 30.5 Pages of the 1635 Psalter showing the four parts of the Old Hundredth, with the melody in the tenor. Courtesy of the Church of Scotland.

of psalms. Although by the time of the famous 1635 psalter the number of these common metre tunes had risen to thirty-one, as time went on the number in actual use dwindled greatly.

The health of the music of Church and nation was not helped by the departure of the court to London and the consequent neglect and ultimate demise of the Chapel Royal, nor by the political and religious turmoil of the second half of the seventeenth century. Various attempts were made by the civil and religious authorities to maintain standards of church music. An early example was the Act of Parliament of 1579, which proposed a 'tymous remeid' through the re-establishment of song schools with qualified masters in the main burghs.[27] This had its effect and there is evidence of a resulting re-emergence of choirs to sing parts in church and to lead the singing. In Aberdeen, for example, a 'double seat' was built in the parish church for song school pupils,[28] and in Stirling in 1621 the pulpit and lectern were remodelled so that 'commodious seating' might be provided for 'the maister of the sang schuil and his bairnis'.[29] These and later measures, however, did not have a lasting effect.

SACRED AND SECULAR

The depth of the tedium into which church music had fallen, and the irritation it caused among worshippers, is evidenced by the fact that when reform came it happened virtually overnight.[30] Revival seems to have started in Monymusk in Aberdeenshire in whose parish church Sir Archibald Grant had in 1748 formed a choir. His pioneering work with the local choir came to have national significance through the appointment as choirmaster of Thomas Channon, an English soldier based at Aberdeen. The instructions given to him are evidence of the kind of problems being experienced at the time. Channon was to recover the old and introduce the best of the new. He was to teach the tune without excessive ornamentation. Harmonies were to add grandeur and solemnity. Under his baton, a craze for choir singing swept Aberdeen, culminating in a demonstration one Sunday morning in 1755 in the kirk of St Nicholas. In spite of being condemned as 'a newfangled profanation' and rated by one commentator as a worse disaster than the 1745 rebellion,[31] the new ways were warmly and widely embraced. Such was the enthusiasm that it became common for 'singing lofts' to be installed in churches to house the burgeoning number of choir members[32] (in Huntly three, to house separately those who sang the melody, the tenor and the bass). It was clear that many found the new tempi undignified – thus Burns' wry instruction in The Ordination to 'skirl up "the Bangor"'.

Soon afterwards, eight 'classes' in psalm singing were established in Edinburgh, regulated by a committee of the great and the good who also sanctioned any new tunes sung in the parish kirks. For its part, Glasgow opened a 'free school ... to encourage and promote the improvement of church music'. It is noteworthy that the teachers appointed by each city came respectively from Durham and Manchester. This enthusiasm for learning about and improving church music was very much of its time. Although the eighteenth century had begun under the shadow of intolerance, with trials for heresy and witchcraft and the purging from parishes of those who could not embrace Presbyterianism, the strong system of schooling and the progressiveness of the universities had contributed to what was later dubbed 'the Scottish Enlightenment'. With several of its pioneers coming from Scotland's manses, this movement broke new ground in law, sociology and philosophy, while at the same time producing the classic order and symmetry of Edinburgh's New Town.[33] At the time of Channon's demonstrations, universities were reaching out into their communities with popular lectures to eager audiences.[34]

In spite of the intermingling in practice of church and culture, there has been at times a strong sense of separation between what took place in church and what in the common life of the people. Suspicion tended to linger that music was more at home with the profane side of human nature, and there are many examples of kirk sessions (the body of oversight in each parish church) attempting to regulate local music making. A later example was when violin-player Dall Munro, a Skye catechist in the early 1800s, who had been accustomed to taking his own instrument with him on his travels, was converted by revivalist preachers and successfully persuaded the people of Skye to make a bonfire of their fiddles.[35] Against this must be balanced the considerable number of manses which made significant contributions to the encouragement of music and song. The compilers of the William Stirling part-book which contained, as well as psalms, Scottish courtly dances and vocal music, have been identified as the minister of Ancrum (d. 1685) and his father, minister of Baldernock,[36] while the Rev. Patrick Macdonald of Kilmore and his brother – their father, minister at Durness, was a prolific composer – published the first collection of Highland airs (1784).[37] Even in those parts where more puritan attitudes prevailed it was clear that it was possible for the fullness of the folk tradition to survive and flourish, evidenced by the plentiful harvest of songs gathered by collectors in the 1940s and 1950s: 200 songs from the natives of the Catholic islands of Barra and Eriskay (total population of 2,500), but just as many noted down from the Presbyterian-dominated island of Raasay (population of 300).[38] Either the Church was less influential than believed or, more likely, accommodations were made in practice.

One place where the separation between the holy and profane persisted was in the weekly choir practice where, throughout the eighteenth century, it was deemed unsuitable to sing the sacred words. The 'practice verses' which resulted ranged from the moralistic:

> Our bodies are but brickle barks
> which sweem the seas of fame;
> and if by sloth we miss our aim,
> we'll sink in seas of shame

to the musical:

> Come, let us sing the tune of French,
> the second measure low;
> the third ascendeth very high,
> the fourth doth downward go.

Clearly some verses were composed on the way home from choir practice, or later on as the ceilidh gained momentum:

> O Bangor's notes are unco high
> an' try the lassies sair;
> they pech an' grane an' skirl an' screech,
> till they can sing nae mair.[39]

The opposite could also happen, and in cases where piety triumphed over pleasure secular songs were given religious words, as when in the following century in some places the traditional words of the waulking songs of the Highland women stretching the tweed were replaced by more godly texts.[40]

Church and culture overlapped also in that music whose origins were in worship also served as the currency of social gatherings. Black's *Lessons on the Psalms* would have been used in the song schools for instrumental tuition but they would also have been heard in the drawing room.[41] It was not uncommon to find them in the manuscript collections of the seventeenth century, typically with three lively melodies embellishing the psalm tune, cheek by jowl with 'musick fyne', courtly songs, continental dances, and popular (even at times bawdy) songs.[42] This unselfconscious mixing of repertoires continued in later gatherings which clearly combined a religious with a social purpose. Groups who met to practise psalms were in the habit of letting their hair down with a healthy diet of 'canons, airs, and catches',[43] often appended to the psalm books. Bodies such as the Sacred Music Institution founded in Glasgow in 1796 (and meeting in Mrs Lamont's tavern!) and the Union of Precentors (Glasgow, 1811) not only would widen the Sunday repertoire but would enable people to hear and to sing the more substantial works of Handel, Pergolesi, Haydn, Beethoven and others.

Another feature of the eighteenth-century soundscape, both sacred and secular, was the upsurge of interest in 'Scottishness', dating from the surrender of a separate Scottish Parliament in 1707 and intensified by the events surrounding the Jacobite uprising of 1745. It has been suggested that at a time of massive change, 'primitive' societies were valued as preserving virtues the new society was deemed to have lost.[44] Spurious though much of these sentiments were, it was at this time that valuable collections were made of the songs which had throughout co-existed with the psalms. It was no doubt inspired by such as William Thomson's *Orpheus Caledonius* (1725, 1733) that some church music editors saw in national song a means of adding to the rather restricted repertoire as well as of promoting the influence and edification of religion. As early as 1683, William Geddes's *Saints' Recreation* had found potential new psalm tunes in such as 'The bonny broom' and 'We'll all go pull the heather'.[45] Over a century later, in their *Christian Songs Versified for the Help of the Memory*, the Glasite sect made use of the melodies of such songs as 'Bonny Jean', 'The Birks o' Invermay' and 'Gala Water'.[46]

In the nineteenth century, church and culture remained closely interrelated, and the influence of religious values was strong. Many initiatives of a social or political nature even borrowed their structures from the church, and sometimes its accompaniments. The Chartist movement, which had a deeply religious ethos in Scotland, organised itself at one point into 'churches',[47] while the Rechabites and other temperance movements found in specially written 'hymns' on the evils of drink a powerful way of affirming and propagating their message. Later, the socialist movement was to have its own version of the Sunday Schools, incorporating a ceremony very like christening,[48] as well as its choirs (the Glasgow Orpheus Choir, famous in the 1930s for its renderings of 'Crimond' and 'All in an April Evening', was one).

ORGANS AND PRECENTORS

The nineteenth century also brought changes in the way vocal music was accompanied, although the Episcopal Church had begun to install organs at least as early as 1730.[49] Organs in Scotland are recorded from 1437 and it seems that by the time of the Reformation their use in church and court was widespread. Although the use of the organ in worship – and to a great extent the instruments themselves – did not survive these changes, the reason, at least in the earliest period, may have been less deliberate destruction than that new patterns had made them redundant. Prior to the Reformation, organs had been used not so much to accompany singing as to embellish and decorate worship,[50] and were usually played not by a permanent organist but by one of the singers as required. Possibly to have used organs to accompany congregational singing would have been a new use of the instrument but this was either not thought of or the opportunity not taken.[51] If at the time of the Reformation no specific position was taken (it would seem) over organs, certainly during the tense situation in the second half of the seventeenth century, organs, along with other reminders of Anglican or Roman Catholic ways, such as statues, vestments and the observance of holy days ('ceremonies'), were specifically targeted. Typical was the proposal of the kirk session of Holyroodhouse (1643) that the organ now lying idle – 'an unprofitable instrument, scandalous to our profession' – be sold and the proceeds given to the poor.[52]

Thus, although accompanied psalm singing seems to have continued outside worship (as at the arrival of Mary, Queen of Scots), no instruments were to be used in worship (confirmed by the General Assembly of 1638) and the psalms were led by the 'uptakers', later called 'raisers' and then 'precentors'. These selected the tune and led the singing. Precentors were often masters in the song school, sometimes school master, sometimes also session clerk, but in times of the decline of the importance given to music they might also double as grave digger or beadle. In the main they were figures of great authority, and inevitably stories grew up round them. One precentor, who prided himself on his memory and eschewed the guidance of books, is recorded as having begun Psalm 107, which includes vivid scenes of a storm at sea, but before the psalm was ended found himself leading the congregation in the ballad of Sir Patrick Spens lying 50 fathoms deep beneath the waves 'with the Scots Lords at his feet'.[53]

Even if there were some precentors who were incompetent, who only knew one tune, or had but an indifferent voice, many more steadily ensured that the psalm was 'raised' and some made outstanding contributions to both Church and society. Three nineteenth-century precentors are commemorated on a plaque on Edinburgh's Calton Hill (Fig. 30.6). The fact that two of them became famous opera singers and one was in demand for his concerts of Scottish song on both sides of the Atlantic is another example of the fruitful relationship between church music and culture. These had their equals among those who continued to make their careers in the Church. Notable was R A Smith (d. 1829), precentor first of Paisley Abbey and then of St George's, Edinburgh, whose choir practices were so popular that observers had to have tickets of admission. In Edinburgh, with the support of his ministerial colleague Dr Andrew Thomson, composer of the famous tune to 'Ye gates, lift up your heads on high' and father of the first Reid

Figure 30.6 This plaque at the foot of Calton Hill, Edinburgh, gives no clue that it commemorates three famous Scottish professional singers who began their careers as precentors in Edinburgh churches. Photograph by the author.

Professor of Music at the University, Smith brought his choir to a pitch of technical perfection and greatly expanded the possible musical repertoire in worship.[54]

Smith's refining of choral singing was soon extended, of necessity, to that of congregations which were still at that time, according to the Scottish Guardian, the home of 'ignorance and bad taste ... Some shout at the utmost pitch of their voices – some wheeze with their breath – some sing through their nose, others with clenched teeth'. Thomas Hately (d. 1867), the Precentor of the Free Church General Assembly, attracted huge numbers to his classes round the country in the history and practice of psalm-singing, with nine hundred recorded on one occasion in Greenock. In the north east, William Carnie drew some two thousand to a lecture on psalmody, resulting in the formation, in 1870, of the Choir of the Thousand Voices which held a weekly practice of psalms, hymns and anthems. The desire of these reformers was not only to improve the quality of singing but the quality of the music sung. Smith's book had included none of the earliest tunes. These later precentors, and particularly Joseph Mainzer, saw to it that the great melodies of the 1564 psalter were given their voice again, and at the same time the Rev. Neil Livingston published (1864) an edition of the most famous Scottish psalter of all, that of 1635. One tool in this renewal of church music which

Figure 30.7 The precentor's pulpit in St Martin's church near Perth is partially obscured by his successor, an organ, in leading the praise. © Crown copyright: RCAHMS.

swept the country was the 'tonic sol-fa' method, particularly associated with John Curwen, when each note was designated by its place in the scale, a system which remained one of the standard methods for choirs to read music within living memory, whether in church or school or burgh choir.[55]

One practice which died hard, that of the 'lining out' by the precentor of each line of the psalm, may only have entered the repertoire on the adoption of the Westminster Directory for Public Worship (1645), a document largely prepared by the dominant reforming wing in England but accepted by the General Assembly with modifications. Along with this went the custom of each individual ornamenting the melody in his/her own way. In time, this gave way to four-part harmony and to styles of singing which allowed the meaning of the text to prevail, but the musical and poetic forms of Gaelic culture, together perhaps with geographical isolation, has meant that this style has not only survived but flowered in the Highlands and Islands, producing a sound so exotic and striking that some have sought Coptic origins.[56]

The transition from precentor to organist did not go smoothly in the Presbyterian Churches, and a full-scale pamphlet war was still being waged as late as 1830 with such titles as 'A new stop to the organ' (against!). *The Psalmody of the Free Church of Scotland* 1845 frowned on the 'foreign aid of instruments or of a trained band [of singers]'. After the installation of a reed organ in Greyfriars', Edinburgh, in 1863, the mould was finally broken and organs became more common – and blessed by assemblies (the Church of Scotland in 1864, the United Presbyterians in 1872 and the Free Church in 1883).[57]

PSALMS AND HYMNS

From the earliest Reformation psalters, canticles (other passages from Scripture suitable for singing) had been printed alongside the psalms, but over a long period the feeling grew that 'the solemn praises of a New Testament Church are too much limited when confined entirely to ... Old Testament composures'.[58] Finally, in 1781, a collection of Translations and Paraphrases in Verse of Several Passages of Sacred Scripture was published for trial throughout the Church of Scotland. This was a step too far for some and a considerable proportion of the Corstorphine membership, decrying 'prelatic leanings', left to form their own congregation. Some of these paraphrases, such as 'O God of Bethel' with its couplet, 'Through each perplexing path of life/our wand'ring footsteps guide' (from Genesis), and 'Come let us to the Lord our God/ with contrite hearts return' (Hosea), touched the communal nerve and still appear with regularity in new hymn collections.

In this collection there were added five hymns that were not based on Scriptural passages, one by Isaac Watts whose earlier inspired reworkings of the psalms had allowed them to break free from the straitjacket of the Old Testament to create powerful statements of the Christian faith. It was inevitable that interest would grow in making new utterances in the mould of the psalms which would not only enshrine the doctrines of the New Testament but record the faith of contemporary Christians. Some denominations, like the Relief Church, the Episcopal Church and the Methodists, had been using 'freely composed' hymns for some time. Revivalist movements, such as the Moody and Sankey campaigns of the 1870s and the Tent Missions of the north-east, familiarised people with alternatives.[59] In the Gaelic regions, poetry and song on religious themes had persisted through the Reformation, but the growth of evangelicalism in the eighteenth century found in the hymn both a mode of biblical instruction and evangelical appeal.[60] Barrel organs, their heyday between 1750 and 1850, and generally imported from England, found new audiences for such tunes as 'Hemsley' ('Lo! he comes') and 'Miles Lane' ('All hail the power'). It was not, however, until the middle of the nineteenth century that the three main Presbyterian strands of the Church brought out their first hymn books.

Until then, variety was sought in other ways. The plethora of psalters, published as the eighteenth century gave way to the nineteenth, if they succeeded in their purpose of increasing the musical repertoire of the Church, also brought problems. As well as a multitude of psalm tune 'soundalikes' of little distinction, there developed a predilection for the ornate and the decorative. 'Repeater' tunes were hugely popular, where phrases were passed between pairs of voices or where lines were repeated to delay and enhance a resounding climax, too often with some sacrifice of sense. Oratorios were ransacked for melodies to be 'torn from their context and tortured into the shape of a psalm-tune'.[61] What was worse was that the original Scottish psalm tunes were largely forgotten. These collections were generally not officially approved but enjoyed great popularity. At the Union of 1929, referred to earlier, the tune sung so enthusiastically by the crowd (*Eastgate*) was a repeater tune which appeared in neither Church's approved psalter.

While the debates were raging over hymns and organs in the Presbyterian Churches, the Episcopal Church had long since returned to a richer musical fare in worship, with hymns, canticles, anthems and sung liturgies.[62] The recovery of choral singing in Roman Catholic services had also taken place, attracting many Protestants.[63] Churches were influenced by each other's practices. It comes as a surprise to many that the Church of Scotland, the Free Church and the United Presbyterian Church published prose psalters using Anglican chants round the turn of the twentieth century, the first of which in its preface commenting that 'the practice of chanting has become so general that little detailed explanation of its principles is necessary'. Other publications provide evidence also of the use of canticles and liturgical texts like the Sanctus and the Doxology. These volumes were often prepared for the use of single congregations, like R A Smith's *Sacred Harmony of the Church of Scotland* for St George's, Edinburgh (1828), and *The Scottish Book of Praise* for Park Parish Church, Glasgow (1876), which unusually printed all its material across two facing pages so that the text fell under the music.

Not everyone embraced hymns, and some branches of Presbyterianism still sing only psalms in church. Isabella MacLean recalls being warned, during a visit to a Free Presbyterian household around 1910, 'Pa doesn't mind songs, but he won't allow hymns', and also recalls the ceilidh where 'tent hymns' provided some of the fare and how a man 'who had been a precentor in the Free Kirk all his grown-up life and wouldn't have imperilled his soul by singing a hymn in church sang as heartily as the rest'.[64] It has been suggested, however, that the characteristic 'chanting' delivery of some ministers of these churches, found mainly in the Highlands and Islands, in prayer and preaching is itself a form of music.

THE TWENTIETH CENTURY

The twentieth century opened on a different musical landscape, one with a recognised place for the professional musician. The Glasgow Athenaeum School of Music (later the Royal Scottish Academy of Music and Drama) had opened in 1890, music degrees could be taken in Edinburgh University from 1893, Scotland was producing composers who were acclaimed at home and abroad. The links between church and culture continued. One of Scotland's leading composers of his time, Sir Alexander Mackenzie (d. 1935), from teaching in a Church-run education college, went on to become Principal of the Royal Academy of Music in London, while many of the tutors at the Athenaeum were also notable church musicians. This momentum has continued, assisted by the consistent promotion of Scottish music by the BBC and by such organisations as the McEwen Bequest, administered through the University of Glasgow and set up in memory of another notable Scottish composer, Sir John B McEwen (d. 1948), to promote the performance of chamber music.[65] This flowering was recognised in the Church Hymnary: Third Edition (1973) in its unique commissioning of tunes from some eight contemporary composers, native to or working in Scotland.

The second half of the century witnessed a 'hymn explosion', arising from movements which sought a greater interaction between the Christian faith and the common life of the world, when such matters as scientific discovery, the rhythms of working life, and care for the planet became the subjects of hymns. In this, the Dunblane Music Group, meeting in the ecumenical Scottish Churches' House in Dunblane, acted as a stimulus and a monitor with an influence throughout the United Kingdom. In the closing decades, the musicians and writers of the Iona Community, such as John L Bell and Graham Maule, began their production of hymns, songs and settings of short liturgical items, which were both biblically literate and rich in contemporary imagery and experience. A feature of these has been the recovery of traditional Scottish melodies (the mid part of the century had seen a revival of folk song following the work of such collectors as Marjorie Kennedy-Fraser – the daughter of one of the precentors commemorated on the Calton Hill.)

In 1998, in *Common Ground*, a hymn book was published which was prepared by all the Churches in Scotland together, the first time there has been a common collection since before the Reformation. Significantly, as well as local material, the book contained a proportion of songs from the 'world church', valued both as introducing Scottish congregations to the experience and insight of churches serving in radically different contexts and as helping to renew Scottish traditions of unaccompanied singing, including singing in parts. The strong presence of material from the Roman Catholic Church in the collection reflected that Church's response to guidance from the Second Vatican Council (1962–5) that more use should be made of local music and that there should be greater participation on the part of the congregations. This was not the only example of music building bridges between the Churches in a century noted for ecumenical initiatives. The Scottish Churches Organist Training Scheme (SCOTS) was established in 1997 by the Churches with the Scottish Federation of Organists and the Royal School of Church Music to train and encourage church musicians in their early stages.

The same church music explosion brought with it much writing in popular styles, not all of it as contemporary as was claimed. Some of it was good of its kind while much else lacked challenge in the content of both music and words, the latter tending to recall earlier formulations of a more

Figure 30.8 John L Bell of Iona, writer and composer of hymns for the twenty-first century. Photograph by Paul Turner.

individualistic kind out of kilter with the gospel-in-daily-life emphasis of the times, or consisting of simple biblical iterations rather different from the 'worked through' biblical insight of the traditional hymn writers. Commercially viable, these songs were widely disseminated in print and on compact disc. Discriminating use, however, and more imaginative musical arrangement enabled many of these to contribute well to the widespread search for worship which both had wider appeal and was better connected to the spirit of the times. The principle of 'disposability' came to be accepted in this period in which certain (indeed, most) new hymns would be seen as offering nourishment to the Church for a while before being allowed to give way to others. The supplementing of 'official' hymn books by such collections as *Songs of God's People*, with an intended life of ten years or so, was another expression of this principle.

Parallel to these developments was the rise in Christian rock bands, several of them Scottish, reaching their audiences mainly through concerts, recordings, and events like the Greenbelt music and arts festival rather than through worship. In the USA their umbrella organisation commanded 12 per cent of all compact discs sold, three times the size of the classical market.

Some saw themselves as providing alternatives to the regular rock scene, others sought to declare the Gospel message openly, while others set out to reflect on human society from the perspective of Christian faith. Characteristic of modern church song of all kinds was that it might be accompanied by instruments other than the organ, or by no instruments at all. Increasingly, the organ is being supplemented by other keyboard instruments, often electronic, and by groups of instruments, some in the particular form known as the 'praise band'. The eclectic nature of the repertoire was reflected also in the facilities provided by the electronic media. *HymnQuest* was only one of the systems which gathered together as many known hymns as possible so that people might be able to choose themes and styles as well as hear the melodies to which they are set.

While at the turn of the twenty-first century hymn singing was no longer central to people's lives in the way that the psalms once were, nevertheless hymns still had significance for people who were not regular churchgoers, for example, in offering reference points on communal occasions such as weddings or funerals. The very popular televised *Songs of Praise* programme, one of the world's longest-running television programmes ever, created a canon of hymns and religious songs which cut across denominational boundaries. Another BBC programme, on both radio and television, was the annual *Carols from King's* (King's College, Cambridge), evidence of the place Christmas carols retained in people's affection. Sometimes, in *Songs of Praise* programmes, 'secular' songs were included which had lyrics that made sense also in a religious context.[66] This cross-over between the secular and the sacred was seen also in the use of Elton John's 'Candle in the wind' at the funeral of Princess Diana in 1997. At funerals, when people did not have their own repertoire of hymns, such songs as 'My way' or 'My heart will go on', the love theme from the film *Titanic*, were often requested.

It is likely that the fourth edition of the *Church Hymnary* will not only restore the bulk of the psalms to the repertoire of the Church (the previous edition had included only a selection), offered in a variety of settings as well as metrical, but also propose a canon of hymns and songs which, while acknowledging popular taste, will match the psalms in the quality of their utterance and the depth of their insight. It is indicative of the sharing of repertoire between the Churches that it will also contain parts of mass settings intended for congregational participation and written by one of Scotland's leading composers, James MacMillan, himself a Catholic, whose concert works characteristically engage both with social and political issues while deriving inspiration from the music and text of the mass. At the close of the century, church music in Scotland was in as good a state of health as it ever has been.

NOTES

1. Kerr, 1886, 77; Prothero, 1903, 241; Glover, 1927, 33–4; Reith, 1933, 342.
2. Devine, 1999, 313.

3. Patrick, 1949, 213.
4. Preece, 2000, 43.
5. Now in St Serf's Parish Church, Dunning.
6. Preece, 2000, 46.
7. Purser, 1992, 22–8.
8. Purser, 1992, 42.
9. Preece, 2000, 55–7.
10. Preece, 2000, 49.
11. Preece, 2000, 29.
12. Cant, 1950, 26ff.
13. Preece, 2000, 84–5.
14. Farmer, 1947, 76.
15. McRoberts, 1976, 105.
16. Cant, 1950, 3.
17. Preece, 2000, 83.
18. Reese, 1940, 193ff.
19. Cant, 1950, 126.
20. Preece, 2000, 91–2.
21. A full account of his work is found in Preece, 2000, Part II and in Ross, 1993.
22. Farmer, 1947, 109.
23. Munro, G J. The Scottish Reformation and its consequences. In Preece, 2000, 274, 298.
24. See Elliott, 1960. There is an example also in Preece, 2000, 284.
25. Munro, 2000, 277.
26. Munro, 2000, 276.
27. *The Acts of the Parliament of Scotland*, III, 174, quoted in Patrick, 1949, 109.
28. *Burgh Accounts of Aberdeen*, quoted in McMillan, 1931, 82.
29. Patrick, 1949, 135.
30. Patrick, 1949, 150ff.
31. Farmer, 1947, 266; Patrick, 1949, 154.
32. Patrick, 1949, 160.
33. Devine, 1999, 64–6.
34. Devine, 1999, 80.
35. McCaughey, T P. Protestantism and Scottish Highland culture. In Mackey, 1995, 203.
36. Stell, 1999, 183ff.
37. McCaughey, 1995, 190.
38. MacLean, 1975, 103.
39. Patrick, 1949, 164ff.
40. McCaughey, 1995, 189.
41. Munro, 2000, 298.
42. Stell, 1999, *passim*.
43. Farmer, 1947, 269.
44. Devine, 1999, 242.
45. Farmer, 1947, 193.
46. Farmer, 1947, 271.
47. Devine, 1999, 279.
48. Smout, 1997, 262.
49. Farmer, 1947, 276.
50. Preece, 2000, 98.
51. Inglis, 1991, 33.
52. Inglis, 1991, 80.
53. Patrick, 1949, 129ff.
54. Patrick, 1949, 199ff.
55. Patrick, 1949, 198–9.

56. See, for example, Gaelic Psalms from Lewis, Scottish Tradition 6, CDTRAX 9006 (Greentrax), from the archives of the School of Scottish Studies, University of Edinburgh.
57. Farmer, 1947, 366ff.
58. Minute of the Presbytery of Paisley, 22 April 1747, quoted in Maclagan, 1889, 182.
59. MacLean, 1998, 76.
60. MacDonald, K D. Hymnology, Gaelic. In *DSCHT*, 424.
61. Patrick, 1949, 186.
62. Maclean, A. Episcopal worship. In Forrester and Murray, 1984, 109.
63. Dilworth, M. Roman Catholic worship. In Forrester and Murray, 1984, 125.
64. MacLean, 1998, 76.
65. Elliott and Rimmer, 1973, 70.
66. Barr, 2001, 108.

BIBLIOGRAPHY

Barr, Andrew. *Songs of Praise: The Nation's Favourite*, London, 2001.
Cant, R G. *The College of St Salvator*, St Andrews, 1950.
Collinson, Francis. *The Traditional and National Music of Scotland*, London, 1966.
Devine, T M. *The Scottish Nation 1700–2000*, London, 1999.
Elliott, K, Rimmer, F. *A History of Scottish Music*, London, 1973.
Elliott, K, ed. *Fourteen Psalm-Settings of the Early Reformed Church in Scotland*, London, 1960.
Farmer, H G. *A History of Music in Scotland*, London, 1947.
Forrester, D, Murray, D, eds. *Studies in the History of Worship in Scotland*, Edinburgh, 1984. A revised edition was published in 1996.
Glover, T R. *Saturday Papers*, London, 1927.
Inglis, Jim. *The Organ in Scotland before 1700*, Schagen, 1991.
Johnson, David. *Music and Society in Lowland Scotland in the Eighteenth Century*, London, 1972.
Kerr, John. *The Psalms in History and Biography*, Edinburgh, 1886.
Mackey, James P, ed. *An Introduction to Celtic Christianity*, Edinburgh, 1995.
Maclagan, Douglas. *The Scottish Paraphrases*, Edinburgh, 1889.
MacLean, C I. *The Highlands*, Inverness, 1975.
MacLean, Isabella G. *Your Father and I*, ed. C. MacLean, East Linton, 1998.
McMillan, W. *The Worship of the Scottish Reformed Church, 1550–1638*, London, 1931.
McRoberts, David. *The Medieval Church of St Andrews*, Glasgow, 1976.
Patrick, Millar. *Four Centuries of Scottish Psalmody*, London, 1949.
Preece, Isobel Woods. *Our Awin Scottis Use*, ed. S. Harper, Glasgow, 2000.
Prothero, R E. *The Psalms in Human Life*, London, 1903.
Purser, John. *Scotland's Music*, Edinburgh, 1992.
Reese, Gustave. *Music in the Middle Ages*, London, 1940.
Reith, George M. *Reminiscences of the United Free Church General Assembly*, Edinburgh, 1933.
Ross, D J. *Musick Fyne: Robert Carver and the Art of Music in Sixteenth-Century Scotland*, Edinburgh, 1993.
Smout, T C. *A Century of the Scottish People*, London, 1997.
Stell, Evelyn F. 'Sources of Scottish instrumental music 1603–1707', PhD thesis, University of Glasgow, 1999.

31 Christian Art in Scotland

MURDO MACDONALD

The importance of St Ninian in the development of Christianity in Scotland is echoed in some of the earliest Christian art available to us. From near the centre of his cult at Whithorn comes a cross incised on a roughly dressed upright slab of stone, known as the 'St Peter' Stone.[1] In form it is simple, clearly the product of a religious community not able, and indeed perhaps not willing, to call on great material resources. Yet its beauty of proportion and value as an object of contemplation gives us sight of the cultural and spiritual depth of the artist and his fellows. This elegant work from the seventh century, or perhaps earlier, is a worthy starting point for a consideration of Christian art in Scotland. By the time it was carved, Scotland was entering a remarkable period of aesthetic achievement, which would reflect what Arnold Toynbee called 'the originality of this far western Christendom'.

Thanks to the efforts of the followers of St Ninian, St Columba, and other pioneers of the new religion, from an artistic perspective Scotland was part of an early Christian cultural project which extended across the entire landmass of the modern country, as far north as Shetland and linking south-west into Ireland and south-east into England. While from a political point of view one might analyse this period in terms of struggles between Picts, Britons, Gaels and Northumbrians, from the perspective of Christian art one can see it in terms of exchange of ideas, techniques, influences and artists. It is when one notes a unity of spiritual purpose combined with an ethnic diversity that one can begin to understand why the Christian art of this period has such vitality. Consider, for example, the illuminated set of gospels known as the Book of Durrow, made in the late seventh century. It is not certain where it was made, but Iona – at the heart of the Gaelic-speaking continuum of western Scotland and Ireland – is high on the list of likely locations. What is known is that this illuminated manuscript was made within one of the great centres of the Church established by St Columba. Yet in style it can be linked with Pictish carved work, most particularly the seventh- or eighth-century Papil stone from Shetland (Fig. 31.1), which has as part of its design an animal which relates strongly to one drawn in the Book of Durrow. Similarly the interlace of the eighth-century Pictish cross slab at Glamis can be related to interlace found in the Lindisfarne Gospels, made in Northumbria at the Columban foundation of Lindisfarne at the end of the seventh century.

Figure 31.1 Papil Stone, Shetland. Courtesy of the Trustees of the National Museums of Scotland, NMS X.IB 46.

The Book of Durrow has been described as 'a compendium of European art'[2] and this comment draws attention to the international vision of these seventh-century artists. Sculpted works from the next century, notably the St Andrews Sarcophagus[3] and the Ruthwell Cross,[4] display a similar link to European art, although at the same time, the originality of the Pictish St Andrews Sarcophagus in particular, must be stressed.[5] The Ruthwell Cross adds another dimension to our understanding of this period. Unlike the Book of Durrow, which was made by Gaelic speakers writing in Latin, the Ruthwell Cross was made by Northumbrians and has a vernacular Anglo-Saxon text in runic script (see Fig. 15.1). This text quotes from the mystical poem, *The Dream of the Rood* (c700), which is now acknowledged as a key reference point for the beginnings of Scottish literature. Its sense of spiritual focus can be matched but not exceeded and it shows that from its beginnings Christian thought in Scotland was communicated at a high level of poetical art, in the local language. Seventh-century Gaelic poetry in praise of Columba, from the time of the Book of Durrow, underlines this.

The Book of Kells (c800), again most likely produced on Iona, can be noted as the high point of this period of art in Scotland. Carl Nordenfalk wrote of Lindisfarne and Iona as 'two poles between which an electric arc of unusual luminous intensity sprang to life'[6] and 'luminous intensity' is the best short description of the Book of Kells that one could wish for. This manuscript takes to its heart the opening of St John's Gospel, remaking the words *'in principio erat verbum'* into a work of art where meaning, belief and image unite (Fig. 31.2). From the same period one finds the great sculpted crosses of Iona, which remind one again of the expertise, both technical and artistic, of stone carvers in Scotland at this time, an expertise which stretched from Iona to St Andrews and from Ruthwell to Shetland.[7]

This unified period of Celtic and Anglo-Saxon Christian art came to a close at the time of the making of the Book of Kells. Viking raiders sacked Lindisfarne in 793, ushering in a period during which no religious foundation was safe. Such disruption resulted in Iona being abandoned as the centre of the Columban *familia* in 849 and the movement of works of art for safekeeping, both to mainland Scotland and to Ireland. A work which may well have been one of the treasures that was moved from Iona at that time is the Monymusk Reliquary,[8] or *Brec Bennoch* of St Columba, a numinous work which is thought to have held a relic of the saint himself. It is a wooden box clad in decorative metalwork of silver, copper alloy and bronze, and this metalwork has been identified as Pictish. Some six hundred years after it was made, it is reputed to have been worn about the neck of a cleric at the battle of Bannockburn in 1314.

This period saw the decline of the art of the Celtic Church and new impetus and direction given to the development of Christian art in Scotland, linking directly with contemporary European developments, through the influence of the Anglo-Saxon Queen Margaret (c1046–93) and her Anglo-Norman-inclined sons, in particular the English-educated David I (c1080–1153). Frustratingly little work survives, but Margaret herself may

Figure 31.2 Book of Kells, opening page of St John's Gospel, c800. Courtesy of The Board of Trinity College Dublin, TCD MS 58 fol. 282r.

well have been the owner of the so-called Celtic Psalter,[9] an Irish or Scottish work from the tenth or eleventh century, which refers back in its decoration to the high period of Celtic and Anglo-Saxon illumination.[10] Another notable survival that relates back to this earlier tradition is the Book of Deer (see Fig. 2.7),[11] made at the Aberdeenshire monastery of the same name in the ninth or tenth century. In its mannered development of figures it echoes the style of figurative carving on contemporary stone crosses in Scotland and Ireland. As Viking raids gave way to Viking settlement and integration, significant influence on art came from this quarter also. In the northern isles in due course this combined with Anglo-Norman influence in the remarkable structure of the cathedral of St Magnus in Kirkwall. In the Highlands and Islands one finds further evidence of the piety of this Norse kingdom in the care with which the bishops of the Lewis chess pieces (twelfth century) are represented.[12]

The introduction of Anglo-Norman influence into the Scottish Church was expressed in grand form in the Romanesque and subsequently Gothic architecture of abbeys and cathedrals throughout the land. Although often ruined or heavily restored, these buildings testify to this achievement. From a post-Reformation viewpoint it is hard to conceive of Scotland as a land rich in the Christian art which would have been the natural accompaniment of that architecture. Yet, as John Higgitt has put it, 'the medieval Church in

Scotland was comparable with the Church elsewhere in Europe in its wealth of ornament and imagery'.[13] Such wealth is hinted at by the surviving tracery of, for example, Melrose Abbey and St Michael's church, Linlithgow. Religious art, including stained glass and illuminated manuscripts, would have been a commonplace. Indications of what once was can be found in a variety of media. An illuminated initial of the charter of Kelso Abbey, dating from 1159, not long after the death of David I, is skilfully executed. Other twelfth-century works include the notable bell shrines of west Highland origin in copper alloy.[14] Further evidence of fine metalwork is found in seal matrices such as the thirteenth-century brass seal matrix of Brechin Cathedral, which displays artistry and workmanship of a high order. International links are underlined by the remarkable illuminated manuscript, the Murthly Hours, made in Paris probably in the 1280s.

Pilgrims' metal badges and tokens have also been recovered, reminding one of the role which local and foreign shrines played in the religious life of Scotland. A recent, illuminating book on the subject, laments that 'these marvellous shrines ... were almost completely destroyed within a short period of time around 1559 and 1560, due to the reforming zeal of the Protestant cause'.[15] The author nevertheless does a remarkable job in bringing alive to the reader, not just a sense of what has been lost in artistic terms, but the wider sense of the role of that art for the pilgrim. It helps one to understand pre-Reformation Christianity in Scotland as having at its heart a network of pilgrimage routes from Whithorn in the south and Iona in the west to St Andrews in the east and Kirkwall in the north. This gives one's appreciation of surviving works of art a different dimension, for many of them were not just for the edification of the local congregation, but also for the enlightenment of the pilgrim. This then is the context of, for example, the fifteenth-century sculpted frieze of the life of St Miren in Paisley Abbey; or of the *Coigrich*, the silver-gilt crozier shrine of St Fillan of Glendochart, which is a fine example of fifteenth-century west Highland work.

Figurative sculpture survives in an architectural context from as early as the twelfth century, for example around the Romanesque doorway of the church at Dalmeny near Edinburgh. Examples surviving from the thirteenth century include a group of bosses carved with faces and foliage in Glasgow Cathedral and a beautiful head of Christ from St Andrews. The latter is all that remains of a statue, and in the context of that ruined cathedral is a poignant reminder of what was lost during the Reformation. The fifteenth-century carvings at Rosslyn chapel display the virtuosity of Scottish architectural masons. Very few carvings survive, however, that can be thought of as the kind of independent works of art that one can take for granted in many parts of continental Europe. A heavily damaged, but still notable, fifteenth-century sculpture of virgin and child at Melrose Abbey is a rare survival. Probably the work of a French or Flemish sculptor, it emphasises again the international links of Scottish art at this time. Some effigies also survive from this period, for example those from a tomb in the old church at Houston, Renfrewshire, which are believed to represent Sir Patrick

Houston and his wife Agnes Campbell and – notable for its quality – the head of Bishop Wardlaw's effigy at St Andrews. Further indication of a vigorous tradition of stone carving during the fifteenth and sixteenth centuries can be found in the decoration of sacrament houses, for example at St Marnock's church at Fowlis Easter near Dundee, built in the mid-fifteenth century. A later example of fine quality, dating from 1551, can be found at Deskford in Banffshire. A number of large panels in the church at Fowlis give us a rare insight into pre-Reformation painting in Scotland. A broadly executed but sensitively conceived crucifixion scene (Fig. 31.3) contrasts with a mystical vision of the risen Christ, and both images have a clear resonance with contemporary Flemish art. The Fowlis works were made in the second half of the fifteenth century, and they only survive because they were painted over during the Reformation (see also Figs. 2.6, 3.1). Indeed, an instruction for this over-painting is vividly conveyed to us in the Minutes of the Synod of Fife in May 1612 in which it was ordained 'that the paintrie quhilk is upon the pulpitt and ruid-laft, being monuments of idolatrie, sal be obliterate be laying it over with grein color'.[16]

Other reminders of the close link to Flanders are provided by survivals such as the Flemish book of hours of Brechin Cathedral,[17] and a

Figure 31.3 Detail of the crucifixion painting. Fowlis Easter collegiate church, second half of the fifteenth century.

superb altarpiece painted for Trinity Collegiate Church in Edinburgh. As the authors of a study on the altarpiece remind us 'in the mid-fifteenth century contacts between Scotland and the Netherlands were both close and frequent' on levels political, dynastic, commercial and cultural. The Trinity Altarpiece dates from the 1470s.[18] It provides a pan-European commercial context for the appreciation of pre-Reformation religious art in Scotland for it was painted by one of the finest artists in Europe at the time, Hugo van der Goes, and commissioned by Edward Bonkil, a Scottish merchant based in Bruges. This is part of a pattern of patronage, for the same artist's most well-known work, *The Adoration of the Shepherds* (1473–8)[19] was commissioned by another businessman based in Bruges, the Florentine banker Tommaso Portinari.[20] Contemporary sources also note the making in Flanders of church furnishings for the abbeys of Melrose and Pluscarden.[21] Michael Lynch has noted that 'the fifteenth century, which is too easily cast off as the period of a church in the first throes of terminal decline, was in fact an age of great spiritual awakening'.[22] The paintings at Fowlis and Trinity College are an indication of this, as is the somewhat later – and equally remarkable – Fetternear Banner.[23] This rare textile survival dates from 1520 and was made for the use of members of the cult of the Holy Blood, a cult which came to Scotland, like so much else, via Bruges.

Throughout the period from the twelfth to the sixteenth century, a distinctive art was developing in the Highlands. This west Highland school provides an intriguing counterpoint to the art of the Lowlands.[24] Where the art of the Lowlands during this period can be considered as a part of the European mainstream, this Highland work, although related, is more individual both in form and iconography. The quality of Highland metal work has been noted above; from the fourteenth century a major stone-carving tradition was established on Iona and the surrounding area. In style and content it brings together Celtic, Norse and Romanesque influences. Its surviving products are primarily in the form of grave-slabs, effigies and freestanding crosses, and these were produced up to the time of the Reformation throughout the western Highlands and the Isles. A good example of a freestanding cross is Maclean's Cross on Iona, not least for its formal integration of knotwork with Romanesque-derived foliage patterns. Grave-slabs made for warriors and chiefs are equally impressive, and have an added interest for their frequent representation of that high status product of Norse-originated technology, the Highland galley. One such representation can be found in a late work of this school, the tomb of Alexander Macleod in the church of St Clement at Rodel in Harris. This work was made in the mid-sixteenth century, perhaps begun about 1527, making it contemporary with high-quality wood carvings from the east of the country, notably an Annunciation from a late-Gothic set of carvings made for Cardinal Beaton (Fig. 31.4). These panels were made within a few decades of the Reformation in Scotland, indeed the man who commissioned them, Cardinal Beaton, is most commonly remembered for his role in the killing of the reformer George Wishart in 1546, and for his own assassination in the same year. The point

Figure 31.4 The Annunciation from the Beaton Panels. Courtesy of the Trustees of the National Museums of Scotland, NMS H.KL 222.1.

must, however, be made that the relative scarcity of pre-Reformation Scottish art cannot simply be ascribed to the enthusiasm of iconoclasts. That was one of a wider set of difficulties which included regular English invasions from the thirteenth to the seventeenth century. These invasions were particularly disastrous for the Border abbeys. Such destruction was compounded by the removal from Scottish society of the remaining body most able to commission, repair and maintain works of art, the royal court, which relocated to London with the union of the crowns in 1603.

Whatever the unfortunate effect of this combination of factors on art, positive aspects of the period must also be noted. One finds, for example, a wonderful aesthetic achievement in silver communion cups and baptismal basins created in the seventeenth century. This is puritan design of a high order, and however much one may abhor the destruction of art inherent to this period, one can celebrate these works. Religious imagery did, however,

Figure 31.5 Detail of King David playing his harp from the Dean House Panels. Courtesy of the Trustees of the National Museums of Scotland, NMS H.KL 70.

survive the Reformation in Scotland in the form of illustration integrated with decorative schemes of one sort or another. An example dating from 1636 can be seen at the church of Grandtully in Perthshire. As Deborah Howard notes, 'this is no Puritan scheme'.[25] However, it is worth considering that illustration of biblical scenes was more acceptable in post-Reformation Scotland than is usually assumed. Another example, this time in a secular building, is found in a series of painted panels from Dean House in Edinburgh. This includes an image of David playing the harp (Fig. 31.5).[26] This sort of work reminds one that iconoclasm cannot be identified with loss of religious imagery in any simple sense. As Philip Benedict has noted, the acceptability of religious imagery in a Calvinist context can be related to its location, not to its subject matter. He notes that 'Calvin ... stressed the distinction between private homes and churches, allowing biblical scenes in private households'.[27] These schemes of illustration may thus have been permitted, perhaps even when placed within a family chapel as at Grandtully, as aids to pedagogy rather than objects of idolatory.[28] These decorative schemes are part of the wider sixteenth-century tradition of exuberantly curvilinear decorative subjects painted on wooden ceilings, carried out in castles and mansions around the county.[29] Contemporary with these works, and similarly vigorous in execution, are examples of graveyard wall tombs, notably those of Greyfriars' churchyard in Edinburgh. Not least of the points of interest here is the Nasmyth tomb dating from 1614, which integrates scripture quoted in English with a late-Renaissance sculptural language. Such quotations from scripture can also be seen in ceiling paintings dating from 1638 in the Skelmorie Aisle at Largs, carried out for Sir Robert Montgomery. The tomb here is an impressive triumphal arch and other aisle tombs of significance include, early in the seventeenth century, those designed by the London-based sculptor Maximilian Colt at Scone and Dunbar. Cognate native Scottish work can be seen, for example in Kinnoull old church, Perth. From late in the century, the Baroque Queensberry Monument at Durisdeer in Dumfriesshire is of considerable interest. It is the work of another distinguished London-based sculptor, John Nost.[30] As one approaches the eighteenth century such tombs become increasingly secular in character.

In the wake of the Reformation, Scottish painting remade itself effectively in the Protestant mode of the individual portrait in oil on canvas. Such portraiture is not simply an art form that avoids religion. Its link with Reformation thinking is fundamental for it emphasises, by implication at least, the spiritual autonomy of the individual before God. Indeed one of its most characteristic forms in Scotland was the image of the Protestant scholar. This can be seen as early as 1581 in a portrait of George Buchanan by the Netherlandish artist Arnold Bronckhorst.[31] Bronckhorst helped to establish the post-Reformation portrait tradition in Scotland, paving the way for native artists such as George Jamesone in the first half of the seventeenth century. From Jamesone there is an unbroken development to Allan Ramsay and Henry Raeburn in the latter part of the eighteenth century, at which point one can consider Scottish art as again able to take its place alongside the best of the age. But this period from the late sixteenth to the late eighteenth century sees a notable lack of those same painters depicting scenes from scripture. With the exception of the early seventeenth-century decorative painting already mentioned, such painting thus drops out of Scottish art for two centuries.

It is in this context that the outpouring of paintings of religious subjects in the nineteenth century is of all the more interest. This begins to feed through from history painting, with a visual re-engagement with the Reformation period as a time of ideological struggle. The tensions within Christianity emerge as the grand narrative of Scottish history, yet the emphasis for these artists is as much on the struggle as on the competing positions, and Mary, Queen of Scots, is shown with the same sympathy, often by the same artists, as are dying Covenanters. These artists were keenly aware of literary works as quarries for their art, not only the works of Burns, Scott and Hogg, but also in the histories of David Hume, Gilbert Stuart and William Robertson, as well as sources such as Knox's *History of the Reformation in Scotland*. Burns and Scott set an important example in their sympathies for both Jacobite and Covenanter and this dual sympathy acts as a template for the artists of the nineteenth century. As early as 1765, James Boswell had commissioned from Gavin Hamilton a painting of an event from Mary's life. The painting, *The Abdication of Mary Queen of Scots*, was not completed until 1776.[32] The theme was explored again by Hamilton's followers Alexander Runciman in 1786 and David Allan in 1796. Allan's proposal for a cycle of works on the life of Mary, Queen of Scots, set an agenda for the painters of the next century, not least with respect to the key confrontations between John Knox and Mary. Outstanding among these painters was a talented son of the manse, David Wilkie, and it is hard to overestimate the example that Wilkie set for the development of religious art in Scotland. He used his skills to depict major historical events such as *The Preaching of John Knox before the Lords of Congregation, 10 June 1559*[33] which dates (as a sketch) from 1822. Scenes of domestic piety were of equal importance to Wilkie notably his evocation of Burns's *Cotter's Saturday Night*[34] from 1837. From two years later come a sketch of *Samuel in the Temple*[35] and his unfinished *John Knox Dispensing the Sacrament at Calder House* (Fig. 31.6).[36] Wilkie's travels in Italy and Spain added a further dimension in which he shows a keen interest in the practices of the Roman Catholic Church. In the period immediately before his untimely death in 1841, he was travelling in the Holy Land, with the intention of seeing at first hand the locations of biblical events. There he produced a wonderful series of sketches such as *Hebrew Women Reading the Scriptures at Jerusalem*.[37] Two of Wilkie's friends extended the range of this religious art further. David Roberts revolutionised public perception of the Holy Land through a major series of lithographs published in the 1840s. William Allan recorded scenes from the life of Mary Queen of Scots, and also painted *The Signing of the National Covenant in Greyfriars Kirkyard*.[38] This was probably painted in 1838 to mark the bicentenary of the event.[39]

Allan's *Greyfriars* painting ushers in the decade of the Disruption of the Church of Scotland. For artists, the debates of the period were a further spur to activity. This was a period of reassessment of how art could be used in the service of a Protestant Christianity. The Covenanting theme was pursued not only by William Allan but also by Thomas Duncan in works such as *The Death of John Brown of Priesthill* (1844).[40] It had been explored in the previous decade by George Harvey, in acutely observed scenes of open-air baptism and preaching such as *The Covenanters' Preaching* (1830).[41] And it was to remain a powerful strand of Scottish art well into the latter part of the century, for example in Robert Herdman's *After the Battle: a Scene of Covenanting Times* (1870).[42] Full-scale paintings of events from the Bible were also becoming more common, particularly in the work of Robert Scott

Figure 31.6 David Wilkie, *John Knox Dispensing the Sacrament at Calder House*. Courtesy of the National Gallery of Scotland, NGS 323.

Lauder, for example his *Christ Teacheth Humility* from 1847.[43] In works such as these, Lauder displays his profound knowledge of the European tradition and this is equally true of his Episcopalian contemporary, William Dyce. Dyce made religious work of great distinction throughout his career from his *Woman of Samaria* (1833)[44] and *Joash Shooting the Arrow of Deliverance* (1844)[45] to his unexpected and beautiful evocation of Christ in a Highland setting, *Christ as the Man of Sorrows* (1860) (Fig. 31.7).[46]

As part of the aesthetic reassessment which took place in the 1840s one must also remember the shift in perception of Scottish Romanesque and Gothic architecture made possible by the publication of R W Billings' *Baronial and Ecclesiastical Antiquities of Scotland*.[47] Billings replaced the notion of picturesque ruins in a landscape setting with an impressive visual appreciation of their architectural integrity. The high quality of his images is reflected in the fact that they are still frequently used today as an alternative to photographs in books on Scottish architecture. This interest in architecture was matched by a revival of interest in stained glass. 1845 saw the publication by James Ballantine of *A Treatise on Painted Glass*. As Michael Donnelly has pointed out, 'this slim volume was the first Scottish text on the subject in modern times'.[48] Donnelly notes that Ballantine's early commissions in Scotland came from the Episcopalian Church and from English patrons. In 1856 he was disappointed not to be awarded the major contract for stained

Figure 31.7 William Dyce, *Christ as the Man of Sorrows*. Courtesy of the National Gallery of Scotland, NGS 2410.

glass in Glasgow Cathedral, but the fact that this contract was awarded to the Royal Bavarian Glassworks demonstrates the international dimensions of the debate around the re-establishment of stained glass in Scotland's Presbyterian churches. That same year Ballantine was commissioned to contribute to the restoration of Greyfriars church in Edinburgh and this commission marks the beginning of a sustainable native stained glass tradition. Others such as Daniel Cottier followed Ballantine's lead, as did Ballantine's son and grandson.

It should not be forgotten that the Disruption had a seminal influence on another art, namely photography, for, thanks to the suggestion of David Brewster, David Octavius Hill and Robert Adamson used the newly invented medium as a method of recording the features of those who participated in the formation of Free Church in Edinburgh in 1843.[49] Hill eventually finished his painting of this event, *The Signing of the Deed of Demission*, in 1866 (see Fig. 5.5).[50]

A final development that one can note in the 1840s was the increasing availability of high quality book illustrations, for example a fine set for Bunyan's *Pilgrim's Progress*, drawn by David Scott. These, engraved by the artist's brother, William Bell Scott, were published in 1850 and they reflect a period of strong growth in popular illustration, religious and otherwise. An example of this wide distribution of religious art of high quality can be found in *The Comprehensive Family Bible* published by Blackie in 1852.[51] The significance of the illustrations is emphasised in the preface: 'The Engraved

Illustrations include Historical Subjects, carefully selected from Ancient and Modern Masters; but the larger portion consists of views of Mountains, Rivers, Lakes and other Natural Scenery; together with Cities and Towns, existing or in ruins; Temples, Tombs &c.; the whole having direct reference to Bible Incidents and History, and more especially illustrating the Fulfilment of Prophecy'. The title page of each testament is both drawn and engraved by William Bell Scott, and Scott also provided the design for the family record page, which is positioned at the end of the Old Testament. The artists illustrating events include Guido Reni, Nicolas Poussin, Leonardo da Vinci, and Rembrandt.[52] Such works complement topographical illustrations by predominantly Scottish artists including J A Houston, Sam Bough, and the brothers W L and R P Leitch. These steel engravings are of high quality, a number of the engravers used being those who had helped to develop this method of printmaking while working with J M W Turner in the 1830s.[53]

Art in a religious context was also widely available through the rapidly developing medium of wood engraving. Notable here was the magazine *Good Words* edited by Norman Macleod.[54] In 1862, for example, one finds the list of illustrators including many of the best young Scottish artists of the period, in particular students of Robert Scott Lauder at the beginning of their careers. They include John MacWhirter, William McTaggart, W Q Orchardson, John Pettie and Tom Graham.[55] Another artist involved was Jemima Blackburn. The following year a major contribution was made by the English Pre-Raphaelite, John Everett Millais: a series of images to complement Thomas Guthrie's modern readings of the parables.[56] The visual ethos of the magazine is reflected in an obituary of a Free Church member and mill-manager which contains the words 'he had an immense reverence for the great art critic of our age [Ruskin], and a boundless sympathy with the high moral tone of his writings.'[57] In short, where *Good Words* went, the opportunity to appreciate good art went also.

The period from 1840 to 1860 saw significant sculptural projects, among them a remarkable group of statues of figures from Reformation and Covenanting times sited in the Valley Cemetery beside the church of the Holy Rood in Stirling. The work of Alexander Handyside Ritchie, these figures include Henderson, Melville, Knox, Guthrie, Renwick and Erskine, and a figure group in marble of the Solway martyrs.[58] Among the significant paintings of this period were, as well as those by Dyce already mentioned, works which explored the roots of the Reformation such as William Bell Scott's picture (from 1854) of Luther's older friend, the painter Albrecht Dürer, on the balcony of his house in Nuremberg.[59] Scott was also notable for writing one of the earliest biographies of Dürer in English, a book complete with etchings by Scott himself.[60] Scott's work is complemented by Joseph Noel Paton's *Dawn: Luther at Erfurt* (1861).[61] In the 1870s and 1880s Paton was among the artists who provided designs for Ballantine's stained glass at Dunfermline Abbey. Similarly, Robert Herdman, noted above for a Covenanter image, provided designs for glass in St Giles' in Edinburgh.

In contrast to the pioneering experiments by Wilkie early in the nineteenth century, it will be clear that by the end of the century artists were exploring Christian material in all the major media from painting through sculpture to stained glass. In Glasgow in 1882 James Guthrie's *A Highland Funeral*[62] (see Fig. 22.8) took further the tradition of social realism asserted by Wilkie in his *Cotter's Saturday Night*. As a kind of mid-term between these two works one can note John Phillip's *Baptism in Scotland* from 1850[63] and a kind of conclusion was given to this type of work in 1891 by J H Lorimer's *Ordination of the Elders*.[64] At the end of the 1880s David Gauld painted his remarkable *St Agnes*[65] ushering in a period in Scottish art dominated by a distinctively Scottish transformation of Arts and Crafts principles. If any one artist from late in the century can symbolise this with respect to religious art it is Dublin-born Phoebe Anna Traquair.[66] She made an extraordinary contribution in a wide variety of media. Her work epitomises the philosophy of William Morris with respect to the social and public uses of art. Morris was also a significant influence on her style. Traquair's contributions ranged from large-scale mural schemes[67] to the intimacy of etchings to accompany lectures on Dante by her friend, the Free Church minister Alexander Whyte.[68] Among her finest work is a set of four large embroidery panels, *The Progress of a Soul*,[69] which display an artistic ability combined with a powerful faith (Fig. 31.8).[70] A number of artists, including Traquair, contributed illustrations to Bunyan. Notable in this regard was an edition of *Pilgrim's Progress*[71] with etched images by William Strang, one of the outstanding figures of the revival in that art. Strang's contribution also included scenes of the New Testament and of Milton's *Paradise Lost*.[72] On a more popular level, Strang's contemporary William Hole made a set of eighty watercolour images of the life of Christ, of wide availability when published as a book with an introduction by George Adam Smith.[73] William Hole was a prolific illustrator and much of this work is of religious relevance, among it illustrations of genre scenes for Ian Maclaren's best-selling novel, *Beside the Bonnie Briar Bush*.[74] He also contributed to the growing interest in the history of the Celtic Church through his major mural scheme for the newly established Scottish National Portrait Gallery. One of the panels here shows *St Columba's Mission to the Picts* (1898). Interest in the saint had been enhanced by the fact that the thirteen-hundredth anniversary of the saint's death fell in 1897. A number of other artists painted major works relating to the life and works of Columba during that decade and later. Perhaps the most surprising is the synthesis of Church history and experimental landscape painting made by William McTaggart. His *Coming of St Columba*[75] from 1895 shows the saint's arrival in the artist's native Kintyre. Such works built on a tradition of works with a Columban or Iona theme which dates back – from an artistic rather than an antiquarian perspective – to William Bell Scott and David Roberts. Subsequent to McTaggart, the use of Iona as a place to paint by the Colourists Peploe and Cadell is well known, but a more specifically religious focus to that landscape is to be found in the work of D Y Cameron.

The aspect of Celtic revivalism in Scottish religious art can be noted

Figure 31.8 Phoebe Anna Traquair, *The Progress of the Soul: The Entrance*. Courtesy of the National Gallery of Scotland, NGS 1865 A.

in the interest in Celtic crosses and interlace on monuments from the mid-nineteenth century onwards.[76] West Highland work began to be given its due in 1881 with the posthumous publication of James Drummond's *Sculptured Monuments in Iona & The West Highlands*. Drummond was a distinguished artist and a notable administrator, who had earlier painted works such as *Sabbath Evening* (1847).[77] He was a friend of Noel Paton, and one should note that Paton's work also has a relevance in the context of Celtic revival. Following Dyce, he explored Arthurian legend in visual form, for example *Beato Mundi Corde* from 1890.[78] Arthurian tales of Celtic Christian quest make up a strong dimension of the art of Scotland at this time, receiving attention from among others Jessie M King, for example, her remarkable painted cover from 1903 for a copy of *The High History of the Holy Graal*.[79]

Another explorer of Arthurian legend was John Duncan. He was the key artist of Celtic-revival religious painting in the early twentieth century, painting a number of works relating to saints of the Celtic Church and to Iona, notably *St Bride* (1913) (Fig. 31.9),[80] *The Adoration of the Magi* (1915),[81] and *St Columba and the White Horse*.[82] Duncan came from Dundee and as a young man he may well have been inspired by the restoration of the nearby

Figure 31.9 John Duncan, *St Bride*. Courtesy of the National Gallery of Scotland, NGS 2043.

Figure 31.10 Hew Lorimer, *Our Lady of the Isles*, South Uist. Courtesy of the Trustees of the National Museums of Scotland, NMS SLA C29977.

parish church of Fowlis Easter, and its pre-Reformation paintings.[83] He made paintings for churches of various denominations, including the Stations of the Cross for St Peter's church in Edinburgh (now lost). This Roman Catholic church was built to designs by Sir Robert Lorimer in 1906 and Lorimer's architecture provides the setting for much of the most significant Scottish religious art of the early twentieth century. Most remarkable of all is the Scottish National War Memorial, at the heart of Edinburgh Castle. In marked contrast to the restrained abstraction of Lutyens' Cenotaph in London, Lorimer's shrine to those who died in World War I depends on the participation of the leading sculptors and stained-glass artists of the day. Among the sculptors were Alexander Carrick, Phyllis Bone, Pilkington Jackson and Alice and Morris Meredith Williams. The stained glass was by Douglas Strachan, whose work is among the finest examples of Scottish art of the period. In something of an historical irony, Strachan made stained glass for another Lorimer project, the restoration of St John's, Perth, where Knox had preached his sermon against idolatory in 1559. Notable stained glass was also made from the 1930s onwards by William Wilson and, indeed, by John Duncan.

Having begun this chapter with work from well over a thousand years ago, I conclude with a work created within fifty years of the time of writing, Hew Lorimer's *Our Lady of the Isles* (1959) (Fig. 31.10). This gigantic work in granite, some 25 feet in height, has been aptly described by Douglas Hall as 'hugely dignified'.[84] It is to be found on South Uist and it is perhaps the finest work of Roman Catholic art made by a Scottish artist since the Reformation. Hew Lorimer was Robert Lorimer's son, but under the influence of Alexander Carrick he became a sculptor rather than an architect. He studied with the English stone carver Eric Gill. Like Gill he converted to Roman Catholicism, and, as with Gill, his faith deeply informed his work.

In conclusion, perhaps contrary to what might be imagined, religious art in Scotland – whatever its losses and its transformations over the years – is an area of great vitality and extraordinary interest.

NOTES

1. Museum, Whithorn.
2. Laing and Laing, 1992, 153.
3. Cathedral museum, St Andrews.
4. Ruthwell church, Dumfriesshire.
5. See James, 1998.
6. Nordenfalk, 1977, 26.
7. See, for example, Fisher, 2001; Foster, 1998.
8. NMS.
9. Edinburgh University Library.
10. Higgitt, 1990, 30–1.
11. Cambridge University Library.
12. NMS; and the British Museum.
13. Higgitt, 1990, 28.
14. The Kilmichael Glassary Bell Shrine and the Guthrie Bell Shrine are both in the NMS.
15. Yeoman, 1999.
16. See Apted and Robertson, 1962.
17. Dundee University Library.
18. Thompson and Campbell, 1974, 55–7.
19. Ufizzi, Florence.
20. Thompson and Campbell, 1974, 52.
21. Thompson and Campbell, 1974, 53.
22. Lynch, 1993, 28.
23. NMS.
24. For a comprehensive account of this school, see Steer and Bannerman, 1977.
25. Howard, 1995, 190.
26. This panel is now in the NMS. See also the ceiling painted for Andrew Lumsden in what is now known as Provost Skene's House in Aberdeen. It dates from 1626 and includes a number of scenes from the life of Christ including a crucifixion.
27. Benedict, 1999, 32.
28. My thanks, for discussion of this point, to Charles McKean.
29. Duncan Macmillan has noted that this type of work seems to be surprisingly specific to Scotland, and seems to have influenced Scandinavian painting.

Macmillan, 1990, 56.
30. A useful overview, 'Family aisles and monuments', is provided in Howard, 1995. See also Howarth, 1991; Graham-Campbell, 1982.
31. Scottish National Portrait Gallery, Edinburgh.
32. Hunterian Gallery, Glasgow. See Smailes and Thomson, 1987, 106–8.
33. Sketch, 1822, Petworth House, Sussex.
34. Glasgow Museums and Art Galleries.
35. National Gallery of Scotland, Edinburgh.
36. National Gallery of Scotland, Edinburgh.
37. Reproduced in *The Wilkie Gallery*, London, 1849–50.
38. City Art Centre, Edinburgh.
39. Howard, 2001, 83. This important reassessment of Allan also contains commentary on his *Murder of Archbishop Sharpe* painted in 1821.
40. Glasgow Art Galleries and Museums.
41. Glasgow Art Galleries and Museums.
42. National Gallery of Scotland, Edinburgh.
43. National Gallery of Scotland, Edinburgh.
44. City Museum and Art Gallery, Birmingham.
45. Kunsthalle, Hamburg.
46. National Gallery of Scotland, Edinburgh.
47. Published by William Blackwood and Sons, Edinburgh, 1845–52.
48. Donnelly, 1997, 18.
49. An interesting commentary is 'Artists and evangelicals', ch. 9 of Macmillan, 1986. See Macmillan, 2000, ch. 10, 'Genre, history and religion'. For an account of Hill and Adamson's work, see, *inter alia*, Michaelson, 1970; Stevenson and Ward, 1986; Stevenson, 2002.
50. Free Church College, Edinburgh. A short account of the painting is given in Anderson, 1994.
51. For drawing my attention to this work thanks are due to my parents J S and C W Macdonald.
52. The other artists involved are Hubner, Tenniel, Warren, Begas, Baroccio, Selous, Duncan, Coypel, Schadow, Schopin, Penley, Dolci, Foster and Teblin.
53. They include Miller, Forrest, Cousen, Willmore, Richardson, and Le Keux.
54. *Good Words* has been described as one of the three outstanding illustrated magazines produced in Britain at this time (Goldman 1994, 37).
55. Macleod, 1862.
56. Macleod, 1863. Like much of the art in this magazine these works by Millais were engraved by the London firm of Dalziel.
57. Macleod, 1873, 199.
58. My thanks to Elspeth King of the Smith Art Gallery, Stirling, for this information.
59. *Albrecht Dürer at Nürnberg*, National Gallery of Scotland, Edinburgh.
60. William Bell Scott, 1869.
61. National Gallery of Scotland, Edinburgh.
62. Glasgow Museums and Art Galleries.
63. Aberdeen Art Gallery and Museums.
64. National Gallery of Scotland, Edinburgh.
65. National Gallery of Scotland, Edinburgh.
66. For a full account of her work see Cumming, 1993.
67. Notably for the Catholic Apostolic Church, the song school of St Mary's Cathedral, and the Mortuary Chapel of the Sick Children's Hospital, all in Edinburgh.
68. Edinburgh, privately printed, T & A Constable, 1890.
69. National Gallery of Scotland, Edinburgh.
70. 1899–1902. National Gallery of Scotland, Edinburgh.

71. London, John C. Nimmo, 1895.
72. London, Routledge, 1905.
73. London, Eyre and Spottiswoode.
74. For example his etchings for Maclaren, I. *Beside the Bonnie Briar Bush*, London, 1896.
75. National Gallery of Scotland, Edinburgh.
76. In 1888, the young Charles Rennie Mackintosh followed the by then well-established fashion for gravestones in the form of Celtic crosses by designing one for the Glasgow Necropolis. An early example of a design using the Celtic cross is to be found on the cover of Daniel Wilson's *Archaeology and Prehistoric Annals of Scotland*, published in 1851. Notable jewellery, based on a number of Celtic crosses, was made by the Ritchies of Iona early in the twentieth century. See MacArthur, 2003.
77. City of Edinburgh Museums and Galleries.
78. McManus Gallery, Dundee (Dundee City Art Gallery).
79. NLS.
80. National Gallery of Scotland, Edinburgh.
81. Private Collection.
82. Carnegie Dunfermline Trust.
83. See 'The Parish Church of Fowlis Easter and its Proposed Restoration', *Dundee Advertiser*, 6 September 1888. My thanks to Lesley Lindsay for locating this material.
84. Hall, 1991, 114.

BIBLIOGRAPHY

Allen, J R and Anderson, J. *The Early Christian Monuments of Scotland*, Edinburgh, 1903.
Allen, J R. *Celtic Art in Pagan and Christian Times*, Edinburgh, 1904.
Anderson, J. *Scotland in Early Christian Times*, 2 vols., Edinburgh, 1881.
Anderson, W. *A Guide to the Free Church of Scotland College and Offices*, Edinburgh, 1994.
Apted, M. *The Painted Ceilings of Scotland 1550–1650*, Edinburgh, 1966.
Apted, M R and Robertson, W N. Late fifteenth century church paintings from Guthrie and Foulis Easter, *PSAS*, 95 (1961–2), 262–79.
Athill, P, ed. *William Strang*, Sheffield, 1981.
Bain, G. *Celtic Art: The Methods of Construction*, Glasgow, 1951.
Bendiner, K. David Roberts in the Near East: Social and religious themes, *Art History*, 8 (1985), 158–77.
Benedict, P. Calvinism as a culture? Preliminary remarks on Calvinism and the visual arts. In Finney, 1999.
Billings, R W. *Baronial and Ecclesiastical Antiquities of Scotland*, Edinburgh, 1852.
Bowe, N G and Cumming, E. *The Arts and Crafts Movements in Dublin and Edinburgh*, Dublin, 1998.
Brooke, D. *Wild Men and Holy Places: St Ninian, Whithorn and the Medieval Realm of Galloway*, Edinburgh, 1994.
Caldwell, D H, ed. *Angels, Nobles and Unicorns: Art and Patronage in Medieval Scotland*, Edinburgh, 1982.
Caldwell, D H. In search of Scottish art: Native traditions and foreign influences. In Kaplan, 1990, 45–60.
Campbell, M. *David Scott*, Edinburgh, 1990.
Cassidy, B, ed. *The Ruthwell Cross*, Princeton, 1992.
Caw, J L. *William McTaggart: A Biography and an Appreciation*, Glasgow, 1917.
Chadwick, N. *Celtic Britain*, London, 1963.
Cheape, H. The world of a nineteenth-century artist in Scotland [Noel Paton], *ROSC*, 3 (1987), 77–90.

Chiego, W J, Miles, H A D and Brown, D B. *Sir David Wilkie of Scotland*, Raleigh, 1987.
Christian, J and Stiller, C. *Iona Portrayed: The Island through Artist's Eyes, 1760–1960*, Inverness, 2000.
Collingwood, W G. *The Early Crosses of Galloway*, Dumfries, 1925.
Cumming, E. *Phoebe Anna Traquair*, Edinburgh, 1993.
Dalgetty, A B. *History of the Church of Foulis Easter with Illustrations of the Medieval Paintings*, Dundee, 1933.
Donnelly, M. *Scotland's Stained Glass*, Edinburgh, 1997.
Drummond, J. *Sculptured Monuments in Iona and the West Highlands*, Edinburgh, 1881.
Errington, L. *The Artist and the Kirk*, Edinburgh, 1979.
Errington, L. *Masterclass: Robert Scott Lauder and his Pupils*, Edinburgh, 1983.
Errington, L. *Robert Herdman*, Edinburgh, 1988.
Errington, L. *William McTaggart*, Edinburgh, 1989.
Errington, L. Celtic elements in Scottish art at the turn of the century. In Christian, J, ed. *The Last Romantics: The Romantic Tradition in British Art*, London, 1989.
Errington, L. Sir Walter Scott and nineteenth century painting in Scotland. In Kaplan, 1990, 121–35.
Errington, L. *Tribute to Wilkie*, Edinburgh, 1985.
Fawcett, R. *Scottish Architecture from the Accession of the Stewarts to the Reformation 1371–1560*, Edinburgh, 1994.
Fisher, I. Early Christian Archaeology in Argyll. In Ritchie, G, ed. *The Archaeology of Argyll*, Edinburgh, 1998, 181–204.
Finney P C, ed. *Seeing Beyond The Word: Visual Arts and the Calvinist Tradition*, Grand Rapids, 1999.
Fisher, I. *Early Medieval Sculpture in the West Highlands and Islands*, Edinburgh, 2001.
Foster, S M. *Picts, Gaels and Scots*, London, 1996.
Foster, S M, ed. *The St Andrews Sarcophagus: A Pictish Masterpiece and its International Connections*, Dublin, 1998.
Goldman, P. *Victorian Illustrated Books 1850–1870: The Heyday of Wood-Engraving*, London, 1994.
Gow, I. The Scottish National War Memorial. In Pearson, 1991, 105–9.
Graham-Campbell, D. *Scotland's Story in her Monuments*, London, 1982.
Hall, D. The twentieth century. In Pearson, 1991, 111–26.
Henderson, G. The Seal of Brechin Cathedral. In O'Connor and Clarke, 1983, 399–415.
Henderson, G. *From Durrow to Kells: the Insular Gospel-Books 650–800*, London, 1987.
Henderson, I. *The Picts*, London, 1967.
Higgitt, J. Art and the Church before the Reformation. In Kaplan, 1990, 28–44.
Higgitt, J. *The Murthly Hours: Devotion, Literacy and Luxury in Paris, England and the Gaelic West*, London, 2001.
Howard, D. *Scottish Architecture: Reformation to Restoration 1560–1660*, Edinburgh, 1995.
Howard, J, ed. *William Allan, Artist and Adventurer*, Edinburgh, 2001.
Howarth, D. Sculpture and Scotland 1540–1700. In Pearson, 1991, 27–37.
James, E. The Continental context. In Foster, 1998.
Kaplan, W, ed. *Scotland Creates*, London, 1990.
Kemplay, J. *John Duncan: A Scottish Symbolist*, San Francisco, 1994.
Laing, L and Laing, J. *Art of the Celts*, London, 1992.
Laing L and Laing, J. *The Picts and the Scots*, Stroud, 1993.
Lynch, M. Scottish culture in its historical perspective. In Scott, P H, ed. *Scotland: A Concise Cultural History*, Edinburgh, 1993.
MacArthur, E M. *Iona Celtic Art*, Iona, 2003.
Macdonald, M. *Scottish Art*, London, 2000.
Macleod, N, ed. *Good Words for 1862*, London, n.d.
Macleod, N, ed. *Good Words for 1863*, London, n.d.

Macleod, D, ed. *Good Words for 1873*, London, n.d.
Macmillan, D. *Painting in Scotland: The Golden Age*, London, 1986.
Macmillan, D. *Scottish Art, 1460–1990*, Edinburgh, 1990.
Macmillan, D. *Scottish Art 1460–2000*, Edinburgh, 2000.
McNeill, P and MacQueen, H, eds. *Atlas of Scottish History to 1707*, Edinburgh, 1996.
McRoberts, D. The Fifteenth-Century Altarpiece of Fowlis Easter Church. In
Michaelson, K. *David Octavius Hill and Robert Adamson*, Edinburgh, 1970.
Meehan, B. *The Book of Kells*, London, 1994.
Meehan, B. *The Book of Durrow*, Dublin, 1996.
Morrison, J. *Painting the Nation*, Edinburgh, 2003.
Noel-Paton, M H and Campbell, J P. *Joseph Noel Paton 1821–1901*, Edinburgh, 1990.
Nordenfalk, C. *Celtic and Anglo-Saxon Painting*, London, 1977.
Richardson, J S. *The Medieval Stone Carver in Scotland*, Edinburgh, 1964.
O'Connor, A and Clarke, D V, eds. *From the Stone Age to the 'Forty-Five*, Edinburgh, 1983.
Oddy, R. The Fetternear Banner. In O'Connor and Clarke, 1983, 416–26.
Pearson, F, ed. *Virtue and Vision; Sculpture and Scotland 1540–1990*, Edinburgh, 1991.
Pearson, F. *William Wilson*, Edinburgh, 1994.
Pointon, M. *William Dyce, 1806–1864: A Critical Biography*, Oxford, 1979.
Radford, C A R and Donaldson, G. *Whithorn and Kirkmadrine*, Edinburgh, 1953.
Ramsay, M P. *Calvin and Art: Considered in Relation to Scotland*, Edinburgh, 1938.
Ritchie, A. *Viking Scotland*, London, 1993.
Ritchie, A. *Iona*, London, 1997.
Scott, W B. *Memoir of David Scott RSA*, Edinburgh, 1850.
Scott, W B. *Albert Durer: His Life and Works*, London, 1869.
Simpson, D. *The Celtic Church in Scotland*, Aberdeen, 1935.
Smailes, H and Thomson, D. *The Queen's Image: A Celebration of Mary, Queen of Scots*, Edinburgh, 1987.
Smith, B. *D. Y. Cameron*, Edinburgh, 1992.
Spearman, R M and Higgitt, J, eds. *The Age of Migrating Ideas: Early Medieval Art in Northern Britain and Ireland*, Edinburgh, 1993.
Steer, K, Bannerman, J and Collins, G H. *Late Medieval Monumental Sculpture in the West Highlands*, Edinburgh, 1977.
Stevenson, S and Ward, J. *Printed Light: The Scientific Art of William Henry Fox Talbot and David Octavius Hill with Robert Adamson*, Edinburgh, 1986.
Stevenson, S. *The Personal Art of David Octavius Hill*, New Haven and London, 2002.
Stratford, N. *The Lewis Chessmen and the Enigma of the Hoard*, London, 1997.
Tanis, J R. Netherlandish Reformed traditions in graphic art. In Finney, 1999.
Thompson, C and Campbell, L. *Hugo van der Goes and the Trinity Panels in Edinburgh*, Edinburgh, 1974.
Thomson, D. *Painting in Scotland 1570–1650*, Edinburgh, 1975.
Weaver, L. *The Scottish National War Memorial*, London, 1927.
White, C. *The Enchanted World of Jessie M. King*, Edinburgh, 1989.
Wilson, D. *The Archaeology and Prehistoric Annals of Scotland*, Edinburgh, 1851.
Yeoman, P. *Pilgrimage in Medieval Scotland*, London, 1999.

Index

A
Aaron (citizen of Caerleon) 16
Aaron of Lincoln 256
abbeys 583–4
   appointments to 105
   building of 577, 583
   Deer 110, 364
   Dunfermline 57, 578, 583, 584, 665
   English invasions and 584, 659
   Holyrood 51, 446, 583, 584, 586
   Inchcolm 630
   Kelso 584, 587, 656
   Melrose 50, 584, 656, 658
   Newbattle 256
   Pluscarden 583, 658
   property post-Reformation 110
   Westminster 57
   see also Iona; Paisley Abbey
Abbot of Unreason 382
Abbotsford 421
Abdul Aziz Bin Fahd, prince 290
Abercorn 439, 578
Aberdeen
   altars 89, 580
   alternative healing 321, 326
   bishops 577
   Camphill Community 318
   Catholic community 171, 176
   choirs 64, 464, 638, 639
   church buildings 577
   church conversion 602
   Church of Scotland Summer Mission 457, 458
   cinemas 427
   Congregational church 602
   cremations 468
   Free Church College 405
   Gordon Evangelistic Mission 500
   Haliblude passion play 337
   Jewish community 257, 259, 268, 270, 271
   King's College 7, 66, 616, 620
   Marischal College 257, 541
   maritime museum 602
   Middlefield 506
   mosque 290
   Paganism 317
   praying society 517
   Quaker converts 237
   Queen's Cross church 464, 618, 622
   St John's church 81
   St Machar 34, 47
   St Mary's church 173
   Scottish Episcopal Church 199
   song schools 638
   synagogues 270
   Theosophical Society 318
   Torry, episcopalian community 212
   Unitarian church 246
   see also St Machar's Cathedral; St Nicholas' church
Aberdeen, Lady 397, 401
Aberdeen Book Agent and Colportage Society 397
Aberdeen Breviary 34, 47, 382
Aberdeen Ecclesiological Society 591
Aberdeen presbytery 138
Aberdeen synod 519
Aberdeen University 107, 108, 110, 116, 582
Aberdeenshire
   Catholic community 112, 171, 176
   Haddo House 401
   Lutheranism 62
   Scottish Episcopal Church 202, 213, 216
   see also Deer; Monymusk
Aberdour 34
Aberlady parish minister 445–6
Aberlour 206, 225
Abernethy 577
Abernethy Trust 512
Abraham (Jewish money lender) 256
abstinence societies *see* temperance movement
Abu Bakr Siddiq 287
Act of Union
   Church of Scotland and 9, 103, 104, 146, 531, 532, 536
   Church Patronage (Scotland) Act and 533
   music and 641

Presbyterian church government and 9, 112, 191, 531, 532
universities 541
Action of Churches Together in Scotland 235, 313
Acts of the Parliaments (1540), illustration in published text 79, 81, 83
Adams, Robert 404
Adamson, Robert 664
Adelaide Place Baptist church, Glasgow 609
Adler, Dr Herman 267
Adomnán, abbot of Iona 15, 34
   Life of St Columba 26–30, 32, 379, 380
      Columba's dealings with Picts 22, 23, 29
      Columba's kindness to crane 44
      Columba's voice 380, 630
adultery 113, 118, 128, 130, 446, 544
   see also sexual conduct
Advisory Committee on Artistic Questions 592
Aed mac Ainmirech, king of Uí Néill 28
Aedán mac Gabráin, king of Dál Riata 27–30, 363–4, 380
Africa 214, 284, 286, 520–2, 524, 525, 528–9
   foreign missions 138, 517, 520–2, 524, 525, 528–9
   see also North African community
African Church, Malawi 528
African languages 521, 524, 525
agape (shared meal) 239
'Age of Saints' 15–41
Agricola, Gnaeus Julius 16, 17
Ahl al-Hadith movement 291
Ahmadiyya, the 288
Aikenhead, Thomas 130
Ailither, abbot of Clonmacnoise 27
Ailred, abbot of Rievaulx 25, 46, 47
Airdrie 173, 183
Airth, Sir Walter 86
Airthrey 46
Albany Trust 609
Alcuin 26
Alcuin Club 208
Aldhelm, bishop of Wessex 24
Alexander I, king of Scots 46, 576–7
Alexander, Jake 34
Alford 409
Ali, 4th caliph 287
Ali ibn Abi Talib 287
All Saints' church, Inverary 223

Allan, David 662
Allan, Rev. Tom 416–17, 458, 507, 508–9, 539
Allan, William 662
altars 89, 580
   burial near 75
   Dundee 65, 72–3, 580
   Edinburgh 64, 65, 70, 89, 90, 580
   Haddington 83
   Leith 73
   Perth 72, 580
   portable 85–6
   see also Trinity Altarpiece
alternative healing 318, 321–3, 325, 326–7
Alva, St Serf at 46
Am Fasgadh museum, Kingussie 223
Amadeus, Alexander 256
America 258, 259, 366
American influence
   alternative beliefs 237, 317, 323, 328
   Christian Science 252, 317
   Church of Scotland communion practice 472
   Church of the Nazarene 248
   Churches of Christ 242
   Congregationalism 240, 241
   Episcopalianism 192, 197, 223
   Jehovah's Witnesses 252, 254
   Mormons 252, 253
   Seventh Day Adventists and 252, 253
   Unitarians and 246
   see also Graham, Billy; Moody, Dwight; Universalist Church of America
Amnesty International 529
An Comunn Gaidhealach 373
Anabaptists 237, 244
   see also Baptists
Ancient Order of Hibernians 181
Ancient Orthodox Church 254
Ancrum, minister of 640
Anderson, Andrew 396
Anderson, David 64
Anderson Free church, Glasgow 605
Anderson, George 356
Anderson, James 444
Anderson, Sir Robert Rowand 210, 591, 592, 595
Andersonian University 261, 520
   see also Strathclyde University
Anderston, Scottish Episcopal Church 204, 206
Andrew of Wyntoun 47, 381

Anglican Communion of Churches 205,
    224, 225, 227
Anglicans *see* Church of England;
    Episcopalianism
Angus 34, 504, 598–9, 603, 619, 625
Angus, John 637
Annan Committee of Enquiry 295
Anthroposophical movement 318
anti-Catholicism
    Episcopalians 200, 211
    Presbyterians 182–3, 313, 422, 495, 497
    see also sectarianism
Antiburghers 133, 138, 534
Antonine Wall 16, 17
Appin 197, 204, 212, 228, 366, 453
Applecross 34
Arab community 286
    see also Muslim community
Arabic language 297, 300
Arbroath Town Mission 489
Arbuthnott church, manuscripts 53
Arbuthnott, Robert, portable altar 85
Arbuthnott, Sir Robert, prayer book 63, 71,
    76, 81, 83, 87, 89
Arcane School 326
Architectural Heritage Society of Scotland
    599, 601
architecture 575–97
    Church of Scotland 9
    Edinburgh mosque 290
    Gothic 577, 579, 580, 585, 655, 663
    Gothic revival 588–9, 591, 592, 594–5
    Norman 580, 592
    Romanesque 578, 579, 591, 595, 655,
        663
    St Mungo Museum of Religious Life
        and Art 307
    Scottish Episcopal Church 209–10
    see also Architectural Heritage Society
        of Scotland; church buildings;
        Scottish Churches Architectural
        Heritage Fund
Ardchattan 587
Argyll, 10th duke of 223
Argyll
    alternative beliefs 317, 321
    Catholic community 173, 176
    Church of Scotland 367, 369, 370, 492
    Gaelic New Testament 367
    stone circles 575
    see also Campbeltown; Dál Riata
Argyll synod 367, 369, 370
Arisaig 157, 173, 369

Arles, Council of 16
Armagh 44, 151
Arminianism 241, 244
aromatherapy 321, 322
Arran 460
art 652–74
    Buddhist 308
    Calvinism and 309, 660–1
    Catholic community 61, 63, 66, 71,
        74–5, 80, 670
    Celtic 630, 652–4, 658
    Church of Scotland 596–7
    Hindu 308
    Holland and 658
    Islamic 308
    Jewish community 275, 308
    monarchs and 63
    Norse influence 655, 658
    Pictish 318, 652, 654
    St Mungo Museum 307–8
    saints in 47–53
    Scottish Episcopal Church 217
    see also embroidery; illustrations;
        jewellery; murals; paintings;
        sculptures; stained glass; woodcuts
Articles Declaratory, Church of Scotland
    130, 141, 532, 535
Artisans' Hall, Glasgow 497
arts 390, 599, 601–2, 609
    see also architecture; art; films;
        literature; music
Ashenheim, Louis 258
Asher, Asher 258
Asian community 255, 284, 285, 299, 315
    see also Bangladeshi community; Indian
        community; Malaysian community;
        Pakistani community
Assemblies
    Christian Brethren 247
    see also General Assembly of the
        Church of Scotland; General
        Assembly of the Free Church of
        Scotland
Assemblies of God 250
Assembly Hall, Free Church of Scotland
    135
assisted suicide 548
Associate (Burgher) Presbytery and Synod
    134
Associate Presbytery, The 133
Associated Presbyterian Churches of
    Scotland and Canada 140, 375, 537
Athanasius 16, 379

Athelstaneford, battle of 381
Atholl 264
Atlas, Rabbi 268
Auchengeich mission station, Glasgow 501
Auchinleck church, Ayrshire 603
Auchterarder, Barony church 609
Auchterarder United Free Church 138
Auld Kirk
    Ayr 622
    Eyemouth 603
Auld Lichts 134, 138, 140
Authorised Version *see* Bible, Authorised Version
Ayr
    Auld Kirk baptismal enclosure 622
    Baptist church 610
    Catholic community 173
    chaplains' stipends 64
    elders' and deacons' election 128
    Jewish community 258, 270, 271
    obits and prayers 65, 87, 90
    Paganism 317
    St John's church 64
    see also Glasgow and Ayr synod
Ayrshire
    Catholic community 171, 173, 179
    church conversions 601, 603
    Church of Scotland Summer Missions 458
    church relocated to Japan 605
    Jewish pedlars 266
    Lithuanian community 179
    St Quivox church 560, 563
    see also Cunninghame; Irvine, Ayrshire; Kyle and Carrick
Ayrshire Christian Union 499

B

badminton 221, 322, 465, 594
Bahai faith 298
Bailey, Alice 323, 326, 328
Balanqual, Walter 444
Baldernock, minister of 640
Balfour Declaration 274
Balfour, Robert 587
Ballachulish 203, 229
Ballantine, James 663, 664, 665
Ballentine, William 151–2, 155
Balornock, Glasgow 506
Balquhidder church, Perthshire 625
Banchory 34
Banff 61, 73, 242

Banffshire
    Catholic community 143, 152, 157–8, 170
    Episcopalian orphanage 206, 225
    Our Lady of Grace well 151–2, 155
    sacrament house and sculptures 87, 657
    see also Cullen
Bangladeshi community 284
banners 70, 88, 658
baptism
    Assemblies of God 250
    Baptists 243–4
    Catholic Church 62, 86, 144, 147, 159
    Christian Brethren 247
    Church of Scotland 244, 472, 474, 582, 583, 622
    Church of the Nazarene 248
    Churches of Christ 242
    Churches of God 248
    Congregationalists 238
    Elim Pentecostal Churches 251
    Glasites 239
    Jehovah's Witnesses 254
    Mormons 253
    Salvation Army 249
    Scottish Episcopal Church 194, 197, 204, 211, 221, 225, 229
        painting of 208, 217
    Seventh-Day Adventists 253
    Unitarian churches 246
baptismal basins 582, 583, 586, 622, 659
    see also baptismal fonts
baptismal enclosure, Ayr Auld Kirk 622
baptismal fonts 472, 582, 583, 592, 622
    Hopkirk 598
    sculptures 72, 81, 86, 87
    see also baptismal basins
baptismal pews 618
Baptist Home Missionary Society for Scotland 491
Baptist Union of Scotland 244, 252, 506
Baptists 11, 243–4
    Christian Brethren and 247
    church buildings
        Ayr 610
        Edinburgh 239, 244, 598
        Glasgow 244, 268, 589, 609
        Paisley 589
        Tiree 371
    Churches of Christ and 242
    early congregations 239, 482, 518
    evangelism 244, 374
    foreign mission work 519, 520

Gaelic language and culture 371, 373, 374, 375
Gaelic-speaking clergy 374, 375
Haldane brothers and 240
home mission work 243, 371, 373, 491, 499, 506
membership 244, 531
Seventh-Day Adventists and 253
see also Anabaptists
Bar Kichba sports club 276
Barbour, John 47, 337, 381
Barclay church, Edinburgh 589
Barclay, Rev. William 418, 425, 460
Barony church
    Auchterarder 609
    Glasgow 494, 605
Barr, Elizabeth 138
Barra 34, 157, 373, 374, 640
barrel organs 645
Barrie, Sir James M 389, 390, 409, 410
Barth, Karl 622
Basis and Plan of Union, Church of Scotland 532, 535
Bassendyne Bible 340
Bathgate 445
Battlefield, Glasgow 264
Baxter, Nicholas 78
Baxter, Richard 366
Baxter, Stanley 466
baxters, Edinburgh 65
baxters' missal 63
Bayne, Anna and Hay 526
BBC 221, 295–305, 411, 414–19, 547, 647
    see also radio broadcasting; television broadcasting
Bearsden 452
Beatles, the 427
Beaton, David, cardinal 83, 106, 658
Beaton, James, cardinal 106
Beauly, Old Kilmorach church 601
Beaverbrook, William Maxwell Aitken, 1st Baron 423
Beckles, bishop 205
Bede 20, 21–6, 29
Bedell, William 367
Bedlam Theatre 606
Begg, Ian 307
beggars 116, 443
Belford church, Edinburgh 599
Belgium 158, 188, 264
Bell, John L 647
bell shrines 656
Bellie church, Fochabers 586

bells see church bells; handbells; mort bells
Bellshill, Lanarkshire Christian Union 500
Belmont church, Glasgow 606
Beltane Fire Festival 317
Ben MacDhui 319
Benedict, Philip 660
Benedictine order 56, 149, 158, 576
benefaction boards 627
benefices 107, 108, 110, 128, 256
Bengal, foreign mission work 523
Bengali language 526
Bernard, Emily and Amy 527
Bernard, Dr Letitia 527
Bernham, David de, bishop of St Andrews 578–9
Bernicia 22, 23
Berwickshire 86, 114, 579
Beveridge, William Henry, 1st Baron 416
Bhowanipur College 526
Bible 9–10, 236
    Assemblies of God view of 250
    Authorised Version 394, 411, 425
        Gaelic Bible and 368, 369
        Geneva Bible and 395
        language of 346, 352, 353, 367
    Baptist view of 244
    Bassendyne Bible 340
    Catholic 159, 335, 338, 340, 356, 369
    Christian Brethren view of 247
    Christian Science view of 252
    Church of Scotland and 340, 395, 442, 456, 484, 637
    Church of Scotland worship and 9–10, 395, 438, 439, 456
    Church of the Nazarene and 248
    Comprehensive Family Bible, The 664
    Congregationalists and 242
    copying at Iona 32
    Daily Study series 418
    education and 10, 394, 395, 396, 484
    Elim Pentecostal Churches 251
    Expositor's Bible, The 407
    familiarity with 159, 213, 229, 350, 396, 425, 439
    Geneva Bible 340, 354, 395
    historical authenticity challenged 405, 425, 538–9
    illustrations 662, 664–5
    in English 344, 347, 356–7, 395
        Authorised Version 346, 352, 353, 367
        Catholic 159, 356
        importation of 340

in Korean 528
in Latin 159, 335, 338, 340
in Scots 340, 342, 346, 352–5, 356, 357–9
in seTswana 521
Jehovah's Witnesses and 254
Knox, John and 340, 395
literacy and 367, 394, 396, 484
literature and 338, 346, 379
Methodists view of 245
Moffatt translation 356–7, 411
Mormons' view of 253
mystery plays and 337
New English Bible 357
printing and distribution 395–6, 397, 398, 413, 424–5, 492, 637
    Bible societies 396, 485, 490–1
    home mission work 484, 485, 486, 487, 488, 489
    in Gaelic 397, 490–1
quotation in sculpture and ceiling paintings 661
Revised Standard Version 357
Salvation Army and 249
Scottish Episcopal Church and 203, 224, 226, 229, 367
see also canticles; fundamentalism; Gaelic Bible; psalms
Bible classes 229, 454, 455, 457, 459–60, 467
    see also Glasgow Bible College
Bible societies 396, 398, 484–5, 490, 492, 496, 511
    national headquarters 604
Bible Training Institute 604
Bible Women 504
bigamy 113
Biggar collegiate church 85, 90
Biggar presbytery 437, 439–40
Billings, R W 663
Birnie church 578
Birr, Synod of 34
Birre, Robert 90
bishops
    Aberdeen 577
    Catholic 105, 172
        Catholic schools 179, 180
        Celtic Church 575
        Church courts 117
        discouragement of Bible reading 159
        obligation to preach 344
        politics and 174, 181

        Scottish mission 149, 151, 155, 157, 158, 159, 161
        Second Vatican Council and 185–6
        see also vicars-apostolic
    Church of Scotland 110, 112, 128, 130–1, 438, 439
        see also superintendents
    Councils of Arles and Rimini 16
    Presbyterian view of 131, 141, 216
    Scots Confession and 582
    Scottish Episcopal Church 10, 191–2, 194, 218, 223
        audience with George IV 198
        blessing by 204
        confirmation by 197
        Lambeth Conference 205
        newspapers and 200, 216
        origins 194–5, 208–9, 227
        Reformation 4th centenary celebrations 217
        Seabury's consecration 192, 205, 217
    see also Episcopalian clergy
Bishops' Report, the 216, 423
Black, A & C 396
Black Acts 111
Black Isle 34, 212, 318–19, 587, 603, 608
Black Isle Bible Society 492
Black, John, *Lessons on the Psalms* 640–1
Blackadder, Rev. D 484
Blackadder, Robert, archbishop of Glasgow
    Glasgow Cathedral 50, 56, 75, 80, 88
    prayer book 67, 71, 83
Blackburn, Jemima 665
Blackfriars Church, Glasgow 398
Blackhall, Andro 637
Blackie, Professor 541
Blackie (publisher) 396, 664
Blackwood 396
Blaikie, Rev. W G 493, 494, 522
Blair, Hugh 9, 385, 447
Blanc, Hippolyte 589
Blantyre, Malawi 522, 525, 528
blasphemy 128, 130, 547
Blind Hary 381–2
Board of National Mission, Church of Scotland 512
Bodley, G F 595
Body and Soul bookshop, Edinburgh 321, 329
Boece, Hector 7, 381
Bogue, David 240, 517
Bombay 496, 517, 520, 523, 526, 527

Bombay Harbour Mission 496
Bombay University 520, 523
Bonaparte, Louis Lucien, prince 352, 353
Bone, Phyllis 669
Bo'ness, Carriden church 591
Boniface, bishop of Mainz 26
Bonkil, Edward 658
Bonnington Mill 243
Bonnybridge 319
Book of Armagh 44, 46
Book of Common Order 344, 346, 365, 478
Book of Common Prayer 196, 197, 199, 201, 209, 218, 477
    'Churching of Women' service 204
    diminution of importance 225, 227
    evangelical Episcopalians 198, 201
    funeral services 204, 477
    language 346
    qualified congregations 192, 194, 196
    see also Common Prayer (Scotland);
        Scottish Book of Common Prayer;
        Scottish Liturgy
Book of Deer 52, 53, 364, 655
Book of Durrow 652, 654
Book of Kells 50, 654
Book of Pluscarden 65
Book of the Dean of Lismore 365
books 81, 159, 338–9, 384–5, 394–411
    censorship 115
    copying at Iona 32
    death preparations 78
    Episcopalian 219
    illustrations 67, 664–5, 666, 668
    in Gaelic 365, 366
        see also Gaelic Bible; Gaelic
            literature; Gaelic psalters
    in schools 115, 158
    in Scots 338–9
    ministers and 446, 447
    monarchy and 63
    read during church services 81, 85
    St Patrick's church, Edinburgh 177–8
    see also Bible; Books of Hours;
        catechisms; gospel books;
        hagiographies; hymns; literacy;
        literature; missals; prayer books;
        psalters; Scottish Liturgy; sermons;
        tracts
Books of Hours 63, 71, 81, 657
Booth, William 248, 249, 410
Bòrd na Gàidhlig 375
Borders
    Catholic community 143, 175
    see also Berwickshire; Dumfriesshire;
        Galashiels; Jedburgh; Melrose
        Abbey; Peebles; Traprain Hoard
Boroughmuir High School 606
Borrowman, Alex 357, 358
Boston, Thomas 386, 404
Boswell, James 386, 662
Bothans collegiate church 89
Bothwell parish, mission station 501
Bough, Sam 665
Bourignon, Madame 194
Bowden church 592, 598
Bower, Walter, abbot of Inchcolm 47, 381, 382
Bowmore church, Islay 587
Boys Brigade 221, 276, 458–60, 561
Bragg, Melvyn 420
Brahma Kumaris 317
Breadalbane, home mission 482
Breadalbane Folklore Centre 319
Brec Bennoch (Monymusk Reliquary) 639
Brechin, round tower 577
Brechin Cathedral 581, 583, 627, 656, 657
Brechin Society for Missions, Schools and Tracts 484
Brelwi movement 291
Brewster, David 664
Bridei son of Derilei, king of the Picts 34
Bridei son of Meilochon, king of the Picts 22–3, 29, 629
Bridgeton, Glasgow 218, 507
Bridgeton Village Mission 494
Bridie, James 185, 390
Bristo Baptist church, Edinburgh 239
British and Foreign Bible Society 398, 484, 485, 490
British Broadcasting Corporation *see* BBC
British South Africa Company 528
broadcasting 295–305, 413, 414–21, 425–7, 428, 547
    Asian community and 299
    Church of Scotland and 296, 297, 298, 300, 417
    Hindu community and 298, 301, 302, 303–4, 426
    hymns 414, 416, 421, 457
    Indian community and 299
    Islamic faith and 302, 426
    Jewish community and 298, 301, 303
    Muslim community and 297, 298, 300, 301, 303, 304
    see also radio broadcasting; television broadcasting

Brodick, Arran 460
Bronckhorst, Arnold 661
Broughton church
    Edinburgh 609
    Peeblesshire 603
Brown, Dr Callum 455, 460, 492
Brown, David 257
Brown, George Mackay 390, 391
Brown, Gordon 475
Brown, John 90
Brown, Rev. R F 618
Brown, W E 183
Browne, Robert 238
Brownists *see* Congregationalist churches
Bruce, F F 247
Bruce, Marion 89
Bruce, Sir William 586
Brude *see* Bridei
Bruges 633, 658
Bryce, David 591
Bryce, James 519
Buccleuch church, Dalkeith 599
Buccleuch, duke of 208, 437
Buchan *see* north east Scotland
Buchan, James 419
Buchan, John 410, 422, 423
    John Buchan Centre 603
Buchanan, Claudius 518–19
Buchanan, Dugald 366, 368
Buchanan, George 113, 384, 385, 661
Buchanan, Rev. Robert 493–4, 495
Buddhists and Buddhism 301–2, 303, 304, 308, 310, 315, 426
buildings *see* architecture; church buildings; mosques; sacrament houses; synagogues; temples
bull horns 630
Bunney, Herrick 465
Bunyan, John 366, 664, 666
Burgess Oath 133
burgh councils, pre-Reformation 63–4
Burghers 133, 138, 518, 534
burial sites
    Jewish community 259–60, 261, 268–9
    Malcolm III and Queen Margaret 57
    near converted churches 603
    pre-Reformation 75–6
    St Ninian 22, 23, 24, 25, 55
    Scottish Episcopal Church 206
    Stewart, James, earl of Moray 476
    see also funerals; kirkyards; tombs
Burleigh, J H S 405
Burleigh, Michael 427

Burn, William 581, 587
Burns, Robert 387, 445
    artists and 662
    'Cottar's Saturday Night' 396, 637, 662, 666
    'Holy Fair' 387, 469, 471
    'Holy Willie's Prayer' 351
    'Inventory' 393
    'Ordination' 387, 639
    paintings and 662
    Scots language and 348, 354
    Skinner (Tullochgorum) and 195
    'Tam O Shanter' 387
Burntisland, Fife 209, 584–5, 618
Burra Isle 501
    see also Papil Stone
Burrell Collection, Glasgow 307
Bute 34, 460, 499, 509, 595
Bute, 2nd marquis of 491
Bute, 3rd marquis of 178
Butterfield, William 595
Buxton, Sir Thomas Fowell 525

C

Cabell, Alexander, canon 89
Cadder, Glasgow 32, 501
Caddy, Eileen and Peter 324
Cadell, Francis 666
Caer Clud 317
Caird, Rev. John 400
Caithness 244, 491
Caithness, earl of 150–1
Calcutta, foreign mission work 519, 520, 523, 526, 527
Calderwood Lodge 271
Caledonii 17
Callander, Maclaren leisure centre 322
Callanish, Lewis 317, 575
Calton Hill, Edinburgh 317, 605, 642, 647
Calvinism 8, 106, 141, 236, 237, 253
    art and 309, 660–1
    education and 385, 524
    Evangelical Union rejection of 241
    Scottish Enlightenment and 385
    Scottish Episcopal Church and 194, 200, 386
    Shorter Catechism and 394
    view of state's role 113
    Westminster Confession 140
Cambuslang 240, 500
Cameron, Rev. Alexander 372
Cameron, D Y 666

682 • SCOTTISH LIFE AND SOCIETY

Cameron, Donald 198
Cameron, Henry Paterson 355, 356
Cameronians 133, 138, 518
Campbell, Agnes 657
Campbell, Alexander 242
Campbell, Angus Peter 390
Campbell, Archibald, bishop of Aberdeen 195
Campbell, Daniel 587
Campbell, Donald, archbishop of Glasgow 172
Campbell, Rev. Dugald 367, 369
Campbell, Rev. Duncan MacGregor 372
Campbell, John 397
Campbell, Rev. John Gregorson 372
Campbell, Dr John Lorne 373
Campbell, John Mcleod 208, 398, 400
Campbell, Neil, 10th duke of Argyll 223
Campbeltown 173, 540
Camphill Community 318
Canada 173, 278, 355, 358, 480
Canal Boatman's Institute, Glasgow 497
Candida Casa, Whithorn 22, 23, 24
Canna, St Edward's church 609
canon law 118
    see also Code of Canons, Scottish Episcopal Church
Canongate (publisher) 425
Canongate Boys' Club 507
Canongate Hall, Edinburgh 534
Canongate kirk, Edinburgh 586
Canonmills Baptist church, Edinburgh 244
canticles 209, 637, 645, 646
Cape Colony
    foreign mission work 517, 520–1, 522, 524
    see also South Africa
Capella Nova 465
Carbost, Skye 371
Cardonald church 591
Careston church bell 619
Carey, Kenneth, bishop 223
Carey, William 519
Carfin, shrine 184
Carfin Pilgrimage Centre 318
Carloway, Lewis 375
Carluke, Lanarkshire Christian Union 500
Carlyle, Alexander 386, 437
Carlyle, Thomas 388–9, 394
Carmichael, Alexander 371
Carmichael, Katherine, lady of Cambusnethan 86
Carmichael, Robert 239

Carmunnock Bible and Missionary Society 485
Carnegie, Andrew 593
Carnie, William 643
Caroline Divines 194
Carrick, Alexander 669, 670
Carriden church, Bo'ness 591
Carrington church, East Lothian 609
Carroll, Paul Vincent 185
Carron, river 444
Carrubers Close mission, Edinburgh 248, 489
Carswell, John, superintendent of Argyll 365, 367
carvings *see* sculptures
Carvor, Robert 633, 635
catechisms
    Catholic 72, 78, 81, 90, 159, 342
    Church of Scotland
        education and 115, 156, 347, 483, 484
        ministers' duties 439, 442, 470
        omitted from communion preparation 471, 472
    language 340, 342, 367, 394
    Scottish Episcopal Church 197, 202, 224
    see also Larger Catechism; Shorter Catechism
catechists
    Church of Scotland 486–7, 491, 639
    colporteurs as 493
    Highlands and Islands 194, 371, 480, 483, 486–7, 491, 639
    Scottish Episcopal Church 194, 197
    SPGH 240, 482
    SSPCK 480, 483, 491
Cathedral of the Isles, Bute 595
cathedrals *see* names of individual cathedrals
Catholic and Protestant mixed marriages 216
Catholic Apostolic church 214, 607
Catholic Choral Society, Glasgow 178
Catholic Church 236, 315, 550
    baptism 62, 86, 144, 147, 159
    Bible 159, 335, 338, 340, 356, 369
    canon law 118
    catechisms 72, 78, 81, 90, 159, 342
    censorship by 115
    church attendance 153, 173, 177, 184, 187
    church government 143, 145, 151, 162, 172, 173, 175–7

courts 117
early Christianity 15–41, 575
ecumenism and 422
education and 105, 175, 188
Gaelic Bible 369
Gaelic language 364–5, 366, 369, 372–3, 374
Highlands and Islands 112, 147, 155, 160, 162, 173, 372
home missions 144–5, 146, 149, 157, 175, 495
land and property 105, 108, 110, 143
languages 185–6, 356
Latin services 58, 66, 186, 336, 356, 382
literature and 382
liturgy 61, 86, 159–60, 185–6, 337, 382, 630
  see also masses
Lowland Vicariate 162, 173
marriages, Penal Laws and 147
membership in Scotland 162–3, 170–3, 177, 182, 186, 466–7
monarchy and 105, 154, 548
parishes 184–5
post-Reformation 8, 11, 62–3, 112, 143–90, 536, 548
prayer 67, 88–91, 147–8, 153, 159
pre-Reformation 19–36, 60–91, 104–5, 364–5
preaching 77–8, 344
Protestant Reformation and 105–6, 108, 110, 143, 365, 435, 582
psalms 147
radio broadcasting 416
rosary 81, 147, 153
St Mungo Museum displays 307, 309
Scottish Mission 145, 147–9, 151, 154, 155, 157–63, 174
  see also vicars-apostolic
Scottish Parliament (–1707) and 62, 145–6
Scottish-Asian Christian Fellowship 255
seminaries 155, 157–8, 162, 177
temperance movement 495
welfare responsibilities 105, 115, 173, 175, 177, 184–5
  see also Celtic Church; Counter-Reformation, the; Protestant Reformation, the
Catholic church buildings 157–8, 161–2, 172–3, 575–81, 588, 594, 595–6

Catholic clergy 172, 173, 184, 187, 188
  education 145, 148, 154, 157–8, 162, 177, 434
  Gaelic culture and 372–3
  Gaelic-speaking 144–5, 364–5, 374
  Highlands and Islands 144–5, 149
  home missions 144–5, 495
  Irish 151, 172, 173, 174, 175, 188
  music and 630, 631, 632, 634, 635
  Penal Laws and 147–9
  pre-Reformation 61, 105, 434, 580
  Protestant Reformation and 106, 108, 143, 365, 435
  see also Benedictine order; benefices; bishops, Catholic; Dominican order; Franciscan order; Jesuit order
Catholic community
  Aberdeenshire 112, 171, 176
  Argyll 173, 176
  art 61, 63, 66, 71, 74–5, 80, 670
  Ayrshire 171, 173, 179
  Banffshire 143, 152, 170
  Borders 143, 175
  Catholic/Protestant marriages 216
  Clyde Valley 171, 173
  drama 185
  Dundee 171, 173, 176, 178
  Edinburgh 163, 171, 172–3, 174, 176, 180–1
  education 156–9, 173, 179–80, 183, 186
    schools 156–9, 179–80, 184, 216, 540, 541, 548
  funerals 86–7
  Gaelic scholarship 372–3
  Galloway 176
  Glasgow 163–4, 171, 172–3, 176, 177, 178, 182
    education 179, 183
    politics 174, 178
  Greenock 163
  Highlands and Islands 144–5, 146, 151, 159–60, 163, 170, 181
    schools 156–7, 158, 179
  Irish immigrants 11, 163–4, 171–2, 173, 175, 315, 495
  Lanarkshire 173–4, 179
  literature 61, 78, 80, 86, 185, 382
  Lithuanian immigrants 178–9
  Lothians 175
  Lowlands 147, 151, 163
  music 187, 646, 647, 649
  New Abbey 143
  Newton Stewart 173

north east Scotland 61–2, 143, 170, 178, 179
Orkney 155, 162
Paisley 172, 176
persecution 145–55
Polish immigrants 182, 315
politics 153–4, 160, 161, 174–5, 178, 179, 180–4
Stirlingshire 175
temperance movement 174, 177, 178, 495
women in 153, 157, 180
World Wars I and II and 181–3
Catholic Eucharistic Congress 183
Catholic Schools Society 174
Catholic Social Guild 183
Catholic Truth Society 183
Catholic Union 183
Catholic Young Men's Societies 177, 183
Catholicism *see* Catholic Church; Catholic community
Catstane, Kirkliston, Midlothian 19–20
Cecil, Robert, 3rd marquess of Salisbury 528
Celtic art 630, 652–4, 658
Celtic Church 6–7, 575–6, 577, 630
  see also St Columba; St Ninian
Celtic Football Club 309
Celtic Psalter 655
Celtic revivalism 666, 668
'Celtic' spirituality 313, 319
Celtic Trails Scotland 321
Cenél Conaill 26, 27
censorship 115
Central Halls, Methodist Church 245
Central Mosque, Glasgow 290
Ceolfrith, abbot of Jarrow 21–2, 23, 24, 25, 26
Ceres church, communion table 587
Chalmers Close, Edinburgh 605
Chalmers, Peter MacGregor 591, 592, 595
Chalmers, Thomas
  Bible societies 398
  communion service 588
  Disruption and 387
  foreign missions 523
  home missions 485–6, 493, 494, 495, 496, 497, 500, 509
  Morningside church, Edinburgh 606
  publications 398
  sermons 398
  see also St John's parish, Glasgow
Chambers, Robert 389, 405

Chambers, William 389, 591
Chanda diocese, Scottish Episcopal Church links with 214
Channon, Thomas 639
chants 630, 646
Chapel Royal 633, 638
chapels *see* church buildings
Chapman, Walter 65
charismatic groups 236, 251–2
charitable trusts 542–3
Charles I, king of Great Britain and Ireland 434, 436, 585
Charles, Prince 427
Charlotte Baptist Chapel, Edinburgh 244
Charteris, Dr Archibald 462, 560, 561, 563
Chartist movement 641
Chepman, Walter 396
Chesterholm 19
Chewa language 525
children, medical treatment 545, 548
children's books 406
children's hostel, Glasgow 264
Children's Special Service Mission 499
China, foreign mission work 527–8
China Inland Mission 527
Chinese College, Malacca 527
Chinese community 11, 315
Chirwa, Orton 529
Chirwa, Dr Vera 529
Chitambo's village, Zambia 521
Chiume, Kanyama 529
Choir of the Thousand Voices 643
choirs 463–5, 561, 593, 638–9, 642–4
  Aberdeen 64, 464, 638, 639
  Dunkeld Cathedral 632
  Edinburgh 206, 463, 464, 642–3
  Glasgow 463, 464, 641
  Granton 568
  home missions 506
  Monymusk 639
  Paisley Abbey 463, 465
  practice verses 352, 640
  Richmond 568–9
  Scottish Episcopal Church 206, 219
  socialist movement and 641
  song schools and 638
  Stirling 638
  Sunday schools 457, 465
  see also song schools
choral societies, Jewish community 275
Chovevei Zion 274
Christ *see* Jesus Christ
Christ Church, Bridgeton, Glasgow 218

INDEX • 685

Christian, definition 235, 236
Christian, duke of Holstein 63
Christian Aid Collection 229
Christian Brethren 247–8, 496
Christian Centre Ministries 607
Christian College, Madras 526
Christian Science 252, 317
Christian Social Union 214
Christianity
   acculturation 6
   alternative beliefs and 327
   diversity 315
   early 15–41, 575–6
   Picts and 21, 22–3, 29, 36, 575, 629, 666
   religious broadcasting 295–8, 302, 303, 304
Christians Linked Across the Nation (CLAN) Gathering 252
Christie, William 246
Christmas 6–7
   Church of Scotland 9, 472, 504, 546
   Free Church of Scotland 504
   Scottish Episcopal Church 213, 218, 223
Chryston mission station, Glasgow 501
Church Accommodation Committee, Church of Scotland 494
Church and Missionary Department, Church of Scotland 466
Church and Nation Committee, Church of Scotland 9, 390, 422, 423, 536
church architecture *see* architecture
church attendance 11, 313, 416, 426, 452
   Catholic Church 153, 173, 177, 184, 187
   Church of Scotland 141, 145–6, 156, 455, 466–7
   compulsory 145–6, 153, 156, 177
   mission services 489, 508
   monarchs 63
   Scottish Episcopal Church 211, 218
church bells 68, 75, 90, 91, 598, 619–20
   see also bell shrines; handbells
church buildings 55–7, 575–97
   Baptist
      Ayr 610
      Edinburgh 239, 244, 598
      Glasgow 244, 268, 589, 609
      Paisley 589
      Tiree 371
   Catholic 157–8, 161–2, 172–3, 575–81, 588, 594, 595–6
   Catholic Apostolic church 607
   Celtic Church 575–6, 577
   Christian Science church 252
   Church of Scotland 9, 582–94, 596–7, 599, 613, 616–18
      Church Extension policy and 506
      Disruption and 10
      financial responsibility for 533, 587, 594
      First Book of Discipline 582, 619, 622
   conversion 12, 268, 598–615
   Free Church of Scotland 135, 588, 589
   furnishings 616–28, 658
   home missions and 221, 493–4, 506
   pilgrimage and 580
   protection 614–15
   Protestant Reformation and 581–5
   relocation 604–5
   Scottish Episcopal Church 202, 205–6, 209–10, 211, 212, 215
      as community centres 211, 228–9
      Home Mission Campaign 221, 506
      interiors 227
      meeting houses 191, 197
   singing lofts 639
   see also abbeys; architecture; mosques; synagogues; *and* names of individual buildings
Church Extension policy, Church of Scotland 505–6, 511, 593–4
church government
   Act of Union and 9, 112, 191, 531, 532
   Assemblies of God 250
   Catholic Church 143, 145, 151, 162, 172, 173, 175–6
   Christian Brethren 247
   Church of Scotland 107, 109, 111–12, 113, 127–42, 532–6
      see also bishops, Church of Scotland; courts; *First Book of Discipline*; kirk sessions; presbyteries; *Second Book of Discipline*; synods
   Church of the Nazarene 248
   Congregational Union of Scotland 241
   Evangelical Union 241
   Free Church of Scotland 135, 136
   Methodism 244
   Salvation Army 248–9
   Scottish Episcopal Church 197, 205, 224–5
   Scottish Parliament (–1707) and 132, 532
   United Presbyterian Church 136
   see also courts; Episcopalianism; kirk sessions; Presbyterianism; presbyteries; synods

church halls 453, 465, 612
  see also mission halls
Church Hill Theatre, Edinburgh 601
Church House, Glasgow 507
church lighting 589, 627
  see also stained glass
Church Missionary Society 199
Church of England
  charismatic movement and 251
  Church of Scotland and 141, 214, 216, 417
  ecumenical dialogues 214, 216
  foreign mission work 519
  Harvest Thanksgiving 218
  liturgy 132–3, 200, 205
  monarchy and 548–9
  Protestant Reformation and 103–4
  St Thomas church, Edinburgh 498
  Scottish Episcopal Church and 192, 198, 205, 208, 209, 213, 218
  see also Book of Common Prayer
Church of England clergy 434, 435, 440
Church of England in Scotland ('English Episcopal Church') 198, 200, 205
Church of Ireland 202, 247, 367
Church of Jesus Christ of the Latter-Day Saints (Mormons) 235, 252, 253
Church of North India 214
Church of Scotland 8–9, 106–18, 130, 315, 550
  Act of Union and 9, 103, 104, 146, 531, 532, 536
  alternative beliefs and 317, 321, 329
  anti-Catholic report 182–3, 313
  Articles Declaratory 130, 141, 532, 535
  baptism 244, 472, 474, 582, 583, 622
  Baptists and 243
  Basis and Plan of Union 532, 535
  Bible and 340, 395, 442, 456, 484, 637
    in services 9–10, 395, 438, 439, 456
  Board of National Mission 512
  broadcasting and 296, 297, 298, 300, 417
  catechisms
    education and 115, 156, 347, 483, 484
    ministers' duties 439, 442, 470
    omitted from communion preparation 471, 472
  catechists 486–7, 491, 639
  censorship by 115
  charismatic movement and 251, 252

Christmas 9, 472, 504, 546
Church Accommodation Committee 494
Church and Missionary Department 466
Church and Nation Committee 9, 390, 422, 423, 536
church attendance 141, 145–6, 156, 455, 466–7
church buildings 9, 582–94, 596–7, 599, 613, 616–18
  Church Extension policy and 506
  Disruption and 10
  financial responsibility for 533, 587, 594
  First Book of Discipline 582, 619, 622
church choirs 463–5
Church Extension policy 505–6, 511, 593–4
church government 107, 109, 111–12, 113, 127–42, 532–6
  see also courts; Episcopalianism, Church of Scotland; First Book of Discipline; Second Book of Discipline
church halls 453
Church of England and 141, 214, 216, 417
civil authority and 111, 112, 117, 134, 141
  see also monarchy
Committee on the Instruction of Youth 455
Communion 439, 469–72, 562–3, 582–3, 587–8, 623, 624
Communion cups 472, 569, 596, 660
Communion tables 622–4
  Burntisland church 585
  Communion served from 471, 588, 592, 623
  Communion taken seated at 469, 470, 471, 582–3, 587–8, 617, 622–3
Congregationalists and 238, 239, 242, 482
courts 117, 118, 127, 130–1
  see also General Assembly of the Church of Scotland; kirk sessions; presbyteries; synods
disestablishment called for 134, 208
dramatic clubs 466
ecumenical activity 214, 216

education role 244, 395, 426, 435, 538, 540, 541
   catechisms 115, 156, 347, 483, 484
   education of ministers 9, 385, 435, 437, 441, 444, 446–7
   First Book of Discipline 107–8, 114–15, 128, 540
   home mission work 484
   parishes and 113, 443–4, 532, 533, 540
elders 549
   as moderators 137
   as visitors 486
   Communion and 472, 623, 624
   election of 128
   funeral services and 477
   ministers' appointment 438
   women 138, 462
Episcopalianism 109, 111, 112, 131, 194, 367, 440, 441
   church buildings 585–6
   Communion 469
   First Book of Discipline 107
   last Episcopal incumbent 194
   Second Book of Discipline 111
   see also bishops, Church of Scotland
evangelism and evangelicals 386, 485–7, 507, 509, 522–3
   Disruption and 532, 533
   Haldane brothers 240, 482
   opposition to 482, 518
   Summer Mission 457–8
finances 110–11, 128, 435, 447, 533
   church buildings 533, 587, 594
   First Book of Discipline 107, 108, 110, 128, 436
Foreign Mission Committee 519, 520
foreign missions 517, 518, 519–20, 522–3, 524, 526–7, 528
funerals 476–8, 546
Gaelic language and culture 365, 367, 370, 372, 373, 374, 375
Gaelic psalter 369, 370
Gratis Sabbath school movement 483–4
Great Charter 111–12
Haldane brothers and 240
high church movement 591
Highlands and Islands 144, 482, 492
   Argyll 367, 369, 370, 492
   catechists 486–7, 491, 639
   church buildings 587
   presbytery meetings 439, 443
   schools 365–6

Home Board 504, 506, 507, 509
Home Mission Committee 494, 501
home missions 483, 484, 486–7, 495, 501–2, 504–9
Iona Community 506
James VI, king of Scots and 112, 439, 441, 585, 623
land and property 107, 108, 110–11
language and 339–47, 349, 357
   Gaelic 365, 370, 373, 374, 375
law and 531, 532–6
   see also Act of Union
literature and 386, 390
marriage 474–6
media and 423
membership 141, 452, 507–8
   decline of 10–11, 12, 455, 466, 509, 549
Methodism and 244–5
monarchy and 109, 111, 112–13, 131, 536, 549
   accession oath 132, 536
   appointment of bishops 110, 112, 130, 131, 439
   calling of General Assemblies 112, 130
   imprisonment of ministers 441
   see also names of individual monarchs
moral governance 114, 128, 130, 141, 443, 539
music 593, 625, 629, 637–46, 647
'New Age' religions and 316
New Charge Development 506
parishes 105, 113, 439–40, 443–4, 532, 533, 540
   welfare responsibilities 113, 115, 135–6, 532, 533, 542
patronage and 112, 437, 438, 441, 443, 445
   secessions 133, 134, 240, 532, 533–4
politics and 113, 385, 440–1, 536
prayer 435, 438, 442, 476, 477–8
Probationers' Society 487, 488
Quakers and 238
readers 435, 620, 625
St Andrew Press 418
schools 115, 156, 365–6, 370, 395, 426, 435, 540
   catechisms in 115, 156, 347, 483
   travelling people 504
Scottish Episcopal Church and 192, 198, 208, 216, 591

Scottish Parliament (–1707) and 106–7, 108–10, 113–14, 128, 130, 131, 532
Seaside/Summer Mission 457–8
secession from 133–41, 240, 313, 349, 441, 518, 532, 533–4
Second Book of Discipline 104, 111, 112
secularisation and 10–11
sermons 10, 344, 438, 470
Social Welfare Committee 536
Socinianism and 246
Sunday schools 454–7, 483–4, 486, 518, 612
superintendents 107, 109, 110
synods 111, 130, 131, 132, 136, 437
Union and Readjustment Committee 502
union with other churches 141, 535
union with United Free Church 138, 141, 216, 535, 537, 629, 646
universities and 9, 114–15, 156, 385, 395, 426, 541–2
First Book of Discipline 107–8
ministers and 9, 435, 437, 439
welfare responsibilities 113, 115, 135–6, 532, 533, 542
ministers and ministers' wives 441, 443, 444
poor relief 113, 115, 128, 136, 395, 435, 443, 532, 542
Westminster Confession and 132, 140, 346, 532, 534, 535
witchcraft and 129–30
Woman's Guild 138, 461–3, 561, 612
women 137–8
church seating 618
elders 138, 462
foreign mission work 526–7
ministers' wives 441–2, 446, 462, 611, 612
ordination 462
Youth Fellowship 460–1
see also Calvinism; Disruption, the; General Assembly of the Church of Scotland; kirk sessions; Protestant Reformation; Scots Confession
Church of Scotland ministers 107, 109, 110, 114, 433–51, 563
Anabaptist beliefs 243
appointment 112, 133, 240, 438, 439, 441, 445
see also patronage
as teachers 437, 438
baptisms and 472, 474

careers 437–40, 564
catechisms and 439, 442, 470
charismatic movement and 251, 252
Church of England clergy and 434, 435, 440
Communion preparations 470
congregations and 442–5, 534
education of 9, 385, 435, 437, 441, 444, 446–7
elders and 438
election 128
finances 111, 445, 462
benefices 107, 108, 110
stipends 114, 433, 434, 435–7, 533, 587
funerals and 477, 546
Gaelic-speaking 366, 375, 438
General Assembly and 434, 442, 444, 446, 447, 486
home missions 486–7, 491, 493, 494, 501
House of Commons and 548, 550
imprisonment 441
in literature 349–50, 387
judicial and other posts forbidden 117–18
kirk sessions and 137, 438, 439, 443, 612
last Episcopalian 194
libraries 446
literacy and 396
marriage ceremonies and 475
marriage of 441–2
media and 423
misdemeanours 442, 445–6, 447
missionary ministers 487, 491, 494, 501
parishioners and 114, 115, 443–4
politics and 440–1
presbyteries and 437, 438, 439, 443, 444, 447, 487
quoad sacra parishes 533
Revolution (1688) and 440, 441, 444–5, 446
Royal Bounty Committee 486
Scots Confession and 110
Scots language and 349–50
Scottish Enlightenment and 9
Seaside/Summer Mission 457, 458
sexual conduct 442, 446
social origins 435–7
status 387, 389–90, 433–51, 611, 613
teachers and 443–4
time spent in parishes 439–40
universities and 435, 437, 439
welfare responsibilities 441, 443, 444

women as 138
see also bishops, Church of Scotland; Episcopalian clergy; glebes; manses; sermons
Church of Scotland ministers' wives 441–2, 446, 462, 611, 612
Church of the Nazarene 248
church organs 465, 498, 593, 597, 625, 641–5, 646
Church Service Society 477, 591
Church Sisters 504
Church without Walls, A, radio series 463
'Church without Walls' report 508
Church Women's Mission Association 221
Churches Agency for Interfaith Relations in Scotland 311
Churches of Christ 242
Churches of God 247–8
churchgoing *see* church attendance
'Churching of Women' service, Scottish Episcopal Church 204
Cimang'anja language 525
cinemas 427, 599
see also films
Circulating Tract Society, Haddington 485
Citizens' Theatre 185
civil courts *see* courts
civil marriages 476
Clach an t-Sagairt 147
Clackmannan church 589
Clackmannanshire 17, 34, 46
Claim of Right 109
CLAN (Christians Linked Across the Nation) Gathering 252
Clark, Dr Lynn Schofield 427
Clark, T & T 396, 404–5
Clarke, Dr Adam 483
Clarke, Tristram 195
Clarkson, Thomas 284
Classical Gaelic 365, 367, 368, 370
Claverhouse, Graham of, Viscount Dundee 629
Clear, Claudius 407
Clearances, the 173
Clement, bishop of Dunblane 580
clerks plays 382
Clonmacnoise, St Columba's visit to 27
Cluny Hill Hotel, Forres 325
Clyde Valley, Catholic community 171, 173
Clydebank, Dalmuir Old church 604
co-operation between churches *see* ecumenical activity
Coaltown of Wemyss, mission station 502

Coatbridge 173, 500
Coats Memorial Baptist church, Paisley 589
Cobden Street mission station, Glasgow 501
Cockburn, Henry, Lord Cockburn 348
Cocker, W D 424
Code of Canons, Scottish Episcopal Church 198, 201
Cogitosus 22
Cohen, Isaac 257
Coia, Jack 595
Coigrich 656
Coldingham Priory, Berwickshire 86, 579
Coleridge, Samuel Taylor 9
collection ladles 625
College (Blackfriars) Church, Glasgow 398
College of Bishops 194, 227
College of Justice 117, 118
collegiate churches, creation of 580
Collins (publisher) 396
colporteurs 397, 492–3
see also pedlars
Colt, Maximilian 661
Colvend church 591
Colville, John 500
Commissary Courts 118
Committee for Gaelic Missions, United Secession Church 491
Committee of Articles, Scottish Parliament (–1707) 131
Committee on the Instruction of Youth, Church of Scotland 455
Common Prayer (Scotland) 209
see also Scottish Book of Common Prayer
Commonwealth, the (17th century) 237, 238
Communion
  Assemblies of God 250
  Burntisland church 585
  Christian Brethren and 247
  Church of Scotland 439, 469–72, 562–3, 582–3, 587–8, 623, 624
  Churches of Christ 242
  Churches of God 247–8
  Congregationalists 239
  Elim Pentecostal Church 251
  Episcopalian churches 197, 469
  James VI, king of Scots 623
  Jehovah's Witnesses 254
  Knox, John and 469, 617
  Mormons 253
  open-air 372
  Quakers' view of 237, 238
  radio broadcast of 416

Richmond Craigmillar and Granton
    parish churches 568–9
Salvation Army 249
Scottish Episcopal Church 196–7, 202,
    209, 218, 220–1, 225, 227
service in Scots 357
Seventh-Day Adventists 253
Unitarian churches 246
see also Eucharist
Communion cups 209, 472, 569, 596, 660
Communion tables
  Church of Scotland 622–4
    Burntisland church 585
    Communion served from 471, 588,
        592, 623
    Communion taken at 469, 470, 471,
        582–3, 587–8, 617, 622–3
  Episcopalian churches 197
Comper, Sir Ninian 210
Comrie church 609
Conal mac Comgaill, king of Dál Riata 23,
    27
Concordat of Leith 110
Condon, Michael 174
Confession of Faith see Negative
    Confession; Scots Confession;
    Westminster Confession
Confucius 527–8
Congo Free State 521
Congregation for the Propagation of the
    Faith (*Congregazione de Propaganda
    Fide*) 145, 148, 149, 151, 154, 159,
    495
Congregational Federation 242
Congregational Union of England 242
Congregational Union of Scotland 240, 241,
    242, 482, 491
Congregationalist churches 11, 237, 238–42,
    518, 520
  Aberdeen 602
  Baptists and 243
  Church of Scotland and 238, 239, 242,
      482
  Church of the Nazarene and 248
  Dundee 239, 242
  Gaelic and 373, 375
  Glasgow 248, 599
  Highlands 482
  Sunday schools 483–4
  Tiree 371
Connachar, John 198
Connolly, James 178
conscientious objection 547–8

conservation areas 615
Constantine I (Roman emperor) 16
Constantine III (Roman emperor) 17
Constantine, St 34
Constantinople 398
Constantius, *Life of Germanus* 18
constitution, Church of Scotland 107, 532,
    535
  see also First Book of Discipline; Second
      Book of Discipline
Constitutional Associate Presbytery 134
Contin 404
Convention of Estates, Protestant
    Reformation and 106
Conveth kirk, St Laurence of Canterbury
    and 34
Cooper, Rev. James 591
Cooper, James Fenimore 410
Coptic Church 254–5, 604, 644
Cormac (monk) 29
Cormack, John 183
coronation oaths 109, 132, 337, 536
Corpus Christi altar, Linlithgow 89
Corpus Christi celebrations 64, 70, 80, 86,
    382
Corstorphine church, music 645
Cosgrove Care 264, 265
Cosgrove, Rev. Dr I K 268
Cottier Community Centre 599
Cottier, Daniel 664
Council of Arles 16
Council of Laodicea 252
Council of Nicaea 16
Council of Rimini 16
Council of Sardica 16
Council of Trent 118, 543, 633
Councils of Churches 242, 245, 250
Counter-Reformation, the 103, 112
Coupar, Sir Andrew 64
Court of Session 533, 534, 548, 587
Court of Teinds 533
courts 117–18, 130–1, 135–6, 533, 534, 535,
    537
  see also General Assembly of the
      Church of Scotland; kirk sessions;
      presbyteries; synods
Covenanters 133, 137, 441, 444, 445, 629,
    660, 662
  see also First Covenant, the; National
      Covenant
Cowcaddens church, Glasgow 602
Cowgate, Edinburgh, Hammermen's
    Chapel 51

Cowie 212
Cowley Fathers of Oxford 498
craft guilds *see* guilds
Craig, Very Rev. A C 418, 428
Craig church, Angus, conversion to domestic use 603
Craig church conversion, Angus 603
Craig Phatric, Inverness-shire 29
Craiglockhart college, Edinburgh 180
Craigton, Glasgow, Jewish burial site 268
Crail, Cross of 67
Crail collegiate church 83
Crailing church 592, 598
Cramond, Edinburgh 452, 460–1, 585
Crathie Church 400, 624
Crawford church conversion, Lanarkshire 603
Crawford, Rev. Dugald 366
Crawfordjohn church, Lanarkshire 603
cremation 229, 468, 478, 546
Crieff, St Ninian's church 604
Cripple League, Glasgow United Evangelical Association 497
Crockett, Samuel Rutherford 409, 410
Crofters' War 181
Cromar, Tom 453
Cromarty church, Black Isle 587, 608
Cromarty Ladies Missionary and Bible Society 492
Cromarty Religious Association 492
Cromwell, Oliver, army of 237, 243
Cronin, A J 185
Crosby, Bing 427
Cross of Crail 67
Crosshill synagogue 270
Crown, the *see* monarchy
crucifix pendants 72
Crucifixion, the, depictions of 66, 70, 71, 72, 79–80, 83, 87–8
Cruden 211
Cul Dreimne or Cul Drebene, battle of 27, 380
Cullen, Banffshire 62, 81, 89–90
Cullen, Paul, cardinal 175
Culross, Fife 32, 42, 46
Cumbernauld 215, 506
Cumbraes 34, 457, 458
Cumbraes Missionary and Bible Association 485
Cumméne, abbot of Iona 22, 26
Cumnock Society for Religious Purposes 485
Cunninghame 34

Cupar, Fife, Baptists in 243
Cupar Old church, Fife 587
curriculum, parish schools 115, 156
Currie, Rev. James 453
Curwen, John 644
cutty stools 618

D

Daiches, Salis, rabbi 269, 277
Daily Study Bible series 418
Dairsie church, Fife 585
Dál Riata 27, 363–4
    see also Aedán mac Gabráin, king of Dál Riata; Conal mac Comgaill, king of Dál Riata
Dali, Salvador 308
Dalkeith, Buccleuch church 599
Dalkeith presbytery, Church of Scotland mission stations 502
Dalmeny parish church 578, 656
Dalmuir Old church, Clydebank 604
Dalry, Ayrshire, Catholic community 173
Dalry House 206, 216
Damnonii 17
Dance of Death sculpture, Roslin chapel, Midlothian 74–5
Darby, J N 247
Darwin, Charles 389, 405, 538
Daughters of Charity 175
David de Bernham, bishop of St Andrews 578–9
David I, king of Scots 46–7, 57, 576–7, 578, 583, 598, 654
David II, king of Scots 65
Davidson, Randall Thomas, archbishop of Canterbury 208
Davitt, Michael 181
Day, Colin 417
deaconesses 137, 138, 504, 511
deacons 128, 135–6, 137, 138, 223, 486
Dean House, Edinburgh 660
Deer 34, 110, 364, 655
    see also Book of Deer
'democratic intellect', the 9
Dempster, Dr John A H 395, 396
denominations, definition 235
Deobandi movement 291
Derry, Columban monastery at 27
Deskford, Banffshire, sacrament house and sculptures 87, 657
devil, the, depiction of 80
devolution 227, 423
    see also Scottish Parliament (1999-)

Dewar, Rev. Daniel 372
Dey, Rev. James 418
Diana, Princess 649
Diarmait (St Columba's attendant) 30
Dicalydonae 17
Dickens, Charles 404, 410
Dickson, David 240
Dieckhoff, Fr Henry 373
Dilworth, M 160
Dingwall Ladies' Association for Aiding Bible and Missionary Societies 492
Dinwiddie, Rev. Dr Melville 296, 414, 415, 416, 420, 428
Diocletian 16
Directory for the Publick Worship of God 132, 346, 472, 474, 644
Disciples of Christ 242
Disruption, the 9, 10, 531, 532–4, 538, 540
    Annals of the Disruption 404
    art and 662–5
    Auld Light Burghers 138
    Chalmers, Thomas and 387
    evangelicals 370, 387, 532, 533
    foreign mission work 520, 522, 527
    patronage and 133, 532, 533–4
    see also Free Church of Scotland
divorce 113, 118, 225, 544
domestic uses, church conversions 602–3
Dominican order 158, 183–4
Donaldson, Alexander 89
Donaldson, Gordon 114, 226, 229, 426
Donaldson, Islay Murray 410
Donaldson, M E M 219
Donaldson, William 403, 409
Donegal, St Columba and 26
Donnelly, Michael 663
Dornoch, St Finnbarr 34
Dornoch Cathedral 424, 580–1, 583
D'Orsey (Episcopalian clergyman) 204
Douai 158, 159, 356
Douglas church, Lanarkshire, stained glass 51, 83, 84
Douglas, Gavin 78
Douglas, Hugh 417
Douglas, Sir James 381
Douglas, John 582
Douglas, marquis of 150–1
Douglas, Neil 246
Douglas-Home, Sir Alec 216
Doune church conversion, Perthshire 603
Dove Hill Relief Congregation, Glasgow 485
Dowanhill church, Glasgow 599

Dowanhill College, Glasgow 180
Dowden, John, bishop of Edinburgh 208, 209
Doyle, Sir Arthur Conan 318
drama 346, 390
    Ane Satyre of the Thrie Estaitis 338, 383–4, 633
    Catholic community 185
    Church of Scotland dramatic clubs 466
    clerks' plays 382
    craft guilds 72, 80
    Douglas 386, 447
    grammar schools 385
    Haliblude 337
    Jewish community 275
    mystery plays 337–8
    radio broadcasts 416
    Seaside/Summer Missions 458
    Sunday schools 457
    see also Corpus Christi celebrations; pageants; theatres
Drane, J 329
Druids 317
Druim Cett, Ireland 28
Drumclog 629
Drummond, Henry 401
Drummond, James 401, 668
Dryburgh Abbey 584
Dublin Street Baptist church, Edinburgh 244, 598
Dufay, Guillaume 633
Duff, Alexander 519, 520, 523, 524, 526
Dull church, Perthshire 604
Dumbarton Bible and Missionary Society 485
Dumfries 147, 252, 326
Dumfries parish church, Holy Rood altar 89
Dumfriesshire 315, 335, 608, 654, 661
Dumyat, Clackmannanshire 17
Dunadd 28
Dunbar 209, 661
Dunbar, Gavin, bishop 62
Dunbar, William 66, 80, 85, 87, 339
Dunbartonshire, hostel for Jewish refugee children 264
Dunblane, Scottish Churches House 597, 647
Dunblane Cathedral 580, 583, 591–2
Dunblane Music Group 647
Duncan, A A M 21, 23
Duncan, Helen 318
Duncan, John 668–9

Duncan, Thomas 662
Dundee
   altars 65, 72–3, 580
   baxters' missal 63
   burial sites 75–6, 268
   Catholic community 171, 173, 176, 178
   church maintenance 64
   Congregationalists 239, 242
   Corpus Christi pageant 70
   election of elders and deacons 128
   Gate Fellowship 251–2
   Glasite church 239, 587
   Indian community 284
   Irish Catholic immigrants 173
   Jewish community 257, 259, 268, 270, 271
   'kail kirk' 239
   Mid-Craigie 506
   mosques 288, 290
   Muslim community 288
   Presbyterianism 127
   St Andrew's church 239, 587
   St Mary's church 63, 580, 581, 583
   St Mary's Mothercraft Centre 216
   St Paul's church 595
   St Salvator's church 595
   synagogues 259, 270
   Unitarian church 246
   see also Fowlis Easter church
Dundee East Church Female Domestic Mission 495
Dundee Juvenile Bible and Missionary Society 484
Dundee Missionary Society 517
Dundrennan Abbey 579
Dunfermline
   alternative healing fairs 326
   Benedictine order 576
   Henryson and 382
   Jewish community 258, 270, 271
   pilgrimage centre 57
   St Leonard's church 591
   St Margaret's church 595
Dunfermline Abbey 57, 578, 583, 584, 665
Dunfermline and Kinross presbytery, Church of Scotland mission stations 501
Dunfermline Missionary and School Society 484
Dunglass, collegiate church 580
Dunkeld 50, 61, 176
Dunkeld Cathedral 583, 632
Dunkeld Missionary Society 485

Dunn, Jim 311
Dunn, Samuel 483
Dunnett, Rev. Arthur 504, 505
Dunning, St Serf at 46
Duns Scotus 381
Dupplin Cross 630
Dürer, Albrecht 665
Durham Cathedral 57, 577
Durie, John 629, 637
Durisdeer, Queensberry Monument, Dumfriesshire 661
Durness, minister of 640
Durobrivae 16
Duror of Appin, Episcopalian community 212
Durrow, monastery of 26–7, 32
Dyce, William 663, 665, 668
Dysart, St Serf at 46

E
Eaglesham, Jewish golf club 276
Earl Marischal 110
Earlsferry 56
East India Company 240, 284, 519
East Kilbride, Scottish Episcopal Church 215
East Lothian, Carrington church 609
East Renfrewshire Council 265
'East-West' alternative healing group, Aberdeen 321
Easter 83, 160, 213, 218, 229, 424, 472
   see also Holy Week
Easter Ross Ladies' Association for Aiding Bible and Missionary Societies 492
Eastern District, Catholic Church 173
Eckford church 592
ecumenical activity 155, 235, 236, 313, 422, 550
   Assemblies of God 250
   Baptists 244
   Catholic Church 422
   Church of Scotland/Church of England dialogues 214, 216
   Congregational Union of Scotland 242
   Crieff conference centre 604
   Elim Pentecostal Churches 251
   Glasgow 268, 311, 317
   home missions 504
   Jewish community 268
   Methodist Church 245
   music and 647
   Paganism and 317
   Quakers 238

Salvation Army 250
Scottish Churches House 597, 647
Scottish Episcopal Church 214, 216, 223, 224–5, 229
Seventh-Day Adventists 253
see also Glasgow Inter-Faith Committee; St Mungo Museum of Religious Life and Art; Tell Scotland movement
Edderton church 608
Eddy, Mary Baker 252
Eden Christian Fellowship, St Andrews 251–2
Edgar, king of Scots 46
Edinburgh
    altars 64, 65, 70, 89, 90, 580
    alternative healing fairs 326
    anti-Catholic riots 183
    Baptists in 239, 243, 244, 598
    baxters 65
    Bible societies 398, 485
    Body and Soul bookshop 321, 329
    Buddhism 315
    Burgess Oath 133
    burial sites 259, 661
    Calton Hill 317, 605, 642, 647
    Canongate Hall 534
    Catholic Apostolic church 607
    Catholic community 163, 171, 172–3, 174, 176, 180–1
    choirs 206, 463, 464, 642–3
    Christian Science church 252
    church buildings 586, 589
    church conversions 599, 601, 604, 605–7, 609
    church demolition 598, 599
    Church Hill Theatre 601
    Church of England 498
    Church of Scotland ministers' time in parishes 439–40
    Commissary Court 118
    Corpus Christi pageant 70
    Craiglockhart teacher training college 180
    Cramond 452, 460–1, 585
    Dean House 660
    Episcopal Theological College 218, 225
    Episcopalian congregations 192, 197, 198, 199, 200, 201, 225
    Filmhouse, the 599
    Flanders and 633
    Gaelic chapel 163, 446
    Gaelic preaching 376
    Gaelic Schools Society 490–1
    George Watson's school 461
    German congregations 254
    Glasites 239, 599
    Granton church 560–4, 568–70
    Greyfriars kirk 464, 585, 645, 664
    Greyfriars kirkyard 660, 661
    Hammermen's Chapel 51
    Highland Tolbooth 599
    Holy Trinity church 606–7, 609
    Holyrood Abbey 51, 446, 583, 584, 586
    home missions 248, 265, 487–8, 489
    hospitals 91, 115, 444
    Hub, the 599
    Indian community 284
    Jewish community 256–7, 260, 273, 277
        burial sites 259
        Christian missions 265
        education 266, 271, 274
        Israel and 274
        literary societies 274–5
        Medical Mission 265
        peddling 266
        politics 272, 274
        synagogues 259, 261–2, 269, 271, 272, 277, 315
        welfare organisations 262, 263, 265
        youth groups 277
    Jewish Representative Council 272
    Lady Glenorchy's chapel 137
    Methodist Church Central Hall 245
    mosques 11, 288, 290
    music 642, 646
    Muslim community 288
    Nasmyth tomb 661
    Netherbow Arts Centre 390, 599
    Old St Paul's church 219
    Old Town Information Centre 599
    Our Lady of Loretto image 62
    Paganism 317
    paintings 658, 660, 669
    politics 180–1, 272, 274
    praying societies 479, 483, 485
    psalm singing classes 639
    public penance 66
    Quaker converts 237
    Queen's Hall 599
    Richmond Craigmillar church 564–71
    Ross House 262
    St Barbara's light, Kirk o' Field church 70
    St George's church 606, 642, 646
    St George's Free church 591

St Giles' Day procession 86
St John's church 201, 204, 219
St Mary in the Fields church 90
St Mary's Cathedral 206, 209, 214, 228, 594, 595
St Patrick's church 177–8
St Thomas church 498
Salisbury Centre 321
Sciennes School 271
Scottish Enlightenment and 639
sculptures 76–7, 89
Sick and Burial Society 177
stained glass 51, 664, 665
Total Abstinence Society 177
Trades Union Congress 214
Trinity College 63, 580, 605, 658
Unitarian church 246
United Presbyterian Church seminary 524
United Presbyterian Mission 494
West Register House 606
Youth Fellowship 460–1
see also Holyroodhouse, Palace of; Leith; St Giles' church
Edinburgh Association of Ladies for the Advancement of Female Education in India 527
Edinburgh Bible Society 398, 485
Edinburgh Castle 63
Edinburgh Christian Mission 24
Edinburgh City Mission 487–8
Edinburgh Festival performance, St Mark's gospel 358
Edinburgh Festival Society 599
Edinburgh Gaelic Schools Society 490
Edinburgh Gratis Sabbath School Society 483
Edinburgh Hebrew Congregation 272, 273
Edinburgh Jewish Board of Guardians 263
Edinburgh Medical School 257
Edinburgh Missionary Society 480
Edinburgh presbytery 191, 519
Edinburgh Religious Tract and Book Society 397, 485
Edinburgh, Treaty of 106
Edinburgh University
    Chair of Celtic 372
    Dominicans 183–4
    Gaelic-speaking clergy 372
    Indian students 284
    Jewish Society 272
    Jewish students 256, 257, 258, 266
    legacies 444
    music 646
    prayers at graduation 12
    Students' Association 606
    Thomas Chalmers Chair of Theology 542
Edmonston, Archibald, of Duntreath 85
education 539–42
    alternative beliefs and practices 326
    alternative therapies 322
    Bible and 10, 394, 395, 396, 484
    Calvinism and 385, 524
    catechisms 115, 156, 347, 483, 484
    Catholic Church and 105, 175, 188
    Catholic clergy 145, 148, 154, 157–8, 162, 177, 434
    Catholic community 156–9, 173, 179–80, 183, 186
    schools 156–9, 179–80, 184, 216, 540, 541, 548
    church buildings converted for 605–6, 609
    Church of Scotland's role 244, 395, 426, 435, 538, 540, 541
    catechisms 115, 156, 347, 483, 484
    education of ministers 9, 385, 435, 437, 441, 444, 446–7
    First Book of Discipline 107–8, 114–15, 128, 540
    home mission work 484
    parishes and 113, 443–4, 532, 533, 540
    Episcopalian clergy 201, 203, 206, 218, 225, 374
    Faith Mission personnel 499
    Findhorn Community 325
    foreign mission work 519, 520, 523, 524, 525–7
    Gaelic language 365–6, 368
    holistic 321
    home mission work 483–4, 493
    Jesuit order's role 175
    Jewish community 258, 263, 266, 270–2, 274, 278
    lay influence on 389
    literature and 382–3
    missionaries 498
    music 639, 642, 646–7
    Muslim community 284, 288, 290, 291, 293
    prohibition of education abroad 158
    seceding churches and 136
    Shorter Catechism and 347, 393–4, 426

women 158, 488, 520, 526–7
see also literacy; schools; song schools; Sunday schools; teachers; universities
education committee membership 541
Edward III, king of England 57
Edward the Confessor, king of England 57
Edwards, Jonathan 240
Eeles, F C 204, 208
effigies *see* sculptures
Effingham, countess of 397
Eid prayers 288
Eigg, St Donnan 34
Eileach an Naoimh 575
elders
    Church of Scotland 549
        as moderators 137
        as visitors 486
        Communion and 472, 623, 624
        election of 128
        funeral services and 477
        ministers' appointment 438
        women 138, 462
    Episcopalian congregations 197
    Evangelical Union 241
    Seventh-Day Adventists 253
    United Free Church of Scotland 137
    see also kirk sessions
Elgin, praying society 517
Elgin Academy 540
Elgin presbytery 540
Elim Pentecostal Churches 251
Eliot, George 410
'élite' and 'popular' religion 60
Elizabeth I, queen of England 106
Elizabeth II, queen of the United Kingdom 609
Elizabeth, Queen (Queen Mother) 223
Elliot, Dr Alison 137
Elliot, James 587
Elliott, William 586
Elphinstone, William, bishop of Aberdeen 62
Elphinstone, William, bishop of Aberdeen, *Aberdeen Breviary* 34, 47, 382
embroidery 66, 227, 597, 627, 666
emigrants in transit 259, 263
emigration 155, 163, 173, 182, 213, 259, 278
Engels, Friedrich 538
English Baptist Order 244
English congregations, Scottish Episcopal Church *see* qualified clergy and congregations
English Episcopal Church 198, 200, 205
English immigrants 210, 211, 228, 315
English language 335–62
    Bible 344, 347, 356–7, 395
        Authorised Version 346, 352, 353, 367
        Catholic 159, 356
        importation of 340
    Episcopal services in Appin 204
    foreign mission work 519, 523–4
    Gaelic and 374–5, 376
    in schools 365–6, 368
    Jewish education in 263, 270, 274
    literacy and 347
    prayer in 335, 350–2
    Scots language and 335–62, 385
engraving 664, 665
Enlightenment *see* Scottish Enlightenment
Enzie, the, Catholic community 143
Episcopal Fund, Scottish Episcopal Church 198
Episcopal Synod, Scottish Episcopal Church 205, 207
Episcopal Theological College, Edinburgh 218, 225
Episcopalian churches 191–234
    Assemblies of God and 250
    church government 127, 197
    Communion 197, 469
        see also Eucharist
    influence 537–8, 550
    see also Church of England; English Episcopal Church; Scottish Episcopal Church
Episcopalian clergy 194–6, 197–8, 201, 208–9, 218, 223–4, 225–6, 227
    community care and discipline 211, 219–21
    dress 202
    education 201, 203, 206, 218, 225, 374
    Gaelic-speaking 194, 201, 203, 209, 218, 374
    Jacobitism 10, 191, 194, 198
    last in Church of Scotland 194
    origins 195, 201, 208, 209, 227
    persecution 191, 441
    qualification 191–2, 195–6, 198
    Revolution (1688) and 444–5
    stipends 201, 203, 215, 223, 224
    women 223, 224, 225, 226
    see also bishops, Scottish Episcopal Church; qualified clergy and congregations

Episcopalian community
    anti-Catholicism 200, 211
    Appin 197, 204, 212, 228
    books 219
    Edinburgh 192, 197, 198, 199, 200, 201, 225
    evangelicals 198, 200, 201, 205, 225, 227, 498
    Glasgow 198, 199, 200, 203–4, 211, 229
    Greenock 199
    Highlands and Islands 8, 197, 202, 212
    Inverness 197, 203
    Jacobitism and 8, 10, 191, 192, 194, 198, 200
    Kirkcaldy 199
    literature and 386
    marriage 197–8, 204, 211, 221, 225, 226, 229
    non-jurors 8, 191, 192, 197, 198, 200, 206
    north east Scotland 197, 198, 199, 202–3, 212
        clergy from 195, 201, 209
        funeral customs 204, 212
        'high church' theology 200
        marriage customs 204
        mysticism 194
    orphanage 206, 225
    Paisley 198, 199
    persecution 191–8, 201
    Perthshire 197
    Peterhead 197, 212
    politics 208, 216, 221, 229
    teacher training college 206, 216
    temperance movement 211
    World Wars I and II and 218, 220
    see also qualified clergy and congregations; Scottish Episcopal Church
Episcopalianism
    abolition by Scottish Parliament (–1707) 132
    Church of Scotland 109, 111, 112, 131, 194, 367, 440, 441
        church buildings 585–6
        Communion 469
        First Book of Discipline 107
        last Episcopal incumbent 194
        Second Book of Discipline 111
        see also bishops, Church of Scotland
    James VI, king of Scots and 112, 585
    Presbyterianism and 127, 191, 194, 367, 440–1

    see also bishops; Scottish Episcopal Church
Eric Liddell centre 609
Eriskay 373, 640
Erroll, earls of 85, 150–1
Erskine, Ebenezer 237
Erskine, John 443, 665
Erskine, Thomas, of Linlathen 208, 214
Eskdalemuir, Samye Ling centre, Dumfriesshire 315
Eskford church 598
Ettrick Bay, Bute 460
Eucharist 32, 77, 85–6, 87, 89, 160, 187, 207
    see also Communion
European Regional Development Fund 307
Evangelical Union 240–1, 247
evangelism and evangelicals, Christian 11, 235–6, 386, 396, 403, 479, 517–18, 522–3
    Assemblies of God 250
    Baptists 244, 374
    broadcasting and 416, 417, 419
    Catholic see Scottish Mission
    Christian Brethren 247
    Church of Scotland 386, 485–7, 507, 509, 522–3
        Disruption and 370, 387, 532, 533
        Haldane brothers 240, 482
        opposition to 482, 518
        Summer Mission 457–8
    Church of the Nazarene 248
    Congregationalists and 240, 242
    Elim Pentecostal Churches 251
    English Episcopal Church 200, 205
    Episcopalian community 198, 200, 201, 205, 225, 227, 498
    Faith Mission 324, 499
    Free Church of Scotland and 499
    Gaelic culture and 366, 371–2, 373
    Gaelic-speaking areas 367–8, 370–2, 490, 491–2
    Haldane brothers 240
    literacy and 396
    literature and 386, 403
    mission halls 248
    music and 371, 372, 498, 499, 500, 645
    other faiths and 308
    praying societies 240, 479, 485
    Protestant anti-Catholicism 495
    radio broadcasting 416
    Salvation Army 249
    social policy and 491–3

temperance movement and 493
tract distribution 396–7, 398, 485, 486
see also foreign missions; Graham, Billy; home missions; Moody, Dwight L; revivalism; Tell Scotland Movement
Ewing, bishop 208
Ewing, Rev. Grenville 480, 482, 483, 485
Ewing, Greville 242
Exclusive Brethren 247
Eyemouth, Auld Kirk 603
Eyre, Charles Petre, archbishop of Glasgow 172, 175, 176, 177, 178

F

Faber, *Facts and Assertions respecting Popery* 200
fabrics 627
Fahd, king of Saudi Arabia 290
fairs 116, 326–7
Faith Mission 324, 499, 509
'Faith to Faith' seminars 311, 312
'Faithfully Yours' exhibition 311
Falconer, Rev. Ronald 296, 297, 414, 416–17, 418, 420, 428
Falkirk 258, 264, 266, 270, 271, 466–7
Falkland church, communion table 587
Falkland Palace 73, 88
family life 543–5
see also marriage; sexual conduct
Farquhar, John Nicol 526
Farquharson, Archibald 371
Farquharson, John 482
Farr church, Sutherland 603
Ferguson, John 181
Ferguson, Ron 461, 510
Ferguson, W 441
Fergusson, Adam 385
Fergusson, Robert 348, 386
Fetternear banner 70, 88, 658
Fife
    Burntisland 209, 584–5, 618
    church conversions 603
    Church of Scotland mission stations 501–2
    Coptic Orthodox Church 254–5, 604
    Culross 32, 42, 46
    Cupar 243, 587
    Dairsie church 585
    Jewish pedlars 266
    Kemback church 584
    Newmills manse conversion 613
    St Serf and 34, 42, 46

    Westbank Natural Health Centre 318, 321, 327
    see also Kirkcaldy
Fife, earls of 56
Fife synod, Fowlis Easter church art 657
Filmhouse, the, Edinburgh 599
films 414, 427
Findhorn Foundation and Community 318, 320, 321, 323, 324–6, 327, 328
Finlay, Vera 242
Finney, Charles 241
Finnie, Charles 248
Finnio (British bishop and missionary) 25
First Book of Discipline 104, 107–8, 110, 114, 118, 128
    adultery 118, 128
    church buildings 582, 619, 622
    education 107–8, 114–15, 128, 540
    Episcopalianism 107
    finances 107, 108, 110, 128, 436
    funerals 476, 477
    stipends 436
    universities 107–8
First Church of Christ Scientist 252
First Covenant, the 106
First Direct bank 321
First Secession 133
Five Wounds, cult of 89, 633
Flanders 633, 657, 658
Fleming, Lord 85
Fleming, Dr Archibald 414
Fleming, John (minister of Cockpen) 448
Fleming, John (minister of Colinton) 447–8
Fletcher, David 445
Fochabers 586, 603
folk literature 382, 385, 387, 391, 410
folk music 373, 640, 647
Folklore Institute of Scotland 373
football 309, 416, 427, 447, 460, 465
Forbes, Alexander Penrose, bishop of Brechin 207, 211
Forbes, G H 209
Forbes, John Hay, Lord Medwyn 204
Forbes, John, of Abercorn 439
Forbes, Rev, Peter 175
Forbes, Robert, bishop 192, 197
Forbes, Sir William 198, 204
Foreign Mission Committee
    Church of Scotland 519, 520
    Free Church of Scotland 519
foreign missions 517–29
    Africa 138, 517, 520–2, 524, 525, 528–9
    Baptists 519, 520

Bengal 523
Bombay 496, 517, 520, 523, 526, 527
Calcutta 519, 520, 523, 526, 527
Cape Colony 517, 520–1, 522, 524
Chalmers, Thomas 523
China 527–8
Church of England 519
Church of Scotland 517, 518, 519–20, 522–3, 524, 526–7, 528–9
Disruption and 520, 522, 527
education 519, 520, 523, 524, 525–7
Free Church of Scotland 519, 520, 522, 523, 524, 527, 528
India 517, 519, 520, 522–4, 526–7
Jamaica 520
Jesuit order 522, 527, 528
languages 519, 523–4
Nigeria 138, 520
politics and 528
Relief Church 517, 518
Rosie, Thomas 496
Scottish Episcopal Church 199
Secession Church 517
teachers 527
Tswana people 525
United Presbyterian Church 520, 523, 524, 527, 528
women 138, 520, 526–7
see also British and Foreign Bible Society
Form of Presbyterial Church Government, The 132
fornication *see* adultery; sexual conduct
Forres 324, 325
Forrester, Marjorie 465
Forret, Thomas 344
Fort William 219, 221
Fort William College 519
Forteviot 575
Foundry Boys' Mission, Glasgow 497
Fowlis Easter church
    Holy Innocents' event 633
    offering plate 72
    paintings 51–3, 657, 658, 669
        Crucifixion 79–80, 83, 657
        Jesus 66, 71–2, 657
        Mary and Jesus 71, 81
        St Catherine 52, 67, 68, 79
        St John 52, 84
    sculptures 81, 83, 657
Fowrross, David 71
Fox, George 237
Franciscan order 79, 144–5, 175

Frankfurt 127
Fraser, Brian 459
Fraser, Rev. James 368
Fraser, James, of Brea 246
Fraser, Marjory Kennedy 372, 647
Fraserburgh, Episcopal bishop's blessing 204
'Free Church Cases' 537
Free Church College 405
    see also Smith, William Robertson
Free Church of Scotland 135, 138, 453, 534–5, 537
    Annals of the Disruption 404
    Assembly Hall 135
    Christmas treats 504
    church buildings 135, 588, 589
    church government 135, 136
    church organs 625
    duchess of Gordon joins 200
    evangelism 499
    finance 135
    Foreign Mission Committee 519
    foreign missions 519, 520, 522, 523, 524, 527, 528
    formation 133, 135, 387, 534, 664
    Gaelic language and culture 370, 371, 372, 375
    Glasgow hall 255
    Home Mission Committee 494
    home mission work 493, 499, 502, 506
    Ladies Highland Association 503–4
    Lanarkshire Christian Union and 500
    Lochcarron 130
    ministers' payment 135
    music 499, 593, 643, 645, 646
    Scottish-Asian Christian Fellowship 255
    Scottish Ecclesiological Society and 591
    Scottish Episcopal Church and 208
    secessions from 140
    United Presbyterian Church and 134, 140, 537
    see also Disruption, the; United Free Church of Scotland
Free Church of Scotland (Continuing) 140, 375, 537
Free Presbyterian Church of Scotland 136, 140, 371, 375, 409, 537, 589
freemasonry 218, 275
Fresh Air Fund, Glasgow United Evangelical Association 497
Friends of the Western Buddhist Order 315
Fullarton, Margaret 87

fundamentalism 236
    Baptists 244
    Christian Brethren 247
    Church of the Nazarene 248
    Elim Pentecostal Church 251
    Mormons 253
    St Mungo Museum displays and 311
    Salvation Army 249
    Seventh-Day Adventists 253
funerals
    Book of Common Prayer 204, 477
    Catholic pre-Reformation 86–7
    church bells 75, 620
    Church of Scotland 476–8, 546
    hymns 477, 649
    Scottish Episcopal Church 198, 204, 211, 212, 221, 229
    Stewart, James, earl of Moray 476
    see also burial sites; cremation
furnishings, church buildings 596–7, 616–28, 658

G

Gadamer, H G 311
Gaelic Bible 367–9, 374
    Authorised Version and 368, 369
    British and Foreign Bible Society 490
    Catholic Church 369
    Classical Gaelic and 365, 367, 368
    distribution 397, 491
    evangelism and 370
    New Testament 367, 369, 490
    revision 373, 376
    Scottish Episcopal Church 203, 367
    Scottish Parliament (–1707) and 395
    SSPCK and 367–8, 490
Gaelic Camp 375
Gaelic Chapel, Edinburgh 163, 446
Gaelic church, Inverness 609
Gaelic Episcopal Society 203, 374
Gaelic language and culture 363–78, 490
    Baptists and 371, 373, 374, 375
    catechisms 367
    Catholic Church 364–5, 366, 369, 372–3, 374
    Church of Scotland 365, 367, 370, 372, 373, 374, 375
    Classical Gaelic 365, 367, 368, 370
    Congregationalist churches and 373, 375
    Dál Riata 27, 363–4
    education 365–6, 368
    English and 374–5, 376
    evangelism and 366, 371–2, 373
    Free Church of Scotland 370, 371, 372, 375
    Glasgow 203–4, 376
    Jacobitism 365, 366
    James VI, king of Scots 365
    journalism 373
    St Mungo Museum 307
    schools 365–6, 368, 490
    Scottish Episcopal Church and 203–4, 212, 225, 374
    sermons 364, 366–7, 371, 372
    SSPCK attitude to 490
    teachers 366, 370
    see also Gaelic literature; Gaelic music; Highlands and Islands
Gaelic literature 364–5, 366, 369, 373–4
    see also Gaelic Bible; Gaelic poetry; Gaelic prayer books; Gaelic psalters
Gaelic music 371, 372, 644, 645
Gaelic poetry 366, 371, 372, 373, 374, 645, 654
Gaelic prayer books 203, 212
Gaelic psalters 365, 367, 369–70
Gaelic scholarship 372–3
Gaelic School Societies 369, 370, 490–1
Gaelic Society 199, 200, 225
Gaelic-speaking areas
    catechists 194, 371
    Episcopalian congregations 203–4
    evangelicals 367–8, 370–2, 490, 491–2
    lay missionaries 502
    revivalism 370, 371, 372
    see also Highlands and Islands
Gaelic-speaking clergy 374–5, 376
    Baptist 374, 375
    Catholic 144–5, 364–5, 374
    Church of Scotland 366, 375, 438
    Edinburgh University 372
    Episcopalian 194, 201, 203, 209, 218, 374
    missionaries 370
Gaitens, Edward 185
Galashiels 206, 219, 220, 221, 239, 317
Galloway 17, 20, 23, 176, 364
    see also Whithorn
Galston Bible and Missionary Society 484
Galt, John 348, 386, 387, 388, 433–4
Garden, Dr George 194
Garnethill Synagogue 267, 268, 270, 272, 277
    children's hostel beside 264

Jewish Board of Guardians' base 263
Scottish Jewish Archives Centre 275
Garry, Flora 356
Garvald church 592
Gate Fellowship, Dundee 251–2
Gatehouse-of-Fleet, Kirkcudbrightshire 323, 603
Gateway Theatre 390
Gau, John 62, 342
Gauld, David 666
Geddes, Alexander 161
Geddes, Jenny 617
Geddes, John, bishop 161, 162, 163
Geddes, William 641
General Assembly of the Church of Scotland
   Act of Union and 9
   articulation of Church views 550
   bishops' appointment 128
   Board of National Mission 512
   boards and committees 136
   broadcasting of 417
   calling of 112, 130
   Campbell, John McLeod, deposed 400
   catechisms 131–2, 470
   Chapel Act 532, 533, 534
   church membership 549
   Communion services 471, 472, 588, 624
   Directory for the Publick Worship of God 132, 472, 644
   education and 114, 115
   evangelicals and 532
   first gathering 128
   foreign missions 518, 519, 520, 523, 528–9
   funeral services 477
   Gaelic Psalter 370
   Glas, John, deposed 239
   home missions condemned by 483
   Iona Community and 506
   James VI, king of Scots and 112
   lay preaching 240, 518
   Lord High Commissioner 536, 548
   Mass celebrations condemned by 85, 86
   media coverage of 423
   meeting place 135
   ministers and 434, 444, 446, 447, 486
   ministers' marriages and 442
   moderators 137, 138
   monarchy and 112, 130, 536
   music 642, 645
   patronage and 133
   plays banned 346
   Prayers for Divine Service authorised by 478
   Revolution (1688) and 441
   Royal Bounty Committee 486–7, 491
   schools 370
   Scottish Parliament (1999-) and 423, 530
   Secession and 133
   Second Book of Discipline approved by 111
   Socinianism and 246
   Stewart monarchs 112, 536
   Sunday schools condemned by 483, 518
   vernacular Bible and 340
   Veto Act 532, 533–4
   Westminster Assembly documents approved by 132
   Winning, Thomas, addresses 186
   women's ordination 138
   see also Disruption, the
General Assembly of the Free Church of Scotland 135, 140, 493, 534, 643, 645
General Assembly of the United Free Church of Scotland 138, 502
General Assembly of Unitarian and Free Christian Churches in Britain 246
General Associate (Antiburgher) Synod 134
General Associate Synod 133, 518
General Register of Sasines 113
General Synod, Scottish Episcopal Church 198, 224, 225
Genesis, Scots translation of book of 355, 356
Geneva 106, 127, 236, 340
Geneva Bible 340, 354, 395
George III, king of Great Britain and Ireland 192, 447
George IV, king of Great Britain and Ireland 198
George of Spalding 64
George Watson's school, Edinburgh 461
German congregations 254
Germanus, bishop of Auxerre 18
Germany 258, 259, 328, 462
   see also Frankfurt
Gibbon, Lewis Grassic 9, 390
Gibbs, James 581
Gibraltar 397
Gibson, Philip 65
Giffnock 264, 265, 270, 271, 276
   see also Jewish Community Centre
Giffnock and Newlands Synagogue 270

Gifford, Yester church 587
Gilbert de Moravia, bishop of Dornoch 580
Gildas 16, 17, 20, 25
Gill, Eric 670
Gillies, John 398
Gilliland, Billy 465
Gillis, James, bishop 173, 174
Girls' Brigade 458
Girvan, Ayrshire, Summer Mission 458
Gladstone, William Ewart 181, 213, 214
Glamis, Pictish cross slab 652
Glas, John 238–9
  see also Glasites
Glasgow
  Ahmadiyya, the 288
  alternative healing fairs 326
  Andersonian University 261, 520
  anti-Catholicism 183, 497
  Artisans' Hall 497
  Asian community 255, 285
  Balornock 506
  Baptist churches 244, 268, 589, 609
  Blackfriars Church 398
  Burgess Oath 133
  Burrell Collection 307
  Cadder 32, 501
  Canal Boatman's Institute 497
  Catholic Choral Society 178
  Catholic community 163, 171, 172–3, 176, 177, 178, 182
    education 179, 183
    Irish immigrants 163–4, 171, 172
    politics 174, 178
  Catholic Schools Society 174
  choirs 463, 464, 641
  Christ Church 218
  Christian Science church 252
  church badminton team 465
  church buildings 589, 591, 595
  church conversions 599, 602, 603, 604, 605, 606
  church demolition 598
  church relocation 604
  cinemas 427
  College (Blackfriars) Church 398
  Congregationalist churches 248, 599
  Dove Hill Relief Congregation 485
  Dowanhill College 180
  ecumenical activity 268, 311, 317
  Episcopalian community 198, 199, 200, 203–4, 211, 229
  Free Church of Scotland hall 255
  Gaelic speakers 203–4, 376

German congregations 254
Glasite church 239
Glenduffhill 269
Govan 34, 229, 274, 506, 510
Greenhead 494–5
guilds 65
Hebrew College 271, 272
Henry Wood Hall 599
home missions 493, 494, 497, 501, 506, 507
  Glasgow City Mission 248, 488–9, 512
hospitals 263
Indian community 284
Islamic Centre 290
Jewish community 258–60, 261, 277–8
  burial sites 259–60, 261, 268–9
  business and employment 258–9, 266, 274
  divisions within 261, 263, 267, 268, 272
  ecumenical activity 268
  education 258, 263, 266, 270–2, 278
  establishment 256, 257, 315
  immigrants in transit 259
  orphanage and children's hostel 264
  politics 268, 272–4
  size 257, 258, 259, 260, 261, 264, 270, 277
  social and cultural activity 274, 275, 276, 277, 278
  synagogues 259, 261, 267–9, 270, 271, 277
  university education 258, 266
  welfare 259, 262, 263–5, 266, 278
Kelvinside 460, 507, 589
Laurelpark School 606
Maryhill 268, 497
'Meet Your Neighbour' event 311, 317
Merryflatts 263
music 639, 641, 646, 647
Muslim community 281, 288, 290
Netherlee 460, 464
Orthodox Church cathedral 254
Paganism 317
Pollok 453, 604
Pollokshields 270, 604
Riddrie 268
Sacred Music Institution 641
St David's parish 486
St Eloy statue 67, 80
St George's (Tron) church 586, 601

INDEX • 703

St John's parish 486, 588, 623
St Kentigern (St Mungo) and 32, 33
St Margaret's church
    Knightswood 622
    Newlands 221, 595
St Mungo Museum of Religious Life and Art 293, 306–12, 318
Scottish-Asian Christian Fellowship 255
Shechita Board 268
Sherbrooke St Gilbert's church 622
Spiritualism 318
Sunday schools 454, 456, 494
Theosophical Society 318
Tron church 586, 601
Tron theatre 601
Union of Precentors 641
Unitarian church 246
United Presbyterian Church 494–5
Youth Fellowship 460
see also Gorbals, Glasgow
Glasgow and Ayr synod 396
Glasgow and Galloway, bishop of 227
Glasgow and West Coast Mission 496, 503, 504
Glasgow Athanaeum School of Music 646, 647
Glasgow Auxiliary Society, London Missionary Society 484, 485
Glasgow Bible College 498
Glasgow Bible Society 398
Glasgow, bishop of 256
Glasgow Buddhist Centre 315
Glasgow Cathedral 56, 68, 580, 583, 591
    Blackadder, Robert and 50, 56, 75, 80, 88
    choir 463
    Lady aisle 75, 88
    Lady Chapel 75, 80
    paintings 75
    pilgrimage centre 56, 580
    sculptures 50, 71, 73, 75, 80, 88, 656
    stained glass 663–4
    statues of saints 50
    see also St Mungo Museum of Religious Life and Art; Society of Friends of Glasgow Cathedral
Glasgow Catholic Schools Society 174
Glasgow City Council 307
Glasgow City Mission 248, 488–9, 512
Glasgow Committee for Colportage 397
Glasgow Development Agency 307
Glasgow, earl of 484
Glasgow Hebrew Benevolent Loan Society 263
Glasgow Hebrew Burial Society 268–9
Glasgow Hebrew Congregation 261
Glasgow Hebrew Philanthropic Society 262
Glasgow Inter-Faith Committee 296–7, 298
Glasgow Jewish Board of Guardians 263, 264–5, 268
Glasgow Jewish Institute 277
Glasgow Jewish Institute Players 275
Glasgow Jewish Representative Council 272–3, 309
Glasgow Jewish Sick Visiting Association 265
Glasgow Jewish Welfare Board 264, 265
Glasgow Maccabi 276
Glasgow Missionary Society 517, 520, 524, 527
Glasgow Necropolis 259–60
Glasgow Orpheus Choir 641
Glasgow Police Office 488
Glasgow presbytery, mission stations 501
Glasgow Religious Tract Society 485
Glasgow School Board 271
Glasgow Sharing of Faiths Group 311
Glasgow Summer School of Theology 409
Glasgow United Evangelical Association 497–8
Glasgow University 107, 108, 582
    African student 524
    Catholic community and 183
    church building conversions 605
    Henryson and 382
    Indian students 284
    Islamic Studies 293
    Jewish Society 272
    Jewish students 258, 266
    McEwen Bequest 647
    Struther legacy 444
    vagabond scholars 116
Glasgow University Press 396
Glasgow Working Men's Evangelistic Association 497
Glasites 239, 587, 599, 641
glass *see* stained glass
glebes 436, 442, 613
Glen Garry, emigration from 163
Glen More 29
Glenalmond, Trinity College 206, 213
Glenbranter, Navvy Mission 505
Glencoe 219–20, 319, 505
Glencraig mission station, Fife 501
Glendaruel, Kilmodan church 587

Glendochart, St Fillan 34, 656
Glenduffhill, Glasgow, Jewish burial site 269
Glenlivet 155, 157, 158
Glenorchy, Lady 137-8
Glenorchy church 587
Glenrothes 215, 506
Goes, Hugo van der 658
Goethe, Johann Wolfgang von 389
Golombok, Zevi 275
Goodfellow, James 494
Gorbals, Glasgow
   Baptist church building conversion 268
   Episcopalianism 229
   Jewish community 261, 263, 264, 265, 266, 277
   synagogues 267-8, 270
   Muslim community 281
Gorbals Medical Mission 497
Gordon, Charles 173
Gordon, duchess of 200, 397
Gordon, Elizabeth 87
Gordon, Esme 585
Gordon Evangelistic Mission, Aberdeen 500
Gordon, George, 1st duke of Gordon and 4th marquis of Huntly 153
Gordon, George, 1st marquis of Huntly 149
Gordon, Harry 452
Gordon, J F S 209
Gordon, Patrick 89
gospel books 50, 53
Gosse, Edmund 410
Gothic architecture 577, 579, 580, 585, 655, 663
Gothic revival architecture 588-9, 591, 592, 594-5
Gottlieb, Rabbi 268
Govan, Glasgow 34, 229, 274, 506, 510
Govan, Sheena 324, 328
Govan, William 524
Govanhill, Jewish community 264
Graemsay mission station, Orkney 501
Graham, Billy 417, 464, 508-9, 539
Graham, James, 1st marquis and 5th earl of Montrose 154
Graham, James Gillespie 589, 595
Graham, John, of Claverhouse, Viscount Dundee 629
Graham Street synagogue, Edinburgh 269, 271
Graham, Tom 665
Graham, William, of Kincardine 85

grammar schools 115, 156, 157, 158, 365, 382, 385
Grampian Television 419, 420
Grandtully church, Perthshire 660, 661
Grant, Sir Archibald 639
Grant, I F 223
Grant, Dr Lachlan 373
Grant, Peter 371
Granton parish church, Edinburgh 560-4, 568-70
Grantown, Baptist mission 491
Gratis Sabbath school movement 483-4
gravestones 75, 658
   see also sculptures
graveyards see burial sites; kirkyards
Gray, Alasdair 425
Gray, Andrew 64
Gray, Bishop 175
Gray, Gordon, cardinal 172
Gray, John 89
Gray, Rev. Dr Nelson 419, 420
Great Charter, Church of Scotland 111-12
Great Invocation, the 326
Great Synagogue, Glasgow 268
Greenbank, Edinburgh, Youth Fellowship 460, 461
Greenbaum, Avrom 275
Greenbelt music and arts festival 648
Greenhead, Glasgow, United Presbyterian Church 494-5
Greenock
   Asian community 284
   Catholic community 163
   church badminton team 465
   Episcopalian congregation 199
   Irish Catholic immigrants 163
   Jewish community 257, 268, 269-70, 271
   music 643
   St Ninian's church 453
   Town Mission 487
Greenock Auxiliary Society, London Missionary Society 485
Greenock Missionary Society 517
Greenside church, Edinburgh 589
Gregory XV, Pope 145
Greig, Sir Thomas 61
Greyfriars kirk, Edinburgh 464, 585, 645, 664
Greyfriars kirkyard, Edinburgh 660, 661
Groam House museum 318-19
Gude and Godlie Ballatis, The 339, 343, 356, 477, 636

Guild, the *see* Woman's Guild
guilds 64–5, 67, 72, 75, 80, 86, 585
    see also Corpus Christi celebrations
Gujerati language 523
Gunn, Neil 390
Gutenberg, Johannes 413, 426, 428
Guthrie church, Angus 625
Guthrie collegiate church 84
Guthrie, James 666
Guthrie, Thomas 665
Guyon, Madame 194
gypsies 116

H
Haddington 73, 83, 86, 88, 485, 592
Haddo House, Aberdeenshire 401
Hadrian's Wall 16, 17
Hædde, bishop of Wessex 24
Haggart, Alistair, bishop 223, 227
hagiographies 15, 42–7, 72
    language 44, 337
    Queen Margaret 46
    St Columba 26, 43–4, 379–80
        see also Adomnán, abbot of Iona,
        Life of St Columba
    St Kentigern (St Mungo) 32–4, 42, 44, 46
    St Ninian 22, 25, 47, 53
    St Serf 34, 42, 44, 46
Haldane, James 240, 244, 480, 482, 518
Haldane, Robert 240, 244, 480, 482, 491, 518
Halima (follower of Livingstone) 522
Hall, Douglas 670
Hamburg, Jewish immigrants from 259
Hamilton 173–4, 321, 500
Hamilton, Alexander 439
Hamilton, Gavin 662
Hamilton, J Arnott 592
Hamilton, Sir James, of Priestfield 446
Hamilton, John, archbishop of St Andrews 66, 342, 344
Hamilton, Patrick 106, 635
Hamilton presbytery, Church of Scotland mission stations 501
hammermen craft 65, 67, 72, 80
Hammermen's Chapel, Edinburgh 51
Hammond, Payson 499
Hanafi and Hanbali schools, Islamic law 287
handbells 630
Hannay, dean 617
Hardie, Keir 181
Hardy, Helen L 504

Hardy, Thomas 410
Hare Duke, Michael, bishop 223
Harper, Michael 251
harps 630
Harris 8, 658
Harris, George 246
Harrison, William 160
Harvest Thanksgiving 218
Harvey, George 662
Harwood, Ronald T Pengilley 254
Hary (Blind Hary) 381–2
Hateley, Thomas 643
Hatt, dean 211
Hattersley, Roy 410
Hay, George (architectural historian) 617
Hay, George, bishop and Vicar Apostolic 155, 161, 162–3
Hay, John, Baron Yester 89
Hay, William, 3rd earl of Erroll 85
Hazlitt, William 398
healing 57, 67, 86, 319, 525, 580, 630
    prayers and 46, 72, 73
    wells and shrines 57, 67, 86, 580
    see also alternative healing; medical treatment
healing stones, St Fillan 319
Healthworks, Stirling 322
hearses (candleholders) 627
Hebrew Benevolent Loan Societies 263, 266
Hebrew College, Glasgow 271, 272
Hebrew language, Jewish education in 271, 272, 274
Hebrides *see* Western Isles
Heenan, John Carmel, canon 417
Helensburgh, Christian Science society 252
Helmsdale, Sutherland, manses 613
Henderson, Alexander 665
Henderson, G D 395, 408
Henderson, George, bishop of Argyll and the Isles 216, 221
Henderson, George (co-translator of Bible) 353
Henderson, Hamish 229
Henderson, Ian 223, 453
Henderson, John 210
Hendrie, Herbert 596
Henry I, king of England 46
Henry II, king of England 46
Henry VIII, king of England 103, 584
Henry, prince (son of James VI) 385
Henry Wood Hall, Glasgow 599
Henryson, Robert 65, 79, 382–3
Hepburn, Ade 89

Hepburn, Anna, Dame 446
Herdman, Robert 662
'hereditary Episcopalians' 210–11, 219, 228
Heritage Lottery Fund 615
Herman, Arthur 396
Herries family 143
Herron, Dr Andrew 136
Hetherington, Alastair 418, 419
Hexham 243
Higgitt, John 655–6
High Commission, courts of 130–1
High Kirk, St Giles' church, Edinburgh 583, 591
Highet, John 313
Highland Development League 373
Highland Missionary Society 491
Highland Mysterworld, Glencoe 319
Highland Roads and Bridges, commissioners of 587
Highland Tolbooth, Edinburgh 599
Highland Vicariate, Catholic Church 160, 162, 173
Highlands and Islands
   alternative healing 321
   art 658
   Bible societies 490, 492
   Caledonii 17
   catechists 194, 371, 480, 483, 486–7, 491, 639
   Catholic Church 112, 147, 155, 160, 162, 173, 372
   Catholic clergy 144–5, 149
   Catholic community 144–5, 146, 151, 159–60, 163, 170, 181
      schools 156–7, 158, 179
   Church of Scotland 144, 482, 492
      Argyll 367, 369, 370, 492
      catechists 486–7, 491, 639
      church buildings 587
      ministers' stipends 587
      presbytery meetings 439, 443
      Royal Bounty Committee 486
      schools 365–6
   Congregationalism 482
   home missions 370, 490–2, 502
      Catholic Church 144–5, 146, 149, 157
      Free Church of Scotland 499
      SSPCK 480, 487
   Lewis chess pieces 655
   literacy 159
   Lowlands and 8, 104, 151, 163, 490
   meditation group 326
   music 644, 646
   newspapers 373
   Parliamentary churches 587
   schools
      Catholic 156–7, 158, 179
      Gaelic language and 365–6, 368, 490
      Ladies Highland Association 503
      Scottish Episcopal Church 203
      SSPCK 156, 158, 366, 368, 480
   Scottish Episcopal Church 8, 197, 202, 203, 212
   sculptures 658, 668, 670
   SSPCK 156, 158, 366, 368, 480, 487
   Sunday observance 546
   Theosophical Society 318
   see also Gaelic language and culture; Gaelic music; Gaelic-speaking areas; Western Isles
Hill, David Octavius 664
Hill, George 523, 526
Hillhead Baptist church, Glasgow 244, 589
Hillhead church, Glasgow 591
Hillman, Samuel, Rabbi 268
Hindus and Hinduism 11, 283, 302, 303–4, 315, 538
   art 308
   broadcasting and 298, 301, 302, 303–4, 426
   foreign missions 519, 523, 526
   neo-Hindu groups 317
   St Mungo Museum 308, 310
Hinton St Mary 16
Historic Scotland 614, 615
Hitler, Adolf 427
Hodder-Williams, Sir Ernest and Lady 410
Hoddom church, St Kentigern and 32
Hodgkinson, Arthur 226
Hogg, A G 526
Hogg, James 8, 348, 350–1, 386, 388, 662
Hogmanay 6, 9, 497, 546
Hole, William 666
holistic education 321
holistic healing 318
Holland 106, 188, 258, 328, 658
Holloway, Richard, bishop of Edinburgh 223, 227, 229, 422, 423
Holmes, Timothy 522
Holocaust, the 309
Holy Blood, cult of 70, 88, 89, 90, 633, 658
Holy Cross, cult of 89–90
Holy Innocents' Day 633
Holy Rood, cult of 83, 89, 90

Holy Trinity church
    Edinburgh 606–7, 609
    St Andrews 580, 583, 587, 592
    Soutra 605
Holy Week 218–19
    see also Easter; Passion, the
Holy Wounds of Jesus altar, Aberdeen cathedral 89
    see also Five Wounds, cult of
Holyrood Abbey, Edinburgh 51, 446, 583, 584, 586
Holyrood Free church, Edinburgh 609
Holyroodhouse, Palace of 51, 63, 642
Holzhandler, Dora 308
Home Board, Church of Scotland 504, 506, 507, 509
Home, D D 318
Home, earl of 51
Home, Rev. John 386, 447
Home Mission Appeal 506
Home Mission Campaign 221
Home Mission Committee
    Church of Scotland 494, 501
    Free Church of Scotland 494
Home Mission Crusade 506
Home Missionary Society 480, 482
home missions 470–516
    Angus 504
    Baptists 243, 371, 373, 491, 499, 506
    Bible distribution 484, 485, 486, 487, 488, 489
    Bible societies 484, 496, 511
    Breadalbane 482
    Catholic Church 144–5, 146, 149, 157, 175, 495
    Chalmers, Thomas 485–6, 493, 494, 495, 496, 497, 500, 509
    choirs 506
    church buildings and 221, 493–4, 506
    Church of Scotland 483, 484, 486–7, 495, 501–2, 504–9
    ecumenical activity 504
    Edinburgh 248, 265, 487–8, 489
    education 483–4, 493
    Free Church of Scotland 493, 499, 502, 506
    Glasgow 248, 493, 494, 497, 501, 506
    Highlands and Islands 490–2
        Catholic Church 144–5, 146, 149, 157
        Free Church of Scotland 499
        Gaelic-speaking missionaries 370
        lay missionaries 502

    SSPCK 480, 487
    Jewish community 265
    media 501, 510
    Open Brethren 499
    Orkney 492
    prayer 482, 488, 489, 499, 506, 511
    Scottish Episcopal Church 221, 498, 506
    Sunday schools 487, 489, 504, 518
    Tayside 504
    temperance movement 493
    tract distribution 484, 485, 486, 487, 488, 489
    visiting 485–9, 492, 494, 501, 506, 509, 510
    welfare work 492–5
    women 492, 494, 495, 503–4, 511
    young people 506–7
Home, Patrick, of Polwarth 85
homosexual partnerships 545, 549
Honeyman, John 581
Hong Kong 527
Hopkirk church 598
Horne, Janet 130
hospitals 91, 115, 263, 444
hostels, Jewish community 263, 264
House of Commons 548, 549
    see also Westminster Parliament
House of Lords 534, 543
    see also Westminster Parliament
housing, church conversions to 602–3
Houston, Renfrewshire 656
Houston, J A 665
Houston, Sir Patrick 656–7
Howard, Deborah 660
Howard, Elizabeth 153
Hownam church 598
Hromadka, Josef 417
Hub, the, Edinburgh 599
humanism 312, 323, 383, 386
Humanist Society of Scotland 312
Hume, David 386, 538, 662
Hungary, Jewish refugees from 264
Huntly, 1st marquis of 149
Huntly, 4th marquis of 153
Huntly church, singing loft 639
Hutcheson, Francis 385
Hutcheson, Margaret 86–7
Hutchinsonianism 194
Huxley, J S 413, 426
Huxley, T H 538
hymns 643, 645–6, 647–9
    broadcasting 414, 416, 421, 457
    Church Hymnary 396, 647, 649

evangelism and 498, 499, 500, 645
funeral services 477, 649
Gaelic 364, 371, 373
Gude and Godlie Ballatis, The 343, 636
Methodist Church 645
Moody and Sankey rallies 498
psalms and 352, 637
Relief Church 645
Scottish Episcopal Church 209, 645
temperance movement 641
see also psalms

I

Ian G Lindsay & Partners 592
illustrations
    Acts of the Parliaments (1540) 79, 81, 83
    Bible 662, 664-5
    books 67, 664-5, 666, 668
    gospel books 53
    see also paintings; woodcuts
imagery *see* art
imams 287-8, 290, 303
immigration 11, 426
    English 210, 211, 228, 315
    Findhorn Community 325
    Indian 283, 538
    Italian 11, 178, 179, 315
    Jewish 257-9, 260-1, 263-4, 277
    Lithuanian 178-9
    Muslim 281, 283, 288, 290, 291, 300
    Polish 182, 258, 315
    Russian 258, 265, 274
    see also Irish immigrants
incest 113, 545
Inchcolm, abbey 630
Inchcolm Antiphoner 630-1
Inchinnan 34, 57
Inchkeith, St Serf at 46
Inchmahome, St Colmán (Colmoc) and 36
Independents *see* Congregationalist churches
India 317, 517, 519, 520, 522-4, 526-7
    see also Bombay; Church of North India; East India Company
Indian community 284
    broadcasting and 299
    immigration 283, 538
    Islam and 281, 286, 288, 290, 291
    Scottish-Asian Christian Fellowship 255
    see also Hindus and Hinduism; Sikhs and Sikhism

Indian languages 291, 523-4, 526
'Inglis' 337, 339, 344
    see also Scots language
Innes, George 197
Innes, Rev. George 202
Innocent VIII, Pope 114
Institutional Writers 538
inter-faith work *see* ecumenical activity
International Bible Students Association *see* Jehovah's Witnesses
International Christian College 498
International Mission Conference 526
International Society for Krishna Consciousness 317
Inverary, All Saints' church 223
Inveresk 437, 502
Inverness
    alternative healing fairs 326
    Baptist mission 491
    church conversions 609
    Episcopalian community 197, 203
    Gaelic church 609
    Jewish community 258, 268, 270, 271
    Quaker converts 237
    St Michael and All Angels church 604
    television broadcast from 418
    Theosophical Society 318
Inverness presbytery 492
Inverness-shire 29
Inverness Society for Educating the Poor in the Highlands 492
Invertiel church, Kirkcaldy 604
Iona 32, 36, 654
    Aedán mac Gabráin's accession 27
    Benedictine monasticism 56
    Book of Durrow and 652
    Book of Kells and 654
    church buildings 55-6, 575, 587
    painting and 666, 668
    pilgrimage centre 55-6, 656
    St Columba and 22-3, 26, 27, 30, 32, 219, 363
    sculptures 47, 50, 56, 630, 654, 658
    see also Adomnán, abbot of Iona
Iona Community 506, 509, 647
    see also MacLeod, George
Iqra Academy 293
Iranian community, Islam and 284, 286, 288
Iraqi community, Islam and 288
Ireland 6, 28, 218, 367
    see also Armagh; Church of Ireland;

Derry; Donegal; Irish Catholic
clergy; Irish immigrants; John of
Ireland; St Columba; St Ninian;
Tailtiú
Ireland, John 338
Irish Catholic clergy 151, 172, 173, 174, 175,
188
Irish Home Rule movement 181
Irish immigrants
Catholic 11, 163–4, 171–2, 173, 175, 315,
495
Church of Scotland and 422
Navvy Mission 505
Episcopalian 206, 210
Irish language, hagiographies 44
irregular marriage 118, 475–6, 543
Irvine, Ayrshire 87, 240, 434
see also Perceton and Dreghorn church
Isaiah, Scots translation of book of 354,
355
Islam 281–94, 549, 550
Africa and 284, 286
art 308
Asian community 284, 285
Bangladeshi community 284
blasphemy and 547
broadcasting 302, 426
clergy 287–8, 303
conversion to 293, 300
Indian community and 281, 286, 288,
290, 291
Iranian community and 284, 286, 288
Iraqi community and 288
law schools 287
Malaysian community 284, 286
mysticism and 302
North African community 284, 286
Pakistani community 281, 283, 284, 285,
286, 288, 290
St Mungo Museum 308, 311
Turkish community 284, 286
see also Muslim community
Islamic Centre, Glasgow 290
Islamic Society of Britain 292
Islamic Studies, Glasgow University 293
Islay 365, 587, 603
Isle of May, St Ethernan (Adrian) 34
Isle of Whithorn, chapel 55
Israel 265, 273, 274, 278, 291
Italian immigrants 11, 178, 179, 315
Italy 397
ITV 419–20, 421
Iverach, James 400

J

Jack, R D S 410
Jackson, Pilkington 669
Jacobitism
Catholics and 160, 161
Episcopalians and 8, 10, 191, 192, 194,
198, 200
Gaelic language and 365, 366
music and 641
Jacobus de Voragine 83, 337
Jamaica, foreign mission work 520
James III, king of Scots 63
James IV, king of Scots 55, 56, 63, 65, 68, 85,
338
James V, king of Scots 62, 63, 65, 632–3, 635
James VI, king of Scots
accession to English throne 112
accession to Scottish throne 108
Authorised Version of Bible 346
Buchanan, George and 385
Church of Scotland and 112, 439, 441,
585, 623
Communion 623
Episcopalianism 112, 585
Gaelic language and 365
literature 114, 383, 385
Negative Confession 111
witchcraft and 114
James VII, king of Scots 109, 586
James VIII and III (the 'Old Pretender') 194
see also Jacobitism
James, P D 425
James, Robert 89
Jamesone, George 627, 661
Jamesone, Mary 627
Jamie, Kathleen 390
Jansenism 159
Japan, church relocation to 605
Jedburgh, monastery church 63, 89
Jedburgh Abbey 584
Jeffreys, George 251
Jehovah's Witnesses 235, 252, 254
Jenkins, Robin 389–90
Jesuit order 145, 146, 149
education and welfare role 175
foreign missions 522, 527, 528
schools and seminary 158
welfare work 175
Jesus Christ
Beatles compared to 427
images of 70, 71–2, 84, 656
impersonated in radio broadcast 416
in Eucharist 87, 89

paintings of 66, 71–2, 81, 657
sculptures of 61, 71, 81, 83, 87, 656
see also Crucifixion, the; Passion, the
jewellery 72, 75
Jewish Blind Society 265
Jewish Boards of Guardians 259, 263,
    264–5, 266
Jewish Care Scotland 265
Jewish community 11, 256–80, 315
    art 275, 308
    broadcasting and 298, 301, 303
    burial sites 259–60, 261, 268–9
    Christian missions to 265
    drama 275
    ecumenical activity 268
    education 258, 263, 266, 270–2, 274, 278
        schools 264, 271, 272
        students 256, 257, 258, 266
    hostels 263, 264
    immigration 257–9, 260–1, 263–4, 277
    literary societies 274–5, 276
    music 275
    newspapers 275
    orphanage 264
    pedlars 266
    politics 268, 272–4
    Sabbath observance 262, 268
    St Mungo Museum 308, 310
    sculpture 275
    synagogues 259, 261–2, 267–70, 271,
        272, 277, 315
    welfare work 259, 262–5, 266, 278
    women 304
    youth groups 276, 277
    see also Holocaust, the; Judaism
Jewish Community Centre, Giffnock 270,
    271, 277
Jewish Dispensary 265
Jewish Hospital Fund and Sick Visiting
    Association 265
Jewish Housing Association 264
Jewish Institutes 277
Jewish Lads' Brigade 276
Jewish Literary Societies 274–5
Jewish Old Age Home for Scotland 264
Jewish Renewal Committee 272
Jewish Representative Councils 272, 309
Jewish Societies, universities 272
Jewish Teenage Centre 272
Jewish Welfare Board 264, 265
Jewish Youth Forum 277
Jocelin of Furness 32, 33, 46
John Buchan Centre 603

John, Elton 649
John Kerr Memorial Church, Edinburgh
    598
John Knox House museum 599
John of Crumme 83
John of Fordun 47, 381
John of Ireland 87
John Paul II, Pope 186
Johnson, Robert 635
Jolly, Alexander, bishop of Moray 201, 204
Jordanhill College of Education 457
Joss, Willie 466
journalism *see* media; newspapers
'Journey of the Spirit' guided tour 321
Jubilee Fund appeal, Glasgow Jewish
    Board of Guardians 263
Judaism 6, 302, 308, 426, 550
    see also Jewish community
Judgement Day 76, 80, 84, 89, 90
Julius (citizen of Caerleon) 16
Jumaah (Prayer Day) 301
Jung, Carl 61

K
Kaffraria diocese, Scottish Episcopal
    Church and 214
'kail kirk', Dundee 239
Kalabari art 308
Kaunda, Kenneth 521, 522
Keiss, Caithness, Baptists in 244
Keith, George, 4th Earl Marischal 110
Keith Missionary Association 485
Kello, John 446
Kelman, James 390
Kelso 407
Kelso Abbey 584, 587, 656
Kelso North church 589
Kelso Old church 586–7
Kelvin, William Thomson, 1st Baron Kelvin
    of Largs 214
Kelvinside, Glasgow 460, 507, 589
Kelvinside (Botanic Gardens) church,
    Glasgow 604
Kelvinside Hillhead church, Glasgow 591
Kemback church, Fife 584
Kemp, Andro 637
Kemp, William 83
Kennedy, Dr, of Dingwall 404
Kennedy, James, bishop of St Andrews 633
Kennedy, Rev. John 372
Kennedy, Walter 71
Kilarrow church, Islay 587
Kilbirnie church 618

Kilbrandon, Lord 229
Kilchoman, Islay 365, 603
Kilchousland, Kintyre, St Constantine 34
Killin, Breadalbane Folklore Centre 319
Kilmany church, communion table 587
Kilmarnock, Ayrshire, Catholic community 173
Kilmartin, Argyll 317
Kilmaurs church, Ayrshire 601
Kilmodan church, Glendaruel 587
Kilmorich church 587
Kilmory, London Missionary Society Glasgow Auxiliary Society 484
Kilmuir Easter Bible Society 492
Kilpatrick, pilgrimage centre 57
King James Bible *see* Bible, Authorised Version
King, Jessie M 668
Kingarth, Bute, St Blane 34
King's College
    Aberdeen 7, 66, 616, 620
    Cambridge 649
King's Cross, Arran 460
Kingussie 223, 491
Kinneff 130, 617
Kinneil, St Serf at 46
Kinnoull old church, Perth 661
Kintyre 28, 34
Kirk o' Field church, Edinburgh, St Barbara's light 70
Kirk, Rev. Robert 367, 368, 369, 370
kirk sessions 117, 128, 132, 135, 443
    beggars and vagabonds 116
    Boys' and Girls' Brigades leaders 458
    church bells and 619
    Communion non-attendance sanction 472
    Cromarty 587
    education responsibilities 533
    Episcopalian congregations 197
    Evangelical Union 241
    Holyroodhouse 642
    ministers and 438, 439, 443, 612
    moderators 131, 137, 443
    moral governance 114, 128, 130, 141, 443
    music and 639, 642
    St Andrew parish, approval of play 346
    Scottish culture and 141
    seceding churches 135
    security of offerings 625
    Sunday observance enforcement 116
    welfare work 115, 135–6, 533, 542
    witchcraft investigation 130
    see also elders
Kirkcaldy, Fife 199, 239, 254–5, 604
Kirkcaldy presbytery, Church of Scotland mission station 502
Kirkcudbrightshire, Gatehouse-of-Fleet 323
Kirkliston, Midlothian 19–20, 578
Kirkmadrine, Galloway 20
Kirkurd, Peeblesshire 444, 446
Kirkwall, Orkney 492, 656
Kirkwall Cathedral 51, 56
Kirkwood, Rev. James 367
kirkyards
    fairs in 116
    see also burial sites; Greyfriars kirkyard
Knapdale 28
Knights of St John, Torphichen preceptory 383
Knights Templar 319, 604
Knox, John 395
    antipathy to Mass 469
    Bible and 340, 395
    Buchanan, George and 384
    Communion and 469, 617
    History of the Reformation in Scotland 104, 662
    imprisonment 106
    language 340, 342, 344
    Lyndsay, Sir David and 384
    paintings of 662
    pulpit 620
    St John's church, Perth 669
    Scots Confession and 107
    sculpture of 665
    sermon at funeral of Regent Moray 476
    stipend 436
    Wishart, George, and 343
    women and 137
    see also John Knox House museum
Knoydart 163, 173
König, Karl 318
Koran *see* Qur'an, the
Korean language, Bible in 528
Kraemer, Hendrik 526
Kramer, Heinrich 114
Kuruman missionary station, Cape Colony 521
Kuwait 288, 290
Kyle and Carrick, Gaelic language 364

L

Ladies Highland Association 503–4
Lady aisle, Glasgow Cathedral 75, 88

Lady Chapel
    Dunblane Cathedral 580
    Glasgow Cathedral 75, 80
    Roslin chapel 74–5
Lady Glenorchy's chapel, Edinburgh 137
Laigh Kirk, St Giles' church, Edinburgh 583
laird's lofts 592–3, 618
lairs *see* burial sites
Lallans Society 357
Lamb, Andrew, bishop of Brechin 627
Lamb, Christian 91
Lamb, Thomas 446
Lambeth Conference 205
Lambhill Evangelistic Mission 497
Lambhill Gospel Band 497
Lammington church, Lanarkshire 601
Lamont, Mrs 641
Lamont, Rev. Donald 374
Lanark, Corpus Christi pageant 70
Lanarkshire
    Catholic community 173–4, 179
    church conversions 601, 603
    Jewish pedlars 266
    mission halls 500
    St Bride's church, Douglas 83, 84
    Scots language prayers 349
    see also Airdrie; Cambuslang; Carfin; Hamilton; Motherwell
Lanarkshire Christian Union 499–500
Langside Synagogue, Glasgow 270
languages 335–62
    Book of Common Prayer 346
    catechisms 340, 342, 367, 394
    Catholic Church 185–6, 356
    Church of Scotland and 339–47, 349, 357
        Gaelic 365, 370, 373, 374, 375
    foreign missions 519, 523–4
    grammar schools 365
    hagiographies 44, 337
    Knox, John, and 340, 342, 344
    pre-Reformation church services 66, 68, 335, 337
    Protestant Reformation's effect on 339–40, 341–4, 346–7
    St Mungo Museum displays 307
    Scots Confession 343, 344
    see also African languages; Arabic language; English language; Gaelic language and culture; Hebrew language; Indian languages; Irish language; Korean language; Latin;
Mandarin language; Scots language; Yiddish language
Larger Catechism 131–2, 346
Largs, Skelmorlie Aisle 661
Larner, Christina 130
Latin 335, 338–9, 356, 360, 364
    Catholic Bible 159, 335, 338, 340
    Catholic Church services 58, 66, 186, 336, 356, 382
    hagiographies in 44
    in schools 365
    literature in 381, 384, 386
Latin Mass Societies 185
Latourette, K S 517
Laud, William, archbishop of Canterbury 585
Lauder church 586
Lauder, Robert Scott 662–3, 665
Lauderdale, duke of 586
Laud's Liturgy 196, 346, 585
Laurelpark School, Glasgow 606
Laurencekirk 34, 46
Laurie, Canon 219
law 530–59
    Church of Scotland and 531, 532–6
    Islamic 287
    see also Act of Union; canon law; courts; legal system; Penal Laws; Scottish Parliament (–1707); sumptuary laws
Law, James 447
Law, William 404
Lawrie, Thomas 445
lay baptism 472, 474
lay missionaries 137, 501, 502, 505, 511
lay preaching 240, 500, 518
Le Feuvre, Amy 410
League of the Cross 177, 495
lecterns 625
legal system 117–18, 531–2
    see also courts; law
Legerwood parish church 578
Legge, James 527–8
Leiper, William 599
Leitch, W L and R P 665
Leith
    Baptists in 243
    Christian Science church 252
    emigrants passing through 259
    Episcopalian chapel 197
    Methodist Central Hall 245
    Norwegian church 601
    Preceptory of St Anthony's 87

St Barbara's altar 73
St Ninian's manse 613
see also Water of Leith
Leith Auxiliary Society, London Missionary Society 485
Leith, Concordat of 110
Leith School of Art 601
Lekpreuik, Richard 343
Leneman, Leah 130
Lenman, Bruce 194
Lennon, John 427
Leo VIII, Pope 536
Leo XIII, Pope 181
Leonard, Tom 360
Leonardo da Vinci 665
Leopold II, king of the Belgians 521
Leslie, Alexander 148
Leslie, Elizabeth 85
Leslie, laird of 86
Lesmahagow, St Machutus 34
Leuchars parish church 578
Levi, Joseph 260
Levington, John 91
Levison, Leon 265
Lewis 8, 212, 317, 375, 575, 655
Lewis, Bernard 282
Liberton church 589
Liddell, Eric 528
    see also Eric Liddell centre
Life Boys 458
Lincluden 50, 580
Lindisfarne 652, 654
Lindisfarne Gospels 652
Lindsay, Sir David *see* Lyndsay or Lindsay, Sir David
Lindsay, Ian G, & Partners 592
Linlithgow 50, 68, 89, 582, 591, 635, 656
Linlithgow presbytery, foreign mission work 519
Linton church 592
Linyanti, Africa 522
Lisenheim, Moses Henry 259
listed buildings 614–15
literacy
    Bible and 367, 394, 396, 484
    English language associated with 347
    evangelism and 396
    Glasgow and Ayr synod 396
    Highlands and Islands 159
    ministers and 396
    pre-Reformation 63, 413
    Protestant Reformation and 383
    Sunday schools and 483
    tract distribution and 397
    see also education
literary societies 274–5, 276
literature 379–92, 401, 403
    Anglo-Saxon 335
    Bible and 338, 346, 379
    Catholic community 61, 78, 80, 86, 185, 382
    children's books 406
    Church of Scotland and 386, 390
    education and 382–3
    Episcopalian community 386
    evangelism and 386, 403
    in schools 115
    in Scots 337–9, 340, 342–4, 348, 352, 381–7
    James VI, king of Scots and 114, 383, 385
    Kailyard school 409–10
    Latin 381, 384, 386
    Mary, mother of Jesus in 338
    ministers in 349–50
    the Passion in 337, 338
    Protestant Reformation and 383–5
    see also Bible; books; drama; folk literature; Gaelic literature; language; poetry
Lithuanian Catholic immigrants 178–9
Little, Robert 482
liturgies
    Catholic Church 61, 86, 159–60, 185–6, 337, 382, 630
    Church of England 132–3, 200, 205
    radio programme about 302
    Scottish Episcopal Church 192, 196, 200, 201–2, 218
    modern 224, 226, 227, 228
    see also Book of Common Order; Book of Common Prayer; Directory for the Publick Worship of God; Laud's Liturgy; missals; Salisbury or Sarum Use; Scottish Liturgy
Livingston 224
Livingston, Rev. Neil 643
Livingstone, David 407, 520, 521–2, 523, 525, 528
Livingstonia, Malawi 522, 528
Lloyd George, David 410
LMS *see* London Missionary Society
Local Ecumenical Projects 224
local preachers, Methodist 245
Loch Awe, Navvy Mission 505
Loch Earn, St Fillan 34

Loch Fyne, Portavadie 460
Loch Lomond, St Kessog 34
Loch Ness monster 319
Lochaber, Episcopalianism 197
Lochbroom, communion table 587
Lochcarron, Free Church 130
Lochinver church, Sutherland 604
Lochleven, St Serf at 46
Lochore mission station, Fife 501
Lockhart Memorial church, Edinburgh 609
London Jewish Board of Guardians 259
London Missionary Society 517, 525
    foreign missions 520, 524, 526
    Liddell, Eric and 528
    Scottish support for 484–5, 518
    teacher missionaries 527
Lonmay chapel, Scottish Episcopal Church 197
Lord Chancellor's office 548
Lord High Commissioner, General Assembly of the Church of Scotland 536, 548
Lord's Day observance *see* Sabbath observance
Lord's Day Observance Society 415
Lords of the Isles 365
Lord's Supper celebrations *see* Communion
Loretto, shrine 67, 72
Lorimer, Hew 670
Lorimer, J H 666
Lorimer, R L C 358
Lorimer, Sir Robert 210, 592, 669, 670
Lorimer, W L 357–8, 359
Lorn presbytery 501
Lothian Road church, Edinburgh 599
Lothians 17, 32, 175, 364, 439–40
    see also East Lothian; Midlothian
Louk, Elizabeth 89
Louk, Sir Henry 89
Loukopia, Galloway 23
Love, James 517
Lovedale Institution 524
Lovers of Zion 274
Low, William, canon 211, 213
Lowland Vicariate, Catholic Church 162, 173
Lowlands
    art 658
    Baptists in 244
    Catholic community 147, 151, 163
    church buildings 592
    Church of Scotland ministers' stipends 437

Gaelic language 363, 364
Highlands and Islands and 8, 104, 151, 163, 490
languages 339
Theosophical Society 318
Lowth, Robert 368
Lubavitch Organisation 271
Luing mission station 501
Lullingstone, Kent 16
Lurie, Rabbi 268
Luss, St Kessog 34
Luther, Martin 103, 105, 413
Lutheran churches, Scottish Episcopal Church and 224
Lutheranism 62, 236
Lutyens, Sir Edwin Landseer 669
Lyall, David 570
Lyell, Sir Charles 538
Lynch, James, co-adjutor vicar-apostolic 175
Lynch, Michael 658
Lyndsay or Lindsay, Sir David 67, 79, 86, 338, 383–4, 385, 633

M
McBain, Alexander 218
MacBain, Dr Alexander 372
McBain, John 216, 218
Macbeth 381
Maccabi Complex, Giffnock 265
MacCaig, Duncan 446
McCartney, Crear 635
McCoist, Ally 427
MacColl, Malcolm, canon 213–14, 218
MacColla, Fionn 389
McConnachie, Catherine 138
McCormick, John 297
McCowen, Alex 358
MacDiarmid, Hugh 355–6, 357, 389, 390
MacDonald, Fr Allan 373
MacDonald, Rev. Donald N 374
Macdonald, Duff 525
MacDonald, Finlay J 373–4
MacDonald, George 394, 403
MacDonald, Rev. John 370, 371, 372
MacDonald, Joseph, archbishop of Edinburgh 178, 184
MacDonald, Mary 371
MacDonald of Keppoch 150
Macdonald, Rev. Patrick 640
MacDonald, Rev. Roderick 374
Macdonald, Very Rev. Sandy 458
Macdonald, Stuart 129

MacDougall, Duncan 371
MacEachen, Rev. Fr Ewen 369, 373
McEwen Bequest 647
McEwen, Sir John B 647
McEwen, Robert 494–5
MacFarlan, Rev. Donald 457
MacFarlane, Rev. Alexander 366, 368, 370
MacFarlane, Patrick 366
MacGill, Patrick 185
McGowan, Andrew 141
MacGrath, John 390
MacGregor, Duncan 365
MacGregor, James, dean of Lismore 365
Macinnes, Alexander, dean 219–20
Macinnes, Rev. Duncan 372
Macinnes, Duncan, bishop 218
MacKay, Brian, vicar 365
MacKay, Rev. Girvan 374
Mackay, James, Lord Mackay of Clashfern 140
Mackenzie, Agnes Mure 218, 219
Mackenzie, Sir Alexander 647
MacKenzie, Rev. Colin N 374
Mackenzie, Compton 185, 219
Mackenzie, Ian 296
Mackenzie, Rev. Ian 418–19
Mackenzie, John 521
Mackey, Professor 542
Mackie, Neil 465
Mackinnon, Donald 227, 372
Mackintosh, Charles Rennie 589
Mackintosh, Donald, archbishop of Glasgow 182
McLagan, Rev. James 372
MacLaine, Rev. Alexander 368
Maclaren, Ian 409, 666
Maclaren leisure centre, Callander 322
McLauchlan, Rev. Thomas 372
McLean, Archibald 239
Maclean, Arthur John, bishop 209
Maclean, Campbell 417
MacLean, Colin 500
MacLean, Rev. Donald 372
Maclean, Dorothy 324
MacLean, Isabella G 394, 404, 406, 646
MacLean, Rev. Malcolm 373
MacLean, Sorley 390
Maclean's Cross, Iona 658
McLeish, Henry 290
McLellan, Rt Rev. Andrew 295
MacLennan, Rev. Malcolm 372
Macleod, Alexander 658
Macleod, Alexander, of Dunvegan 83

Macleod, Donald 140, 400, 403
MacLeod, Rev. George 416, 418, 423, 506, 509
MacLeod, Rev. Kenneth 372
MacLeod, Rev. Malcolm 373
MacLeod, Rev. Norman 370, 372
MacLeod, Norman 403, 407
MacLeod, Rev. Norman 494
MacLeod, Norman 665
MacLeod, Rev. Dr Roderick 374, 502
McLuskey, Rev. Fraser 452
McMarquess, John 369
Macmillan, Very Rev. Gilleasbuig 453
MacMillan, James 187, 422, 649
MacMurray, John 416
McNabb, Vincent 183
MacNicol, Nicol 526
MacPhail, Dr Matilda 527
Macpherson, Rev. John 398
McPherson, Robert 419
MacQueen, Professor 24
McTaggart, William 665, 666
MacVicar, Rev. Kenneth 453
MacWhirter, John 665
Madonna (singer) 424
Madras 526, 527
Madrid, Scots College 158
Maeatae 17
magazines *see* periodicals
Magnus Barelegs, king of Norway 56
Magnus Maximus 17
Magnusson, Anna 463
Magnusson, Mamie 461, 462
Magnusson, Sally 421
Maguire, John, archbishop of Glasgow 172, 178
Main Street Synagogue, Gorbals, Glasgow 268
Mainzer, Joseph 643
Maitland, John, duke of Lauderdale 586
Maitland, Sir Richard, of Lethington 86
Makey, W 434, 436
Makololo chieftaincy, Livingstone and 522
Malacca, Chinese College 527
Malawi 522, 525, 528–9
Malaysian community, Islam and 284, 286
Malcolm III, king of Scots 57, 576, 630
Malcolm IV, king of Scots 605
Malcolm, Lord Fleming 85
Maliki school, Islamic law 287
Malison, Alexander 89
Malison, Gilbert 89
Malta 397

Mandarin language 307, 527
Manning, Henry E, archbishop of
    Westminster 175, 177
manses 433, 474, 475, 506, 587, 611–13
    see also glebes
Marathi language 523
Margaret, queen (consort of James III) 63
Margaret, queen (consort of James IV) 63
Margaret, queen (saint and consort of
    Malcolm III) 381, 576, 577, 630
    art and 654–5
    burial place and shrine 57
    endowment of pilgrims' ferry 56
    gospel books 53
    hagiography 46
    Seventh-Day Adventists and 252
Marian Players, the 185
Marie de Lorraine, queen (consort of James
    V) 632
Marine, Theodore 257
Marischal College, Aberdeen 257, 541
Marists 175
maritime museum, Aberdeen 602
Marnock manse 613
marriage 543–4, 547
    Catholic, Penal Laws and 147
    Church of Scotland 474–6
    Church of Scotland ministers' 441–2
    civil ceremonies 476
    Episcopalian community 197–8, 204,
        211, 221, 225, 226, 229
    homosexual partnerships and 545, 549
    matrimonial law 118
    mixed Catholic/Protestant 216
    Mormons 253
    Muslim community 288, 292–3, 309
    Scots language service 357
    see also irregular marriage
marriage pews 618
Marshall, Bruce 185
Martin Cross, Iona 630
Martin, Richard 282–3
martyrdom 67
    Aaron and Julius (citizens of Caerleon)
        16
    Hamilton, Patrick 106
    Ogilvie, John 146
    St Alban 16
    St Catherine 68
    St Magnus 56
    Solway martyrs 629, 665
    Wishart, George 106, 658
    women Covenanters 137

Marx, Karl 538
Mary (mother of Jesus) 62, 81, 83, 84–5, 382
    altars dedicated to 75, 89
    depictions of 70, 71, 81, 83, 84–5, 582
        paintings 62, 71, 81, 84
        sculptures 61, 73, 81, 656
        stained glass 51, 84
        statues 62, 66, 91
        woodcuts 71
    Scots language prayers to 338
Mary I, queen of England 103
Mary King's Close, Edinburgh 76–7
Mary Magdalene 382
Mary of Guise 62, 71, 105, 106, 383
Mary, queen of Scots
    abdication 108
    greeted by psalm singing 629, 642
    jewellery 72, 75
    paintings of 661, 662
    prayers for 65, 90–1
    Protestant Reformation and 106, 107,
        108, 109, 112, 469
    Treaty of Edinburgh and 106
Maryhill, Glasgow 268, 497
Mason, Elizabeth 63
Masonic Lodges see freemasonry
mass media see media
mass stones 147
masses 88–9, 153, 469, 580
    General Assembly on 85, 86
    in secret 147, 161
    Latin 186, 382
    music 147, 635, 649
    proscription of 107, 109, 145
    St Catherine's 68
    St Gregory's 53
    St Michael's 635
    vernacular 185
    see also obits
Matar, Nabil 283
Matheson, Duncan 397–8
Matilda, queen of England 46
Maule, Graham 647
Mavisvalley mission station 501
Maximus see Magnus Maximus
Maxwell family 143
Maxwell, Ian 522, 523
Maxwell-Stuart, P G 128
Mayfield church, Edinburgh 589
Mayo, Isabella Fyvie 401
Mayo, J A 414
media 413–29, 501, 510
    see also books; broadcasting; films;

newspapers; periodicals; sound recording
Medical Mission to the Jews, Edinburgh 265
medical profession 257, 258, 266, 267, 284, 527, 548
Medical School, Andersonian University 261
medical treatment 32, 265, 545, 548
   see also healing; hospitals; welfare
meditation groups 315, 324, 326, 327, 328
mediums 318, 323
Medwyn, Lord 204
'Meet Your Neighbour' multi-faith event 311, 317
meeting houses
   Glasites, Edinburgh 599
   Scottish Episcopal Church 191, 197
Meiklefolla, Scottish Episcopal Church 202, 213
Melrose Abbey 50, 584, 656, 658
Melville, Andrew 111, 435, 665
Menard, Charles 595
Menteith, Robert 446
Menzies, Rev. Robert 366
Menzies, Thomas, of Pitfodels 64
merchant guilds *see* guilds
Merryflatts, Glasgow 263
Methodist Church 244–5, 248, 482, 483, 645
Methodist Conference of Great Britain 245
metrical psalms *see* Psalms
Metrical Psalter *see* Psalter
Miathi 23
Mid-Craigie, Dundee 506
Middlefield, Aberdeen 506
Midlothian 19–20, 264, 578
   see also Roslin chapel
migration 10–11, 184, 241
   see also emigration; immigration
Millais, John Everett 665
Millar, Euphemia 527
Millar, J H 409
Millennial Dawn *see* Jehovah's Witnesses
Miller, Hugh 405–6
Miller, Lydia 406–7
Miller, William 253, 526
Million Shilling Fund 506
Millport, Church of Scotland Seaside/Summer Mission 457, 458
Milne, Sandy 213
Milton, John 666
Milton of Campsie, Ancient Orthodox Church 254

ministers *see* Church of Scotland ministers
missals 63, 81, 83
mission halls 248, 265, 497, 500
   see also church halls
Mission of Friendship 506
mission stations 501–2, 503
missionaries *see* Finnio (British bishop and missionary); foreign missions; home missions; lay missionaries; St Columba; St Ninian
missionary ministers 487, 491, 494, 501
Missionary Society *see* London Missionary Society
missions *see* evangelism and evangelicals; Faith Mission; foreign missions; home missions; Navvy Mission; Radio Missions; Salvation Army; Scottish Mission
Mitchell, Antony, bishop of Aberdeen and Orkney 209
Mitchell, David Gibb 355
Mitchell, George 497
Mitchison, Naomi 390
Mitchison, R 130, 441
Moderates and Moderatism 8, 385, 396, 447, 518, 532
moderators 131, 137, 138, 443
Moffat, Mary 521
Moffat, Robert 520, 521
Moffatt, James 356–7, 411
Moidart 173
Moltmann, Jürgen 563
monarchy
   art commissioning 63
   books and 63
   Buchanan, George and 384
   Catholic Church and 105, 154, 548
   church attendance 63
   Church of England and 548–9
   Church of Scotland and 109, 111, 112–13, 131, 536, 549
      accession oath 132, 536
      appointment of bishops 110, 112, 130, 131, 439
      calling of General Assemblies 112, 130
      imprisonment of ministers 441
   church property annexation 110–11
   coronation oaths 109, 132, 337, 536
   Eucharist celebration 85
   hospitals and 115
   music and 632–3
   prayers for 65

Scottish Episcopal Church and 10, 192, 194, 202
see also Jacobitism; names of individual monarchs
monasteries
   Deer 364, 655
   Derry 27
   Durrow 26–7, 32
   Jedburgh 63, 89
   music and 630
   Newbattle Abbey 256
   Protestant Reformation and 105, 108, 110, 584
   St Serf's, Culross 42, 46
   see also abbeys; Iona; priories
Monteith, Robert, of Carstairs 175
Montgomery, Sir Robert 661
Montrose 239, 246, 477
Montrose, 1st marquis and 5th earl of 154
Monty Python team 427
Monymusk, Aberdeenshire, church choir 639
Monymusk Reliquary 639
Moody and Sankey Mission 497, 498
Moody, Dwight L 248, 393, 498, 499, 500, 509
Morar 160, 163, 173
Moravia, Gilbert de, bishop of Dornoch 580
Moray 245, 586, 603
   see also Findhorn Foundation and Community
Moray, earl of 108, 476
Moray House Institute 606
Moray-Knox church, Edinburgh 599
Moray Rudolf Steiner School 325
Moray synod, foreign missions 519
Morison, James 241
Mormons 235, 252, 253
Morningside church, Edinburgh 606, 609
Morningside High church, Edinburgh 601
Morrice, William 406
Morris, William 666
Morrison, Richard 413
mort bells 620
Morton, Ralph 506–7
Mosias, Moses 257
mosques 11, 288–91, 302
Mothers' Unions, Scottish Episcopal Church 221
Motherwell, Catholic see 176
Moubray, Andrew 90
Moulin church, Perthshire 609

Mozambique 525
Muchalls 204, 211, 212
Muhammad, prophet 287, 288
Muir, Edwin 390
Muir of Ord, Scottish Episcopal Church school 203
Muirchu 22
Mukden, China 528
Mull 371
Mulliner, Samuel 253
Multilateral Conversations 242, 245
Muluzi, Dr Bakili 529
Mumbai see Bombay
Munro, Dall 639
mural fabrics 627
murals 607, 666
Murchison, Rev. Dr Thomas Moffat 373, 374
Murdoch, John, bishop 174, 175
Murphy, Richard 500
Murray, Charles 393, 394, 409
Murray, James A H 353
Murray, John 89
Murray, John (publisher) 407
Murthly Hours 656
Museum of Atheism, St Petersburg 312
museums 223, 312, 318–19, 599, 602, 603–4
   see also St Mungo Museum of Religious Life and Art
music 629–51
   BBC and 302–3, 647
   Catholic Church masses 147, 635, 649
   Catholic clergy and 630, 631, 632, 634, 635
   Catholic community 187, 646, 647, 649
   Celtic Church 630
   church conversions for 599, 602
   Church of Scotland 499, 593, 625, 629, 637–46, 647
   Dunblane Music Group 647
   ecumenical activity and 647
   education 639, 642, 646–7
   evangelicals and 371, 372, 498, 499, 500, 645
   Free Church of Scotland 499, 593, 643, 645, 646
   Glasites 641
   healing and 630
   Jacobitism and 641
   Jewish community 275
   monarchy and 632–3
   monasteries and 630
   Moody and Sankey Mission 498

Presbyterian churches 593
Protestant Reformation and 636–8
recordings of 414, 427
Richmond Craigmillar and Granton parish churches 568–9
St Columba and 630
schools 206, 639, 646, 647
Scottish Episcopal Church 209, 641, 645, 646
Scottish Music Centre 602
Sunday school performances 457
teachers 639, 642
United Presbyterian Church 593, 646
see also barrel organs; choirs; church organs; folk music; Gaelic music; hymns; psalms; rock music; song schools
musicians, treated as vagabonds 116
Muslim clergy 287–8, 303
Muslim community 11, 281–94, 315, 538, 549
broadcasting and 297, 298, 300, 301, 303, 304
education 284, 288, 290, 291, 293
immigration 281, 283, 288, 290, 291, 300
marriage 288, 292–3, 309
prayer 287, 288, 301, 303
St Mungo Museum displays 310
schools 293
secularisation and 291
sermons 291
women 293, 303, 311
see also Islam
Muslims, portrayal in 'The Dance of the Sevin Deidly Synnis' 80
Musselburgh 67, 72, 499
Musselburgh chapel, Scottish Episcopal Church 197
Muthill, Episcopalian clergy 201
Myers, Joseph Hart 257
Myllar, Andrew 396
Myln or Mylne, Alexander, bishop of Dunkeld 61, 86, 632
Myrtle, Harriet 406
mystery plays 337–8
mysticism 194, 302, 320, 409

N
Nairn 517, 604
Nairnshire Society for the Propagation of the Gospel 484
Naismith, James 89
Napier University, Edinburgh 605–6

Nasmyth tomb, Edinburgh 661
National Bible Society of Scotland headquarters 604
National Covenant 131, 133, 585, 660
see also Covenanters
National Trust for Scotland 609
Navvy Missions 505
Nazarene Church 248
Nechton, king of the Picts 21–2, 23, 24, 25, 26
Negative Confession 111
Neil, Thomas 65
Nelson (publisher) 396
Nennius 18
Netherbow Arts Centre, Edinburgh 390, 599
Netherlands see Holland
Netherlee, Glasgow 460, 464
Netherlee and Clarkston Synagogue 270
Neuburger, Julia, rabbi 308
New Abbey, Catholic community 143
'New Age' religions 7, 236, 316, 318, 321–7, 328
see also Paganism
New Charge Development, Church of Scotland 506
New English Bible 357
New Kilpatrick, Glasgow 464
New Lights 134, 140, 534
New North church, Edinburgh 606
New Testament see Bible
New Year celebrations
Iranian 288
see also Hogmanay
Newark Lodge 264
Newbattle Abbey 256
Newbattle church 618, 625
Newburn church, Fife 603
Newhaven church, Edinburgh 609
Newington, Edinburgh, United Presbyterian Mission 494
Newington St Leonard's church, Edinburgh 599
Newlands, Peter 89
Newmills, Fife, manse 613
newspapers 200, 216, 275, 373, 403, 423, 465
see also periodicals
Newton Dee, Aberdeen, Camphill Community 318
Newton, John 518
Newton Mearns 264, 270, 271
Newton Stewart, Catholic community 173
Nicaea, Council of 16

Nicene Creed 224
Nicoll, Catherine 409
Nicoll, Rev. Harry 408–9
Nicoll, Mildred 409
Nicoll, William Robertson 407–11
Nicolson, John 483
Nicolson, Thomas, bishop and vicar-apostolic 151, 155, 158–9
Niddrie *see* Richmond Craigmillar church
Niebuhr, Reinhold 452
Nielsen, J 284
Niemoller, Martin 417
Nietzsche, Friedrich 538
Nigeria 138, 520
Nigg church 608
Nisbet, James 586
Nisbet, Murdoch 340
non-jurors, Scottish Episcopal Church 8, 191, 192, 197, 198, 200, 206
Nordenfalk, Carl 654
Norman architecture 580, 592
Norse influence, art 655, 658
North African community, Islam and 284, 286
North Berwick, witchcraft case 114
North church, Aberdeen 602
North-East Coast Mission 496
north east Scotland
    alternative beliefs and practices 318
    Catholic community 61–2, 143, 170, 178, 179
    Episcopalian community 197, 198, 199, 202–3, 212
        clergy from 195, 201, 209
        funeral customs 204, 212
        'high church' theology 200
        marriage customs 204
        mysticism 194
    music 643
    Protestant Reformation and 110
    Tent Missions 645
    see also Aberdeen; Aberdeenshire; Banffshire; Deer; Fraserburgh; Nairn; Peterhead
North Kelvinside, Glasgow 507
North Kessock, Scottish Episcopal Church school 203
North Morningside church, Edinburgh 609
Northcott, Cecil 521
Northern District, Catholic Church 173
Northern Missionary Society 492
Northern Rhodesia 528
Northumbria 24, 36, 335, 652, 654

Norwegian church, Leith 601
Nost, John 661
Nôtre Dame cathedral 446
Novantae 17, 20
novels 348, 387, 388, 390, 403, 404, 666
Ntaoeka (follower of Livingstone) 522
Nyanja language 525
Nyasaland *see* Malawi
Nyasaland African Congress 528, 529

O
oaths
    allegiance and abjuration 8, 10, 191, 198
    court trials 545
    see also Burgess Oath; coronation oaths; non-jurors
Oatlands, Glasgow, railway mission 497
Oban 219, 221, 228, 589
obits 64, 65, 67–8, 70, 78, 90, 91
    candles at 75
    dedication 89
occultism 323, 326
O'Connell, Daniel 174
O'Donnell, Manus 379–80
O'Donnell, William 367
offering plates 72, 625
Officials Courts 117, 118
Ogilvie, John 146
Ogilvy, Alexander 87
Ogston, David 357
oil industry, chapel of 581
oils, blessing of 159
Old Kilmorach church, Beauly 601
Old Kirk *see* Auld Kirk
Old Lights 134, 138, 140
Old St Paul's church, Edinburgh 219
Old Town Information Centre, Edinburgh 599
'One Scotland, Many Cultures' campaign 315
Ong, Walter J 411
Onward and Upward Association 401, 410
Open Brethren 247, 499
oral literature 382, 385, 387, 391, 410
Orange Order, Scottish Episcopal Church and 206, 208
Orchardson, W Q 665
Order of Knights Templar 319, 604
Order of the Ascending Spirit 321
organs *see* barrel organs; church organs
orientalists 282–3
Origen 16

INDEX • 721

Original Secession Church 138, 141, 237
Orkney
    Baptist mission 491
    Catholic community 155, 162
    education committee membership 541
    Episcopalian community 212
    home mission work 492
    Kirkwall 51, 56, 492, 656
    mission stations 501
    Picts and 29
    St Magnus 51, 56
    St Peter's church, Sandwick 608
    Stenness 575
    United Secession Church 492
orphanages 206, 225, 264
Orthodox Church cathedral, Glasgow 254
Orthodox churches 224, 236, 315
Orwell church 604–5
Ossian 447
Otto, Julius Conrad 256
Our Lady of Grace well, Banffshire 151–2, 155
Our Lady of Loretto church, Clydebank 604
Our Lady of Loretto image, Edinburgh 62
overseas dioceses, Scottish Episcopal Church links with 214
overseas missions *see* foreign missions
Owen, John 238
Oxford Movement 205, 206–7, 211, 594
Oxford Street Synagogue 268
Oxford University Press 396
Oxnam church 598

P
Pae, David 403
Pagan Federation 317
pagan material, St Mungo Museum 310
paganism 317, 371
    see also 'New Age' religions
pageants 70, 72, 80, 86, 633
Paget, Sir William 67
paintings 61, 657–8, 660–3, 666, 668–9
    Bible quotation in 661
    Covenanters in 660, 662
    Edinburgh 658, 660, 669
    Fowlis Easter church 51–3, 657, 658, 669
        Crucifixion 79–80, 83, 657
        Jesus 66, 71–2, 657
        Mary and Jesus 71, 81
        St Catherine 52, 67, 68, 79
        St John 52, 84

        Glasgow Cathedral 75
        Jesus in 66, 71–2, 81, 657
        Mary (mother of Jesus) in 62, 71, 81, 84
        Perth burgh church 63
        saints in 52–3, 63, 67, 68, 79, 84, 666, 668
        Scottish Episcopal Church baptism 208, 217
        Seabury's consecration 217
    see also illustrations
Paisley
    Catholic community 172, 176
    church buildings 589
    Episcopalian community 198, 199, 206
    Glasite church 239
    St Mirren 34
    Sunday school society 483–4
Paisley Abbey
    choir 463, 465
    church building 584, 592
    precentor 642
    retable of two Marys 70
    St Miren frieze 656
    St Mirin chapel 50–1
    St Mirin retable 67
    statues of saints 81
    Sunday school funding 483
Paisley and Eastern Renfrewshire Bible Society 484
Paisley Missionary Society 517
Paisley Society for Gaelic Missions 491
Pakistani community
    Islam and 281, 283, 284, 285, 286, 288, 290
    Scottish-Asian Christian Fellowship 255
Palace Collieries mission station 501
Palestinian-Israeli conflict 291
palm crosses 218–19, 223
Palmer, Thomas 246
Panton, Miss 201
Papa Stour mission station, Shetland 501
Papa Westray mission station, Orkney 501
Papil Stone, Shetland 652
Paraphrases, in post-Reformation worship 9, 498
Pardo, David 257
Paris 7, 158
parish missions *see* home missions
parish schools *see* schools
Parish Society, St David's, Glasgow 486
parishes
    beggars and vagabonds in 116

Catholic church 184–5
Church of Scotland 105, 113, 439–40, 443–4, 532, 533, 540
  welfare responsibilities 113, 115, 135–6, 532, 533, 542
  Church of Scotland ministers and 114, 115, 439–40, 443–4, 533
  land and property registration 113
  post-Reformation 113, 114, 434, 494, 506, 532–3, 534
  pre-Reformation 105, 577, 578–9, 582, 598
  see also kirk sessions; *quoad sacra* parishes
Park church, Glasgow 599, 646
Parkhead Congregational Church, Glasgow 248
Parliamentary churches, Highlands and Islands 587
Parliaments *see* Act of Union; Scottish Parliament (–1707); Scottish Parliament (1999-); Westminster Parliament
Parnell, Charles Stewart 181
parricide 113
Partick, Glasgow 454
'Partners in Mission' consultation 229
Passion, the 72, 73, 81, 87–8, 89, 337, 338
Paterson, Don 390
Paterson, Janet 75
Paterson, Thomas Whyte 355
Paton, Joseph Noel 665, 668
Patrick, bishop of Moray 83
patronage
  abolition of 134
  Church of Scotland and 112, 437, 438, 441, 443, 445
  secessions and 133, 134, 240, 532, 533–4
Patten, William 67
pedlars 266
  see also colporteurs
Peebles 19, 64, 89–90, 439–40
Peebles collegiate church 71, 90
Peebles, David 637
Peeblesshire 444, 446, 603
Pehthelm, bishop of Whithorn 22, 23, 24, 25, 26
Pelagius 16, 18
Penal Laws 146–53, 154, 159–60, 161, 162, 197, 198
Pengilley Harwood, Ronald T 254
penitential seats 618

Penny Savings Bank, St Patrick's, Edinburgh 177
Pentecost, Church of Scotland Communion at 472
Pentecostal Church of Scotland 248
Pentecostal Church of the Nazarene 248
Pentecostal Churches 250–5
Peploe, Samuel John 666
Perceton and Dreghorn church, Ayrshire 605
periodicals 185, 397, 401, 403, 404, 407
  see also newspapers
persecution 145–55, 191–8, 201, 309, 441
  see also Penal Laws
Perth
  altars 72, 580
  Asian community 284
  Baptists 243
  Burgess Oath 133
  Christian Science society 252
  Corpus Christi celebrations 70, 80
  election of elders and deacons 128
  Episcopalian services 197
  Glasite church 239
  hammermen craft 65, 72, 80
  Kinnoull old church 661
  manses 613
  Paganism 317
  riot (1559) 106
  St Barbara's altar 72
  St Bartholomew painting 63
  St John's church 357, 580, 583, 592, 669
  St Ninian's Cathedral 595
Perth Missionary Society 517
Perthshire
  Balquhidder church 625
  church and community 453
  Doune church, conversion 603
  Dull church 604
  Episcopalian community 197
  Grandtully church 660, 661
  Lady Glenorchy's chapel 137
  Moulin church 609
  Strathfillan 34, 137
  Tibbermore church 608
Peterhead, Episcopalian community 197, 212
Peterhead Missionary Society 484
Petheid, Mariote 90
Pettie, John 665
pew rents or sittings *see* seat rents
Philip, John 520, 523, 524, 525
Phillip, John 666

Phillips, Rev. E P 268
photography 664
Physgill 55
'Pictish War' 17
Picts 7, 364
    art 318, 652, 654
    Christianity and 18, 21, 22–3, 29, 36, 575, 629, 666
    Germanus' victory against 18
    Groam House museum 318
    Orkney and 29
    Romans and 17
    St Columba and 22–3, 29, 629, 666
    St Serf and 34
    sculptures 364, 575, 652, 654
pilgrimage 53, 55–7, 67, 151–2, 155, 577
    art and 656
    Carfin Pilgrimage Centre 318
    church buildings and 580
    radio programme about 304
    routes 598, 656
    see also shrines
pilgrims' ferry, endowment of 56
Pilkington, F T 589
Piper Alpha disaster 619
Piping Centre 602
Pitcairne, Archibald 386
Pitlochry, school for travelling people 504
Pitsligo church 618, 622
Pius X, Pope 184
plays *see* drama
Plockton church 587
Pluscarden Abbey 583, 658
Plymouth Brethren *see* Christian Brethren
poetry 86, 348, 390
    'Dream of the Rood, The' 335, 654
    in Gaelic 366, 371, 372, 373, 374, 645, 654
    in Latin 384
    in Scots 348, 381–2
    James VI and 385
    MacDiarmid, Hugh 355–6, 390
    Ossian 447
Polish immigrants 182, 258, 315
politics
    Catholic community 153–4, 160, 161, 174–5, 178, 179, 180–4
    Church of Scotland 113, 385, 440–1, 536
    Episcopalian community 208, 216, 221, 229
    foreign missions and 528
    Government administration 531
    James VI, king of Scots 385

    Jewish community 268, 272–4
    'One Scotland, Many Cultures' campaign 315
    psalm-singing and 629
    see also Jacobitism; monarchy
Pollok, Glasgow 453, 604
Pollokshields, Glasgow 206, 264, 453, 604
Pollokshields Synagogue, Glasgow 270
Polwarth, Edinburgh, John Kerr Memorial Church 598
poor relief *see* medical treatment; welfare work
Poors Aisle, Cromarty church 587
Pope, Rev. Alexander 372
'popular' and 'élite' religion 60
Port na h-Aifrinne 147
portable altars 85–6
Portavadie, Loch Fyne 460
Portinari, Tommaso 658
Portmahomack, Tarbet Old church 603
Portnahaven church, Islay 587
Portree, Skye, Free Presbyterian Church 140
Porvoo Agreement 224
Potterow mosque, Edinburgh 11
Poussin, Nicolas 665
practice verses, psalms 352, 640
praise bands 649
Pratt, John 209
prayer 12
    at funerals 198, 204, 229, 476, 477–8
    Bailey's 'Great Invocation' 326
    Catholic Church 67, 88–91, 147–8, 153, 159
    Church of Scotland 435, 438, 442, 476, 477–8
    death preparations 77, 78
    for Mary, queen of Scots 65, 90–1
    for monarchy 65
    for the dead 60, 64, 65, 70, 75, 80, 87, 90
    healing by 46, 72, 73
    home missions 482, 488, 489, 499, 506, 511
    in English 335, 350–2
    in Latin 338
    in Scots 338, 344, 346, 349, 360
    in vernacular 435
    Lyndsay's view of 86
    music and 646
    Muslim community 287, 288, 301, 303
    payment to poor for 68, 87, 90
    St Kentigern and 42
    Scottish Episcopal Church 196, 198, 202, 204, 227, 229

Sunday schools 456, 457
   to Mary (mother of Jesus) and 62, 81, 338
   to St John 83–4
   World Day of Prayer 229
   see also obits; rosary, the
Prayer Book Society (Scotland) 225
prayer books
   Arbuthnott's 63, 71, 76, 81, 83, 87, 89
   Blackadder's 67, 71, 83
   in Gaelic 203, 212
   Scottish Episcopal Church 191, 192, 194, 196, 199, 201, 203, 212, 213, 215
   taken to church 85
   see also Book of Common Order; Book of Common Prayer; Scottish Book of Common Prayer
prayer desks 625
praying societies 240, 479, 483, 485, 517
precentors 370, 593, 641–5, 647
Preceptory of St Anthony's, Leith 87
Presbyterian Churches 250, 550, 593
   see also names of individual Churches
Presbyterianism 8–9, 127–42
   alternative beliefs and 315–16, 319, 328
   church government 109, 111–12, 127–42, 191, 531, 532, 535
   Episcopalianism and 127, 191, 194, 367, 440–1
   view on bishops 131, 141, 216
Presbyterians
   anti-Catholicism 182–3, 313, 422, 495, 497
   Scottish Episcopal Church and 204, 207–8
presbyteries 111, 127, 132, 434
   Highlands and Islands, meetings 439, 443
   ministers and 437, 438, 439, 443, 444, 447, 487
   moderators 131, 137, 138
   quoad sacra charges 533
   school inspections 115, 540
   seceding churches 135
   supervision of universities 541
   see also Associate (Burgher) Presbytery; Constitutional Associate Presbytery; Relief Presbytery
Prestwick, Paganism 317
priests *see* Catholic clergy; Episcopalian clergy
printing *see* books
priories, Protestant Reformation and 105

Probationers' Society 487, 488
processional banners 70, 88
Propaganda see Congregation for the Propagation of the Faith (*Congregazione de Propaganda Fide*)
Protestant Reformation 7–10, 103–23, 217, 236–7
   Catholic Church and 105–6, 108, 110, 143, 365, 435, 582
   church buildings and 581–5
   Church of England and 103–4
   language and 339–40, 341–4, 346–7
   literacy and 383
   literature and 383–5
   Mary, queen of Scots and 106, 107, 108, 109, 112, 469
   monasteries and 105, 108, 110, 584
   music and 636–8
   north east Scotland 110
   Scottish Parliament (–1707) and 106–7, 108–10, 145
   see also Episcopalianism; Presbyterianism
Protestants
   marriage to Catholics 216
   St Mungo Museum displays 309
Proverbs, Scots translation 355
Provincial Synod, Scottish Episcopal Church 205, 214
'psalm in reports' settings 637
psalm singing 629, 639, 642, 644
psalms 9, 394, 498, 629–30, 637–41, 645–6, 649
   Celtic Church 630
   Free Church services 499
   Gude and Godlie Ballatis, The 343, 636
   hymns and 352, 637
   practice verses 352, 640
   precentors and 642–5
   Scots translation of 353, 354, 355, 357
   Scott, Sir Walter on 629
   Scottish Episcopal Church services 209
   substituted for Catholic Daily Office 147
   Sunday schools and 483
   see also hymns; psalters
psalters 370, 396, 637–8, 643, 645–6
   Cathach or Battler 380
   Celtic Psalter 655
   copied by St Columba 30, 380
   in Gaelic 365, 367, 369–70
   Rous' version 346, 347, 348
Ptolemy 17, 19, 20, 23

publishing *see* books; newspapers; periodicals; tracts
Pugin, A W 589
Pugin spire, Highland Tolbooth, Edinburgh 599
pulpit falls 597
pulpit steps 210
pulpits 582, 584, 587, 591, 592, 620–2, 625
    Burntisland church 585
    church organs and 593
    crosses behind 597
    Lauder church 586
    St George's Free church 591
    St Nicholas' church, Aberdeen 617
    St Salvator's chapel, St Andrews 620
    Ullapool church 604
    woodwork 596
Pune, India, foreign mission work 526
Punjabi language, St Mungo Museum 307
Puritanicus 479–80, 511

Q

Quakers 237–8, 492
qualified clergy and congregations, Scottish Episcopal Church 191–2, 194, 195–6, 197, 198
Queen, Isaac 257
Queen Street church, Edinburgh 609
Queen's Cross church
    Aberdeen 464, 618, 622
    Glasgow 589
Queen's Hall, Edinburgh 599
Queens Park Synagogue, Glasgow 270
Queensberry Monument, Durisdeer 661
Queensferry 56
Quest (youth society), Greenbank, Edinburgh 461
Quincy, Robert de 256
quoad sacra parishes 533
Qur'an, the 287, 291, 293

R

Raasay, folk music 640
Raby, John 483
radio broadcasting 221, 295–305, 413, 414–17, 418, 420, 426–7
    Billy Graham's Crusade 508
    Carols from King's 649
    A Church without Walls 463
    'Faith of a Woman' 303–4
    Fireside Sunday School 457
    In Good Faith 298
    'Insights on Offer' 301
    Islamic worship 302
    'Making a Meal of it' 301
    on Boys' Brigade 459
    on Hindu music 302–3
    on liturgies 302
    on pilgrimage to Whithorn 304
    Thought for the Day 297, 298, 419
Radio Missions 416–17, 420, 510
Raeburn, Sir Henry 204, 661
railway mission, Glasgow 497
Rajneesh movement 317
Ramadan 288
Ramsay, Alexander 85, 445
Ramsay, Allan 348, 386, 661
Ramsay, C Lilias 252
Ramsay, E Mary 252
Ramsay, Edward Bannerman, dean of Edinburgh 200, 201, 203
Ramsden, Omar 209
Ramshorn church, Glasgow 605
Rangers Football Club 309
Ratcliffe, Alexander 183
Rattray, Thomas, bishop of Dunkeld 195
Rea, Rev. Ernest 299–300
readers, Church of Scotland 435, 620, 625
Rechabites 641
records, Scottish Episcopal Church 212–13
Reekie, Stella 297
Reform Synagogue 270
Reformation, the *see* Protestant Reformation
Reformed Church of England and Wales 242
Reformed Presbyterian Church of Scotland 133, 138, 140
Reid, Thomas 9, 10, 385–6
Reiki 322
Reith, John 411, 414, 420, 428
Reith lectures 298
relics
    Edward the Confessor 57
    St Andrew 56, 57, 381
    St Columba 56, 86, 380, 639
    St Magnus 56
    St Ninian 53, 55
Relief Church 484, 517, 518, 534, 645
Relief Presbytery 133, 134
Relief Synod 134
Religious Society of Friends (Quakers) 237–8, 492
Religious Tract and Book Society of Scotland 397, 410, 492–3
relocation, church buildings 604–5

Rembrandt 665
Rendell, Ruth 425
Renfrewshire 265, 656
    see also Paisley
Reni, Guido 665
Renwick, James 665
'repeater' tunes 646
Representative Church Council, Scottish
    Episcopal Church 205, 208, 214
Restalrig, pilgrimage centre 57, 67
Restoration Settlement 191
revivalism 239–40, 241, 244, 249
    Assemblies of God 251
    Gaelic-speaking areas 370, 371, 372
    hymns 645
    Seventh-Day Adventists 253
    see also Celtic revivalism; Moody and
    Sankey
Revolution (1688) 440, 441, 444–5, 446
Revolution Settlement 133, 146, 385, 585
Rhodes, Cecil 528
Rhodesia 528
Rhydderch Hael, king of Britons of
    Dumbarton 29–30, 32
Richard, Cliff 427
Richardson, Robert 633
Richelieu, cardinal 446
Richmond Craigmillar church, Edinburgh
    564–71
Richmond Street synagogue, Edinburgh
    269
Riddell, Henry Scott 353
Riddrie, Glasgow, Jewish burial site 268
Rifkind, Malcolm 274
Rimini, Council of 16
Ritchie, Alexander Handyside 665
Ritchie, Rocco 424
Robert I, king of Scots 55, 381
Robert, bishop of St Andrews 577
Robert MacLehose and Co Ltd 396
Roberts, David 662, 666
Robertson, William 385, 386, 447, 595, 662
Robin Hude 382
Robinson, John Arthur Thomas, bishop
    425
Robinson, Mairi 343, 344
Robson, Joseph Philip 353
rock gongs 630
rock music 648–9
Rodel, Harris 658
Rognvald, earl 578
Rolland, William 64
Roman Britain 16–17

Roman Catholicism see Catholic Church;
    Catholic clergy; Catholic
    community
Romanesque architecture 578, 579, 591, 595,
    655, 663
Romanesque art 658
Rome, Scots College 158
Rood see Holy Rood
rosary, the 81, 147, 153, 338
Rosebery, earl of 448
Rosemarkie, Black Isle 34, 318–19
Rosie, Thomas 496
Roslin chapel, Midlothian 319, 580
    sculptures 51, 74–5, 79, 80, 81, 87, 656
Ross, Alexander 210
Ross, Andrew 407
Ross, Anthony 183
Ross House, Edinburgh 262
Ross, John 528
Ross presbytery 492
Ross, Rev. Thomas 370
Rossend Castle, laird of 585
Rosskeen Religious Association 492
Rosslyn chapel see Roslin chapel,
    Midlothian
Rothesay, Bute, Faith Mission 499, 509
Rous, Francis 346
Roxburgh Place synagogue, Edinburgh 269
Roy, Kenneth 418–19
Royal Academy of Music 647
Royal Bavarian Glassworks 664
Royal Bounty Committee, General
    Assembly of the Church of
    Scotland 486–7, 491
Royal National Mission to Deep Sea
    Fishermen 497
royal saints 46–7
    see also Margaret, queen (saint and
    consort of Malcolm III)
Royal School of Church Music 647
Royal Scottish Academy of Music and
    Drama 646
Royal Scottish National Orchestra 599
Ruddiman, Thomas 386
Runciman, Alexander 662
Rushdie, Salman 291, 547
Russell, bishop 200, 208
Russell, Charles T 254
Russia 258, 265, 274, 312
Russian Jewish Relief Fund 263
Ruth, Scots translation of book of 353,
    357
Rutherford Free church, conversion 603

INDEX • 727

Rutherford, Joseph Franklin ('Judge Rutherford') 254
Rutherford, Samuel 386
Ruthwell Cross, Dumfriesshire 335, 654

S

Sabbath observance 116–17, 141–2, 447, 546
   Jewish community 262, 268
Sacks, Dr Jonathan, chief rabbi 298
sacrament houses 81, 87, 657
Sacred Music Institution 641
Sahaja Yoga 317
St Adomnán *see* Adomnán, abbot of Iona
St Adrian 34
St Aidan parish church 86
St Alban 16
St Aloysius church, Glasgow 595
St Andrew 51, 56–7, 381
St Andrew kirk session 346
St Andrew Press 418
St Andrews
   Christians Linked Across the Nation Gathering 252
   church buildings 577–8
   Eden Christian Fellowship 251–2
   election of elders and deacons 128
   Holy Trinity church 580, 583, 587, 592
   Paganism 317
   Presbyterianism 127
   Queen Margaret and 576
   St Rule's church 56, 578
   St Salvator's chapel 620
   St Salvator's College 68, 633
   St Salvator's collegiate church 84–5, 631–2
   sculptures 654
   Theosophical Society 318
   Wolfenbüttel manuscript 631
St Andrew's and St George's church, Edinburgh 586
St Andrews Cathedral
   church building 56–7, 578, 579, 583
   depiction of head of Jesus 71, 656
   pilgrimage centre 56–7, 580, 656
   St Andrew relics 56, 57
   statues 51, 66
St Andrew's church
   Dundee 239, 587
   Edinburgh 586
   Glasgow 211, 602
St Andrews Psalter 637
St Andrews Sarcophagus 654
St Andrews University 7, 107, 108, 116, 582
St Anne 71
St Anne's altar, St Giles' church, Edinburgh 64
St Anthony, painting in Fowlis Easter church 52
St Anthony's, Preceptory of, Leith 87
St Athanasius 16, 379
St Baldred 34
St Barbara 67, 70, 72, 73
St Bartholomew, painting of, Perth burgh church 63
St Baya 34
St Blane 34
St Boniface (Cuiretán) 34
St Brendan of Birr 27
St Bride, painting of 668
St Bride's church
   Douglas, Lanarkshire 83, 84
   East Kilbride 596
   Edinburgh 601
St Brigid 50, 382
St Catherine 52, 67, 68, 79, 89
St Christopher 51, 65
St Clement's church, Rodel, Harris 658
St Colmán 36
St Colman's church, Portmahomack 603
St Colmoc 36
St Columba 25, 26–32, 363
   death of 30, 32, 56, 379
   early Christian art 652
   Gaelic poetry in praise of 654
   hagiography 26, 43–4, 379–80
      see also Adomnán, abbot of Iona, *Life of St Columba*
   Iona and 22–3, 26, 27, 30, 32, 219, 363
   Latin and 335
   music and 630
   paintings of 666, 668
   Picts and 22–3, 29, 629, 666
   psalter copied by 30, 380
   relics 56, 86, 380, 639
   St Finnian and 20, 26
   St Kentigern and 33
   sculpture of 50
   shrine 56
St Columba's church, Glenrothes 594
St Columba's-by-the-Castle, Scottish Episcopal congregation 206, 211
St Constantine 34
St Conval 34, 57
St Cuthbert's altar, Edinburgh 65
St Cuthbert's church, Edinburgh 464, 589
St David's parish, Glasgow 486

St Devenic 34
St Donnan 34
St Drostan 34
St Duthac 34, 56, 577
St Duthus church, Tain 608
St Edward's church, Canna 609
St Eloy 67, 80
St Enoch 382
St Ethernan 34
St Fergus 36, 50
St Fillan 34, 42, 43, 44, 319, 656
St Finbar of Cork 25
St Finnbarr 34
St Finnian 20, 26
St Finnian of Clonard 25
St Finnian of Moville 25
St Fothad's mission station 501
St Fotin 36
St Francis 153
St Francis de Sales 153
Saint Francis-in-the-East, Glasgow 507
St Gabriel's altar, St Giles' church, Edinburgh 70
St George's church, Edinburgh 606, 642, 646
St George's Free church, Edinburgh 591
St George's in the Field church, Glasgow 603
St George's (Tron) church, Glasgow 586, 601
St George's West church, Edinburgh 454, 464
St Gilbert's church, Glasgow 622
St Giles' church, Edinburgh
  acoustics 592
  altars 580
  burial of earl of Moray 476
  Chirwa memorial service 529
  choir 463
  collegiate church 580
  Communion services 471
  Durie, John and 629
  High Kirk 583, 591
  Holy Blood altar 89
  Holy Blood confraternity 70, 88
  Holy Blood prebend 90
  Holy Rood prebend 90
  Laigh Kirk 583
  music and 465
  non-religious uses 609
  St Anne's altar 64
  St Gabriel's altar 70
  St Ninian's altar 90
  St Salvator's prebend 90
  sculptures 89
  seat rents 454
  stained glass 665
  subdivision of 583, 609
  violence against ministers 445
  West Kirk 583
St Giles' church, Elgin 618
St Giles' Day procession, Edinburgh 86
St Gregory, Mass of, picture of 53
St James' altar, Haddington parish church 83
St James' church
  Cruden 212
  Paisley 589
St James' (Pollok) church, Glasgow 604
St James the Less 50, 52
St John 52, 83–4
  see also Knights of St John
St John the Evangelist 50, 51, 83–4, 89
St John's church
  Aberdeen 81
  Ayr 64
  Ballachulish 218
  Edinburgh 201, 204, 219
  Forfar 223
  Perth 357, 580, 583, 592, 669
St John's Gospel, Book of Kells and 654
St John's parish, Glasgow 486, 588, 623
St John's Street church, Edinburgh 606
St Kenneth 36
St Kentigern (St Mungo) 32–4, 42, 44, 46, 56, 382
St Kentigerna 36
St Kessog 34
St Kevoca 36
St Laurence of Canterbury 34, 46
St Leonard's church
  Dunfermline 591
  Edinburgh 599
  St Andrews 622
St Leonard's hospice, St Andrews 580
St Lis see Senlis, Simon de, earl of Northampton and Huntingdon
St Luke, stained glass window, Douglas Kirk 51
St Luke's church
  Broughty Ferry 589
  Edinburgh 604
St Machan 36
St Machar 34, 43, 47, 382
St Machar's Cathedral, Aberdeen 89, 381, 583, 620, 625

INDEX • 729

St Machutus 34
St Madoc 36
St Maelrubai 34
St Magnus 51, 56
St Magnus Cathedral, Orkney 578, 583, 655
St Magnus church, Tingwall 75
St Maioca 36
St Margaret *see* Margaret, queen (saint and consort of Malcolm III)
St Margaret of Antioch 72–3
St Margaret the Virgin, altar, Dundee 72–3
St Margaret's church
    Dunfermline 595
    Knightswood, Glasgow 622
    Newlands, Glasgow 221, 595
St Mark, stained glass window 51
St Mark's chaplainry, Dundee 65
St Mark's gospel 358, 424–5
St Marnoc 36
St Marnock's church *see* Fowlis Easter church
St Martin 20, 382
St Martin, church of 22, 23, 24, 25, 55
St Mary in Childbirth prebend, Peebles 71
St Mary in the Fields church, Edinburgh 90
St Mary of Loretto, shrine 72
St Mary Wynd hospital, Edinburgh 91
St Mary's Cathedral, Edinburgh 206, 209, 214, 228, 594, 595
St Mary's church
    Aberdeen 173
    Bo'ness 595–6
    Cullen 62
    Dundee 63, 580, 581, 583
    Edinburgh 586
    Glasgow 595
    Haddington 88, 592
    Motherwell 509
St Mary's Mothercraft Centre, Dundee 216
St Matthew, stained glass window 51
St Matthew's gospel, Scots translation 353
St Matthias, painting of 52
St Maura 34
St Medan 36
St Michael 50, 83, 582, 635
St Michael and All Angels church, Inverness 604
St Michael's altar, Dundee 65
St Michael's church, Linlithgow 89, 582, 591, 635, 656
St Miren, Paisley Abbey 50–1, 67, 656
St Mirren 34
St Monan 36
St Mungo *see* St Kentigern
St Mungo Museum of Religious Life and Art 293, 306–12, 318
St Mungo's church, Alloa 589
St Nicholas' church, Aberdeen 581, 583
    altars 89, 580
    choir 464, 639
    church bells 620
    church maintenance 64
    furnishings 616–17, 618–19, 622, 625, 627
    missal binding 83
    pulpit 617
    St Ninian's Day 68
St Ninian 6, 20–6
    art 652
    dedications to 34
    depictions of 52–3
    hagiography 22, 25, 47, 53
    relics 53, 55
    shrines 26, 53, 55
    statues of 51
    tomb 22, 23, 24, 25, 55
    see also Whithorn
St Ninian's altar, St Giles' church 90
St Ninian's Cathedral, Perth 595
St Ninian's Cave, Whithorn 55, 153
St Ninian's church
    Crieff 604
    Greenock 453
St Ninian's Day 68, 153
St Ninian's manse, Leith 613
St Olai Kirke, Helsingør 36
St Oswald's church, Edinburgh 606
St Palladius 34
St Patrick 18, 20, 44, 46, 57, 307
St Patrick's church, Edinburgh 177–8
St Patrick's well, Kilpatrick 57
St Paul 50, 51, 52
St Paul's church
    Dundee 595
    Glenrothes 595
    see also Old St Paul's church, Edinburgh
St Peter 20, 52
St Peter's church
    Edinburgh 669
    Galashiels 219, 221
    Peterhead 209
    Sandwick, Orkney 608
St Peter's Stone 20, 652
St Petersburg, Museum of Atheism 312
St Quivox church, Ayrshire 560, 563

St Regulus 36
St Rule's church, St Andrews 56, 578
St Salvator's altar, St Nicholas church, Aberdeen 89
St Salvator's chapel, St Andrews 620
St Salvator's church, Dundee 595
St Salvator's College, St Andrews 68, 633
St Salvator's collegiate church, St Andrews 84–5, 631–2
St Salvator's prebend, St Giles' church 90
St Saviour's Child Garden 219
St Sebastian, sculpture of 51
St Serf 15, 33, 34, 42, 44, 46
St Serf's monastery, Culross 42, 46
St Simon, painting of 52
St Stephen's church, Edinburgh 586
St Ternan 36, 53
St Thenew 382
St Thomas 50, 52, 527
St Thomas church, Edinburgh 498
St Thomas the Apostle and St Margaret the Virgin altar, Dundee 72–3
St Thomas's church, Keith 594
St Triduana 36, 57, 67
St Vincent de Paul welfare societies 177
St Vincent Street church, Glasgow 589
St Winnin 36
St Winnoc 36
saints 15–59, 66–7, 80, 83–5
    burial near altars of 75
    Catholic liturgy 382, 630
    depictions of 47–53, 66, 67, 582
    Lyndsay's views on cult of 67, 86
    paintings of 52–3, 63, 67, 68, 79, 84, 666, 668
    royal saints 46–7
        see also Margaret, queen (saint and consort of Malcolm III)
    sculptures of 47–51
    singing abilities 630
    statues of 48–51, 67, 80, 81, 83, 88, 307, 582
    see also names of individual saints
Saline church, Fife 603
Salisbury, 3rd marquess of 528
Salisbury Book of Hours 53
Salisbury Centre, Edinburgh 321
Salisbury church, Edinburgh 599
Salisbury Road synagogue, Edinburgh 269, 277
Salisbury or Sarum Use 630
Salmon, Rev. C A 499
Salter, Joseph 284

Salvation Army 248–50
Samye Ling centre, Eskdalemuir 315
Sandeman, Robert 239
Sandemanians 239, 244
Sandford, Daniel, bishop of Edinburgh 198
Sandilands, Sir James 383
Sandymount, Glasgow, Jewish burial site 269
Sanford, John 563
sang schools see song schools
Sankey, Ira D 497, 498, 500
Sanskrit 526
Sardica, Council of 16
Sarmento, Jacob de Castro 257
Sarwar, Mohammed 293
Saudi Arabia 288, 290
Saxons 18
Sayers, Dorothy L 416
Scalan seminary, Banffshire 157–8
Scanlan, archbishop 185
scheduled monuments 614, 615
schismatics 194
schools 91, 115, 539–41
    books in 115, 158
    Catholic community 156–9, 179–80, 184, 216, 540, 541, 548
    church conversion 606
    Church of Scotland 156, 395, 426, 435, 540
    catechisms 115, 156, 347, 483
    curriculum 115, 156
    First Book of Discipline 107, 540
    Greenock 487
    Highlands and Islands 365–6, 370
    ministers and 443–4
    presbytery inspections 115, 540
    Sunday schools and 483
    travelling people 504
    Edinburgh 271, 461
    foreign missions 524
    Glasgow 606
    Islamic law 287
    Jesuit order 158
    Jewish community 264, 271, 272
    Ladies Highland Association 503
    languages in 365–6, 368, 490
    literature in 115
    music 206, 639, 646, 647
    Muslim community 293
    Scottish Episcopal Church 203, 205, 211, 212, 214, 216, 221, 228
        Code of Canons provision for 198
        Trinity College 206, 213

seceding churches 136
sex education debate 422
SSPCK 156, 158, 366, 368, 480
travelling people 504
United Industrial School 173
see also education; grammar schools; song schools; teachers
Schotz, Benno 275, 595
Sciennes School, Edinburgh 271
Scientology 317
SCIFU (Scottish Churches Initiative for Union) 245
Scone, tombs 661
Scorsese, Martin 427
Scot, John 343
Scot, Reginald 114
Scotch Baptists 239, 242, 244
Scoto-Catholic movement 591
Scots Colleges 158
Scots Confession
   Articles Declaratory and 141
   bishops and 582
   language 343, 344
   legislation and 108, 109, 112, 534, 537
   ministers and 110
   Negative Confession 111
   preparation and adoption 104, 107
   Scottish Parliament (–1707) and 104, 108, 109
   true Church characteristics 127
Scots language 336–62
   Bible in 340, 342, 346, 352–5, 356, 357–9
   books in 338–9
   Burns, Robert, and 348, 354
   Communion service 357
   English and 335–62, 385
   hagiographies 337
   literature in 337–9, 340, 342–4, 348, 352, 381–7
   marriage service 357
   ministers and 349–50
   prayer in 338, 344, 346, 349, 360
   sermons in 338, 339, 344, 346, 349, 355, 357, 382
   services in 357
Scots Language Society 357
Scott, Adam 350–1
Scott, Andrew, bishop 174
Scott, David 664
Scott, David Clement 525
Scott, Gilbert 594, 595
Scott, Hew 438, 439, 442, 446

Scott, Sir Walter 404, 438
   depiction of Scotland 8, 388, 389
   Episcopalian and Presbyterian affiliation 204
   evangelism and works of 386
   language in works of 348, 350, 351–2
   on psalms 629
   paintings and 662
   see also Abbotsford
Scott, Walter Francis, 5th duke of Buccleuch 208
Scott, William Bell 664, 665, 666
Scott-Moncrieff, George 185
Scottish-Asian Christian Fellowship 255
Scottish Bible Society 398
Scottish Book of Common Prayer 215, 218, 223, 227
   see also Common Prayer (Scotland)
Scottish Catholic Historical Association 185
Scottish Churches Architectural Heritage Fund 615
Scottish Churches Committee 615
Scottish Churches House, Dunblane 597, 647
Scottish Churches Organist Training Scheme 647
Scottish Churches Renewal Movement 251
Scottish Coast Mission 496
Scottish Congregational Church 242, 243
Scottish Council of Jewish Communities 273
Scottish Ecclesiological Society 591
Scottish Enlightenment 9, 328, 385, 386, 447, 538, 639
Scottish Episcopal Church 10–11, 191–230, 536, 550
   American influence 192, 197, 223
   anti-Catholicism 200, 211
   architecture 209–10
   art 217
   baptism 194, 197, 204, 211, 221, 225, 229
   painting of 208, 217
   Bible 203, 224, 226, 229, 367
   Bible classes 229
   burial sites 206
   Calvinism and 194, 200, 386
   catechisms 197, 202, 224
   catechists 194, 197
   Catholic Apostolic congregations join 214
   Catholic/Protestant mixed marriages 216
   charismatic movement and 252

choirs 206, 219
Christmas 213, 218, 223
church attendance 211, 218
church buildings 202, 205–6, 209–10, 211, 212, 215
    as community centres 211, 228–9
    Home Mission Campaign 221, 506
    interiors 227
    meeting houses 191, 197
    relocation 604
church government 192, 197, 205, 224–5
Church of England and 192, 198, 205, 208, 209, 213, 218
Church of Scotland and 192, 198, 208, 216, 591
'Churching of Women' service 204
clergy *see* bishops, Scottish Episcopal Church; Episcopalian clergy
Code of Canons 198, 201
College of Bishops 104, 227
Communion 196–7, 202, 209, 218, 220–1, 225, 227
Easter 213, 218, 229
ecumenical activity 214, 216, 223, 224–5, 229
elders 197
English immigrants and 210, 211, 228
Episcopal Fund 198
Episcopal Theological College 218, 225
evangelicals 198, 200, 201, 225, 227, 498
finance 206, 215, 221, 223, 224, 226
foreign missions 199
Free Church of Scotland and 208
funerals 198, 204, 211, 212, 221, 229
Gaelic Bible 203, 367
Gaelic language and culture 203–4, 212, 225, 374
Gaelic-speaking clergy 194, 201, 203, 209, 218, 374
hanging sign 217
'hereditary Episcopalians' 210–11, 219, 228
home missions 221, 498, 506
hymns 209, 645
Irish immigrants and 206, 210
Jacobitism and 8, 10, 191, 192, 194, 198, 200
kirk sessions 197
liturgy 192, 196, 200, 201–2, 218
    modern 224, 226, 227, 228
    see also Book of Common Prayer; Laud's Liturgy; Scottish Liturgy

marriage 197–8, 204, 211, 221, 225, 226, 229
meeting houses 191, 197
membership 192, 200, 212, 214, 215–16, 221, 223, 227–9
monarchy and 10, 192, 194, 202
Mothers' Unions 221
music 209, 641, 645, 646
mysticism 194
newspapers and 200, 216
non-jurors 8, 191, 192, 197, 198, 200, 206
Orange Order and 206, 208
Orthodox churches and 224
overseas links 214
prayer 196, 198, 202, 204, 227, 229
prayer books 191, 196, 203, 212, 213, 215
    see also Book of Common Prayer
Presbyterians and 204, 207–8
psalms 209
qualified clergy and congregations 191–2, 194, 195–6, 197, 198
records 212–13
Representative Church Council 205, 208, 214
schools 205, 211, 212, 214, 216, 221, 228
    Code of Canons provision for 198
    Highlands and Islands 203
    Trinity College 206, 213
seat rents 206
sermons 194, 202–3, 209, 214
Social Responsibility Committee 216
stained glass 663
status 536
Stewart monarchs 10, 192, 194
Sunday schools 216, 219, 604
synods 198, 205, 207, 214, 224, 225
teachers 192, 206, 216
welfare responsibilities 198, 199
Woman's Conference 208
women 219, 221, 223, 224, 226, 422
    Churching of Women service 204
    Episcopalian clergy 223, 225, 226
worker priests 225
see also Episcopalian community
Scottish Episcopal Church Society 198, 199
Scottish Executive 315, 422, 531
Scottish Federation of Organists 647
Scottish Film Council 599
Scottish Football Association 416
Scottish Interfaith Council 311, 313, 317
Scottish Jewish Archives Centre 275

Scottish Jewish Standing Committee 273
Scottish Labour Pary 181
Scottish Liturgy, Scottish Episcopal Church
    192, 196, 201–2, 205, 208, 209, 215,
    218
  1911 revision 209
  1970 revision 226, 227
  1982 revision 227
Scottish Mission, Catholic Church 145,
    147–9, 151, 154, 155, 157–63, 174
  see also vicars-apostolic
Scottish Missionary Society 517, 520
Scottish Music Centre 602
'Scottish national chalice' 209
Scottish National Portrait Gallery 666
Scottish National War Memorial 669
Scottish Office 531
Scottish Parliament (–1707)
  Acts of the Parliaments (1540) 79, 81,
    83
  Catholic Church and 62, 145–6
  church government and 132, 532
  Church of Scotland and 106–7, 108–10,
    113–14, 128, 130, 131, 532
  clergy's right to vote 113
  Committee of Articles 131
  education abroad prohibited 158
  Episcopalianism abolished by 132
  Gaelic Bible and 395
  pre-Reformation support for religion
    63, 113
  Protestant Reformation and 106–7,
    108–10, 145
  Scots Confession and 104, 108, 109
  song schools 638
  vernacular Bibles and psalters 395,
    637
  Westminster Assembly documents
    approval 132
  Westminster Confession and 132, 532,
    534
Scottish Parliament (1999-) 531, 536
  General Assembly of the Church of
    Scotland and 423, 530
  Malawi and 529
  opening ceremony 629
  prayers 12
  sex education debate 422
  Wright, Kenyon, canon 227
Scottish prayer book (1637)
  Scottish Episcopal Church 196
  see also Common Prayer (Scotland);
    Scottish Book of Common Prayer

Scottish Redundant Churches Trust 607–8,
    615
Scottish Society for the Propagation of
    Christian Knowledge
  Gaelic Bible and 367–8, 490
  Gaelic language and culture 490
  missionaries and catechists 480, 483,
    491
  missionary ministers 487
  schools 156, 158, 366, 368, 480
  Sunday schools 483
Scottish Synod
  Methodist Church 245
  United Reformed Church of the United
    Kingdom 243
Scottish Television 419, 420
Scottish Tourist Board 307
Scottish Unitarian Association 246
Scribner, Robert 60
Scrimgeour, John 63
Scrimgeour, Marion 85
Scripture Union Scotland 499, 512
scriptures
  Hindu 302
  see also Bible
Scroggie, John 500
sculptures 61, 654, 656–7, 661, 665, 668
  Annunciation, the 658
  Banff market cross 61, 73
  baptismal fonts 72, 81, 86, 87
  Book of Deer and 655
  Deskford sacrament house 87, 657
  Falkland Palace 73
  Fowlis Easter church 81, 83, 657
  Glasgow Cathedral 50, 71, 73, 75, 80,
    88, 656
  Highlands and Islands 658, 668, 670
  Iona 47, 50, 56, 630, 654, 658
  Jesus 61, 71, 81, 83, 87, 656
  Jewish community 275
  Knox, John 665
  Mary (mother of Jesus) 61, 73, 81,
    656
  Mary King's Close, Edinburgh 76–7
  Passion, the 72, 73
  Pictish 364, 575, 652, 654
  Roslin chapel 51, 74–5, 79, 80, 81, 87,
    656
  Ruthwell cross 335
  St Andrews 654
  St Giles' church, Edinburgh 89
  St Peter's Stone 652
  saints 47–51

Scottish National War Memorial 669
Seton, Lord 68
Tealing church, Angus 599
see also statues
Seabury, Samuel, bishop 192, 205, 217
Seaman's Bethel, Glasgow 497
Seaman's Chapel, Glasgow 488, 497
Seaman's Friend societies 496
Seaside Mission, Church of Scotland 457–8
seat rents 206, 454, 486, 587
seating 616–19, 623
secession
    from Church of Scotland 133–41, 240, 313, 349, 441, 518, 532, 533–4
    from Free Church of Scotland 140
Secession Church 134, 484, 517, 518
    see also Original Secession Church; United Secession Church
Second Book of Discipline 104, 111, 112
Second Vatican Council 185–6, 188, 356, 647
Secretary of State for Scotland 531
sectarianism 309, 311, 313, 422, 497, 509, 549
    see also anti-Catholicism
sects 235–55
secularisation 4, 6, 10–11, 160, 188, 390, 530
    building conversions 613
    Catholic Church and 183, 188
    Labour Party and 183
    Muslim community and 291
    Pae's novels and 403
secularisation thesis 313, 315
Ségéne, abbot of Iona 26
Sekeletu, chief of the Makololo 522
Selgovae 17, 19
Sellars, James 591
seminaries, Catholic 155, 157–8, 162, 177
Senlis, Simon de, earl of Northampton and Huntingdon 46
Serampore, India, Baptist mission 519
Seres, Robert (elder and younger) 75
sermons 438–9, 589, 591
    at funerals 476
    Blair, Hugh 447
    Chalmers, Thomas 398
    Church of Scotland 10, 344, 438, 470
    Ewing, Grenville 482
    Glasgow City Mission 488, 489
    in *Ane Satyre of the Thrie Estaitis* 338
    in English 335
    in Gaelic 364, 366–7, 371, 372
    in Scots 338, 339, 344, 346, 349, 355, 357, 382

London Missionary Society 525
Methodists 483
Moody and Sankey Mission 498
Muslim community 291
printing and distribution 355, 357, 366–7, 371, 398, 400–1
Scotichronicon and 382
Scottish Episcopal Church 194, 202–3, 209, 214
Tract Societies 485
Seton, Lord, sculpture of 68
Seton collegiate church 68, 72, 86, 87, 88, 580
Seton, George, 5th baron 72
Seton, Mary 75
seTswana language, Bible in 521
Seventh-Day Adventists 252, 253–4
Severus, Lucius Septimius 16, 17
sexual conduct 65, 67, 78–9, 113, 130, 545
    Church of Scotland ministers 442, 446
    see also adultery; bigamy
Seymour, Edward, 1st earl of Hertford and duke of Somerset 67
Shafi'I school, Islamic law 287
Shaikh, Asma 309
Shakespeare, William 381, 425
Shannon, Rev. Bill 458
Shapiro, Rabbi 268
Sharpe, George 248
Shaw, Duncan 114, 530–1
Shaw, Geoff 460–1
Shaw, George, abbot of Paisley 81
Shaw, Margaret Fay 373
Shaw, Rev. Neil 373
Shechita Board, Glasgow 268
Sheed, Frank 183
Sheppard, Dick 414
Sherbrooke St Gilbert's church, Glasgow 622
Shetland
    Church of Scotland mission stations 501, 502
    early Christian art 652
    education committee membership 541
    Methodist Church 245, 482
    Papil Stone 652
    Quaker preachers 492
    St Ninian, dedications to 34
Shettleston, Glasgow, Jewish burial site 269
shi'ite Islam 286, 287–8
Shinwell, Emanuel 274
Shorter Catechism 140, 346, 411, 470
    Calvinism and 394

education and 347, 393–4, 426
General Assembly of the Church of
    Scotland approval 131–2, 470
printing, publication and distribution
    394, 398
Shotts, Lanarkshire Christian Union 500
shrines 56–7, 61, 67, 72, 656
    Carfin 184
    healing 57, 67, 86, 580
    St Ninian 26, 53, 55
    see also bell shrines; pilgrimage
Sibbald, James 53
Sick and Burial Society, Edinburgh 177
Siddiqui, Dr Mona 293
Sikhs and Sikhism 11, 283, 302, 310, 315,
    426
Simons, Michael 274
Simpson, Archibald 581
Simpson, David 369
Sinclair, Catherine 214
Sinclair, Sir David, of Swynbrocht 75
Sinclair, Sir John 372, 386, 434, 448
Sinclair, Margaret 184
Sinclair, Sir William 244
singing see choirs; psalm singing; song
    schools
singing lofts 639
Single Carritch see Shorter Catechism
Sinn Fein 218
Sinton, Rev. Thomas 372
Sissons, Peter 466
Sixtus IV, Pope 85
Skateraw 212, 507
Skelmorlie Aisle, Largs 661
Skene, W F 214
Skerries mission station, Shetland 501
Skinner, John, bishop of Aberdeen 198, 200,
    201
Skinner, John ('Tullochgorum') 195
Skye 140, 371, 375, 491, 639
Slade (group) 427
Slamannan, Falkirk, Jewish convalescent
    home 264
Slater, Oscar 268
Slessor, Mary 138, 527
Smail, Tom 251
Small, Christopher 394–5
Smeaton mission station, Inveresk 502
Smith, Adam 538
Smith, Alexander 148
Smith, Alice Thiele 404
Smith, bishop 159
Smith, Sir George Adam 408, 666

Smith, Iain Crichton 390
Smith, James 586, 587
Smith, Rev. John 368, 370
Smith, Joseph 253
Smith, R A 642–3, 646
Smith, Sir William 459
Smith, William Pirie 404
Smith, William Robertson 10, 404, 405, 425,
    539
Smith, William Wye 355, 356, 358
Smout, T C 501
social care see welfare work
social life
    churches and 452–68, 563
    Jewish community 274–7
    see also church halls
Social Responsibility Committee, Scottish
    Episcopal Church 216
Social Welfare Committee, Church of
    Scotland 536
Society for Propagating the Gospel at
    Home (SPGH) 240, 482, 500, 518
    see also Nairnshire Society for the
    Propagation of the Gospel
Society for the Prevention of Cruelty to
    Animals 259
Society of Friends of Glasgow Cathedral
    307
Society of Friends (Quakers) 237–8, 492
Society of Probationers of the Church of
    Scotland 487, 488
Socinus, Faustus 245–6
Soga, Rev. Tiyo 524
Solemn League and Covenant 131, 133
Solway martyrs 629, 665
Somerset, duke of 67
Somerville, Rev. A N 499
Song of Solomon, Scots translation of 353,
    355
song schools 115, 632, 638, 639, 640, 642
    see also choirs
Sorensen, Rev. Alan 420
sound recording 414, 427
South Africa 520–1, 524, 525
    see also British South Africa Company;
    Cape Colony
South Carolina 237
South Clerk Street synagogue, Edinburgh
    269
South Portland Street Synagogue, Glasgow
    268
South Uist 373, 374, 670
Southend, Kintyre 28

Southern General Hospital, Glasgow 263
Southern Rhodesia 528
Soutra, Holy Trinity church 605
Spalding Club 208
Spangler, David 323
Spark, John 444
Spark, Muriel 390
Speaight, Robert 416
Spence, Alan 390
Spence, David 444
Spens, Sir Patrick 642
Spicer, Andrew 617
Spiritualism 318
spirituality 320, 321, 323, 327, 425, 428
Spottiswoode Club 208
Spottiswoode, John, archbishop of St
    Andrews 585
Sprenger, Jacob 114
Springburn, Glasgow, railway mission 497
Sprott, G W 591
Spurgeon, C H 404
Sri Chinmoy 317
SSPCK *see* Scottish Society for the
    Propagation of Christian
    Knowledge
stained glass 51, 596–7, 656, 663–4
    Cottier Community Centre 599
    Douglas 51, 83, 84
    Dunfermline Abbey 665
    Glasgow Cathedral 663–4
    Greyfriars kirk, Edinburgh 664
    Holyrood Abbey, Edinburgh 51
    St Giles' church, Edinburgh 665
    St Michael's church, Linlithgow 635
    Scottish Episcopal Church 663
    Scottish National War Memorial 669
    studios for 601
Stanley, Arthur, dean of Westminster 400
Statistical Account of Scotland 349, 386, 434,
    447, 448
statues 61, 67, 185, 582, 665
    Mary (mother of Jesus) 62, 66, 91
    saints 48–51, 67, 80, 81, 83, 88, 307, 582
    Shiva Nataraja 308
    see also sculptures
Stead, Tim 618
Stead, W T 401
Steiner, Rudolf 318
    see also Moray Rudolf Steiner School
Stenness, Orkney 575
Stevenson, Eric 597
Stevenson, R L 388, 389, 394
Stewart, Rev. Alexander 368

Stewart, Andrew 417
Stewart, Charles Edward, Prince 198
Stewart, J S 418
Stewart, James, earl of Moray 108, 476
Stewart, James (Lovedale Institution
    principal) 524
Stewart, James, of the Glen 198
Stewart, Rev. John 368
Stewart monarchs
    Catholicism and 154
    General Assemblies of the Church of
        Scotland 112, 536
    music and 632–3
    Scottish Episcopal Church and 10, 192,
        194
    support for church activities 63
    see also Jacobitism; names of individual
        monarchs
Stilicho, Flavius 17
stipends
    Church of Scotland ministers 114, 433,
        434, 435–7, 533, 587
    Episcopalian clergy 201, 203, 215, 223,
        224
    pre-Reformation clergy 64, 436
    see also benefices; patronage
Stirling 64, 317, 322, 326, 638, 665
Stirling Missionary Society 517
Stirling presbytery 57
Stirling, William 640
Stirlingshire 175, 266
Stobo parish church 578
Stockbridge church, Edinburgh 599
stone circles 317, 575
stone crosses *see* sculptures
Stonehaven jail, painting of Episcopalian
    baptism 208, 217
Stonehouse, Lanarkshire Christian Union
    500
stones *see* healing stones; mass stones;
    Papil Stone; St Peter's Stone
Stornoway, Episcopalian community 212
Storrar, Will 570
Strachan, Douglas 596, 669
Strang, William 666
Stranraer 317, 318
Strasbourg 127
Strathaven, Lanarkshire Christian Union
    500
Strathbogie, Catholic community 143
Strathclyde 17, 29–30, 32
Strathclyde Community Relations Council
    297

Strathclyde University 605
    see also Andersonian University
Strathfillan, Perthshire 34, 137
Strathin, Wat 64
Strathmiglo, Westbank Natural Health
    Centre 318, 321, 327
Streather, Paul 418
Strong, Rowan 211
Struther, William 444
Stuart family (Terregles/Traquair),
    Catholicism 143
Stuart, Gilbert 662
Stuart, Rev. James 368
Stuart, Jamie 358–9, 360
Stuart, Rev. John 368
Stuart, John, 2nd marquis of Bute 491
Stuart, John, 3rd marquis of Bute 178
Sufism 291
summer missions 457–8, 499
sumptuary laws 116
Sunday observance see Sabbath observance
Sunday school society, Paisley 483–4
Sunday schools 483–4
    catechisms 411, 483
    choirs 457, 465
    Church of Scotland 454–7, 483–4, 486,
        518, 612
    Congregationalist churches 483–4
    Falkirk 467
    Fireside Sunday School programme 457
    General Assembly of the Church of
        Scotland condemns 483, 518
    Glasgow 454, 456, 494
    home missions 487, 489, 504, 518
    Inverness 604
    Ladies Highland Association 504
    literacy and 483
    manse as meeting place 612
    performances 457
    prayer 456, 457
    prizes 404, 410
    Scottish Episcopal Church 216, 219, 604
    socialist movement and 641
    SSPCK 483
    teachers 456, 486, 489, 494
sunni Islam 286–7
superintendents, Church of Scotland 107,
    109, 110
Sutherland 603, 604, 613
Sutherland, Alexander 75
Sutherland, duchess of 581
Sutherland, earl of 150–1
Swan, Annie S 409, 410

Swave, Peter 63
Symington, Lanarkshire 349
Symington church 592
Symmachus, Pope 20
synagogues 259, 261–2, 267–70, 271, 272,
    277, 315
Synod of Birr 34
Synod of the Associated Presbyterian
    Churches 537
    see also Associated Presbyterian
        Churches of Scotland and Canada
Synod of Whitby 7
synods
    Church of Scotland 111, 130, 131, 132,
        136, 437
    foreign missions 518
    Free Church of Scotland 136
    Methodist Church 244, 245
    Scottish Episcopal Church 198, 205,
        207, 214, 224, 225
    seceding churches 135
    United Reformed Church of the United
        Kingdom 243
    see also Associate (Burgher) Presbytery
        and Synod; General Associate
        (Antiburgher) Synod; General
        Associate Synod; Relief Synod;
        United Associate Synod

T
tableaux vivants 72
Tablighi-jama'at 291
T'ai Chi 322
Tailler, Simon 631
Tailtiú, County Meath 27
Tain 34, 56, 577, 608
Tait, Archibald Campbell, archbishop of
    Canterbury 208
Talmage, Dr 404
Talmud Torah 270, 271, 276
Tankel, Henry 298
Tarbet Old church, Portmahomack 603
Tarbuck, Jimmy 427
Taylor, Sir Thomas 417
Tayside, home mission work 504
teacher training colleges 180, 206, 216, 524
teachers
    Catholic schools 157, 179, 180, 183, 548
    Church of Scotland schools 115, 366,
        435, 540, 541
    ministers and 443–4
    ministers as 437, 438, 446
    foreign missions 526, 527

Gaelic language and 366, 370
Gaelic School Societies 370, 490
Henryson 382
Islamic Studies 293
Jewish 266, 267, 271, 272
music 639, 642
Scottish Episcopal Church 192, 206, 216
Sunday schools 456, 486, 489, 494
Westminster Confession and 540
Tealing church, Angus 598–9
Tealing parish 238
teinds 110
    see also Court of Teinds
television broadcasting 414, 417–21, 425, 426–7, 547, 649
Telford, Thomas 587, 604
'Tell Scotland' movement 417, 507, 508, 509, 539
Teltown, County Meath 27
temperance movement
    Catholic community 174, 177, 178, 495
    Episcopalian community 211
    home missions 493
    hymns 641
    periodicals 404
Templars 319, 604
Temple church, East Lothian, conversion to domestic use 603
Temple, William 209
temples 303, 315
Templewood, Argyll 575
Tennant, William 348
Tent Hall, Glasgow United Evangelical Association 497–8
Tent Missions, north east Scotland 645
Terregles, Catholic community 143
Tertullian 16
theatres 185, 390, 601, 605, 606, 610
    see also drama
Theological Institute 225
theological schools and colleges 218, 225, 409
Theosophical Society 318
Thirlestane Castle 586
Thirty-nine Articles of Religion, the 198, 223
Thomas á Kempis 366
Thompson, Christina 527
Thomson, Alexander 589
Thomson, Dr Andrew 642–3
Thomson, D P 457, 507, 509
Thomson, Derick 373
Thomson, John 447

Thomson, William 641
Thomsone, David 444
Thornwood mission station, Bothwell parish 501
Three Hours Service 218
Thurgot, prior of Durham 46
Thurso, Baptist mission 491
Tibbermore church, Perthshire 608
Tillycoultry, St Serf at 46
Tindale see Tyndale, William
Tingwall, Shetland, St Magnus church 75
Tirchán 46
Tirconnell, barony of 26
Tiree 371
Titwood church, Pollokshields, Glasgow 604
'To Serve Thee Better' campaign 224
Tobermory, Baptist mission 491
Todd, Margo 141
Todd, Dr Stewart 625
Tokyo, church relocation to 605
tombs 75, 661
    Macleod, Alexander, of Dunvegan 83
    St Kentigern (St Mungo) 56
    St Ninian 22, 23, 24, 25, 55
    see also burial sites
Torphichen preceptory, Knights of St John 383
Torry, Episcopalian community 212
Torry, Patrick, bishop 197
Total Abstinence Society, St Patrick's, Edinburgh 177
Toulmin, S E 571
Town Mission, Greenock 487
Townhead, Glasgow 497, 501
Toynbee, Arnold 652
Tract and Colportage Society of Scotland 397
tract societies 397, 484, 485, 492, 511
    see also Jehovah's Witnesses
Tractarian (Oxford) Movement 205, 206–7, 211, 594
tracts
    evangelicals and 396–7, 398, 485, 486
    home missions 484, 485, 486, 487, 488, 489
    James VI's 385
    literacy and 397
    pre-Reformation 61
    publishing industry 492
Trades Union Congress, Edinburgh 214
Traill, James, bishop of Down and Connor 196

Transcendental Meditation 317
transportation, church buildings 604–5
Traprain Hoard 18–19, 20
Traquair, Catholic community 143
Traquair, Phoebe Anna 607, 666
Treaty of Edinburgh 106
Treaty of Union *see* Act of Union
Trent, Council of 118
Trinity Altarpiece 658
Trinity church *see* Holy Trinity church
Trinity College
    Edinburgh 63, 580, 605, 658
    Glenalmond 206, 213
Trinity Collegiate Church, Edinburgh, paintings 658
Trinity Congregational Church, Glasgow 599
Trinity Hospital, Edinburgh 444
Tron church
    Edinburgh 599
    Glasgow 586, 601
Tron theatre, Glasgow 601
True Jesus Church 604
Truth and Unity Movement 225
Tswana people, foreign missions 525
Tudor, Margaret (consort of James IV) 63
Tulideff, Andrew 64
Tullybody, St Serf at 46
Turkish community, Islam and 284, 286
Turner, J M W 665
Turriff, Episcopalian clergy 201
Twaddle, Alison 462, 463
Tweeddale, marquis of 587
Tyler, Evan 396
Tyndale, William, New Testament 340, 344, 395
Tyndrum, Navvy Mission 505
Tynninghame, St Baldred 34

U

Uí Néill 27, 28
Uinnio (Finnio) 25
Ulaid 28
Ullapool church 603–4
Ulster 27, 28, 171
Union and Readjustment Committee, Church of Scotland 502
Union of Congregational Churches 240
Union of Precentors 641
Union of South Africa 520–1
Union of the Parliaments *see* Act of Union
Unitarians 235, 245–6
United Arab Emirates 288, 290

United Associate Synod, Secession Church 134
United Free Church of Scotland 140, 453, 535, 537
    Book of Common Order 478
    deacons 138
    elders as moderators 137
    mission stations 502
    ordination of women 138
    Scottish Ecclesiological Society and 591
    union with Church of Scotland 138, 141, 216, 535, 537, 629, 646
    see also Free Church of Scotland
United Industrial School 173
United Jewish Israel Appeal 271
United Presbyterian Church 134, 140, 453, 534
    church government 136
    church organs 593, 625, 645
    Edinburgh seminary 524
    foreign missions 520, 523, 524, 527, 528
    Free Church and 134, 140, 537
    Glasgow 494–5
    McEwen, Robert 494
    mission stations 502
    music 593, 646
    see also United Free Church of Scotland
United Presbyterian Mission, Edinburgh 494
United Reformed Church of the United Kingdom 242–3
United Secession Church 241, 491, 492
United Synagogue of Glasgow 268, 272
United Working Men's Christian Mission 497
Universalist Church of America 246
Universalists *see* Unitarians
universities
    Buchanan, George and 384
    Catholic students 183
    church conversions 605–6
    Church of Scotland and 9, 114–15, 156, 385, 395, 426, 541–2
    First Book of Discipline 107–8
    ministers and 9, 435, 437, 439
    Disruption and 10
    foundation 7, 381
    Indian students 284
    Jewish community and 258, 266, 272
    Muslim students 284, 290
    Pakistani students 284
    Protestant Reformation and 8
    Scottish Enlightenment and 639

vagabond scholars 116
Westminster Confession and 541
women's education 520
see also names of individual universities
Unreason, Abbot of 382
Urdu language 291, 307
Ure, David 448
Urquhart, Inverness-shire 29
USA *see* America
usury 113

V
vagrants 116, 443
Valley Cemetery, Stirling 665
vernacular, the
    Bibles and psalters in 340, 395, 637
    masses in 185
    prayer in 435
    see also African languages; English language; Gaelic language and culture; Indian languages; Scots language
Verturiones 17
vicars-apostolic 173, 174, 176, 495
    see also Hay, George; Lynch, James; Nicolson, Thomas
Victoria Infirmary, Glasgow 263
Victoria, Queen 400, 609
Vikings, influence on art 655, 658
Vincent of Beauvais 47
Virgin Mary *see* Mary (mother of Jesus)
Voltaire 9
Votadini 17, 19, 20
Vulgate *see* Latin, Catholic Bible

W
Waddell, P Hately 354–5, 358
Waldef, abbot of Melrose 46
Walker, William 202
Wallace, Robert 447
Wallace, William 381–2
Waller, H 522
Wallyford mission station, Inveresk 502
Walton, O F 410
Ward Chapel, Dundee 242
Wardlaw, Henry, bishop of St Andrews 657
Wark, David 447
Watch Tower Bible and Tract Society *see* Jehovah's Witnesses
Water Newton, Cambs 16
Water of Leith 243, 444

Watson, Dr John 409
Watson, T L 589
Watts, Isaac 366, 645
Waugh, Alec 517
Waverley Station 580, 605
Webb, Pauline 295
Webster, Rev. Alexander 394, 447
Webster, Gordon 596
Wedderburn, John and James 343, 383
Wedderburn, Robert, vicar of Dundee 65, 383, 477
weddings *see* marriage
welfare work
    Catholic Church 105, 115, 173, 175, 177, 184–5
    Church of Scotland 113, 115, 135–6, 532, 533, 542
        ministers and ministers' wives 441, 443, 444
        poor relief 113, 115, 128, 136, 395, 435, 443, 532, 542
    home missions 492–5
    Jewish community 259, 262–5, 266, 278
    Richmond Craigmillar church 565–6, 567
    Salvation Army 249–50
    Scottish Episcopal Church 198, 199
    seceding churches 136
Wellington church, Glasgow 464, 589
wells and shrines, healing 57, 67, 86, 580
Welsh, Rev. David 534
Wesley, John 138, 244, 248
West Church of St Nicholas *see* St Nicholas' church, Aberdeen
West Coast Mission 496–7
West Kirk, St Giles' church, Edinburgh 583
West Register House, Edinburgh 606
Westbank Natural Health Centre 318, 321, 327
Wester Coates church, Edinburgh 604
Westerkirk church, Dumfriesshire 608
Western District, Catholic Church 172, 173, 175, 176–7
Western Isles
    Catholic Church 112, 155, 173, 372
    early church buildings 576
    education committee membership 541
    Gaelic-speaking clergy 375
    mission stations 503
    Sabbath observance 546
    West Coast Mission 496–7
    see also names of individual islands
Westminster Abbey 57

Westminster Assembly of Divines 131–2, 346, 347, 393
Westminster Confession 131, 140, 346
   Church of Scotland 132, 140, 346, 532, 534, 535
   education in tenets of 156
   Free Church divisions over 140
   Scottish Parliament (–1707) and 132, 532, 534
   teachers and 540
   universities and 541
Westminster Parliament 130, 132–3, 134
   see also House of Commons; House of Lords
Westminster Standards 371
Westray, Orkney, Baptist mission 491
Whaling, Frank 317
Wheatley, John 178, 181
Wheeler & Sproson 594
White, John 87, 422, 423
White, Very Rev. John 506
White, Joseph Blanco 200
Whitekirk, East Lothian, pilgrimage centre 57
Whiteriggs, Airdrie 183
Whithorn 20, 25, 26, 36, 382, 575
   Candida Casa 22, 23, 24
   pilgrimage centre 53, 55, 304, 656
   St Martin's church 22, 23, 24, 25, 55
   St Ninian's Cave 55, 153
   St Peter's Stone 20, 652
   see also St Ninian
Whithorn Cathedral 55, 583
Whithorn Priory, statue of St Ninian 51
Whitley, Rev. Harry 454, 461, 462, 465
Whittinghame Farm School 264
Whyte, Alexander 666
Whyte, I D, and Whyte, K A 442
Whyte, Thomas 447
Wilberforce, William 525
Wilkie, David 662, 666
Wilkie, William 447
William I, king of Scots 256
William of Touris 80
William Stirling part-book 640
Williams, Alice and Morris Meredith 669
Williamson, Henry 246
Wilson, A N 425
Wilson, John 520, 523–4, 526
Wilson, Margaret (missionary) 520, 526
Wilson, Margaret (Solway martyr) 629
Wilson, Thomas 187

Wilson, William 596, 669
Winchester, Simon 396
Winning, Thomas, cardinal 186, 422, 423
Winram, John 582
Winton, earl of 150–1
Winzet, Ninian 342
Wishart, Adam 90
Wishart, Agnes 65
Wishart, George 78, 106, 343, 658
Wishart, John 90
Wishaw, Lanarkshire Christian Union 500
Wiston Lodge, Biggar 461
witchcraft 114, 128–30, 385, 546–7
Withrington, Donald J 405
Wode, Thomas 637
Wolfenbüttel manuscript 631
Woman's Conference, Scottish Episcopal Church 208
Woman's Guild, Church of Scotland 138, 461–3, 561, 612
women
   alternative beliefs and 325, 326, 327–8
   Catholic community 153, 157, 180
   Christian Brethren 247
   Church of Scotland 137–8, 462, 467, 526–7, 618
      ministers' wives 441–2, 446, 462, 611, 612
   education 158, 488, 520, 526–7
   'Faith of a Woman' programmes 303–4
   foreign missions 138, 520, 526–7
   Hindu community 303–4
   home missions 221, 492, 494, 495, 503–4, 511
   Jewish community 304
   Knox's views on 137
   Livingstone's African followers 522
   Lyndsay's views on 79
   martyrdom 137, 629, 665
   medical profession 527
   Muslim community 293, 303, 311
   ordination 549
      Church of Scotland 138, 462
      Congregationalist churches 242
      Scottish Episcopal Church 223, 224, 225, 226
      Unitarian churches 246
      United Free Church of Scotland 138
   Scottish Episcopal Church 204, 219, 221, 223, 224, 225, 226, 422
   Seventh-Day Adventists 253

sexual purity 67, 79
Unitarian churches 246
United Free Church of Scotland 138
witchcraft accusations 130
wood carvings *see* engraving; sculptures
woodcuts 67, 71, 79
Wordsworth, Charles, bishop of St Andrews 207
worker priests, Scottish Episcopal Church 225
World Association for Christian Broadcasting 418
World Council of Churches 409
World Day of Prayer 229
Wormald, J 114
Wright, Alexander 253
Wright, Kenyon, canon 227
Wright, Roderick, bishop of Argyll and the Isles 423
Wright, Rev. Dr Ronald Selby 415, 507
Wyclif, John, New Testament translation 340
Wynd Mission, Glasgow 493

X
Xhosa people, mission work among 522, 524

Y
Yarrowkirk 19
Yester church, Gifford 587
Yetholm church 598
Yiddish language 270, 275, 277
Young, Edward 366
Young Irelanders 174
Young Muslims 291–2
young people 177, 183, 276, 277, 375, 467, 506–7
  see also schools; Sunday schools
Young, William 66
Young Women's Groups 463
Youth Fellowship, Church of Scotland 460–1

Z
Zambia 521, 522, 525
Zionism 274, 275, 277